# Comprehensive Treatise of Electrochemistry

*Volume 2: Electrochemical Processing*

# COMPREHENSIVE TREATISE OF ELECTROCHEMISTRY

Volume 1    THE DOUBLE LAYER
Edited by J. O'M. Bockris, Brian E. Conway, and Ernest Yeager

Volume 2    ELECTROCHEMICAL PROCESSING
Edited by J. O'M. Bockris, Brian E. Conway, Ernest Yeager, and Ralph E. White

Volume 3    ELECTROCHEMICAL ENERGY CONVERSION AND STORAGE
Edited by J. O'M. Bockris, Brian E. Conway, Ernest Yeager, and Ralph E. White

Volume 4    ELECTROCHEMICAL MATERIALS SCIENCE
Edited by J. O'M. Bockris, Brian E. Conway, Ernest Yeager, and Ralph E. White

# Comprehensive Treatise of Electrochemistry

## Volume 2: Electrochemical Processing

Edited by

**J. O'M. Bockris**
Texas A & M University
College Station, Texas

**Brian E. Conway**
University of Ottawa
Ottawa, Ontario, Canada

**Ernest Yeager**
Case Western Reserve University
Cleveland, Ohio

**Ralph E. White**
Texas A & M University
College Station, Texas

PLENUM PRESS · NEW YORK AND LONDON

Library of Congress Cataloging in Publication Data

Main entry under title:

Electrochemical processing.

(Comprehensive treatise of electrochemistry; v. 2)
Includes bibliographies and index.
1. Electrochemistry—Industrial applications. I. Bockris, John O'M. II. Series.
QD552.C64 vol. 2 [TP255]             541.3'7 [660.2'97]             80-24836
ISBN 0-306-40503-2

© 1981 Plenum Press, New York
A Division of Plenum Publishing Corporation
233 Spring Street, New York, N.Y. 10013

All rights reserved

No part of this book may be reproduced, stored in a retrieval system, or transmitted, in any form or by any means, electronic, mechanical, photocopying, microfilming, recording, or otherwise, without written permission from the Publisher

Printed in the United States of America

# Contributors

**Fritz Beck,** University of Duisburg, Gesamthochschule, FB 6 Elektrochemie, D 4100 Duisburg 1, West Germany

**Donald L. Caldwell,** The Dow Chemical Company, Freeport, Texas 77541

**J. E. Colman,** John E. Colman & Associates, Ltd., 1485 Shamrock Lane, Oakville, Ontario L6L 1R1, Canada

**V. A. Ettel,** J. Roy Gordon Research Laboratory, INCO Metals Company, Sheridan Park, Mississauga, Ontario L5K 129, Canada

**William B. Frank,** Aluminum Company of America, Alcoa Laboratories, Alcoa Center, Pennsylvania 15069

**Warren E. Haupin,** Aluminum Company of America, Alcoa Laboratories, Alcoa Center, Pennsylvania 15069

**James P. Hoare,** Electrochemistry Department, Research Laboratories, General Motors Corporation, General Motors Technical Center, Warren, Michigan 48090

**Anton J. Hopfinger,** Department of Macromolecular Science, Case Western Reserve University, Cleveland, Ohio 44106

**N. Ibl,** Eidgenössische Technische Hochschule Zürich, Technisch-Chemisches Laboratorium, CH-8092 Zürich, Switzerland

**Klaus Köster,** Institute of Chemical Technology, Technical University of Darmstadt, Darmstadt, West Germany

**Mitchell A. LaBoda,** Electrochemistry Department, Research Laboratories, General Motors Corporation, General Motors Technical Center, Warren, Michigan 48090

**P. W. T. Lu,** Advanced Energy Systems Division, Westinghouse Electric Corporation, Box 10864, Pittsburgh, Pennsylvania 15236

**Kenneth A. Mauritz,** T. R. Evans Research Center, Diamond Shamrock Corporation, Painesville, Ohio 44077

**Christoph J. Raub,** Forschungsinstitut für Edelmetalle und Metallchemie, D7070 Schwäbisch Gmünd, West Germany

**S. Srinivasan,** Department of Energy and Environment, Brookhaven National Laboratory, Upton, New York 11973

**B. V. Tilak,** Hooker Chemical Corporation, Research Center, Long Road, Grand Island, New York 14072

**H. Vogt,** Technische Fachhochschule Berlin, Fachbereich Verfahrenstechnik, D-1000 Berlin 65, West Germany

**Hartmut Wendt,** Institute of Chemical Technology, Technical University of Darmstadt, Darmstadt, West Germany

**R. Woods,** CSIRO, Institute of Earth Resources, Division of Mineral Chemistry, Port Melbourne, Victoria 3207, Australia

# Preface to Comprehensive Treatise of Electrochemistry

Electrochemistry is one of the oldest defined areas in physical science, and there was a time, less than 50 years ago, when one saw "Institute of Electrochemistry and Physical Chemistry" in the chemistry buildings of European universities. But, after early brilliant developments in electrode processes at the beginning of the twentieth century and in solution chemistry during the 1930s, electrochemistry fell into a period of decline which lasted for several decades. Electrochemical systems were too complex for the theoretical concepts of the quantum theory. They were too little understood at a phenomenological level to allow their ubiquity in application in so many fields to be comprehended.

However, a new growth began faintly in the late 1940s, and clearly in the 1950s. This growth was exemplified by the formation in 1949 of what is now called The International Society for Electrochemistry. The usefulness of electrochemistry as a basis for understanding conservation was the focal point in the founding of this Society. Another very important event was the choice by NASA in 1958 of fuel cells to provide the auxiliary power for space vehicles.

With the new era of diminishing usefulness of the fossil fuels upon us, the role of electrochemical technology is widened (energy storage, conversion, enhanced attention to conservation, direct use of electricity from nuclear–solar plants, finding materials which interface well with hydrogen). This strong new interest is not only in the technological applications of electrochemistry. Quantum chemists have taken great interest in redox processes. Organic chemists are interested in situations where the energy of electrons is as easily controlled as it is at electrodes. Some biological processes are now seen in electrodic terms, with electron transfer to and from materials which would earlier have been considered to be insulators.

It is now time for a comprehensive treatise to look at the whole field of electrochemistry.

The present treatise was conceived in 1974, and the earliest invitations to authors for contributions were made in 1975. The completion of the early volumes has been delayed by various factors.

There has been no attempt to make each article emphasize the most recent situation at the expense of an overall statement of the modern view. This treatise is not a collection of articles from *Recent Advances in Electrochemistry* or *Modern Aspects of Electrochemistry*. It is an attempt at making a mature statement about the present position in the vast area of what is best looked at as a new interdisciplinary field.

| | |
|---|---|
| *Texas A & M University* | J. O'M. Bockris |
| *University of Ottawa* | B. E. Conway |
| *Case Western Reserve University* | Ernest Yeager |
| *Texas A & M University* | Ralph E. White |

# Preface to Volume 2

This volume brings together some dozen processes well known to the electrochemist and treats them according to their various degrees of importance.

The production of hydrogen is one of the more important processes, particularly with respect to the prospects of a hydrogen economy. No one would doubt, however, that the most commercially important electrochemical processes at the present time are the production of aluminum and of chlorine. Each of these processes has a separate chapter devoted to it.

We have treated inorganic and organic synthesis in separate chapters as topics alone, but thereafter proceeded to the electrofining and electrowinning of metals, which we clearly distinguish from electroplating.

Electrochemical machining is an area of growing importance which is tending to replace older methods of handling materials, and may be extended to some semiconductors.

The theory of membranes is a subject that is not in the central field of electrochemistry, but it is nevertheless an important one, not only with respect to the possibility of interpreting biological facts, but also with regard to batteries and other electrochemical machines.

Electrochemical painting is one of the greater achievements that contributed to the success of electrochemical technology during the 1960s, and after due consideration, we thought that it was worth devoting an entire chapter to it.

The last chapter in this volume involves an intriguing area which has blossomed out to be a part of electrochemistry; namely, the processes by which the various minerals are separated from each other by means of flotation. The realization of the electrochemical nature of such processes promises fair weather for increased electrochemical research in the area, and eventually the tailoring of suitable flotation agents.

*Texas A & M University*     J. O'M. Bockris
*University of Ottawa*     B. E. Conway
*Case Western Reserve University*     Ernest Yeager
*Texas A & M University*     Ralph E. White

# Contents

**1. Electrolytic Production of Hydrogen**
*B. V. Tilak, P. W. T. Lu, J. E. Colman, and S. Srinivasan*

| | |
|---|---|
| 1. Rationale for Electrolytic Production of Hydrogen | 1 |
| 2. General Aspects of Water Electrolysis Technology | 3 |
|    2.1. Types of Water Electrolyzers | 3 |
|    2.2. Mass and Energy Balance for a Water Electrolyzer | 9 |
|    2.3. Typical Process Flow Sheet for a Water Electrolysis Plant | 12 |
| 3. Alkaline Water Electrolyzers | 13 |
|    3.1. Thermodynamics | 13 |
|    3.2. Electrode Kinetics | 16 |
|    3.3. Commercial Electrolyzers—Design Features and Operating Characteristics | 33 |
|    3.4. Novel Alkaline Electrolyzers | 49 |
|    3.5. Projected Advances in Alkaline Water Electrolysis Technologies | 57 |
| 4. Solid Polymer Electrolyte (SPE) Water Electrolyzers | 61 |
|    4.1. Thermodynamics | 61 |
|    4.2. Electrode Kinetics | 61 |
|    4.3. Electrode Configurations | 63 |
|    4.4. Electrolytic Cells—Design Aspects | 64 |
|    4.5. Comparison of Operating Parameters of Various Cells | 78 |
| 5. Economics of Hydrogen Production | 79 |
| 6. Futuristic Concepts for Electrolytic Production of Hydrogen | 84 |
|    6.1. Use of Anode Depolarizers | 84 |
|    6.2. Water Vapor Electrolysis | 86 |
|    6.3. Thermochemical–Electrochemical Hybrid Cycles | 90 |
|    6.4. Photoelectrolysis of Water | 91 |
| 7. Heavy Water—A Useful By-Product | 92 |
| *Appendix* | 96 |
| *References* | 97 |

## 2. Production of Chlorine
*Donald L. Caldwell*

| | |
|---|---|
| 1. Introduction | 105 |
|   1.1. Significance | 105 |
|   1.2. Scope | 105 |
|   1.3. Major End Uses | 106 |
| 2. Historical Survey | 107 |
|   2.1. Nonelectrolytic Processes | 107 |
|   2.2. Early History of Electrolysis | 108 |
|   2.3. The Diaphragm Cell | 108 |
|   2.4. The Mercury Cell | 109 |
| 3. Electrolytic Decomposition of Sodium Chloride | 110 |
|   3.1. Manufacturing Processes | 110 |
|   3.2. Cell Voltage | 112 |
|   3.3. Current Efficiency | 117 |
|   3.4. Energy Efficiency | 120 |
|   3.5. Energy Consumption | 120 |
| 4. Cell Components | 121 |
|   4.1. Anodes | 121 |
|   4.2. Cathodes | 127 |
|   4.3. Cell Separators | 132 |
| 5. Cell Technology | 138 |
|   5.1. Diaphragm Cells | 138 |
|   5.2. Mercury Cells | 149 |
|   5.3. Membrane Cells | 151 |
| 6. Chloralkali Plant Auxiliaries | 156 |
|   6.1. General | 156 |
|   6.2. Direct Current Electric Power | 157 |
|   6.3. Brine Purification | 158 |
|   6.4. Chlorine Processing | 158 |
|   6.5. Hydrogen Recovery | 159 |
|   6.6. Caustic Soda Processing | 159 |
| 7. Future Trends | 160 |
| *References* | 162 |

## 3. Inorganic Electrosynthesis
*N. Ibl and H. Vogt*

| | |
|---|---|
| 1. Introduction | 167 |
|   1.1. Scope | 167 |
|   1.2. History and Current Outlook of Inorganic Electrosynthesis | 168 |
| 2. Chlorate | 169 |
|   2.1. Industrial Significance | 169 |
|   2.2. Development of Theory and Technology; Main Features of the Present State of the Art | 170 |
|   2.3. Fundamentals | 173 |

|  |  |  |
|---|---|---|
| 2.4. Discussion of the Factors Relevant for Industrial Chlorate Electrolysis | | 189 |
| 2.5. Industrial Cells | | 196 |
| 3. Hypochlorite | | 201 |
| 3.1. Industrial Significance | | 201 |
| 3.2. Development of Theory and Technology | | 203 |
| 3.3. Reaction Fundamentals | | 203 |
| 3.4. Industrial Cells | | 205 |
| 4. Perchloric Acid and Perchlorate | | 208 |
| 4.1. Methods of Preparation | | 208 |
| 4.2. Industrial Significance | | 209 |
| 4.3. Development of Theory and Technology | | 210 |
| 4.4. Reactions and Mechanisms | | 210 |
| 4.5. Industrial Cells | | 214 |
| 5. Bromate | | 216 |
| 6. Iodate and Periodate | | 219 |
| 6.1. Iodate | | 219 |
| 6.2. Periodate | | 220 |
| 7. Peroxodisulfate | | 221 |
| 7.1. Reaction Fundamentals | | 221 |
| 7.2. Industrial Cells | | 224 |
| 8. Hydrogen Peroxide | | 226 |
| 8.1. Hydrogen Peroxide by Anodic Oxidation | | 226 |
| 8.2. Hydrogen Peroxide by Direct Cathodic Reduction | | 227 |
| 8.3. Hydrogen Peroxide by Indirect Reduction | | 229 |
| 8.4. Development of Technology | | 229 |
| 9. Perborate | | 230 |
| 10. Permanganate | | 231 |
| 10.1. Industrial Significance | | 231 |
| 10.2. Reaction Fundamentals | | 231 |
| 10.3. Development of Technology | | 232 |
| 10.4. Industrial Cells | | 232 |
| 11. Chromic Acid | | 234 |
| 11.1. Industrial Significance | | 234 |
| 11.2. Development of Technology | | 234 |
| 11.3. Reaction Fundamentals | | 234 |
| 11.4. Industrial Cells | | 235 |
| 12. Sodium Sulfate Electrolysis | | 236 |
| *Auxiliary Notation* | | 237 |
| *References* | | 237 |

## 4. Electro-Organic Syntheses
*Klaus Köster and Hartmut Wendt*

|  |  |
|---|---|
| 1. Introduction | 251 |
| 1.1. Historical Survey | 251 |
| 1.2. Electro-organic Synthesis and Electrochemical Engineering | 252 |
| 1.3. Mediated Electrochemical Synthesis | 253 |

| | | |
|---|---|---|
| 1.4. | Future Development | 253 |
| 1.5. | Scope of this Chapter | 253 |
| 2. Direct Electrochemical Oxidation and Reduction of Organic Molecules | | 253 |
| 2.1. | Introduction and Survey | 253 |
| 2.2. | Molecular Orbital (MO) Representation and Energy Demand for the Anodic Radical Cation and Cathodic Radical Anion Formation from Unsaturated Hydrocarbon Molecules | 254 |
| 2.3. | Hückel Molecular Orbital (HMO) Representation of the Electrochemical Oxidation and Reduction of Double Bonds between Carbon and Heteroatoms | 258 |
| 2.4. | Reactivity of Radical Cations and Radical Anions Generated from Unsaturated Hydrocarbon Molecules | 259 |
| 2.5. | Typical Chemical Reactions of Radical Cations and Radical Anions Generated Electrochemically from Unsaturated Hydrocarbons | 261 |
| 2.6. | Electrochemical Generation of C Radicals and Their Reactions | 277 |
| 2.7. | Summary of Reaction Types for the Direct Electrochemical Conversion of Organic Substances | 283 |
| 2.8. | Influence of Electrosorption for Product Composition and Yields of Electro-organic Synthetic Processes | 285 |
| 2.9. | Solvent Electrolyte Systems and Electrolyte Materials Used in Electro-organic Synthesis | 289 |
| 3. Mediated Electrochemical Conversion of Organic Substrates | | 290 |
| 3.1. | Description of Mediated Electro-organic Synthesis | 290 |
| 3.2. | Anodic Mediator System | 290 |
| 3.3. | Cathodic Mediator System and Typical Mediated Homogeneous and Heterogeneous Cathodic Conversions | 291 |
| 4. Semitechnical and Technical Electrochemical Organic Synthesis | | 291 |
| 4.1. | General Remarks | 291 |
| 4.2. | Anodic Processes | 292 |
| 4.3. | Cathodic Processes | 294 |
| *References* | | 295 |

## 5. Electrometallurgy of Aluminum

*Warren E. Haupin and William B. Frank*

| | | |
|---|---|---|
| 1. Hall–Heroult Cell for Alumina Reduction | | 301 |
| 2. Electrolyte | | 303 |
| 2.1. | Composition and Physical Properties | 303 |
| 2.2. | Ionic Constitution of the Electrolyte | 305 |
| 2.3. | Dissolution of Alumina | 306 |
| 3. Electrode Reactions | | 306 |
| 3.1. | Cathode Reactions | 306 |
| 3.2. | Anode Process | 308 |
| 4. Current Efficiency | | 314 |
| 4.1. | Factors Controlling Reoxidation Rate | 317 |
| 4.2. | Other Losses in Current Efficiency | 317 |
| 5. Energy Considerations | | 318 |

6. Electromagnetic Effects . . . . . . . . . . . . . . 321
7. Producing High-Purity Aluminum . . . . . . . . . . 321
8. Electrolysis of Aluminum Chloride . . . . . . . . . 322
*References* . . . . . . . . . . . . . . . . . . 324

# 6. Electrolytic Refining and Winning of Metals
*V. A. Ettel and B. V. Tilak*

1. Introduction . . . . . . . . . . . . . . . . . . 327
2. Electrochemical Principles of Electrorefining and Electrowinning . . . 328
    2.1. Electrochemical Selectivity . . . . . . . . . . . 328
    2.2. Addition Agents . . . . . . . . . . . . . . 330
    2.3. Mass Transport in Refining and Electrowinning Cells . . . . 331
3. Technological Principles of Refining and Electrowinning . . . . . 335
    3.1. Soluble Anodes . . . . . . . . . . . . . . 335
    3.2. Insoluble Anodes for Electrowinning . . . . . . . . 338
    3.3. Cathodes . . . . . . . . . . . . . . . . 340
    3.4. Electrolytic Cells . . . . . . . . . . . . . . 343
    3.5. Electrical Circuits . . . . . . . . . . . . . 345
    3.6. Electrolyte Circuits . . . . . . . . . . . . . 347
    3.7. Slime Handling . . . . . . . . . . . . . . 348
    3.8. Material Handling . . . . . . . . . . . . . 349
    3.9. Electrolyte Mist in Electrowinning . . . . . . . . . 349
4. Operating Practices . . . . . . . . . . . . . . . 350
    4.1. Copper Refining . . . . . . . . . . . . . . 350
    4.2. Lead Refining . . . . . . . . . . . . . . . 356
    4.3. Nickel Refining . . . . . . . . . . . . . . 357
    4.4. Silver Refining . . . . . . . . . . . . . . 360
    4.5. Refining of Other Metals . . . . . . . . . . . 362
    4.6. Zinc Electrowinning . . . . . . . . . . . . . 363
    4.7. Copper Electrowinning . . . . . . . . . . . . 367
    4.8. Nickel Electrowinning . . . . . . . . . . . . 371
    4.9. Cobalt Electrowinning . . . . . . . . . . . . 374
    4.10. Electrowinning of Other Metals . . . . . . . . . 375
*References* . . . . . . . . . . . . . . . . . . 377

# 7. Electroplating
*Christoph J. Raub*

1. Introduction . . . . . . . . . . . . . . . . . . 381
2. Present-Day Technology of Electroplating with Metals . . . . . 381
3. Electrolytically Produced Metallic and Nonmetallic Layers . . . . 383
4. Electroforming . . . . . . . . . . . . . . . . 384
5. Solutions Used in Electroplating . . . . . . . . . . . 384
6. Alloy Plating . . . . . . . . . . . . . . . . . 385

7. Properties of Deposits . . . . . . . . . . . . . . . 388
8. Inhibitors . . . . . . . . . . . . . . . . . . 390
9. Throwing Power . . . . . . . . . . . . . . . . 391
10. Leveling . . . . . . . . . . . . . . . . . . 393
11. Further Work on Alloys . . . . . . . . . . . . . 393
12. Effect of Complexing . . . . . . . . . . . . . . 395
13. Summary . . . . . . . . . . . . . . . . . . 396
Suggested Reading . . . . . . . . . . . . . . . . 396
References . . . . . . . . . . . . . . . . . . . 397

## 8. Electrochemical Machining
### James P. Hoare and Mitchell A. LaBoda

1. Introduction . . . . . . . . . . . . . . . . . . 399
2. Corrosion Process Fundamentals . . . . . . . . . . . 400
   2.1. Local Cell Corrosion . . . . . . . . . . . . . 400
   2.2. Bimetallic Corrosion . . . . . . . . . . . . . 401
   2.3. Anodic Corrosion . . . . . . . . . . . . . . 401
3. Electropolishing . . . . . . . . . . . . . . . . 403
4. Jet Etching . . . . . . . . . . . . . . . . . . 413
5. Electrochemical Grinding . . . . . . . . . . . . . 418
6. Principles of Electrochemical Machining . . . . . . . . 423
   6.1. The Gap and Feed Rate . . . . . . . . . . . . 423
   6.2. Temperature Effects . . . . . . . . . . . . . 426
   6.3. Pressure Effects . . . . . . . . . . . . . . 429
   6.4. Hydrogen Bubble Effects . . . . . . . . . . . 429
   6.5. Effects of Corrosion Products . . . . . . . . . . 433
   6.6. Metal Removal Rate Considerations . . . . . . . 434
   6.7. Effects of Stray Currents . . . . . . . . . . . 443
   6.8. Addition Agent Studies . . . . . . . . . . . . 445
   6.9. Surface Finish Considerations . . . . . . . . . . 446
7. Electrochemical Machining Operations . . . . . . . . . 447
   7.1. External Shaping . . . . . . . . . . . . . . 447
   7.2. Die Sinking . . . . . . . . . . . . . . . . 449
   7.3. Plunge-Cutting . . . . . . . . . . . . . . . 450
   7.4. Turning . . . . . . . . . . . . . . . . . 451
   7.5. Trepanning . . . . . . . . . . . . . . . . 451
   7.6. Internal Grooving . . . . . . . . . . . . . . 451
   7.7. Wire Cutting . . . . . . . . . . . . . . . 451
   7.8. Deburring . . . . . . . . . . . . . . . . 451
   7.9. ECM Machines . . . . . . . . . . . . . . . 453
8. The Electrochemistry of Electrochemical Machining . . . . 456
   8.1. New Electrolyte Studies . . . . . . . . . . . . 456
   8.2. ECM Precautions with Oxidizing Salt Electrolytes . . . 458
   8.3. Polarization Studies . . . . . . . . . . . . . 462
   8.4. Transpassive Dissolution of Anodic Films . . . . . . 465
   8.5. Surface Brightening . . . . . . . . . . . . . 467

CONTENTS

|   |   |   |
|---|---|---|
| 8.6. Solution Flow Effects | | 468 |
| 8.7. Other ECM Electrolytes | | 471 |
| 8.8. Mixed Electrolytes | | 475 |
| 8.9. High-Strength, High-Temperature Alloys | | 483 |
| 8.10. The ECM of Other Metals | | 490 |
| 8.11. Pulsed ECM Methods | | 496 |
| 8.12. Consensus | | 497 |
| 9. Electrochemical Machining Applications | | 497 |
| 9.1. Stem Drilling | | 500 |
| 9.2. Fixture Electrochemical Machining (Cell Type) | | 503 |
| 9.3. ECM Broaching | | 507 |
| 9.4. Electrolytic Grinding (ELG, ECG) | | 508 |
| Auxiliary Notation | | 511 |
| References | | 511 |

## 9. Theory of the Structure of Ionomeric Membranes
### Anton J. Hopfinger and Kenneth A. Mauritz

|   |   |
|---|---|
| 1. Introduction | 521 |
| 2. Cluster Formation Model | 523 |
| 2.1. General Mechanism of Cluster Formation | 523 |
| 2.2. The Dry State | 524 |
| 2.3. Exposure to Water | 525 |
| 2.4. Exposure to Aqueous Ionic Solutions | 526 |
| 2.5. Application of the Theory to Nafion | 527 |
| 3. Cluster Property Model | 529 |
| 4. Summary | 534 |
| References | 535 |

## 10. Electrodeposition of Paint
### Fritz Beck

|   |   |
|---|---|
| 1. Introduction | 537 |
| 2. Historical Development | 540 |
| 3. Anodic EDP Process | 541 |
| 3.1. General Considerations | 541 |
| 3.2. Static Properties of the Bath and of the Film | 543 |
| 3.3. Electrocoagulation as the Primary Process | 545 |
| 3.4. Thickness Growth of Film as the Secondary Process | 552 |
| 4. Cathodic EDP Process | 559 |
| 5. Deposition of Dispersions and Latices | 560 |
| 6. Chemistry of Resins and of the Bath | 561 |
| 7. Pretreatment of the Substrate | 562 |
| 8. Practical Aspects of the EDP Process | 562 |
| 9. Follow-Up Steps after Electrodeposition | 563 |
| 10. Electrochemical Alternative Processes | 564 |
| References | 566 |

## 11. Mineral Flotation
*R. Woods*

| | |
|---|---|
| 1. Introduction | 571 |
| 2. Adhesion of Particles to Gas Bubbles | 572 |
|    2.1. Thermodynamics of Bubble Attachment | 572 |
|    2.2. Kinetics of Bubble Attachment | 575 |
| 3. Flotation of Oxide Minerals | 576 |
|    3.1. Surface Charge | 576 |
|    3.2. Collector Adsorption | 577 |
|    3.3. Influence of Inorganic Ions in Solution | 578 |
|    3.4. Interaction between Hydrocarbon Chains | 579 |
|    3.5. Influence of Neutral Organic Molecules | 581 |
| 4. Flotation of Sulfide Minerals | 582 |
|    4.1. Mixed Potential Mechanism of Collector Adsorption | 582 |
|    4.2. Anodic Oxidation of Sulfide Minerals | 583 |
|    4.3. Anodic Oxidation of Collectors | 586 |
|    4.4. Cathodic Reduction of Oxygen at Sulfide Surfaces | 589 |
|    4.5. Modulation of Sulfide Flotation | 590 |
|    4.6. Investigations of the Flotation of Mineral Particles as a Function of Potential | 592 |
| *References* | 594 |
| *Index* | 597 |

# Notation

| | | | |
|---|---|---|---|
| $a_\pm$ | mean activity; $a_i$, $a_j$ activities of species $i$, $j$ | $E_{CB}$ | energy of conduction band |
| $c$ | concentration (molar); velocity of light (cm s$^{-1}$) | $E_F$ | Fermi level |
| | | $E_H$ | measured potential on the hydrogen scale in the *same* solution |
| $C_1$, $C_2$, etc. | differential capacities of regions 1, 2, etc. | $E_{NHE}$ | measured potential on the scale of the normal hydrogen electrode |
| $cn$ | coordination number | | |
| $d$ | thickness, e.g., of a film, or of a dielectric | $F_{ss}$ | energy of surface states |
| $D$ | diffusion coefficient | $E_{VB}$ | energy of valence band |
| $D_{x_\psi}$ | dissociation energy for molecule $x_\psi$ | $\mathscr{E}$ | electrostatic field |
| $\mathbf{D}$ | dielectric displacement | $f_\pm$ | rational activity coefficient (mean) |
| $e$ | electron charge | $F$ | Faraday constant |
| $E$ | potential (cf. electrode, on metal–solution potential difference, in kinetics) | $g$ | interaction parameter, in non-Langmuir isotherms |
| $E_{cal}$ | measured potential on the scale of the normal calomel electrode | $g_{ij}(r_{ij})$ | radial distribution function (of distance $r_{ij}$); pair correlation function |

| | | | |
|---|---|---|---|
| $G, H, S$ | free energy enthalpy, and entropy (per mole) | $n_p^0$ | concentration of holes in bulk |
| $h$ | Planck's constant | $N_A$ | concentration of charge acceptors |
| $i$ | current density | $N_D$ | concentration of charge donors |
| $I_0$ | intensity of light | | |
| $I$ | current moment of inertia | $N_{SS}$ | concentration of surface states |
| $J$ | flux; quantum number for rotation | $P$ | pressure (Pa), e.g., $P_{O_2}$, presence of a gas, $O_2$; momentum |
| $k$ | with subscript, rate constants | | |
| $k_s$ | salting out (Setschenow) coefficient | $P(E)$ | probability (for state of energy E) |
| $k$ | Boltzmann constant | $q, Q$ | partition function |
| $K$ | thermodynamic equilibrium constant | $Q_i$ | charge for some species, $i$, e.g., on a surface |
| $K_1, K_2$, etc. | integral capacities of regions 1, 2, etc. | | |
| $m$ | concentration (molal); mass of particle | $r_i$ | radius of an ion |
| | | $r_{ij}$ | distance between particles $i, j$ |
| $M$ | molarity; $N$ no longer used; number of particles | $R$ | molar gas constant; resistance |
| | | $t$ | time |
| $n$ | solvation number; quantum number for vibration | $T$ | absolute temperature (K); with subscript, nmr relaxation times ($T_1, T_2$) |
| $n_{CB}$ | density of electronic states in the conduction band | $U$ | internal energy |
| | | $v$ | velocity (usually of a reaction); mobility of ion under 1 V cm$^{-1}$ charge |
| $n_e$ | concentration of electrons | | |
| $n_e^s$ | concentration of electrons at the surface | $V$ | volume; partial molar volume |
| $n_e^0$ | concentration of electrons in bulk | $x, y, z$ | coordinate system; distances |
| | | $y_\pm$ | stoichiometric activity coefficient (mean, molar) |
| $n_p$ | concentration of holes | | |
| $n_e^s$ | concentration of holes at the surface | $\neq$ | activated state (used as superscript) |

## Greek Symbols

| | |
|---|---|
| $\alpha$ | light absorption coefficient; transfer coefficient; specific expansibility |
| $\beta$ | charge-transfer symmetry factor; specific compressibility |
| $\gamma$ | surface tension |
| $\gamma_{\pm}$ | stoichiometric activity coefficient (mean) molal |
| $\delta$ | diffusion-layer thickness; barrier thickness |
| $\Delta_i^{i,b}\varphi$ | potential inside a metal ($i = m$), semiconductor ($i = $ sc), or insulator ($i = $ ins) |
| $\Delta_1{}^i\varphi$ | potential drop at the inner Helmholtz plane $\varphi$ ($i = M$, sc, ins, etc.) |
| $\Delta_b{}^2\varphi$ | potential in the diffuse (Gouy) double layer |
| $\Delta_2{}^i\varphi$ | potential in the Helmholtz layer ($i = M$, sc, or ins) |
| $\Gamma_i$ | surface excess of species $i$ |
| $\varepsilon$ | permittivity; quantum efficiency |
| $\zeta$ | zeta potential |
| $\eta$ | overpotential; viscosity |
| $\theta$ | fractional surface coverage; relative permittivity; dielectric constant |
| $\kappa$ | conductivity; Debye–Hückel parameter |
| $\Lambda_{\pm,c}$ | molar ionic conductivity at concentration $c$ |
| $\Lambda_c$ | molar conductivity at concentration $c$ |
| $\Lambda_\infty$ | molar conductivity at infinite dilution |
| $\Lambda_{\pm,\infty}$ | molar ionic conductivity at infinite dilution |
| $\mu$ | electric dipole moment; or chemical potential |
| $\mu_e$ | mobility of electrons |
| $\mu_p$ | mobility of holes |
| $\mu^0$ | standard chemical potential |
| $\tilde{\mu}$ | electrochemical potential |
| $\nu$ | stoichiometric number; frequency of vibration (s$^{-1}$) |
| $\tilde{\nu}$ | wave number (cm$^{-1}$) |
| $\rho$ | density of space charge, resistivity |
| $\rho(E)$ | volume charge density |
| $\rho_i(E)$ | density of states ($i = M$, sc, or ins) |
| $\sigma$ | surface charge density in distribution; charge in double-layer region (subscripted) divided by area |
| $\sigma_e$ | capture cross section of electrons |
| $\sigma_m$ | charge on metal surface, divided by area |

| | | | |
|---|---|---|---|
| $\sigma_p$ | capture cross section of holes | | molar function of $x$ |
| $\tau$ | relaxation time | $\varphi$ | inner potential |
| $\phi$ | double-layer potential (subscripted for indication of region) | $\Delta\varphi$ | Galvani potential |
| | | $\chi$ | surface potential |
| | | $\Delta\chi$ | surface potential difference |
| $\phi_x$ | apparent molar function of $x$; with subscript $\bar{x}$, partial | $\psi$ | outer potential |
| | | $\Delta\psi$ | Volta potential |
| | | $\omega$ | angular frequency |

# 1
# Electrolytic Production of Hydrogen

**B. V. TILAK, P. W. T. LU, J. E. COLMAN, and S. SRINIVASAN**

## 1. Rationale for Electrolytic Production of Hydrogen

In order to meet the anticipated enhanced demands for hydrogen as a chemical feedstock, as a process gas, and as a clean fuel, the development of techniques for bulk hydrogen generation from nonfossil primary energy sources is vital. With the continued increase in costs and dwindling availability of natural gas and oil, which are the major sources of hydrogen in most countries, the production of hydrogen from coal, or by water electrolysis using electricity derived from hydroelectric, nuclear, solar, geothermal, or fusion energy, will become quite attractive in the near future. Since the energy crisis of 1973, methods have been actively pursued for the production of electricity and/or heat (forms of energy which are not easily storable or transportable over long distances) from renewable sources. Even if such methods are found, there will still be a need for portable fluid fuels which will have to be manufactured on a large scale from the above-mentioned resources. Portable

---

**B. V. TILAK** • Hooker Chemical Corporation, Research Center, Long Road, Grand Island, New York 14072. **P. W. T. LU** • Advanced Energy Systems Division, Westinghouse Electric Corporation, Box 10864, Pittsburgh, Pennsylvania 15236. **J. E. COLMAN** • John E. Colman & Associates, Ltd., 1485 Shamrock Lane, Oakville, Ontario, L6L 1R1 Canada. **S. SRINIVASAN** • Department of Energy and Environment, Brookhaven National Laboratory, Upton, New York 11973.

fuels will be most essential for transportation applications. Hydrogen, methanol, or ethanol are the most likely candidates for fluid fuels. Water electrolysis is the only proven technology for production of hydrogen from nonfossil fuel primary energy sources. Other methods such as thermochemical, photochemical, or biochemical are in the infant research stage and from an engineering and economic standpoint show little prospects of success for production of hydrogen on a commerical scale. Due to the intermittent nature of some of the renewable energy sources (solar, wind), energy storage in some other form becomes essential. Here again, electrolytic production of hydrogen is most attractive. $H_2$ is ideal, whereas methanol or ethanol will be an alternative fuel for future transportation applications. In the nuclear, solar, or fusion era, the most convenient method of methanol production will be by the gas phase catalytic reaction of $CO_2$ (from carbonate rocks, the atmosphere, or the ocean) and electrolytic hydrogen. Ethanol can be easily manufactured from any form of biomass.

At the present time, hydrogen is used in large quantities by the chemical and agricultural industries—1 quad (i.e., $10^{15}$ BTU) of hydrogen per year in the U.S.A.—for the production of ammonia and methanol, hydrocracking of petroleum, hydrogenation of fats and oils, and as a reducing agent in the metallurgical and semiconductor industries (Figure 1). Since the energy crisis of 1973, it became increasingly apparent that there is merit in a "hydrogen economy" and by using hydrogen as a fuel (energy conversion in fuel cells, aircraft engines, energy storage, and transportation), vast quantities of hydrogen will have to be produced by water electrolysis.

Hydrogen derived from coal needs to be purified for several chemical and fuel applications. The cost of purification may raise its cost to approximately

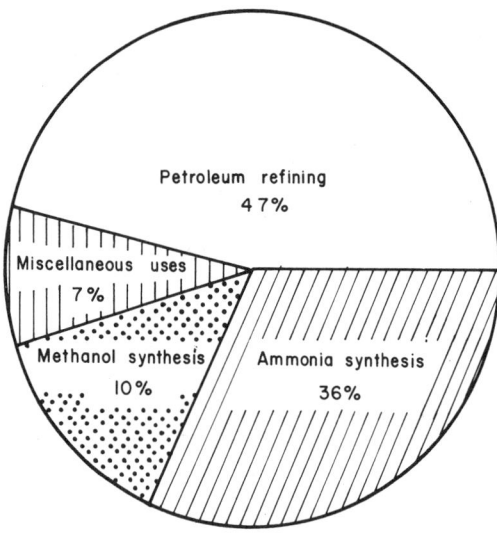

Figure 1. Hydrogen consumption pattern in U.S.A. in 1974.

the same level as that of electrolytic hydrogen. In certain cases where relatively small quantities of hydrogen and/or oxygen are required, on-site electrolysis of water will again be more economical than its production by other methods (e.g., water–gas reaction) at a central site, if the cost of storage and transport of the gases to the site of application is included.

## 2. General Aspects of Water Electrolysis Technology[4-7,39,42]

Electrolysis is one of the best-known and simplest methods for producing pure hydrogen, either on a small or large scale, from an abundantly available source—water. Water electrolysis plants (i) operate with few moving parts, (ii) require little space, (iii) are nonpolluting, and (iv) can be maintained with semiskilled labor. Further, the products of the reaction, $H_2$ and $O_2$, are physically separated during the evolution at the electrodes. Several large-scale electrolytic plants have been constructed and operated for several years (for example, water electrolyzers in the Aswan Dam require 150 MW of power and produce hydrogen at the rate of 35,000 $m^3$/hr).

### 2.1. Types of Water Electrolyzers

Electrolyzers fall into two major categories: monopolar, tank-type cells (see Figure 2) or bipolar, filter press cells (see Figure 3). The primary distinction between these two types of cells is the electrical polarity or polarities of the electrodes, and the manner in which these electrodes are connected together in a container full of electrolyte.

In a monopolar cell (Figure 2), all faces of an electrode have only one polarity performing a single electrochemical process, either positive, which discharges oxygen, or negative, which discharges hydrogen. A number of positive anode electrodes interleaved with negative cathode electrodes can then be installed into a single gas-tight tank, with all the positive electrodes connected together in parallel, and similarly, all the negative electrodes connected together in parallel. The form of the electrode pack is normally a number of vertical parallel plates. Thus, quite large monopolar cells can be built containing a very high electrode surface. As illustrated in Figure 2, suitable diaphragms such as asbestos cloth, are inserted around either the positive or negative electrodes to separate the oxygen and hydrogen gases. Circulation within the cell is by simple gas lift. A monopolar cell therefore can have very large electrical currents with a low voltage equal to one electrode pair of one cathode and one anode, which is roughly 2 V.

In a bipolar cell (Figure 3), all electrodes, except the terminal electrodes, have one side with a positive polarity and the opposite side with a negative polarity. The terminal electrodes which lead the current into and out of the

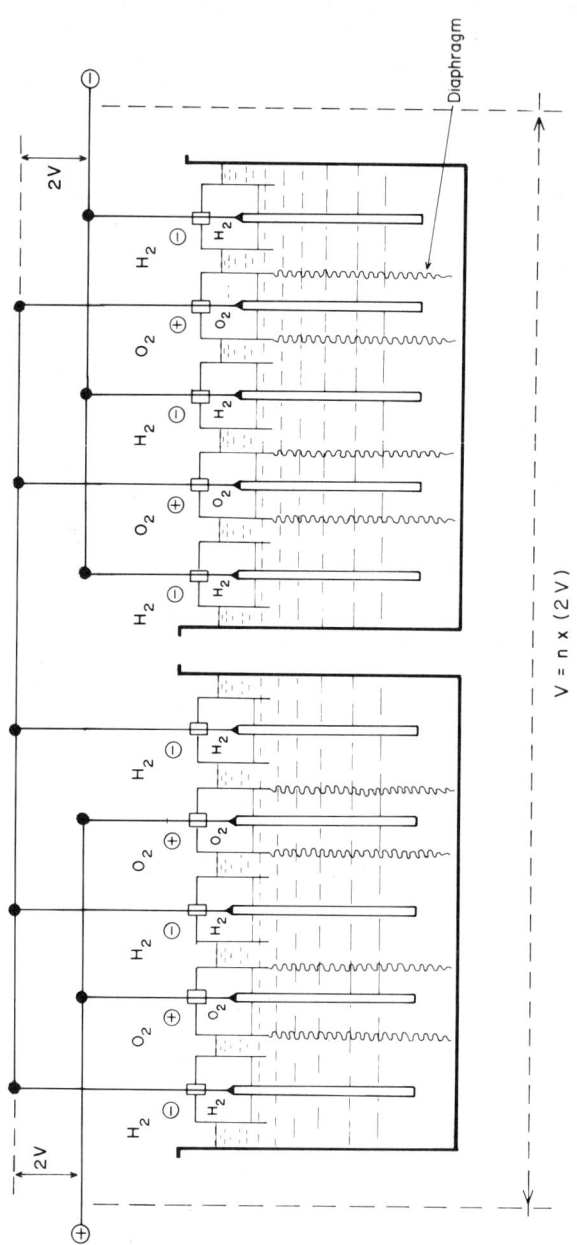

Figure 2. Unipolar tank-type cell construction. ($n$ refers to the number of unit cells).

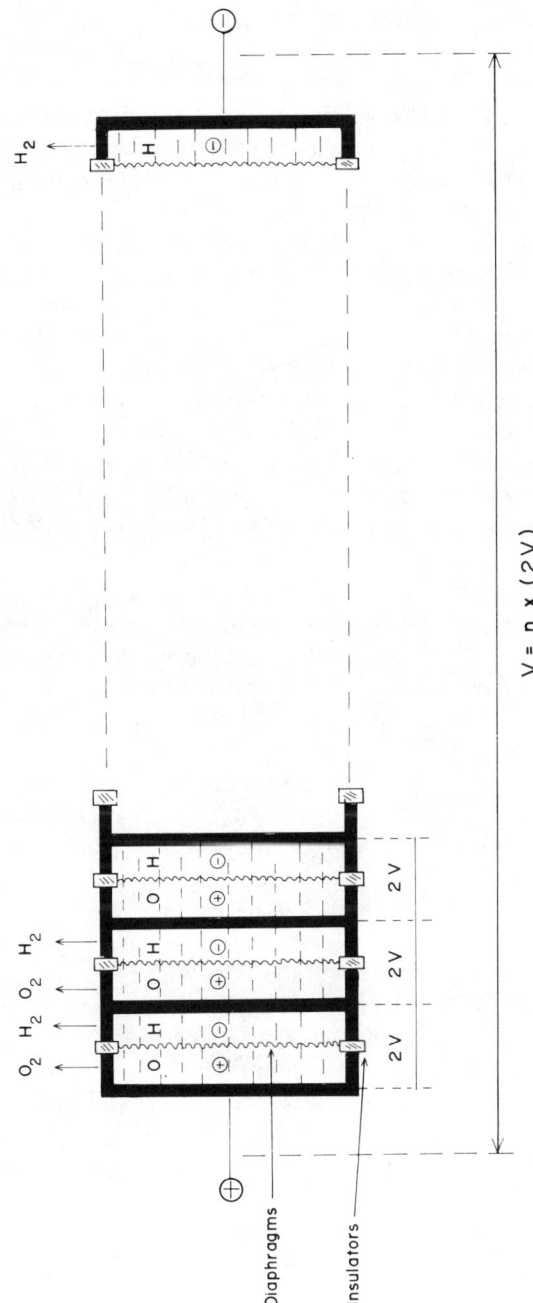

Figure 3. Bipolar filter-press-type cell construction. ($n$ refers to the number of unit cells.)

bipolar cell are in fact monopolar. (A more suitable term for bipolar cells might be multipolar cells.) As in monopolar cells, the electrode pack is normally a number of vertical parallel plates. Thus a large number of electrodes can be placed side by side, with diaphragms in between to separate the gases, and with insulating frames between the electrodes in a filter press type of assembly. The assembly is held together by several longitudinal tie bolts, which serve to press the insulating frames complete with electrodes, diaphragms, and gaskets together, to prevent liquor leaks. Each electrode is therefore electrically in series with (through the electrolyte), but insulated from its neighbors.

The electrical current therefore flows in series through a number of sandwiches of bipolar electrode, anolyte, diaphragm, catholyte, bipolar electrode, etc., with electrolyte being pumped to each individual anolyte or catholyte chamber, and a mixture or slurry of gas and electrolyte leaving each anolyte or catholyte chamber. Current therefore flows from one end of the filter press assembly to the other end. Since there may be up to several hundred intermediate bipolar electrodes, with each pair of electrodes requiring about 2 V, in a filter-press-type assembly, the overall voltage may be quite high. Compared to monopolar cells, bipolar cells are of comparatively high voltage with a low current.

In order to obtain the desired hydrogen and/or oxygen output, the required number of modules of monopolar cells can be arranged in series electrically (see Figure 4), or the required number of modules of bipolar cells can be arranged in electrical series–parallel (see Figure 5), depending upon the system's current and voltage, so as to match the output of a suitable rectifier or rectifiers.

The advantages and disadvantages of monopolar and bipolar cells are summarized in Table 1. The advantages of a monopolar cell can be reduced to the fact that it is a simple tank, rugged, has few parts, and is therefore easily manufactured and maintained. The disadvantages can be reduced to one simple statement, a relatively high voltage compared to bipolar cells. The relatively high voltage is a result of long current paths from the terminals on top of the cell tank to the electrodes, and the large interelectrode gaps.

The advantages of a bipolar cell essentially resolve the major disadvantage of the monopolar cell. Bipolar cells are noted for their relatively low unit cell voltages, resulting in lower power consumptions per unit of gas produced, even at relatively high current densities. Low voltages are due solely to very short

Figure 4. Electrical arrangement (series) of monopolar cells.

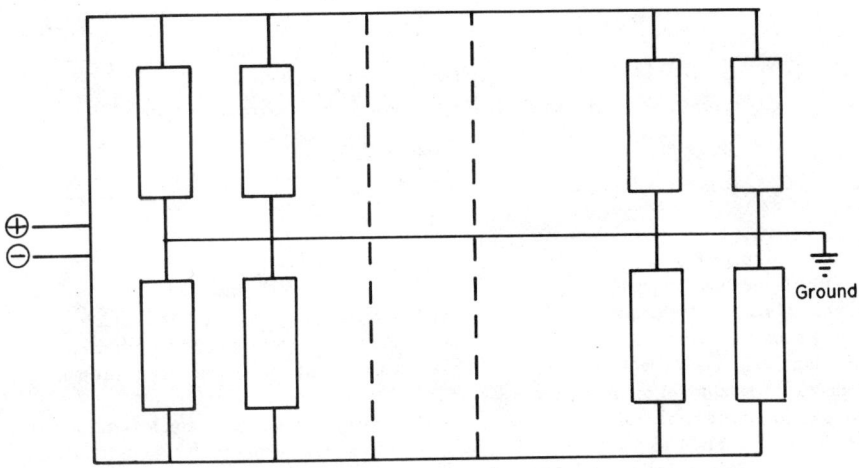

Figure 5. Electrical arrangement (series–parallel) of bipolar cells.

current paths in the electrodes, and the ease with which very narrow inter-electrode gaps are achieved. The disadvantages of a bipolar cell are due to the relative complexity of the system, the accurately machined parts, numerous gasketed joints, pumps and filters, and relatively complex fabrication and maintenance techniques.

The historical result of these advantages and disadvantages is that all early water electrolysis cells were of the simple tank variety. However, in recent years as power costs have increased, and as modern technologies have developed and decreased in cost, particularly plastic technologies, the bipolar cells have steadily replaced most of the monopolar cells. All commerical cells are now of the bipolar type with only one notable exception (see Section 3 for details), and use steel, stainless steel, or nickel cathodes, and nickel anodes in 25–35% KOH and operate in the temperature range 60–90°C. Electrolytic hydrogen is generated at 100–300 mA cm$^{-2}$ at cell voltages of 1.9–2.2 V. This corresponds to an overall energy efficiency of 70–80%, after consideration of energy requirements of auxiliaries.

Electrolysis of water to produce hydrogen and oxygen at high pressures is often essential. From an electrochemical point of view, the cell performance is improved under these conditions because of the reduction of the size of the gas bubbles which results in lowered cell voltage. Furthermore, to minimize activation overpotentials at both electrodes, it is necessary to increase the operating temperature above 100°C and consequently pressurization of the cells is essential. On the economical side, operation of an electrolyzer under pressure can eliminate the need for a compressor. (It may be noted that most hydrogenation processes (e.g., production of ammonia), are carried out at elevated pressures.) The only large-scale commercial pressure water electrolyzer is manufactured by Lurgi.[1]

Table 1
Comparison of Monopolar and Bipolar Hydrogen–Oxygen Cells

| Monopolar | Bipolar |
|---|---|
| *Advantages* | |
| Simple and rugged design. | Lower unit cell voltages. |
| Relatively inexpensive parts. | Higher current densities. |
| Simple fabrication techniques. | Intercell busbars greatly reduced. |
| Few gasketed surfaces. | Rectifier costs more easily optimized. |
| Individual cells easily checked. | Can readily operate at higher pressures and temperatures. |
| Cells easily isolated for maintenance. | Pressure operation eliminates compressors. |
| No parasitic currents in system. | Easier to control entire system for temperature and electrolyte level. |
| Minimum disruption to production (say, by single cell failure) for maintenance problems. | Fewer spare parts required. |
| Cells easily maintained on site. | Individual cell frames can be very thin, thus providing a large gas output from a small piece of equipment. |
| No pumps or filters required. | Fallout from military and aerospace programs in fuel cells as well as hydrogen oxygen generation has greatly assisted bipolar cell development. |
| Simple internal gas lift circulation. | Mass production of plastic cell components could result in lower capital costs. |
| | Potential to operate at very high current densities. |
| | Electrical arrangement of electrolyzers can allow a ground potential where the gases and electrolyte leave the system, or electrolyte enters the system. |
| *Disadvantages* | |
| Difficult to achieve small interelectrode gaps. | Sophisticated manufacturing and design techniques required. |
| Heavy intercell busbars. | Parasitic currents lower current efficiency. |
| Inherently higher power consumption from potential drop in cell hardware. | External pumping, filtration, cooling, and gas disengaging system required. |
| Cell pressures and temperatures limited by mechanical design. | Malfunction of a unit cell difficult to locate. |
| Each cell requires operator attention for temperature, electrolyte level, and gas purity. | Repair to a unit cell requires entire electrolyzer to be dismantled (in practice). |
| Sludge and corrosion products collect within cell. | Higher disruption to production for maintenance problems. |

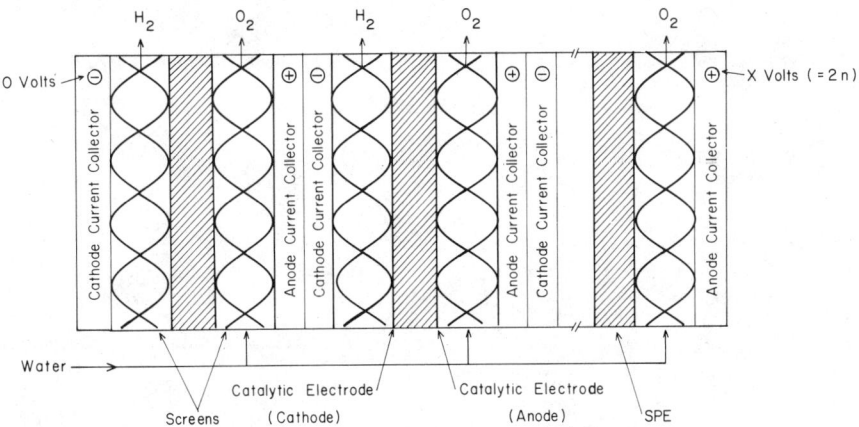

Figure 6. Design of multicell for "solid polymer electrolyte water electrolyzer." ($n$ refers to the number of unit cells.)

A novel and unconventional type of water electrolyzer is being developed at the General Electric Company.[2] With the technology achieved in the Gemini hydrogen–oxygen fuel cell, General Electric Company sought to carry out the reverse reaction (i.e., water electrolysis) in a cell of similar design. In this cell, a Nafion membrane is used as the electrolyte layer. It is essentially a sulfonated Teflon, and is a proton conductor when the membrane is in a wetted acid-form condition. Platinum particles, impregnated on one side of the membrane together with carbon, serve as the cathode electrocatalyst. The anode contains a proprietary electrocatalyst, similarly impregnated. The cells are arranged in a bipolar arrangement (Figure 6).

Section 3 contains detailed descriptions of the present status of alkaline water electrolyzers and advances being made to improve this technology from the points of view of lowering capital costs and approaching 100% efficiency (thermal). Progress being made in the development of the solid polymer electrolyte water electrolysis technology is presented in Section 4. Economics and some advanced concepts for electrolytic production of hydrogen are discussed in Sections 5 and 6, respectively.

One of the potential by-products during electrolytic generation of hydrogen is heavy water. Recent developments in this related area are outlined in Section 7.

### 2.2. Mass and Energy Balance for a Water Electrolyzer

The components of a mass and energy balance for a simple water electrolyzer are illustrated in the process flowsheet in Figure 7 (note in the following text the term *electrolyzer* includes the electrolytic cell and the

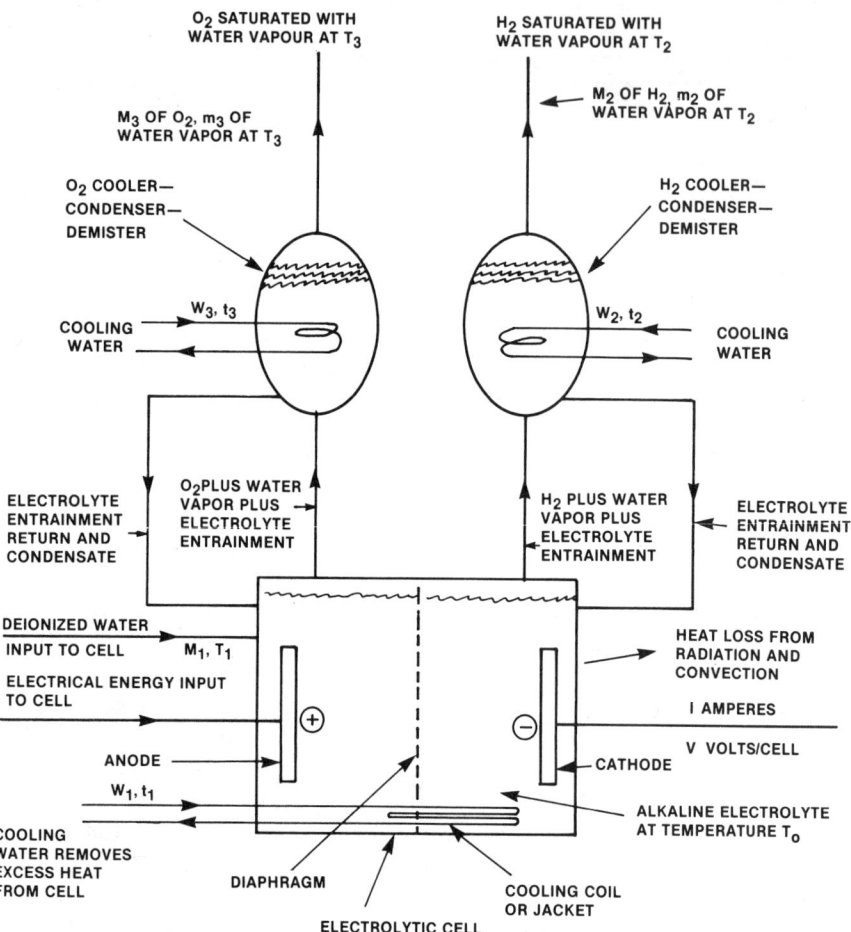

Figure 7. Mass and energy balance for a simplified $H_2$–$O_2$ cell. $W$ is the weight of cooling water flowing at temperature $t$, $M$ and $m$ the weight of water entering the cell or water and gas leaving the cell, $T$ the temperature of the electrolyte in the cell or of the fluids or gases entering or leaving the cell, $I$ the current entering the cell, and $V$ the voltage across the cell.

cooler–condenser–demisters, as depicted in Figure 7). Although a simple water electrolyzer such as a monopolar cell is illustrated, the components required for a more complex multipolar cell are very similar.

In order to appreciate the components of a mass and energy balance, it is essential to understand the functions of each unit process in the entire process. The electrolytic cell produces a steady stream of hydrogen and oxygen gas bubbles evolving from the cathode and anode, respectively, by the decomposition of water:

$$2H_2O \rightleftharpoons 2H_2 + O_2 \qquad (1)$$

## ELECTROLYTIC PRODUCTION OF HYDROGEN

As noted later (see Section 3 for details), the decomposition of water at sensible rates of reaction and at reasonable temperatures requires more energy input to the cell than is thermodynamically required. Thus, for the mass balance, the electrolytic cell requires a continuous input of deionized water to equate the water that is decomposed to hydrogen and oxygen, and the water vapor present in these gases that are saturated with respect to water at the prevailing exit gas temperature and pressure of the entire system. For an energy balance, the input electrical energy to the electrolyzer must equal the energy leaving in the form of hydrogen and oxygen gas, the water vapor in these gases, the energy lost from the cell because of radiation and convective cooling, and the excess energy leaving via the water in the cooling coils or jackets.

Deionized water is essential for water electrolyzers. Since the electrolyzer is a closed system, impurities such as chlorides, calcium, and magnesium must be removed from the water to prevent these from accumulating and causing problems during operation.

The hydrogen and oxygen gases rise as a steady stream of very small bubbles in the electrolytic cells. As these bubbles break free from the surface of the electrolyte, they carry some amount of entrained electrolyte, and are also saturated with respect to water at the prevailing operating temperature and pressure in the electrolytic cell. Therefore, both gas streams pass through a cooler–condenser–demister, all functions often being performed in a single vessel. Thus, for the mass balance, the gases entering the cooler–condenser–demister have the entrained electrolyte removed by the demister, and most of the water vapor condensed by cooling. The entrained electrolyte and condensed water vapor are returned to the electrolytic cell. For an energy balance, the enthalpy of the inlet gases to the cooler–condenser–demister must equal the enthalpy of the exit gases plus the enthalpy of liquid returned to the cell, plus heat removed in the cooling water. The gases leaving the cooler condenser–demister are of course saturated with water vapor at the prevailing temperature and pressure of these gases. The relationships for a mass balance of the electrolyzer may be expressed as follows:

Weight of deionized water, $M_1$
  = weight of oxygen gas $M_3$, and hydrogen gas $M_2$, plus weight of water vapor, $m_3 + m_2$, to saturate these gases at temperatures $T_3$ and $T_2$, respectively, leaving the electrolyzer system,
  = (weight of oxygen and hydrogen gas plus weight of water vapor calculated from the concentration of the electrolyte at the cell temperature $T_0$ plus entrained electrolyte) − (weight of condensed water vapor, plus entrained electrolyte returned to the electrolytic cell).

The weights of water vapor to saturate the hydrogen and oxygen gas can be calculated from vapor pressure curves of water (for gases leaving the cooler–condenser–demister) and of alkaline solutions for concentration of alkali employed (for gases leaving the electrolytic cell). The weight of oxygen and hydrogen can be calculated from the input current to the electrolytic cell and Faraday's law. The energy balance for the active electrolyzer may be expressed as follows:

Electrical energy input to cell = $V \times I$
 = Energy to decompose water to hydrogen and oxygen gas at temperature $T_2$ and $T_3$, respectively, plus enthalpy of water vapor $m_3$ and $m_2$, saturating these gases at temperatures $T_2$ and $T_3$, plus heat loss from the electrolytic cell by radiation and convection, plus heat leaving in the cooling coils or jackets of the electrolytic cell and the cooler–condenser–demister, less enthalpy of the entering deionized water.

Similar formulations can be written exclusively for the electrolytic cell and the cooler–condenser–demisters.

The energy to decompose water can be calculated from the heat of formation of water from hydrogen and oxygen gas with appropriate corrections for the actual operating temperature and pressure of the electrolytic cell. The enthalpy of water vapor saturating the gases can be obtained from a standard text on the thermodynamic properties of steam. The heat loss from the cell by convection and radiation can be calculated by techniques described in Reference 3.

### 2.3. Typical Process Flow Sheet for a Water Electrolysis Plant

A typical process flow sheet illustrating the essential components of a water electrolysis plant is shown in Figure 8.

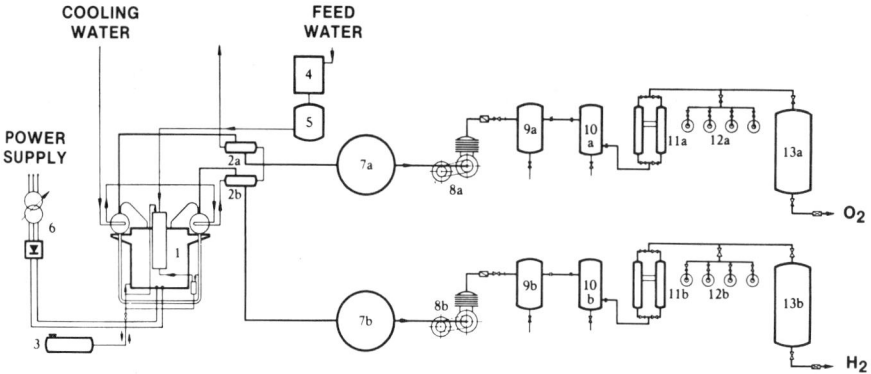

Figure 8. Process flow sheet of a water electrolysis plant. 1, electrolyzer; 2a, $O_2$ cooler and scrubber; 2b, $H_2$ cooler and scrubber; 3, electrolyte storage tank; 4, feedwater deionizer; 5, feedwater tank; 6, electric power transformer rectifier system; 7a, $O_2$ gasometer; 7b, $H_2$ gasometer; 8a, $O_2$ compressor; 8b, $H_2$ compressor; 9a, intermediate storage tank for $O_2$; 9b, intermediate storage tank for $H_2$; 10a, purification unit for $O_2$; 10b, purification unit for $H_2$; 11a, $O_2$ drier; 11b, $H_2$ drier; 12a, $O_2$ cylinder filling station; 12b, $H_2$ cylinder filling station; 13a, $O_2$ storage tank; 13b, $H_2$ storage tank.

The major component in the process is the electrolytic cell, item 1 in Figure 8. Note that a Brown Boveri bipolar cell is shown in Figure 8, whereas the mass and energy balance of Figure 7 illustrates a typical tank-type monopolar cell (see Table 1 for the differences between monopolar and bipolar cells). The major flow sheet differences between the systems of Figures 7 and 8 are that the circulation of electrolyte in the latter case is by a pump, and cooling of the electrolyte is by a cooler external to the electrolytic cell.

Deionized water is fed continuously to the electrolytic cell from the deionizer, item 4, and the feed-water storage tank, item 5. A makeup supply of alkali electrolyte is stored in the electrolyte storage tank, item 3. Electric power enters the line or stack of electrolytic cells from the electric power transformer and rectifier system, item 6. The transformer and rectifier system has suitable control functions to adjust the current to the cells and hence, the gas production rate. Water—to remove excess heat from the system, to cool the gases, and remove most of the water vapor from the gases—is fed into the various separators and coolers.

A mixture of saturated gas and electrolyte leaves the cell and is separated and cooled in two vessels along the top of the electrolytic cell, item 1. From these two vessels the gases enter oxygen and hydrogen coolers and scrubbers, items 2a and 2b, respectively. These coolers and scrubbers further reduce the water vapor content of the gases and remove any entrained electrolyte. From the oxygen and hydrogen scrubbers and coolers, the cool and relatively dry gases pass onto surge chambers, the oxygen and hydrogen gasometers, items 7a and 7b. If the oxygen is not required as a product, it can be vented to atmosphere after the oxygen cooler and scrubber, and all subsequent equipment, items 7a, 8a, 9a, 10a, 11a, 12a, and 13a, omitted.

From the gasometers, items 7a and 7b, the gas is compressed by compressors, items 8a and 8b, and stored in high-pressure intermediate storage tanks, items 9a and 9b. The following operations are required to purify the gases in purification units, 10a and 10b (note: the purification unit for hydrogen may be a catalytic deoxifier), and to dry the gases free of water vapor in units, items 11a and 11b. Finally, the gases may be bottled at the cylinder filling stations, items 12a and 12b, or stored in the final high-pressure storage tanks, items 13a and 13b.

## 3. Alkaline Water Electrolyzers

### 3.1. Thermodynamics

Hydrogen is produced by the electrolysis of aqueous KOH (usually 25–30 wt %) solutions and can be represented by the overall reaction

$$H_2O \rightarrow \tfrac{1}{2}O_2 + H_2 \tag{2}$$

which requires 2 faradays of electricity to produce 1 g mol of $H_2$ at 0°C and 760 torr (or 6.95 cm$^3$ of $H_2$/A min or 15.6 SCF/1000 A hr, at 68°F and 14.7 psi, assuming the current efficiency to be 100%). The component reactions describing the overall reaction (2) are the discharge of $H_2$ at the cathode and $O_2$ at the anodes as shown through the following equations:

$$2H_2O + 2e^- \rightarrow H_2 + 2OH^- \tag{3}$$

$$2OH^- \rightarrow O_2 + 2H^+ + 4e^- \tag{4}$$

The thermodynamic decomposition voltage corresponding to the Gibbs free energy change of reaction (2) of 56.69 kcal (g mol)$^{-1}$ [4-8] (at 25°C, 1 atm pressure) is 1.229 V. Reaction (2) is endothermic—the heat of the reaction ($\Delta H$) being 68.32 kcal (g mol)$^{-1}$ at 25°C and 1 atm pressure. This difference between the free energy change and enthalpy change arises from the entropy changes in the overall process and must be balanced by either supplying or removing heat from the system (and not by conversion to electricity). Thus, below a cell voltage of 1.48 V (but above 1.23 V), corresponding to $\Delta H/nF$, the electrolysis cell would act as a refrigerator absorbing heat from the surroundings, and above 1.48 V, heat is generated and must be removed, for an isothermal operation of the cells. In fact, the thermoneutral voltage ($E_{tn}$) defined as[7]

$$E_{tn} = \frac{\Delta H}{nF} \tag{5}$$

for the cell will be some value greater than 1.48 V, because of the requirement that heat be supplied to raise the temperature of the feed-water to that of the electrolyte, and to evaporate water which is carried off with the product gases.[9] It may be noted that all practical cells operate above a cell voltage of 1.48 V owing to the various losses in the cell, and hence heat removal becomes a key engineering feature in designing water electrolysis cells.

The heat losses reflect energy losses in the electrolysis system, and they may be quantitatively described by comparing the cell voltage with the higher-heating-value voltage $V_{HHV}$ defined[9] as

$$V_{HHV} = [H(H_2)_T + H(O_2)_T - H(H_2O)_{25°C}]/nF \tag{6}$$

where $H$ refers to the enthalpy of the species noted in parenthesis. Thus, a water electrolysis cell operating at 2.5 V, for example at 25°C, has an energy efficiency of $(1.48/2.5) \times 100 = 59.2\%$. The energy efficiency can be improved by operating the cells at higher temperatures, since the theoretical minimum of electrical energy (i.e., the thermodynamic decomposition voltage $E_{rev}$) is lower at high temperatures, while the temperature dependence of the higher-heating-value voltage is small and $E_{tn}$ changes[8] only to a small extent from 1.48 V at 25°C to 1.49 V at 1000°C (1832°F). Furthermore, the kinetics

# ELECTROLYTIC PRODUCTION OF HYDROGEN

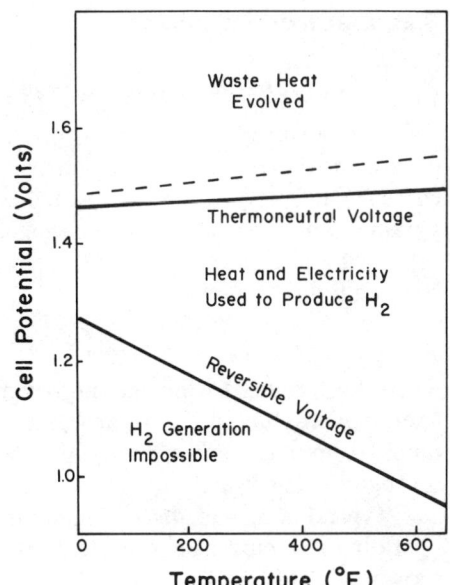

Figure 9. Idealized operating conditions for water electrolysis. Solid line for thermoneutral voltage was calculated using Eq. (5) and the dotted line for the higher-heating-value voltage using Eq. (8).

of the electrode reactions and of ion transport are accelerated at elevated temperatures.

The temperature dependence of $E_{rev}$ and $V_{HHV}$ may be calculated from Eqs. (6) and (7), with results presented in Figure 9. $E_{rev}$ varies from 1.229 V at 25°C (77°F) to 1.040 V at 264°C (507.2°F). The high-heating-value voltage shows only a relatively small temperature coefficient, increasing from 1.481 V at 25°C to 1.536 V at 264°C:

$$E_{rev} = 1.5184 - 1.5421 \times 10^{-3} T$$
$$+ 9.523 \times 10^{-5} T \ln T + 9.84 \times 10^{-8} T^2 \tag{7}$$

$$V_{HHV} = 1.415 + 2.169 \times 10^{-4} T + 1.52 \times 10^{-8} T^2 \tag{8}$$

$T$ in Eqs. (6) and (7) is expressed in degrees Kelvin. Equation (8) can be derived by substituting Eq. (9) (describing $\Delta H$ variations with temperature) into Eq. (10), which takes into account the energy requirements involved in heating the feed-water from its temperature ($T_f$) to $T$.

$$\frac{\Delta H_T}{nF} = 1.5475 - 3.2312 \times 10^{-4} T + 4.820 \times 10^{-7} T^2 - 4.82 \times 10^{-10} T^3 \tag{9}$$

$$V_{HHV} = \frac{\Delta H_T}{nF} + \frac{1}{nF} \int_{T_f}^{T} \bar{C}p(H_2O) \, dT \tag{10}$$

where $\bar{C}p(H_2O)$ refers to the heat capacity of water.

## 3.2. Electrode Kinetics

### 3.2.1. Components of Cell Voltage

The power consumption for operating water electrolysis cells is directly proportional to the cell voltage and inversely to the current efficiency. Since the current efficiency for alkaline water electrolysis is very close to 100%, the parameter of importance in evaluating various designs of the water electrolysis cells is the magnitude of the cell voltage, which is composed of the various contributions[10] shown by

$$E = E_{rev} + \eta_a + \eta_c + \eta_\Omega + \eta_{hw} \tag{11}$$

where $E_{rev}$ refers to the thermodynamic decomposition voltage, $\eta_a$ to the anodic overpotential, $\eta_c$ to the cathodic overpotential, $\eta_\Omega$ to the electrolyte ohmic drop between the anode and the cathode, and $\eta_{hw}$ to the ohmic drop through the hardware.

Typical values of the components of the water electrolysis cell voltage operating at a current density (c.d.) of 150 mA/cm² (140 A/ft²) at 75°C are shown in Table 2. These values of the components of cell voltage can be calculated for a given cell configuration from the various contributions shown in Eq. (11) as shown in the following sections.

### 3.2.2. Thermodynamic Decomposition Voltage

The variation of the reversible cell voltage can be calculated from the Nernst equation, Eq. (12), as[9]

$$E_{rev,T,P} = E_{rev,T,P=1} + \frac{RT}{nF} \ln \frac{a_{H_2}(a_{O_2})^{1/2}}{a_{H_2O}} \tag{12}$$

where $a$ denotes activity, $R$ the gas constant, and

$$E_{rev,T,P=1} = 1.5184 - 1.5423 \times 10^{-3} T$$
$$+ 9.524 \times 10^{-5} T \ln T + 9.84 \times 10^{-8} T^2 \tag{13}$$

[valid between 298 K (25°C) and 528 K (250°C)].

Table 2
Typical Components of Water Electrolyzer Cell Voltage
(c.d., 150 mA/cm²; Temperature, 75°C)

| | |
|---|---|
| $E_0$ | = 1.19 V |
| $\eta_a$ | = 0.30 V |
| $\eta_c$ | = 0.30 V |
| $\eta_\Omega$ | = 0.25 V (anode–cathode gap ~ 4 mm) |
| $\eta_{hw}$ | = 0.11 V |
| Cell voltage | = 2.15 V |

**ELECTROLYTIC PRODUCTION OF HYDROGEN**

Substituting $P - P_{H_2O}$, where $P_{H_2O}$ refers to the vapor pressure of water over the electrolyte, for the activities of $H_2$ and $O_2$, and the ratio of $P_{H_2O}$ to the vapor pressure of pure $H_2O$ ($P°_{H_2O}$) at temperature $T$ for the activity of $H_2O$ in Eq. (12) results[9] in

$$E_{rev,T,P} = E_{rev} + 4.309 \times 10^{-5} T \ln \frac{(P - P_{H_2O})^{1.5} P°_{H_2O}}{P_{H_2O}} \quad (14)$$

where

$$\ln P°_{H_2O} = 37.043 - \frac{6275.7}{T} - 3.4159 \ln T$$

$$\ln P_{H_2O} = 0.016214 - 0.13802\, m + 0.19330\, m^{1/2} + 1.0239 \ln P°_{H_2O}$$

and $m$ refers to electrolyte molality.

Thus, for 25% KOH ($m = 5.94$ mol per kg of solvent) at 363 K, the influence of pressure would result in correcting the $E_{rev}$ values by $-10$ mV at 1 atm pressure and by 113 mV at 100 atm.

### 3.2.3. Hydrogen Overvoltage

The cathode overpotential ($\eta_c$) contribution to the measured cell voltage may be expressed, in general, by the Tafel equation[11]:

$$\eta_c = 2.303 \frac{RT}{\alpha F} \log \frac{i}{i_0} \quad (15)$$

where $\alpha$ refers to the transfer coefficient and $i_0$ to the exchange c.d. of the reaction, which is a function of the nature of the electrode material. The mechanistic and electrocatalytic aspects of the hydrogen evolution reaction have been extensively investigated and discussed,[12–14] and the kinetic data pertinent to alkaline solutions have been compiled in Table 3.[15] These studies generally exhibited[16,17] periodic variations of hydrogen overpotential with atomic number, work function, etc., both in alkaline and acidic media—Ni, Pd, and Pt with $d^8s^2$, $d^{10}s^0$, and $d^9s^1$ electronic configurations, respectively, exhibiting minimum values and Zn, Cd, Hg with $d^{10}s^2$ electronic configuration showing maximum values (see Figure 10).[16]

Since nickel is a reasonably stable material in alkaline solutions (unlike Fe), and relatively inexpensive, extensive investigations[15–22] were carried out on various nickel-based materials (e.g., alloys, intermetallic compounds, interstitials, etc.). Polarization data on smooth nickel electrodes[8] (presented in Figure 11), show marked variation of the kinetic parameters (see Table 3) with temperature with $i_0$ changing from $2 \times 10^{-4}$ A/cm$^2$ at 80°C to $4 \times 10^{-2}$ A/cm$^2$ at 264°C and $\alpha = 0.5$. The changing Tafel slopes with overpotential and temperature suggest the electrochemical desorption mechanism (see Section 4.2.1 for details) to be operative on Ni in alkaline solutions at high

Table 3
Kinetic Parameters for the $H_2$ Evolution Reaction from Alkaline Solutions

| Metal | Concentration | Temperature, °C | $-\log i_0$, A/cm$^2$ | Tafel slope, mV | Activation energy, kcal/mol | Mechanism[a] | Reference[b] |
|---|---|---|---|---|---|---|---|
| Ag | 1 N NaOH | 30 | 6.5 | 120 | | | 1 |
| | 0.1 N NaOH | 25 | 6.5 | 122 | | | 2 |
| | 1.1 or 0.3 N NaOH | 21 | 5.35 | 43–109 | | | 2 |
| | 6 N NaOH | 25 | 6 | 120 | | | 29 |
| Al | 6 N NaOH | 25 | 4.52 | 140 | | | 29 |
| Au | 0.1 N NaOH | 25 | 5–7 | 71–120 | | | 3 |
| | 0.1 N NaOH | 25 | 5.4 | 120 | | | 4 |
| Sb | 6 N NaOH | 25 | 6 | 150 | | | 29 |
| Bi | 0.1 M $Na_2B_4O_7$ + 1 M NaOH + 1 M $HClO_4$ | 25 | 10.3 | 110 | | | 32 |
| Carbon | 40% NaOH | 40 | 4.534 | 148 | | | 5 |
| | 40% NaOH | 60 | 3.982 | 167 | | | 5 |
| | 40% NaOH | 80 | 3.856 | 160 | | | 5 |
| Electrokohle | 27% NaOH | 25 | 8.333 | 90 | | | 6 |
| Lichtenberg | 27% NaOH | 80 | 4.231 | 130 | | | 6 |
| Graphite | 4% NaOH | 20 | 5.522 | 186 | | | 7 |
| | 23.8% NaOH | 20 | 5.936 | 171 | | | 7 |
| | 4% NaOH | 80 | 5.965 | 142 | | | 7 |
| | 23.8% NaOH | 80 | 5.130 | 162 | | | 8 |
| | 40% NaOH | 60 | 4.071 | 140 | | | 8 |
| | 40% NaOH | 80 | 3.679 | 140 | | | 9 |
| | 44.3% NaOH | 20 | 4.541 | 148 | | | 10 |
| | 10% NaOH | 90 | 3.613 | 155 | | | 10 |
| | 20% NaOH | 90 | 4.45 | 100 | | | 10 |
| | 20% NaOH | 90 | 1.86 | 285 | | | 10 |
| | 30% NaOH | 90 | 4.727 | 110 | | | 10 |

# ELECTROLYTIC PRODUCTION OF HYDROGEN

| | | | | | | |
|---|---|---|---|---|---|---|
| Electrokohle | 40% NaOH | 90 | 5.053 | 95 | | 10 |
| | 50% NaOH | 90 | 5.053 | 95 | | 10 |
| | 22% NaOH | 20 | 4.303 | 138 | | 11 |
| | 22% NaOH | 50 | 4.225 | 124 | 3.93 | 11 |
| | 22% NaOH | 90 | 3.743 | 119 | | 11 |
| | 43% NaOH | 70 | 3.616 | 122 | | 11 |
| | 43% NaOH | 90 | 3.519 | 117 | 2.77 | 11 |
| Tokoku carbon | 20.09% NaOH | 60 | 5.67 | 118 | | 11 |
| | 20.09% NaOH | 70 | 5.621 | 118 | 3.26 | 11 |
| | 20.09% NaOH | 80 | 5.549 | 118 | | 11 |
| | 37.4% NaOH | 60 | 5.336 | 119 | | 11 |
| | 37.4% NaOH | 70 | 5.145 | 119 | 7.13 | 11 |
| | 37.4% NaOH | 80 | 5.073 | 119 | | 11 |
| | 43.2% NaOH | 70 | 3.745 | 149 | 10.8 | 11 |
| | 43.2% NaOH | 80 | 3.550 | 151 | | 11 |
| | 43.2% NaOH | 90 | 3.327 | 150 | 13.1 | 11 |
| Société carbon | 43.8% NaOH | 70 | 4.121 | 140 | 24.5 | 11 |
| | 43.8% NaOH | 80 | 3.68 | 153 | | 11 |
| | 43.8% NaOH | 90 | 3.475 | 162 | 12.0 | 11 |
| Nippon carbon | 42.1% NaOH | 70 | 3.020 | 149 | 10.8 | 11 |
| | 42.1% NaOH | 80 | 2.826 | 155 | | 11 |
| Co | 0.1 M KOH | 25 | 1.87 | 186 | Single crystals | 31 |
| | 6 N NaOH | 25 | 4.3 | 140 | | 29 |
| | 1 N NaOH | 25 | 2.26 | 145 | | 30 |
| Cu | 0.01 N NaOH | 38 | 6–7 | 114 | | 3 |
| | 0.1 N NaOH | 25 | 7 | 120 | | 4 |
| | 6 N NaOH | 25 | 5.3 | 160 | | 29 |
| Cd | 6 N NaOH | 25 | 6.398 | 160 | | 29 |
| Cr | 6 N NaOH | 25 | 7 | 120 | | 29 |
| Fe | 0.1 N NaOH | 25 | 6 | 120 | | 3 |
| | 0.1 N NaOH | 25 | 6.046 | 120 | | 12 |
| | 0.1 N NaOH | 25 | 5.82 | 120 | | 4 |
| | 6 N NaOH | 25 | 4 | 150 | | 29 |

Table 3 *(continued)*

| Metal | Concentration | Temperature, °C | $-\log i_0$, A/cm² | Tafel slope, mV | Activation energy, kcal/mol | Mechanism[a] | Reference[b] |
|---|---|---|---|---|---|---|---|
| Mg | 6 N NaOH | 25 | 5.523 | 350 | | | 29 |
| Mo | 0.1 N NaOH | 25 | 6–7 | 80–116 | | | 3 |
|    | 6 N NaOH | 25 | 5 | 140 | | | 29 |
| Mn | 6 N NaOH | 25 | 7.52 | 120 | | | 29 |
| Nb | 1 N NaOH | 25 | 7.5 | 140 | | | 13 |
|    | 1 N NaOH | 25 | 6.2 | 140 | | | 14 |
| Ni | 0.1 N NaOH | 20 | 6.4 | 101 | | | 15 |
|    | 2 N NaOH | 26–28 | 4.9 | 100 | | | 16 |
|    | 0.1 N NaOH | 30 | 5.3 | 89 | | | 17 |
|    | 4 N NaOH | 25 | 5 | 89–93 | | | 18 |
|    | 0.5 N NaOH | 25 | 6.1 | 96 | | 2 | 19 |
|    | 1.0 N NaOH | 25 | 5.1 | 90 | | | 20 |
|    | 0.1 N NaOH | 25 | 5 | 92 | | | 12 |
|    | 0.1 N NaOH | 25 | 5.35 | 130 | | 2 | 4 |
|    | 6 N NaOH | 25 | 5 | 140 | | | 29 |
|    | 0.1 N KOH | 0–20 | | | 9–12 | | 33 |
|    | (High-surface-area materials 0.1–30 m²/g) | | | | | | |
|    | 50% KOH | 80 | 3.959 | 140 | | 8–14 kcal | 21 |
|    | 50% KOH | 150 | 3.745   2.022 | 54   280 | | | 21 |
|    | 50% KOH | 208 | 3.097   1.699 | 70   320 | | 3 | 21 |
|    | 50% KOH | 264 | 3.032   1.699 | 66   200 | | | 21 |
| Pb | 0.5 N NaOH | 25 | 6.47 | 129 | | 5 | 19 |
|    | 6 N NaOH | 25 | 5.398 | 250 | | | 29 |
| Pd | 0.1 N NaOH | 25 | 5.0 | 125 | | 1 | 3 |
|    | 0.1 N NaOH | 25 | 3.60 | 105 | | | 4 |
| Pt | 0.1 N NaOH | 25 | 4.17 | 114 | | | 22 |

# ELECTROLYTIC PRODUCTION OF HYDROGEN

| Material | Solution | Temp | a | b | i | — | Ref |
|---|---|---|---|---|---|---|---|
| Ir | 0.5 N NaOH | 25 | 4.06 | 117 | | | 19 |
| | 0.1 N NaOH | 25 | 3.4 | 105 | | 3 | 4 |
| Rh | 0.1 N NaOH | 25 | 3.26 | 125 | | 3 | 4 |
| | 0.1 N NaOH | 25 | 4.0 | 118 | | | 4 |
| | 0.01 N NaOH | 25 | 4.85 | 120 | | | 3 |
| Sn | 6 N NaOH | 25 | 5.52 | 230 | | | 29 |
| Ta | 6 N NaOH | 25 | 6.523 | 150 | | | 29 |
| Ti | 6 N NaOH | 25 | 6 | 140 | | | 29 |
| W | 0.5 NaOH | 25 | 6.6 / 7.53 | 80 / 100 | | 2 | 19 |
| | 0.1 M NaOH | 30 | 7.66 / 6.29 | 58 / 93 | | | 23 |
| | 0.5 M NaOH | 30 | 7.34 / 6.22 | 61 / 93 | | | 23 |
| | 1.0 M NaOH | 30 | 7.299 / 6.14 | 62 / 95 | | | 23 |
| | 1.0 M NaOH | 30 | 6.6 / 5.56 | 59 / 98 | | | 23 |
| Zn | 6 N NaOH | 25 | 6.398 | 210 | | | 29 |
| Zr | 6 N NaOH | 25 | 6.523 | 220 | | | 29 |

## Alloys and compounds

| Material | Solution | Temp | a | b | i | — | Ref |
|---|---|---|---|---|---|---|---|
| Au–0.05% Cu | 0.05 N NaOH | 20 | 8.229 | 6 | 53 | 122 | 24 |
| | | 30 | 7.638 | 5.854 | 55 | 115 | |
| | | 38 | 7.301 | 5.699 | 53 | 104 | |
| | | 48 | 6.854 | 5.495 | 52 | 120 | |
| | 0.10 N NaOH | 20 | 7.699 | 5.745 | 52 | 120 | 24 |
| | | 30 | 7.444 | 5.553 | 53 | 120 | |
| | | 38 | 7.149 | 5.398 | 55 | 122 | |
| | | 48 | 6.886 | 5.276 | 48 | 110 | |
| Stainless steel 304 | 0.1 M NaOH | 25 | 5.37 | 134 | | | 12 |
| | 1.0 M NaOH | 25 | 5.09 | 133 | | | |
| | 3.0 M NaOH | 25 | 5.14 | 121 | | | |
| | 5.0 M NaOH | 25 | 5.71 | 115 | | | |
| | 10 M NaOH | 25 | 5.38 | 105 | | | |
| | 6 M NaOH | 25 | 5 | 120 | | | |
| Ni | 0.1 N NaOH | 25 | 5.4 | 130 | | | 29 |
| Ni–12 at % Pd | 0.1 N NaOH | 25 | 4.82 | 115 | | 2 | 29 |
| Ni–35 at % Pd | 0.1 N NaOH | 25 | 4.5 | 130 | | 3 | 4 |
| Ni–68 at % Pd | 0.1 N NaOH | 25 | 3.49 | 130 | | | 4 |

Table 3 (continued)

| Metal | Concentration | Temperature, °C | $-\log i_0$, A/cm$^2$ | Tafel slope, mV | Activation energy, kcal/mol | Mechanism[a] | Reference[b] |
|---|---|---|---|---|---|---|---|
| Pd | 0.1 N NaOH | 25 | 3.46 | 110 | | | 4 |
| Ni | 0.1 N NaOH | 25 | 5.35 | 130 | | | |
| Ni–25 at % Pt | 0.1 N NaOH | 25 | 5.0 | 155 | | | |
| Ni–50 at % Pt | 0.1 N NaOH | 25 | 4.02 | 170 | | | |
| Ni–63 at % Pt | 0.1 N NaOH | 25 | 3.89 | 120 | | | |
| Pt | 0.1 N NaOH | 25 | 3.40 | 120 | | | |
| Pt–< 5% V | 0.1 N NaOH | 25 | 5.0 | 158 | | | 4 |
| Pt–< 5% Mo | 0.1 N NaOH | 25 | 5.75 | 165 | | | 4 |
| TiC | 0.1 N NaOH | 25 | 6.52 | 140 | | | 4 |
| TaC | 0.1 N NaOH | 25 | 4.7 | 300 | | | 4 |
| B$_4$C | 0.1 N NaOH | 25 | 5.7 | 135 | | | 4 |
| TiC(97%) + Ni(3%) | 42.2% NaOH | 70 | 2.667 | 141 | 7.16 | | 11 |
| | 42.2% NaOH | 80 | 2.538 | 145 | | | |
| | 42.2% NaOH | 90 | 2.359 | 153 | 10.5 | | |
| Fe | 0.1 N NaOH | 25 | 5.82 | 120 | | | 4 |
| Fe–75% Ni | 0.1 N NaOH | 25 | 5.46 | 115 | | | 4 |
| Ni | 0.1 N NaOH | 25 | 5.35 | 130 | | | 4 |
| AlSb | 0.1 N NaOH | 25 | 1.76 | 215 | | Extrapolated from Ref. 25 | |
| TiAl$_3$ | 0.1 N NaOH | 25 | 3.2 | 110 | | | |
| FeAl$_3$ | 0.1 N NaOH | 25 | 0.8 | 130 | | | |
| CuAl$_2$ | 0.1 N NaOH | 25 | 1.34 | 200 | | | |
| NiAl$_3$ | 0.1 N NaOH | 25 | 0.68 | 90 | | | |
| Zn | 9 N KOH | 25 | 8.824 | 124 | | | 26 |
| Zn–2% Hg | 9 N KOH | 25 | 9.569 | 116 | | | |
| –0.2% Pb | 9 N KOH | 25 | 10.097 | 98 | | | |
| Zn–0.8% Pb | 9 N KOH | 25 | 11.222 | 86 | | | |
| –0.05% Fe | 9 N KOH | 25 | 7.155 | 158 | | | 26 |

| | | | | | |
|---|---|---|---|---|---|
| Zn-8% Cd | 9 N KOH | 25 | 7.824 | 154 | |
| Zn-0.2% Pb | 9 N KOH | 25 | 8.699 | 137 | 26 |
| Zn-0.8% Pb | 9 N KOH | 25 | 8.886 | 134 | |
| Zn-2% Pb | 9 N KOH | 25 | 8.208 | 172 | |
| Zn-0.05% Mn | 9 N KOH | 25 | 7.292 | 138 | 26 |
| Zn-0.5% Mn | 9 N KOH | 25 | 7.125 | 140 | |
| Zn-2% Hg –0.2% Pb | 9 N KOH | 25 | 9.222 | 125 | 26 |
| Zn-0.8% Pb –0.05% Fe | 9 N KOH | 25 | 9.046 | 125 | 26 |

[a] Mechanism:
1. Slow discharge–fast electrochemical desorption
2. Slow discharge–fast recombination
3. Fast discharge–slow electrochemical desorption
4. Fast discharge–slow recombination
5. Coupled discharge–electrochemical desorption
6. Coupled discharge–recombination

[b] References:

1. M. A. V. Devanathan, J. O'M. Bockris, and W. Mehl, *J. Electroanal. Chem.* **1**, 143 (1959).
2. T. Yamazaki and H. Kita, *J. Res. Inst. Catalysis Hokkaido Univ.* **13**, 77 (1965).
3. N. Pentland, J. O'M. Bockris, and E. Sheldon, *J. Electrochem. Soc.* **104**, 182 (1957).
4. R. J. Mannan, Ph.D. thesis, University of Pennsylvania, 1967.
5. M. M. Jaskic, D. R. Jovanovic, and I. M. Csonka, *Electroch.m. Acta.* **13**, 2077 (1968).
6. G. I. Volkov, *Proisvodstvo, Chlora. i Kausticeskoj Sody Metodom Elektroliza S. Rtunym Katodom*, Chimijo, Moscow (1968), p. 87.
7. T. Eriksson, thesis, Royal technical University, Stockholm, 1968.
8. F. Hine, M. Okada, S. Yoshizawa, and S. J. Okada, *J. Electrochem. Soc. (Japan)* **27**, E46 (1959).
9. V. P. Pancesnaja, L. I. Struk, and V. G. Prichodcenko, *Porcskovaja Metallurgia* **54**, 21 (1967).
10. R. Burian (see Reference 11).
11. I. Roussar, J. Hostomsky, S. Rajasekharan, and V. Cezner, *Coll. Czech. Chem. Commun.* **39**, 1 (1974).
12. R. N. O'Brien and P. Seto, *J. Electrochem. Soc.* **117**, 32 (1974).
13. A. L. Rotinyan and N. M. Kozhevnikova, *Zh. Fiz. Khim.* **37**, 1818 (1963).
14. M. J. Joncich, L. S. Stewart, and F. A. Posey, *J. Electrochem. Soc.* **112**, 717 (1965).
15. J. O'M. Bockris and E. C. Potter, *J. Chem. Phys.* **20**, 614 (1952).
16. M. A. V. Devanathan and M. Selvaratnam, *Trans. Faraday Soc.* **56**, 1820 (1960).
17. A. C. Makrides, *J. Electrochem. Soc.* **109**, 977 (1962).
18. J. L. Weininger and M. W. Breiter, *J. Electrochem. Soc.* **111**, 707 (1964).

19. J. O'M. Bockris and S. Srinivasan, *Electrochim. Acta* **9**, 31 (1964).
20. A. Matsuda and T. Ohmori, *J. Res. Inst. Catalysis Hokkaido Univ.* **10**, 203 (1962).
21. M. H. Miles, G. Kissel, P. W. T. Lu, and S. Srinivasan, *J. Electrochem. Soc.* **123**, 332 (1976).
22. I. A. Ammar and S. Darwish, *J. Phys. Chem.* **63**, 983 (1959).
23. I. A. Ammar, S. Darwish, and R. Salim, *J. Electroanal. Chem.* **52**, 443 (1974).
24. I. A. Ammar and S. Riad, *J. Phys. Chem.* **62**, 660 (1958).
25. A. I. Golubev and M. N. Ronzhim, *Zashch Metal* **1**, 199 (1965).
26. T. S. Lee, *J. Electrochem. Soc.* **122**, 171 (1975).
27. I. B. Barmashenko, Ekh. Ignatenko, and I. P. Zakharchenko, *Zh. Prikl. Khim.* **38**, 2827 (1965). (Porous Ni–Fe electrodes in 2.7 N NaOH in the temperature range 20–80°C were studied. Kinetic data are not available and need extrapolation.)
28. I. B. Barmashenko and V. I. Shapaval, *J. Appl. Chem. USSR* **32**, 827 (1959). (Porous Fe–Ni electrodes in 15% NaOH at 20 and 80°C—kinetic data not available.)
29. M. D. Zholudev and V. V. Stender, *J. Appl. Chem. USSR* **31**, 711 (1958).
30. L. Peraldo Bicelli, C. Romagnani, and M. Rosania, *J. Electroanal. Chem.* **63**, 238 (1975).
31. I. V. Kudryashov, E. S. Burmistrov, and V. L. Kirlis, *Electrokkhimiya* **6**, 737 (1970).
32. D. E. Williams and G. A. Wright, *Electrochim. Acta* **21**, 851 (1976).
33. O. S. Abramzon, S. P. Chernyshov, and A. G. Pshenichnikov, *Sov. Electrochem.* **12**, 1520 (1976).

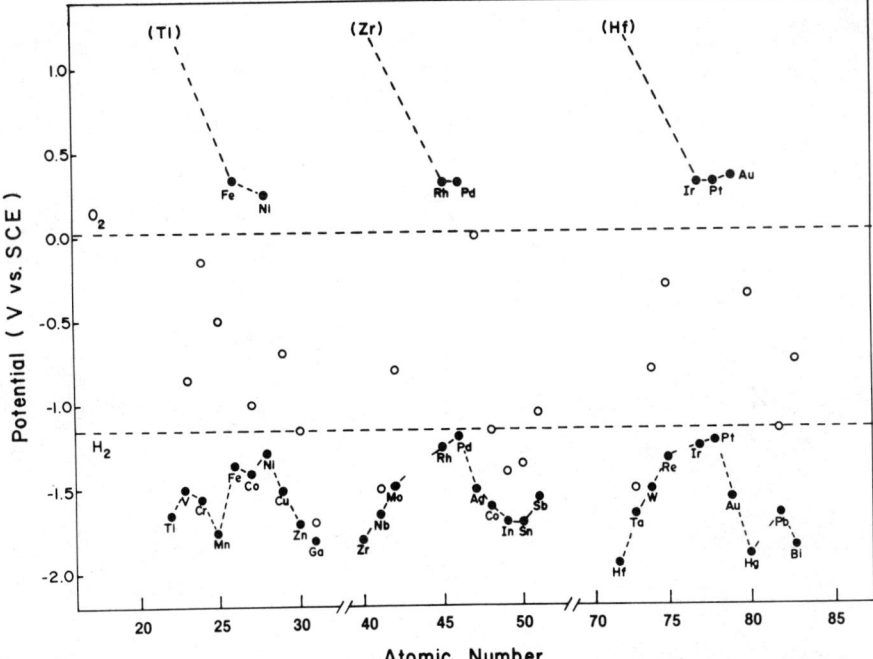

Figure 10. Potential ($V$ vs. SCE) variations with atomic number in 30% KOH at 80°C and a current density of 2 mA/cm$^2$ (with permission from *J. Electrochem. Soc.*).

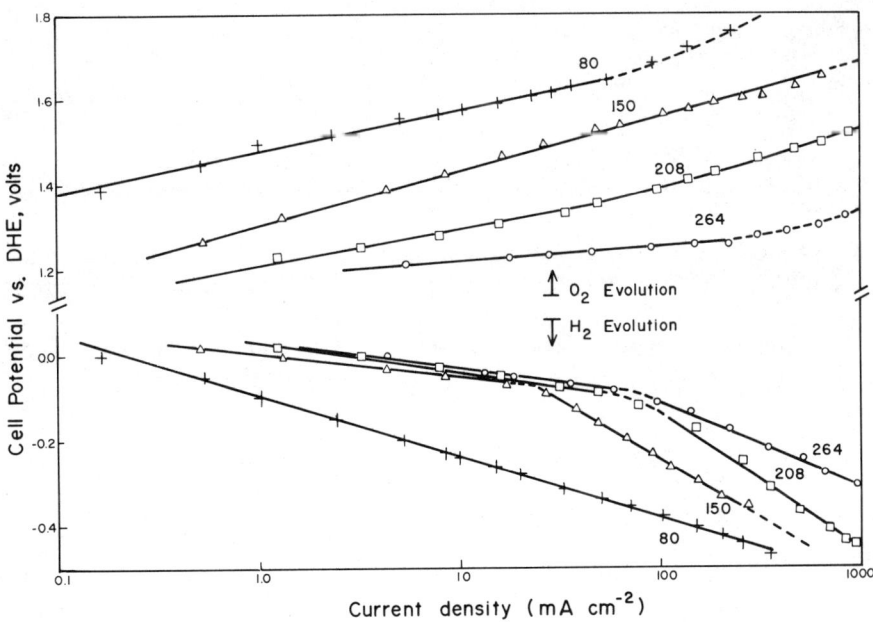

Figure 11. Polarization data for $H_2$ and $O_2$ evolution reactions on polished nickel electrodes in 50 wt % KOH at 80, 150, 208, and 264°C (with permission from *J. Electrochem. Soc.*).

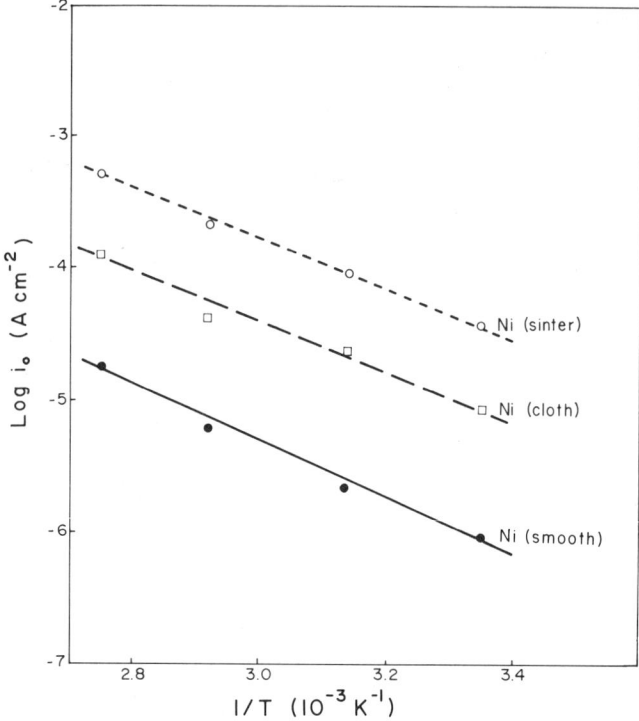

Figure 12. Arrhenius plots for exchange current (smooth electrodes) or apparent exchange current (cloth and sinter electrodes) for the hydrogen evolution reaction on Ni electrodes in 34% KOH (with permission from the authors and Pergamon Press).

temperatures. Comparison[18] of the $i_0$ values on various forms of Ni (see Figure 12) shows that high $i_0$, and hence low overpotential, can be achieved by effectively utilizing the high surface area of porous electrode structures.

One of the problems observed during the usage of Ni-based materials for water electrolysis is the loss of activity as indicated by the time variation[10,18,21,23] of the cathode potential by about 100–200 mV (see Figure 13). While this phenomenon appears reversible in the sense of being able to reactivate the cathodes by current interruption, two possible reasons[21] may be attributed to this variation:

(a) adsorption of metallic and/or organic impurities on the electrode surface blocking the effective sites for the hydrogen evolution reaction.
(b) hydrogen permeation into nickel and subsequent change of mechanism from electrochemical desorption to rate-controlling discharge resulting in higher cathode overpotential.

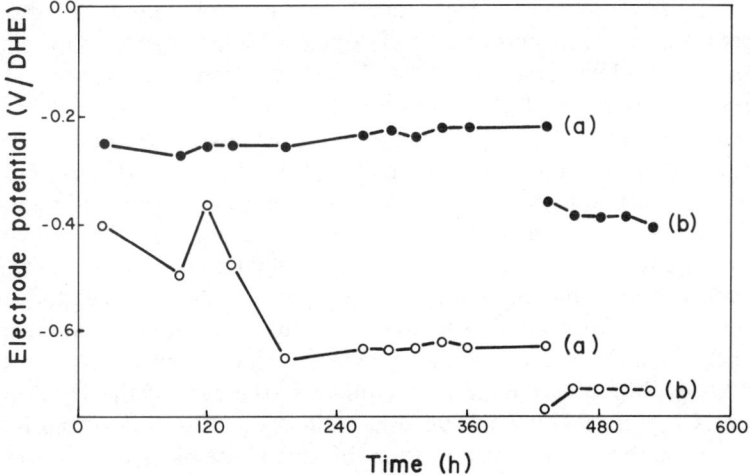

Figure 13. Cathode overpotential variations with time using Ni and nickel boride cathodes during water electrolysis in 30% KOH solutions at 90°C (with permission from *J. Electrochem. Soc.*). ○, Ni screen; ●, Ni boride; (a) 0.333 A/cm$^2$; (b) 0.833 A/cm$^2$.

Of the various materials examined so far, high-surface-area nickel boride,[21] Ni–S, Ni–Al, Ni–Zn, Ni sinter impregnated with cobalt molybdate and iron molybdate[20] appear promising from the viewpoint of achieving low overvoltage and minimizing the losses in catalytic activity with time. It may be noted here that the time-variation phenomenon[10] is also noticed with Fe cathodes presumably for the same reasons as noted above.

### 3.2.4. Oxygen Overvoltage

The oxygen overpotential component of the water electrolysis cell voltage is of the same form as Eq. (15), but the mechanistic aspects of the oxygen evolution reaction from alkaline solutions have not been as extensively[24,25] examined as the hydrogen evolution reaction except during the last decade. Studies on the electrocatalytic effects involved during the oxygen evolution reaction are limited by the restricted choice of materials due to anodic dissolution and complications arising from the formation of surface oxide films—often having poor electronic conductivity—on which oxygen evolution always takes place.

The mechanism of $O_2$ evolution is complex compared to the pathways suggested for the hydrogen evolution reaction, and the most generally accepted mechanism[8,24,25] involves the following steps:

$$OH^-_{ads} \rightleftharpoons OH_{ads} + e^- \qquad (16)$$

$$OH^- + OH_{ads} \rightleftharpoons O_{ads} + H_2O + e^- \qquad (17)$$

$$O_{ads} + O_{ads} \rightleftharpoons O_2 \qquad (18)$$

where one of the charge transfer steps is rate controlling. Observed dependence of transfer coefficients and Tafel slope variations suggest a slow electron transfer step at low temperatures and a slow recombination step at high temperatures on nickel electrodes in 50 wt % KOH solutions.

Recent studies[16,18,20,26–33] on the electrocatalytic activities of various metals, alloys (or intermetallics), and pure and mixed oxide surfaces (see Table 4 for the kinetic parameters for the oxygen evolution reaction and Figures 11 and 14) indicate that "nickel-based" materials are probably the best catalyst surfaces for the oxygen evolution reaction in alkaline solutions. Thus, noble metal additions are not mandatory to achieve low overvoltages for the $O_2$ evolution reaction from alkaline media. It is of interest to note that there is little difference[18] in the polarization behavior of smooth and porous structures during $O_2$ evolution (see Figure 15), unlike in the case of the $H_2$ evolution reaction (see Figure 12). This suggests ineffective utilization of the internal surface area of the porous body, presumably due to the blockage of the surface by the gas bubbles arising from the bubble morphology effects involved in gas evolution kinetics. The optimized electrode structures reported so far for $O_2$ anodes include sintered Ni electrodes impregnated with "nickel oxide," anodized nickel surfaces, NiOOH, high-surface-area $NiCo_2O_4$ electrodes, and 50 Ni–50 Fe alloys.[31]

As in the case of the $H_2$ cathodes, Ni anodes also exhibit potential variation[30] with time (see Figure 16) due to progressive transformation of trivalent nickel to tetravalent nickel oxide—a relatively poor electronic

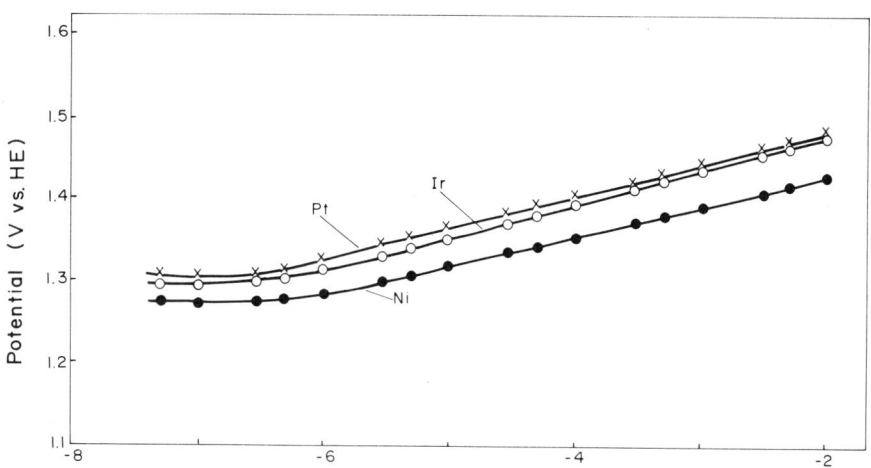

Figure 14. Comparison of the polarization characteristics on smooth Ni, Pt, and Ir electrodes in 34% KOH eq. at 90°C during $O_2$ evolution (with permission from the authors and Pergamon Press).

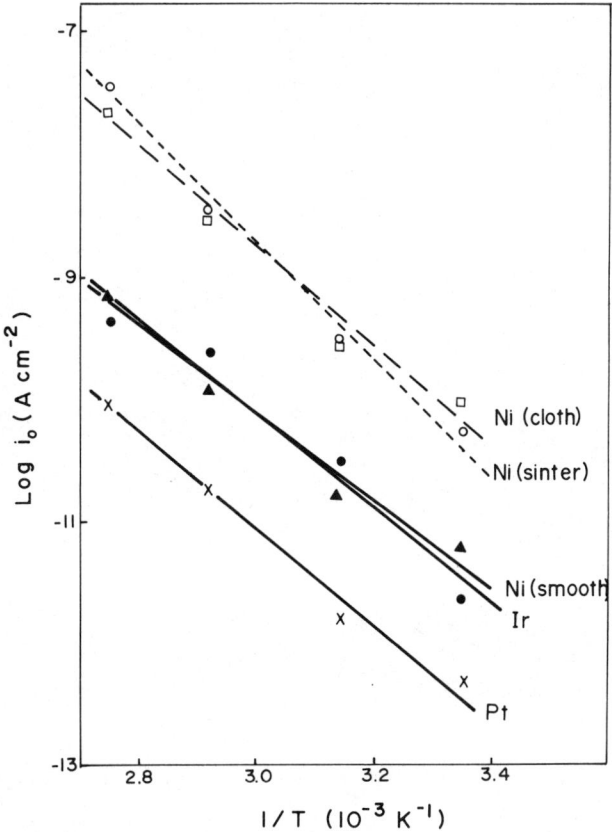

Figure 15. Arrhenius plots of exchange current (smooth electrodes) or apparent exchange current (Ni cloth and sinter electrodes) for the oxygen evolution reaction from 34% KOH solutions (with permission from the authors and Pergamon Press).

conductor—at a potential of 1.56 V (vs. $H_2$ electrode). Oxide thickening on Ni was ruled out as an interpretation for these time effects based on ellipsometric studies.[30,32] The gradual oxidation of $Ni^{3+}$ to $Ni^{4+}$ can be retarded significantly by (a) increasing the operating temperature to lower the equilibrium ratio of $Ni^{4+}$ to $Ni^{3+}$ ions in the surface film, (b) maintaining the anode potential below 1.56 V by using high-surface-area electrodes and thereby lowering the "real current density," and (c) introducing a stable secondary cation into the Ni oxide film to preserve the trivalent Ni form as in $NiCo_2O_4$. Long-term studies[21] using $NiCo_2O_4$ anodes have shown relatively good stability (see Figure 16) of the spinel over 150 days of operation and 100 mV lower overpotential compared to a relatively smooth, low-surface-area nickel screen at a c.d. of 330 mA/cm².

## Table 4
### Kinetic Parameters for the Oxygen Evolution Reaction in KOH Solutions

| Material | Temperature, °C | KOH, wt % | Tafel slope, V | | Transfer coefficient[a] | | Exchange current density, A/cm² | | Activation energy, kcal/mol | Reference |
|---|---|---|---|---|---|---|---|---|---|---|
| | | | Low $\eta$ | High $\eta$ | Low $\eta$ | High $\eta$ | Low $\eta$ | High $\eta$ | | |
| Pt | 80 | 30 | 0.046 | | 1.5 | | $1.2 \times 10^{-9}$ | | | 28 |
| Ir | 80 | 30 | 0.055 | | 1.3 | | $5.0 \times 10^{-8}$ | | | 28 |
| Rh | 80 | 30 | 0.067 | | 1.0 | | $3.8 \times 10^{-7}$ | | | 28 |
| Ru | 80 | 30 | 0.067 | | 1.0 | | $3.0 \times 10^{-8}$ | | | 28 |
| Os | 80 | 30 | 0.070 | | 1.0 | | $1.0 \times 10^{-7}$ | | | 28 |
| Pd | 80 | 30 | 0.067 | | 1.0 | | $1.9 \times 10^{-7}$ | | | 28 |
| Ni | 80 | 30 | 0.062 | | 1.1 | | $2.3 \times 10^{-7}$ | | | 28 |
| | 80 | 50 | 0.095 | | 0.74 | | $4.2 \times 10^{-6}$ | | | |
| | 150 | 50 | 0.125 | | 0.67 | | $1.8 \times 10^{-4}$ | | 18 | 8 |
| | 208 | 50 | 0.085 | 0.135 | 1.1 | 0.71 | $6.0 \times 10^{-4}$ | $3.5 \times 10^{-3}$ | | |
| | 264 | 50 | 0.032 | | 3.3 | | $1.0 \times 10^{-3}$ | | | |
| Co | 80 | 30 | 0.126 | | 0.56 | | $3.3 \times 10^{-6}$ | | | 28 |
| Fe | 80 | 30 | 0.191 | | 0.37 | | $1.7 \times 10^{-5}$ | | | 28 |

# ELECTROLYTIC PRODUCTION OF HYDROGEN

| Catalyst | | | | | | | | | Ref |
|---|---|---|---|---|---|---|---|---|---|
| 75 Ni–25 Ir[b] | 80 | 30 | | | | | | 8.8 × 10$^{-7}$ | 28 |
| 50 Ni–50 Ir[b] | 80 | 30 | 0.047 | 0.086 | 1.5 | 1.0 | 3.4 × 10$^{-8}$ | 2.0 × 10$^{-5}$ | 29 |
| 25 Ni–75 Ir[b] | 80 | 30 | 0.064 | | 1.1 | | 6.0 × 10$^{-7}$ | | 29 |
| 75 Ni–25 Ru[b] | 80 | 30 | 0.056 | | 1.3 | | 2.4 × 10$^{-8}$ | | 29 |
| 75 Ni–25 W[b] | 80 | 30 | 0.114 | | 0.61 | | 4.6 × 10$^{-5}$ | | 29 |
| 50 Ni–50 W[b] | 80 | 30 | 0.125 | | 0.56 | | 2.5 × 10$^{-4}$ | | 29 |
| Ni$_3$Ti | 80 | 30 | 0.064 | | 1.1 | | 4.7 × 10$^{-7}$ | | 29 |
| NiTi | 80 | 30 | 0.057 | 0.094 | 1.3 | 0.74 | 6 × 10$^{-8}$ | 2.8 × 10$^{-6}$ | 29 |
| NiTi$_2$ | 80 | 30 | 0.383 | | 0.18 | | 6.5 × 10$^{-5}$ | | 29 |
| TiO$_2$ + 30 mol % Pt Oxide | 80 | | 0.225 | | 0.31 | | 9.5 × 10$^{-6}$ | | 28 |
| TiO$_2$ + 30 mol % Co Oxide | 80 | 30 | 0.193 | | 0.36 | | 2.0 × 10$^{-6}$ | | 28 |
| TiO$_2$ + 30 mol % Ni Oxide | 80 | 30 | 0.259 | | 0.27 | | 1.4 × 10$^{-4}$ | | 28 |
| TiO$_2$ | 80 | 30 | 0.118 | | 0.59 | | 7.3 × 10$^{-9}$ | | 28 |
| TiO$_2$ + 30 mol % RuO$_2$ | 80 | 30 | 0.068 | | 1.0 | | 6.4 × 10$^{-7}$ | | 28 |
| TiO$_2$ + 100 mol % RuO$_2$ | 80 | 30 | 0.066 | | 1.1 | | 2.5 × 10$^{-6}$ | | 28 |
| Ni Co$_2$O$_4$ | 80 | 30 | 0.042 | | 1.67 | 9.2 | 2.9 × 10$^{-8}$ | | 26 |

[a] Transfer coefficient = 2.303 $RT/bF$, where $b$ is the Tafel slope.
[b] In atom percent.

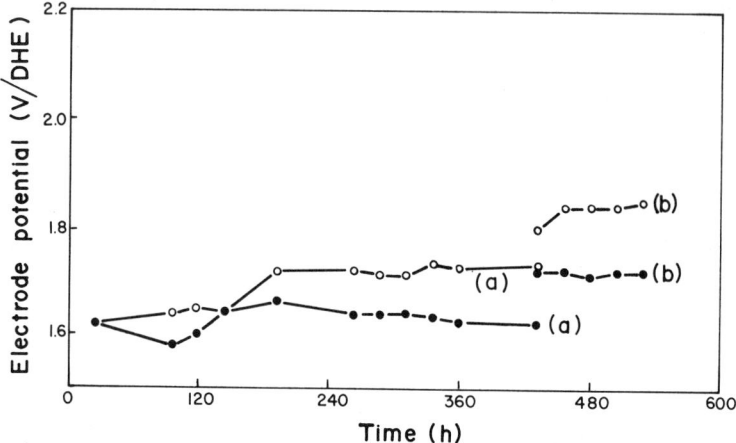

Figure 16. Time dependence of oxygen overpotential on Ni and $NiCo_2O_4$ electrodes during water electrolysis in 30% KOH at 90°C (with permission from *J. Electrochem. Soc.*). ○, Ni screen; ●, $NiCo_2O_4$; (a) 0.333 A/cm$^2$; (b) 0.833 A/cm$^2$.

### 3.2.5. Ohmic Drop across the Anode and Cathode—Diaphragms

The ohmic drop between the anode and the cathode is a substantial contribution to the cell voltage, although smaller than the overvoltage losses. The ohmic drop is composed of the potential drop across the gas + electrolyte mixture and the separator material. While void-fraction effects[7,34] and hence conduction in heterogeneous media[35] coupled with two-phase flow[36] are important in cell design considerations, the major ohmic losses arise from the separator material used to prevent the short circuiting of the anode and the cathode and separation of the gas bubbles from one compartment to the other. Since dissolved gas crossover is serious, the diaphragm should (a) have small pores with capillary pressure greater than the maximum differential pressure applied across the cell, (b) be wettable, (c) prevent preferential collection of gas in the pores, (d) offer high resistance to the electrolyte flow, and (e) have low resistance to the current flow. These requirements[39] impose incompatible restrictions on the design of the diaphragms since small pore size or voltage not only reduces the flow rate but also enhances the electrolyte resistance due to its inverse proportionality to the porosity of the separator material.

Asbestos is the most common material used in alkaline water electrolyzers in the form of woven asbestos cloth sometimes supported by fine wire, or mat or felt of asbestos fibers in pressure electrolyzers. While asbestos is an ideal separator material below 100°C in an alkaline environment, an asbestos diaphragm fails at temperatures over 100°C owing to its enhanced dissolution in strong alkaline solutions. Since operation at high temperatures (>100°C) is a prerequisite for the development of water electrolyzers approaching 100%

*Table 5*
**An Analysis of the Usefulness of Various Types of Materials as Separators for Alkaline Water Electrolysis at Temperatures in the Range 80–150°C**

| Materials—Class or type | Comments |
| --- | --- |
| Asbestos material | Ideal separator under 100°C in alkali |
| Woven material | Asbestos works well; used commercially. Boron nitride—not stable in alkali especially above 100°C. |
| Potassium titanate paper (Teflon binder) | Very good in alkali above 100°C. |
| Nafion | Very good in acid at all temperatures; works well in 20% NaOH at 100°C and above. |
| Nonwoven (felts) | Low-resistance material but limited life at elevated temperature. Also possible higher diffusion of gases. |
| Battery separator | Works well as separator material. Most are of polyethylene base—not suitable for higher temperature. |
| Cationic membrane | With the exception of Nafion, found to be suitable in regard to resistance but as a rule, short life above 100°C. |
| Anionic membrane | Very high resistance and short life. |
| Porous membrane | Teflon base, excellent durability above 100°C. All have high resistance (2–10 times greater than asbestos). |

energy efficiency, several materials[17,23,32,37–40] have been examined (see Table 5) for use as diaphragms and the following appear promising at temperatures >100°C in alkaline solutions.

(a) Teflon-based cation exchange membranes (Nafion) in 20% NaOH solutions at 120–160°C. However, their resistivity is higher in >20% NaOH solutions.

(b) Teflon-bonded potassium titanate. Its resistivity is independent of the nature of the electrolyte and has proven stability for over 5000 hr. However, it appears to embrittle in regions that are not in contact with the electrolyte.

(c) Asbestos treated with a binder (e.g., tin hydrosol) or in NaOH solutions containing sodium silicate.

(d) Polysulfone fabricated in thin fiber form.

### 3.3. Commercial Electrolyzers—Design Features and Operating Characteristics

#### 3.3.1. Monopolar Tank-Type Cells

During the early history of water electrolysis, monopolar tank-type cells enjoyed an exclusive status and there were several manufacturers of these cells, both in Europe and North America. However, in recent years, their numbers have steadily decreased as the competition from the more energy-efficient

*Table 6*
*Manufacturers of Commerical Bipolar Filter Press Cells and their Differences*

| Manufacturer | Pressure of operation | Electrolyte feed, and gas–electrolyte outlet system |
|---|---|---|
| Brown Boveri & Cie | Ambient | Individual pipes external to filter press frames, to and out of each cell chamber. |
| DeNora | Ambient | Individual pipes external to filter press frames, to and out of each cell chamber. |
| Norsk Hydro | Ambient | Ducts within filter press frames with small bore holes from the ducts, to and out of each cell chamber. |
| Lurgi Gesselschaft | 30 atm | Ducts within filter press frames with small bore holes from the ducts, to and out of each cell chamber. |

bipolar cells (not necessarily true for the future) increased, with only one monopolar technology, by the Electrolyser Corporation, Ltd.,[10,42–47] of significance still remaining. See Table 6 for a list of bipolar cell manufacturers.

### 3.3.1.1. The Electrolyser Corporation, Ltd. Technology[10,42–47]

This company, the only leading manufacturer of monopolar tank-type cells today, is a very strong competitor to the more numerous manufacturers of bipolar cells (see Table 6). The cell is widely known as the Stuart cell, after its inventor, A. Th. Stuart, founder of the Electrolyser Corporation, Ltd., which has been in the hydrogen electrolyzer business for almost 50 years. The company has 500 installations of such equipment in approximately 80 different countries. There equipment has gained a reputation for reliability and low maintenance, in spite of varying climatic conditions.

The Stuart cell (see Figure 17) is an all-welded, rectangular, low-carbon-steel tank, nickel plated on the inside for additional corrosion protection and long life. A gas-tight cover, also nickel plated, supports the oxygen-collecting chamber, asbestos cloth separators, and the electrodes. High-conductivity steel is used in the construction of the electrodes; the anodes are plated with a heavy sponge nickel and the cathodes specially treated to generate a more active surface for low cathodic overvoltage. An important feature of the electrodes is a unique construction which provides a large surface area for electrolysis in a minimum volume.

Surrounding each anode is a woven asbestos cloth diaphragm which prevents the mixing of the gases, and guides the generated oxygen to the oxygen chamber located immediately below the cell cover. Hydrogen generated on the cathodes rises between, but outside, the diaphragms to the hydrogen compartment under the cover. In each Stuart cell, the electrodes are suspended, but insulated from the cell cover. The copper terminals to the electrodes are silver plated to achieve a low contact resistance with the silver-plated copper busbar. All the anodes are connected together by

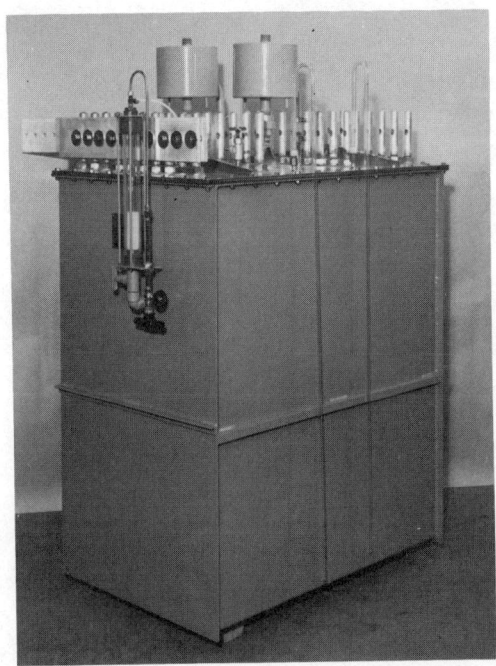

Figure 17. A monopolar tank-type Stuart cell manufactured by the Electrolyser Corporation, Ltd. (with permission from Electrolyser Corporation, Ltd.).

the busbar, as are all the cathodes. By connecting several electrodes in this manner in parallel, a low cell voltage of ~1.9 V is achieved.

Although the electrodes are at a potential of ~1.9 V apart, the cell tank, cover, diaphragm suspension system, and gas chambers are allowed to float electrically. Thus, the only electrical insulation in the cell is from the electrodes to the tank, and between the tank and the floor.

The temperature of the cells is controlled by regulating water flow to a cooling jacket on one side of the cell. The alkaline electrolyte level is controlled automatically by an external float control chamber and valve. Both the $H_2$ and $O_2$ gases, largely separated from the electrolyte within the cell tank, enter individual gas cooler–condenser–demisters (see Figure 7) immediately above the cells. All entrained electrolyte and condensed water vapor returns to the cell tank by gravity flow. Circulation of the electrolyte within the cell is by the gas lift principle only. Thus there are no moving parts in the Stuart cell.

The operating characteristics of the Stuart cell are shown in Table 7. Briefly, these cells operate at ambient pressures and at 70°C using 28% KOH as the electrolyte. To achieve a reasonable cell voltage, a current density (based on the area of the diaphragm between the electrodes) of $1340 \text{ A/m}^2$ is employed to operate these cells.

A typical commercial installation of Stuart cells is shown in Figure 18. The cells are connected together electrically in series. Large gas headers located

## Table 7
### Typical Cell-Operating Parameters

| Manufacturer: | The Electrolyser Corp., Ltd | Brown Boveri & Cie | Norsk Hydro A.S. | De Nora S.P.A. | Lurgi GmbH |
|---|---|---|---|---|---|
| Cell type: | Monopolar tank | Bipolar filter press | Bipolar filter press | Bipolar filter press | Bipolar filter press |
| Operating pressure: | Ambient | Ambient | Ambient | Ambient | 450 psig |
| Operating temperature: | 70°C | 80°C | 80°C | 80°C | 90°C |
| Electrolyte: | 28% KOH | 25% KOH | 25% KOH | 29% KOH | 25% KOH |
| Current density, A/M²: | 1340 | 2000 | 1750 | 1500 | 2000 |
| Cell voltage, V: | 1.90 | 2.04 | 1.75 (after 1 yr operation) | 1.85 (increases to 1.95 after 2 yr) | 1.86 |
| Current efficiency, %: | >99.9 | >99.9 | >98 | ~98.5 | 98.75 |
| Oxygen purity, %: | 99.7 | ≥99.6 | 99.3–99.7 | 99.6 | 99.3–99.5 |
| Hydrogen purity, %: | 99.9 | ≥99.8 | 98.8–99.9 | 99.9 | 99.8–99.9 |
| Power consumption, DC-kWh per normal m³ $H_2$[a]: | 4.9 | 4.9 | 4.3 | 4.6 | 4.5 |

[a] *Note*: Current required for 1 nm³ hydrogen = 2393 A hr.

# ELECTROLYTIC PRODUCTION OF HYDROGEN

Figure 18. A typical plant built by the Electrolyser Corporation, Ltd. (with permission from Electrolyser Corporation, Ltd.).

above the cells collect the $H_2$ and $O_2$ gases, which pass through water seals (a safety measure to equalize pressures in the cell and thereby prevent accidental increases in pressure) before entering gasometers (see Figure 8). Other headers, located at floor level between rows of cells, feed cooling water and deionized water to each cell.

### 3.3.2. Bipolar Filter Press Cells†

All bipolar filter press cells are designed on the principles of the plate and frame filter press for liquid–solids separation, and although there are many manufacturers, the differences in technology are only minor.[3a] Commercial bipolar filter press cells differ significantly in only two aspects. These are the pressure of operation and the method of feeding the electrolyte into the anolyte and catholyte chambers and extracting the mixture of generated gases and electrolyte from each chamber. Table 6 lists the major manufacturers of bipolar electrolyzers and their differences. Figures 19–21 (Brown Boveri cell)

† See Appendix for a description of Krebskosmo Technology.

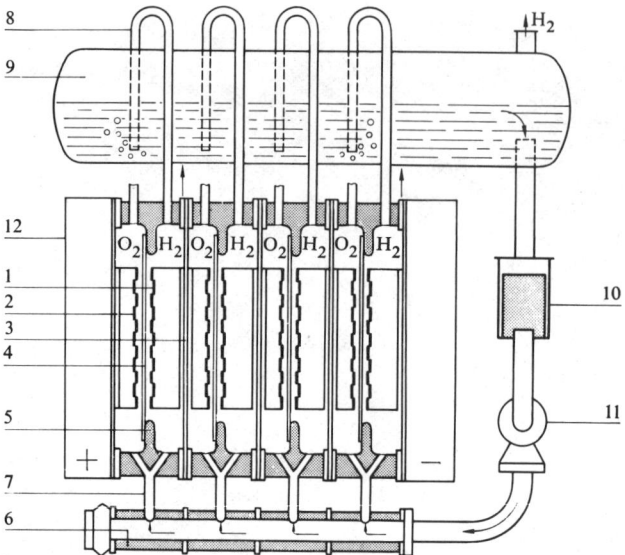

Figure 19. Schematic cross section of a BBC electrolyzer (with permission from Brown Boveri & Cie). 1, cathode; 2, anode; 3, carrier electrode; 4, diaphragm; 5, cell frame; 6, electrolyte distributing system; 7, cell feed tube; 8, gas outlet; 9, gas separator (8 and 9 drawn for the hydrogen side only); 10, electrolyte filter; 11, electrolyte pump; 12, press plates.

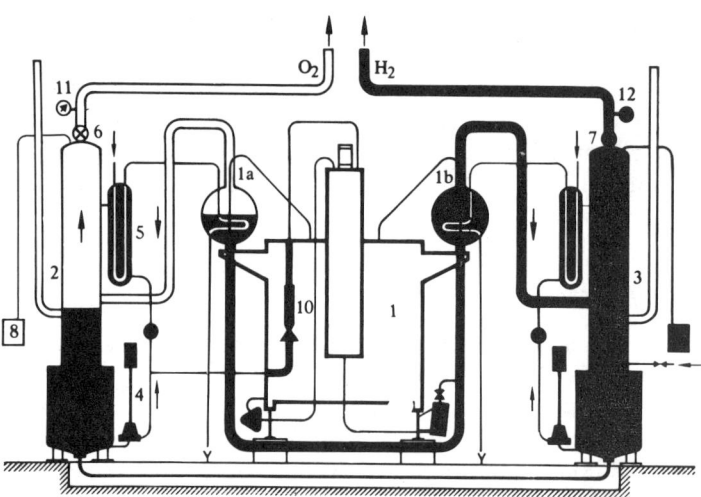

Figure 20. Flow diagram of a Brown Boveri electrolyzer (with permission from Brown Boveri & Cie). 1, electrolyzer; 1a, gas separator $O_2$; 1b, gas separator $H_2$; 2, gas scrubber $O_2$; 3, gas scrubber $H_2$; 4, circulating pump for gas scrubber; 5, heat exchanger; 6, $O_2$ valve; 7, $H_2$ valve; 8, $O_2$ gas analyzer; 9, $H_2$ gas analyzer; 10, rotameter; 11, $O_2$ gas thermometer; 12, $H_2$ gas thermometer.

# ELECTROLYTIC PRODUCTION OF HYDROGEN

Figure 21. End view of Brown Boveri electrolyzer (with permission from Brown Boveri & Cie).

illustrate the typical method of feeding electrolyte into each cell chamber and removing the mixture of gas–electrolyte out of each cell chamber by individual pipes external to the filter press frames. Figures 28–30 (Lurgi Gesellschaft cell) depict the typical method of feeding and removing electrolyte and gases to and from *each* cell chamber by ducts within the filter press and small bore holes from these ducts into each cell chamber.

### 3.3.2.1. Brown Boveri & Cie Technology[48,49]

The first industrial bipolar electrolyzer was designed around 1900 by Schmidt. A slightly modified version of this design was commercialized by Maschienenfabrik Oerlikon under the name Oerlikon–Schmidt electrolyzer. In 1967, Brown Boveri joined with Maschienenfabrik Oerlikon, preserving the name of the Oerlikon electrolyzer system in the literature.

Since 1902, more than 800 of these electrolyzers have been built, the largest installation being at Aswan, Egypt (replacing the older plant of a different origin). This plant, when completed in late 1980, will consist of 144 electrolyzers with a nominal rating of 162-MW electrical input and a hydrogen generating capacity of 32,400 $m^3$/hr.

As with most bipolar cells, the Brown Boveri system is highly modularized, so that standardized elements such as cell frames, electrodes, and diaphragms allow construction of cells with a wide variation in generating capacity. Standard units producing from about 5 to 300 $m^3$/hr are currently manufactured. Any number of these electrolyzers can then be assembled in a suitable series–parallel electrical arrangement to supply the total desired hydrogen output. Each electrolyzer may contain up to 80 unit cells or cell frames. Thus, the electrolyzer voltage will be approximately equal to the number of cells multiplied by the unit cell voltage of $\sim 2$ V.

Each cell frame contains only the diaphragm and the electrodes (see Figure 19). A forced electrolyte circulating system feeds the alkaline solution via a header into each half of the cell frame. The forced circulating system contains a metal screen filter to remove solid impurities which may otherwise plug the pipes to or from each cell frame. Circulation through the cells is entirely by a pump, unlike the gas lift circulation in monopolar cells.

A mixture of electrolyte and the respective gas (only the hydrogen outlet is shown in Figure 19) leaves each half of every cell frame through its external gas outlet, and then through a dip pipe into the gas separator which also acts as a cooler of the separated electrolyte to maintain the cell at a constant temperature. The dip pipes act as hydraulic safety seals and pressure regulators. Both gas separators are connected together by a U tube which balances the pressures between the hydrogen and oxygen side of the diaphragms. It is from this U tube that the electrolyte flows through a mixing vessel from where the circulating pump sucks it through the filter and back to the cells. From the gas separators, the gases enter gas scrubbers and coolers (see Figures 8 and 20).

Table 7 lists the typical operating data for the Brown Boveri electrolyzer. The construction of a Brown Boveri electrolyzer is shown in Figure 19. A woven asbestos diaphragm is stretched in the cell frame, which in turn is pressed between insulating frame gaskets on both sides of the center plate. Pre-electrodes are mounted on both sides of the center plate. The anode is nickel-plated steel whereas the cathode is plain steel supplied with or without an activated surface. The activated surface of the cathode is reported to last only a few years. Several such frame and electrode elements can be mounted together between a pair of press plates at the end of the bipolar assembly. The press plates are held together by several tie rods stretching the length of the bipolar assembly shown in Figure 21, with the tension on the tie rods carefully set to hold the cell frames, gaskets, and electrodes together with a pressure sufficient to prevent electrolyte leakage.

The inlet and outlet pipes from each cell frame are of pure nickel and lengths of plexiglass are used to insulate these pipes and thereby minimize parasitic currents and permit visual inspection of the circulating electrolyte. The horizontal tank gas separators are on the top and both sides of the cell frame assembly, and are constructed of a corrosion-resistant chrome–nickel–molybdenum steel. The filter which removes solid particles, including any coarse asbestos fibers, from the system is of a fine nickel mesh. It is significant to note that all the components in this electrolyzer subject to anodic stress and/or in contact with the alkaline electrolyte are protected by a special nickel plating.

## ELECTROLYTIC PRODUCTION OF HYDROGEN

Figure 22. Two fully automatic BBC electrolyzers built in Kuwait—type EBN—220–40, generating 150 m$^3$ of hydrogen per hour (with permission from Brown Boveri & Cie).

Brown Boveri manufacture several sizes of electrolyzers (see Figure 22), all factory preassembled to minimize field construction time and costs. The largest electrolyzer unit, type EBK-385-80, is capable of generating 300 m$^3$ of H$_2$ per hour, and has 80 unit cells between the press plates. This unit operates at about 9000 A and 164 V and has overall dimensions of 4.9 m high, 3.2 m wide, and 7.1 m long and weighs (when full of 25% KOH) about 66 metric tons. Figure 23 illustrates the layout of the first 56 electrolyzers built in Aswan, Egypt in 1973 and is typical of all bipolar filter-press-type plants.

### 3.3.2.2. Norsk Hydro A.S. Technology[4,50–52]

Norsk Hydro has been manufacturing H$_2$ for over half a century and was one of the pioneers to use cheap hydropower to generate H$_2$ to manufacture ammonia-based fertilizers. By the early 1960s, it had the largest H$_2$ plants in the world (see Table 8, which illustrates plants by other manufacturers), with a peak capacity of 100,000 normal m$^3$ of H$_2$ per hour. However, some of this capacity has been converted to H$_2$ from hydrocarbons. In recent years, Norsk Hydro has been offering its cell technology for sale, and a total of about 11 plants have been sold, mostly in Scandinavia and Europe.

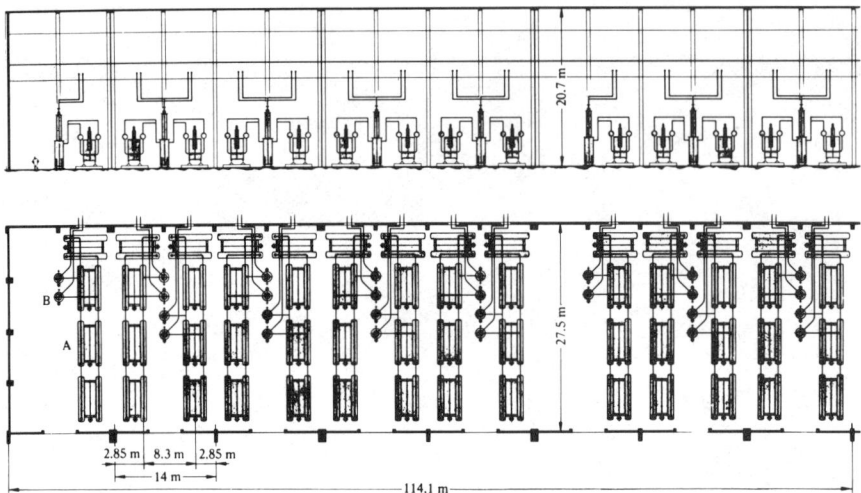

Figure 23. Layout of a large water electrolysis plant in Aswan, Egypt with a rated output of 12600 m³ H₂/hr. A: Group of electrolyzers 4 X EBK 385-70; B: Gas scrubber GA 1000 (with permission from Brown Boveri & Cie).

The Norsk Hydro electrolyzers are of the bipolar filter press type. Unlike the Brown Boveri and DeNora cells, the electrolyte feed, and gas–electrolyte outlet are through ducts within the filter press frames with small bore holes from the ducts to and out of each cell chamber (compare Figures 19 and 24). A series of identical gas-producing cells are firmly pressed together between two solid end frames and held together by four spring-loaded tie bars.

The individual cell construction presented in Figure 24 shows the electrodes, gaskets, and a diaphragm sandwiched together to form two separate compartments for the generation of $H_2$ and $O_2$ gases. The main support electrode is common to two neighboring cells, with perforated plate fore electrodes fastened by rivets or screws to each side of the main electrode. Both

*Table 8*
*Hydrogen Plants by Various Manufacturers*

| Plant location | Electrolyzer manufacturer | Capacity, normal m³ $H_2$/hr |
|---|---|---|
| 1. Nangal, India | DeNora | 30,000 |
| 2. Aswan, Egypt | Brown Boveri | 22,000 (to be 33,000 in 1980) |
| 3. Ryukan, Norway | Norsk Hydro | 27,900 |
| 4. Ghomfjord, Norway | Norsk Hydro | 27,100 |
| 5. Trail, Canada | Trail | 15,200 |
| 6. Cuzco, Peru | Lurgi | 4,500 |
| 7. Huntsville, Alabama | Electrolyser Corporation | 535 |

**ELECTROLYTIC PRODUCTION OF HYDROGEN**  43

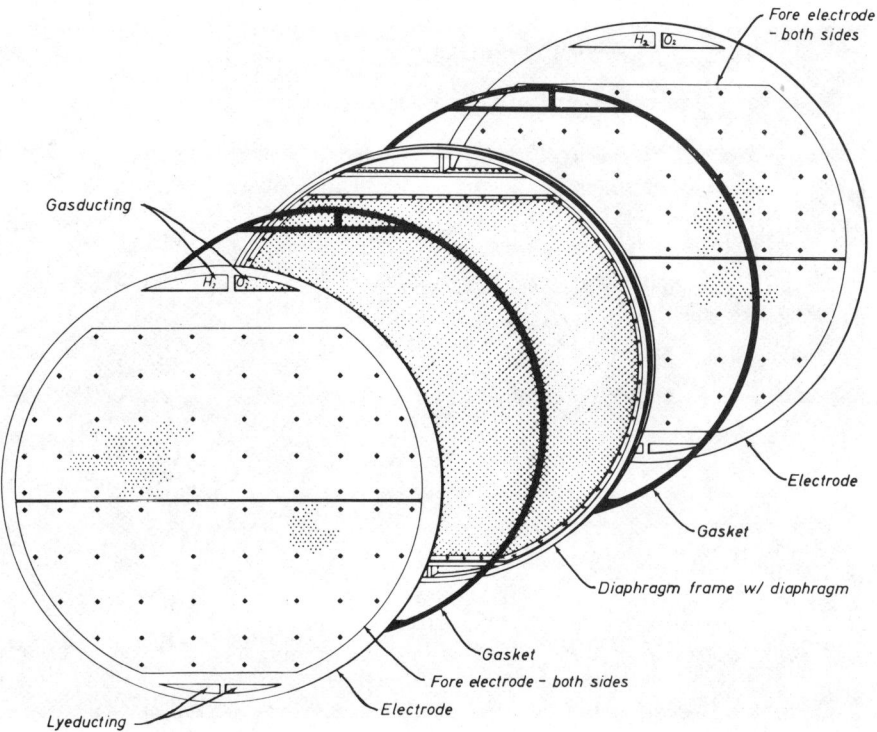

Figure 24. Norsk Hydro electrolyzer—exploded view of one electrolyzer cell (with permission from Norsk Hydro A.S.).

fore electrodes are mounted at a distance from the main support electrode to provide free passage of the gases and to reduce the cell voltage.

The diaphragm is of an asbestos cloth, 2 mm thick, reinforced with nickel thread. On each side of the diaphragm frame there are small passages into one of the two gas ducts at the top rim of the diaphragm ring. The gaskets are of a special rubber quality. The upper gas–electrolyte ducts terminate in two gas separators at the front of the electrolyzer.

Sandwiching a number of these electrodes, gaskets, and diaphragms together forms an electrolyzer (see Figures 25 and 26). A typical electrolyzer has about 235 unit cells, each with a cross-sectional area of the electrodes of 2.1 m$^2$, carries a load of 3600 A (ranges from 1000 to 4000 A) and generates about 350 normal m$^3$ of H$_2$ per hour. Such an electrolyzer weighs about 59 metric tons when full of electrolyte, and is about 11 m long and 2.5 m high. All metal parts are of mild steel, protected where necessary by a nickel coating, or by cathodic protection to prevent corrosion.

Typical operating data for a Norsk Hydro electrolyzer are presented in Table 7. The cell voltage is low owing to the active nickel coating, which,

Figure 25. A typical Norsk Hydro electrolytic hydrogen generation plant—back view (with permission from Norsk Hydro A.S.).

however, degrades after a number of years (see Figure 27). In locations where power costs are relatively high, the electrodes can be readily recoated. As do all bipolar filter press cells, the Norsk Hydro system requires an external gas separator, electrolyte coolers, filters, and pumps.

### 3.3.2.3. Lurgi Gmbh Technology[53–55]

Twenty years have passed since Lurgi put its first pressure electrolyzer, adopting the process of the Zdansky–Lonza cell from a Swiss company by an exclusive license agreement, into operation for the generation of $H_2$ and $O_2$. Since that time, many electrolyzers have been built throughout the world.

The principal differentiating feature of the Lurgi electrolyzer is that it operates at 450 psig, thus saving the cost of gas compressors, reducing equipment size, and lowering the cell voltage. The other noteworthy feature is that the electrolyte feed and gas–electrolyte outlet system (similar to the Norsk Hydro cell) is through ducts within the filter press frame with small bore holes from the ducts to and out of each cell chamber (see Figures 28 and 29). In all other respects the Lurgi cell is very similar to the Brown Boveri and the Norsk Hydro systems (e.g., gas separators, electrolyte coolers, and pumps).

Figure 26. A typical Norsk Hydro electrolytic hydrogen generation plant—front view (with permission from Norsk Hydro A.S.).

As can be seen in Figures 28 and 29, nickel-plated wire and activated gauze electrodes are laid directly on the 3-mm-thick asbestos board diaphragms. The electrodes are then pressed against the pure asbestos diaphragm by the embossed cell partition wall. The embossed or dimpled nickel-plated steel plates fixed into the cell frames provide electrical contact between the wire mesh anodes and cathodes of the adjoining cells, and also permit the $H_2$

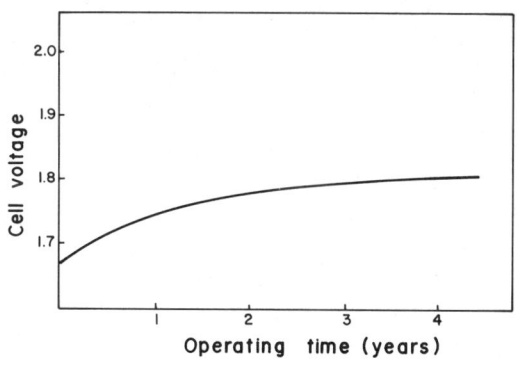

Figure 27. Cell voltage of Norsk Hydro cells. Operating conditions: load, 1750 A/m$^2$; temperature, 80°C; electrolyte, 25% KOH.

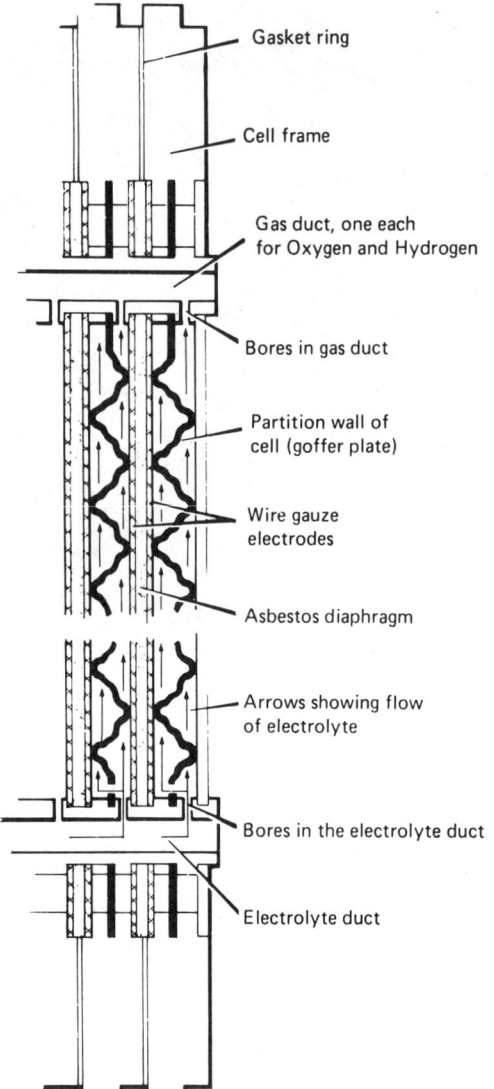

Figure 28. Sectional view of Lurgi electrolyzer cell [with permission from Her Majesty's Service (Reference 55)].

and $O_2$ gas bubbles to rise behind the electrodes. Each cell frame is separated and sealed by a Teflon-covered gasket.

A number of cell modules (e.g., cell partition wall, anode, cathode, asbestos board diaphragm) can then be assembled together and held between end press head plates and a system of six tie rods. Figure 30 illustrates a side and end view of a Lurgi electrolyzer and shows the electrolyzer with horizontal gas cooler separators on the top of the cell. Figure 31 depicts a typical Lurgi

# ELECTROLYTIC PRODUCTION OF HYDROGEN

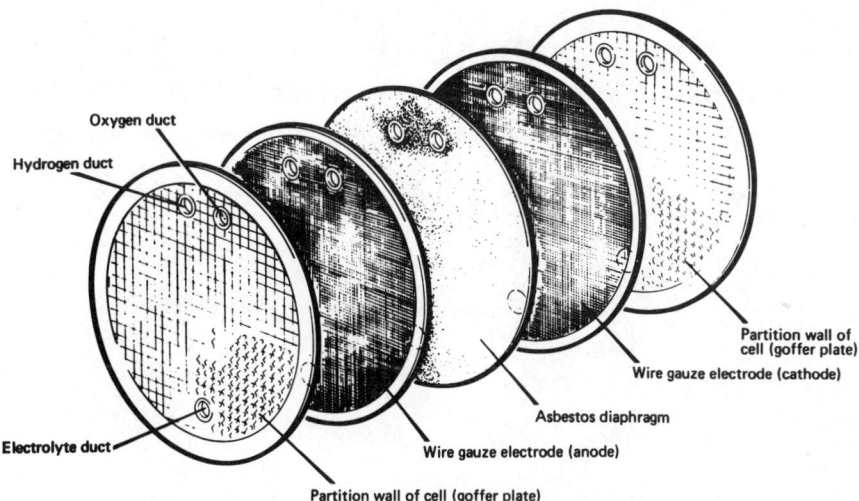

Figure 29. Exploded view of electrodes, partition plate, and diaphragm of Lurgi electrolyzer cell [with permission from Her Majesty's Service, (Reference 55)].

electrolyzer installation. Note that the electrolyzer is supported by the lower two tie rods on insulators.

Electrolyzers of a wide range of capacities are supplied. The largest unit, with a current capacity of about 6000 A and 350 unit cell modules, can produce about 800 normal $m^3$ of $H_2$ per hour. Typical operating conditions for Lurgi electrolyzer are presented in Table 7.

### 3.3.2.4. DeNora S.P.A. Technology[56-58]

DeNora has built water electrolyzers throughout the world, the most famous being at Nangal in Northern India (see Table 8). The DeNora electrolyzer, being of the filter press type, consists of a number of bipolar elements assembled between two end plates by means of tie rods (see Figure 32).

The bipolar element consists of a 5-cm-thick steel frame surrounding a bipolar plate to which expanded planar metal electrodes are connected on both sides. The expanded metal electrodes, both anode and cathode, are nickel-plated mild steel and activated using a proprietary sulfide process. This activation lasts only for about two to three years (see Table 7). However, they can be removed and reactivated. Use of expanded metal not only increases the available electrode area compared to flat plates, but also facilitates the withdrawal of gases from the interelectrode gap.

A unique feature of the DeNora electrolyzer is the use of a double-woven specially treated asbestos diaphragm. These are in physical contact with each other with the space in between vented to the atmosphere. During normal

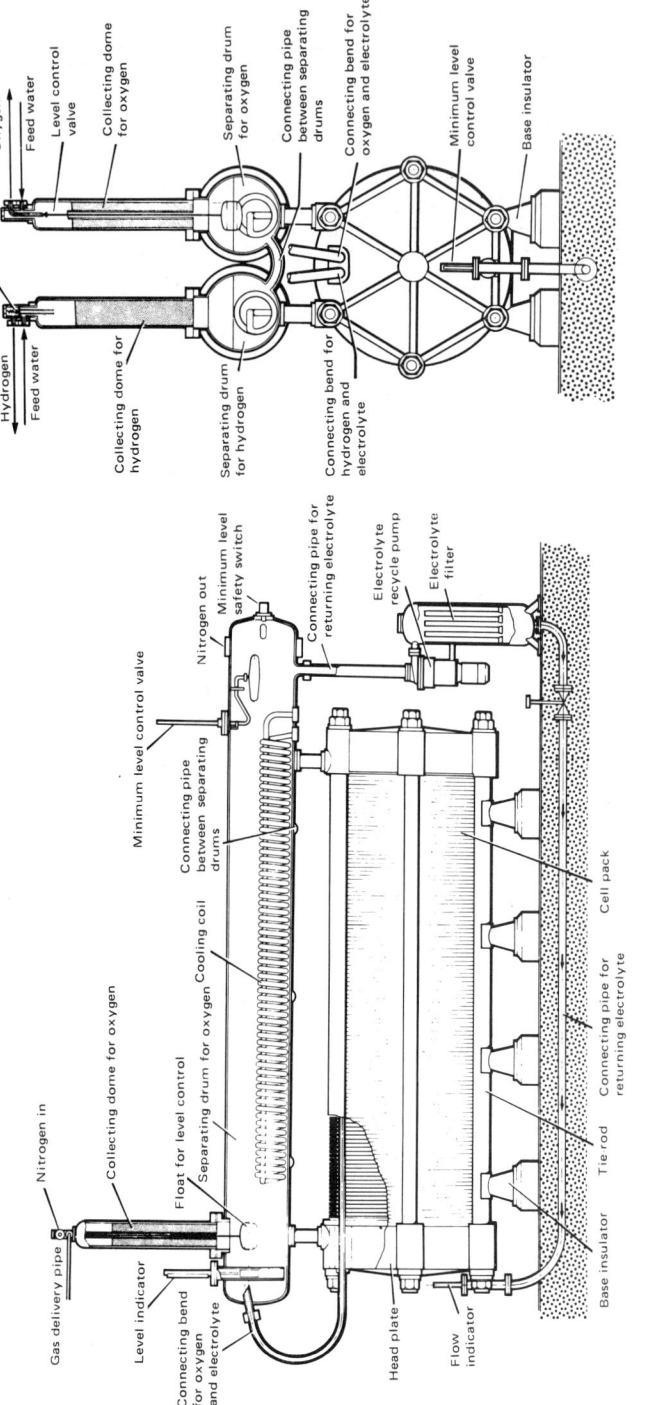

Figure 30. Sectional (A) and end (B) views of Lurgi electrolyzer [with permission for Her Majesty's Service (Reference 55)].

Figure 31. Typical Lurgi electrolyzer installation (with permission from Lurgi Umwelt und Chemokechnik GmbH).

operation, the diaphragms are pressed against one another, but any penetration of gas bubbles results in the formation of a larger bubble between the diaphragms, which is vented to the outside, and thus cannot intermix with the opposing gas. Hence, the gas purity from the DeNora cell is greater than 99.9%, and the possibility of formation of explosive gas mixtures is greatly reduced.

Product gases discharging from the bipolar filter press block enter separator drums and bubble compartments located on top of the electrolyzer. The separator drums remove entrained electrolyte from the rising product gases, and the bubble compartments maintain uniform gas pressure in the electrolyzer system.

DeNora electrolyzers are supplied in standard sizes of 2500, 4500, and 10–12,000 A, with the number of bipolar elements varying from 40 to about 100. The largest electrolyzer plant built for Nangal in India had 60 units each of 108 cells. Each electrolyzer is 50 m wide, 5.5 m high, and 15 m long.

## 3.4. Novel Alkaline Electrolyzers[4,32,59,60]

In recent years considerable research and development have been devoted to improving the efficiency and lowering the cost of alkaline water electrolyzers. Many advanced concepts have been proposed (see Section 5 for

Figure 32. Typical DeNora water electrolyzer (with permission from Panclor SA).

details). The worldwide effort and volume of literature has become considerable, as alternatives to the present fossil fuel economy are investigated of which the hydrogen economy is the most promising.

Only two advanced concepts have been developed so far to the commercial or semicommercial stage, although the units currently available have a very low hydrogen output. The two systems developed are manufactured by Life Systems and Teledyne Energy Systems.

### 3.4.1. Life Systems, Inc. Technology[61,62]

Life Systems has been developing this technology for the past ten years. Their principal efforts have been devoted to the generation of oxygen for space

# ELECTROLYTIC PRODUCTION OF HYDROGEN

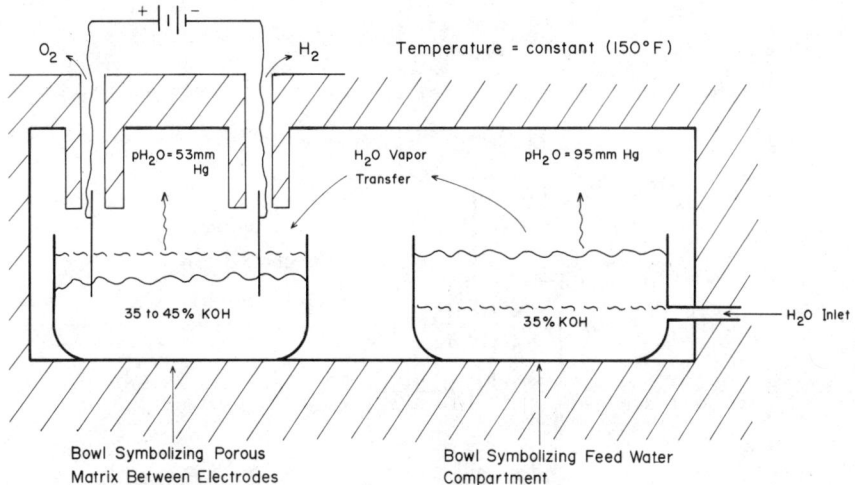

Figure 33. Principle of Life Systems static feed-water electrolysis cell (with permission from Life Systems, Inc.).

applications, but since late 1976 they have extended their technology to chemical energy and hydrogen energy systems.

Unlike the conventional alkaline water electrolyzers where the electrodes are immersed in the electrolyte, the Life Systems cells rely upon a static feed-water concept.

In the static feed-water system, the water to be electrolyzed is fed to the cell electrolyte as a vapor. Figure 33 illustrates the concept, and shows a thermally insulated box in which two bowls of electrolyte are placed. When power is applied to the electrodes, water in the cell electrolyte is consumed. As a result, the concentration of the cell electrolyte increases, causing its vapor pressure to drop below that of the feed compartment electrolyte.

Figure 34 illustrates the principle of the practical cell configuration. Each cell is divided into four main compartments: a water feed compartment, a hydrogen gas compartment, an oxygen gas compartment, and a liquid coolant compartment. Compartment separation and liquid–vapor phase separation is achieved by the capillary action provided by liquid-filled asbestos sheets (matrices). Catalyzed, porous electrodes support the cell matrix forming a composite electrolysis site. Plastic screens similarly support the water feed matrix. A number of these cells are assembled in a bipolar filter press configuration as shown in Figure 35.

Materials of construction of these cells include injection-molded polysulfone cell frames, activated porous nickel electrodes, Life Systems reconstituted asbestos for the feed-water matrix, and cell electrolyte matrix. End plates and other external components are of 316 stainless steel. Various

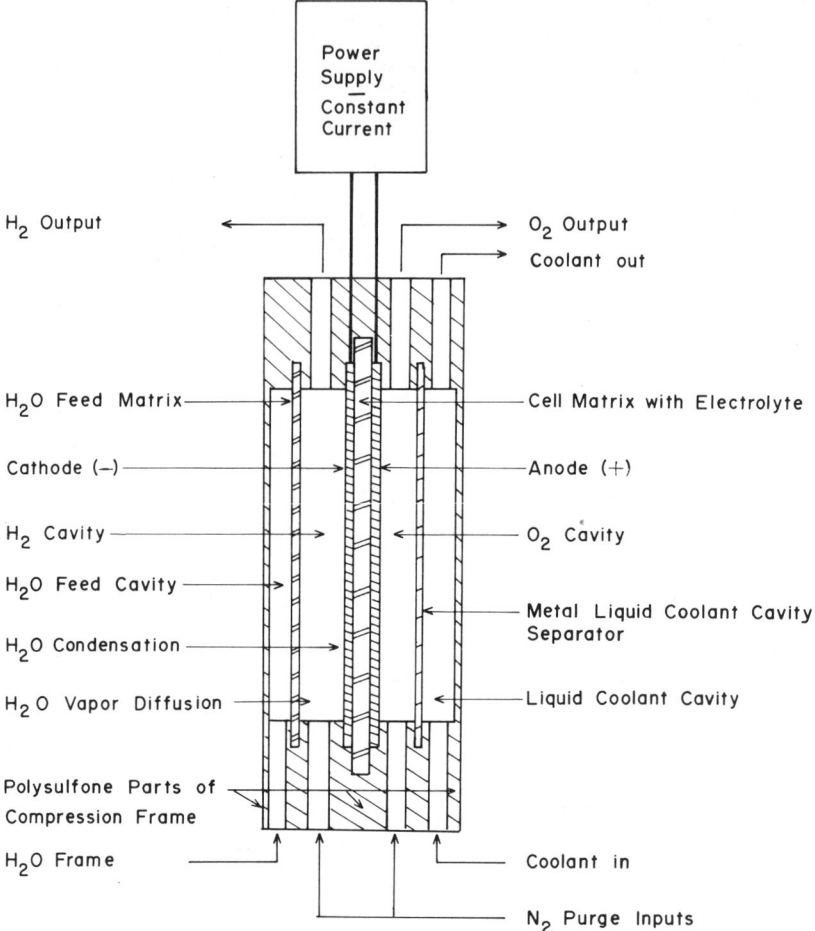

Figure 34. Life Systems cell configuration (with permission from Life Systems, Inc.).

nonmetallics employed in cell construction are ethylene–propylene O-rings, Teflon tape, and polypropylene screens to support the matrices.

The present design capability of the Life Systems cell is as follows: maximum pressure, 600 psig; maximum temperature, 220°F; maximum current density, 600 ASF; power consumption, ~4.4 kWh per normal $m^3$ of $H_2$ (at 600 ASF and ambient pressure).

The major advantages of the Life Systems static feed-water cell are that (i) the product gases do not need to be separated from the electrolyte, (ii) semipure feed water may be used, (iii) catalytic electrodes and the electrolyte remain uncontaminated since the water comes to the cathode as a pure vapor, and (iv) the interelectrode gap may be small and is free of gases.

# ELECTROLYTIC PRODUCTION OF HYDROGEN

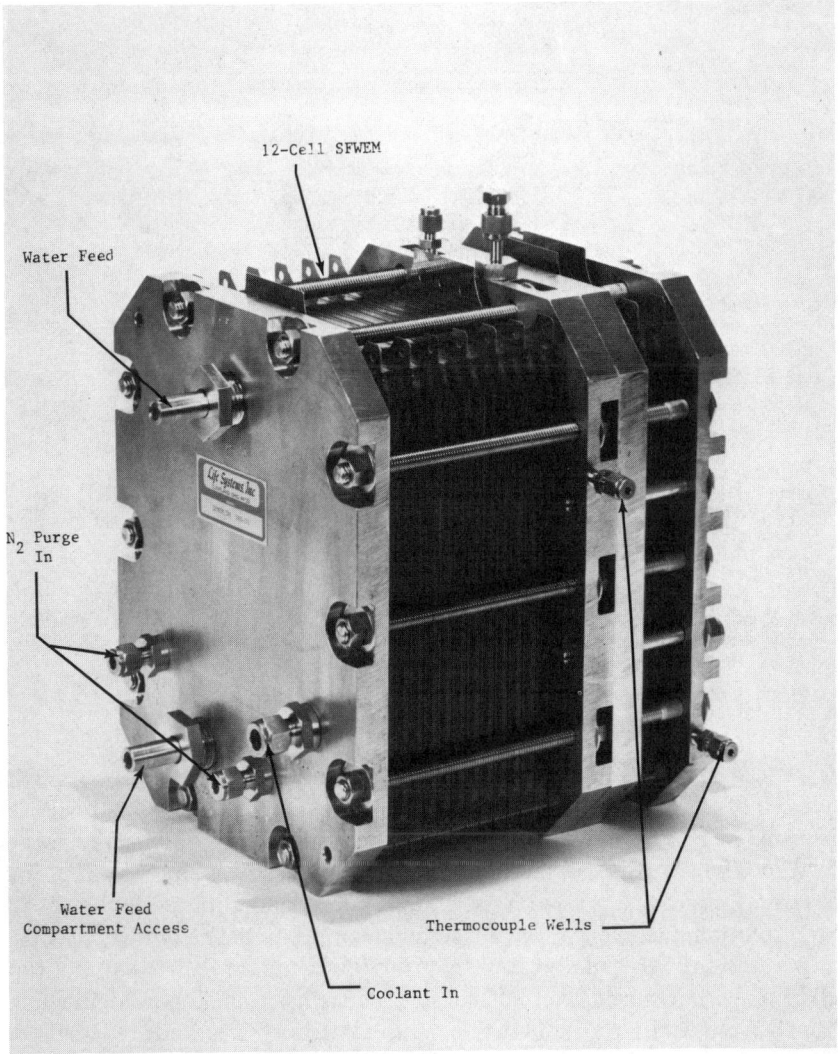

Figure 35. 12-cell water electrolysis module with electrolytic dehumidifier (with permission from Life Systems, Inc.).

### 3.4.2. Teledyne Energy System's Technology[38,63–65]

In mid-1972, a new approach to the design of water electrolyzers was introduced by Teledyne, designated as the HG series of hydrogen generators. The capacity of these cells is quite small, varying from 0 to 60 l/hr of hydrogen at STP. The product line was extended in 1973 with the HS series of hydrogen generators, with capacities varying from 1200 to 12,000 l/hr of hydrogen at

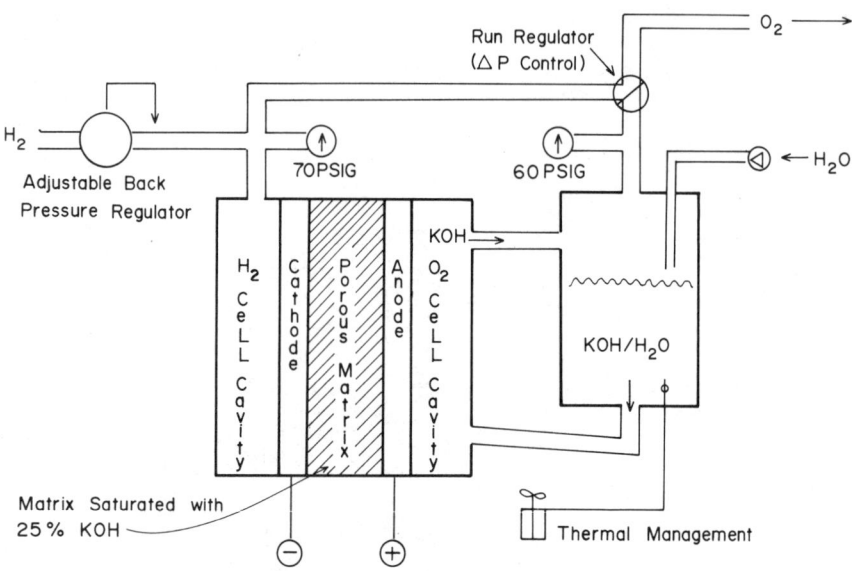

Figure 36. Principle of Teledyne's single irriguous electrolysis cell (with permission from Teledyne Energy Systems).

STP. Over 400 of the HG systems and 100 of the HS systems have been sold so far.

The approach to the cell design is quite unique, termed the *single irriguous style cell of water electrolysis,* in which only one electrode is flooded with the electrolyte (see Figure 36). Basically, the electrolysis cell contains the porous electrodes which are separated spatially and electrically by a piece of porous asbestos board or paper. With the asbestos separator saturated with a 25% KOH solution, the application of a suitable voltage to the electrodes results in the generation of $H_2$ and $O_2$ gases which flow from and through the appropriate cell cavities. The porous wetted asbestos matrix has a small pore diameter of about 0.7 $\mu$m, which precludes gas flow directly from the cathode to the anode. This property of the asbestos matrix allows the cell to operate with the hydrogen gas pressure higher than the oxygen gas pressure, and thereby ensures that all the free liquid remains within the anolyte compartment of the cell. Thus, with the hydrogen generated in a liquid-free compartment, only one electrolyte–gas separator, electrolyte tank, cooler, and pump is required.

In the commercial HS series of Teledyne cells, the 29.2-cm-diameter electrodes are woven screens of nickel 200 wire. The screen is 60 mesh with a 0.028-cm-diameter wire. Adjacent and contiguous to the electrode is a flow spacing agent to define the major dimensions of the fluid cavities. This expanded nickel member provides the electrical contact from the bipolar plate to the electrodes.

# ELECTROLYTIC PRODUCTION OF HYDROGEN

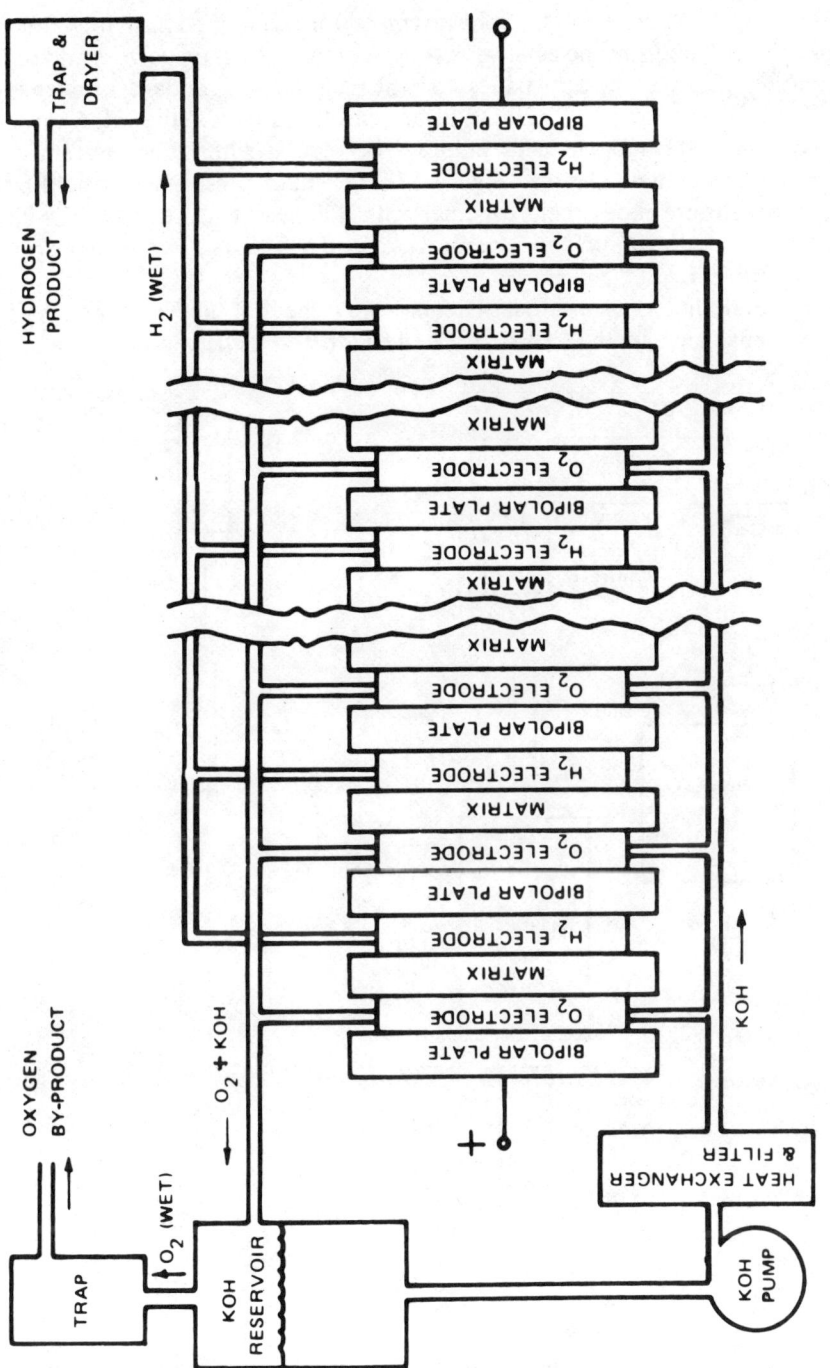

Figure 37. Schematic of Teledyne electrolysis module (with permission from Teledyne Energy Systems).

The metal members and the asbestos matrix are contained in an injection-molded P-1700 polysulfone cell frame which contains the fluid porting and the distribution passages and provides for the cell seals. Standard ethylene–propylene EPT O-rings or Teflon seals are employed to seal the cell frames.

Both the HG and HS cells employ the bipolar filter press system of assembling individual cells (see Figure 37). The filter press stack is held by stainless steel end plates, held together with a series of tie rods fitted with belleville springs to maintain the compressive load on the seals between the cell frames.

Figure 38 illustrates a simple schematic flow sheet of the Teledyne single irriguous approach to the cell and its effect on the overall gas generation

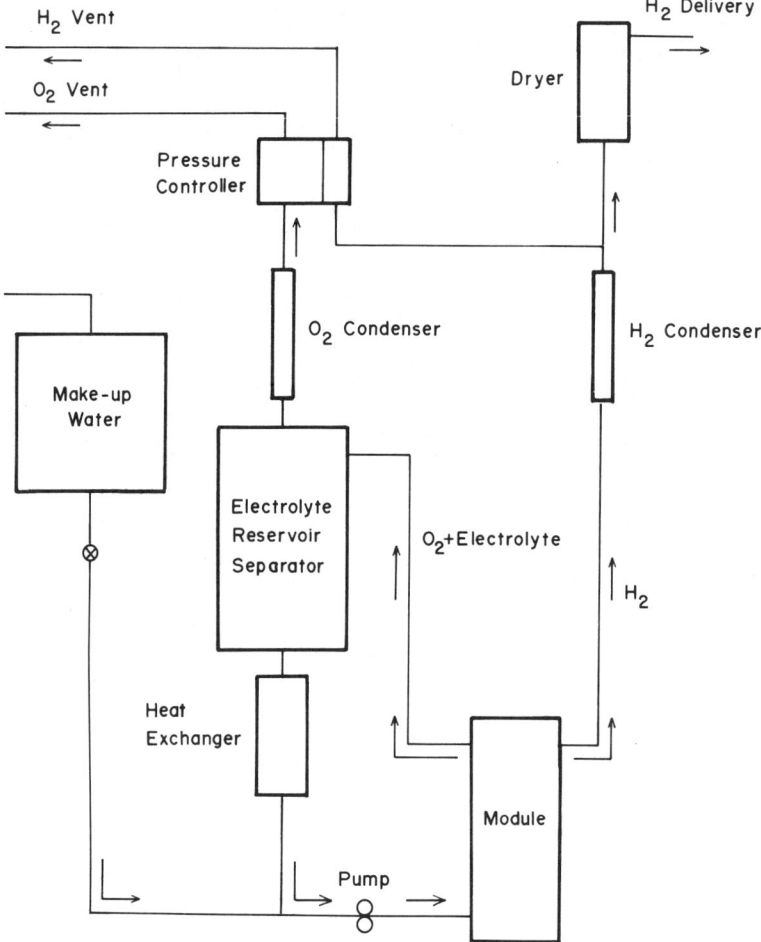

Figure 38. Schematic of Teledyne Electra cell HS series gas generator systems (with permission from Teledyne Energy Systems).

system. The entire gas generation system is contained within a cabinet 0.84 m deep by 1.88 m wide by 1.63 m high, with a total weight of about 1350 kg (see Figures 39a and 39b). The overall characteristics of the Teledyne system are as follows:

> hydrogen pressure: 75 psig
> electrolyte temperature: 32–66°C
> system power consumption†: ~6.0 kWh per normal m³ of $H_2$

## 3.5. Projected Advances in Alkaline Water Electrolysis Technologies

In view of increasing energy and manufacturing costs, all the leading manufacturers whose technologies were discussed earlier (see Sections 3.3 and 3.4) are concentrating their efforts to improve their technologies[32,105,129] in terms of lowering the power consumption and capital costs (see Section 5), with the ultimate goal being aimed at achieving an energy efficiency of ~100% by operating the electrolyzers at cell voltages close to the thermoneutral voltage (see Section 3.1 for details). To minimize and/or avoid extensive development costs towards new equipment, the general approach appears to be in the direction of modifying current alkaline electrolyzer technology by incorporating "advanced" anodes, cathodes, separators, and optimizing the operating conditions and cell designs.

As discussed in Section 3.2, the major energy losses in operating a water electrolyzer are associated with cathodic and anodic overpotential, and ohmic drop across the separator and cell hardware (see Table 2). Hence, extensive research and development efforts are focused towards developing activated anodes and cathodes, thin separators to lower the ohmic drop, and cell designs that provide a narrow gap between the anode and the cathode. Thus, to achieve[129] a cell voltage of 1.6 V at 400 mA/cm², considerable developmental efforts are required to lower the overpotential losses to ~0.25 V (compared to the present value of ~0.7–0.8 V at the same current density) and the total ohmic drop to about 0.16 V. The former can be accomplished by using "activated" electrodes at moderately elevated temperatures (e.g., 140°C or less). The ohmic drop can be reduced by using (a) thinner separators of thicknesses in the range 0.1–0.5 mm with low resistance (compared to ~4-mm thickness of the asbestos separator with a resistance of ~1.0–1.3 $\Omega\,cm^{-2}$) and (b) smaller anode–cathode gap. In addition to lowering the manufacturing costs, the search and testing of materials with long-term stability particularly at higher temperatures and lower cost, is in progress. Most of the research efforts in the above areas by the various water electrolyzer manufacturers is proprietary, and hence only limited information available in the public domain is outlined here.

† Includes auxiliary power, equivalent $H_2$ purification losses for 99.9996% pure product and ac to dc power consumption losses.

1. SOLID STATE BRIDGE RECTIFIER
2. WATER QUALITY INDICATOR
3. SHUTDOWN REGULATOR
4. PRESSURE TRANSDUCER
5. $O_2$ PRESSURE GAUGE
6. PREPRESSURE REGULATOR
7. $H_2$ PRESSURE GAUGE
8. PRESSURE SWITCHES
9. BACK PRESSURE REGULATORS
10. ELECTROLYSIS MODULE
11. MOLECULAR SIEVE DRYERS
12. CHOKE
13. POWER TRANSFORMER
14. CONTROL BOX

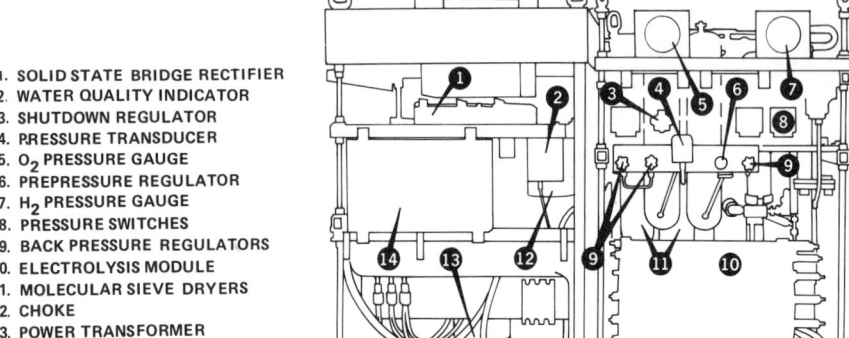

Figure 39a. Schematic of Teledyne Electra cell HS series hydrogen generator system (with permission from Teledyne Energy Systems).

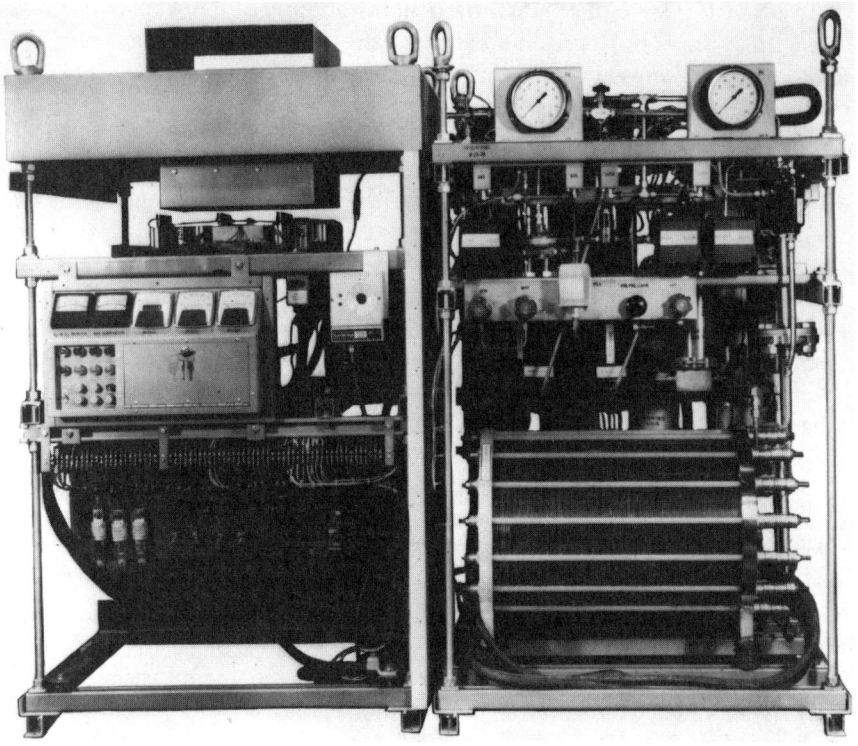

Figure 39b. Typical Teledyne Electra cell HS series hydrogen generator system described in Figure 39a (with permission from Teledyne Energy Systems).

Electrolyser Corporation[10,46] is actively engaged in developing its monopolar technology to produce hydrogen at the same costs as are projected for SPE technology. The projection for 1983 includes achieving a cell voltage of 1.74 V (@ 134 mA/cm$^2$) compared to the present value of 2.04 V, at an energy efficiency of 87%, by reducing the overvoltage contribution by 75 mV and the ohmic contribution by 50%. Ni-boride-based materials for cathodes and $NiCo_2O_4$, prepared by the freeze-dry technique, for anodes are being tested[10] for potential application in its future generation cells.

Brown Boveri & Cie[49] is aiming at enhancing the electrolyzer efficiency by electrocatalysis and by increasing the current density from 2 to 3.4 kA/m$^2$, which will allow higher capacities with single modules. By the end of 1978, BBC projected building electrolyzer modules with hydrogen generating capacities of 500 m$^3$/hr.

In-house developments of Norsk Hydro technology[51] include new diaphragms, improved cell designs for operation at higher loads, and in a joint venture with Electricite de France, work continues on improved active electrode coatings. Results from plant operation with new active coatings on electrodes show cell voltages as low as 1.71 V (at a load of 3600 A at 80°C and 25% KOH) after one year of operation.

To achieve the economic optimum, Lurgi Gmbh[1] is concentrating on reducing both the power consumption and the capital costs. To reduce the power requirements, the operating temperature is being raised from 90°C to 130–150°C. Improved asbestos separators with an order of magnitude lower resistance than the present asbestos diaphragms have been developed to withstand the higher operating temperatures. In addition, via improved electrode activation, the power consumption is reduced without loss in cell performance characteristics over a period of two years. To lower capital costs, the cell diameter will be increased to 1.6 m to achieve high volume production. It is projected that the energy consumption could be lowered to 100 kWh/MSCF (3.531 kWh per normal m$^3$ of $H_2$) at an operating pressure of ~450 psig and a temperature of 90°C—the current density being 2 kA/m$^2$.

DeNora S.P.A.[19,40] is actively engaged in developing novel ruthenium and rhenium deposited (by galvanic or plasma-spraying techniques) on Fe electrodes for use as cathodes, and Ni (or Co), stainless-steel-coated Ni, or Ni–Fe as anodes to lower the unit energy consumption. The low loading levels ($\leq 50$ g/m$^2$ for anodic coatings and $\leq 20$ g/m$^2$ for cathodic coatings) of catalysts on the substrates make them attractive from a cost viewpoint. Other coatings, which consist of Ni or Fe sulfides, exhibit low overpotentials even at 10 kA/m$^2$ and long-term stability with respect to coating wear and voltage variation. The projected power consumption at a current density of 5 and 10 kA/m$^2$ is 4.27 kWh/m$^3$ of $H_2$ and 5.12 kWh/m$^3$ of $H_2$, respectively.

Teledyne Energy Systems (TES), under contract with Brookhaven National Laboratory and sponsorship of the Department of Energy, has been studying[38,63,130] the various parameters relevant to lowering the cost of electrolytic production of hydrogen. A moderately sized Applied Research and Industrial Electrolysis System (ARIES) was built for testing purposes at temperatures up to 125°C and optimization studies. Developments in the major areas of attention[38,63,130] are noted below:

(a) *Metal Corrosion*. For the KOH-wetted fluid lines, tanks, etc., stainless steel 316 was found to offer better corrosion resistance than stainless steel 304 at high operating temperatures.

(b) *Materials within System Components*. Areas of concern within the system include the bearings and seals of the circulating electrolyte pumps, valve seats and seals, and the electrolyte filter element materials for operation at >80°C. To minimize the downtime due to failure of the above components, materials offering superior chemical resistance are being tested. Promising materials for seals and elastomers would include Perfluoroalkoxy (PFA), Teflon (PTFE), or Kalrez; asbestos- or potassium-titanate-fiber-reinforced polysulfone or polyphenylene sulfide for bearing and structural applications; filter elements prepared from polyphenylene sulfide or polybenzimidazole (PBI).

(c) *Materials for Cell Frame*. Several materials were tested and polyphenylene sulfide appears satisfactory for operation up to 150°C, polysulfone up to 125°C, and polyarylsulfone and polyethersulfone at temperatures lower than 100°C. Asbestos- or potassium-titanate-fiber-reinforced thermoplastics would offer even greater strength possibilities.

(d) *Separators*. Various thermoplastic separators which will allow a differential pressure of up to 4 atm with a fiber diameter of 5 $\mu$m and wettable, are being examined to replace the presently used asbestos diaphragms. Teflon-bonded potassium titanate and polybenzimidazole which are stable up to 100°C appear promising in preliminary tests.

(e) *Electrodes*. Increasing the temperature of operating cells with conventional nickel wire screen electrodes was shown to raise the voltage efficiency from 65% to 77.5% at 125°C. For further improvements, increasing the electrode surface area and investigations with both anode and cathode electrocatalysts were conducted.

The tests with several types of anodes indicated (a) $Co_2NiO_4$ was not significantly effective as an $O_2$ evolution electrocatalyst and (b) only a slight improvement was observed when the anode roughness factor with and without $Co_2NiO_4$ was increased from about 4 to values in excess of $10^4$.

Changes to the cathode were of greater significance. Of the four cathodes tested at 100°C (operating pressure 4 atm), $Ni_2B$-coated nickel screen, proprietary cathode TES-C-110 with Pt (<1 mg/cm$^2$) showed a cell voltage of 1.88 and 1.84 V, respectively (450 mA/cm$^2$), while nickel screen and the plain TES-C-110-0 (high-surface-area screen) showed a cell voltage of 2.05 and 1.965 V, respectively. Operating the TES-C-110/Pt cathodes at 125°C allowed 1.70 V (450 mA/cm$^2$) equivalent to 87% voltage efficiency.

Thus, operating advanced electrocatalytic and novel materials at high current densities and at elevated temperatures is the general approach for reducing the power consumption of the TES electrolysis modules from 27.7 kWh/lb of $H_2$ (5.5 kWh per normal $m^3$ of $H_2$) to 21 kWh/lb of $H_2$ (4.2 kWh per normal $m^3$ of $H_2$).

## 4. Solid Polymer Electrolyte (SPE) Water Electrolyzers

### 4.1. Thermodynamics

As discussed previously (see Section 3), most conventional water electrolyzers employ ~25% KOH(aq) as the electrolyte. However, more recently, General Electric Company has developed a unique solid polymer electrolyte (SPE) water electrolysis technology[66-69] wherein a solid sheet of Nafion perfluorosulfonic acid membrane serves the purpose of the "electrolyte." Teflon-bonded catalysts are pressed onto the surfaces of the SPE membrane to generate an intimate electrode–electrolyte contact. During electrolysis, the electrochemical reaction taking place at the anode of an SPE cell is

$$6H_2O(l) \rightarrow 4H_3O^+ + 4e^- + O_2(g) \qquad (19)$$

and at the cathode, the hydroxonium ions are discharged to produce $H_2$ gas as

$$4H_3O^+ + 4e^- \rightarrow 4H_2O(l) + 2H_2(g) \qquad (20)$$

the overall reaction being the same as represented by Eq. (2). Hence, the thermodynamic aspects of water electrolysis are identical to those formulated for alkaline water electrolyzers and discussed in Section 3.1.

### 4.2. Electrode Kinetics

#### 4.2.1. Hydrogen Evolution Reaction

Unlike the case of alkaline water electrolyzers where $H_2O$ molecules are discharged to generate $H_2$ gas, the hydrogen evolution reaction occurs from the discharge of hydroxonium ions since the hydrated SPE used in the water electrolysis cell is highly acidic, with a pH equivalent of a 10 wt % $H_2SO_4$ solution.[70] The electrode kinetic and electrocatalytic aspects of this reaction have been extensively investigated[11,12,14,25,71] and discussed; the generally accepted mechanism in acidic media being the discharge of hydroxonium ions leading to the formation of adsorbed atomic hydrogen followed by their removal by either recombination [Eq. (22)] or electrochemical desorption [Eq.

(23)] as given by the following equations:

$$S + H_3O^+ + e \rightarrow S\text{-}H + H_2O(l) \tag{21}$$

$$2S\text{-}H \rightarrow 2S + H_2(g) \tag{22}$$

$$S\text{-}H + H_3O^+ + e \rightarrow H_2O(l) + H_2(g) \tag{23}$$

(S refers to the active site on the cathode surface.)

In SPE water electrolysis cells, platinum is currently used as the electrocatalyst on the cathodes—the probable mechanism of $H_2$ evolution on Pt in acid solutions[11,25,72] being fast discharge followed by rate-determining recombination [Eq. (22)]. On the highly activated platinum black used in the SPE cell, the hydrogen evolution reaction is probably controlled[73] by diffusion of molecular hydrogen away from the electrode surface.

### 4.2.2. Oxygen Evolution Reaction

The anodic evolution of oxygen is a highly irreversible process, requiring considerable overpotential. In acidic media, as encountered in SPE cells, only noble metals and alloys are suitable for use[12,25,71,74,75] as electrocatalysts for the oxygen evolution reaction. This reaction has been extensively studied on platinum electrodes (or more precisely, on oxidized platinum electrodes);[76] and a large number of probable mechanisms have been suggested.[77-80] Taking into account the results of various theoretical and experimental investigations, the generally accepted mechanism for the anodic evolution of oxygen on oxide-covered platinum surfaces under the Langmuir conditions of adsorption of intermediates is as follows[12,25,71,74,75]:

$$S + 2H_2O(l) \rightarrow S\text{-}OH + H_3O^+ + e^- \tag{24}$$

$$2S\text{-}OH \rightarrow S + S\text{-}O + H_2O(l) \tag{25}$$

$$2S\text{-}O \rightarrow 2S + O_2(g) \tag{26}$$

in which the primary transfer process [i.e., Eq. (24)] is the rate-determining step. Studies of the kinetics of the oxygen evolution reaction from acid media have also been carried out on other noble metals and noble metal alloys.[12,24,71,74,75] As inferred from the observed electrode kinetic parameters, a reaction mechanism similar to the one for platinum electrodes [i.e., Eqs. (24)–(26)] was reported for gold,[81] palladium,[81] and ruthenium[82] in the entire potential range, and for iridium[83,84] and rhodium[83] in the high overpotential regions. Dual mechanisms were found for the oxygen evolution reaction on iridium and rhodium electrodes.[82,83] At low overpotentials, the proposed rate-determining step for oxygen evolution on these two electrodes is

$$S\text{-}OH + 2H_2O \rightarrow S\text{-}O\cdots H\cdots OH^- + H_3O^+ \tag{27}$$

following the primary electron transfer reaction [Eq. (24)].

## 4.3. Electrode Configurations

In the SPE water electrolysis technology, a solid sheet of Nafion perfluorosulfonic acid membrane of 10–12 mils in thickness is currently employed as the electrolyte as well as the separater material. As illustrated schematically in Figure 40, Teflon-bonded catalysts in the form of fine particles are pressed, using a proprietary manufacturing procedure, onto each face of the SPE membrane to form a structurally stable membrane–electrode assembly.[66,68–70,85] Charge carriers in the (wet) membrane are hydroxonium ions ($H^+ \cdot xH_2O$), which migrate through the solid electrolyte by passing from one sulfonic acid group to the adjacent one. The sulfonic acid groups are chemically bound to the perfluorocarbon backbone and do not move; thus the concentration of hydrated ions remains constant within the SPE membrane.[86]

While operating an SPE water electrolyzer, deionized pure water is circulated at a sufficiently high flow rate (to remove the waste heat) over the anode where it is decomposed electrochemically, producing oxygen gas, hydroxonium ions, and electrons. As shown in Figure 40, the overall reaction at the anode can be represented by Eq. (19). The hydroxonium ions move through the SPE membrane and then recombine with electrons, which pass via the external circuit, to form hydrogen gas at the cathode. $H_2$ and $O_2$ gases are generated at a stoichiometric ratio at any desired pressure. Depending upon

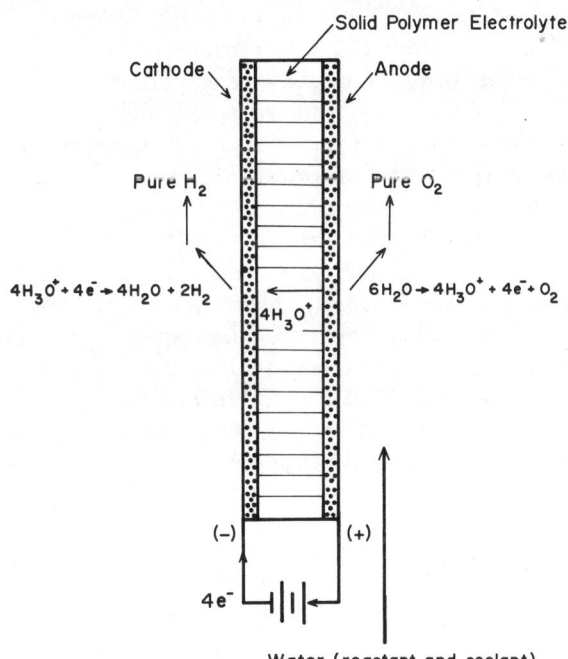

Figure 40. Schematic diagram of an electrode–membrane assembly.

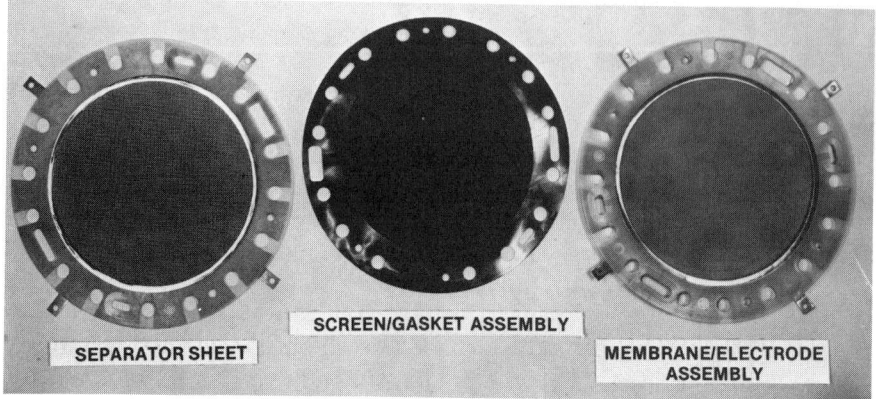

Figure 41. Cell components for an SPE water electrolyzer (active electrode area approximately 335 cm$^2$) (with permission from General Electric Company).

the operating temperature and the differential pressure across the SPE membrane, a small quantity of water, coupled with the hydroxonium ions, penetrates through the membrane and is finally discharged with the hydrogen gas evolved at the cathode. At 100°C, for example, there are approximately 3.5 to 4.0 water molecules transported with each hydrogen ion.[70]

Figure 41 illustrates[68,69] the membrane–electrode assembly for the SPE water electrolyzers wherein the electrocatalysts pressed on to the SPE membrane are in the form of fine particles exhibiting extremely high specific surface areas.[68,69] The interelectrode spacing is approximately equivalent to the thickness of the membrane, which is in general not more than 0.03 cm (i.e., 12 mils). It may be noted that at the present time the Lurgi-pressurized alkaline water electrolyzers are constructed of single cells of 1-cm thickness.[29] Due to the maximization of active surface areas of the electrodes and the minimization of interelectrode spacing, the SPE water electrolyzers are capable of operation at current densities of over 1 A/cm$^2$, which is at least five times higher than the production rate of hydrogen in conventional alkaline water electrolyzers with a cell voltage of 1.85–2.0 V at 80°C and of about 1.6 V at 150°C. However, the electric resistance at the SPE–electrode junction contributes[87] a significant loss to the overall efficiency of an SPE water electrolyzer. This ohmic loss within the junction can be reduced by optimizing the bonding between catalyst and SPE membranes.

### 4.4. Electrolytic Cells—Design Aspects

#### 4.4.1. Cell Configuration

The SPE water electrolyzers, as shown in Figures 40–43 are designed and constructed on the basis of a special filtration-type configuration.[6] Prior to

Figure 42. A 27-cell electrolysis module for use in a space craft life support system (with permission from General Electric Company).

1975, SPE water electrolyzers were generally composed of three basic components[66,68,69]: membrane–electrode assembly, separate sheet and screen–gasket assembly, as shown in Figures 40 and 41. On each side of the membrane–electrode assembly, a screen–gasket assembly was incorporated to form either the hydrogen or oxygen compartments. The screen package, composed of multilayer expanded metal screens, serves as a membrane support to allow operation at high differential pressures across the membrane and also provides electronic conductivity to electrocatalysts on the SPE membrane. A flat silicone rubber gasket attached to the screen package (see Figure 41) was used for sealing the gas compartments and manifolding the feed-water and evolved gases between cells. The separator sheet separated the hydrogen compartment of one cell from the oxygen compartment of its adjacent cell. The bipolar current collector was placed between two adjacent cells in the stack by means of the screen packages contacting the common metallic separator sheets. The cells were in turn stacked between rigid end plates as illustrated in Figure 42, which shows a 27-cell electrolysis module developed for use as an oxygen generator in a spacecraft life support system. With a view of eliminating corrosion problems, the metallic components in the cells were normally fabricated of either titanium alloys or niobium. Spring-loaded tie rods, located in the cell gasket area, were employed to maintain adequate gasket and cell compression.

In the past three to four years there has been considerable progress in the optimization of cell configuration.[85,88] Figure 43 depicts the schematic representation of two consecutive SPE water electrolysis cells. Foil-backed

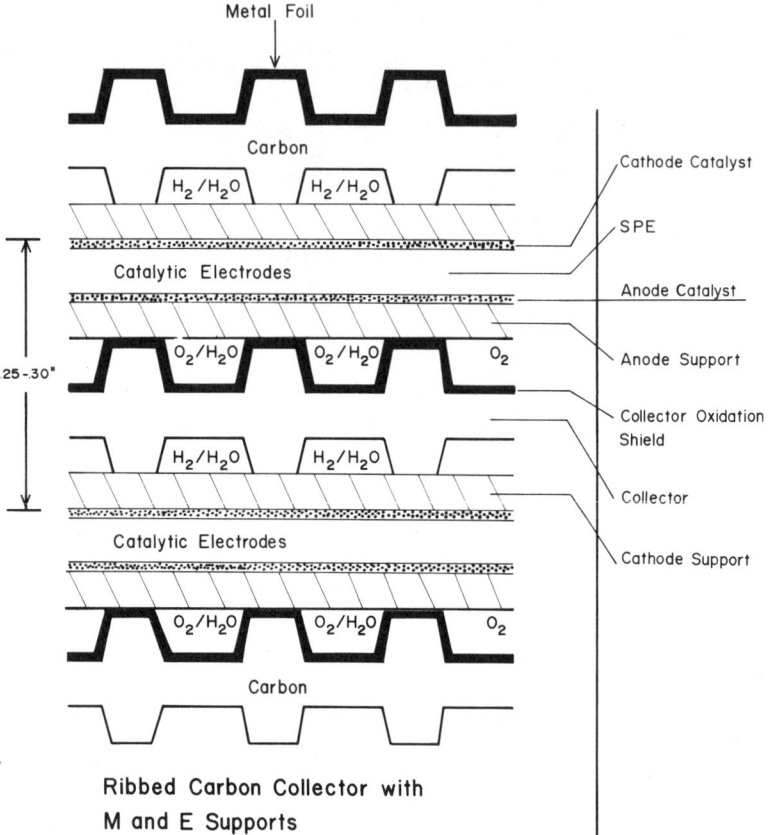

Figure 43. Schematic representation of two SPE single cells connected in series (with permission from *J. Appl. Electrochem.*).

molded carbon plates have been developed to replace the expensive transition metal screens as the current collectors in the SPE water electrolyzers. The molded carbon current collectors, as shown in Figure 44, are fabricated by pressing a mixture of carbon and phenolic resin on a thin ribbed titanium foil. The incorporation of the molded carbon current collectors results in a gasketless sealing configuration which not only eliminates the need for costly silicone rubber gaskets, but also provides a more reliable seal, allowing a leak-tight operation of up to 35–50-atm gas pressures. It may be noted that in the improved cell configuration, the SPE membrane itself acts as a gasket between the sealing surfaces of two consecutive current collectors. As seen in Figure 43, the titanium side of a molded collector, which exhibits a high corrosion resistance at any anodization potentials, is exposed to the oxygen chamber of an SPE cell. As a result, the carbon side of this collector is incorporated into the hydrogen chamber of its adjacent cell.

OXYGEN SIDE (ANODE)   HYDROGEN SIDE (CATHODE)

Figure 44. Foil-backed molded carbon current collector (laboratory size) (with permission from General Electric Company).

The SPE membrane is normally devoid of dimensional stability, particularly at high temperatures. Therefore, electrode supports are required to prevent the creep of membrane–electrode assemblies into the flow fields within the molded current collectors. The criteria for the selection of electrode supporting materials are (a) high electric conductivity, (b) adequate porosity to allow the fluids to pass to and from the electrode surfaces, and (c) sufficient dimensional stability to resist the membrane creep. At present, a thin sheet of porous carbon fiber paper is used as an electrode support on the cathode side, as shown in Figure 43.[85,88] On the anode side, the electrode support is made of either perforated titanium foil or porous titanium plate, which is fabricated by acid etching or a powder metallurgy process, respectively.[89] A number of electrolysis cells, arranged in series electrically and with parallel fluid feed and discharge, as illustrated in Figure 43, are assembled between pneumatically loaded end plates to form an electrolysis module. The end plates are currently constructed of steel.

Table 9 shows the components of a SPE water electrolysis cell voltage operating at a current density of 2 A/cm$^2$.[90] It can be seen that the ohmic losses contribute a substantial portion to the overall cell voltage. These voltage losses due to ir drops can be minimized by optimizing the cell configuration, and extensive efforts are being made to achieve further improvement in the cell design of SPE water electrolyzers. Apart from the optimization of cell configuration, the basic technology development program[32,70,85,88,89] at the General Electric Company is aimed at (a) high-activity electrocatalysts, (b) low-cost current collectors, (c) high-temperature–high-pressure capability, and (d) low-cost SPE, in parallel with a hardware scale-up effort. The progress made up to the present time (1979) and the needed research and development activities in each of these areas are presented in the following sections.

Table 9
Measured and Projected Cell Potential of an SPE Water Electrolyzer Operating at 2 A/cm$^2$

| Parameter | Existing technology in 1974 (82°C) | Projected potentials for 1980 (120°C) | Projected potentials for 1985 (150°C) |
|---|---|---|---|
| $E_0$ | 1.18 | 1.15 | 1.12 |
| $\eta_c$ | 0.04 | 0.04 | 0.04 |
| $\eta_a$ | 0.38 | 0.30 | 0.20 |
| $\eta_\Omega$ | 0.45 | 0.20 | 0.19 |
| $\eta_{hw}$ | 0.05 | 0.05 | 0.05 |
| $E_{cell}$ | 2.10 V | 1.74 V | 1.60 V |

### 4.4.2. Investigation of Electrocatalysts for SPE Water Electrolyzers

The slowness of the oxygen evolution reaction results in significant efficiency loss in water electrolyzers.[25,74,75] Attempts have been made to find more reversible oxygen catalysts. As has been pointed out in the literature,[83,84,87,91–93] iridium and noble metal alloys exhibit higher electrocatalytic activities than platinum for the oxygen evolution reaction from acidic media. As shown in Figure 45, iridium has about 100 times the exchange current density as that of platinum.[91] Although ruthenium is the most active noble metal for the oxygen evolution reaction,[94,95] the high electrocatalytic activity of this metal is marred by its rapid corrosion in the potential range of oxygen evolution.[94,96] Although ruthenium oxide ($RuO_x$) prepared by thermal decomposition technique exhibits[97] higher corrosion resistance than

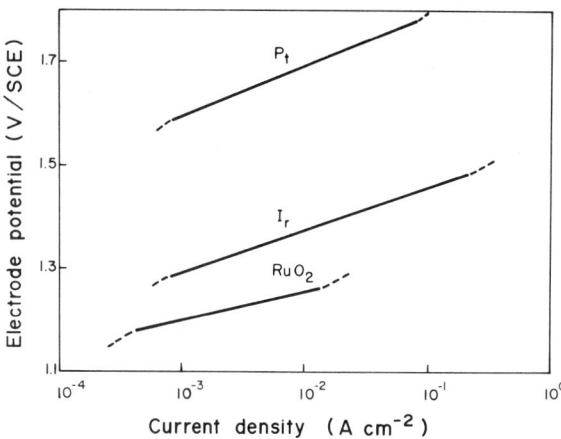

Figure 45. Tafel plots for oxygen evolution on Pt, Ir, and $RuO_2$ in 1 $M$ $H_2SO_4$ at 80°C (with permission from Miles and Srinivasan).

metallic ruthenium, significant performance degradation of $RuO_2$ was observed presumably because of the gradual accumulation of $RuO_2$ film on the catalytic surface of $RuO_x$ particles by a dissolution–precipitation mechanism.[97]

At present, the chemical composition of anode catalysts used in commercial SPE water electrolyzers is still proprietary. The main purpose in the catalytic electrode development is to achieve high-activity and low-cost anode catalysts. Binary and ternary alloys of Pt, Ir, Ru, and other transition metals (e.g., Ti, Ta, Nb, Zr, W, Hf), as well as the mixed oxide catalysts have been investigated for oxygen evolution in SPE water electrolyzers. As seen from Figure 46, a mixed oxide catalyst, designated as E-50 at General Electric, exhibits much lower oxygen overpotentials than a Pt-black anode. The catalyst E-50, having a BET surface area in the range 40–160 $m^2/g$, is probably prepared by the reduction of Pt–Ir mixed oxides. The performance of this catalyst for oxygen evolution is currently considered as a base case in the development of alternate anode materials.

According to some estimates,[97,98] an improvement of at least 70 mV in oxygen overpotential is required in order to meet the performance objective. It has been identified that the catalyst WE-3, which exhibits low oxygen overpotential and high performance stability as illustrated in Figure 47, is most promising for use as an anode catalyst in SPE water electrolyzers.[85,98] In an endurance test lasting 3000 hr, the SPE cell using WE-3 anode showed an average cell voltage of ~37 mV better than the base case. It is believed that the catalyst WE-3 is probably composed of reduced mixed oxides containing Ru, Ir, and Ta, which is much less expensive than the Pt–Ir mixed oxides. Therefore, a significant reduction in the anode fabrication cost can be anticipated by using a WE-3 anode.

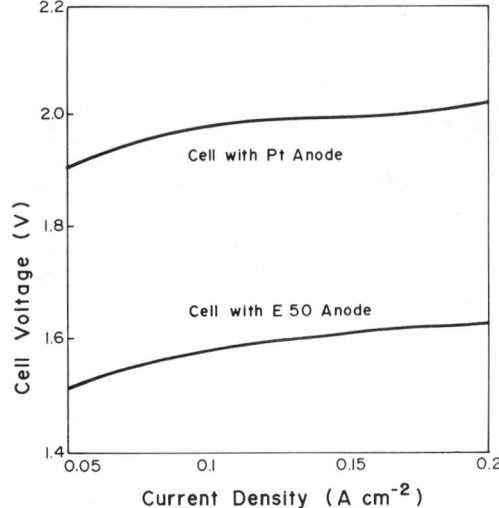

Figure 46. Comparison of performance of two SPE water electrolyzers using E-50 and Pt black as anode catalysts (at 49°C) (with permission from General Electric Company).

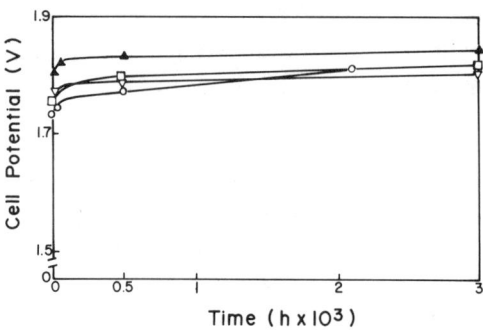

Figure 47. The endurance test of various anode catalysts at approximately $1 \text{ A/cm}^2$ and 82°C (with permission from General Electric Company). Anode and cathode loadings, $4 \text{ g/ft}^2$. □: Cell, 05-1; anode, WE-2; cathode, Std Pt. ○: Cell, 125-1; anode, WE-1; cathode, Std Pt. ◇: Cell, 17-2; anode, WE-3; cathode Std Pt. ●: Cell, baseline; anode, WE-4; cathode, Std Pt.

With a view to achieving further cost improvement, efforts have been made to reduce the noble metal loading in anode catalysts through the incorporation of non-noble metals.[70,85,88,89,98,99] According to LaConti, Fragala, and Boyack,[70] anode noble metal loadings have been successfully lowered from $4 \text{ mg/cm}^2$ (in 1976) to approximately $1 \text{ mg/cm}^2$, without loss in the electrode performance. At the present time (1979), a series of anode catalysts with loadings at a level of $0.25 \text{ mg/cm}^2$ are being evaluated at General Electric. In short-term tests, some of these low-loading catalysts exhibited decent performance characteristics for oxygen evolution, but information regarding their long-term stability is still not available.

Figure 48 illustrates the cost reduction and performance improvements of two promising anode catalysts, WE-3 and E-100, in comparison to the base case E-50.[85,89] With catalyst WE-3, one is able to achieve 40–70% of the cost reduction goal, along with an improvement of ~50 mV in the electrode performance. An SPE cell using E-100 anode, which probably consists of iridium oxide, fabricated by a dry process, shows promise to realize more than 90% of the cost reduction. At present, however, the long-term stability of catalyst E-100 (loading, $0.25 \text{ mg/cm}^2$) for oxygen evolution is still doubtful. The continuing experimental work on the reduction of noble metal loadings in anode catalysts is aimed at a desired goal of $0.18 \text{ mg/cm}^2$.[32]

In the research and development of hydrogen catalysts for SPE water electrolyzers, focus has been mainly on the reduction of noble metal loadings through the use of graphite or transition metal silicide supports. Prior to being fabricated into an electrode structure, graphite or silicide in the form of powder is sputtered with platinum to increase both the electronic conductivity and the electrocatalytic activity. At a current density of $\sim 1 \text{ A/cm}^2$, the SPE water electrolysis cells, which were constructed using platinum sputtered on graphite substrates as the cathodes (with loadings ranging from 0.25 to $1.0 \text{ mg/cm}^2$), showed a cell potential that is only 30 mV higher than the base case cell with a cathode composed of pure Pt black of a loading of $2 \text{ mg/cm}^2$.[70] However, the SPE water electrolyzers, using silicide as a diluent of noble metals on the cathode, exhibited an extremely poor performance, presumably because of the

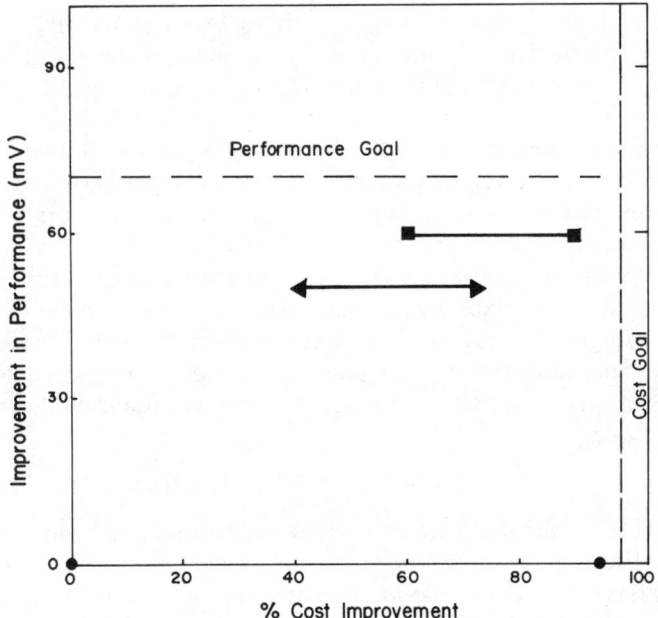

Figure 48. Cost reduction versus performance improvements for new anode catalyst WE-3 and E-100 (with permission from General Electric Company). ●, E-50; ■, E-100; ▲, WE-3.

high resistance of this material. More recently, Pt–Pd alloys at 0.25 mg/cm$^2$ loading have also been evaluated as cathode catalysts in SPE water electrolyzers. At $\sim$1 A/cm$^2$ and 82°C, the performance of an SPE cell incorporated with a Pt–Pd cathode, is about 20 mV worse than that of a base case cell voltage. The ultimate goal in the cathode development is to reduce its loading of noble metals to a level of $\sim$0.02 mg/cm$^2$ with no significant sacrifice in the electrode performance.

### 4.4.3. Current Collector Development

In SPE water electrolyzers, a bipolar current collector serves mainly as (a) a conductor for electron transfer from the oxygen electrode of a cell to the hydrogen electrode of the adjacent one, (b) a separator of the hydrogen and oxygen gases evolved in the cathode and anode chamber, respectively, and (c) a mechanical support of the SPE membrane under operational differential pressures. It also provides the flow fields for the feed-water and evolved gases. Based on the present (1979) technology, about 47% of the capital cost of SPE water electrolyzers is for fabrication of the current collectors.[89] The criteria for the selection and evaluation of reliable and economical colector materials include (a) high electronic conductivity, (b) capability of being impervious to hydrogen and oxygen, (c) low contact resistance with electrocatalysts, (d) adequate corrosion resistance under operating conditions, (e) low fabrication costs, and (f) prolonged lifetime.

In the earlier work on SPE water electrolysis technology, transition metals or alloys that exhibit high corrosion resistance in acidic media were used to construct the current collectors. Of these niobium and Ti–0.2-wt%-Pd alloy, coated with a thin layer of platinum, showed contact resistance as low as gold,[100] and are thus most suitable for use as current collectors on the anode side. In the hydrogen chamber, a small quantity of hydrogen evolved may diffuse into the crystal lattice of the collector materials. Thus, hydrogen embrittlement takes place on cathode current collectors which results in the performance degradation of SPE water electrolyzers. Of some potential candidates that have been investigated as cathode current collectors in SPE cells, the resistance to hydrogen embrittlement was found to decrease in the following order[70]:

$$\text{graphite} > \text{Zr} > \text{Nb} > \text{Ti} = \text{Ta}$$

Since 1976, molded carbon current collectors have been developed at General Electric. The molded current collector is manufactured by incorporating a mixture of carbon powder and phenolic resin onto an *in situ* formed titanium foil shield. Small laboratory-sized current collectors as shown in Figure 44, have been evaluated in an SPE water electrolyzer for over 12,000 hr with satisfactory performance. An SPE water electrolysis cell using a molded carbon current collector has exhibited excellent performance up to current densities of ~5.5 A/cm$^2$, as seen from Figure 49.[98] The application of molded collectors in SPE cells has resulted in substantial reduction in the cost of electrolyzer production from \$260/ft$^2$ for the 1975 technology to the present \$64/ft$^2$.[89,133]

### 4.4.4. High-Temperature Capability

Most of commercial SPE water electrolyzers operate at ~80°C. As shown in Figure 50, temperature plays an extremely important role in improving the

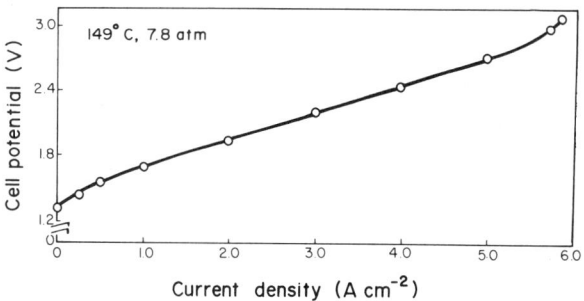

Figure 49. Performance characteristics of an SPE water electrolysis cell using molded current collector (with permission from General Electric Company).

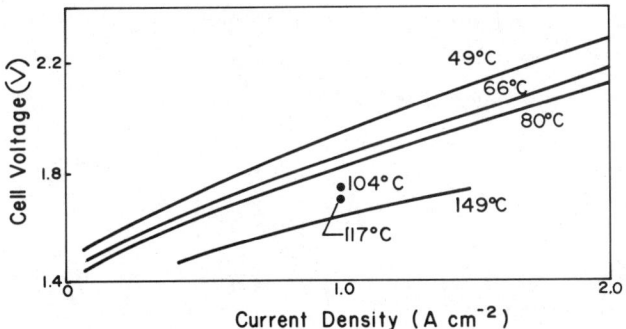

Figure 50. Effect of temperature on the performance of an SPE water electrolyzer (active cell area: approximately 46 cm$^2$) (with permission from General Electric Company).

performance of an SPE water electrolyzer. Thus, at a normally operating current density of 1 A/cm$^2$, an improvement of ~150 mV in the cell voltage can be achieved by raising the operating temperature from 80 to 149°C. Evaluation of various cell components is currently being conducted to define the life characteristics of SPE water electrolyzers at 149°C.[70]

There are disadvantages in operating the cells at high temperature, which include (a) enhanced parasitic losses arising from the diffusion of product gases across the SPE membrane, (b) additional SPE support requirements to prevent the membrane creep, (c) difficulties in finding chemically and mechanically stable materials for use as gaskets, sealants, and cell frames, and (d) enhancement of hydrogen embrittlement effects on the cathode side of a current collector. According to some estimates, the parasitic diffusion loss of hydrogen generated in the cathode chamber is as high as 8% if an SPE water electrolyzer operates at 149°C and at ~40 atm. The technical drawback on gasket and sealant materials has been overcome by the development of a "gasketless" sealing configuration. As discussed in Section 4.4.1, this "gasketless" sealing technique, in which the SPE membrane serves as a gasket-type seal, is capable of withstanding gas pressures up to 40 atm at elevated temperatures. At the General Electric Company, as SPE water electrolyzer using the Nafion® 120 membrane and molded carbon current collectors has operated for more than 5000 hr at a constant current density of ~1 A/cm$^2$ and at 149°C.[70,85,88,89] This cell shows an average voltage efficiency over 85% and a slight degradation in its performance with time.

### 4.4.5. Advanced Studies on SPE Membranes

Nafion® 120, which is currently used as the solid electrolyte in the SPE water electrolyzers, is a copolymer of polytetrafluoroethylene (PTFE) and polysulfonylfluoride vinyl ether containing pendant sulfonic groups,[86] the resulting chemical structure of Nafion membrane being

$$\{(CF_2CF_2)_x(CF_2CF)_y\}$$
$$\begin{array}{c} | \\ O \\ | \\ CF_2 \\ | \\ CF_3-CF \\ | \\ O \\ | \\ CF_2 \\ | \\ CF_2 \\ | \\ SO_3H \end{array}\Bigg]_z$$

Selected physical and chemical properties of Nafion® 120 membranes are given in Table 10. After equilibrating at 100°C in pure water, the SPE (wetted) membrane contains ~28 wt % of water and thus has reasonably low resistance for the transfer of hydroxonium ions (being ~0.46 $\Omega$ cm$^2$ at 25°C). In order to optimize the performance and lifetime of SPE water electrolyzers, each Nafion membrane is treated, prior to being impregnated with electrocatalysts on its surfaces, by introducing a small quantity of catalysts into the bulk of the membrane itself. Hydrogen peroxide and certain ions such as Fe$^{2+}$ that are probably present in the feed-water have been confirmed to attack Nafion structure, leading to considerable degradation in the performance of the cells with time.[70] Therefore, purification techniques for the deionization and the contamination control of feed-water are essential. In addition, the projected reduction of membrane thickness from the conventional 0.03 cm to the order of 0.015 cm would result in a substantial improvement of ~250 mV in the cell voltage at ~2 A/cm$^2$.[70] However, the parasitic losses arising from the gas permeation through the SPE membrane also increase with the reduction in the membrane thickness.

Table 10
Physical and Chemical Properties of Nafion 120 Membrane at 25°C[a]

| | |
|---|---|
| Equivalent weight | 1200 |
| Ionic (H$_3$O$^+$) resistance | 0.46 $\Omega$ cm$^2$ |
| Tensile at break | 2500 psi |
| Elongation at break | 150% |
| Mullen burst strength | 150 psi |
| Water content | 28 wt % |
| Ion exchange capacity | 0.83 meq/g dry SPE |
| H$_2$ permeability | $5.6 \times 10^{-4}$ cm$^3$ cm cm$^{-2}$ hr atm |
| O$_2$ permeability | $3.0 \times 10^{-4}$ cm$^3$ cm cm$^{-2}$ hr atm |
| Hydrodynamic H$_2$O permeability | $2.7 \times 10^{-5}$ cm$^3$ cm cm$^{-2}$ hr atm |
| Electro-osmotic permeability | $7.5 \times 10^{-4}$ cm$^3$ C$^{-1}$ |

[a] The membrane ~0.03 cm (wet) in thickness was equilibrated at 100°C in pure water.

The present cost (about \$25/ft$^2$) of Nafion membranes is far too expensive for the SPE water electrolysis system to be economical for industrial and utility applications. According to some reports, the production cost of Nafion membranes will not be lower than \$20/ft$^2$ in the near future. From an economic viewpoint, it is of crucial importance to develop low-cost SPE membranes. Attempts have been made at General Electric to evaluate several radiation-grafted ion exchange membranes such as trifluorostyrene and fluorocarbon for use as SPE in water electrolyzers.[85,88,89,99] All the candidate members lack long-term stability under water electrolysis operating conditions. Up to now (1979), the developmental efforts towards the identification of low-cost SPE membranes have not shown any positive results.[99]

Recently, Asahi Chemicals in Tokyo, Japan developed a perfluorocarboxylic acid membrane, which appears to be quite competitive with Nafion membranes for use as an ion-exchange membrane in the chloralkali cells. General Electric has contacted this membrane manufacturer to obtain the SPE for evaluation in water electrolyzers, but no information is available regarding its performance in SPE cells. Essentially, the ultimate goal in developing promising alternate membranes for SPE water electrolysis systems is to reduce the membrane cost to about 1/8 of the current price of Nafion membranes.[89]

### 4.4.6. Scale-Up of SPE Water Electrolysis Systems

Small SPE water electrolyzers for use as pure hydrogen suppliers for gas chromatographs and other laboratory apparatus have been commercialized since 1973.[85] As shown in Figure 51, these electrolysis units are composed of single cells with an active electrode area of 46 cm$^2$. For the large-scale production of hydrogen, water electrolyzers with such small electrodes are neither practical nor economical. Starting in 1976, General Electric became quite active in developing large-scale SPE water electrolyzers. The cell scaling-up program[88,89] is being conducted in two steps: an initial effort to scale-up the cell area from 46 cm$^2$ to ~2300 cm$^2$ cell development during 1979.

In the initial scaling-up program, an SPE water electrolysis module with a 50-kW power input was designed and constructed. As shown in Figure 52, this module consists of two stacks of SPE cells on either side of a center bus plate that serves as a negative electrode. The feed-water enters and water–gas mixtures leave each stack through manifold plates located adjacent to the center bus plate. The SPE cells within the stack are electrically in series with parallel fluid feed and discharge. The membrane–electrode assembly incorporated into each SPE cell is shown in Figure 53.[85,88,89]

At the end of each stack, an end plate containing a pneumatically loaded copper diaphragm that is sealed to the end plate cavity using an elastomeric sealant is incorporated (see Figure 52). The end plates (about 12 cm in thickness) are made of steel and act as the positive electrode terminals. This design for the end plates is capable of withstanding a compression pressure up

Figure 51. A small SPE water electrolysis module with an active area of approximately 46 cm$^2$ (with permission from General Electric Company).

to ~48 atm and a gas-generation pressure to ~40 atm. Figure 54 shows the actual end plates assembled in the 50-kW SPE water electrolysis module.[88] The evaluation of this module is planned for a start-up in 1979. More recently, General Electric has started the work on the design and assembly of a 200-kW water electrolysis module by increasing the number of SPE cells with an active electrode area of ~2300 cm$^2$.

According to General Electric,[85,88,89] the scaling-up program is aimed at the construction of a 5-MW water electrolysis system during 1983. The projected performance characteristics of this full-scale demonstration system are presented in Table 11.[85,88,89,98] Like the conventional alkaline water

Table 11
Projected Performance Characteristics of a 5-MW SPE Water Electrolysis System

| | | |
|---|---|---|
| Hydrogen pressure | 6.8 atm | 40 atm |
| Operating temperature | 149°C | 149°C |
| System energy efficiency | 90% | 85% |
| Electrolyzer voltage efficiency | 93% | 88% |
| Cell potential at 1 A/cm$^2$ | 1.58 V | 1.63 V |
| Active cell area | ~9300 cm$^2$ | ~9300 cm$^2$ |
| System lifetime | >20 years | >20 years |
| Electrolyzer lifetime | 40,000 hr | 40,000 hr |

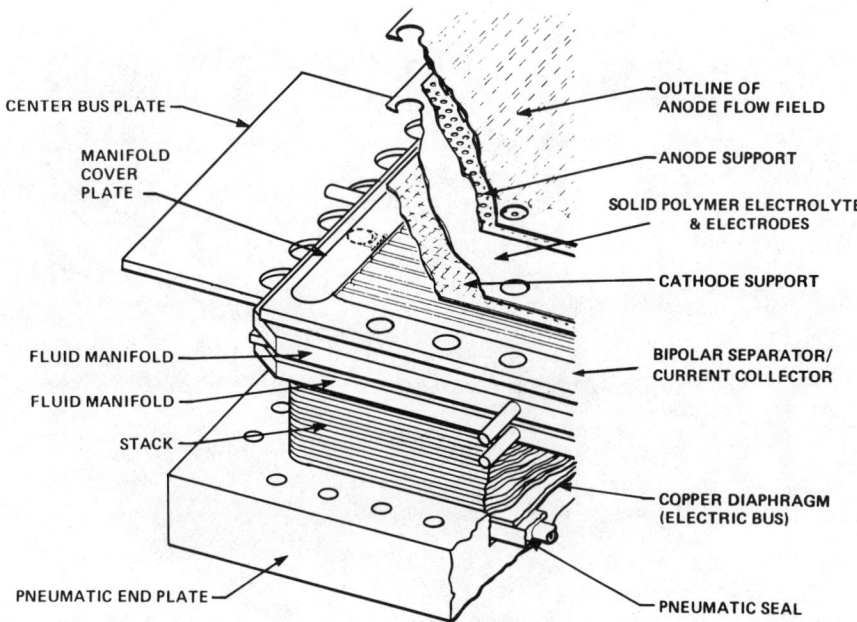

Figure 52. Detailed configuration of a 50-kW water electrolysis module (with permission from General Electric Company).

Figure 53. Appearance of a large-scale membrane–electrode assembly with an active cell area of approximately 2300 cm$^2$ (with permission from General Electric Company).

Figure 54. Actual end plates assembled in a 50-kW water electrolysis module (with permission from General Electric Company).

electrolyzers, the cell potential of an SPE water electrolyzer increases with rising pressure, essentially because of the power required for compressing the hydrogen gas,[5] resulting in decreased voltage efficiency as the operating pressure is increased.

The ultimate goals shown in Table 11 for the development program on SPE water electrolysis technology are a result of a design study of a 58-MW SPE water electrolysis plant, performed for Brookhaven National Laboratory in 1975.[85,89,101] Figure 55 shows the 58-MW electrolysis system, suitable for utility applications. It consists of storage cylinders for metal hydride to the right of the electrolysis module and a 26-MW hydrogen–air fuel cell unit (for converting $H_2$ back to electricity) at the rear of the building. According to General Electric,[89] this electrolysis plant is subject to modification to produce hydrogen for use as a large-scale chemical feedstock or as a supplement to natural gas.

## 4.5. Comparison of Operating Parameters of Various Cells

In comparison to the alkaline water electrolyzers, the key advantages claimed for the SPE system are[66,68,69]:

Figure 55. Artist's concept of 58-MW SPE water electrolysis plant (with permission from General Electric Company).

(a) Higher current density capacity at relatively low cell potentials, resulting in lower capital cost, size, and weight for the electrolysis modules and less power requirement for unit of hydrogen produced;

(b) Operation at higher pressures (up to ~204 atm), eliminating the need for a compressor when highly pressurized hydrogen is required;

(c) Use of a solid electrolyte with charge carriers, chemically bound to the polymer chain, resulting in no corrosive species which could carry over into downstream system components.

On the basis of 1978 electrolysis technology, the performance of SPE water electrolyzers, compared with that of conventional alkaline cells, is presented in Figure 56, which also depicts the cell performance expected in 1983.[85] Table 12, which summarizes the operating parameters of SPE water electrolyzers and alkaline water electrolyzers, shows the advantages of SPE technology over others in being able to operate at high current densities with high voltage efficiency.

## 5. Economics of Hydrogen Production [102–107]

One of the problems in calculating the cost of hydrogen is the treatment of the investment capital and the manner in which it is apportioned to the unit of gas generated at an appropriate interest rate.

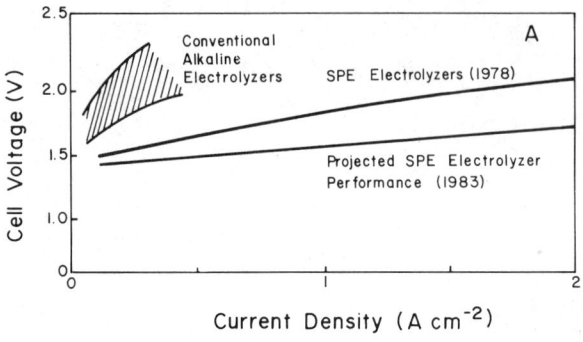

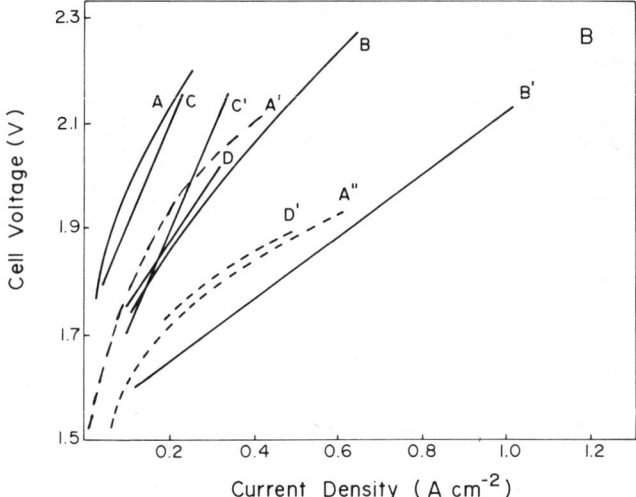

Figure 56. (A) Comparison of SPE water electrolysis cell performance with alkaline water electrolyzers (with permission from General Electric Company). (B) Projected cell performance of some alkaline water electrolyzers. Electrolyser Corp.[46]: A (1978), A' (1980), A" (1983); DeNora S.p.A.[19]: B (present), B' (with improved catalysts); BBC, Oerlikon[6]: C (nonactivated), C' (activated); Teledyne[6]: D (current), D' (projected near term).

A method that has received wide acceptance in the chemical industry is the *profitability index* or discounted cash flow methods. The profitability index is defined as that interest rate, continuously compounded, at which a sum of money equal to the capital investment in the project would have to be invested in an annuity fund in order for the fund to make payments equal to and at the same time as the receipts from the project. The receipts from the project include, of course, the return of the original capital investment plus the continuously compounded interest rate.

## Table 12
### Comparison of Operating Parameters of Various Water Electrolyzers

| Manufacturer | Electrolyte | P, atm | T, °C | Current density, mA/cm$^2$ | Cell potential, V | Voltage efficiency,[a] % |
|---|---|---|---|---|---|---|
| General Electric[b] | Nafion | 40 | 82 | 1080 | 1.83 | 81 |
| General Electric[b] | Nafion | 40 | 149 | 1080 | 1.70 | 87 |
| Electrolyser | 28 wt % KOH | 1 | 70 | 134 | 1.90 | 78 |
| Teledyne | 25 wt % KOH | 6.8 | 82 | 430 | 1.84 | 80 |
| Lurgi | 25 wt % KOH | 30 | 95 | 190 | 1.84 | 80 |
| Bamag | 26 wt % KOH | 1 | 80 | 250 | 1.92 | .77 |
| Norsk Hydro | 25 wt % KOH | 1 | 80 | 175 | 1.75 | 83 |
| Life Systems | 35 wt % KOH | 1 | 93 | 650 | 1.84 | 80 |

[a] Based on the higher heating value of hydrogen.
[b] On the basis of the 1978 technology.[(98)]

It may be shown that the annual capital recovery factor for the above definition is expressed by the following relationship:

$$\frac{R}{C} = \left\{ \left[ \frac{i(1+i)^n}{(1+i)^n - 1} \right] \right\} \quad (28)$$

where $R/C$ is the annual capital recovery factor ($R$ being annual payments), $C$ the original capital investment, $i$ the annual compounded investment rate, and $n$ the life of project in years.

Thus, for example, if a project costs $1,000,000, the life of the project is ten years, and the interest rate desired before taxes is 25%, the annual payments would have to be

$$R = \$1,000,000 \frac{0.25(1 + 0.25)^{10}}{(1 + 0.25)^{10} - 1} \quad (29)$$

or

$$R = \$280,000 \quad (30)$$

Table 13 provides capital recovery factors for several interest rates and project lives.

The chemical industry generally regards the economic life of a project as no more than ten years, and requires interest rates on capital before taxes to be no lower than 25%. This apparently high interest rate is normally expected to account for taxes, and to offset the risks inherent in the present Western business world that tend to reduce profitability (e.g., recessions, labor problems, plant operating or maintenance problems, obsolescence either within the plant or the product end use process, inflation, etc.).

Table 13
Capital Recovery Factors for Several Interest Rates and Project Lives

| Project life, years \ Interest rate, % | 10 | 20 | 25 | 30 | 40 |
|---|---|---|---|---|---|
| 5  | 0.264 | 0.334 | 0.372 | 0.411 | 0.491 |
| 10 | 0.163 | 0.239 | 0.280 | 0.323 | 0.414 |
| 15 | 0.131 | 0.214 | 0.259 | 0.306 | 0.403 |
| 20 | 0.117 | 0.205 | 0.253 | 0.302 | 0.400 |
| 25 | 0.110 | 0.202 | 0.251 | 0.300 | 0.400 |

The components of cost to produce $H_2$ are as follows:
- capital recovery (including compounded interest)
- power consumption
- utilities (e.g., feed-water, cooling water, steam, compressed air, nitrogen)
- maintenance supplies and labor
- operating labor
- taxes and insurance
- management and overheads

Components within the capital cost should include the following:
- electrolyzer module and auxiliaries
- power transformers, rectifiers, and power lines
- equipment transportation costs
- land and land services such as sewers, drainage, city water, etc.
- dedicated utility services such as cooling towers and pumps, compressors, nitrogen purge system, feed-water treatment, and deionized system
- buildings and services such as lighting and heating
- spare parts excluding maintenance supplies
- site construction costs
- engineering and project management costs
- financing costs
- training and plant start-up costs

The actual complete capital cost of an electrolyzer installation is difficult to assess from the published literature, since definitions of what is included are *never* quite the same.

Other factors that must be considered in any treatment of the cost to produce hydrogen are the following:
- plant capacity or operating factor (i.e., time in any year that the plant actually operates)
- power conversion efficiency
- current efficiency of electrolysis cells

## ELECTROLYTIC PRODUCTION OF HYDROGEN

Analysis of the cost estimates shows that the cost of 1 normal m³ of $H_2$ is given by the very simple expression

$$\$ \text{ per normal m}^3 \text{ of } H_2 = \left[\left\{\frac{2.393\, C_1}{8760\, fn_1}(R + a + b + c + d)\right\}\right]$$
$$+ \left[\left\{\frac{P}{n_2} \cdot C_2(1+e)\right\}\right] \quad (31)$$

where
- $C_1$ = total cost of turnkey operating $H_2$ plant expressed in \$/kA of input energy/unit cell,
- $C_2$ = power cost in \$/kWh,
- $P$ = power consumption to produce 1 normal m³ of hydrogen in kWh,
- $R$ = capital recovery factor to recover original capital investment,
- $n_1$ = current efficiency in the electrolyzer cells (varies from 0.98 to 0.995 but usually is 0.99),
- $n_2$ = power conversion efficiency (normally about 0.97),
- $f$ = plant operating factor (normally about 0.95),
- $a$ = the fraction of capital cost required to cover maintenance labor and materials (normally about 0.03),
- $b$ = the fraction of capital cost required to cover taxes and insurance (normally about 0.02),
- $c$ = the fraction of capital cost required to cover operating labor (varies widely but assumed to be about 0.02),
- $d$ = the fraction of capital cost required to cover management and overhead costs (varies very widely but assumed to be about 0.02),
- $e$ = the fraction of total power cost required to cover utilities (varies widely but assumed to be about 0.05),
- 2.393 = the number of kAh required to produce 1 normal m³ of hydrogen,
- 8760 = the number of hours in a year,
- kA = the kiloamps entering the plant,
- $N$ = the number of monopolar cells receiving kiloamps or total number of bipolar unit cells in plant (=number of electrolyzers × number of unit cells/electrolyzer).

Inserting assumed values for $a$, $b$, $c$, $d$, $e$, $n_1$, and $n_2$, Eq. (31) reduces to

$$\$ \text{ per normal m}^3 \, H_2 = 0.0002905\, C_1(R + 0.09) + 1.083\, PC_2 \quad (32)$$

For an average case in which
- $C_1$ = \$400/kA = installed capital cost,
- $R$ = capital recovery factor for 10-year project life and return on investment of 25% is 0.280,
- $P$ = power consumption of 4.5 kWh per normal m³ of $H_2$,
- $C_2$ = cost of power of 20 mils/kWh or \$0.02/kWh,

the cost of 1 normal m³ of $H_2$ would be \$0.14 per normal m³ of $H_2$. This simplified treatment of hydrogen cost is valid for most engineering purposes since the capital cost plus the power cost represents a very substantial portion (>90%) of the total cost of producing hydrogen.

It should be noted that in all electrochemical cell systems, there is a correlation between capital cost and power consumption. This is a result of the pronounced effect of current density. A decrease in the current density will decrease the power consumption and costs, but increases the capital cost, and vice versa. Thus, if capital costs and power costs are introduced in Eqs. (31) and (32) as a function of current density, then the optimum current density can be estimated from the expression

$$\text{\$ per normal m}^3 \text{ of } H_2 = 0.0002905\, F_1 A (R + 0.09) + (1.083\, F_2 A C_2) \tag{33}$$

Thus, for any particular power cost $C_2$, and required capital recovery factor $R$, the optimum current density for any specific application may be calculated.

$F_1 A$ and $F_2 A$ in Eq. (33) refer to the capital costs and power costs, respectively, expressed as a function of current density.

It is claimed that the total installed capital cost of various technologies is very approximately as follows in 1979 U.S. dollars.
- For monopolar cell technology: $600/kA[46]
- For bipolar cell technology: $650–$750/kA[46,51,105]
- For SPE cell technology: $250/kA[107] (1985 projection)
  (Note that no commercial plants have been built in 1979 based on this latter technology.)

Assuming an average capital cost of $500/kA, the components of the operating cost by the methods described above for a 20,000 normal m$^3$ of hydrogen per hour is as shown in Table 14. These estimates were arrived at based on an average unit cell voltage of 1.9 V, a power cost of 20 mils/kWh, and other assumptions as given in the definitions above.

## 6. Futuristic Concepts for Electrolytic Production of Hydrogen

### 6.1. Use of Anode Depolarizers

Oxygen is generally an unwanted by-product in the electrolytic production of hydrogen. This may not be so in the future since in the steel industry, both hydrogen and oxygen are required, and pure oxygen is desirable for sewage treatment plants. In the future, hydrogen–oxygen alkaline fuel cells may have a role as emergency power supplies. However, in the context of production of electrolytic hydrogen as a chemical feedstock or as a fuel, during load leveling in nuclear power plants or using intermittent power sources (photovoltaics, wind, solar steam power plants), it is essential to reduce the electric input into

Table 14
Operating Costs for an Average 20,000 normal m³ per hour Hydrogen Plant
(Capital Cost: $23,930,000 1979 U.S. Dollars)

| Component costs | U.S. Dollars/year | U.S. Dollars per normal $m^3$ $H_2$ |
|---|---|---|
| Annual capital recovery for 10-year plant life and 25% compound interest | 6,700,000 | 0.0403 |
| Power cost at 20-mil/kW hr and plant-operating factor of 0.95 | 15,761,000 | 0.0947 |
| Maintenance labor and materials at 3% of capital cost | 718,000 | 0.0043 |
| Plant insurances and taxes at 2% of capital cost | 479,000 | 0.0029 |
| Plant-operating labor at 2% of capital cost | 479,000 | 0.0029 |
| Management and overhead costs at 2% of capital cost | 479,000 | 0.0029 |
| Plant utilities at 5% of power cost | 788,000 | 0.0047 |
| Total: | 25,404,000 | 0.1527 |

water electrolyzers where the problems are with the oxygen electrode reaction owing to its high reversible potential and its high degree of irreversibility. By the use of anode depolarizers, it should be possible to lower significantly the voltage input for electrolytic hydrogen production. For example, sulfurous acid may be oxidized to sulfuric acid at the anode. This reaction is the electrochemical step in the Westinghouse thermochemical–electrochemical hybrid cycle for hydrogen production.[108] The anodic and cathodic partial reactions and the overall reaction are represented by

$$\text{Anodic:} \quad H_2SO_3 + H_2O \rightarrow H_2SO_4 + 2H^+ + 2e^- \quad (34)$$

$$\text{Cathodic:} \quad 2H^+ + 2e^- \rightarrow H_2 \quad (35)$$

$$\text{Overall:} \quad H_2SO_3 + H_2O \rightarrow H_2SO_4 + H_2 \quad (36)$$

The thermodynamic reversible potential for this reaction is only 0.17 V as compared to 1.23 V for water electrolysis. The sulfur dioxide required for this reaction may be made available from stack gases. It is also possible that the sulfuric acid, if produced in high enough concentrations, could be a useful by-product (say, for the fertilizer industry).[109] Some problems are encountered with the sulfurous acid oxidation reaction. Firstly, migration of sulfurous acid to the cathode causes its reduction to sulfur at the electrode and hence its poisoning. Thus, a membrane, impermeable to this species, is required between the anode and cathode. Secondly, there is a significant increase in the open-circuit potential with increase in acid concentration. Effectively, the cell potential will be of the order of 0.8–1.0 V at practical current densities ($\sim$200 mA cm$^{-2}$). While this means a reduction in electric energy consumption of 40–60% compared with conventional or advanced water electrolyzers, the cost of sulfurous acid will have to be taken into account in the computation of cost of electrolytic hydrogen if it is not available at negligible costs.

Other promising anode depolarizers are low-cost carbonaceous materials (e.g., cellulose)[110]. In a recent study[111] a coal slurry was used as the anodic reactant in a cell containing 3.7 $M$ $H_2SO_4$ (at room temperature) using platinum gauze electrodes with hydrogen evolution occurring at the cathode. The half-cell and overall reactions are

$$\text{Anode:} \quad C(s) + 2H_2O \rightarrow CO_2 + 4H^+ + 4e^- \quad (37)$$

$$\text{Cathode:} \quad 2H^+ + 2e^- \rightarrow H_2 \quad (38)$$

$$\text{Overall:} \quad C(s) + 2H_2O(l) \rightarrow 2H_2(g) + CO_2(g) \quad (39)$$

Though the oxidation rate was low (a few mA cm$^{-2}$) at cell potentials of about 0.8 V, it was suggested that by carrying out the reaction at 200–600°C, a significant enhancement in rates would be possible. There was a decrease in current with time, and this phenomenon was attributed to the surface oxidation of the coal since an aged sample which was removed from the electrolyte, heated in air to 200°C, and returned to the cell showed higher oxidation rates. Taking into acount the chemical energy of the coal, the overall efficiency for hydrogen production can be increased by about 25% and the electric energy input reduced by 30–50%.

### 6.2. Water Vapor Electrolysis

From the thermodynamic and electrode kinetic points of view, there are significant advantages of electrolyzing water at elevated temperatures, say, about 1000°C. Firstly, owing to a decrease in the free energy change of the reaction [Eq. (1)] with temperature (see Section 3.1, Figure 9), the theoretical energy requirements at 1000°C are less than at ambient temperature by about 20%. Secondly, activation overpotential losses at high temperature are greatly reduced as compared with the corresponding values at temperatures at which aqueous electrolyte water electrolysis is carried out. In Figure 9, it can be seen that $V_{HHV}$ for reaction (1) is practically constant at all temperatures. Since at high temperatures the cell potentials for water electrolysis are expected to be considerably less than the higher-heating value voltage, $V_{HHV}$, high-grade heat will have to be supplied to the cell to maintain the desired operating temperature, and hence, the cell performance.

Following the work in which the reverse reaction was carried out in a solid oxide electrolyte fuel cell, there was some interest in developing water vapor electrolyzers using such electrolytes. A regenerative life support system was developed in which carbon dioxide and water were reduced to oxygen, hydrogen, and carbon.[112] This electrochemical cell consisted of sintered

Pt–ZrO$_2$ electrodes, containing 70% Pt and 30% ZrO$_2$ by volume and yttria-stabilized zirconia as the solid electrolyte. Attempts were made at the General Electric Company to develop solid oxide electrolyte water electrolyzers.[113] One of the major problems encountered was finding a material to connect the cells in series. This material must meet stringent requirements: (i) high electronic conductivity, (ii) ability to withstand differences in partial pressures (15–20 orders of magnitude) across the material, (iii) thermal expansion matching characteristics with other cell components, and (iv) chemical compatibility with other cell components.

Pioneering work on the design, fabrication, and testing of solid electrolyte water electrolyzers, fuel cells, and electrically regenerative fuel cells was carried out at Brown Boveri & CIE (BBC) in Heidelberg, Germany.[114] These cylindrical cells containing yttria-stabilized zirconia as the electrolyte have a diameter of 25 mm, height of 12 mm, and wall thickness of 0.5 mm and employ nickel for the hydrogen electrode, and tin-doped indium oxide or perovskite-type oxides (LaNiO$_3$, LaMnO$_3$) for the oxygen electrode. Typical cell-potential–current-density relations in single cells during fuel cell and electrolysis operation are shown in Figure 57. Such types of electrically regenerative fuel cells may be attractive for load-leveling applications in electric utility power plants. Multicells have also been fabricated and tested at BBC. Materials such as cobalt chromite, lanthanum chromite doped with strontium and nickel, and lanthanum manganite doped with strontium have been evaluated as interconnection materials.

Dornier Systems in West Germany is investigating the prospects for large-scale hydrogen production by water electrolysis (Project HOT ELLY),

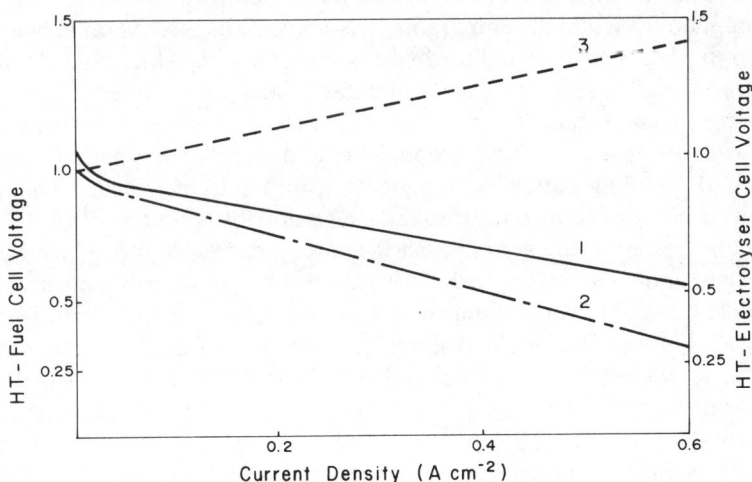

Figure 57. Potential–current-density relations for a BBC high-temperature (HT) solid electrolyte fuel cell (curves 1 and 2) and water electrolysis cell (curve 3) at 1000°C (with permission from the author of Reference 114).

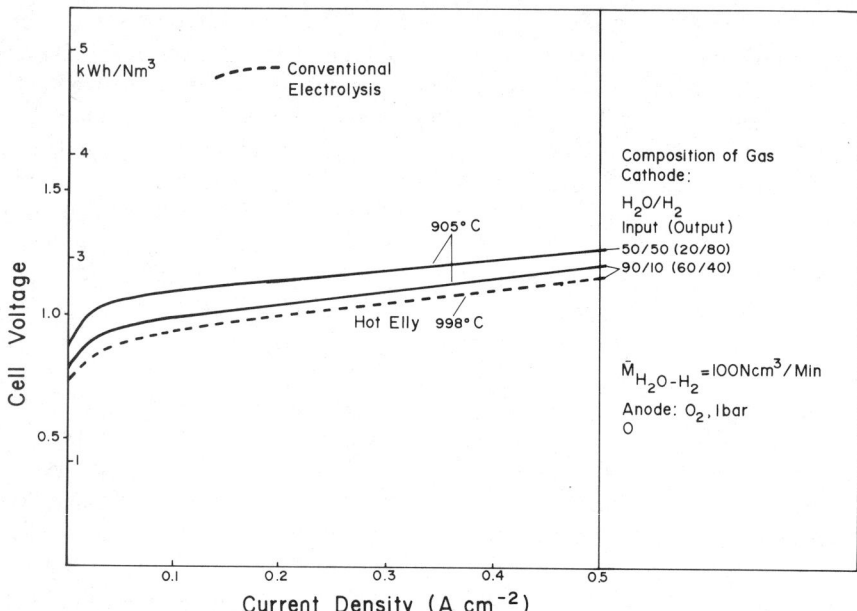

Figure 58. Potential–current-density relations for water electrolysis in the Dornier System solid electrolyte cell using platinum electrodes (with permission from the authors of Reference 115 and J. Electrochem. Soc.).

which will be essential for a future hydrogen energy concept.[115] Single cells have been designed, fabricated, and tested. The potential–current-density relation with noble metal electrodes is shown in Figure 58. With electricity and heat supplied from a high-temperature gas-cooled reactor, overall efficiencies of close to 50% are projected for hydrogen production. This value should be compared with 30–35% for low-temperature water electrolyzers.

In the distant future, when fusion will be the predominant energy source, there will still be a need for transportable and storable synthetic fuels. The concept of coupling fusion power plants with solid electrolyte water electrolyzers is being examined at Brookhaven National Laboratory (BNL).[116] A conceptual design of the system is shown in Figure 59. Heat produced in the fusion reactor blanket is supplied to the endothermic water vapor electrolyzers and to the power conversion unit where the electricity for the electrolyzers is produced. The number of electrolyzers is a function of the blanket temperature, the temperature change within each electrolyzer, the outlet hydrogen concentration, and the upper temperature limits of the conventional steam generators.

In the solid electrolyte, electrochemical cells which were developed in the sixties for fuel cells and water electrolysis, high electrolyte ohmic losses were observed owing to a large anode–cathode gap. In the solid electrolyte fuel cell program, renewed at Westinghouse in 1976, major strides have been made in

## ELECTROLYTIC PRODUCTION OF HYDROGEN

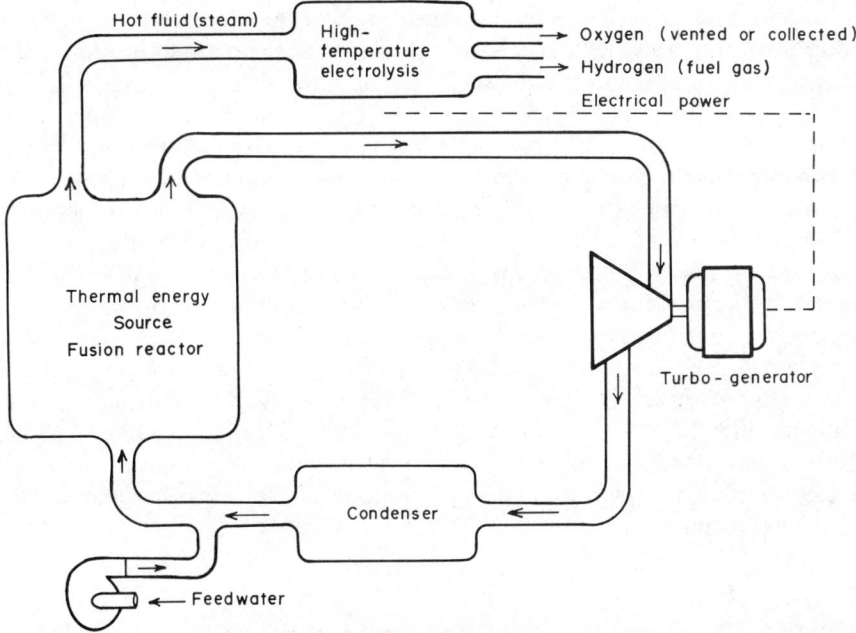

Figure 59. Concept of fusion reactor coupled with solid electrolyte water electrolyzer–General HTE process with conventional power cycle.

the development of thin layer cells.[117] The electrodes, electrolyte, and interconnection layers (on the order of 20 μm in thickness) are chemically vapor deposited in turn on a porous support tube. A schematic of the single cell and multi-cell are shown in Figure 60. Single and multicells have been constructed

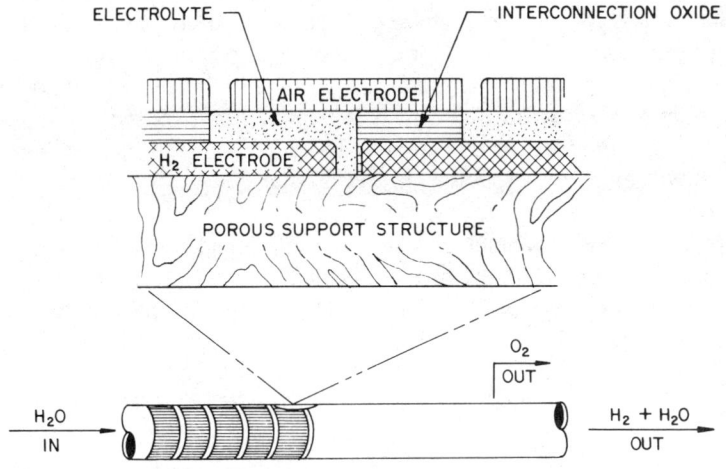

Figure 60. Schematic of single cell and multicell in Westinghouse solid electrolyte fuel cell.

and tested, and in such cells the ohmic losses in the electrolyte have been reduced considerably. In this cell, nickel is used as the anode and tin-doped indium oxide (impregnated with praeseodymium oxide) as the cathode, yttria-stabilized zirconia for the electrolyte, and lanthanum chromite doped with magnesium (to improve electronic conductivity) and aluminum (to match the thermal expansion characteristics with other cell components) for the interconnection materials. In the BNL project, it was proposed that this type of a cell design may be most compatible for high-temperature water electrolyzers coupled with fusion power plants. Calculations of the overall efficiency of a 2000-MW thermal fusion reactor producing hydrogen in a high-temperature electrolyzer have been made. According to this scheme, nine electrolyzers will operate at an inlet temperature of 1380°C and three at a lower concentration to produce hydrogen at a concentration of 95% (i.e., $H_2/H_2O + H_2$); such a plant would produce $2.7 \times 10^8$ scf/day of hydrogen at an overall efficiency of 52.5%. The main problems which would be encountered in developing such a fusion reactor–electrolyzer plant would be in finding stable materials and in the fabrication techniques.

### 6.3. Thermochemical–Electrochemical Hybrid Cycles

In the past few years, many thermochemical cycles have been proposed based on thermodynamic calculations to produce hydrogen on a large scale by the decomposition of water. The main problems connected with such schemes for hydrogen production are as follows: (i) Consideration must be made of the kinetics of the intermediate reactions since the slowness of any of the intermediate steps will enhance the heat requirements; (ii) in serveral of the thermochemical cycles proposed, the overall reaction occurs in a number of intermediate steps which also involve circulation of fluids and separation of fluids (e.g., gas–gas); (iii) at least one of the steps requires a relatively high temperature and corrodible chemicals are involved; (iv) the efficiency of the overall system is Carnot limited and it is very difficult to attain more than two-thirds of the theoretical efficiencies. Because of these factors, practical efficiencies will not be expected to be higher than conventional water electrolysis.

There are a few thermochemical–electrochemical hybrid cycles, which appear more promising than the direct thermochemical route. One of these is the Westinghouse sulfur cycle.[118] The partial reactions for this scheme may be represented as follows:

$$\text{Electrochemical:} \quad SO_2 + 2H_2O \rightarrow H_2SO_4 + H_2 \quad (40)$$

$$\text{Thermal:} \quad H_2SO_4 \rightarrow H_2O + SO_2 + \tfrac{1}{2}O_2 \quad (41)$$

$$\text{Overall:} \quad H_2O \rightarrow H_2 + \tfrac{1}{2}O_2 \quad (42)$$

A brief analysis of the electrochemical step involved in the above scheme was presented in Section 6.1. To maximize the efficiency of this thermoelectrochemical process for hydrogen production, it is essential to attain high concentrations of sulfuric acid in the electrochemical step. Otherwise, the thermal energy requirements for concentration and decomposition of sulfuric acid will be high. Conversely, aiming at high sulfuric acid concentrations in the cell results in high open-circuit potentials and hence high cell potentials for the electrochemical step. Furthermore, in contrast to the electrochemical step where the gas separation problem is overcome, it is not the case with the thermal step. Oxygen must be separated from the gas mixture containing steam, oxygen, sulfur dioxide, and unreacted sulfuric acid. In addition, sulfur dioxide must also be recovered from recycling to the electrolyzer. Progress is being made at Westinghouse to improve the performance of the electrochemical reaction.[108] Overall efficiencies of 45% for hydrogen production projected for the Westinghouse cycle are based on the assumption that the cell potential for the electrochemical cell will not exceed 0.6 V. One of the attractive features of this reaction pathway is that only two intermediate reactions are involved. Surprisingly thermochemical cycles have been proposed which occur in more than six intermediate steps!

Another interesting thermochemical–electrochemical hybrid cycle, which has been proposed by the investigators at the Joint Research Center of the Commission of European Communities at Ispra, Italy, consists of the following reactions[119]:

Electrochemical: $2HBr \rightarrow H_2 + Br_2$ (43)

Chemical: $Br_2 + SO_2 + 2H_2O \rightarrow HBr(g) + H_2SO_4(aq)$ (44)

Thermal: $H_2SO_4 \rightarrow H_2O + SO_2 + \tfrac{1}{2}O_2$ (45)

The complete cycle has been demonstrated at Ispra. The hydrobromic acid electrolysis was carried out with graphite electrodes. By addition of Pt or Pd salts to the electrolyte, the hydrogen overpotential can be considerably reduced. The optimum temperature for the electrochemical step is 100°C. Cell voltages are about 0.8 V at a current density of 200 mA cm$^{-2}$ with a 50% inlet concentration of HBr and 40% HBr and 10% Br$_2$ concentrations at the outlet.

## 6.4. Photoelectrolysis of Water

The photoelectrochemical production of hydrogen was initially demonstrated by Fujishima and Honda,[120] who used a single-crystal TiO$_2$ working anode coupled with a platinum cathode. Polycrystalline TiO$_2$ has also been investigated for use as the oxygen electrode[121] which exhibited electrocatalytic activity comparable to TiO$_2$ single crystal. During operation, photons of energy $h\nu$ higher than the energy gap ($E_g$) of the working electrode, excite hole–electron pairs:

$$h\nu \rightarrow p^+ + e^- \qquad (46)$$

The electrons pass through an external circuit to the cathode, while the holes move to the anode surface, where the following electrochemical reactions occur:

$$H_2O + 2p^+ \rightarrow \tfrac{1}{2}O_2 + 2H^+ \qquad (47)$$

and

$$2OH^- + 2p^+ \rightarrow \tfrac{1}{2}O_2 + H_2O \qquad (48)$$

in acidic and alkaline solutions, respectively. The corresponding reactions on the cathode are

$$2H^+ + 2e^- \rightarrow H_2 \qquad (49)$$

and

$$2H_2O + 2e^- \rightarrow 2OH^- + H_2 \qquad (50)$$

It has been reported[121] that, using $TiO_2$ with an $E_g \approx 2$ eV, the maximum quantum efficiency for the photoelectrolytic process is only 1–2% at $h\nu = 4$ eV. Theoretical analysis on the photoelectrolysis of water has been carried out by Bockris and Uosaki.[122] Depending on the properties of the semiconductor anodes, the calculated quantum efficiencies of photoelectrolysis cells vary from $10^{-2}$% to 10%. Furthermore, Mavroides and co-workers[123] pointed out that the quantum efficiency of a photoelectrolysis cell using $SrTiO_3$ anode, a semiconductor with an electron affinity being 0.2 eV less than that of $TiO_2$, is higher by at least a factor of 10 than that of the cell using $TiO_2$ anode.

According to the work of Bockris,[124] if a quantum efficiency of ~5% for the solar spectrum (not at specific wavelengths) can be practically attained, the production cost of photoelectrolytic hydrogen may be lower than by other methods. (The present quantum efficiency of photoelectrolysis cells is only 0.2%.) However, the energy gaps of existing semiconductor anodes such as $TiO_2$ (3 eV) and $SrTiO_3$ (3.2 eV) are too high for efficient solar-energy conversion. It is of crucial importance to examine other semiconducting materials, which exhibit more favorable characteristics than $TiO_2$ and $SrTiO_3$.

## 7. Heavy Water—A Useful By-Product

As stated in Sections 1 and 5, the cost of hydrogen production by water electrolysis is high relative to its production by the steam-reforming or water–gas reaction. However, if markets can be found for the by-products of water electrolysis, there could be an effective cost reduction of electrolytic hydrogen.

The by-product most often thought about is oxygen; the neglected one being heavy water. The demand for heavy water is increasing in Canada with the development of the natural uranium heavy water moderated CANDU† family of nuclear reactors.[125] It is estimated that by the year 2000, the electricity generation in Canada by this type of reactor will be 80,000 MW. Such reactors require 0.85 kg of $D_2O$ per MW of electricity generated as inventory.

A combined electrolysis catalytic exchange–heave water process (CECE–HWP) is being developed at Chalk River Nuclear Laboratories, Chalk River, Ontario, Canada. The principle of this method is that isotopic separation is involved in two processes—electrolysis and chemical catalytic exchange. The electrolytic hydrogen is passed countercurrently to the feed-water for the electrochemical cells through a catalytic column. An isotopic exchange reaction is effected. In the early work, the catalyst used was platinum on finely divided charcoal. But more recently, a new wet-proofed catalyst has been developed for this application. The reaction involved in the (CECE) separation process is represented by Eq. (51):

$$HD(g) + H_2O(l) \rightleftharpoons H_2(g) + HDO(l) \qquad (51)$$

where the enrichment of deuterium in the electrolyte occurs owing to the kinetic separation factor in the water electrolysis reaction.[126] Reaction (51) occurs in two steps:

$$HD(g) + H_2O(g) \xrightleftharpoons{\text{catalyst}} H_2(g) + HDO(g) \qquad (52)$$

$$HDO(g) + H_2O(l) \rightleftharpoons H_2O(l, g) + HDO(l) \qquad (53)$$

A flow chart of the CECE process is shown in Figure 61. Owing to the countercurrent principle involved, the deuterium content of the water increases down the catalytic column while that of the gas shows the opposite trend. A small-scale laboratory unit was set up and tested at Chalk River. The General Electric solid polymer electrolyte water electrolyzer was used in these experiments. A small combined electrolysis catalytic exchange–heavy water process (CECE–HWP) pilot is located at the Chalk River Nuclear Laboratories (CHRNL). A pilot plant for the CECE–Tritium Recovery Process (CECE–TRP) was constructed under the sponsorship of CRNL and the U.S. Department of Energy at the Mound Laboratory, Miamisburg, Ohio. Both of these plants are in operation and the preliminary results are encouraging. The success of the pilot plant studies should lead to the design, development and demonstration of larger plants (5 → 100 → 400 MW). In the last case (400 MW), the demonstration plant could be sited near a hydroelectric power source. The hydrogen produced could be used for ammonia synthesis. Note that the cost of shipping to locations where it will be used is small compared to the cost of manufacturing heavy water.

† CANDU—Canada Deuterium Uranium.

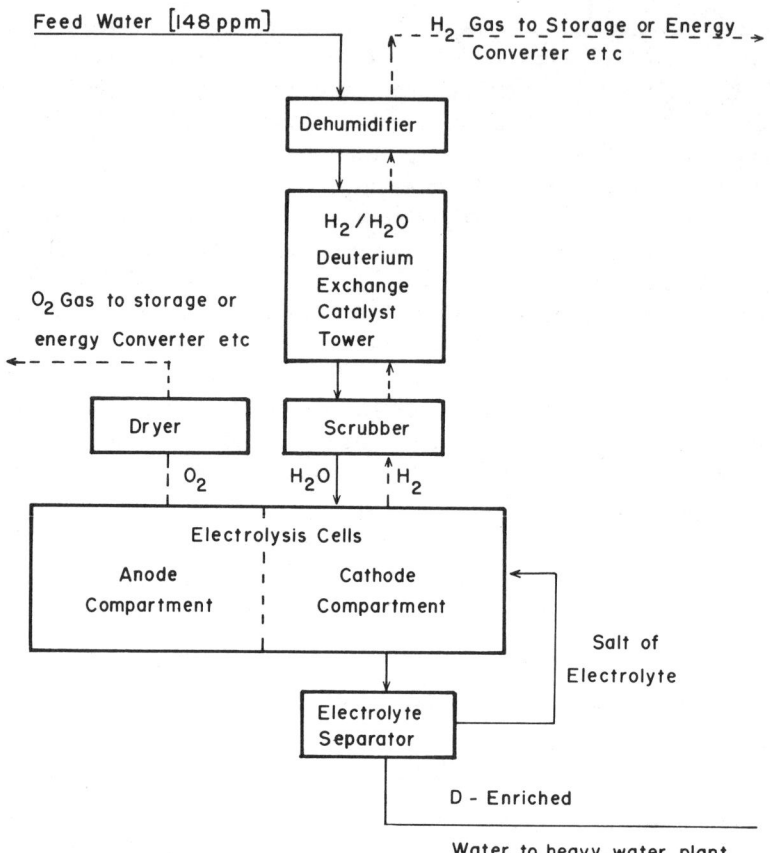

Figure 61. Flow chart of combined electrolysis catalytic exchange–heavy water process (CECE–HWP) (with permission from the author of Reference 125).

Tritium will be required in large quantities for fusion reactors in which the D–T reaction occurs. The CECE–TRP process shows promise of attaining large enrichment factors.

The cost of deuterium in 1977 was quoted at \$214/kg.[125,127,128] The dollar value of heavy water ($D_2O$) per gigajoule (GJ) of hydrogen has been calculated taking into account (i) the fraction of deuterium recovered, (ii) deuterium concentration in the feed-water (which varies from 133 to 148 × $10^{-6}$ D/H + D in Canada), and (iii) price of reactor grade $D_2O$. The calculations show that the net $D_2O$ credit could be as high as \$1/GJ of electrolytic $H_2$. This value may be compared with the oxygen by-product credit which has been estimated at \$0.62/GJ of electrolytic $H_2$. It has also been projected that about 50 kg of $D_2O$ can be recovered per annum for every 1000 metric ton per day ammonia plant. Projected estimates[125,127,128] in Canada indicate that ~1000 kg per annum of $D_2O$ (which is 1/4 of the committed production capacity) could be produced. In the fusion era, the related tritium recovery process may assume increasing importance.

## Acknowledgements

The authors wish to thank Electrolyser Corporation, Ltd., Brown Boveri & Co., Ltd., Norsk Hydro A.S., Lurgi Umwelt und Chemtechnick GmbH, Panclor S.A., General Electric Company, Teledyne Energy Systems, Life Systems Inc., Krebskosmo for cooperation in supplying the technical brochures, photographs, and diagrams and for permission to publish the figures included in this chapter. Thanks are also due to Dr. R. L. LeRoy for sending preprints of his papers. This work was performed under the auspices of the U.S. Department of Energy.

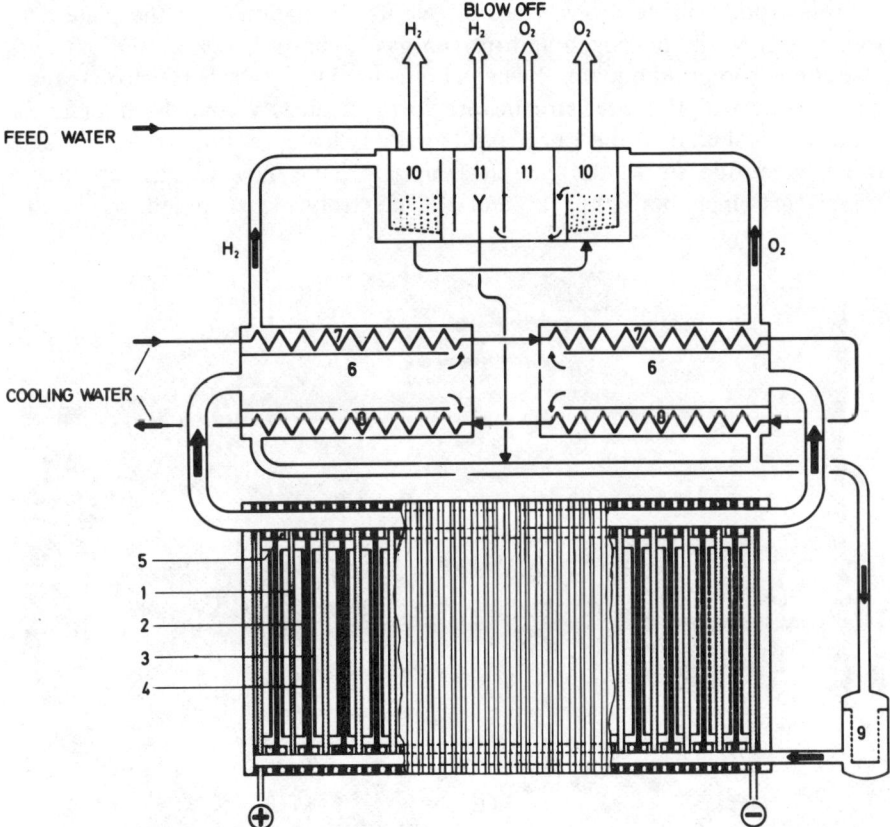

Figure 62. Sectional side view of Krebskosmo water electrolyzer: 1, center electrode; 2, pre-electrode; 3, pre-electrode; 4, diaphragm; 5, spacer; 6, gas separation vessels; 7, gas coolers; 8, electrolyte coolers; 9, filter; 10, pressure equalizing tanks; 11, hydraulic safety seal (with permission from Krebskosmo, Berlin, West Germany).

## Appendix: Krebskosmo Technology[42,131,132]

In 1972, Krebskosmo acquired the well-known and proven Demag water electrolyzer technology. The principle of operation of a Krebskosmo cell is very similar to a Lurgi cell except that the operating pressure is atmospheric and electrolyte circulation is via the gas lift principle, thus requiring no circulating pumps.

As noted in Figure 62, the cells have a typical plate and frame filter press arrangement. The center electrode, cathode pre-electrodes, and the spacers are of nickel-plated steel. Anode pre-electrodes are nickel. Both anode and cathode pre-electrodes are coated with an electrocatalytic layer to reduce the hydrogen and oxygen overvoltages. Asbestos diaphragms are used.

The design of the pre-electrodes allows a large portion of the generated gas to be conveyed to the backside of the electrode, thereby, reducing interelectrode voltage drops. Natural gas lift circulation from the plate and frame cells to the hydrogen and oxygen gas separation vessels and gas and electrolyte coolers eliminates the need for a circulating pump. Electrolyte feed and gas–electrolyte outlet streams are through ducts within the filter press frame, very similar to the Lurgi and Norsk Hydro cells. From the plate and frame electrode stack, the gas–electrolyte streams flow to gas separation vessels in which both the gas and the electrolyte are cooled. From the

Figure 63. Typical installation of a Krebskosmo water electrolyzer plant (with permission from Krebskosmo, Berlin, West Germany).

separators, the hydrogen and oxygen gas streams flow to a pressure equalizing tank and hydraulic safety seal located above the separators (see Figure 63).

As with all filter press bipolar cells, the Krebskosmo cell can be tailor made to the needs of a particular client. However, three standard size cells are available, types WE125, WE250, and WE375, producing up to 60, 135, and 200 normal $m^3$ per hour of hydrogen gas. These three cells require up to 3, 6, and 9 kA at about 1.9 V per unit cell. Typical operating conditions for a Krebskosmo electrolyzer (see Figure 63) would be an operating temperature of 80°C, 28 wt % of KOH, gas purities of 99.9% for hydrogen and 99.5% for oxygen, the power consumption being 4.5 dc kWh per normal $m^3$ of hydrogen.

## References

1. H. Wuellenweber and J. Mueller, Lurgi pressure electrolysis, in *Proceedings of the Symposium on Industrial Water Electrolysis*, Vol. 78-4, S. Srinivasan, F. J. Salzano, and A. R. Landgrebe, eds., Electrochemical Society, Princeton, New Jersey (1978), p. 1.
2. J. H. Russel, Design and development of solid polymer electrolyte water electrolyzers for large-scale hydrogen generation, in *Proceedings of the Symposium on Industrial Water Electrolysis*, Vol. 78-4, S. Srinivasan, F. J. Salzano, and A. R. Landgrebe, eds., Electrochemical Society, Princeton, New Jersey (1978), p. 77.
3. J. H. Perry, ed., *Chemical Engineers Handbook*, 3rd ed., McGraw-Hill, New York (1950), pp. 473–485.
3a. J. H. Perry, ed., *Chemical Engineers Handbook*, 4th ed., McGraw-Hill, New York (1951), pp. 19–62, 19–67.
4. D. P. Gregory, A Hydrogen-Energy System, American Gas Association, Cat. No. L21173 (1973).
5. M. S. Casper, ed., *Hydrogen Manufacture by Electrolysis, Thermal Decomposition and Unusual Techniques*, Noyes Data Corporation, New Jersey (1978).
6. A. P. Fickett and F. R. Kalhammer, Water Electrolysis, in *Hydrogen: Its Technology and Implications*, Vol. 1 of *Hydrogen Production Technology*, K. E. Cox and K. D. Williamson, eds., Chemical Rubber Co., Cleveland, Ohio (1977).
7. J. Evangelista, B. Phillips, and L. Gordon, Electrolytic hydrogen production—an analysis and review, NASA Technical Memorandum No. TMX-71856, December 1975.
8. M. H. Miles, G. Kissel, P. W. T. Lu, and S. Srinivasan, Effect of temperature on electrode kinetic parameters for hydrogen and oxygen evolution reactions on nickel electrodes in alkaline solutions, *J. Electrochem. Soc.* **123**, 332 (1976).
9. R. L. Leroy, C. T. Bowen, and D. J. LeRoy, The thermodynamics of aqueous water electrolysis, *J. Electrochem. Soc.* **127**, 1954 (1980).
10. R. L. Leroy, M. B. I. Janjua, R. Renaud, and U. Leuenberger, Time variations effects in unipolar water electrolyzers and their implications for efficiency improvement, in *Proceedings of the Symposium on Water Electrolysis*, Vol. 78-4, S. Srinivasan, F. J. Salzano, and A. R. Landgrebe, eds., Electrochemical Society, Princeton, New Jersey (1978), p. 63.
11. J. O'M. Bockris and A. K. N. Reddy, *Modern Electrochemistry*, Vol. 2, Plenum, New York (1970).
12. A. J. Appleby, Electrocatalysis, *Mod. Aspects Electrochem.* **9**, 369 (1974).
13. E. W. Brooman and A. T. Kuhn, Correlations between the rate of the hydrogen electrode reaction and the properties of alloys, *J. Electroanal. Chem.* **49**, 325 (1974).

14. A. T. Kuhn, Surveying electrocatalysis, in *Electrochemistry—the Past Thirty and the Next Thirty Years*, H. Bloom and F. Gutman, eds., Plenum, New York (1977), p. 139.
15. B. V. Tilak, in A. J. Appleby's Chapter on Electrocatalysis, *Comprehensive Treatise of Electrochemistry*, Vol. 5, J. O'M. Bockris, E. A. Yeager, and B. E. Conway, eds., Plenum, New York (1982).
16. M. H. Miles, Evaluation of electrocatalysis for water electrolysis in alkaline solution, *J. Electroanal. Chem.* **60**, 89 (1975).
17. S. Srinivasan and F. J. Salzano, Prospects for the hydrogen production by water electrolysis to be competitive with conventional methods, *Int. J. Hydrogen Energy* **2**, 53 (1977).
18. A. J. Appleby, G. Crepy, and J. Jacquelin, High efficiency water electrolysis in alkaline solutions, *Int. J. Hydrogen Energy* **3**, 21 (1978).
19. A. Nidola, P. M. Spaziante, and L. Giuffre, Electrolytic production of $H_2$ from alkaline solutions, in *Proceedings of the Symposium on Water Electrolysis*, Vol. 78-4, S. Srinivasan, F. J. Salzano, and A. R. Landgrebe, eds., Electrochemical Society, Princeton, New Jersey (1978), pp. 102–116.
20. A. J. Appleby and G. Crepy, New developments in alkaline electrolysis technology, in *Proceedings of the Symposium on Water Electrolysis*, Vol. 78-4, S. Srinivasan, F. J. Salzano, and A. R. Landgrebe, eds., Electrochemical Society, Princeton, New Jersey (1978), p. 150.
21. G. Kissel, F. Kulesa, C. R. Davidson, and S. Srinivasan, Selection and evaluation of materials for advanced alkaline water electrolyzers, in *Proceedings of the Symposium on Water Electrolysis*, Vol. 78-4, S. Srinivasan, F. J. Salzano, and A. R. Landgrebe, eds., Electrochemical Society, Princeton, New Jersey (1978), p. 218.
22. M. Prigent, L. J. Mas, and F. Verillon, Electrocatalysis in high performance alkaline water electrolysis, in *Proceedings of the Symposium on Water Electrolysis*, Vol. 78-4, S. Srinivasan, F. J. Salzano and A. R. Landgrebe, eds., Electrochemical Society, Princeton, New Jersey (1978), p. 234.
23. P. J. Moran and G. E. Stoner, Energy losses occurring in alkaline electrolytes, in *Proceedings of the Symposium on Water Electrolysis*, Vol. 78-4, S. Srinivasan, F. J. Salzano, and A. R. Landgrebe, eds., Electrochemical Society, Princeton, New Jersey (1978), p. 169.
24. J. P. Hoare, *The Electrochemistry of Oxygen*, Wiley, New York (1968).
25. J. O'M. Bockris and S. Srinivasan, *Fuel Cells: Their Electrochemistry*, McGraw-Hill, New York (1969).
26. G. Singh, M. H. Miles, and S. Srinivasan, Mixed oxides as oxygen electrodes, in "Electrocatalysis on non-metallic surfaces," *Natl. Bur. Stand. (U.S.) Spec. Publ.* **455**, 289 (1976).
27. S. Gottesfeld and S. Srinivasan, Electrochemical and optical studies of thick oxide layers on iridium and their electrocatalytic activities for the oxygen evolution reaction, *J. Electroanal. Chem.* **86**, 89 (1978).
28. M. H. Miles, Y. H. Huang, and S. Srinivasan, The oxygen electrode reaction in alkaline solutions on oxide electrodes prepared by the thermal decomposition method, *J. Electrochem. Soc.* **125**, 1931 (1978).
29. P. W. T. Lu and S. Srinivasan, Nickel based alloys as electrocatalysts for oxygen evolution from alkaline solutions, *J. Electrochem. Soc.* **125**, 265 (1978).
30. P. W. T. Lu and S. Srinivasan, Electrochemical–ellipsometric studies of oxide film formed on nickel during oxygen evolution, *J. Electrochem. Soc.* **125**, 1416 (1978).
31. J. M. Gras and M. Pernot, Oxygen overvoltages on Ni–Fe 50–50 in KOH 40%, in *Proceedings of the Symposium on Electrode Materials and Processes for Energy Conversion and Storage*, J. D. E. McIntyre, S. Srinvasan, and F. G. Will, eds., Electrochemical Society, Princeton, New Jersey (1977), p. 425.
32. P. W. T. Lu and S. Srinivasan, Advances water electrolysis: technology with emphasis on the use of solid polymer electrolyte, *J. Appl. Electrochem.* **9**, 269 (1979).
33. A. C. Tseung, S. Jasem, and M. N. Mahmood, Oxygen evolution on porous semiconducting oxide electrodes, in *Proceedings of the Symposium on Water Electrolysis*, Vol. 78-4, S.

Srinivasan, F. J. Salzano, and A. R. Langrebe, eds., Electrochemical Society, Princeton, New Jersey (1978), p. 161.
34. C. W. Tobias, Effect of gas evolution on current distribution and ohmic resistance in electrolyzers, *J. Electrochem. Soc.* **106**, 833 (1959).
35. R. E. Meredith and C. W. Tobias, Conduction in heterogeneous systems, *Adv. Electrochem. Electrochem. Eng.* **2**, 15 (1966).
36. G. B. Wallis, *One Dimensional Two-Phase Flow*, McGraw-Hill, New York (1969).
37. W. C. Kincaide and J. N. Murray, Electrolytic hydrogen: a prognosis, in *Proceedings of the First International Energy Agency Water Electrolysis Workshop*, p. 90, BNL Publication No. 21165 (1975).
38. M. R. Yaffe and J. N. Murray, Aqueous caustic electrolyzers at high temperatures, in *Proceedings of the Symposium on Water Electrolysis*, Vol. 78-4, S. Srinivasan, F. J. Salzano, and A. R. Landgrebe, eds., Electrochemical Society, Princeton, New Jersey (1978), p. 132.
39. A. J. Konopka and D. P. Gregory, Hydrogen production by electrolysis: present and future, paper presented at the 10th Intersociety Energy Conversion Engineering Conference, University of Delaware, Newark, August 18, 1975.
40. L. Giuffre, P. M. Spaziante, and A. Nidola, Diaphragms and membranes for the production of hydrogen by electrolysis of alkaline and sulfuric acid solutions, in *Proceedings of the Symposium on Water Electrolysis*, Vol. 78-4, S. Srinivasan, F. J. Salzano, and A. R. Landgrebe, eds., Electrochemical Society, Princeton, New Jersey (1978), p. 190.
41. C. A. Hampel, ed., *The Encyclopedia of Electrochemistry*, Reinhold, New York (1964), pp. 1156–1160.
42. A. T. Kuhn, ed., *Industrial Electrochemical Processes*, Elsevier, Amsterdam (1971).
43. Commerical brochures from the Electrolyser Corporation entitled (i) Electrolytic hydrogen plants and generators and (ii) Electrolytic hydrogen generators for metereological stations.
44. A. K. Stuart, Modern electrolyzer technology in industry, paper presented at the American Chemical Society, Annual National Meeting, Symposium on Non-Fossil Chemical Fuels, Boston, April 9–14, 1972.
45. A. K. Stuart, Electrolyzer technology for $H_2/O_2$ production, paper presented at the Symposium Course on Hydrogen Energy Fundamentals, University of Miami, Miami Beach, Florida, March 3–5, 1975.
46. R. L. LeRoy and A. K. Stuart, Present and future costs of hydrogen production by unipolar water electrolysis, in *Proceedings of the Symposium on Industrial Water Electrolysis*, Vol. 78-4, S. Srinivasan, F. J. Salzano, and A. R. Landgrebe, eds., Electrochemical Society, Princeton, New Jersey (1978), pp. 117–127.
47. Private communication from *The Electrolyser Corporation, Ltd.*, 1979.
48. Commercial brochures from Brown Boveri & Cie entitled (i) Electrolyzers for producing $H_2$, publication No. CH-1S411-290-E; (ii) Electrolyzers type EBN (Oerlikon System, Publication No. CH-1S-411-811-E); (iii) BBC electrolyzers (Oerlikon System), publication No. CH-1S412-241E; (iv) Electrolyzers to produce pure $H_2$ and $O_2$—reference list, publication No. CH-1S411-557-D/E/F/S; (v) Electrolyzer type EBN, drawings No. 1SE 320001, 2, 3, 4, and 7; and (vi) Electrolyzer plants, drawings No. H1SE 320-048, 068, 049, 050, and 051.
49. M. J. Braun, Brown boveri electrolyzers today and in the near future, in *Proceedings of the Symposium on Industrial Water Electrolysis*, Vol. 78-4, S. Srinivasan, F. J. Salzano, and A. R. Landgrebe, eds., Electrochemical Society, Princeton, New Jersey (1978), pp. 16–23.
50. Commercial brochures from Norsk Hydro Notodden Fabrikker, entitled (i) $H_2$ and $O_2$ by electrolysis of water; electrolyzers and electrolysis plants and (ii) Reference list hydrogen electrolyzer plants.
51. K. Christiansen and T. Grundt, Large-scale hydrogen production technology, experience and application, in *Proceedings of the Symposium on Industrial Water Electrolysis*, Vol. 78-4,

S. Srinivasan, F. J. Salzano, and A. R. Landgrebe, eds., Electrochemical Society, Princeton, New Jersey (1978), pp. 24–38.
52. Electrolytic hydrogen for $NH_3$ syntheses—economics becoming more favorable as hydrocarbon prices rise, by The British Sulphur Corporation, Ltd., Nitrogen No. 97, September/October, 1975.
53. Pressure lowers electrolytic hydrogen cost, *Chem. Eng.* (March 7, 1960).
54. H. Wullenweber and J. Mueller, Lurgi pressure electrolysis, in *Proceedings of the Symposium on Industrial Water Electrolysis*, Vol. 78-4, S. Srinivasan, F. J. Salzano, and A. R. Landgrebe, eds., Electrochemical Society, Princeton, New Jersey (1978), pp. 1–15.
55. The explosion at Laporte Industries Ltd., Ilford, April 5, 1975, Her Majesty's Stationery Office, England, ISBN 011 880 3306, 1975.
56. Giant electrolyzers produce hydrogen, *Chem. Eng.* (December 14, 1959).
57. Commerical brochure from Panchlor Chemicals Ltd. (a subsidiary of DeNora), entitled Water electrolysis—a simple way to produce high purity hydrogen.
58. Private communication from Panchlor Chemicals, Ltd., June 25, 1979.
59. E. M. Dukson, J. W. Ryan, and M. H. Smulyon, A technology assessment of the hydrogen economy concept, Record of the 10th Intersociety Energy Conversion Engineering Conference, University of Delaware, Newark, Delaware, August 18–22, 1975.
60. R. L. LeRoy, Hydrogen in Canada's energy future, in Hydrogen in Metals, The Metallurgical Society of CIM, annual volume, 1978, pp. 1–10.
61. F. C. Jensen and F. H. Schubert, Hydrogen generation through static feed water electrolysis, *Hydrogen Energy, Part A*, T. N. Veziroglu, ed. Plenum Press, New York (1975), pp. 425–439.
62. Private communication from Life Systems Inc., February 23, 1979.
63. J. N. Murray, J. B. Laskin, and W. C. Kincaide, The single irriguous approach to water electrolysis, in *Proceedings of the Symposium on Industrial Water Electrolysis*, Vol. 78-4, S. Srinivasan, F. J. Salzano, and A. R. Landgrebe, eds., Electrochemical Society, Princeton, New Jersey (1978), pp. 39–53.
64. J. B. Laskin, Electrolytic hydrogen generators, *Hydrogen Energy, Part A*, T. N. Veziroglu, ed., Plenum Press, New York (1975), pp. 405–415.
65. Commerical brochures from Teledyne Energy Systems, entitled (i) The Teledyne electra cell and (ii) Hydrogen gas generators.
66. L. J. Nuttall, A. P. Fickett, and W. A. Titterington, Hydrogen generation by solid polymer electrolyte water electrolysis, paper presented at American Chemical Society Division of Fuel Chemistry Meeting, Chicago, Illinois, August 1973.
67. W. A. Titterington and A. P. Fickett, Electrolytic hydrogen production with solid polymer electrolyte technology, in Proceedings of the 8th Intersociety Energy Conversion Engineering Conference, Philadelphia, 1973.
68. L. J. Nuttall, A. P. Fickett, and W. A. Titterington, Hydrogen generation by solid polymer electrolyte water electrolysis, in *Hydrogen Energy, Part A*, T. N. Viziroglu, ed., Plenum Press, New York (1975), pp. 441–455.
69. L. J. Nuttall and W. A. Titterington, General Electric's solid polymer electrolyte water electrolysis, paper presented at the Conference on the Electrolytic Production of Hydrogen, City University, London, England, February 1975.
70. A. B. LaConti, A. R. Fragala, and J. R. Boyack, Solid polymer electrolyte electrochemical cells: electrodes and other materials considerations, in *Proceedings of the Symposium on Electrode Materials and Processes for Energy Conversion and Storage*, J. D. E. McIntyre, S. Srinivasan, and F. G. Will, eds., Electrochemical Society, Princeton, New Jersey (1978), pp. 354–374.
71. B. D. McNicol, Electrocatalysis, *Chem. Soc. Spec. Period. Rep.* (1979).
72. J. O'M. Bockris and S. Srinivasan, Elucidation of the mechanism of electrolytic hydrogen evolution by the use of H–T separation factors, *Electrochim. Acta* **9**, 31 (1964).

73. L. Kandler, C. A. Knorr, and C. Schwitzer, The determinative process in the separation of hydrogen on active Pd and Pt cathodes, *Z. Phys. Chem.* **A180**, 281 (1937).
74. M. Breiter, Oxygen overvoltage, some problems in the study of, *Adv. Electrochem. Electrochem. Eng.* **1**, 123 (1961).
75. J. P. Hoare, *The Electrochemistry of Oxygen*, Interscience, New York (1968), Chap. 3.
76. A. K. N. Reddy, M. A. V. Devanathan, and J. O'M. Bockris, Ellipsometric Study of Oxygen Containing Films on Platinum Electrodes, *J. Electroanal. Chem.* **8**, 406 (1964); see also A. K. N. Reddy, M. A. Genshaw, and J. O'M. Bockris, Ellipsometric study of oxygen containing films on platinum anodes, *J. Chem. Phys.* **48**, 671 (1968).
77. J. O'M. Bockris and A. K. M. S. Hug, The mechanism of electrolytic evolution of oxygen on platinum, *Proc. R. Soc. London Ser. A* **237**, 277 (1956).
78. J. O'M. Bockris, Kinetics of activation controlled consecutive electrochemical reactions: anodic evolution of oxygen, *J. Chem. Phys.* **24**, 817 (1956).
79. T. P. Hoar, Mechanism of the oxygen electrode in alkaline solutions, *Proceedings of the 8th Meeting of CITCE, Madrid*, 1956, Butterworths, London (1958), p. 439.
80. A. Damjanovic, A. Dey, and J. O'M. Bockris, Kinetics of oxygen evolution and dissolution on platinum electrodes, *Electrochim. Acta* **11**, 791 (1966).
81. J. J. MacDonald and B. E. Conway, The role of surface films in the kinetics of oxygen evolution at Pd + Au alloy electrodes, *Proc. R. Soc. London Ser. A* **269**, 419 (1962).
82. C. Iwakura, K. Hirao, and H. Tamura, Anodic evolution of oxygen on Ru in acidic solutions, *Electrochim. Acta* **22**, 329 (1977).
83. A. Damjanovic, A. Dey, and J. O'M. Bockris, Electrode kinetics of oxygen evolution and dissolution on Rh, Ir and Pt–Rh alloy electrodes, *J. Electrochem. Soc.* **113**, 739 (1966).
84. D. N. Buckley and L. D. Burke, Part 6, oxygen evolution and corrosion at iridium anodes, *J. Chem. Soc. Faraday Trans. 1* **72**, Part 2, 2431 (1977).
85. L. J. Nuttall, Production and application of electrolytic hydrogen: present and future, in $H_2$: *Production and Marketing*, M. W. Smith and J. G. Santagelo, eds., A.C.S. (1980), p. 191.
86. D. J. Vaughan, *Du Pont Innovation* **4**, 10 (Spring 1973).
87. S. Stuck and A. Menth, Electrochemical studies of noble metal alloys in contact with polymer electrolyte membranes, in *Proceedings of the Symposium on Industrial Water Electrolyzers*, Vol. 78-4, S. Srinivasan, F. J. Salzano, and A. R. Landgrebe, eds., Electrochemical Society, Princeton, New Jersey (1978), pp. 180–189.
88. J. H. Russell, Design and development of solid polymer electrolyte water electrolyzers for large-scale hydrogen generation, in *Proceedings of the Symposium on Industrial Water Electrolyzers*, Vol. 78-4, S. Srinivasan, F. J. Salzano, and A. R. Landgrebe, eds., Electrochemical Society, Princeton, New Jersey (1978), pp. 77–87.
89. J. H. Russell, The development of solid polymer electrolyte water electrolysis for large-scale hydrogen generation, paper presented at the IEEE Power Engineering Society Summer Meeting, Vancouver, Canada, July 15–20, 1979.
90. General Electric Direct Energy Conversion, Solid polymer electrolyte water electrolysis technology development for large-scale hydrogen production, proposal submitted to the American Gas Association, Lynn, Massachusetts, April 22, 1974.
91. Brookhaven National Laboratory, Hydrogen storage and production in utility system, BNL 50472, F. J. Salzano, ed., Upton, New York, 1975.
92. E. W. Brooman and T. P. Hoar, Platinum metal alloys in electrocatalysis, *Platinum Met. Rev.* **9**, 122 (1965).
93. M. H. Miles, E. A. Klaus, B. P. Gunn, J. R. Locker, W. E. Serafin, and S. Srinivasan, The oxygen evolution reaction on Pt, Ir, Ru and their alloys in acid solutions, *Electrochim. Acta* **23**, 521 (1978).
94. J. Llopis and M. Vazquez, Passivation of Ru in hydrochloric acid solution, *Electrochim. Acta* **11**, 633 (1966).

95. L. D. Burke and T. O. O'Meara, Part 2—Behavior at Ru black electrodes, *J. Chem. Soc. Faraday Trans 1* **68**, 839 (1972).
96. J. Llopis, I. M. Tordesillas, and J. M. Alfayate, Anodic corrosion of Ru in hydrochloric acid solution, *Electrochim Acta* **11**, 623 (1966).
97. P. W. T. Lu and S. Srinivasan, Ruthenium oxide anodes for solid electrolyte water electrolyzers: electrochemical and ellipsometric studies, in *Proceedings of the Symposium on Industrial Water Electrolysis*, Vol. 78-4, S. Srinivasan, F. J. Salzano, and A. R. Landgrebe, eds., Electrochemical Society, Princeton, New Jersey (1978), pp. 88–101.
98. L. J. Nuttall, Solid polymer electrolyte water electrolysis development status, paper presented at the 9th Synthetic Pipeline Gas Symposium, Chicago, Illinois, October 31–November 2, 1977.
99. J. H. Russell and L. J. Nuttall, Development status of solid polymer electrolyte water electrolysis for large-scale hydrogen generation, paper presented at the DOE Chemical/Hydrogen Energy Systems Contractors' Review Meeting, Washington, D.C., November 27–30, 1978.
100. R. M. Dempsey and A. B. Laconti, GE Memo Report No. 70-1, General Electric Company, DECP, Lynn, Massachusetts, 1970.
101. SPE water electrolysis technology development for bulk energy storage systems, GE Final Technical Report to Brookhaven National Laboratory, Contract No. 350400S, General Electric, DECP, Wilmington, Massachusetts, November, 1975.
102. H. E. Kroeger, Use of discounted cash-flow method, *Chem. Eng.* (May 16, 1960).
103. J. Linsby, Return on investment: discounted and undiscounted, *Chem. Eng.* (May 21, 1979).
104. C. G. Edge, A practical manual on the appraisal of capital expenditure, Special Study No. 1, Revised Edition, The Society of Industrial and Cost Accountants of Canada, April 1964.
105. D. P. Gregory, K. F. Blurton, and N. P. Biederman, Economic criteria for the improvement of electrolyzer technology, in *Proceedings of the Symposium on Industrial Water Electrolysis*, Vol. 78-4, S. Srinivasan, F. J. Salzano, and A. R. Landgrebe, eds., Electrochemical Society, Princeton, New Jersey (1978), pp. 54–62.
106. M. J. Braun, Advanced alkaline water electrolysis cells: electrode and other material problems, in *Proceedings of the Symposium on Electrode Materials and Processes for Energy Conversion and Storage*, Vol. 77-6, Electrochemical Society, Princeton, New Jersey (1977), pp. 375–381.
107. L. J. Nuttall, Solid polymer electrolyte water electrolysis status for large-scale hydrogen production, in Proceedings of the 14th Intersociety Energy Conversion Engineering Conference, Boston, August 6–7, 1979.
108. P. W. T. Lu and R. L. Ammon, New development in sulfur dioxide depolarized water electrolysis technology, in *Extended Abstracts of the Fall 1979 Meeting of the Electrochemical Society*, October 14–19, Electrochemical Society, Princeton, New Jersey (1979).
109. A. J. Smith and J. D. Hatfield, A new look at the electrolytic production of hydrogen for the manufacture of ammonia, in *Proceedings of the Symposium on Industrial Water Electrolysis*, Vol. 78-4, S. Srinivasan, F. J. Salzano, and A. R. Landgrebe, eds., Electrochemical Society, Princeton, New Jersey (1978), p. 143.
110. F. J. Salzano, private communication, 1979.
111. R. W. Coughlin and M. Farooque, Hydrogen production from coal, water and electrons, *Nature* **279**, 301 (1979).
112. L. Elikan, J. P. Morris, and C. K. Wu, Development of a solid electrolyte system for oxygen reclamation, report prepared under Contract No. NASI-8896 by Westinghouse Electric Corporation for NASA, 1971.
113. H. S. Spacil and C. S. Tedmon, Electrochemical dissociation of water vapor in solid oxide electrolyte cells, thermodynamics and materials: series connected multiple cell stacks, in *Extended Abstracts of Spring Meeting of the Electrochemical Society*, Vol. 68-1, Electrochemical Society, Princeton, New Jersey (1968), p. 194.

114. F. J. Rohr, High temperature solid oxide fuel cells: present status and problems of development, in *Proceedings of the Workshop on High Temperature Solid Oxide Fuel Cells*, May 5–6, 1977, Brookhaven National Laboratory, Upton, New York, 1978, p. 122.
115. W. Doenitz, W. Schmidberger, and E. Steinhell, Perspectives and problems of high temperature electrolysis of water vapor using solid electrolytes, in *Proceedings of the Symposium on Industrial Water Electrolysis*, Vol. 78-4, S. Srinivasan, F. J. Salzano, and A. R. Landgrebe, eds., Electrochemical Society, Princeton, New Jersey (1978), p. 266.
116. H. S. Isacs, J. A. Fillo, V. Dang, J. R. Powell, M. Steinberg, F. Salzano, and R. Benenati, Hydrogen production from fusion reactions coupled with high temperature electrolysis, in *Proceedings of the Symposium on Water Electrolysis*, Vol. 78-4, S. Srinivasan, F. J. Salzano, and A. R. Landgrebe, eds., Electrochemical Society, Princeton, New Jersey (1978), p. 249.
117. A. P. Isenberg, Thin film solid electrolyte fuel cells, in *Proceedings of the Symposium on Electrode Materials and Processes for Energy Conversion and Storage*, Vol. 77-6, Spring 1977 Meeting of the Electrochemical Society, Philadelphia, J. D. E. McIntyre, S. Srinivasan, and F. G. Will, eds., Electrochemical Society, Princeton, New Jersey (1977), p. 682.
118. G. H. Farbman and G. H. Parker, The sulphur cycle hydrogen production process, in $H_2$: *Production and Marketing*, M. W. Smith and J. G. Santagelo, eds., A.C.S. (1980), p. 359.
119. D. VanVelzer, H. Langenkamp, A. Schutz, D. Lalonde, J. Flaen, and P. Feibelmann, Development, design and operation of a continuous laboratory scale plant for hydrogen production by the Mark 13 cycle, in *Proceedings of the 2nd World Hydrogen Energy Conference*, Zurich, Switzerland, August 21–24, 1978.
120. A. Fujishima and K. Honda, Electrochemical photolysis of water at a semiconductor electrode, *Nature* **238**, 37 (1972).
121. J. G. Mavroides, D. I. Tchernev, J. A. Kofalas, and D. F. Kolesar, Photoelectrolysis of water in cells in $TiO_2$ anodes, *Mat. Res. Bull.* **10**, 1023 (1975).
122. J. O'M. Bockris and K. Uosaki, Theoretical treatment of the photoelectrochemical production of hydrogen, *Int. J. Hydrogen Energy* **2**, 123 (1977).
123. J. G. Mavroides, J. A. Kafalas, and D. F. Koselar, Photoelectrolysis of water in cells with $SrTiO_3$ anode, *Appl. Phys. Lett.* **28**, 241 (1976).
124. J. O'M. Bockris, Hydrogen production in a solar energy economy, in *Proceedings of the Symposium Electrode Materials and Processes for Energy Conversion and Storage*, Vol. 77-6, Spring 1977 Meeting of the Electrochemical Society, Philadelphia, J. D. E. McIntyre, S. Srinivasan, and F. G. Will, eds., Electrochemical Society, Princeton, New Jersey (1977), p. 338.
125. M. Hammerli, Heavy water as a valuable by-product of electrolytic hydrogen, in *Proceedings of the 2nd World Hydrogen Energy Conference*, Zurich, Switzerland, August 21–24, 1978, p. 423.
126. J. O'M. Bockris, S. Srinivasan, and D. B. Matthews, Proton transfer across double layers, *Discuss Faraday Soc.* **39**, 239 (1965).
127. M. Hammerli, J. P. Butler, and W. H. Stevens, Peak power and heavy water production from nuclear electrolytic hydrogen and oxygen in Canada, *Int. J. Hydrogen Energy* **4**, 85 (1979).
128. J. P. Butler, J. H. Rolston, and W. H. Stevens, Novel catalysts for isotopic exchange between $H_2$ and liquid water; M. Hammerli, W. H. Stevens, and J. P. Butler, Combined electrolysis catalytic exchange process for $H_2$ isotope separation, in *Separation of $H_2$ Isotopes*, H. K. Ray, ed., ACS Symposium Series No. 68, 1978.
129. A. J. Appleby and G. Crepy, Advanced electrolysis in alkaline solutions, in *Proceedings of the Symposium on Electrode Materials and Processes for Energy Conversion and Storage*, Vol. 77-6, J. D. E. McIntyre, S. Srinivasan, and F. G. Will, eds., Electrochemical Society, Princeton, New Jersey, (1977), p. 382.
130. J. N. Murray and M. R. Yaffe, High efficiency alkaline water electrolysis technology, in *Proceedings of the 14th IECEC Meeting*, Boston, August, 1979, p. 749.

131. E. Hausmann, H. Will, and A. Belloni, Proceedings of the Oronzio denora Symposium on Chlorine Technology, Oronzio deNora Impianti Electrochimici S.P.A., Milano, 1979, p. 69.
132. Commercial brochure from Krebskosmo, Berlin, West Germany, entitled "Water Electrolysis."
133. J. H. Russel, Development of solid polymer electrolyte water electrolysis for large scale $H_2$ production, CONF-791127, Proceedings of the DOE Chemical Energy Storage and $H_2$ Energy Systems Contracts Review, 13–14 November, 1979, Reston, Virginia.

# 2
# Production of Chlorine

**DONALD L. CALDWELL**

## 1. Introduction

### 1.1. Significance

Chlorine and its coproduct, sodium hydroxide (caustic soda), are significant factors in the world economy. They are indispensable intermediates for the chemical industry, and also possess important uses in a variety of other industries. Chlorine is second to aluminum as a consumer of electricity among the electrolytic processes. Direct current (dc) power for chlorine cells accounts for nearly 2% of all electric power generated in the United States.[1]

Caustic soda and chlorine are the largest volume electrolytic chemicals by a wide margin; in 1975 they ranked sixth and seventh in production volume among all U.S. industrial chemicals.[2] Annual global chlorine production capacity was estimated in mid-1978 to be 35.4 million metric tons, including 12.2 million in the United States, 12.7 million in Western Europe, and 3.6 million in Japan.[3] Actual 1978 production in the United States was 10.0 million metric tons.[4]

### 1.2. Scope

The scope of this chapter is restricted to the production of chlorine and caustic soda by electrolysis of aqueous sodium chloride. This process accounts for about 96.5% of all chlorine production.[3] Part of the balance is obtained

---

**DONALD L. CALDWELL** • The Dow Chemical Company, Freeport, Texas 77541.

from by-product hydrochloric acid by electrolytic or nonelectrolytic methods; the remainder occurs as a coproduct in the electrolytic manufacture of potassium hydroxide, sodium, and magnesium.

### 1.3. Major End Uses

#### 1.3.1. The Chlorine–Caustic Balance

The economic health of the chlorine industry requires that a reasonable balance be maintained between consumer requirements for chlorine and caustic soda. Chlorine markets have shown the greater growth; this fact plus the difficulty of storing large quantities of chlorine have led chlorine demand to control the production of caustic soda. Economic pressures in the 1950s and 1960s allowed caustic soda to penetrate markets then dominated by other alkalies such as calcium hydroxide (lime) and sodium carbonate (soda ash). Those uses of caustic soda-as-alkali retain some flexibility, and act as dampers for caustic over- or undersupply.

#### 1.3.2. Chlorine

Chlorine is a highly reactive element. It is capable of supporting combustion, and will react with most elements under specific conditions. Its major industrial uses are first, as a strong oxidizing agent, and second, as a specific chlorinating agent. The largest volume use for chlorine, accounting for 17% of U.S. production in 1978,[4] is in the production of vinyl chloride monomer. Polymerized vinyl chloride, or polyvinyl chloride, is one of the major plastics. The synthesis of other chlorinated organic compounds consumed another 48% of U.S. $Cl_2$ production. Major identified categories include chlorinated ethanes (14%), chlorofluorocarbons (7%), chlorinated methanes (4%), epichlorohydrin (3%), and phosgene (3%). The chlorohydrin process for the synthesis of propylene oxide, an intermediate in the manufacture of urethanes and polyesters, used 8% of U.S. chlorine production in 1978.

Chlorine is used widely as a bleaching agent and as a disinfectant, either in the form of the element or as sodium or calcium hypochlorite. Other inorganic compounds produced from chlorine include hydrogen chloride, nonmetal chlorides such as phosphorous trichloride and sulfuryl chloride, and metal chlorides such as aluminum chloride and titanium tetrachloride. Additional information on chlorine markets may be found in the publication *Chemical Origins and Markets*.[5]

#### 1.3.3. Sodium Hydroxide (Caustic Soda)

Sodium hydroxide has a wide range of industrial applications. Its aqueous solutions are highly alkaline and react with mineral acids and acid gases to form the corresponding sodium salts. Caustic soda reacts with many metal salts to form insoluble hydroxides. Its reaction products with amphoteric metals and their oxides are, however, soluble; both classes of reaction with metals are industrially important. A variety of reactions can occur with organic materials. Organic acids and esters are solubilized by formation of sodium salts. Lignin and hemicellulose are removed from wood pulp in the Kraft process. A particularly valuable class of reaction is known as internal coupling. One such coupling reaction is the dehydrochlorination of propylene chlorohydrin with caustic soda to form propylene oxide, sodium chloride, and water. Additional end use information may be found in *Chemical Origins and Markets.*[5]

Exportation of caustic soda has helped to maintain the chlorine–caustic balance in the major producing countries. About 12% of U.S. caustic production in 1975 was exported, primarily to Latin America. Caustic demand in a developing economy always exceeds chlorine demand. Caustic-consuming basic industries such as mineral processing, paper, glass, and textile manufacture normally precede the development of the chlorine-consuming petrochemical and plastics industries. The chlorine/caustic usage ratio has, in fact, been used as a measure of the maturity of an economy. As nations which now import caustic grow, their internal chlorine demands will grow to the extent that local chloralkali plants will be required; these will reduce the need for imported caustic.

## 2. Historical Survey

### 2.1. Nonelectrolytic Processes

The manufacture of soap by the action of alkali on animal fats or vegetable oils is an ancient process. In 1737, Duhamel du Monceau established that common salt contained the base of natural mineral alkali. Several industrial processes for making alkali from salt were in use before 1800.

The lime–soda process dominated caustic soda production until the late 1940s. The method is limited to dilute solutions, and depends on the low solubility of calcium carbonate in water:

$$Na_2CO_3 + Ca(OH)_2 \rightarrow 2NaOH + CaCO_3\downarrow \qquad (1)$$

Although this process is still used to a certain extent to produce caustic soda for internal uses, for example, in the kraft pulp industry, all caustic marketed today is produced electrolytically.

Chlorine was known to the alchemists at least as far back as the 13th century.[6] Its properties were first investigated systematically by Scheele, who reported his findings in 1774. Bertollet's discovery in 1785 of chlorine's bleaching properties and Macintosh's invention of calcium hypochlorite bleaching powder in 1799 were of great significance to Great Britain. One of

Britain's major industries at that time was the manufacture of textiles. The invention of the power loom had vastly increased production rates; however, bleaching of cloth could only be accomplished by spreading it under the sun in open fields. The development of chlorine bleaching shortened the process from months to a few days and thus was an important factor in Britain's industrial revolution. Chlorine was produced at that time by Scheele's method of reacting manganese dioxide with hydrochloric acid:

$$MnO_2 + 4HCl \xrightarrow{100-110°C} MnCl_2 + Cl_2(g) + 2H_2O \qquad (2)$$

The hydrochloric acid was available as a by-product from the LeBlanc soda ash process. The chlorine was turned into a stable, solid product by passing it over hydrated lime to form calcium hypochlorite, which was the most widely used form for shipping chlorine for over one hundred years.[7] Both products of today's chloralkali industry were thus well-established items of commerce by the early 19th century, manufactured by chemical processes.

## 2.2. Early History of Electrolysis

In 1800, Cruickshank demonstrated the decomposition of salt water by means of an electric current. During the period 1832–1834, Faraday formulated the laws governing the electrolysis of aqueous salt solutions. In 1851, an English patent was granted for the electrolytic production of chlorine from brine. The process did not become commercially feasible until dynamos were developed to the point that large amounts of direct current could be supplied economically. In 1890, the first electrolytic chlorine plant began operation in Griesheim, Germany, using potassium chloride electrolyte. This was followed rapidly by the construction of plants in the United States and Great Britain. The first commercial electrolytic sodium hydroxide was produced in Rumford Falls, New York, in 1893.

One of the most difficult problems facing the infant industry was a means of continuous separation of the electrolysis products, chlorine and sodium hydroxide. In the decade 1883–1893, two fundamentally different cell types were developed to accomplish this separation: the diaphragm cell and the mercury cell. These two designs remain basic to modern chloralkali technology.

## 2.3. The Diaphragm Cell

A diaphragm cell contains a porous solid separator between the electrode compartments. This separator must allow passage of electric current while minimizing migration of hydroxyl ion from cathode to anode compartment. The Griesheim cell utilized a porous cement diaphragm invented by Brauer in

1886. The diaphragm was prepared by mixing portland cement with hydrochloric acid–acidified brine. The steel cell body served as cathode; anodes were carbon or magnetite. The cell was operated batchwise and electrolysis of potassium chloride was carried out for three days until a concentration of about 7% potassium hydroxide was obtained.

The first diaphragm cell operated in Great Britain was the Hargreaves–Bird cell. It had an asbestos cement diaphragm on copper gauze. Back-migration of hydroxyl ion was prevented by adding steam and carbon dioxide to the cathode compartment. This was the first commercial cell with a vertical diaphragm.

The first use of a percolating diaphragm, the basis of all modern diaphragm cells, was made by LeSueur at Rumford Falls in about 1897. Brine was allowed to flow into the anolyte compartment and through the diaphragm by the device of maintaining the electrolyte level higher on the anode than on the cathode side. The slow percolation of electrolyte through the diaphragm countered the migration of hydroxyl ions toward the anode compartment, permitting not only continuous operation but a much higher current efficiency than had previously been obtained. An installation of these cells with essentially the same design remained in operation at the Brown Company in Berlin, New Hampshire at least to the mid-1960s.[8]

Early anodes were of carbon, magnetite, or platinum; the economics of their use was poor, and could not support a major industry. The independent invention of synthetic graphite by Acheson and Castner in the 1890s provided an inexpensive, abundant, reasonably efficient anode material which served the chloralkali industry well for over 75 years.[9]

The next major development took place in 1913, when Marsh designed a cell with interleaved anodes and cathodes. This increased electrode area per unit of cell volume or cell room floor space, reducing capital investment per unit of production. Diaphragms were of asbestos paper wrapped around the cathodes and sealed with cement and putty. The Marsh cell was plagued with leaks and was not entirely successful. In 1928, Stuart of Hooker Electrochemical Company developed a method of depositing asbestos fiber onto the cathode of a Marsh-type cell by immersing it in a slurry and applying a vacuum.[10]

With the invention of the deposited asbestos diaphragm the technology had matured. Although scores of cell types were developed, the basic features of all successful diaphragm cells remained the same for the next 40 years: all had vertical graphite anodes, steel screen cathodes, and percolating diaphragms of vacuum-deposited asbestos.

### 2.4. The Mercury Cell

The second basic type of chlorine cell contains no diaphragm, but achieves separation of electrode products by the use of a mercury cathode. The first practical cells were invented by Castner and Kellner, both of whom applied independently for patents in 1892. The two men ultimately licensed their patents jointly and conducted industrial operations under the name Castner–Kellner.

The mercury cell process actually involves two cells, an electrolyzer and a decomposer. In the electrolyzer chlorine is produced at the anode from chloride ion. At the mercury cathode sodium ion is reduced and forms a sodium–mercury amalgam. The amalgam is passed to the second cell, where the amalgam is reacted with water to form sodium hydroxide, hydrogen, and pure mercury, which is recycled to the electrolyzer.

Castner's cell accomplished the circulation of mercury by imparting a slow rocking movement to the entire cell. In 1894 the Mathieson Alkali Works obtained the U.S. rights. A commercial plant incorporating these cells and Castner's newly developed synthetic graphite anodes was built at Niagara Falls, New York, in 1897. The original rocking cells remained in operation in this plant until 1960.[11]

In 1898, Solvay et Cie began experiments with a stationary cell in which the mercury cathode flowed continuously along the bottom of a slightly inclined, elongated trough. A development of this cell was built in Great Britain in 1902 by the Castner–Kellner Alkali Company. This type of cell, the so-called "Long" cell, became the predominant mercury cell design. The evolution of the mercury cell in the twentieth century has been via cumulative minor improvements, mainly directed toward improved efficiency and increased capacity.[12]

## 3. Electrolytic Decomposition of Sodium Chloride

### 3.1. Manufacturing Processes

There are at present three types of chloralkali cells in commercial use. These are (1) diaphragm cells, (2) flowing-mercury cathode cells, and (3) membrane cells. These cells are illustrated schematically in Figure 1. Their basic difference is the means used to separate the reaction products. The principles of operation of diaphragm and mercury cells have been described briefly in Sections 2.3 and 2.4. The membrane cell is the newest type, and is just coming into widespread use. It differs from the diaphragm cell in that its separator, a cation-exchange membrane, does not permit bulk flow of electrolyte, but only the transport of ions with their associated hydration spheres. In a well-designed cell these ions will be essentially all sodium and hydronium ions. The resulting product is of greater purity and is more concentrated than the effluent from diaphragm cells, providing the economic justification for the membrane cell's greater expense.

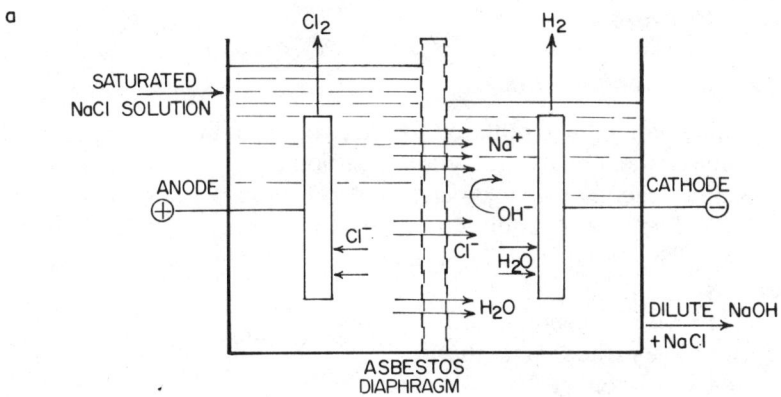

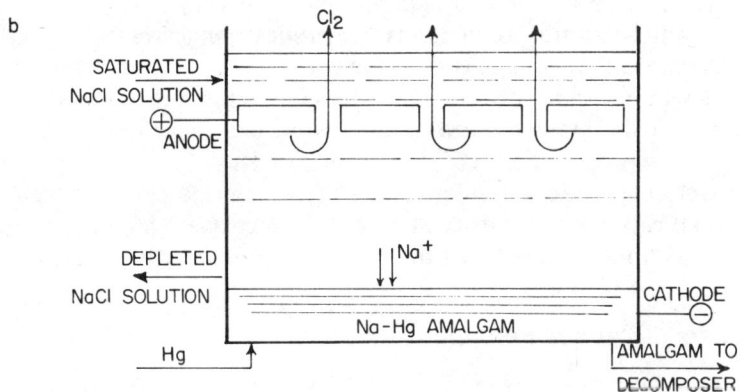

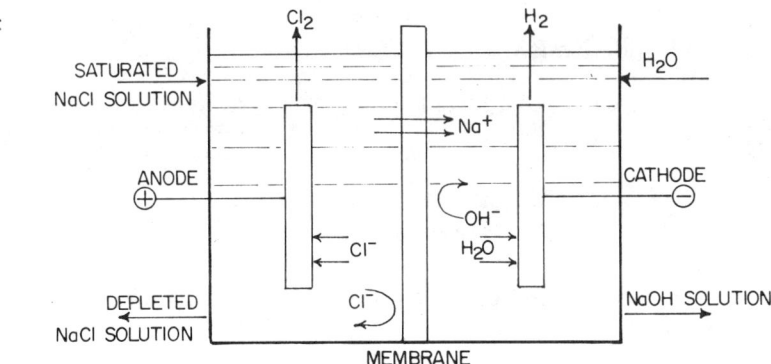

Figure 1. Chloralkali electrolysis cells: (a) diaphragm cell, (b) mercury cell, (c) membrane cell.

## 3.2. Cell Voltage

### 3.2.1. Voltage Components

The minimum voltage required for electrolysis to begin at a given set of cell conditions is the sum of the anodic and cathodic reversible potentials, and is known as the "decomposition voltage." To drive the reaction at an acceptable rate additional voltage is required to overcome cell resistance and electrode overvoltage. The total cell voltage therefore comprises the following components:

1. reversible anode potential ⎫
2. reversible cathode potential ⎬ decomposition voltage
3. anode overvoltage
4. cathode overvoltage
5. ir drop in electrolyte
6. ir drop in separator (diaphragm or membrane cell)
7. ir drop in structural components (electrodes, connectors)

The decomposition voltage is traditionally taken as 2.3 V for diaphragm cells and 3.2 V for mercury cells, although the sums of the true thermodynamic reversible potentials obtained at cell conditions are somewhat lower than these values.

Reversible electrode potentials and ir drop considerations, which are common to all cell types, are discussed below. Electrode overvoltages are a function of electrode material and their discussion is deferred to Section 4.

### 3.2.2. Reversible Anode Potential

The desired chlorine cell anode reaction is the oxidation of chloride ion:

$$2Cl^- \rightarrow Cl_2(g) + 2e^-, \quad E^\circ_{298} = +1.3595 \text{ V} \tag{3}$$

The standard chlorine potential at temperature (K) has been given by Faita, Longhi, and Mussini[13]:

$$E^\circ_T = 1.47252 + (4.82271 \times 10^{-4})T - (2.90055 \times 10^{-6})T \tag{4}$$

at 70°C,

$$\begin{aligned} E^\circ_{343} &= 1.47252 + 0.16650 - 0.34157 \\ &= 1.2965 \text{ V} \end{aligned} \tag{5}$$

The reversible discharge potential is related to the standard potential by the Nernst equation:

$$E(Cl_2) = E^\circ_T + \frac{RT}{2F} \ln P_{Cl_2} - \frac{RT}{F} \ln a_{Cl^-} \tag{6}$$

## Table 1
### Reversible Chlorine Discharge Potentials at 1 atm [15]

| Brine | | $E(Cl_2)$, V | | | |
|---|---|---|---|---|---|
| NaCl, g/liter | $Cl^-$ mol/liter | 50°C | 60°C | 70°C | 80°C |
| 300 | 5.133 | 1.2801 | 1.2741 | 1.2482 | 1.2312 |
| 295 | 5.048 | 1.2805 | 1.2646 | 1.2486 | 1.2317 |
| 290 | 4.962 | 1.2810 | 1.2651 | 1.2492 | 1.2323 |
| 280 | 4.791 | 1.2820 | 1.2661 | 1.2502 | 1.2333 |
| 260 | 4.449 | 1.2840 | 1.2683 | 1.2524 | 1.2356 |
| 240 | 4.107 | 1.2863 | 1.2706 | 1.2547 | 1.2380 |
| 220 | 3.764 | 1.2887 | 1.2731 | 1.2573 | 1.2407 |
| 200 | 3.422 | 1.2914 | 1.2758 | 1.2601 | 1.2436 |
| 180 | 3.080 | 1.2943 | 1.2788 | 1.2632 | 1.2468 |
| 160 | 2.738 | 1.2976 | 1.2822 | 1.2667 | 1.2504 |
| 140 | 2.395 | 1.3013 | 1.2860 | 1.2707 | 1.2544 |

where $R = 1.987$ cal/mol K, $T = K$, $F = 23,060$ cal/V-eq, $P_{Cl_2}$ is the partial pressure of chlorine in atmospheres, and $a_{Cl^-}$ is the activity of chloride ion in mol/l, frequently approximated by the ionic concentration. The partial pressure of chlorine gas is equal to the total pressure less the vapor pressure of water above the solution, which must be corrected for the presence of NaCl in the anolyte. The necessary calculations have been published by MacMullin.[14]

At 70°C, $P_{Cl_2} \approx 0.7$ atm, and 260 g/l NaCl in electrolyte,

$$E(Cl_2) = 1.2965 + 0.0148 \ln(0.7) - 0.0296 \ln(4.44)$$

$$- 1.2965 - 0.0053 - 0.0441 \quad (7)$$

$$= 1.2471 \text{ V}$$

At 25°C, $P_{Cl_2} \approx 0.97$ atm, and 260 g/l NaCl in electrolyte,

$$E(Cl_2) = 1.3595 + 0.0128 \ln(0.97) - 0.0257 \ln(4.44)$$

$$= 1.3595 - 0.0004 - 0.0383 \quad (8)$$

$$= 1.3208 \text{ V}$$

Table 1 is a compilation of reversible chlorine potentials at industrially useful conditions, taken from Currey and Pumplin.[15]

### 3.2.3. Reversible Cathode Potential

The cathode reaction in diaphragm (and membrane) cells is fundamentally different from that in mercury cells. In the former case the reaction is the reduction of water:

$$2H_2O + 2e^- \rightarrow H_2(g) + 2OH^-, \quad E^\circ_{298} = -0.828 \text{ V} \quad (9)$$

In the mercury cell the cathodic reaction is the formation of sodium amalgam:

$$Na^+ + e^- + xHg \rightarrow NaHg_x, \qquad E^\circ_{298} = -1.868 \text{ V} \qquad (10)$$

This is the dominant reaction, even at a 1 V higher standard potential than that of hydrogen formation, because of the high overvoltage of hydrogen on mercury at cell conditions. The difference in potential between diaphragm and mercury cells is equivalent to the potential of the reaction

$$2Na + 2H_2O \rightarrow 2Na^+ + 2OH^- + H_2(g) \qquad (11)$$

which occurs in the decomposition portion of the mercury cell system. No means has as yet been found to utilize this decomposition voltage; the decomposer thus acts as a totally shorted cell.

The temperature coefficient of the hydrogen evolution reaction is conventionally taken as zero. The reversible discharge potential of the cathodic reaction in diaphragm and membrane cells is thus

$$E(H_2) = -0.828 - \frac{RT}{2F} \ln P_{H_2} - \frac{RT}{F} \ln a_{OH^-} \qquad (12)$$

The partial pressure of hydrogen gas over the catholyte is equal to the total pressure minus the vapor pressure of water for that temperature and solution concentration. The vapor pressure of water is lowered by the presence of NaCl and NaOH in the catholyte. The required correction factors are found in MacMullin's paper.[14] His algorithm is used in the calculations below.

At 70°C, 100 g/l NaOH, and 180 g/l NaCl in electrolyte,

$$P_{H_2} = 0.863 \text{ atm}$$

$$\begin{aligned} E(H_2) &= -0.828 - 0.0148 \ln(0.863) - 0.0296 \ln(2.5) \\ &= -0.828 + 0.0022 - 0.0271 \\ &= -0.853 \text{ V} \end{aligned} \qquad (13)$$

At 25°C, 100 g/l NaOH, and 180 g/l NaCl in electrolyte,

$$P_{H_2} \approx 1 \text{ atm}$$

$$\begin{aligned} E(H_2) &= -0.828 - 0 - 0.0257 \ln(2.5) \\ &= -0.851 \text{ V} \end{aligned} \qquad (14)$$

The pressure correction factor at 70°C is thus 2 mV.

### 3.2.4. Electrolyte ir Drop

The diaphragm in a cell of that type is normally drawn directly onto the cathode. The current path is from the anode through the anolyte and diaphragm to the cathode. The electrical resistance across the anode–diaph-

## Table 2
### Resistivities (in Ω cm) of NaCl Solutions vs. Temperature and Concentration

| NaCl, g/liter (at 20°C) | Temperature | | | | | |
|---|---|---|---|---|---|---|
| | 25°C | 60°C | 70°C | 80°C | 90°C | 100°C |
| 320 | 3.987 | 2.251 | 2.000 | 1.798 | 1.634 | 1.497 |
| 310 | 4.000 | 2.263 | 2.012 | 1.812 | 1.643 | 1.506 |
| 300 | 4.024 | 2.286 | 2.035 | 1.832 | 1.665 | 1.525 |
| 290 | 4.058 | 2.316 | 2.058 | 1.853 | 1.686 | 1.549 |
| 280 | 4.102 | 2.347 | 2.085 | 1.879 | 1.712 | 1.573 |
| 270 | 4.151 | 2.383 | 2.117 | 1.907 | 1.738 | 1.598 |
| 260 | 4.208 | 2.420 | 2.153 | 1.940 | 1.767 | 1.622 |
| 250 | 4.269 | 2.461 | 2.192 | 1.976 | 1.799 | 1.652 |
| 240 | 4.336 | 2.503 | 2.232 | 2.014 | 1.835 | 1.685 |
| 220 | 4.508 | 2.608 | 2.317 | 2.093 | 1.908 | 1.753 |
| 200 | 4.737 | 2.747 | 2.449 | 2.208 | 2.012 | 1.851 |

ragm space (the "brine gap") adds a component to the overall cell voltage. The mercury cell contains an analogous brine gap between the anode and the mercury cathode. The electrical resistance of the brine gap arises from (1) the inherent electrical resistance of the solution and (2) the presence of gas bubbles in the solution.

Resistivities of sodium chloride solutions at industrial conditions are given in Table 2. Resistivity is related to the resistance of a conductor by

$$R = \rho \frac{l}{A} \tag{15}$$

where $R$ is the resistance of a conductor of length $l$, cross-sectional area $A$, and resistivity $\rho$, all in consistent units. The gas-free ir drop per centimeter of brine gap is given by

$$E_{brine} = j\rho \text{ V/cm} \tag{16}$$

where $j$ is the current density in A/cm$^2$ and the electrolyte resistivity $\rho$ is in ohm cm.

The effect of gas bubbles on the brine gap resistance is frequently estimated by the relationship of De La Rue and Tobias[16]:

$$\frac{\rho}{\rho_0} = (1 - \varepsilon)^{-3/2} \tag{17}$$

where $\rho/\rho_0$ is the ratio of the electrolyte resistivity with and without gas bubbles, and $\varepsilon$ is the gas void fraction. Hine and Murakami[17] have reported experimental verification of this equation for electrolysis of caustic soda in a vertical cell. Gas void fraction data for chlorine cell anolyte have not been reported in the open literature.

The contribution of the catholyte to the ir drop of a diaphragm cell is normally not considered except as a part of the diaphragm ir drop. The design of some membrane cells may, however, incorporate a catholyte gap between the cathode and membrane. Catholyte ir drop should be considered in such cases. Membrane cell catholyte has low levels of chloride and other impurities, and its resistivity can be closely approximated by that tabulated for pure NaOH solutions by, e.g., Dobos.[18] The gas void fraction studies of Hine et al.[17] should be directly applicable to the membrane cell catholyte gap.

Careful attention is paid in the design of chlorine cells to minimizing brine ir drop. The recent development of metal anodes and polymer-stabilized diaphragms has allowed the width of the brine gap to be decreased. In diaphragm cells natural upward flow of electrolyte is induced in the gap by the rising chlorine bubbles, the so-called "gas lift." This tends to sweep the electrode surface clean of bubbles. Provision must be made in the cell design for the electrolyte to return to the bottom of the anode compartment. If the flow rate in the gap is too low, or if the gap is made too small, blinding of the anode surface with gas can occur, with a consequent voltage penalty. A theoretical treatment of anolyte circulation in diaphragm cells has been published by Kheifets and Gol'dberg.[19]

### 3.2.5. Diaphragm ir Drop

Diaphragms are formulated from electrical insulators such as asbestos and their presence in the current path results in an increased electrical resistance over an equivalent volume of electrolyte alone. The potential drop across the diaphragm is appreciable, amounting to about 10% of the total cell voltage. Hine et al.[20,21] have published experimental studies, reporting the effect on resistance of such factors as compression during preparation and entrapped hydrogen bubbles. The deposited asbestos diaphragm is a partially oriented three-dimensional fibrous mat, and is difficult to characterize physically. Furthermore, its structure changes in operation, its resistance to flow being affected by, e.g., applied current load, anolyte pH, and brine purity. No theoretical model capable of predicting diaphragm ir drop has yet appeared in the open literature.

### 3.2.6. Structural ir Drop

The complete cell circuit includes electrical conductors to and from the rectifier, cell-to-cell connectors (internal or external to the cell), and the

### Table 3
### Resistivities (in $\mu\Omega$ cm) of Chlorine Cell Components at 20°C

| Component | Resistivity |
|---|---|
| Copper | 1.7 |
| Aluminum | 2.6 |
| Steel | 11.9 |
| Lead | 21.9 |
| Titanium | 46.2 |
| Mercury | 95.8 |
| Graphite | 700 |

electrodes themselves, which for maximum efficiency must be designed for uniform current distribution. All these conductors possess finite electrical resistances and thus contribute to the total circuit voltage. Primary consideration to structural resistance must be paid during the design stage; care must also be taken during assembly and operation to minimize all connection resistances. Resistivities of some common chlorine cell materials are given in Table 3. Titanium, now the predominant anode substrate material, has a relatively high resistivity for a structural metal. Current distributors such as Ti-clad copper tubes are commonly used to minimize structural ir drop of titanium anodes. The electrically active surface of the anode is frequently a sheet of expanded mesh. The resistance of this material is most often anisotropic; the pattern of the openings is such that the effective current path length differs for the $X$ and $Y$ dimensions of the sheet. The electrocatalytic coatings placed on titanium anodes are semiconductors, with resistivities $10^3$–$10^4$ times those of metals. These coatings are so thin, however ($\sim 1\ \mu$m), that the ir drop across them amounts at most to a few millivolts.

### 3.3. Current Efficiency

#### 3.3.1. Theoretical Current Consumption

The electrolytic decomposition of aqueous sodium chloride in a chlorine cell produces chlorine gas, hydrogen gas, and aqueous sodium hydroxide:

$$2\text{NaCl(aq)} + 2\text{H}_2\text{O} \xrightarrow{E} \text{Cl}_2\text{(g)} + \text{H}_2\text{(g)} + 2\text{NaOH(aq)} \tag{18}$$

The driving force for this reaction is provided by dc electrical energy. Faraday's law states that 96,487 C (A sec) are required to produce 1 gew (gram-equivalent weight) of electrochemical reaction product. This relationship determines the minimum electrical requirement for chloralkali production:

$$\frac{\text{kAh}}{\text{mt Cl}_2} = \frac{96{,}487 \times 1000}{60 \times 60 \times 35.453} = 755.99 \qquad (19)$$

$$\frac{\text{kAh}}{\text{mt NaOH}} = \frac{96{,}487 \times 1000}{60 \times 60 \times 39.997} = 670.10 \qquad (20)$$

$$\frac{\text{kAh}}{\text{mt H}_2} = \frac{96{,}487 \times 1000}{60 \times 60 \times 1.008} = 26{,}589.23 \qquad (21)$$

where mt is metric tons.

The current efficiency of an electrolytic process is the ratio of the amount produced of the material of interest to the amount theoretically possible. The anode product in a chlorine cell is chlorine; the cathode products are hydrogen and caustic soda. The cell therefore possesses simultaneously three distinct current efficiencies, one for each product. These never equal 100%; losses can occur from chemical and electrochemical side reactions within the cell, from physical loss of products outside the cell, and from leakage of electric current around the cell. Hydrogen efficiencies are rarely measured; most manufacturers settle on either the chlorine efficiency or the caustic efficiency as the "cell efficiency," although the two values coincide only fortuitously.

### 3.3.2. Chlorine Efficiency

The major anodic inefficiency reaction is the oxidation of water:

$$2H_2O \rightarrow O_2(g) + 4H^+ + 4e^- \qquad (22)$$

Under industrial conditions this reaction accounts for 2–4% of the anodic current. Its rate is a function of the anode material, the current density, the temperature, and the NaCl concentration and pH of the anolyte. Since in a diaphragm or membrane cell at steady state this pH, and thus the oxygen production rate, is proportional to the amount of hydroxyl ion entering the anode chamber from the catholyte, this reaction is frequently written as the simple discharge of hydroxyl ion:

$$4OH^- \rightarrow O_2(g) + 2H_2O + 4e^- \qquad (23)$$

although the concentration of hydroxyl ion in chlorine cell anolyte is typically $10^{-11}$ $M$.

Much of the oxygen produced at graphite anodes reacts directly to $CO_2$:

$$2H_2O + C(\text{graphite}) \rightarrow CO_2(g) + 4H^+ + 4e^- \qquad (24)$$

Oxygen-containing anions such as bisulfate may also participate in the anode reactions. Another possible reaction is the anodic oxidation of hypochlorite to chlorate:

$$6OCl^- + 3H_2O \rightarrow 2ClO_3^- + 4Cl^- + 6H^+ + 3/2 O_2(g) + 6e^- \qquad (25)$$

The chlorate concentration in anolyte is typically quite low, but increases with increasing pH; the relative rates of its electrochemical and chemical formation are uncertain.

Chlorine cell anolyte is quite complex chemically. The following reactions are all considered to be at equilibrium:

$$Cl_2(g) \rightleftharpoons Cl_2(aq) \tag{26}$$

$$Cl_2(aq) + Cl^- \rightleftharpoons Cl_3^- \tag{27}$$

$$Cl_2(aq) + H_2O \rightleftharpoons HOCl + H^+ + Cl^- \tag{28}$$

$$HOCl \rightleftharpoons H^+ + OCl^- \tag{29}$$

The chemical formation of chlorate is too slow to reach equilibrium:

$$2HOCl + OCl^- \rightarrow ClO_3^- + 2H^+ + 2Cl^- \tag{30}$$

The equilibrium or rate constants for equations (26)–(30) are known, and have recently been tabulated by Nagy.[22]

The chlorine species dissolved in the electrolyte of mercury and membrane cells is recovered as product; depleted brine is withdrawn continuously from these cells, acidified to drive the chlorine from solution, resaturated with salt, and returned to the cells as fresh feed. Anolyte chlorine is lost in the diaphragm process, however. It is carried through the diaphragm with the bulk flow of electrolyte and is either reduced at the cathode or exits the cell with the catholyte. The chlorine efficiency loss in diaphragm cells due to anolyte solution is primarily a function of anolyte pH and brine flow rate.

### 3.3.3. Caustic Efficiency

Electrolytic inefficiency reactions at the cathode are minor in importance. Hypochlorite flowing through the diaphragm is essentially all reduced by electrochemical and/or chemical means:

$$OCl^- + H_2O + 2e^- \rightarrow Cl^- + 2OH^- \tag{31}$$

$$OCl^- + H_2 \rightarrow Cl^- + H_2O \tag{32}$$

The reduction of chlorate has been a subject of controversy. Hine and Yasuda[23] found no electrochemical reduction, while Veselovskaya et al.,[24] and more recently Nagy[22] report a slow electrochemical reduction on a steel cathode.

The presence of hydrogen in mercury cell gas can result from either a chemical or an electrochemical reaction. In either case the net result is a loss of caustic soda production:

$$2H^+ + 2e^- \rightarrow H_2(g) \tag{33}$$

$$2NaHg_x + 2H_2O \rightarrow 2NaOH + H_2(g) + xHg \tag{34}$$

The major source of caustic loss in diaphragm and membrane cells is the migration of hydroxyl ion from the cathode to the anode compartment, attracted by the positively charged anode. The extent of this back-migration is of course dependent on the physical properties of the separator and is also strongly influenced by the concentrations of $OH^-$ in the catholyte and $Cl^-$ in the anolyte. Caustic efficiency in diaphragm and membrane cells normally runs somewhat lower than chlorine efficiency. A typical diaphragm cell caustic efficiency range is 94–96%.

A mercury cell inefficiency reaction with no counterpart in the other cell types is the so-called "recombination" reaction, which is the reverse of the cell reaction:

$$2NaHg_x + Cl_2 \rightarrow 2Na^+ + 2Cl^- + xHg \tag{35}$$

This reaction always occurs to some extent, and becomes highly significant if the anode–cathode gap is made too small, so that chlorine bubbles can upset the cathodic boundary layer and render the mercury surface accessible to dissolved chlorine.

### 3.4. Energy Efficiency

The energy efficiency of a cell is the product of the voltage and current efficiencies. Voltage efficiency is the ratio of the "decomposition voltage" to the actual working cell voltage.

$$\text{diaphragm cell } \eta_{energy} = \frac{2.3\eta_{current}}{E_{cell}} \tag{36}$$

$$\text{mercury cell } \eta_{energy} = \frac{3.2\eta_{current}}{E_{cell}} \tag{37}$$

The use of energy efficiency as a measure of cell performance has declined in recent years in favor of the energy consumption relationship (kWh/unit $Cl_2$ or NaOH), which is more immediately useful.

### 3.5. Energy Consumption

Calculation of dc energy usage requires knowledge of the cell voltage and current efficiency. The rectifier conversion efficiency is also required if ac energy consumption is to be calculated:

$$\frac{\text{ac kWh}}{\text{mt } Cl_2} = \frac{755.99 E_{cell}}{\eta_{current} \eta_{ac-dc}} \tag{38}$$

$$\frac{\text{ac kWh}}{\text{mt NaOH}} = \frac{670.10 E_{cell}}{\eta_{current} \eta_{ac-dc}} \tag{39}$$

$$\frac{\text{ac kWh}}{\text{mt } H_2} = \frac{26{,}589.23 E_{cell}}{\eta_{current} \eta_{ac-dc}} \tag{40}$$

## 4. Cell Components

### 4.1. Anodes

#### 4.1.1. Chlorine Overvoltage

The chlorine overvoltage at the anode is the difference between the anode's true operating voltage and the reversible chlorine discharge potential at the temperature and electrolyte composition of interest. Overvoltage, frequently called activation overvoltage or overpotential, is a function of the electrode material, the current density, and the temperature and composition of the electrolyte. At current densities of interest for chloride electrolysis overvoltages normally follow the Tafel relationship:

$$\eta = a + b \log j \qquad (41)$$

where $\eta$ is the overvoltage, $a$ and $b$ are constants dependent primarily on the electrode material and secondarily on electrolysis conditions, and $j$ is the current density in $A/cm^2$ of true surface area.

Reported overvoltages for anodes of technological interest differ greatly. The surface area of graphite changes in service, and no definitive pretreatment can exist. Coated metal anodes are quite stable in service, but are difficult to prepare reproducibly. Table 4 presents some representative overvoltages obtained on industrial anodes under service conditions. They were carefully determined, but no claim of universal validity is made. The reference electrode for these measurements was the saturated calomel electrode maintained at

Table 4
Chlorine Overvoltages (in mV) In 300 g/liter NaCl Electrolyte[a]

| Anode material | Temperature °C | Current density, mA/cm$^2$ | | | | | |
|---|---|---|---|---|---|---|---|
| | | 77.5 | 62.0 | 46.5 | 31.0 | 15.5 | 1.6 |
| Graphite | 30 | 229 | 207 | 181 | 158 | 129 | 62 |
| | 50 | 192 | 181 | 167 | 150 | 122 | 46 |
| | 70 | 187 | 174 | 159 | 139 | 108 | 29 |
| | 80 | 173 | 163 | 148 | 130 | 98 | 25 |
| $Co_3O_4$ on titanium | 30 | 63 | 58 | 52 | 46 | 37 | 16 |
| | 50 | 48 | 42 | 37 | 30 | 21 | 7 |
| | 70 | 45 | 39 | 34 | 27 | 19 | 6 |
| | 80 | 46 | 40 | 34 | 27 | 19 | 6 |
| $RuO_2$–$TiO_2$ on titanium | 30 | 61 | 54 | 46 | 38 | 27 | 6 |
| | 50 | 44 | 41 | 36 | 31 | 23 | 6 |
| | 70 | 44 | 40 | 37 | 31 | 24 | 6 |
| | 80 | 50 | 47 | 43 | 36 | 28 | 8 |

[a] Anode potentials measured vs. S.C.E. at 25°C.

25°C. This has been found to be an accurate, reproducible method. The alternative of maintaining the reference electrode at cell temperature has two drawbacks: reference electrode instability and an ambiguity in converting the measurements to the hydrogen scale, arising from the recently reported observation that the temperature coefficient of the reversible hydrogen electrode, far from being zero, is not even linear, but exhibits a change of slope at 60°C.[25]

### 4.1.2. Graphite Anodes

The environment that a chlorine cell anode must withstand is extraordinarily severe. A reasonably stable and economical anode was a prerequisite for the growth of the chloralkali industry. Synthetic graphite, developed independently by Castner and Acheson prior to 1900, proved to be such a material. Graphite remained the best available anode material until the development of coated titanium anodes in the late 1960s.

Graphite anodes have a service life of 6–24 months, being consumed at the rate of 1–3 kg/mt $Cl_2$. Part of the loss is due to electrochemical oxidation, and part is physical wear due to turbulent conditions at the anode face. Commercial diaphragm cell anodes are impregnated with a variety of resins and oils to reduce porosity and thus the mechanical wear rate. Graphite anodes are frequently the limiting factor in the run life of the cells that use them. Their corrosion products, a variety of chlorinated hydrocarbons, contaminate the product streams; solid graphite particles blanket the diaphragm and adversely affect its performance.

The economic impact of graphite wear is significant, and has attracted the attention of many electrochemical technologists over the years. Worthy of mention are the studies of Krishtalik,[26] Vaaler,[27,28] and Hine.[29] Factors found to accelerate graphite wear include high cell temperature, high anolyte pH, and ionic feed brine impurities, especially sulfate.

Transition of the industry from graphite to metal anodes accelerated in the mid-1970s under the influence of sharply increasing energy costs. By early 1979 conversion in the U.S. was nearly complete. The only major exception was Dow, which had favorable cell and power economics unique in the industry. However, in late 1978, Dow announced that it, too, was in the process of converting to metal anodes.[30]

### 4.1.3. Metal Anodes

#### 4.1.3.1. Platinum Group Coatings

Several of the platinum group metals are electrochemically stable under chlorine cell conditions. Platinum anodes were in limited commercial use in the late nineteenth century. They were uneconomical and were rapidly displaced by synthetic graphite.

Another group of metals, the so-called "valve metals" or "film-forming metals" of periodic groups IVB, VB, and VIB are able to withstand the corrosive conditions within a chlorine cell. They are protected by a thin, tightly adherent oxide film, which is incapable of passing electrical current in the anodic direction. The idea of combining the stability and economy of the valve metals with the electrical conductivity of a platinum-group metal coating is very old. Stevens[31] patented a platinum-coated tungsten anode in 1913. These early coated anodes did not find commercial acceptance, in part because the valve metals had a limited availability.

The first valve metal to find widespread industrial markets was titanium. Its strength, light weight, and temperature resistance recommended it to the aircraft industry in the late 1940s. Its present wide availability and low cost have made it the predominant anode substrate material. The idea of a platinum-metal-coated titanium anode was resurrected in the late 1950s by, among others, J. B. Cotton of ICI[32] and H. B. Beer, an independent Dutch inventor.[33]

The inadequacies of graphite anodes are more pronounced in mercury cells than in diaphragm cells. Voltages are higher because of higher current density, and wear rates are higher because of higher temperatures and greater mechanical wear. Mercury cell anodes are horizontal and must be provided with channels, slots, or other means to allow the chlorine to escape from their bottom, active, face. Titanium anodes, which are readily perforated to facilitate gas release, promised simpler designs and better performance. Platinum-coated titanium anodes were given wide publicity in the early 1960s but proved unsuccessful. Platinum wear rates were unacceptably high, and anode–cathode short circuits, always a possibility in cells with mercury cathodes, were found to cause sudden, disastrous loss of coating.

In the mid-1960s Beer and other researchers transferred their attention to the oxides of the platinum group metals, several of which were known to have high, near-metallic conductivities.[34] Best results were obtained with ruthenium dioxide, $RuO_2$. This oxide is isostructural and isomorphous with the rutile form of $TiO_2$. It thus has the ability to form a solid solution with the oxide film on titanium, rendering it conductive. Beer filed for worldwide patent coverage, with the assistance of V. deNora of Oronzio deNora Impianti, Milan, Italy.[35] Patents began issuing in 1968.[36] Beer's U.S. patents covering $RuO_2$ and $RuO_2$–$TiO_2$ solid solution coatings issued in 1973 and 1972.[37] No court tests of the U.S. patents have come to public attention. The German and Japanese equivalents have been challenged, with results unknown to this author in mid-1979.

The solid-solution $RuO_2$–$TiO_2$ anode coating is licensed globally by Diamond Shamrock Technologies, S.A., of Geneva, Switzerland, under the tradename DSA (for dimensionally stable anode). In the 1970s, the DSA has rapidly become the world-standard anode material, possessing long life, very low operating voltage, acceptable efficiency and cost, and reasonable tolerance of a wide range of operating conditions. Life in diaphragm cells is at least eight years. Life in mercury cells is considerably shorter, about two years, but DSAs are not subject to disastrous massive coating loss on shortout as were the earlier platinum coatings.

The exact composition of the commercial DSA coating has been a closely guarded trade secret. It is probable that two types of coating have been used: the original Beer type, with a Ti/Ru mole ratio of approximately 2/1, and a three-component coating patented originally by K. J. O'Leary of Diamond Shamrock Corporation,[38] in which part of the ruthenium is replaced by tin or other non-noble metals. According to the patent literature, the coating composition can affect the anodic inefficiency reaction. One type of coating may favor oxygen formation, while another type may catalyze the formation of chlorate. V. deNora has advocated use of multicomponent coatings to obtain the exact combination of properties desired. One coating recently described[39] contains the oxides of Ru, Ti, Sn, Bi, and Co; it is not known if coatings of this complexity have been used commercially.

A wide variety of preparative techniques are disclosed in Diamond Shamrock's extensive collection of DSA patents. A representative method is described in U.S. Patent No. 3,776,834: the coating precursor is a solution containing $RuCl_3 \cdot 2.5\ H_2O$, $SnCl_2$, butyl titanate, HCl, and butanol. The solution is applied to an expanded titanium mesh substrate and heated in air to a temperature of 450°C for seven minutes. The procedure is repeated ten times; the final oxide coating has a metal mole ratio of $3\,Ru:2\,Sn:11\,Ti$, and has a weight of $1.6\ mg/cm^2$ of anode surface on a $(RuO_2 + SnO_2)$ basis.

Other types of platinum-group coatings have been marketed; they include Pt–Ir alloys, non-solid-solution $RuO_2$, palladium oxides, and platinum "bronzes" (nonstoichiometric platinum oxides of general formula $M_{0.5}^{+}Pt_3O_4$). Vendors known to be active in early 1979 include IMI Marston Ltd., Sigri Electrographit GmbH, Tokyo Denki Kagaku Co., Ltd., and C. Conradty Nürnberg. None of the competitive coatings have as yet made a significant impact on Diamond Shamrock's near monopoly of the merchant metal anode market. One factor in the favorable economics of the DSA, low coating cost, was apparently entirely fortuitous: ruthenium is one of the least-expensive platinum-group metals, is readily available as a by-product of platinum manufacture, and has no major use apart from electrode coatings.

The chemistry and electrochemistry of the $RuO_2$-based electrode have been the subjects of a fairly extensive literature. $RuO_2$ and $IrO_2$ have rutile structures, but high electrical conductivities, and are frequently described as metallic or metal-like conductors.[40] Pure rutile, $TiO_2$, has a room temperature conductivity of only $10^{-13}$ mho/cm,[41] although its conductivity can be increased many orders of magnitude by thermal reduction or doping with a wide variety of elements.[42] The high conductivity of $RuO_2$ and $IrO_2$ was attributed by Rogers et al.[34] to a metal–oxygen $\Pi^*$ band just at the Fermi edge, which is partially filled in $RuO_2$ and $IrO_2$, but empty in $TiO_2$. Riga et al.[42] have recently reported confirmation of the Rogers model by X-ray photoelectron spectroscopy.

Anode coatings are polycrystalline and complex in structure. $RuO_2$ films are nonstoichiometric, containing varying amounts of oxygen and, when prepared from chloride solutions, chlorine.[44] The temperature coefficient of their resistivities is invariably negative, characteristic of semiconductors rather than metallic conductors. Their electrical and electrocatalytic properties are highly dependent on their method of preparation. This complicates evaluation of the works of different laboratories; the situation is worsened by the looseness of the nomenclature. "DSA" and "$RuO_2$" are used almost interchangeably in the literature to describe oxide coatings comprising Ru; Ru and Ti; Ru, Ir, and Ti; and Ru, Ti, and Sn. Confusion of single-metal $RuO_2$ coatings with the bimetal coatings can be especially unfortunate; despite the strictures of trade secrecy, the technical superiority of the $RuO_2$–$TiO_2$ solid solution coating to $RuO_2$ alone or to high-Ru coatings should be known by now to all industrial researchers in the field. This superiority has been confirmed in the open literature by Hine et al.,[45] who studied the physical, chemical, and electrochemical properties of a wide variety of anode coatings.

The DSA of commerce is a highly efficient electrocatalyst for chlorine production. This simple fact may not be apparent to the casual student of the literature. A variety of kinetic parameters have been reported, and several mechanisms have been proposed. A recent comprehensive study by Janssen et al.[46] is instructive. The Tafel slopes for chlorine evolution [$b$ of Eq. (41)] were highly dependent on the temperature of formation of their $RuO_2$–$TiO_2$ electrodes. The minimum slope, 40 mV at 25°C, was found for anodes prepared at 500°C. This Tafel slope corresponds to a transfer coefficient of $0.45 \pm 0.05$ and according to Janssen indicates chlorine formation by the Volmer–Heyrovsky mechanism, consisting of the anodic Heyrovsky reaction

$$Cl_{AD} + Cl^- \rightarrow Cl_2 + e^- \tag{42}$$

together with the cathodic Volmer reaction

$$Cl_{AD} + e^- \rightarrow Cl^- \tag{43}$$

the Heyrovsky reaction being the rate-determining step.

Other workers have reported Tafel slopes of 30 mV,[47] which would correspond to the Volmer–Tafel mechanism, consisting of the anodic Volmer reaction

$$Cl^- \rightarrow Cl_{AD} + e^- \tag{44}$$

followed by the rate-determining Tafel reaction

$$2Cl_{AD} \rightarrow Cl_2 \tag{45}$$

The subject must be considered open. Of prime interest of the chloralkali industry is the fact that the anodic overvoltage in a commercial chlorine cell containing DSAs is of the order of only 50 mV.

For 40 years prior to the introduction of the DSA chlorine cell technology had remained essentially unchanged. The invention of the DSA coincided with, and to a very great extent precipitated, the technological awakening of the industry, enabling it to cope with the successive energy and ecological crises of the 1970s. For the healthy state of today's industry all chloralkali technologists and researchers are indebted to Henri Beer.

### 4.1.3.2. Non-Platinum-Group Coatings

Magnetite, $Fe_3O_4$, saw commercial use as a chlorate anode for some 70 years. Japan Carlit operated magnetite anodes into the 1970s, claiming that the disadvantages of fragility and low conductivity in comparison with graphite were offset by their low wear rates and ability to operate at high temperatures.[48] Magnetite saw limited early use as a chlorine anode until replaced by synthetic graphite; however, its overvoltage for chlorine evolution is extremely high—350 mV at 1 mA/cm$^2$ in 5.3 $M$ NaCl at 25°C.[49]

Magnetite is a member of the large group of compounds classified as spinels. Spinels are oxides having the general formula $A[B]_2O_4$, where the oxygen ions lie in an almost perfect close packed cubic lattice. The A ions occupy tetrahedral interstices of the oxygen lattice and the B ions, octahedral interstices. The so-called inverse structure, $B[AB]O_4$, also exists, and magnetite is in fact an inverse spinel. The A and B ions may possess variable valencies, as long as charge balance is maintained. The most common valence for the A ions is +2; for the B ions, +3. Most spinels are quite tolerant of ionic substitutions, forming solid solutions over a wide range of compositions. The large range of properties obtainable from spinels have attracted the attention of electrode researchers. P. Anthony and A. Martinsons[50,51] of PPG patented a large number of spinel-type compositions as chlorine anode coatings. These have apparently not seen commercial use.

Cobalt oxide spinel coatings ($Co_3O_4$) are the subject of a number of Soviet publications and patents.[52-55] Voltage and efficiency equivalent to $RuO_2$–$TiO_2$ coatings are claimed. The extent of industrial use of these coatings is unknown.

Cobalt oxide spinel coatings have also been patented in the United States.[56,57] This coating is a very effective electrocatalyst for chlorine discharge and can operate at $RuO_2$–$TiO_2$-equivalent voltages even though its bulk resistivity is several orders of magnitude higher than that of $RuO_2$–$TiO_2$. Its catalytic properties are apparently attributable to the presence of surface $Co^{+3}$ ions and the extreme ease of electron transfer by the $Co^{+2} \rightleftharpoons Co^{+3} + e^-$ couple.

Improved coatings based on $Co_3O_4$ have recently been described and patented.[58,59] These coatings have the general formula $M_xCo_{3-x}O_4 \cdot yZrO_2$, where $0 \leq x, y \leq 1$ and M is Cu, Mg, or Zn. They are prepared by applying a solution containing the desired mixtures or thermally decomposable metal salts to a cleaned titanium substrate and baking in air at 250–475°C to form the oxides. The coating step is repeated 6–12 times until the desired coating thickness is obtained. It is claimed that these mixed oxide coatings operate at lower, more stable voltages than $Co_3O_4$. They are metal-deficient semiconductors with conductivities of 0.4–20 mho/cm.

Extensive plant testing of anode coatings of this type has been conducted by Dow Chemical Co. Dow has recently disclosed that its conversion plan for metal anodes includes use of its own coating system as well as that of Diamond Shamrock Technologies.[30]

## 4.2. Cathodes

### 4.2.1. Hydrogen Overvoltage

The hydrogen overvoltage at the cathode is the difference between the operating voltage of the hydrogen-producing cathode and the reversible hydrogen discharge potential in the given electrolyte. At industrial conditions hydrogen overvoltage normally follows the Tafel equation,

$$\eta = a + b \log j \tag{41}$$

where $\eta$ is the overvoltage, $a$ and $b$ are constants characteristic of the solution and the electrode material, and $j$ is the $A/cm^2$ of true surface. The Tafel slope $b$ is approximately 120 mV for most metals at 25°C.

The overvoltage for hydrogen discharge on mercury is very high, which makes the mercury cathode chlorine cell a viable concept. In a typical mercury cell the cathode potential is about $-1.8$ V, approximately 0.9 V more cathodic than the reversible hydrogen discharge potential An average chlorine gas sample from such a cell may contain 0.5% $H_2$, indicating that about 0.5% of the cathodic current is carried by the hydrogen ion, the other 99.5% being carried by the sodium ion. The sodium overvoltage at the mercury cathode is quite small, of the order of 10–20 mV.[11]

The literature of the hydrogen evolution reaction is, of course, very extensive, and will not be reviewed here. Surprisingly, relatively few mechanistic studies have been reported for the highly alkaline solutions of industrial interest. The 1958 study by Zholudev and Stender[60] is still pertinent (Figure 2). Tafel curves were obtained for a number of metals in 6 $M$ NaOH at 25°C. The hydrogen evolution reaction is in general highly irreversible in alkaline solutions. In common with acidic systems, the pure metals with the lowest overvoltages are found to be the Group VIII transition metals.

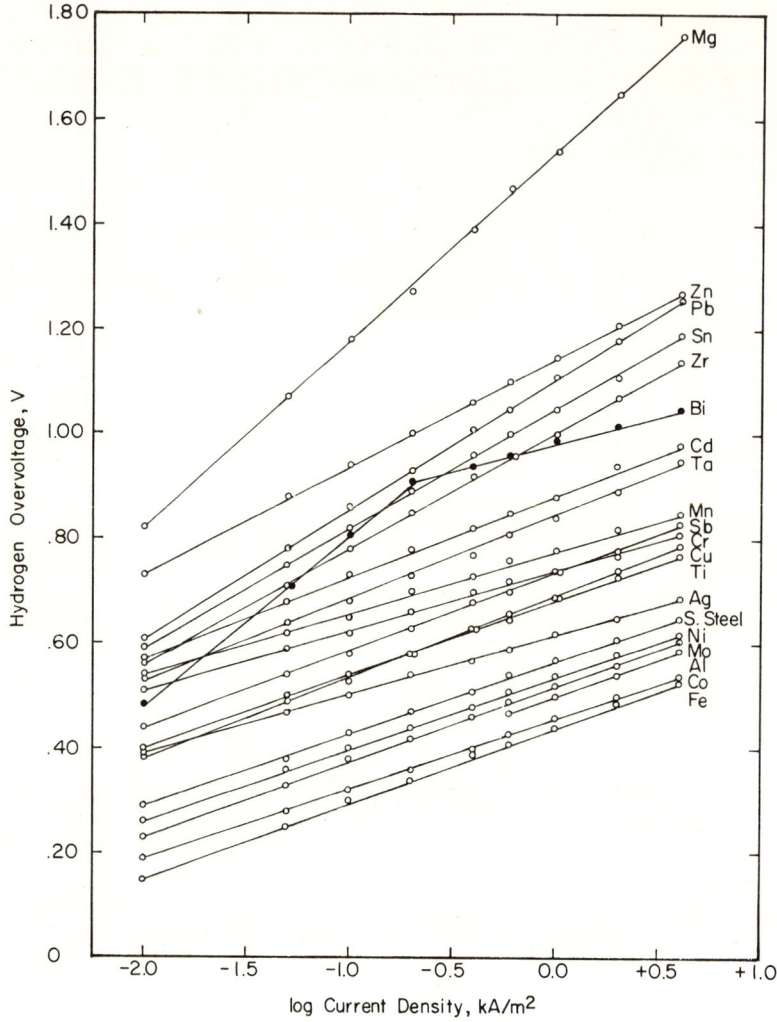

Figure 2. Hydrogen overvoltage for various metals in 6 N NaOH solution at 25°C.[60]

### 4.2.2. Steel Cathodes

The standard cathode material for diaphragm cells is low-carbon steel. Steel has a very favorable combination of properties: a low hydrogen overvoltage, low cost, availability in a wide variety of structural shapes, and durability. It is cathodically protected from corrosion at cell-operating potentials, and with proper care possesses a virtually unlimited service life—Hooker has advertised that some of its cathodes have been in service for over 40 years.

*Table 5*
*Hydrogen Overvoltages (in mV) on Low-Carbon Steel in 100 g/liter NaOH–180 g/liter NaCl Electrolyte*[a]

| Temperature, °C | Current density, mA/cm$^2$ | | | | |
|---|---|---|---|---|---|
| | 155.0 | 77.5 | 46.5 | 15.5 | 1.6 |
| 25 | 371 | 360 | 342 | 308 | 213 |
| 50 | 340 | 315 | 295 | 247 | 131 |
| 70 | 306 | 272 | 246 | 186 | 71 |
| 85 | 296 | 259 | 227 | 161 | 47 |

[a] Cathode potentials measured vs. S.C.E. at 25°C.

Diaphragm cell cathodes are normally formed into pockets, acting as chambers for catholyte and hydrogen. They are foraminous to allow passage of electrolyte from the anode compartment. Some producers favor punched steel sheet; others prefer woven wire screens. Each has points in its favor: punched sheet can be shaped to closer tolerances, permitting closer anode–cathode spacing, while woven wire screen presents a highly irregular surface, promoting adherence of the deposited asbestos diaphragm.

Although steel is a relatively good electrocatalyst for hydrogen evolution, its overvoltage in a diaphragm chlorine cell is typically 300 mV. This value depends on the state of the cathode surface. Some representative overvoltages obtained on an industrial cathode material under service conditions are given in Table 5. The reference electrode for these measurements was the saturated calomel electrode at 25°C.

### 4.2.3. Low-Overvoltage Cathodes

Little serious interest in replacing the steel cathode was shown by the industry until the mid-1970s, when escalating power costs provided sufficient incentive. The mechanical properties and economy of steel make it unlikely that it will be superseded as the base material of the conventional diaphragm cell cathode. Rather, methods are being sought to lower the overvoltage of steel by modifying or coating its surface.

Two approaches can be taken: the true surface area of the cathode can be increased, lowering the true current density and thus decreasing the log $j$ term of the Tafel equation, or a coating with better electrocatalytic properties than steel can be applied.

Early bases for both approaches can be found in the literature. In 1948, Fedom'ev *et al.*[61] studied the overvoltage of various steels in industrially realistic environments. No appreciable differences were found among 15 types of carbon-containing and alloyed steel. However, sandblasting lowered the overvoltage of low-carbon steel by 159 mV at 0.8 kA/m$^2$ (310 g/l NaCl, 75°C). Best results were obtained with electroplated coatings of nickel containing sulfur or selenium. Overvoltage reductions of 380 mV were obtained under the above conditions. The authors concluded that the sulfur or selenium was leached from the cathodes in service, leaving a microporous nickel coating of high surface area.

In 1956, Sasaki and Matsui[62] obtained a patent for electrodes for water electrolysis made by, first, electroplating nickel and zinc components on appropriate electrode base materials, and, second, eluting and removing the zinc by chemical dissolution or anodization, leaving a porous nickel coating of reduced overvoltage. As the cathode reaction in a diaphragm or membrane cell is, in fact, the electrolysis of water, at least one serviceable method of preparing low overvoltage cathodes would appear to be in the public domain. The combination of nickel with the soluble pore former such as aluminum or zinc appears to show promise, and several patents disclosing methods of applying such a coating have appeared since 1976.[63-66]

Catalytic coatings also possess an extensive patent literature. In 1966, Hall and Van Gemert[67] disclosed a variety of molybdenum and tungsten alloys with nickel, iron, and/or cobalt as catalytic coatings for electrolytic cell cathodes. Numerous patents have issued since 1976; a representative sampling is referenced.[68-71] Martinsons and Johnson[72] have patented a method of preparing a coating by cathodic deposition of catalytic materials added to the chlorine cell catholyte. There obviously exists the possibility that catalytic poisons will also deposit at the highly cathodic potentials of a chlorine cell. For this reason, catalytic coatings may prove to have shorter service lives than the high-surface type of coating.

The concept of the low overvoltage cathode has been readily accepted by the chloralkali industry. It has been the subject of several recent technical papers,[73-75] and Diamond Shamrock has announced the construction of a commercial facility for coating cathodes.[75] The absence of a dominating patent analogous to the Beer DSA patents, the ready adaptability of low-overvoltage cathodes to existing cells, and the relative absence of risk involved in their use, could well indicate total conversion of the industry within a decade.

### 4.2.4. Depolarized Cathodes

The oxygen-depolarized cathode, or "oxygen cathode," utilizes a different cathode reaction from the conventional diaphragm or membrane cell:

$$H_2O + (1/2)O_2 + 2e^- \rightarrow 2OH^-, \qquad E^o_{298} = +0.401\ V \qquad (46)$$

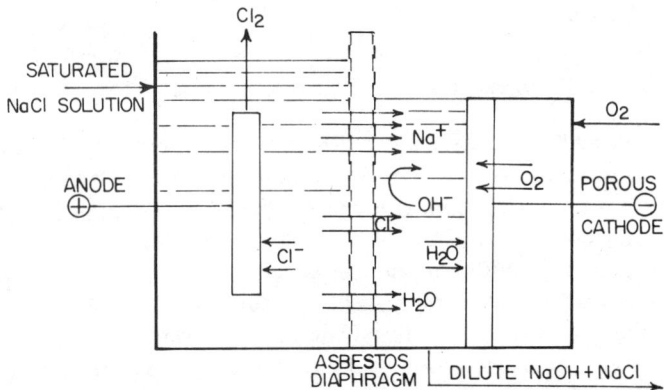

Figure 3. Chloralkali electrolysis cell with asbestos diaphragm and oxygen-depolarized cathode.

The overall cell reaction becomes

$$2NaCl(aq) + H_2O + (1/2)O_2 \xrightarrow{E} Cl_2(g) + 2NaOH(aq), \qquad E_{298} = 0.96 \text{ V}$$

(47)

The cathode reaction is spontaneous, and is in fact the conventional cathode reaction of the alkaline fuel cell. Reduction in theoretical decomposition voltage from that of a hydrogen-producing cell is 58%. Adaptation of the oxygen- or air-depolarized cathode to the diaphragm or membrane cell is not a matter of a simple retrofit but requires a total redesign of electrode and cell. Figure 3 shows a typical schematic design containing a porous solid cathode. The cathode reaction is three phase and for adequate efficiency the cathode porosity must be carefully controlled to facilitate proper transfer of gas and liquid. Fuel cell technology may prove to be inadequate, owing to the higher current density of chloralkali cells and their continuous flow of electrolyte, with consequent possibilities of deposition of catalyst poisons or leaching of active catalyst.

Although obviously a formidable technical challenge, the depolarized cathode has attracted the interest of the chloralkali industry, as would any scheme which promised dramatic power savings. Several patents have issued.[76] In late 1977, Case Western Reserve University and Diamond Shamrock Corporation were awarded a $2.5 million, four-year contract by the U.S. Department of Energy to develop low-cost oxygen electrodes.[77] It is too early to discern the ultimate direction this technology may take. Dotson[78] has recently published a negative evaluation of the practical power savings obtainable with depolarized cathodes. However, it is the only brine electrolysis cell capable in theory of breaking the "2.3 V barrier," and it is certain that industry will not abandon the concept before thoroughly exploring its possibilties.

## 4.3. Cell Separators

### 4.3.1. Asbestos Diaphragms

The diaphragm is a major key to the efficient operation of a diaphragm cell. It determines the overall cell efficiency, the range of caustic strength obtainable from the cell, accounts for an appreciable fraction of the total cell voltage, and, since the introduction of metal anodes, is the life-limiting component of the cell. Although deposited asbestos diaphragms have been in commercial use for 50 years, there is little information in the open literature concerning their physical properties or mode of operation.

Asbestos is a generic term for a variety of hydrated silicate minerals, which are characterized by their fibrous structure. Although there are six different asbestos minerals and several have found use in cell separators, one type, chrysotile, predominates as a chlorine cell diaphragm material. Chrysotile fibers consist of bundles of parallel fibrils. These fibrils are hollow needles with an outer diameter of $\sim 300$ Å and an inner diameter of $\sim 50$ Å, and consist of alternate layers of $Mg(OH)_2$ and $SiO_2$. The number of fibrils in a fiber varies widely; thus the fibers exhibit considerable variance in diameter and length. In preparing commercial diaphragms the fibers are slurried in a suitable liquid medium and sucked by vacuum onto the outer surface of the foraminous cathode. The cathode is then heated to drive off solvents, leaving the diaphragm as a fibrous mat. It is the void space between fibers that provides the electrolyte path, not the tiny pores in the center of the fibrils. It should be apparent that these voids exhibit no geometric regularity and their mathematical simulation would be a formidable task.

Chrysotile asbestos is not chemically stable in chlorine cell electrolytes. $Mg(OH)_2$ is soluble in acid solutions and stable in basic solutions; the reverse is true for $SiO_2$. When a cell is energized, a chrysotile diaphragm will become $SiO_2$ enriched on its acidic, anolyte face, and $Mg(OH)_2$ enriched on its alkaline, catholyte face. The flow through the diaphragm will flush $Mg^{2+}$ ions from the anolyte side toward the catholyte, where they will reprecipitate as $Mg(OH)_2$. This precipitate will constrict the flow channels, decreasing flow rate and efficiency and increasing voltage drop and caustic strength. The diaphragm is said to "tighten." After a period in service, the diaphragm reaches a state of equilibrium with its surroundings and its characteristics stabilize. However, any drastic change in operating conditions will cause the dissolution–reprecipitation process to begin anew.

The function of a diaphragm is to separate the electrolysis products while allowing electrical communication between the cell compartments. An additional requirement in chlorine cells is that the diaphragm must allow bulk flow of electrolyte from anode to cathode compartments. Asbestos is an electrical insulator and probably possesses a charged surface. Such a surface could play a functional role in diaphragm operation by accelerating or retarding ionic migration. This behavior would result in a streaming potential; Hine et al.[20] concluded after a thorough study that the streaming potential of an asbestos diaphragm is negligible at the high ionic strengths prevailing in a chlorine cell. No evidence for any other type of ionic selectivity has been reported. Based on present evidence, the chlorine cell diaphragm functions as a simple physical barrier to flow.

The efficiency and electrical resistance of a diaphragm result from the processes of mass transport within it. Mass transport in an electrolytic solution is the result of three processes; bulk, or convective flow, ionic migration, and diffusion. In a dilute solution the flux of each dissolved species $i$ is given by Newman[79]:

$$\bar{J}_i = c_i \bar{V} - Z_i \mu_i F c_i \nabla \phi - D_i \nabla c_i \qquad (48)$$

The flux $\bar{J}_i$ of species $i$, expressed in mol/cm$^2$ sec, is a vector quantity indicating the direction and number of moles moving per unit time across a 1-cm$^2$ plane oriented perpendicular to the flow of the species. This movement is due to the bulk motion, which has velocity $\bar{V}$. Other driving forces will cause deviation from this average flow: diffusion will occur in the presence of a concentration gradient $\nabla c_i$, and migration of ions of charge $Z_i$ will occur in the presence of a potential gradient $\nabla \phi$. Other terms in Eq. (48) are $F$, Faraday's constant, and $\mu_i$ and $D_i$, the ionic mobility and diffusivity of species $i$. In the limit of infinite dilution, $\mu_i$ and $D_i$ are related by the Nernst–Einstein equation:

$$D_i = \mu_i RT \qquad (49)$$

The simultaneous solution of the ionic flux equations, together with current balance and electroneutrality considerations, should yield concentration profiles of all ionic constituents in the diaphragm, the potential drop, and the current efficiency. The first worker to apply these modern electrochemical concepts to the processes within the diaphragm was Mukaibo.[80] His excellent 1952 study was followed 15 years later by that of Stender, Ksenzhek, and Lazarev,[81] which initiated a series of Russian investigations.[82–87] The original Russian goal was to derive a minimum diaphragm thickness for a given set of electrolysis conditions. Simplifying assumptions were made to allow analytical solutions. Hydroxyl was the only ion considered, and voltage drop was not discussed. This work has become progressively more sophisticated, and most recently the results of numerical solutions of the flux equations have been reported.[87,88] All of this work has been based on dilute electrolyte theory.

Hine et al.[21] have recently obtained experimental confirmation of Mukaibo's theoretical model, some 25 years after it was first proposed. Hine's paper contains much useful data. Kaden and Pohl[89] have reported some porosity measurements obtained by centrifugation. They claim that pore size distribution and percentage participation of pore size classes in the total permeability of the diaphragm can be determined by this relatively simple technique.

In the early 1970s, manufacturers in the United States began using modified diaphragms reinforced with plastic. Asbestos is still the base material of these diaphragms, and determines their performance. Plastic-bonded diaphragms offer improved mechanical and handling properties, resistance to swelling, and longer service lives. Typically, a fluorocarbon resin is added to the asbestos slurry and is admixed with the asbestos when drawn onto the cathode. The cathode is heated, first to evaporate the water, and finally to soften or melt the polymer, which flows sufficiently to coat the asbestos fibers. When cooled the fibers are cemented together, forming a tough, resilient, dimensionally stable diaphragm. Most of the commercially available fluorocarbon resins have been utilized; their physical forms have included solutions, powder dispersions, and fibers. A number of patents have issued, differing in relatively minor details. A representative sample is referenced.[90-92]

### 4.3.2. Porous Plastic Separators

Thin, uniformly porous plastic cell separators would eliminate the well-publicized ecological problems associated with asbestos and hold out the promise of improved cell design and performance: lower ir drop, decreased anode–cathode spacing, and uniform flow properties. Fulfilling the promise has proved elusive; 15 years of research effort has as yet resulted in no commercial success. A number of materials and processes have been described in the patent literature.[93-95] The plastic is normally a fluorocarbon for the required corrosion resistance; porosity is obtained by incorporating a leachable pore-former during fabrication of the sheet. Problem areas range from the theoretical one of establishing the required porosity to the highly practical one of forming and sealing the sheets around the cathode pockets of a typical diaphragm cell. Maintaining adequate pore uniformity and adequate wettability of the fluorocarbon are also challenging tasks. Plastic sheet diaphragms may see their first chloralkali commercialization in flat plate bipolar cells such as Kureha's SK cells rather than in the more conventional pocket cells.[96]

### 4.3.3. Ion-Exchange Membranes

Modest research efforts were made in the 1950s and early 1960s to develop an ion-selective chlorine cell separator. Joint programs by Hooker and Rohm and Haas,[97] and Diamond Alkali and Ionics Corporation,[98] ultimately foundered owing to the instability in the chlorine cell environment of the materials then available. In the late 1960s duPont introduced their line of perfluorinated ion exchange materials, trademarked Nafion, which resulted from their work in fuel cells for the space program.[98] Initial work quickly established that these resins had the necessary chemical stability. Cell performance of their early versions was inadequate. The membranes had an excessive ir drop and their efficiency decreased sharply with increasing caustic strength. Continuing efforts by duPont and chloralkali producers have resulted in Nafion variants with markedly improved properties.

Ion-exchange membranes for chloralkali cell separators are insoluble polyelectrolytes containing negatively charged sites within their structure. These sites attract cations, facilitating their passage through the membrane under the attractive influence of the negative cathode potential. Anions are repelled by the sites. Thus chloride ion is confined to the anolyte compartment; the passage of hydroxyl ion, which is produced at the cathode, is not barred completely, however, owing to its high mobility. Increasing caustic strength increases diffusion forces on the hydroxyl ion, leading to increased penetration to the anolyte compartment with subsequent oxygen discharge at the anode. Minimizing the loss of efficiency due to this back-migration is imperative, and has been sought by modifications of the membrane, cell design, and operating procedures.

The membrane does not permit bulk flow of electrolyte. The only water to pass through the membrane is the water of hydration of the sodium and hydronium ions. This is sufficient to produce a 40–50% solution of NaOH in the catholyte. In practice, water is added to the catholyte compartment to reduce hydroxyl back-migration and increase cell efficiency. The economic optimum caustic strength varies with the membrane and cell design chosen, and has been a matter of considerable controversy.

Nafion membranes are the subject of an excellent and extensive technical literature, primarily by workers at duPont and Diamond Shamrock.[99] Nafion perfluorosulfonic acid membranes are derived from a copolymer of tetrafluoroethylene and a perfluorosulfonylethoxy vinyl ether. In its original form the polymer is thermally stable, hydrophobic, nonionic, and melt processable. Physical and chemical processing of the membrane takes place at this stage. Once it has been hydrolyzed to the ionic form it is physically intractable and chemically inert; the perfluorocarbon structure gives it its chemical resistance, while the appended sulfonic acid groups render it hydrophilic and give it the ability to retard anionic transport. The hydrophilic nature of the polymer varies inversely with the equivalent weight, i.e., with the number of sulfonic acid groups. Practical chloralkali membranes have equivalent weights in the range 1100–1500 g/meq.

Higher-equivalent-weight membranes absorb less water per sulfonic acid group, permitting more efficient anion rejection but resulting in lower electrical conductivity (Figure 4). To take advantage of the increased efficiency obtainable with high-equivalent-weight membranes, duPont developed composite membranes, in which the cathode side of the membrane is a thin barrier layer of high-equivalent-weight polymer. This is backed up by a thicker layer of lower-equivalent-weight material to improve mechanical strength with little

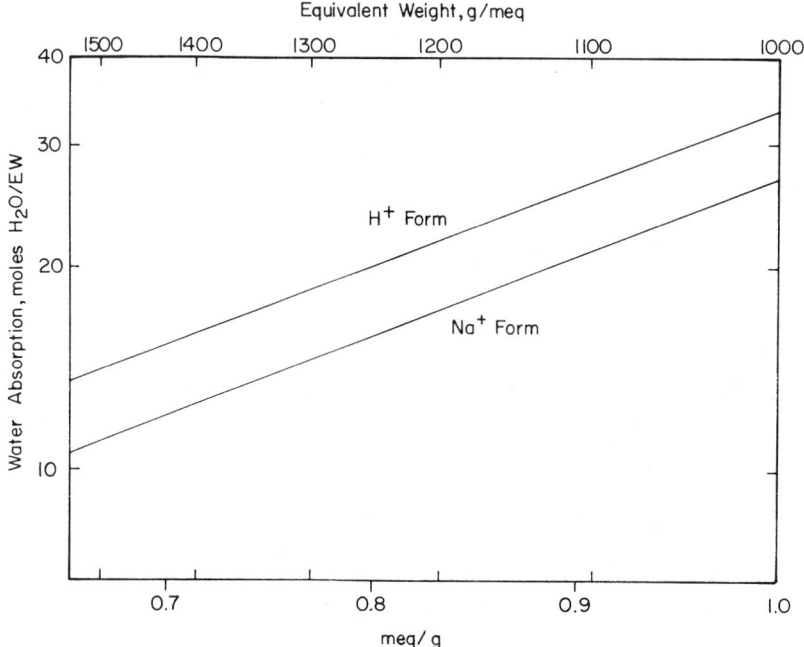

Figure 4. Water absorption of Nafion polymers in sulfonate form.[99]

added electrical resistance. A typical composite membrane comprises 25–40 μm of 1500-equiv-wt polymer on 100 μm of 1100-equiv-wt polymer.

The efficiency of composite membranes is greater than that of single layer membranes up to a catholyte NaOH concentration of about 20%. To further increase efficiency at the higher caustic strengths, duPont developed a new type of Nafion membrane, the "chemically modified" membrane. The concept of the thin anion-rejecting barrier layer was again used. Ion-rejection results from water rejection, so the approach taken was to change the chemistry of the cathodic membrane face to reduce its water-absorptive capability without excessively degrading the membranes' mechanical, chemical, or electrical properties. While in the sulfonyl fluoride form one surface of the polymer is reacted with ethylene–diamine (EDA) to form a surface layer of substituted sulfonamide groups. The EDA concurrently serves to crosslink the polymer.

The EDA-modified sulfonamide membrane allows production of 30% NaOH at current efficiencies greater than 90% (see Figure 5[100]). This performance improvement is not without its price, however; chemically modified Nafion is reported to be less physically robust than its bilayer predecessor, and it is susceptible to chemical attack at the amide groups. Its performance decays with time; Pulver[101] reports current efficiency at 3.1 kA/m$^2$ and 28 wt % NaOH decreased from 93% to 83% over a 360-day period. The choice between the two membrane types would appear to be clear-cut only for those installations with a definite requirement for cell effluent in the 30+% NaOH range.

Patent activity in the area of membrane synthesis has been vigorous. Over 150 patent entities appeared prior to mid-1977. Most of the recent activity has originated in Japan. In 1973 the Japanese government decided to phase out all mercury cell plants in Japan by April 1978. This deadline has been extended, but Japanese researchers certainly exerted every effort to develop a totally new technology in a very short time. Conversion of existing mercury cell plants to membrane cells rather than diaphragm cells is preferred because membrane cells are a much better "fit": the design of a membrane cell chloralkali plant has many features in common with mercury cell plants, and product quality—specifically NaOH purity—is similar.

Ion-exchange membranes and/or cells are presently marketed by two Japanese concerns. Asahi Glass Co., Ltd., is marketing a perfluorocarboxylic acid membrane trade named Flemion.[102] Judging from published cell data[103] this membrane has some unusual features. Current efficiency apparently increases as caustic strength increases from 30% to 40%. It is claimed that water absorption by this membrane decreases markedly with increasing caustic

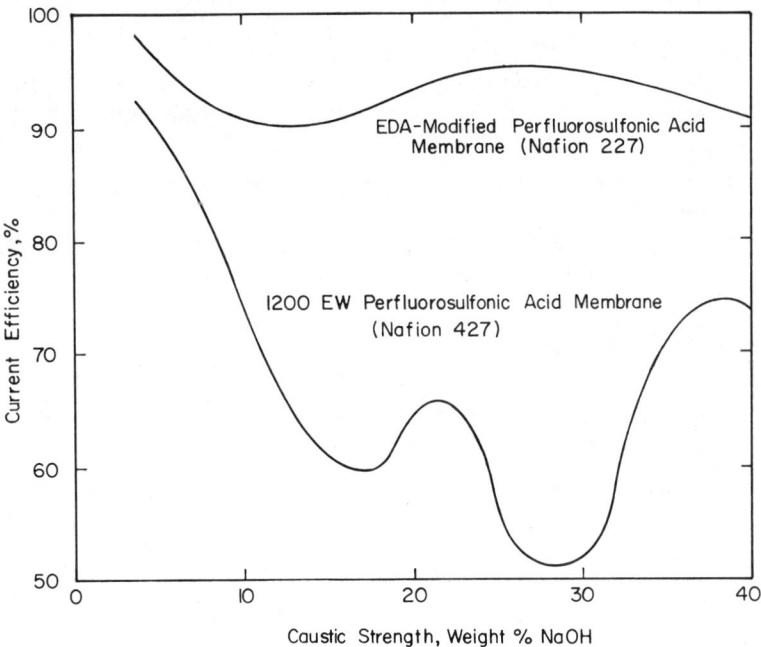

Figure 5. Current efficiency versus caustic strength for Nafion perfluorosulfonic acid membranes.[100]

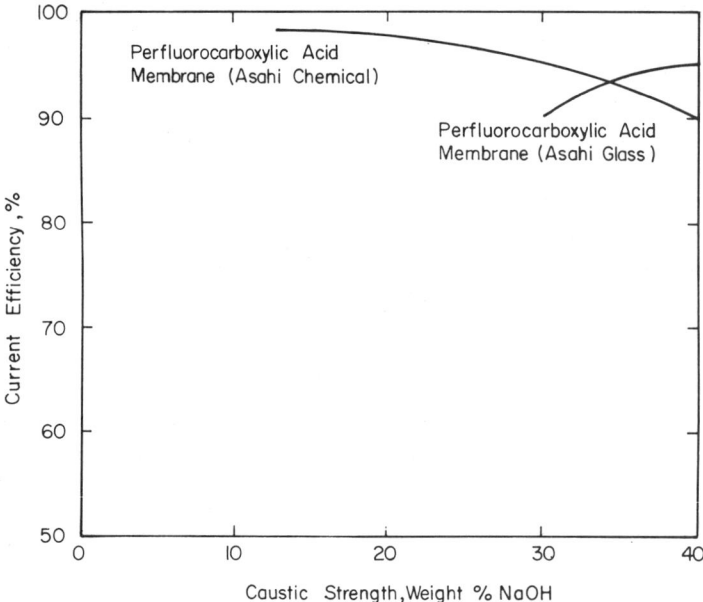

Figure 6. Current efficiency versus caustic strength for perfluorocarboxylic acid membranes.[103,104]

strength, improving its anion-rejecting capability. An increase in hydrophobicity should also lead to increased voltage drop; the published claims appear to substantiate this.

Asahi Chemical Industry Co., Ltd., is marketing cells containing membranes of their own development.[104] Their membrane is also said to be a perfluorocarboxylic acid type. Published data indicate their membrane has characteristics falling between those of perfluorosulfonic acid membranes and the Asahi Glass membrane (Figure 6).

## 5. Cell Technology

### 5.1. Diaphragm Cells

#### 5.1.1. General Comments

Somewhat over half the world's chloralkali production is by the diaphragm cell process.[3] In North America this process has always predominated over the competitive mercury cell process, accounting for about 75% of U.S. chlorine production from 1961 to the present.[4,8] About 90% of the new chlorine capacity installed globally during 1973–1976 utilized diaphragm cells, and most new plants now planned or under construction will use this process.[15]

All modern diaphragm cells are simple, boxlike structures designed to contain a maximum of electrode area in the minimum practical cell volume. This is most commonly achieved by alternating a large number of commonly connected anode blades with cathode pockets which are made a single electrical unit by connections at the sides and/or bottom.

Diaphragm cells can be grouped in two design categories: monopolar and bipolar. Monopolar cells are self-contained units. A chlorine plant normally operates one or more groups of cells, each cell in a group interconnected electrically in series by means of cables or bus bars. The monopolar designs in most widespread use are those of Hooker and Diamond Shamrock. These cell types are discussed in detail below; their operating characteristics are summarized in Table 6. Bipolar cells, sometimes called filter press cells, require no external intercell connections; the anodes of one cell are connected directly to the cathodes of an adjacent cell. These cells are normally operated in units of at least ten cells. The cell units are called electrolyzers or, simply, series. Dow has operated bipolar cells for eighty years. The first bipolar cell to be widely marketed is the Glanor cell, a joint development of PPG Industries, Inc. and Oronzio deNora Impianti Electrochimici. Its operating characteristics are included in Table 6.

Diaphragm cells share many design and operating characteristics. All modern cells employ vertical anodes of graphite or coated titanium, vertical cathodes, and asbestos diaphragms vacuum deposited on the cathodes. Purified, saturated brine (~25% NaCl) is fed to the anolyte compartment from where it percolates through the diaphragm into the catholyte chamber. The percolation rate is controlled by maintaining a higher level of electrolyte in the anolyte compartment to establish a positive adjustable hydrostatic head. A low flow rate through the diaphragm results in a high concentration of caustic in the catholyte, but at the expense of a low caustic current efficiency due to increased back-migration of hydroxyl ion from catholyte to anolyte. The economic optimum flow rate results in the decomposition of about 50% of the incoming salt; the cathode product is a dilute solution containing 8–12% NaOH and 12–18% NaCl.

### 5.1.2. Hooker Cells

The Hooker monopolar cell originated from the 6-kA type S or Stuart cell developed from 1924 to 1934. The type S was gradually scaled up in size and current rating; the latest graphite design, the S4, operates at currents up to 55 kA. Two metal anode cell designs are currently in use: the H-2A, which is rated at 80 kA, and the H-4, a 150-kA design (see Figure 7). These cells contain metal anodes comprised of coated, expanded titanium mesh sheet shaped into a hollow box. Titanium-clad copper rods project vertically through the interior of these boxes, connecting them to the base of the cell and serving to distribute the current to the active anode faces with minimum ir drop.

## Table 6
## Diaphragm Cell Operating Characteristics

| Parameter | Diamond | | | Hooker | | | PPG |
|---|---|---|---|---|---|---|---|
| | MDC-29 | MDC-55 | | H-2A | H-4 | | V-1144 |
| Operating current, kA | 35 | 80 | 75 | 150 | 40 | 80 | 80 | 150 | 72 |
| Current density, kA/m$^2$ | 1.21 | 2.76 | 1.37 | 2.74 | 1.11 | 2.21 | 1.24 | 2.32 | 1.97 |
| Current efficiency, % | 96.5 | 96.5 | 96.5 | 96.5 | 93.0 | 95.4 | 93.8 | 95.6 | 95.5–97.8 |
| Cell voltage (with intercell bus), V | 2.90 | 3.62 | 3.00 | 3.62 | 3.03 | 3.62 | 2.99 | 3.47 | 3.55 |
| dc kWh/metric ton Cl$_2$ | 2310 | 2876 | 2390 | 2870 | 2463 | 2869 | 2410 | 2744 | 2739–2849 |
| Rated Cl$_2$ production, metric tons/day | 1.05 | 2.41 | 2.33 | 4.53 | 1.18 | 2.42 | 2.38 | 4.55 | 2.20 |
| Operating caustic concentration, g/liter NaOH | 120 | | 120 | | 140 | | 140 | | 130–140 |
| Cell dimensions | | | | | | | | | |
| Active anode area, m$^2$ | 29.0 | | 55.0 | | 36.16 | | 64.52 | | 36.5 |
| Length, m | 1.14 | | 1.61 | | 1.87 | | 2.58 | | 5.18 |
| Width, m | 2.21 | | 2.97 | | 2.39 | | 3.11 | | 3.35 |
| Height, m | 2.43 | | 2.58 | | 2.05 | | 2.13 | | 3.66 |
| Cells/electrolyzer | 1 | | 1 | | 1 | | 1 | | 11 |

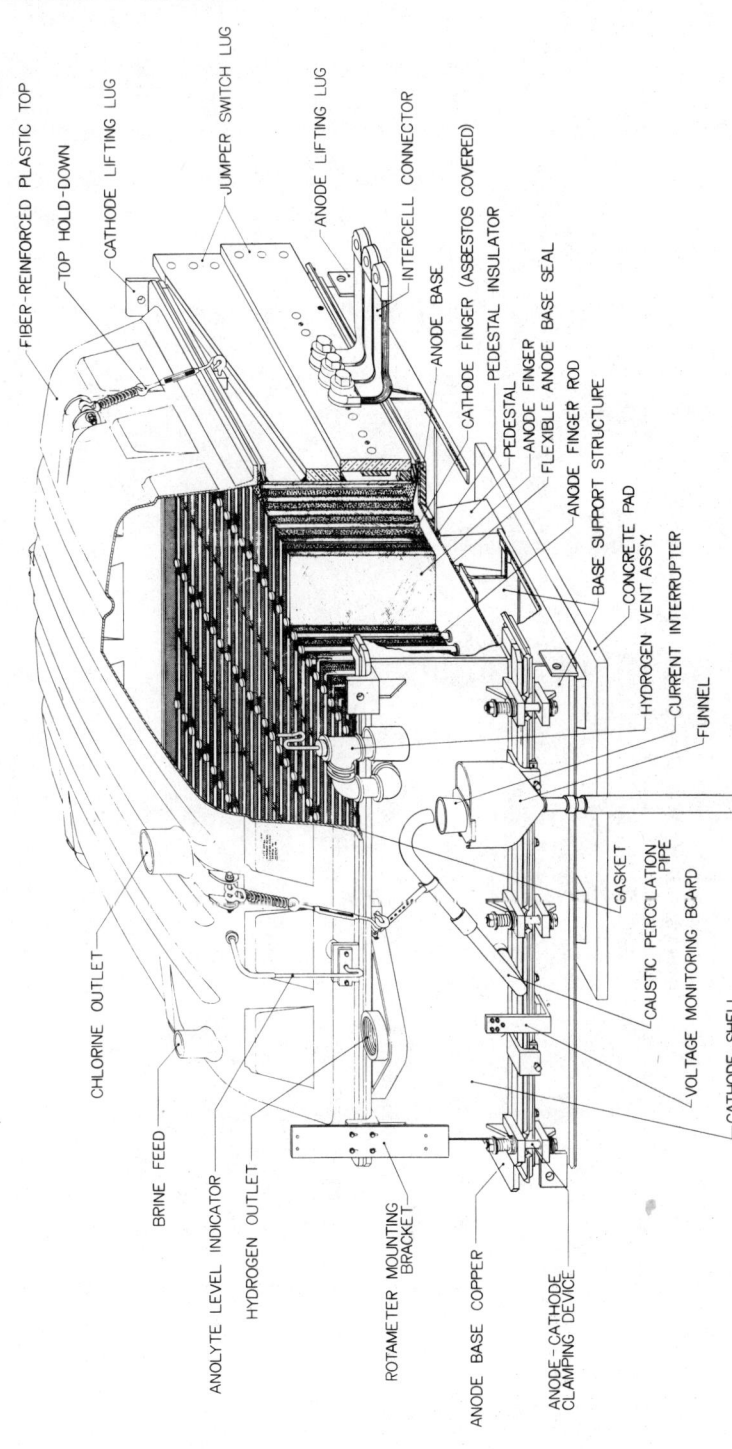

Figure 7. Hooker type H-4 diaphragm cell. Nominal current rating 150 kA. (By permission of Hooker Chemicals and Plastics Corp., Industrial Systems Division.)

## Table 7
### Analysis of Hooker Diaphragm Cell Voltages[105,106]

| Parameter | 1960<br>S-3<br>Graphite anodes<br>Std. asbestos diaphragm<br>Wide gap | 1974<br>H-4<br>DSA<br>Std. asbestos diaphragm<br>Wide gap | 1976<br>H-4<br>DSA<br>HAPP diaphragm<br>Narrow gap | |
|---|---|---|---|---|
| Current density, kA/m² | 1.55 | 2.32 | 1.55 | 2.32 |
| Voltage analysis | | | | |
| Reversible anode potential | 1.32 V | 1.32 V | 1.32 V | 1.32 V |
| Reversible cathode potential | 0.93 | 0.93 | 0.93 | 0.93 |
| Anode overvoltage | 0.33 | 0.03 | 0.03 | 0.03 |
| Cathode overvoltage | 0.27 | 0.30 | 0.27 | 0.30 |
| Anolyte ir drop | 0.49 | 0.49 | 0.27 | 0.34 |
| Diaphragm ir drop | 0.30 | 0.46 | 0.16 | 0.25 |
| Structural ir drop | 0.36 | | 0.17 | 0.26 |
| Anode + contact to base | | 0.11 | | |
| Base | | 0.06 | | |
| Cathode | | 0.09 | | |
| Total cell voltage | 4.00 V | 3.79 V | 3.15 V | 3.43 V |

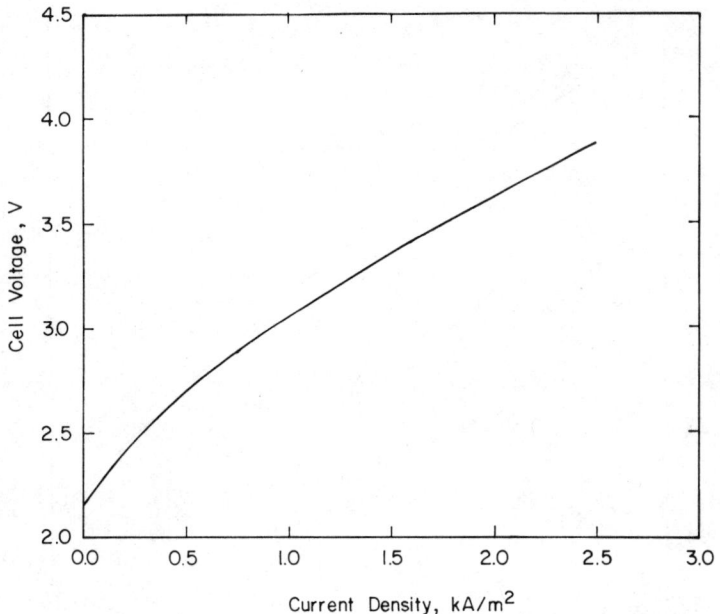

Figure 8. Cell voltage versus current density for Hooker type H-2A and H-4 cells.[105]

The steel cathode pockets are of perforated sheet, and extend across the steel cell body which is insulated from the copper anode base by an elastomeric sheet which also acts as the seal around the anode rods. Diaphragms are resin-bonded asbestos (trade named HAPP) which are vacuum-deposited by conventional techniques. The cell top is of fiberglass-reinforced plastic (FRP) sealed to the cell body with gaskets.

Operating characteristics of the H-2A and H-4 are given in Table 6. Additional useful information has been published by Grotheer and Harke,[105] and Currey.[106] Cell-voltage–current curves for the two cells are shown in Figure 8. Table 7 breaks down the voltage of Hooker cells into individual constituents, showing the gains since 1960. Figure 9 shows the variation of caustic efficiency of the H-4 with caustic strength.

### 5.1.3. Diamond Cells

From 1945 to 1947 the Diamond Alkali Company developed a 20-kA monopolar cell of similar configuration to the type S cell. Their first generation of metal anode cells, the DS series, has been progressively improved since the late 1960s and has now been redesignated as the MDC series (Figure 10). Three sizes are currently marketed, the MDC-20, MDC-29, and MDC-55, the number indicating the anode area in $m^2$. Rated capacity ranges from 20 to

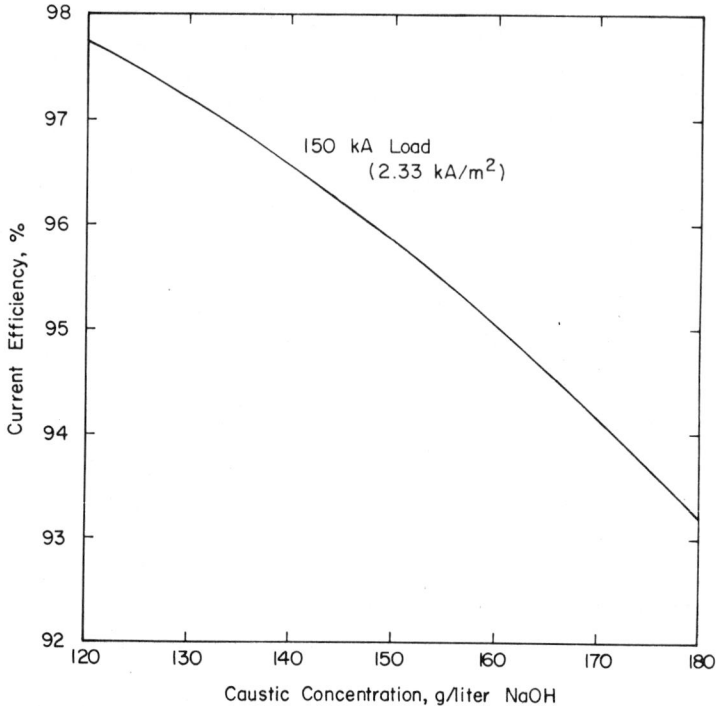

Figure 9. Current efficiency versus caustic concentration for Hooker type H-4 Cell.[105]

150 kA. Most structural features of Diamond cells are similar to those described for Hooker cells: expanded Ti mesh anodes, steel cell bodies with full-width cathode pockets, flat copper base covered by a patented attack-resistant mat,[107] FRP cell tops, and polymer-bonded deposited asbestos diaphragms. The cathode pockets are steel screen rather than punched sheet.

A unique feature of the Diamond cell is the use of an adjustable anode.[108] The surfaces of these anodes are clamped together during assembly of the anode–cathode unit. After adjusting the anodes between the cathode rows, the clamps are removed and the surfaces of the anodes are moved outward to a preset distance from the diaphragm face by the action of spring-mounted current supply members.

Operating characteristics of Diamond cells are given in Table 6. Liederbach[109] has published a detailed breakdown of Diamond cell voltages (Table 8). Current–voltage curves for Diamond cell types are shown in Figure 11.

### 5.1.4. The Dow Cell

Dow is the largest chloralkali producer, accounting for about 1/3 of United States capacity and 1/5 of world capacity. Dow uses its own bipolar

# PRODUCTION OF CHLORINE

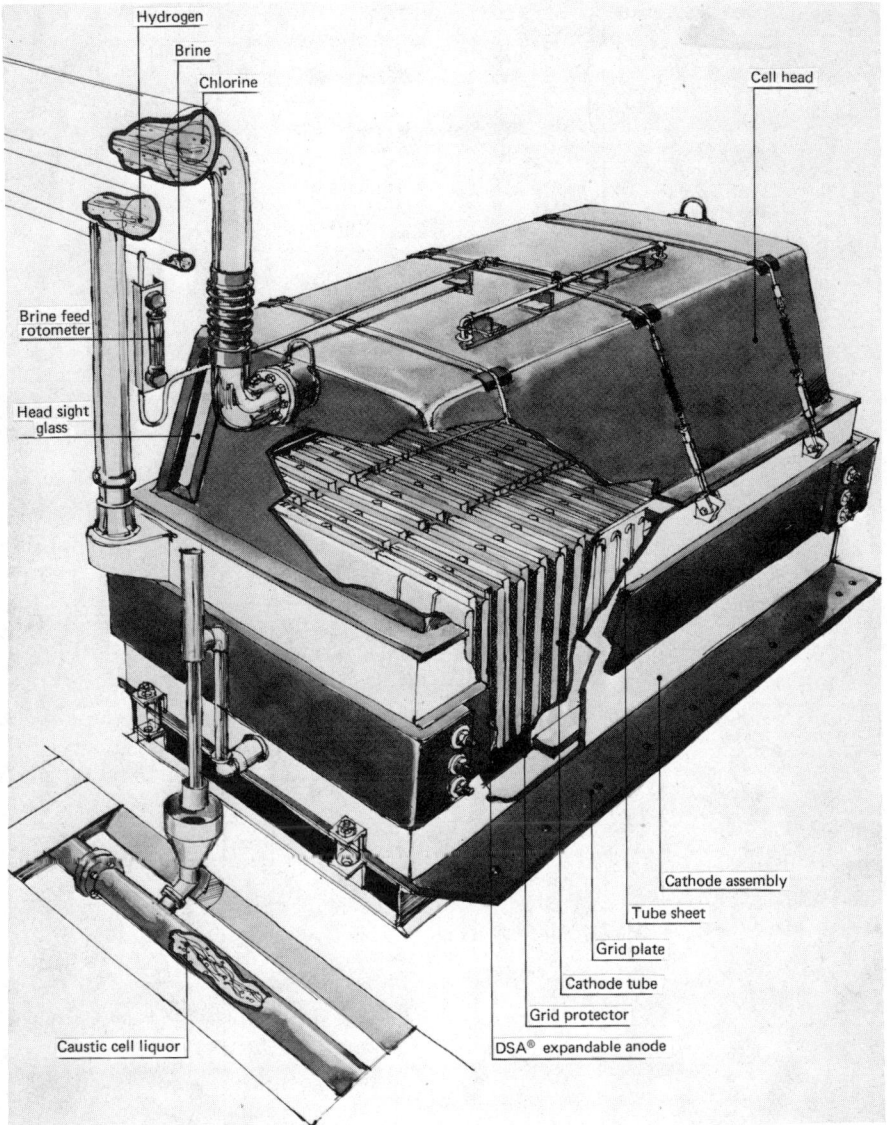

Figure 10. Diamond type MDC-55 diaphragm cell. (By permission of Diamond Shamrock Corp., Electrolytic Systems Division.)

design. It is optimized for low-current-density, high-power-efficiency operation. Since this operating mode requires more cells for a given production capacity than the high current densities favored by the other producers, it is probable that the Dow design is the world's most numerous cell type.

## Table 8
### Analysis of Diamond Diaphragm Cell Voltages[109]

| Year<br>Cell features<br><br>Parameter | 1968<br>Graphite Anode<br>Std. Asbestos<br>Diaphragm | | DSA-Standard<br>Std. Asbestos<br>Diaphragm | | DSA-Standard<br>Mod. Diaphragm | DSA-Expandable<br>Mod. Diaphragm | 1978<br>DSA-Expandable<br>Mod. Diaphragm<br>Coated Cathode |
|---|---|---|---|---|---|---|---|
| Current density, kA/m$^2$ | 1.20 | 2.17 | 1.20 | 2.17 | 2.17 | 2.17 | 2.17 |
| Voltage analysis | | | | | | | |
| Total anode potential | 1.50 V | 1.52 V | 1.29 V | 1.31 V | 1.31 V | 1.31 V | 1.30 V |
| Total cathode potential | 1.10 | 1.16 | 1.10 | 1.16 | 1.16 | 1.16 | 1.01 |
| Anolyte ir drop | 0.50 | 0.87 | 0.23 | 0.44 | 0.52 | 0.32 | 0.32 |
| Diaphragm ir drop | 0.33 | 0.65 | 0.30 | 0.66 | 0.37 | 0.37 | 0.37 |
| Structural ir drop | 0.50 | 0.87 | 0.17 | 0.21 | 0.21 | 0.21 | 0.21 |
| Total cell voltage | 3.93 V | 5.07 V | 3.09 V | 3.78 V | 3.57 V | 3.37 V | 3.21 V |

# PRODUCTION OF CHLORINE

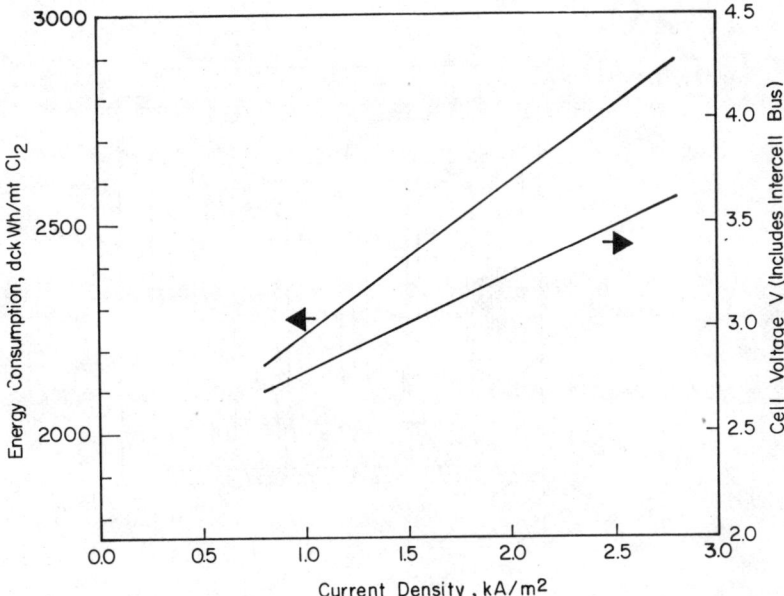

Figure 11. Cell voltage and energy consumption versus current density for Diamond MDC series cells.

Dow does not license its cells, and has revealed no operating data. A representative illustration from the patent literature is shown in Figure 12.[110] In common with most designs, the Dow cell contains parallel-connected anode blades positioned between steel cathode pockets upon which have been deposited asbestos diaphragms. The bipolar units are connected in series of 50 or more cells. The bipolar concept results in compact, economical cells and cell floors; the absence of external cell connectors makes the bipolar cell suitable for outdoor installation.

## 5.1.5. Glanor Electrolyzers

The Glanor electrolyzer is a bipolar cell developed in the early 1970s by PPG and Oronzio deNora. Both vertical (type V) and horizontal (type H) designs have been described[111]; the latter has apparently not been commercialized. A three-cell section of the type V-1144 electrolyzer is shown in Figure 13. Rated at 72–80 kA, it normally contains ten bipolar elements and two end-electrode elements clamped together by tie rods, forming a single assembly of eleven cells.

Each cell has its own brine feed compartment, through which chlorine exits to the collection system. Current is carried into the electrolyzer through the anodic and cathodic end elements. Current flows between cells within the

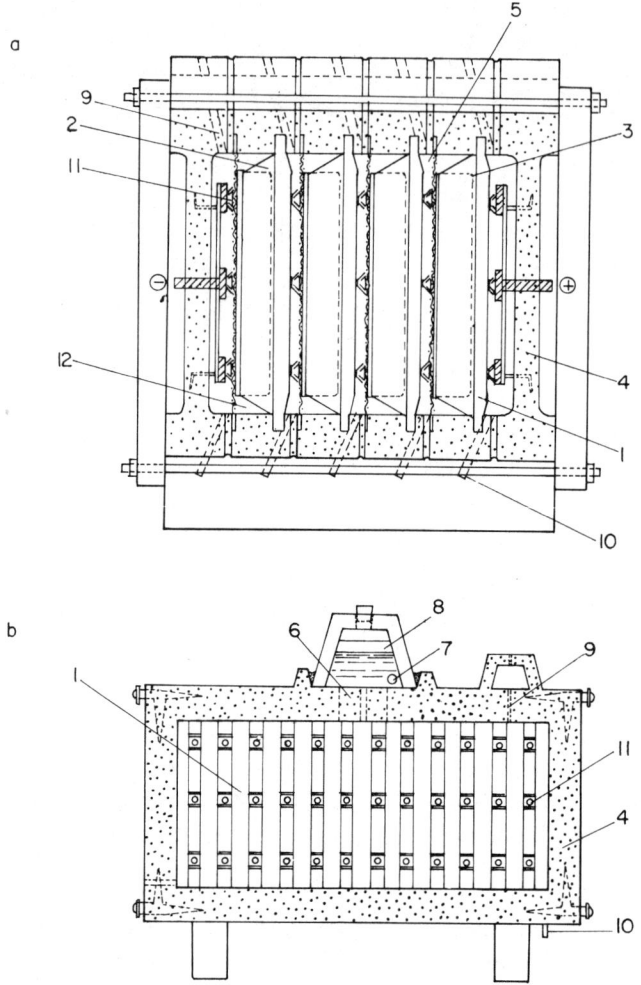

Figure 12. Dow diaphragm cell[110]: (a) longitudinal sectional elevation through four unit cells + two end pieces, (b) transverse sectional elevation: 1, Graphite anode backboard; 2, graphite anode; 3, steel screen cathode pocket with drawn asbestos diaphragm; 4, concrete cell frame; 5, cathode space for caustic liquor + hydrogen; 6, brine inlet + chlorine outlet; 7, intercell brine port; 8, intercell gas partition; 9, hydrogen outlet; 10, caustic outlet; 11, spring clip connector between unit cells; 12, anode space for anolyte + chlorine.

electrolyzer through numerous very short internal connections, made by a special technique to assure very low intercell voltage drops.

The heart of the design is the bipolar electrode. It consists of a steel plate to which the anode fingers (coated Ti mesh) are connected on one side and the cathode fingers (standard steel wire mesh) on the outer. Conventional asbestos

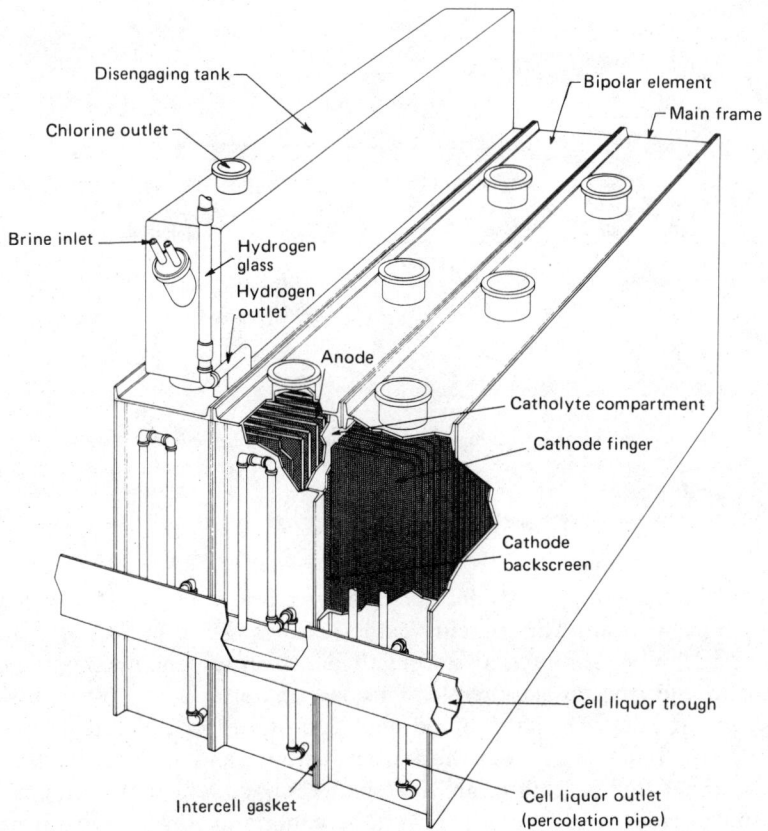

Figure 13. PPG-deNora type V-1144 Glanor® diaphragm electrolyzer (three cells illustrated). (By permission of PPG Industries, Inc.)

diaphragms have been utilized, but the design is adaptable to new diaphragm technology.

The typical circuit is made up of two rows of electrolyzers. A circuit of 14 electrolyzers operating at 80 kA will produce about 375 metric tons of chlorine per day. The floor area required is about 50 × 17 m, exclusive of the rectifier transformer area. Operating characteristics are summarized in Table 6.

### 5.2. Mercury Cells

Modern mercury cells have evolved from the Solvay long cell design of ~1900. A typical cell is shown in Figure 14. As was previously mentioned, the mercury cell actually comprises two cells, an electrolyzer and a decomposer. The major structural components of the electrolyzer are an elongated trough and a gas-tight cover. The trough is normally constructed of steel and its sides

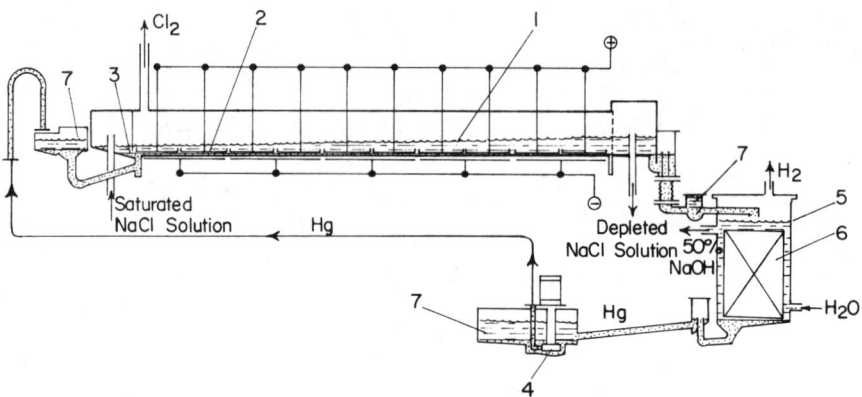

Figure 14. Mercury cathode electrolyzer and decomposer[12]: 1, brine level; 2, metal anodes; 3, mercury cathode (flowing along baseplate); 4, mercury pump; 5, vertical decomposer; 6, graphite packing; 7, wash water.

are lined with hard rubber. Modern electrolyzers are about 10–25 m in length and 1–2.5 m in width. The mercury cathode flows as a thin sheet down the trough. The flow is promoted by sloping the electrolyzer, and making the base as uniform and smooth as possible. The base material is one distinguishing feature among cell designs: machined steel, and rubber, or concrete linings are used. Electrical contact between the mercury cathode and the external circuit is made in the latter case through steel plates imbedded in the lining. The horizontal anodes are graphite, or, more commonly today, mixed-oxide-coated titanium. They are suspended and supported through a rubber-covered steel or multilayered rubber cell cover. The entire assembly is carefully leveled and aligned on insulated pillars.

In operation, purified, saturated brine (~25.5% NaCl) flows down the trough cocurrently with the mercury cathode. Flow rates and current density are chosen to produce an amalgam of 0.25–0.50% sodium at the cathode. The depleted brine leaves the electrolyzer at a concentration of 21–22%. It is stripped of chlorine by air blowing or vacuum treatment, resaturated with solid salt, treated to remove impurities, and recycled to the electrolyzer.

In modern installations the decomposer is usually a cylindrical steel tower. Spheres or irregular pieces of graphite are supported in the body of the tower. Sodium amalgam enters, and hydrogen and sodium hydroxide exit near the top of the tower, while the lower part contains a water inlet and an outlet for the denuded amalgam, which flows to the intake of the mercury pump and is recirculated to the upper end of the electrolyzer.

The decomposer acts as a shorted primary cell in which the graphite is the insoluble cathode and the sodium amalgam is the soluble anode in an electrolyte of sodium hydroxide:

$$2NaHg_x \to 2Na^+ + xHg + 2e^- \tag{50}$$

$$\underline{2H_2O + 2e^- \xrightarrow{\text{graphite}} 2OH^- + H_2} \tag{51}$$

$$2NaHg_x + 2H_2O \to 2NaOH + H_2 + xHg \tag{52}$$

It has not proved possible to recover this energy. In order to match the rate of amalgam decomposition to its rate of formation in the electrolyzer, the reaction energy is allowed to degrade entirely into heat.

About 20 types of mercury cells are in commercial operation today. Characteristics of several of the more significant cell designs are summarized in Table 9.

## 5.3. Membrane Cells

### 5.3.1. General Considerations

The first three commercial membrane cell plants were installed in 1975 by Reed Paper Ltd., of Dryden, Ontario, using Hooker technology; by Asahi Chemical Ind. Co., Ltd., at Nobeoka on Kyushu Island, Japan; and by Asahi Glass Co., Ltd., in Osaka City, Japan. Commercialization activity is greatest in Japan. Plants under construction or in operation in Japan in 1978 had a total production capacity of over 500 metric tons/day of chlorine.[4] In late 1978, Akzo NV. announced a new 700-metric ton/day chlorine plant at Rotterdam, based on Asahi Glass technology. The scheduled completion date is 1983.[112] North American membrane cell plants announced through 1978 tend to be small, in the 20–40-metric ton/day range. Pulp mills are apparently finding such package units to be competitive with diaphragm cells and purchased chlorine and alkali.

Four firms currently active in licensing membrane cell technology are Hooker, Asahi Chemical, Asahi Glass, and Diamond Shamrock, which has a membrane development agreement with E. I. DuPont de Nemours and Company; performance characteristics of the cell types offered by the four vendors are summarized in Table 10. Construction details of the Hooker and Asahi Chemical cells have been publicly disclosed and are given below. Announcements of new plants are being made almost monthly. A good source of current information is *International Electrochemical Progress*; good summaries are contained in the Annual Report of the Electrolytic Industries published in the *Journal of the Electrochemical Society*.

Designers and operators of membrane cells share a common set of problems. These include the following:

1. High operating voltage at moderate-to-high caustic strengths.

## Table 9
### Mercury Cell Operating Characteristics

| Cell type | deNora 21M2 | Krebs-Kosmos 23.2-70 | Krebs–Paris 15KFM-160 | Olin E-812 | UHDE 350-100M |
|---|---|---|---|---|---|
| Rated load, kA | 300 | 300 | 160 | 288 | 345.6 |
| Rated current density, kA/m$^2$ | 12.5 | 12 | 10.4 | 10 | 10 |
| Current efficiency, % | 96.5 | 96 | 97 | 97 | 96 |
| Cell voltage, V | 4.45 | 4.25 | 4.3 | 4.24 | 4.17 |
| Current density, kA/m$^2$ | 10 | 10 | n.a.[a] | 10 | 10 |
| dckWh/metric ton Cl$_2$ | 3490 | 3350 | 3400 | 3305 | 3455 |
| Rated Cl$_2$ production, metric tons/day | 9.19 | 9.14 | 4.92 | 8.87 | 10.53 |
| Anode material | DSA | Metal | Metal | Metal | Metal |
| No. of anodes | 42 | 36 | 24 | 96 | 54 |
| Mercury charge, kg | 2650 | 2750 | 1650 | 3800 | 3680 |
| Cell dimensions | | | | | |
| Cathode area, m$^2$ | 24.0 | 23.2 | 15.4 | 28.8 | 34.6 |
| Cathode length, m | 11.4 | 14.4 | 9.6 | 14.8 | 14.4 |
| Cathode width, m | 2.1 | 1.61 | 1.6 | 1.94 | 2.4 |
| Electrolyzer slope, mm/m | 15 | 18 | n.a. | 10 | 15–20 |

[a] n.a. means data not available.

Table 10
Membrane Cell Operating Characteristics

| Cell type | Hooker MX | Diamond DM-14 | Asahi Chemical | Asahi Glass |
|---|---|---|---|---|
| Operating current, kA | 8.0 | 4.4 | 10.8 | 5.3 |
| Current density, kA/m$^2$ | 3.0 | 3.1 | 4.0 | 2.0 |
| Current efficiency, % | 80–95 | 89 | 93 | 95 |
| Cell voltage, V | n.a.$^a$ | 3.9 | 3.75 | 3.7 |
| dckWh/metric ton NaOH | 2800 | 2950 | 2703 | 2610 |
| NaOH production/cell, metric tons/day | n.a. | 0.14 | 0.36 | 0.18 |
| Caustic concentration, % NaOH | 17–30 | 28 | 22 | 35 |
| Membrane type | Nafion | Nafion | Asahi Chem. | Flemion |
| Electrode area, m$^2$ | 2.7 | 1.41 | 2.70 | 2.64 |
| Cells/electrolyzer | 50 | Variable | 80 | n.a. |
| Electrode configuration | Bipolar | Monopolar | Bipolar | Monopolar |

$^a$ n.a. means data not available.

2. Fragility (and expense) of the membrane imposes design constraints. Successful designs have utilized well-supported flat sheets rather than pockets or other complex structures.

3. Impurities in the brine feed (e.g., $Ca^{2+}$ and $Mg^{2+}$) cause deterioration of membrane performance.

### 5.3.2. The Hooker MX Cell

The MX cell is a bipolar design. The cell system resembles the filter press arrangement of conventional water electrolyzers. (Figure 15). The assemblage of cells is termed a *cell stack*. The number of cells in a stack can vary; the Dryden stack has 26 cells. Each cell contains an anode compartment with DSA and a cathode compartment. One or more flat ion-exchange membranes of similar or dissimilar materials separate the compartments.[113] Hooker has used a dual-membrane configuration, resulting in a three-compartment cell.[97] Membrane size is about 2.7 m$^2$. Individual cells in a stack are connected internally in series. The Dryden plant has 12 cell stacks, each operating at 3000 A and 104 V. Feed and product headers are part of individual plastic cell frames. The rated capacity of the Dryden plant is 50 metric tons/day of chlorine and 9–10% NaOH. This low caustic concentration was required for adequate current efficiency using 1973–1974 membranes. Recent Hooker designs operate in the range 10–15 kA. Current information on the membrane or membranes used is not available. Energy consumption is 2800 kWh/metric ton NaOH at 8 kA.

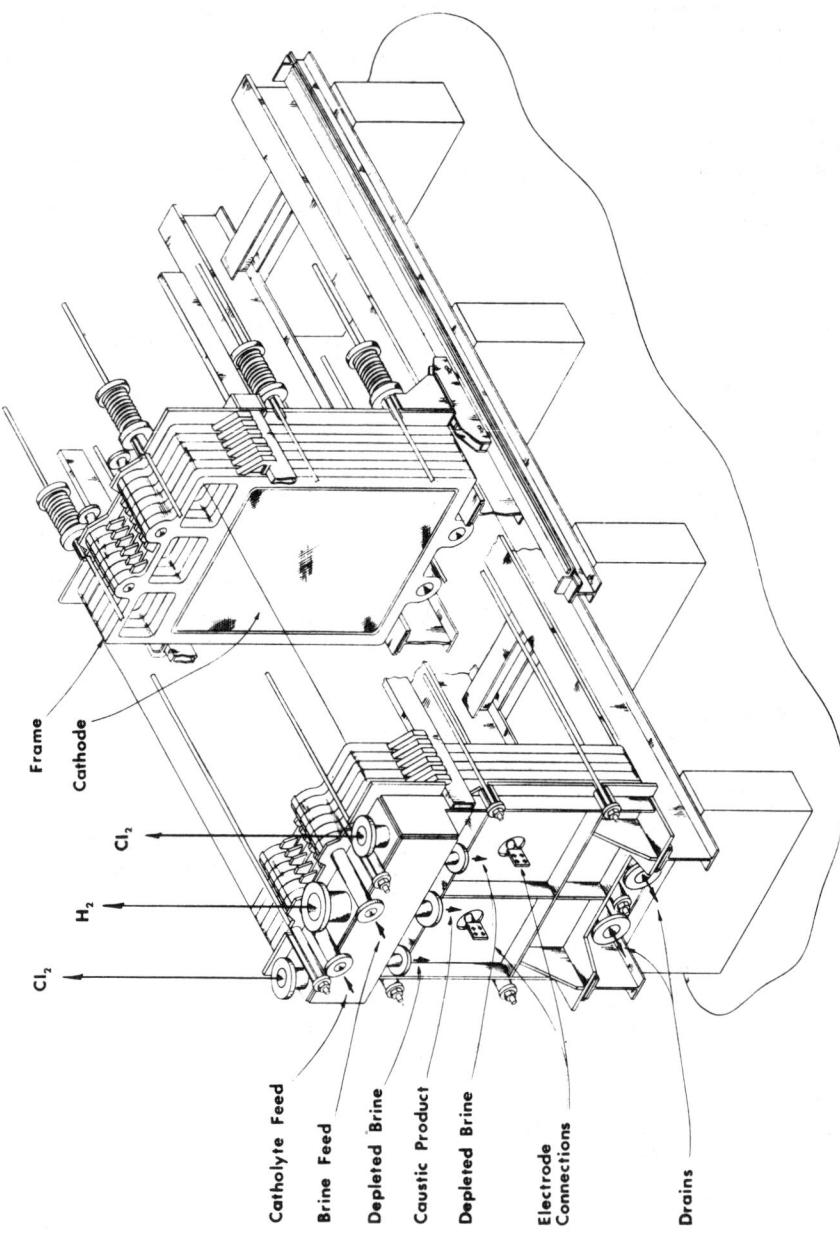

Figure 15. Hooker MX®-II Membrane Cell. (By permission of Hooker Chemicals and Plastics Corp., Industrial Systems Division.)

## 5.3.3. The Asahi Chemical Membrane Electrolyzer[104]

The Asahi Chemical electrolyzer contains 80 bipolar cells. Annual production rate per electrolyzer is 10,070 metric tons of NaOH, based on a cell current of 10.8 kA. Initial operation was with Nafion perfluorosulfonic acid membranes; these have been replaced with Asahi Chemical's own perfluorocarboxylic acid-type membrane. They have also developed their own anode coating, a three-component solid solution containing ruthenium and titanium oxides.

The anode compartment is lined with titanium and the cathode compartment is steel. The cell-to-cell connection is a partition of explosion-bonded titanium on steel. Vertical plates welded at right angles to the partition plate carry current to the electrode surfaces. Holes in the vertical plates permit electrolyte circulation. The electrodes themselves are expanded mesh and are welded to the vertical plates. The membrane is clamped between units using Teflon gaskets. The anode–cathode gap is 2–3 mm. Chlorine and hydrogen are released to the rear of the electrodes and rise up the channels formed by the vertical plates. Asahi Chemical claims this design promotes escape of the cell gases and prevents their becoming trapped between the electrodes and the membrane. Details of cell construction are shown in Figure 16.

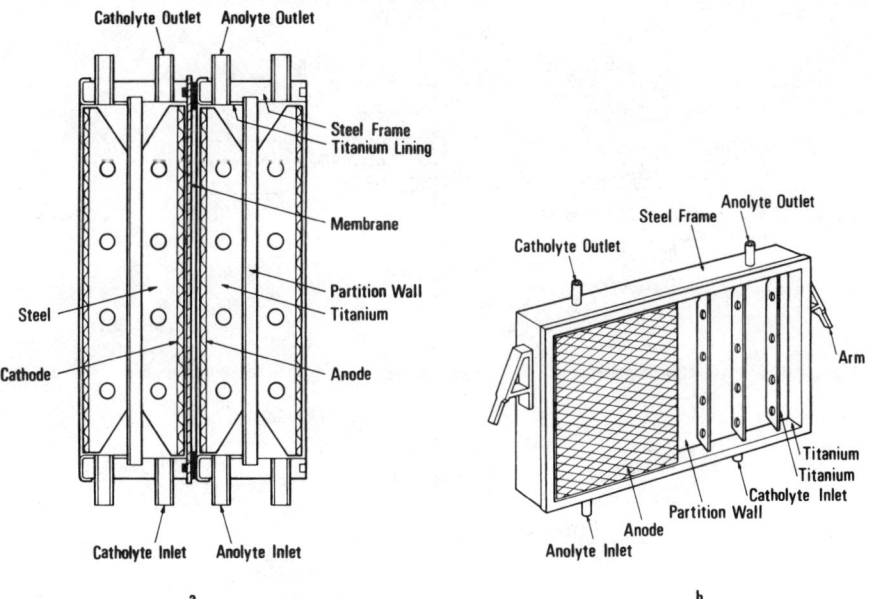

Figure 16. Asahi Chemical membrane electrolyzer[118]: (a) longitudinal sectional elevation (two cells), (b) cutaway view of bipolar cell unit. (By permission of Asahi Chemical Industry Co., Ltd.)

## 6. Chloralkali Plant Auxiliaries

### 6.1. General

Certain ancillary units are common to all chloralkali production facilities, whether they contain diaphragm, mercury, or membrane cells: a means to supply direct current electric power, and equipment for brine purification, chlorine processing, and hydrogen recovery. Diaphragm cell plants for merchant caustic production also require caustic evaporation and salt recovery; membrane cell plants may also require caustic evaporation. A typical diaphragm cell plant diagram is shown in Figure 17; a typical mercury cell plant is illustrated in Figure 18.

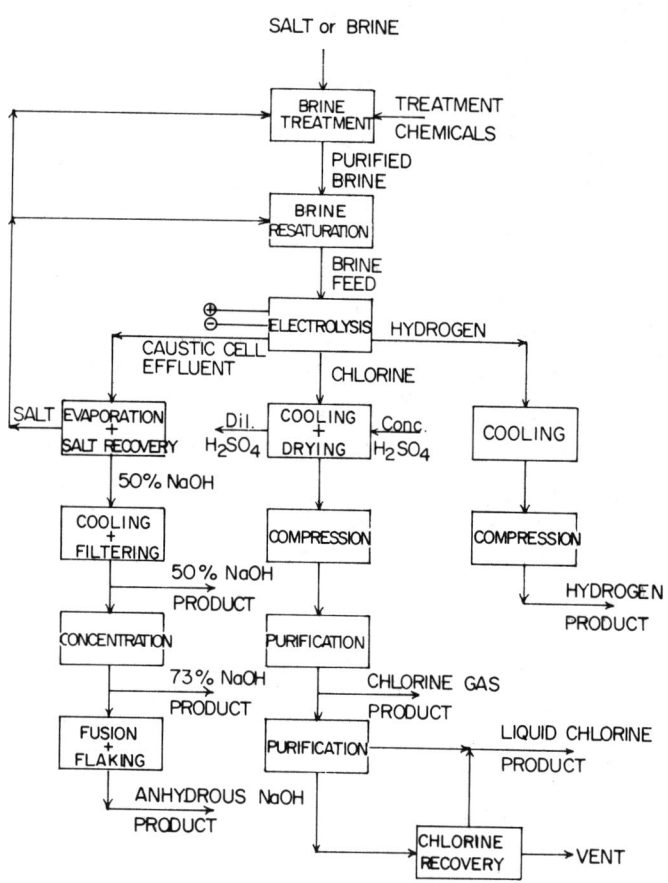

Figure 17. Schematic of chloralkali plant containing diaphragm cells.

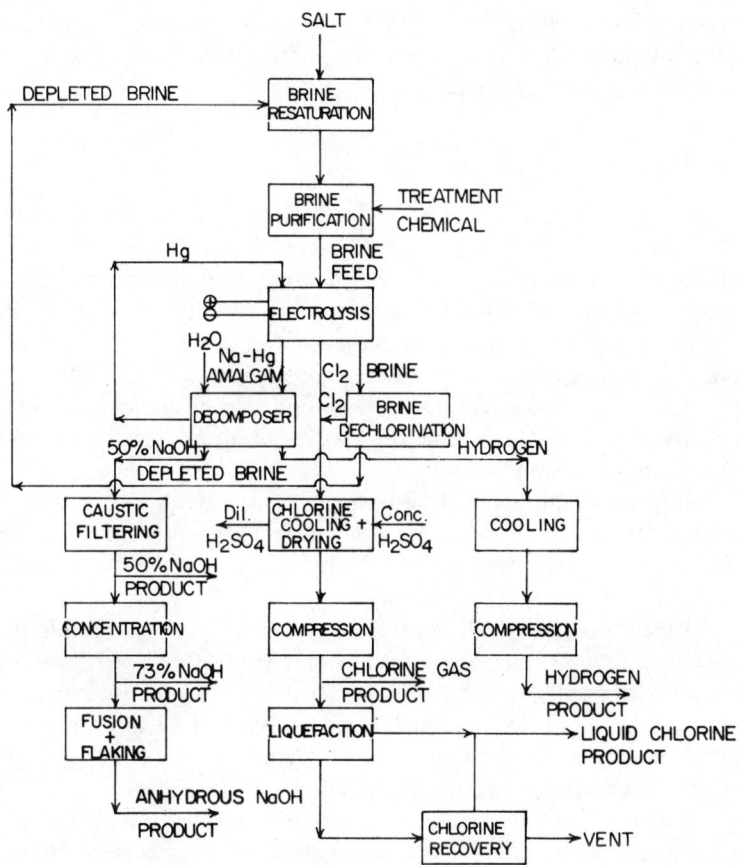

Figure 18. Schematic of chloralkali plant containing mercury cells.

## 6.2. Direct Current Electric Power

Electric power is the largest component in the operating cost of a chloralkali plant. An efficient, reliable source of large direct currents is a necessary part of plant design. High-voltage alternating current is normally reduced to the appropriate circuit voltage by a series of step-down transformers. Rectifiers then convert ac to dc. Solid state silicon diodes are in common use; these achieve rectification efficiencies in the high 90% range at typical circuit voltages. Newly developed thyristors give solid state control of the dc current, eliminating the need for mechanical equipment such as tap changers.

Many large chloralkali producers generate their own electric power. The industry has pioneered the concept of cogeneration, whereby use is made of both ac power and steam produced by the power plants. The energy efficiency of the diaphragm cell process can be greatly improved by using cogenerated steam to evaporate caustic.

### 6.3. Brine Purification

Sodium chloride for brine electrolysis is obtained by solution mining of underground deposits or purchased as solid salt. Diaphragm cell feed brine must be treated for removal of calcium, magnesium, iron, and aluminum salts. These impurities will precipitate in the alkaline layer of the diaphragm, decreasing anolyte flow and reducing the life of the diaphragm. Sulfate ion is detrimental to graphite anodes, but apparently does not affect metal anode operation. Some brines contain ammonium ions or organic nitrogen which can be converted to $NCl_3$ in the cell, with potentially explosive results in the chlorine compression system.

Brine for diaphragm cells is typically treated sequentially with $Na_2CO_3$ (to precipitate $Ca^{2+}$ and $Sr^{2+}$), NaOCl (to convert ammoniacal nitrogen to the volatile $H_2NCl$, which decomposes harmlessly in the cell), and NaOH (to precipitate $Mg^{2+}$ and heavy metals).

Brine for mercury cells must meet stringent purity standards, as even trace concentrations of some heavy metals can lower the hydrogen overvoltage at the cathode sufficiently to produce dangerous amounts of hydrogen in the electrolyzer. Depleted brine from the electrolyzer is dechlorinated, resaturated with solid salt, and returned to the cell. Solid salt is required to maintain the proper water balance around the plant.

The brine dechlorination–resaturation–purification cycle for membrane cells is similar to that for mercury cells. Purity standards are rigid; precipitation of metal hydroxides within the membrane can lead to severe performance degradation or even physical destruction of the membrane.

### 6.4. Chlorine Processing

The chlorine discharged from the electrolytic cells is hot and saturated with water vapor, and contains entrained brine, atmospheric gases, electrolytic oxygen and hydrogen, carbon dioxide, and possible traces of chlorinated hydrocarbons.

The first processing step is normally water condensation in cold water scrubbers or titanium heat exchangers. Further drying is accomplished in multistage countercurrent sulfuric acid contact towers. Chlorinated organics may be removed by washing the gas with liquid chlorine in a countercurrent packed tower.

Chlorine which has been dried and purified may be safely handled in steel piping and equipment. About half of the U.S. production is used directly in the gaseous form. The other half is first compressed and liquefied for storage and/or shipment. Multistage centrifugal compressors are frequently utilized; intercooling between stages is usually necessary to keep the steel processing equipment safely below its upper temperature limit of 150°C. Liquefaction can be accomplished at 18°C and 709–1216 kPa (7–12 atm) or at −40°C at slightly greater than 101 kPa (1 atm) pressure. Present tendency is toward liquefaction at about 203 kPa and 4°C.[2]

Raw chlorine contains noncondensible atmospheric gases plus oxygen, carbon dioxide, and hydrogen from the cell reactions. As chlorine is condensed and removed, the concentration of these gases increases. To maintain hydrogen levels safely below the explosive limit of 5%, air may have to be bled into the compression system between stages. The noncompressible gas, known as "tail gas" or "sniff gas," always contains residual chlorine which must be reclaimed or rendered innocuous. Diamond Shamrock[114] recovers this chlorine by absorption in chilled carbon tetrachloride. Hooker[115] dissolves this chlorine in cooling water in a special tower under about 340 kPa (3.4 atm) pressure. This chlorinated water is then used to cool the hot cell gas in a direct contact cooling tower, releasing the chlorine back into the system.

Chlorine that is not recovered can be converted into sodium or calcium hypochlorite by treatment with dilute caustic soda or calcium hydroxide. The hypochlorite may be used internally, destroyed catalytically[116] to recover salt by the reaction

$$NaOCl \rightarrow NaCl + \tfrac{1}{2}O_2(g) \qquad (53)$$

or treated chemically prior to ultimate disposal.

### 6.5. Hydrogen Recovery

Hydrogen from diaphragm cells is water saturated but otherwise fairly pure. Simple refrigeration systems are normally sufficient to recover usable hydrogen; chemical adsorption may be added. Hydrogen from mercury cells contains entrained caustic soda and mercury, which is removed by scrubbing with sodium persulfate or sodium hypochlorite.

Hydrogen from chloralkali plants is generally not recovered as salable product, but is used in or near the generating plant. Integrated chemical production facilities may use some hydrogen as a reactant, but most chloralkali hydrogen is burned as a low-energy fuel.

### 6.6. Caustic Soda Processing

Caustic soda for commerce is produced in three standard concentrations: 50% water solution, 73% water solution, and anhydrous NaOH. The latter is available in the form of a solid, flakes, or beads. The total quantity of anhydrous and 73% NaOH produced is minor compared with that of 50% solution. Mercury cells produce high-purity 50% NaOH directly from the amalgam decomposers. This product may require filtration to remove graphite particles, but typically contains no more than 10–100 ppm of sodium chloride and 0.3–0.7 ppm of mercury.

Diaphragm cell effluent typically contains 8–12% NaOH and 12–18% NaCl, plus inorganic impurities such as $Na_2SO_4$, $Na_2CO_3$, and $NaClO_3$. Concentration to salable product is accomplished in multiple-effect evaporators. Salt is recovered from each stage. Regular-grade 50% diaphragm cell caustic contains about 1% NaCl and 0.1% $NaClO_3$. This is suitable for most end uses; additional purification can be obtained through ammonia extraction of salt and chlorate, but at considerable increased cost.

Membrane cell effluent is more concentrated than that from diaphragm cells; optimum strength depends on the membrane and cell design, and total plant economics. It typically contains more sodium chloride than that from mercury cells, but this value is low enough that evaporation to 50% NaOH will not cause the solubility of NaCl to be exceeded. Simpler evaporator designs are therefore adequate, reducing total capital and energy requirements of the membrane cell chloralkali plant.

## 7. Future Trends

The worldwide growth rate of both chlorine and caustic usage has averaged 5% per year over the past decade. Future growth will probably continue at about the same rate; at worst, it will match the overall growth of the economy. In addition to gauging the needs of the marketplace, chloralkali manufacturers must continue to direct their attention to two dominant concerns of the 1970s: the rising cost of energy and the necessity of conducting operations with the proper regard for the environment.

The major production technologies for the next two decades will be based on diaphragm cells and membrane cells. Replacement capacity for obsolescent mercury cell plants will be derived from a mixture of the other two cell types. Mercury cell technology made marked progress in the 1950s and 1960s, and even began to increase its share of diaphragm cell-dominated U.S. capacity. Increasing awareness of the environmental hazards of mercury led in 1968 to the strict regulation of plant emissions in the United States. Plants under construction at that time were completed; however, uncertainty as to the possibility of compliance with future standards made new mercury cell installations unattractive risks. No installations of significant capacity have been built since 1970. The trend away from mercury cells is worldwide, led by Japan.

A trend of the 1960s and early 1970s that has apparently peaked is that toward increased cell size. The 150-kA diaphragm cells offered by Hooker and Diamond Shamrock should meet the needs of the largest plants. An exception would be designs adapted to existing mercury cell circuits, such as the 330-kA

Kureha SK cell.[96] Large cells offer economies of labor, instrumentation, and real estate but at the cost of reduced operational flexibility. Savings in cell capital due to a smaller number of cells are to a large extent offset by increased costs inherent in the bulkier handling equipment and the cells themselves. Cell vendors will continue to offer the smaller 35–40-kA cells.

The introduction of metal anode diaphragm cells during the late 1960s allowed producers the option of increasing operating current density, and thus production rate, while maintaining existing circuit voltages. Most chose this option, and new designs were optimized for operation at 2.5 kA/m$^2$ and up, more than doubling the typical graphite cell current density of 1.2 kA/m$^2$. Energy efficiency is inversely proportional to current density. The cost of power is now forcing operating current densities inexorably downward; this trend will continue for the foreseeable future. Cells designed for 2.5-kA/m$^2$ operation cannot be expected to provide an adequate return on invested cell capital when forced to produce at low rates; they contain much more titanium and copper than necessary to handle reduced loads. The next generation of diaphragm cells will be optimized for 1.0–1.5-kA/m$^2$ operation and will emphasize economy of materials.

The possible course of technological progress over the next 20 years could run as follows:

1. Metal anodes will totally supplant graphite. More efficient, cheaper coatings will be sought.

2. Low-overvoltage cathodes will find ready acceptance, and will be in widespread use within five years.

3. The diaphragm itself is now the weak link in the diaphragm cell. A better fundamental understanding of the asbestos diaphragm is needed, whether the future lies with fibrous or polymeric sheet diaphragms. Improvements are needed in uniformity, efficiency, voltage drop, and life.

4. A ten-year maintenance-free cell is possible, and should be the goal.

5. The technology of air- and oxygen-depolarized cathodes is not yet far enough advanced to allow their impact to be predicted, but they should not become significant factors before the late 1980s.

6. While radical changes in external cell geometry are not probable, simplification of cell internals will be sought. One possible direction is indicated by General Electric's anode–membrane–cathode composite material.[117]

7. Membrane cells will replace both diaphragm and mercury cell capacity, beginning with small captive-use plants and medium sized merchant plants. Economic factors will favor diaphragm cells or diaphragm–membrane cell mixtures for large plants.

8. Rate of conversion to membrane cells will be greatly accelerated by membrane cost reductions or increased environmental pressure on asbestos.

9. The first cell manufacturer with a membrane cell design suitable for direct replacement in diaphragm cell circuits will gain a significant competitive advantage.

## Acknowledgments

The author is indebted to the management of the Dow Chemical Company, in particular to Mr. W. A. Rogers, Director of Research and Development, Texas Division, and Mr. F. W. Spillers, Director of Chlor-Alkali Research, Texas Division, for permission to publish this paper. Thanks are due to Dr. M. S. Chao, Dr. M. J. Hazelrigg, and Dr. J. M. McIntyre for supplying previously unpublished data. A special thanks is owed to Mr. W. F. McIlhenny, Associate Scientist, for a critical initial review of the manuscript. Other co-workers participating in the review of this paper included Dr. J. M. McIntyre, Mr. W. J. Lee, Mr. K. A. Poush, and Mr. W. P. Saunders; their helpful comments are gratefully acknowledged.

## References

1. T. R. Beck, in Proceedings of the Workshop on Energy Conservation in Industrial Electrochemical Processes, Argonne National Laboratory Report No. ANL/OEPM-77-1, August 1976, pp. 37–82.
2. J. J. Leddy, I. C. Jones, Jr., B. S. Lowry, F. W. Spillers, R. E. Wing, and C. D. Binger, Alkali and Chlorine Products, in *Encyclopedia of Chemical Technology*, 3rd ed., Vol. 1, John Wiley and Sons, New York (1978), pp. 799–865.
3. Y-C. Yen, Chlorine, Process Economic Program Report No. 61B, Stanford Research Institute, Menlo Park, California, November 1978, p. 3.
4. J. Renner and K. E. Woodard, Jr., Report of the electrolytic industries, presented at the Electrochemical Society Meeting, Boston, Massachusetts, May 1979, p. 4.
5. *Chemical Origins and Markets*, Stanford Research Institute, Menlo Park, California, 1967.
6. L. R. Belohlav and E. T. McBee, in *Chlorine—Its Manufacture, Properties and Uses*, A.C.S. Monograph 154, J. S. Sconce, ed., Reinhold, New York (1962), Chap. 1, pp. 1–9.
7. H. W. Schultze, The chlorine industry—past, present and future, in *Chlorine Bicentennial Symposium*, Electrochemical Society, Princeton, New Jersey (1974), pp. 1–19.
8. M. S. Kircher, in *Chlorine—Its Manufacture, Properties and Uses*, A.C.S. Monograph 154, J. S. Sconce, ed., Reinhold, New York (1962), Chap. 5, pp. 81–126.
9. W. C. Gardiner, Castner, a pioneer inventor in alkali–chlorine, in *Chlorine Bicentennial Symposium*, Electrochemical Society, Princeton, New Jersey (1974), pp. 35–43.
10. K. E. Stuart, U.S. Patent 1,865,152 (1932).
11. R. B. MacMullin, in *Chlorine—Its Manufacture, Properties and Uses*, A.C.S. Monograph 154, J. S. Sconce, ed., Reinhold, New York (1962), Chap. 6, pp. 127–199.
12. D. W. F. Hardie, *Electrolytic Manufacture of Chemicals from Salt*, 2nd ed., The Chlorine Institute, New York (1975), pp. 77–78.
13. G. Faita, P. Longhi, and T. Mussini, Standard potentials of the $Cl_2/Cl^-$ electrode at various temperatures with related thermodynamic functions, *J. Electrochem. Soc.* **114**, 340–343 (1967).
14. R. B. MacMullin, Algorithms for the vapor pressure of water over aqueous solutions of salt and caustic soda, *J. Electrochem. Soc.* **116**, 416–419 (1969).
15. J. E. Currey and G. G. Pumplin, Chloralkali, in *Encyclopedia of Chemical Processing and Design*, Vol. 7, Marcel Dekker, New York (1978), pp. 305–450.
16. R. E. De La Rue and C. W. Tobias, On the conductivity of dispersions, *J. Electrochem. Soc.* **106**, 827–833 (1959).

17. F. Hine and K. Murakami, Bubble effects on the solution ir-drop in a vertical electrolyzer under free and forced convection flow conditions, presented at the Electrochemical Society Meeting, Boston, Massachusetts, May 1978, Abstract No. 281.
18. D. Dobos, *Electrochemical Data*, Elsevier, Amsterdam (1975), p. 85.
19. L. I. Kheifets and A. B. Gol'dberg, The rate of anolyte circulation in diaphragm-type electrolysis cells, *Sov. Electrochem.* (Engl. Transl.) **10**, 1140–1147 (1974).
20. F. Hine and M. Yasuda, Studies on the deposited asbestos diaphragm with a miniature diaphragm-type chlorine cell, *J. Electrochem. Soc.* **118**, 166–173 (1971).
21. F. Hine, M. Yasuda, and T. Tanaka, Mass transfer through the deposited asbestos diaphragm in chlor-alkali cells, *Electrochem. Acta* **22**, 429–437 (1977).
22. Z. Nagy, A mechanistic model for the calculation of material balance for a diaphragm type chlorine caustic cell, *J. Electrochem. Soc.* **124**, 91–95 (1977).
23. F. Hine and M. Yasuda, Studies on the cathodic reaction in the diaphragm-type chlorine cell, *J. Electrochem. Soc.* **118**, 170–173 (1971).
24. I. E. Veselovskaya, E. M. Kuchinskii, and L. V. Morochko, The cathodic reduction of chlorate, *J. Appl. Chem. USSR* (Engl. Transl.) **37**, 85–91 (1964).
25. J. M. McIntyre, Thermal temperature coefficients of the hydrogen electrode, presented at the Electrochemical Society Meeting, Seattle, Washington, May 1978, Abstract No. 541.
26. L. I. Krishtalik, G. L. Melikova, and E. G. Kalinina, Investigation of the effect of electrolysis conditions on the stability of graphite anodes in the chlorine cell, *J. Appl. Chem. USSR* (Engl. Transl.) **34**, 1464–1469 (1961).
27. L. E. Vaaler, Graphite anodes in brine electrolysis, *J. Electrochem. Soc.* **107**, 691–698 (1960).
28. L. E. Vaaler, Graphite–electrolytic anodes, *Electrochem. Technol.* **5**, 170–174 (1967).
29. F. Hine, M. Yasuda, I. Sugiura, and T. Noda, Effects of the active chlorine and the pH on consumption of graphite anode in chlor-alkali cells, *J. Electrochem. Soc.* **121**, 220–225 (1974).
30. *Chem. Eng. (N.Y.)* **86**, 45 (December 18, 1978).
31. R. H. Stevens, U.S. Patent 1,077,894 (1913).
32. J. B. Cotton, E. C. Williams, and A. H. Barber, U.K. Provisional Patent Spec. 22619 (1957); U.K. Patent 877,901 (1961).
33. H. B. Beer, Neth. Patent Appl. 216,199 (1957); U.S. Patent 3,236,756 (1966).
34. D. B. Rogers, R. D. Shannon, A. W. Sleight, and J. L. Gillson, Crystal chemistry of metal dioxides with rutile-related structures, *Inorg. Chem.* **8**, 841–849 (1969).
35. H. B. Beer, Living from invention, *Chem. Ind. (London)*, 491–496 (July 15, 1978).
36. H. B. Beer, S. African Patent 2667/66 (1967).
37. H. B. Beer, U.S. Patents 3,711,385 (1973), 3,632,498 (1972).
38. K. J. O'Leary, U.S. Patent 3,776,834 (1973).
39. V. deNora, Ion selective electrodes, presented at the Electrochemical Society Meeting, Seattle, Washington, May 1978, Abstract No. 458.
40. S. Pizzini and G. Bianchi, Oxides with metallic conductivity, in *The Science of Materials Used in Advanced Technology*, John Wiley and Sons, New York (1973), Chap. 10, pp. 229–241.
41. D. C. Cronemeyer, Electrical and optical properties of rutile single crystals, *Phys. Rev.* **87**, 876–886 (1952).
42. H. P. R. Frederikse, Recent studies on rutile ($TiO_2$), *J. Appl. Phys. Suppl.* **32**(10), 2211–2215 (1961).
43. J. Riga, C. Tenret-Noël, J. J. Pireaux, R. Caudano, and J. J. Verbist, Electronic structure of rutile oxides $TiO_2$, $RuO_2$ and $IrO_2$ studied by x-ray photoelectron spectroscopy, *Phys. Scr.* **16**, 351–354 (1977).
44. G. Lodi, E. Sivieri, A. DeBattisti, and S. Trasatti, Ruthenium dioxide-based film electrodes, *J. Appl. Electrochem.* **8**, 135–143 (1978).
45. F. Hine, M. Yasuda, and T. Yoshida, Studies on the oxide-coated metal anodes for chlor-alkali cells, *J. Electrochem. Soc.* **124**, 500–505 (1977).

46. L. J. J. Janssen, L. M. C. Starmans, J. G. Visser, and E. Barendrecht, Mechanism of the chlorine evolution on a ruthenium oxide/titanium oxide electrode and on a ruthenium electrode, *Electrochem. Acta* **22**, 1093–1100 (1977).
47. G. Faita and G. Fiori, Anodic discharge of chloride ions on oxide electrodes, *J. Appl. Electrochem.* **2**, 31–35 (1972).
48. A. T. Kuhn and P. M. Wright, in *Industrial Electrochemical Processes*, A. T. Kuhn, ed., Elsevier, Amsterdam (1971), p. 533.
49. T. Matsumura, R. Itai, M. Shibuya, and G. Ishi, Electrolytic manufacture of sodium chlorate with magnetite anodes, *Electrochem. Technol.* **6**, 402–404 (1968).
50. P. P. Anthony, U.S. Patent 3,711,382 (1973).
51. A. Martinsons, U.S. Patent 3,711,397 (1973).
52. G. N. Kokhanov, R. A. Agapova, F. I. Mulina, V. V. Avksent'ev, V. L. Kubasov, Yu. V. Dobrov, N. G. Baranova, S. A. Avdeeva, R. I. Kuznetsova, F. V. Kupovich, and Yu. M. Filimonov, USSR Patent 492,301 (1975).
53. M. B. Konovalov, V. I. Bystrov, and V. L. Kubasov, A probe method for the study of the electrochemical characteristics of cobalt oxide anodes, *Sov. Electrochem.* (Engl. Transl.) **11**, 218–220 (1975).
54. M. B. Konovalov, V. I. Bystrov, and V. L. Kubasov, Titanium-base cobalt oxide electrodes, *Sov. Electrochem.* (Engl. Transl.) **12**, 1160–1162 (1976).
55. R. A. Agapova and G. N. Kokhanov, Electrochemical properties of cobalt oxide anodes, *Sov. Electrochem.* (Engl. Transl.) **12**, 1505–1508 (1976).
56. D. L. Caldwell and R. J. Fuchs, U.S. Patent 3,977,958 (1976).
57. D. L. Caldwell and M. J. Hazelrigg, U.S. Patent 4,142,005 (1979).
58. M. J. Hazelrigg and D. L. Caldwell, Cobalt oxide based chlorine cell anodes, presented at the Electrochemical Society Meeting, Seattle, Washington, May 1978, Abstract No. 457.
59. M. J. Hazelrigg and D. L. Caldwell, U.S. Patent 4,061,549 (1977).
60. M. D. Zholudev and V. V. Stender, Overvoltage in the evolution of hydrogen from alkaline solutions, *J. Appl. Chem. USSR* (Engl. Transl.) **31**, 711–715 (1958).
61. N. P. Fedom'ev, N. V. Berezina, and E. G. Kruglova, Cathodes with positive potential of hydrogen formation, *Zh. Prikl. Khim.* **21**, 317–328 (1948).
    457.
62. K. Sasaki and R. Matsui, Japan. Patent 31–6611 (1956).
63. Hooker Chemicals and Plastics Corp., Neth. Patent Appl. 75–07550 (1976).
64. J. R. Brannan and I. Malkin, U.S. Patent 4,024,044 (1977).
65. R. B. MacMullin, German Patent Appl. 2,704,213 (1977).
66. J. R. Brannan, I. Malkin, and C. M. Brown, U.S. Patent 4,104,133 (1978).
67. J. R. Hall and J. T. Van Gemert, U.S. Patent 3,291,714 (1966).
68. S. D. Gokhale, U.S. Patent 3,974,058 (1976).
69. H. H. Hoekje, H. B. Johnson, and R. D. Chamberlin, U.S. Patent 3,990,957 (1976).
70. W. W. Carlin, U.S. Patent 4,010,085 (1977).
71. H. C. Kuo, R. L. Dotson, and K. E. Woodard, U.S. Patent 4,033,837 (1977).
72. A. Martinsons and H. B. Johnson, U.S. Patent 4,105,516 (1978).
73. D. W. Carnell and C. R. S. Needes, Energy-saving catalytically active cathodes for caustic-chlorine production, presented at the Electrochemical Society Meeting, Boston, Massachusetts, May 1979, Abstract No. 260.
74. W. W. Carlin and W. B. Darlington, Activated cathodes for reduced power consumption in electrolytic cells, presented at the Electrochemical Society Meeting, Boston, Massachusetts, May 1979, Abstract No. 261.
75. I. Malkin and J. R. Brannan, Reduction of hydrogen overpotential in a chlorine cell, presented at the Electrochemical Society Meeting, Boston, Massachusetts, May 1979, Abstract No. 262.
76. G. Gritzner, U.S. Patents 4,035,254 and 4,035,255 (1977).

77. *Internat. Electrochem. Progr.* **7**(73), 9 (January 1978).
78. R. L. Dotson, Modern electrochemical technology, *Chem. Eng. (N.Y.)* **85**, 106–118 (July 17, 1978).
79. J. S. Newman, *Electrochemical Systems*, Prentice-Hall, Englewood Cliffs, New Jersey (1973), p. 9.
80. T. Mukaibo, Technical analysis of diaphragm cells for the electrolysis of NaCl solution, *Denki Kagaku* **20**, 482–489 (1952).
81. V. V. Stender, O. S. Ksenzhek, and V. N. Lazarev, Alkali transfer and current efficiency in electrolysis of solutions of chlorides in diaphragm cells, *J. Appl. Chem. USSR* (Engl. Transl.) **40**, 1245–1249 (1967).
82. O. S. Ksenzhek and V. M. Serebrit-skii, Theory of current efficiency in the electrolytic preparation of chlorine and alkali by the diaphragm method, *Sov. Electrochem.* (Engl. Transl.) **4**, 1294–1300 (1968).
83. V. M. Serebrit-skii and O. S. Ksenzhek, Measurement of transport numbers of hydroxyl ions in mixed highly concentrated solutions of alkali and sodium chloride, *J. Appl. Chem. USSR* (Engl. Transl.) **43**, 69–71 (1970).
84. V. M. Serebrit-skii and O. S. Ksenzhek, Theory of current yield during the electrolytic production of chlorine and alkali by the diaphragm method. II, *Sov. Electrochem.* (Engl. Transl.) **7**, 1592–1596 (1971).
85. I. S. Stepanyan, Checking the theory of the unsteady condition for electrolysis of a sodium chloride solution in industrial cells with vertical filtering diaphragms, *Sov. Electrochem.* (Engl. Transl.) **9**, 810–812 (1973).
86. V. L. Kubasov, Estimation of the thickness of the filtering diaphragm of electrolysis vessels for the preparation of chlorine and alkali, *Sov. Electrochem.* (Engl. Transl.) **12**, 72–75 (1976).
87. L. I. Kheifets and A. B. Gol'dberg, Macrokinetics and chlorine cells with filter-action diaphragms. I. The effect of secondary processes on the current yield, *Sov. Electrochem.* (Engl. Transl.) **12**, 1525–1528 (1976).
88. A. B. Gol'dberg and L. I. Kheifets, Macrokinetics chlorine cells with filter action diaphragm. II. Temperature dependence on the current yield, and the limits of the effect of anolyte resaturation, *Sov. Electrochem.* (Engl. Transl.) **12**, 1555–1558 (1976).
89. H. Kaden and A. Pohl, Concerning porosity and pore structure of asbestos diaphragms, *Chem. Tech. (Leipzig)* **30**, 25–28 (1978).
90. J.-A. Leduc, U.S. Patent 3,694,281 (1972).
91. W. B. Darlington and R. T. Foster, U.S. Patent 3,853,721 (1974).
92. R. N. Beaver and C. W. Becker, U.S. Patent 4,093,533 (1978).
93. R. Goldsmith, U.S. Patent 3,281,511 (1966).
94. W. G. Grot, U.S. Patent 3,702,267 (1972).
95. C. Vallance, U.S. Patent 3,930,979 (1976).
96. H. Shibata, Y. Kokubu, and I. Okazaki, The Nobel diaphragm cell: a flexible design for high currents and its performance characteristics at 330 kA, in *Diaphragm Cells for Chlorine Production*, Society of Chemical Industry, London (1977), pp. 53–65.
97. J. E. Currey and J. W. Ahern, Hooker's membrane cell at Reed Paper Ltd.'s Dryden, Ontario plant, presented at the 19th Chlorine Institute Chlorine Plant Manager's Seminar, Montreal, Quebec, February 1976.
98. K. J. O'Leary, Membrane chlorine cell design and operation, in *Diaphragm Cells for Chlorine Production*, Society of Chemical Industry, London (1977), pp. 103–115.
99. Symposium on Fluorocarbon Ion Exchange Membranes, Electrochemical Society Meeting, Atlanta, Georgia, October 1977, Abstract Nos. 436–443.
100. E. H. Price and D. E. Maloney, Nafion perfluorosulfonic acid membranes for the production of chlorine and caustic soda, presented at the 21st Chlorine Institute Chlorine Plant Manager's Seminar, Houston, Texas, February 1978.

101. D. R. Pulver, The Commercial use of membrane cells in chlorine-caustic plants, presented at the 21st Chlorine Institute Chlorine Plant Manager's Seminar, Houston, Texas, February 1978.
102. Y. Oda, M. Suhura, and E. Endo, U.S. Patent 4,065,366 (1977).
103. H. Ukihashi and T. Asawa, Ion exchange membrane for chlor-alkali process, presented at the Electrochemical Society Meeting, Philadelphia, Pennsylvania, May 1977, Abstract No. 247.
104. M. Seko, The ion-exchange membrane chlor-alkali process, *Ind. Eng. Chem. Prod. Res. Dev.* **15**, 286–292 (1976).
105. M. P. Grotheer and C. J. Harke, The development of Hooker's H-2A and H-4 cells, in *Chlorine Bicentennial Symposium*, Electrochemical Society, Princeton, New Jersey (1974), pp. 209–217.
106. J. E. Currey, Recent advances in Hooker chlor-alkali cell technology, in *Diaphragm Cells for Chlorine Production*, Society of Chemical Industry, London (1977), pp. 79–91.
107. R. E. Loftfield and H. W. Laub, U.S. Patent 3,591,483 (1971).
108. E. I. Fogelman U.S. Patent 3,674,676 (1972).
109. T. A. Liederbach, Technical advances in diaphragm chlorine cells, in *Diaphragm Cells for Chlorine Production*, Society of Chemical Industry, London (1977), pp. 41–52.
110. R. M. Hunter, L. B. Otis, and R. D. Blue, U.S. Patent 2,282,058 (1942).
111. V. deNora, Chlorine production using Glanor cells, *Chem. Ing. Tech.* **47**, 141 (1975).
112. *Internat. Electrochem. Progr.* **7**(82), 7. October 1978.
113. S. A. Dahl, Chlor-alkali cell features new ion-exchange membrane, *Chem. Eng. (N.Y.)* **82**, 60–61 (August 18, 1975).
114. R. E. Hulme, U.S. Patent 2,765,873 (1956).
115. T. Hooker and R. H. Miller, U.S. Patent 2,750,002 (1956).
116. D. L. Caldwell and R. J. Fuchs, U.S. Patent 4,073,873 (1978).
117. T. G. Coker, SPE brine electrolyzers, presented at the Oronzio deNora Symposium on Chlorine Technology, Venice Lido, Italy, May 1979.
118. S. Ogawa, Asahi Chemical membrane chlor-alkali process, presented at the Seminar on Developments in Chlor-Alkali Industry, Indian Institute of Chemical Engineers, New Delhi, India, March 1980.

# 3
# Inorganic Electrosynthesis

**N. IBL and H. VOGT**

## 1. Introduction

### 1.1. Scope

This chapter deals with the electrochemical synthesis of inorganic compounds.

Hereinafter, *electrosynthesis* is understood as the formation of compounds directly by electrosynthesis or by subsequent homogeneous reaction. The expansion of the term to subsequent reactions is sensible since the final product often does not appear as the immediate result of an electrode reaction (as in the anodic oxidation of hypochlorite to chlorate, for example). In other cases, homogeneous chemical reactions involving electrode products take place in the vicinity of the electrode (as in hypochlorite formation by hydrolysis of chlorine generated anodically) or at some distance from the electrode, sometimes even outside the cell (as occurs in the synthesis of chlorate by dismutation of hypochlorite and hypochlorous acid, a reaction for which in technical applications it is advantageous to let it take place outside of the electrolytic cell). These cases, too, must be classed with electrosynthesis. A strict conceptional separation of electrosynthesis and electrolysis is inexpedient. In fact, even classical processes of electrolysis can involve electrosynthesis through reactions following the electrode reaction proper (as the formation of sodium hydroxide in alkali chloride electrolysis in diaphragm process).

---

**N. IBL** • Eidgenössische Technische Hochschule Zürich, Technisch-Chemisches Laboratorium, CH-8092 Zürich, Switzerland.   **H. VOGT** • Technische Fachhochschule Berlin, Fachbereich Verfahrenstechnik, D-1000 Berlin 65, West Germany.

Among the compounds generated by electrosynthesis, chlorate and perchlorate are of predominant relevance because they are manufactured industrially on a large scale and by electrosynthesis only. For some other compounds electrosynthesis competes with chemical processes. Particular attention will be devoted to the chlorate electrosynthesis which is the most important industrially and the fundamentals of which are the most complex and the most investigated.

This chapter does not include the manufacturing of elemental inorganics such as chlorine, fluorine, hydrogen and oxygen, and copper and aluminum, which is treated in other chapters. The same applies to NaOH because it is produced simultaneously with chlorine (see Chapter 2 in this volume). The manufacture of manganese dioxide is treated in Vol. 3 of this series, Chapter 6 on primary batteries, with which it is closely linked.

## 1.2. History and Current Outlook of Inorganic Electrosynthesis

Electrosynthesis of inorganic compounds is as old as electrochemistry itself. The fundamentals of all important electrosynthesis processes were laid down at the turn of the century.

Over the years, some inorganic electrosyntheses expanded enormously (as happened for chlorate owing to its use as an intermediate product for the chlorine dioxide generation), others stagnated (e.g., perchlorate), are retrograded, or have even been abandoned (e.g., hydrogen peroxide). On the other hand, some electrosyntheses have recently come into considerable industrial application after having been nearly completely substituted by chemical processes (e.g., hypochlorite).

On a tonnage basis, the electrolytic production of $Cl_2$, NaOH, or Al is much more important than that of the inorganic compounds. Nevertheless an appreciable amount of the latter is currently made by electrolysis on a technical scale.

The research in organic electrosynthesis is for the time being very much more active than in inorganic electrolysis. However, inorganic electrosynthesis processes other than those already mentioned are currently being seriously investigated (for instance, hydrogen peroxide by reduction of oxygen) and might become operational in industry in the near future.

As we have already mentioned the technical applications of inorganic electrosynthesis have fluctuated over the years. It seems likely, however, that, by and large, they will retain their present share of industrial processes and perhaps even expand. An asset of inorganic electrosynthesis is the possibility to contribute to the load leveling of power plants, by operating the electrolysis at a variable current density, designed to make use of the surplus electricity available during certain periods of the day or of the year. In chlorate electrolysis it is easier than in many other electrolytic processes to achieve a variation of current density without impairing the process or causing an excessive additional investment.

## 2. Chlorate

### 2.1. Industrial Significance

Chlorate synthesis is by far the industrially most important inorganic electrosynthesis. A comparable effort in research and technological innovation has not been devoted to any other branch of electrosynthesis. That is partly due to the industrial relevance of the product, partly to the complexity of the electrochemical phenomena.

Today, chlorate is produced exclusively by electrosynthesis. Chemical manufacture is possible and was the only route in the past century. Nowadays, it is only applied to some less common chlorates, which are chemically converted from sodium chlorate, which is itself obtained electrolytically. Sodium chlorate is the most common and most important of all chlorates. The steep increase in chlorate production, shown for some important countries in Figure 1, is mainly due to the extended need in pulp bleaching. In the U.S.A. in 1955, only 29% of the total sodium chlorate produced was used for bleaching purposes, in 1974 it amounted to 78%.[1] This development is representative for Europe, too.

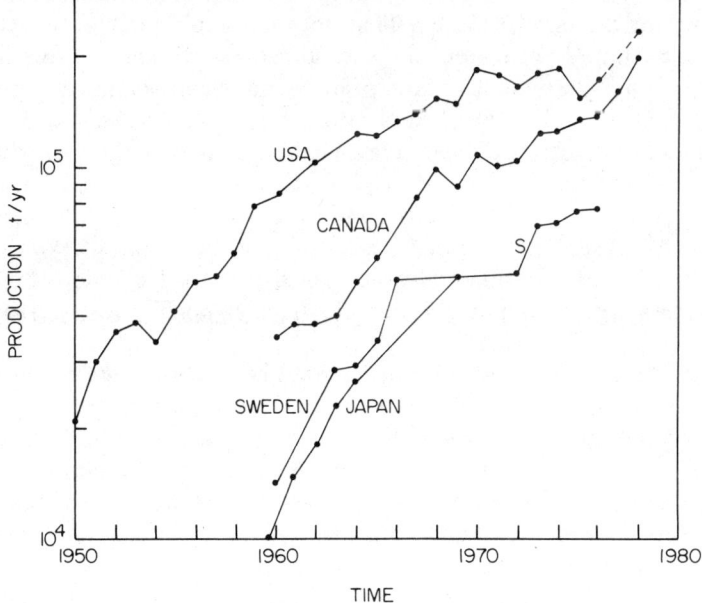

Figure 1. Development of sodium chlorate production by electrosynthesis.

The commercial suppliers of chlorate are not gaining the full benefit of the rapid production increase. Only part of the chlorine dioxide used for bleaching is produced from solid chlorate. The major part is produced in captive plants.[2,3] The first on-site plants for chlorate were erected in the mid-1950s[4] and have been most successful since. In Day–Kesting chlorine dioxide plants[5,6] the chlorate in solution is reduced by hydrochloric acid to form $ClO_2$; the resulting solution, enriched in chloride, is continuously recycled to the electrochemical cells. Introduction of the oxygen bleach of pulp in the last years[7,8] only replaces the first bleach stages and does not affect the upward chlorate production trend.

As compared to the application in pulp industry, the traditional fields of application (sodium chlorate as a total herbicide, potassium chlorate as oxidizing agent in safety matches and for chlorate explosives) have receded into the background. Use of sodium chlorate as an intermediate for perchlorate production has expanded considerably (by a factor of 4.5 in the U.S.A. within the past 20 years[1]), but its share in total chlorate produced has remained constant. At present, worldwide chlorate production capacity amounts to more than 1,100,000 t/year.[315]

## 2.2. Development of Theory and Technology; Main Features of the Present State of the Art

It was stated with good reasons[9] that the most important electrochemical technologies, except for engineering improvements, have remained practically unchanged for the last 100 years. That does not apply to chlorate synthesis, which has been in a permanent innovation process throughout this century, with a first climax in the early years of industrial electrochemical production and important technological changes within the past ten years. The process of development of theory and technology will be sketched in the following.

Electrosynthesis of chlorate is older than currently stated in modern literature. As early as 1802 Hisinger and Berzelius prepared sodium chlorate by electrolysis of sodium chloride,[10] but were not able to show it conclusively. Kolbe[11] confirmed experimentally the electrosynthetic formation of chlorate, and, furthermore, prepared chloric acid from hypochlorous acid by electrosynthesis.

The reactions leading to chlorate were yet utterly unknown, when in 1851 a patent[12] was granted to C. Watt for a chlorate manufacture process[13] which surprisingly shows some essential features that are common in modern industrial plants: the omission of a diaphragm, high temperature of the electrolyte to favor chemical chlorate formation, the separation of chlorate by crystallization leaving the sodium chloride in solution, and the recycling of the resaturated mother liquor to the cells. A practical application is not reported and cannot be expected since no large-scale power sources were available at that time. After the invention of the dynamo, the conditions for industrial chlorate manufacture by electrosynthesis were given, and from 1886 onward plants came on stream in Switzerland, France, Sweden, and the United States. All cells were operated with alkaline electrolyte solutions. An excellent survey of the technology of that time was given by Kershaw.[13]

After the beginning of industrial chlorate electrosynthesis, a widespread interest in the knowledge of the chemical events occurring in chlorate formation developed. Within a few years before and after the turn of the century, several controversial theories were established and sometimes passionately discussed. Based on the work of Oettel,[14,15] Haber and Grinberg,[16] Wohlwill,[17] Foerster, Jorre, and Müller[18,19] studied the main reactions involved in the electrosynthesis of chlorate. Their stoichiometry and important mechanistic features were well established early in this century and were confirmed by more recent work. A detailed discussion, which is still worthwhile reading today, was presented in Foerster's remarkable book *Elektrochemie wässeriger Lösungen* published first in 1905.[20] More recent progress has dealt mainly with mass transport and kinetic aspects.[21–26,71,75,75a,87] A comprehensive review of the whole subject was presented recently by Fioshin.[72]

After the reactions had been clarified, industrial cells were operated with slightly acid electrolytes. In the first industrial cells, attempts were made to solve the problem of cathodic reduction of hypochlorite by use of a diaphragm. Oettel,[15] based on experimental results, propagated the omission of a diaphragm. Landin[27] and Imhoff[28,29] finally proposed the addition of chromic acid or chromate, respectively, to the electrolyte to form a diaphragm on the cathode, the action of which was explained by Wagner[30] in 1954.

Application of these technological features introduced at the time is generally customary today. Favored by the substantial increase in chlorate production within the last 25 years, further substantial improvements were made recently. An important feature is the use of a chemical reactor, separate from the electrolysis cell (Figure 2). It was recognized at the turn of the century[18,31] that the volume of the electrolyte has to be large to ensure that the relatively slow chemical reactions which follow the electrochemical reactions can proceed to a satisfactory extent. At that time it was customary to characterize the quality of a cell design by the term *current concentration*, i.e., the ratio of current to total electrolyte volume of a cell.[31] However, the most favorable residence time spectrum of the electrolyte outside the interelectrode space can only be achieved if the cell is separated from the chemical reactor. Such a separation was first proposed by Schumann-Leclercq.[32]

In the design of the last decade the interelectrode space is kept small and the electrolyte flows with considerable velocity along the electrodes. Hypochlorite generated in the cell reacts mainly outside the cell in the separate reactor. The electrolyte is recycled to the cell after the hypochlorite concentration has dropped to a low value. Optimization of the system with respect to various factors will be discussed in Section 2.4. Current densities as applied in modern cells result in electrolyte flow rates that are much greater than the feed

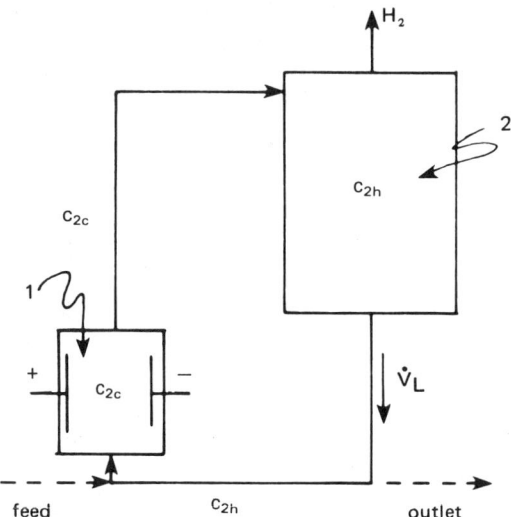

Figure 2. Flow system with separate electrochemical (1) and chemical (2) reactors for electrosynthesis of chlorate.

of fresh solution to the system. Residence time of the electrolyte in the chemical reactor is by far greater than in the cell. Another advantage of this feature is that the losses through reaction of hypochlorite at the cathode are minimized.

For a long time, high electrolyte flow rates were obtained by pumping. Topical cell design dispenses with forced convection and makes use of the high rate of gas bubbles evolved at high current densities to establish a satisfactory self-circulation. Hydrogen evolved at the cathode, together with some oxygen from the anode, departs in the form of small bubbles from the electrode with low steady state rise velocity. The interelectrode space thus becomes substantially enriched with gas, causing a gas lift which makes a pump superfluous[34,35] combined with energy saving and higher regularity of the electrolyte flow.

A precondition for the substantially increased current densities of modern cells was the introduction of noble metal coated anodes in place of the currently applied anodes of graphite, magnetite, and lead dioxide.

Magnetite was the first anode material industrially used in chlorate manufacture.[44] Wear is moderate and the anode behaves virtually dimensionally stable during operation. The material allows higher temperatures than graphite. The disadvantage lies in a low conductivity and high fragility. Anodes of magnetite gained a predominant importance in Japan, where ten years ago more than 50% of the total chlorate was produced with anodes of this material.[36]

In Europe, the preferred anode material for a long time was graphite. It requires a lower capital investment, but undergoes considerable wear[38–41] by formation of $CO$ and $CO_2$ and by becoming brittle, which not only pollutes the electrolyte but also increases the interelectrode distance in the course of operation, thus considerably increasing the ohmic loss in the electrolyte. Several authors have studied in detail the influence of temperature, pH, current density, and potential on the wear of graphite.[37–41,104–107] It increases with increasing temperature; Jakšić[41] found a minimum at pH 6. Impregnation of the graphite with linseed oil reduces the wear substantially.[40–42] A comprehensive survey of the literature on electrode wear can be found in the book by Fioshin.[72]

Lead dioxide[99,100] has a higher conductivity than most metal oxides. The difficulty of getting a pore-free and well-clinging material has been solved in the last years by depositing lead dioxide on graphite and on titanium.

All these anode materials are going to be substituted by coatings of platinum–iridium alloys, ruthenium oxide, or other diverse noble metals or their oxides on titanium. Throughout the world, noble-metal-coated anodes are the only ones used in cells put into commission at present. (For more details see Section 2.5.)

Cathode materials in the first years of electrochemical chlorate manufacture were copper, nickel, and platinum. The proposal to use chromium-plated steel cathodes to prevent cathodic reduction of hypochlorite without a chromate content of the electrolyte was made by Grube and Burkhardt[46] and was repeatedly taken up.[47] A recent investigation[48] evidenced much lower cathodic reduction rates for chromium–nickel alloys than for carbon steel or nickel cathodes. However, the usual cathode material of modern industrial cells is carbon steel (see Section 2.5).

## 2.3 Fundamentals

### 2.3.1. Main Reactions

In the electrolysis of NaCl solutions chlorine is generated at the anode:

$$Cl^- \rightarrow \tfrac{1}{2}Cl_2 + e \quad (1)$$

The cathodic reaction is

$$H_2O + e = OH^- + \tfrac{1}{2}H_2 \quad (2)$$

With a cathode where Na amalgam is formed or with a diaphragm which prevents the mixing of the $OH^-$ ions with the anolyte, the solution remains acid (pH ~ 3) and chlorine is evolved as gas. These are the conditions prevailing in industrial chlorine production. If the $OH^-$ ions are allowed to mix with the anolyte the pH is much higher; the dissolved chlorine undergoes further reactions and chlorate and perchlorate are formed. We will discuss here the reactions involved in the synthesis of chlorate, most of which were formulated by Foerster and Müller toward the beginning of this century.[20,53–55]

*Main reactions in solution*

$$Cl_2 + H_2O \rightarrow HClO + Cl^- + H^+ \tag{3}$$

$$HClO \rightleftarrows ClO^- + H^+ \tag{4}$$

$$2HClO + ClO^- \rightarrow ClO_3^- + 2Cl^- + 2H^+ \tag{5}$$

*Anodic loss reactions*

$$6ClO^- + 3H_2O \rightarrow 2ClO_3^- + 4Cl^- + 6H^+ + \tfrac{3}{2}O_2 + 6e, \quad E_0 = 0.46 \text{ V} \tag{6}$$

or

$$6HClO + 3H_2O \rightarrow 2ClO_3^- + 4Cl^- + 12H^+ + \tfrac{3}{2}O_2 + 6e \tag{6'}$$

$$2H_2O \rightarrow O_2 + 4H^+ + 4e, \quad E_0 = 1.23 \text{ V} \tag{7}$$

$$4OH^- \rightarrow O_2 + 2H_2O + 4e \tag{7'}$$

$$ClO_3^- + H_2O \rightarrow ClO_4^- + 2H^+ + 2e, \quad E_0 = 1.18 \text{ V} \tag{8}$$

*Cathodic loss reactions*

$$ClO^- + H_2O + 2e \rightarrow Cl^- + 2OH^- \tag{9}$$

$$ClO_3^- + 3H_2O + 6e \rightarrow Cl^- + 6OH^- \tag{10}$$

*Loss processes in solution*

$$ClO^- \rightarrow Cl^- + \tfrac{1}{2}O_2 \tag{11}$$

$$Cl_{2\,(diss)} \rightarrow Cl_{2\,(gas)} \tag{12}$$

### 2.3.2. Cathodic Losses and Losses in the Solution

In experiments carried out with steel cathodes Wranglén and Hammar[89] found that the rate† of reaction (9) is proportional to the bulk concentration of hypochlorite. In practice the cathodic losses [reactions (9) and (10)] are kept very small by adding some dichromate to the solution. Its action is due to the formation of a thin solid diaphragm of chromic oxide at the cathode through reduction of the chromate. In the pores of the diaphragm the current density is high and a large potential gradient is built up. This acts in a direction opposite to that of the diffusion of the $ClO^-$ and $ClO_3^-$ ions to the cathode and its high value virtually prevents anions from reaching the cathode.[30] The diaphragm remains very thin and the ohmic losses within it are small.

Reaction (11) is much slower than chlorate formation,[20,56,57,59] but impurities of the solution such as Co, Ni, and Cu may noticeably catalyze the reaction.[58] As Fe and Mg were not found to act catalytically[58] and the

---

† A number of publications dealing with this subject are quoted and discussed in a recent paper by Heal, Kuhn, and Lartey.[170]

# INORGANIC ELECTROSYNTHESIS

concentration of dissolved chlorine is small,[89] both loss processes in the solution (11) and (12) play a negligible role under industrial conditions, unless detrimental corrosion products from anode coatings or impregnations are present.

### 2.3.3. Competition between Chemical and Anodic Chlorate Formation

Let us now turn to the anodic loss reactions.† The anode potential is essentially determined by the current density and by the corresponding overpotential for $Cl^-$ discharge. In technical chlorate electrolysis it is more positive than the standard potentials of reactions (7) and (8), but owing to their high overpotential the latter do not proceed to any appreciable extent. For the production of perchlorate the conditions must be different and the corresponding reactions will be discussed in Section 4.4. Reaction (7) can play a role only if the solution is very alkaline or the $Cl^-$ concentration very low. In chlorate electrolysis the main loss reaction is the anodic oxidation of hypochlorite to chlorate with simultaneous oxygen evolution [reaction (6)]. This electrochemical chlorate formation competes with the chemical formation in the solution through oxidation of the hypochlorite ion by the hypochlorous acid. If all the chlorate produced is formed by chemical reaction (5), 6 faradays are consumed in the oxidation of 1 mol of chloride to chlorate. This represents the maximum current efficiency and is usually taken as 1. If all the chlorate produced is formed by reaction (6) the current efficiency is 0.667 since then 0.75 mol of oxygen (needing 3 faradays) is generated per mol of $ClO_3^-$ produced and one-third of the total current is thus used for the evolution of oxygen. In technical chlorate electrolysis one strives to make chlorate formation through reaction (5) as predominant as possible in comparison to reaction (6). [On the other hand, in hypochlorite production one wishes to make *both* reactions (5) and (6) negligible (see Section 3.3).]

### 2.3.4. Chemical Chlorate Formation

The rate of the chemical chlorate formation [Eq. (5)] has been studied by various authors.[20,33,43,61–68] As is evidenced by the results of a number of investigations,[20,33,61,67] it is given by

$$dc_{ClO_3^-}/dt = f_{HClO}^2 k c_{HClO}^2 c_{ClO^-} \tag{13}$$

where $k$ is the rate constant, $f_{HClO}$ the activity coefficient of HClO, and $f_{HClO}^2$ is approximately equal to the Brønsted kinetic factor of the reaction. Equation

---

† The main anodic reaction (i.e., the discharge of $Cl^-$ ions) is treated in Chapter 2 on chlorine electrolysis and is not discussed here.

(13) is the relationship reported by Foerster[18,20,53–55] and extended by Imagawa.[62d] Using a dissociation constant for hypochlorous acid,

$$K = \frac{c_{\text{ClO}^-} c_{\text{H}^+}}{c_{\text{HClO}}} \qquad (14)$$

Flis and Bynyaeva[67] have shown that Eq. (13) can be conveniently rewritten in terms of the total hypochlorite concentration $c_2$ available for oxidation to chlorate,

$$c_2 = c_{\text{HClO}} + c_{\text{ClO}^-} \qquad (15)$$

Inserting

$$c_{\text{HClO}}^2 c_{\text{ClO}^-} = \frac{(K/c_{\text{H}^+}) c_2^3}{(1 + K/c_{\text{H}^+})^3}$$

and introducing further the activities instead of the concentrations and writing the dissociation constant of hypochlorous acid as

$$K_a = \frac{a_{\text{ClO}^-} a_{\text{H}_3\text{O}^+}}{a_{\text{HClO}} a_{\text{H}_2\text{O}}} \qquad (16)$$

Jakšić et al.[68] derived a modified expression of the dissociation constant

$$K^* \equiv \frac{c_{\text{ClO}^-} a_{\text{H}_3\text{O}^+}}{c_{\text{HClO}}} = K_a \frac{f_{\text{HClO}} a_{\text{H}_2\text{O}}}{f_{\text{ClO}^-}} \qquad (17)$$

Hence, the rate of reaction (5) can be written

$$-\frac{dc_2}{dt} = 3\frac{dc_{\text{ClO}_3^-}}{dt} = 3f_{\text{HClO}}^2 k \frac{K^*/a_{\text{H}_3\text{O}^+}}{(1 + K^*/a_{\text{H}_3\text{O}^+})^3} c_2^3 \qquad (18)$$

Values of $K_a$ are known from experiments[69] and may be approximated in the common temperature range by

$$\log K_a = -7.744 + 0.0075\, T/°C \qquad (19)$$

Evaluation[70] of experimental data of Jakšić et al.[71] shows that $K^*/K_a \approx 14$, if the Foerster scheme, Eq. (5), is taken into account. Data from several authors for $kf_{\text{HClO}}^2$ at various temperatures were collated by Kokoulina and Krishtalik.[33] They are in good agreement except for the rate constant measured by Flis and Bynyaeva,[67] which is substantially lower. Equation (18) makes evident that the reaction rate is highly dependent on the pH value.

Flis and Bynyaeva[67] determined the optimum pH (see below) and found that it is somewhat different from that expected in the case of the classical reaction (5). They concluded that chlorate is also formed along the alternate route

$$2\text{ClO}^- + \text{HClO} = \text{ClO}_3^- + \text{H}^+ + 2\text{Cl}^- \qquad (5a)$$

with a rate constant comparable to that of (5). Recently Taniguchi and Sekine[97] also envisaged the simultaneous formation of chlorate by reactions

(5) and (5a) but regard reaction (5) as the major one. In fact, it is to be expected that the extent to which both reactions participate in the chemical chlorate formation depends on pH. There may also be an influence of $Cl^-$ concentration.

Both Eqs. (5) and (5a) describe global reactions and do not give any information about the rate-determining step. According to the study of Lister[101] the latter is the intermediate formation of chlorite according to

$$2HClO = ClO_2^- + Cl^- + 2H^+ \quad \text{(slow)} \quad (5b)$$

This would correspond to a reaction which is second order with respect to HClO as has been indeed reported by various authors.[61,63,101] It would fit reaction scheme (5) better than (5a). However, Lister himself writes the fast reaction which follows (5b) as

$$ClO_2^- + HClO = ClO_3^- + Cl^- + H^+ \quad \text{(fast)} \quad (5c)$$

We will encounter similar uncertainties regarding various other aspects of chlorate electrolysis. It is remarkable how many relevant facts are not accurately known or controversial in such a time-honored and industrially important process as chlorate electrosynthesis.

Qualitatively, it is obvious that reactions (5) and (5a) can run at high rate only if *both* HClO and $ClO^-$ are present in the solution in sufficient amount, i.e., if the equilibrium of reaction (4) does not lie too much to the right or to the left. This is the case in a medium pH range close to neutrality. However, the actual optimum value depends somewhat upon whether one accepts reaction (5) or (5a). Increase of the temperature strongly accelerates reaction (5). Below 10°C and at pH above 8 it is negligible as compared to reaction (6) but under the conditions of technical electrolysis (40–80°C) it is predominant. Since reactions (5) and (6) are homogeneous and heterogeneous, respectively (i.e., the first one is a volume reaction, the second one a surface reaction), their relative importance depends on the ratio of the solution volume to the anode area and thus on the current concentration (A $m^{-3}$) (see also Section 2.4.3). For this reason technical systems are often built up according to the principle of Figure 2, i.e., they include a chemical reactor with a large hold up connected with an electrolysis cell of comparatively small electrolyte volume, the solution being continuously recycled between the two (see also Section 2.4.5).

### 2.3.5. Anodic Chlorate Formation: Stoichiometry

Let us now turn to the main anodic formation of chlorate. It has been doubted by several authors[62,73-77] that the stoichiometry of this reaction corresponds to that originally proposed by Foerster and Müller [Eq. (6)]. Rius and Llopis[75] concluded from their experiments that the reaction should be written

$$5HClO + 8H_2O \rightarrow 2O_2 + 2Cl^- + 3ClO_3^- + 21H^+ + 16e \quad (20)$$

On the other hand, Shlyapnikov and Filippov[73,78-80] suggested the two following reaction schemes:

$$ClO^- + 2H_2O \rightarrow ClO_3^- + 4H^+ + 4e \tag{21}$$

or

$$2HClO + H_2O \rightarrow HClO_3 + H^+ + Cl^- + 2e \tag{22}$$

Reaction (21) was considered most probable, whereas reaction (22) was regarded as improbable at high temperatures. From the viewpoint of stoichiometry these two reactions are equivalent to the primary anodic chlorate formation envisaged in its time by Foerster[20] as a further possibility besides reaction (6):

$$ClO^- + 2O \rightarrow ClO_3^- \tag{23}$$

With any of these three reactions [(21)-(23)] 6 faradays are needed to make 1 mol of $ClO_3^-$ from 1 mol of $Cl^-$, i.e., the current efficiency as defined earlier is 1. In reality it is much less if the chemical chlorate formation is suppressed. This has then to be explained by assuming that oxygen is evolved by a separate reaction such as (7).

At present, quite a number of experimental data are available from which conclusions regarding the stoichiometry can be drawn. One has to distinguish between a medium (slightly alkaline or acid) and an alkaline pH range. In the following we will discuss mainly the first case which is the one that prevails in industrial practice. Extensive measurements were carried out recently by Jakšić et al.[76,77] in the arrangement shown in Figure 2. It consists of an electrolytic cell and a separate chemical reactor, in which the chemical conversion to chlorate according to Eq. (5) takes place, as is the case in modern chlorate cells.[351] The electrolyte circulates through the system at flow rate $\dot{V}_L$. The conditions were such that cathode losses and the chemical chlorate formation in the electrolytic cell 1 were negligible. Let us now consider two extreme possibilities. We denote by $b$ the current efficiency for chlorate formation ($\varepsilon = b$) in the *complete absence* of chemical conversion of hypochlorite (i.e., if the conversion in the chemical reactor 2 is also nil). If, on the contrary, *all* hypochlorite is converted chemically in reactor 2, $\varepsilon$ increases to 1 (i.e., by the quantity $1 - b$). In that case the total amount of hypochlorite generated and converted per unit time is $I/2F$, according to Eqs. (1) and (3), $I$ being the electrolysis current. In reality, as a simple mass balance for reactor 2 shows only the amount $\dot{V}_L(c_{2c} - c_{2h})$ mol $HClO + ClO^-$ is converted per unit time in the chemical reactor (where $c_{2c}$ and $c_{2h}$ denote the concentrations of $HClO + ClO^-$ in the electrochemical and chemical reactor, respectively, which are assumed perfectly mixed). The increase in current efficiency is therefore only $(1 - b)\dot{V}_L(c_{2c} - c_{2h})/(I/2F)$. We have

$$\varepsilon = b + 2(1 - b)\dot{V}_L(c_{2c} - c_{2h})FI^{-1} \tag{24}$$

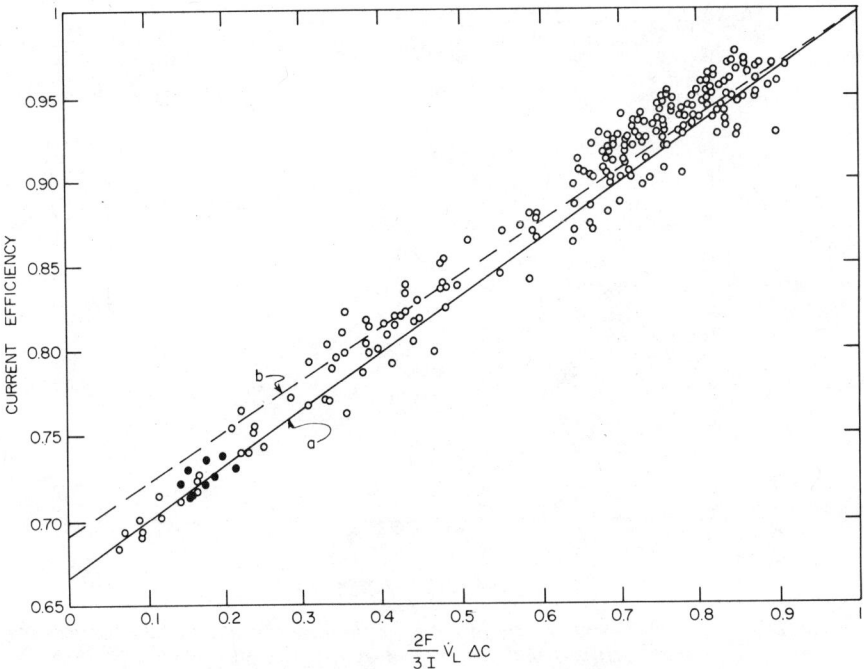

Figure 3. Current efficiencies of chlorate production measured under various conditions in a flow reactor. ○, Jakšić et al.[76]; ●, Jakšić et al.[77] The straight lines represent the theoretical slope according to (a) Foerster, Eq. (6); (b) Rius and Llopis,[75] Eq. (20). $\Delta c = c_{2c} - c_{2h}$.

$b$ is equal to 2/3 for Foerster's reaction (6) and to 0.692 for the stoichiometry of Llopis et al. [Eq. (20)]. The two straight lines of Figure 3 represent Eq. (24) for these two values of $b$. The figure also shows a representative selection of experimental results obtained by Jakšić et al.[76,77] under a variety of conditions. They are in fair agreement with the stoichiometry of Foerster. The same is true for the measurements of Foerster himself and his co-workers.[20] The measurements of Rius et al.[75] are in better agreement with the reaction proposed by them [Eq. (20)] than with Eq. (6). However, the differences between the values predicted by these equations are small (Figure 3). Furthermore, Eq. (24) implies a number of simplifying assumptions (no cathode losses, no $O_2$ evolution through reaction (7) and no chemical chlorate formation in the electrolyte cell, perfectly mixed reactors), and it is difficult to evaluate the accuracy of their fulfilment for the experiments of Figure 3. It is thus hardly possible to decide between the validity of the equations of Foerster and Rius et al. and one may just as well use the simpler Eq. (6). From the above discussion and from all the experimental data available at present, we may

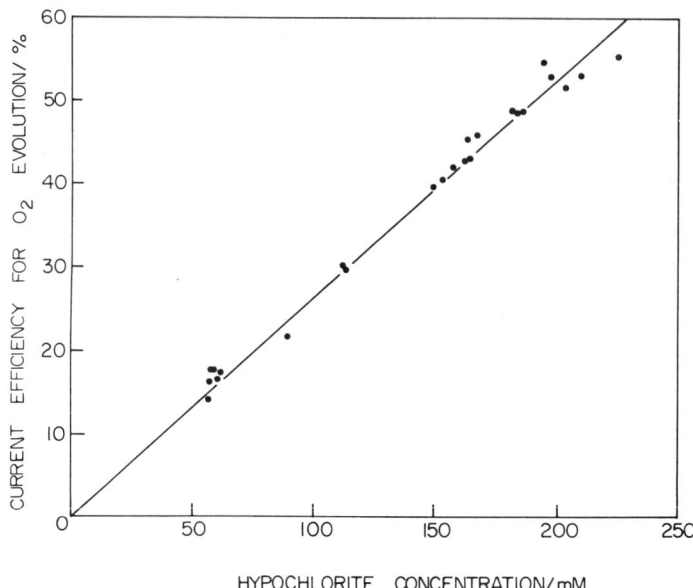

Figure 4. Dependence of current efficiency for oxygen production, $1 - \varepsilon$, on bulk hypochlorite concentration at constant c.d. $j = 40$ mA/cm$^2$, pH = 9–10, $c_{NaCl} = 150$ g/liter and $T = 25°C$.[89]

conclude that at least in the medium pH range the simpler equation of Foerster [Eq. (6)] can be used with an approximation sufficient for practical purposes.

An interesting question is whether the ratio of the rates of oxygen and anodic chlorate formation remains constant, independent of conditions. This appears to be the case, as can be deduced from the experiments of Jakšić et al. (Figure 3) which were carried out at various pH, current densities, and temperatures as well as in the absence and presence of additives such as bichromate or nitrate with Pt or coated titanium electrodes.[76,77] This result is not likely if the chlorate and oxygen were generated in two *separate* anodic reactions such as (21)–(23) and (7). Also the proportionality between the rate of oxygen evolution and hypochlorite concentration (see Section 2.3.6 and Figure 4) points in the same direction. Furthermore, Hammar and Wranglén[89] have found that the ratio of oxygen production to hypochlorite consumption through anodic oxidation corresponds to reaction (6). All these findings suggest that Eq. (6) adequately describes the overall reaction for anodic chlorate formation. They do not give, of course, any information about the details of the mechanism or the rate-determining step.

A different situation is encountered in alkaline solution. The current efficiency for chlorate formation then drops well below 2/3 at high OH$^-$ concentrations (Figure 8). At least some of the oxygen is then formed through a direct discharge, probably of $OH^-$ ions. We will return to this question near the end of Section 2.3.6.

### 2.3.6. Anodic Chlorate Formation: Kinetics and Mechanisms

The kinetics of the anodic oxidation reaction has been studied by various authors.[24,25,81–87] Equation (6) can be written with HClO instead of $ClO^-$ [Eq. (6′)]. Indeed, $H_3O^+$ ions are generated both by the anodic reaction and the hydrolysis of $Cl_2$ [Eq. (3)] so that the diffusion layer is expected to be acidic. However, in concentrated NaCl solutions a buffering of the solution may take place. HClO reacts at potentials close to that of $Cl^-$ discharge. With graphite anodes HClO is oxidized before the $Cl^-$ ions in dilute NaCl solutions but not in concentrated ones.[25] On the other hand $ClO^-$ is oxidized before HClO and thus reacts at potentials much less positive than chlorine generation.[25,60,88] According to Despić et al.[87] the electrochemical oxidation of $ClO^-$ involves as an intermediate step the removal of an electron from the $ClO^-$ ion and the formation of a ClO radical. In the overall reaction mass transport plays an essential role. In a thorough discussion de Valera[21] interpreted from this viewpoint the early measurements of Foerster and Müller[88] and Hammar and Wranglén[89] reinvestigated the problem experimentally using solutions containing various amounts of hypochlorite. They found that the current efficiency for oxygen evolution is proportional to the hypochlorite concentration (Figure 4) and the same is true of the rate of anodic chlorate formation if Foerster's stoichiometry [Eq. (6)] applies. De Valera[21] as well as Hammar and Wranglén[89] came to the conclusion that the anodic chlorate formation is controlled by the diffusion of $ClO^-$ (or HClO) from the bulk solution toward the anode. Figure 5 shows for *concentrated* NaCl solutions (4 $M$) the evolution of the concentrations of chlorate and hypochlorite with time. Curves 1 and 2 correspond to two different initial $ClO^-$ concentrations of 0 and 0.175 $M$, respectively. The concentration of $ClO_3^-$ increases continuously, whereas after some time a steady state is reached for the $ClO^-$. The concentration of the latter in the steady state is independent of whether one starts from low or high $ClO^-$ concentration. Since the experimental conditions (low temperature) were such that the chemical chlorate formation was negligible, the steady state has to be interpreted as being due to equal rates of formation of $ClO^-$ through reactions (3) and (4) and of consumption through the anodic oxidation (6). The rate of chlorate formation (which is given by the slope of the broken line of Figure 5) is seen to vary according to the hypochlorite concentration, in agreement with the mass transport model. However, at *low* NaCl concentrations (0.1 $M$) the rates of chlorate formation were found to be 7–65 times *higher* than expected[24] (Table 1). Ibl and Landolt[24,25] interpreted this by a kinetic model. The hydrolysis of chlorine is a fast reaction. One can thus assume the process to take place essentially very close to the anode, in the inner, virtually convection-free region of the diffusion layer. The hydrolysis

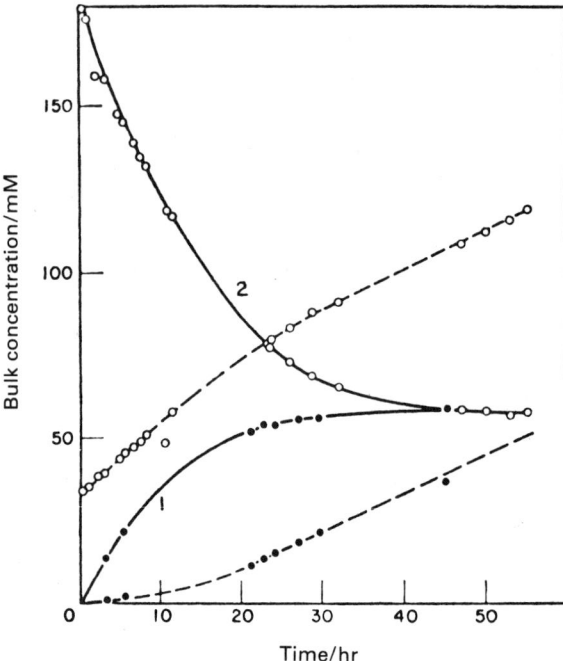

Figure 5. Evolution of bulk concentration of hypochlorite and chlorate with time in the electrolysis of 4 $M$ NaCl solutions (pH 9) with graphite anodes, at $j = 5 \text{ mA/cm}^2$ (Reference 25). ●, No hypochlorite initially present; ○, initial concentration of hypochlorite: 175 m$M$. ——, Hypochlorite; - - -, chlorate.

reaction (3) being of the first order the steady state profile of the chlorine concentration $c_1$ obeys the differential equation

$$D_1 \frac{d^2 c_1}{dy^2} - k_h c_1 = 0 \quad (25)$$

For the HClO concentration $c_2$ we have†

$$D_2 \frac{d^2 c_2}{dy^2} + k_h c_1 = 0 \quad (26)$$

$k_h$ is the rate constant of the hydrolysis reaction (3), $D_1$ and $D_2$ are the diffusion coefficients of $Cl_2$ and HClO, respectively. Since it is not well known how far HClO is dissociated in the diffusion layer $D_2$ must be in fact regarded as an average diffusion coefficient. The negative and positive signs in the Eqs. (25) and (26) reflect that chlorine is consumed, and HClO generated in the diffusion layer. The link between the electrolysis current and the interfacial concentration gradient on one hand, the fact that the concentration of $Cl_2$ drops to zero within the diffusion layer on the other hand, provide the boundary conditions for the integration of Eq. (25). Owing to the low oxidation potential

† In reality, $c_2$ represents the sum of the concentrations of HClO + ClO⁻ which one must regard as a single entity. However, in the above model the solution close to the anode is acid so that $c_2$ is there virtually the concentration of HClO.

Table 1

Chlorate Formation: Comparison of Calculated and Experimental Values

| Cl⁻, mol/liter | $\delta,^a$ cm × 10⁻³ | Flow velocity, cm/sec | $j$, mA/cm² | pH | HClO + ClO⁻, mmol/liter | $N_e^b$, mol/cm² sec × 10⁻⁹ | $N_d$, mol/cm² sec × 10⁻⁹ | $N_h$, mol/cm² sec × 10⁻⁹ | $N_h'$, mol/cm² sec × 10⁻⁹ | $N_e/N_h$ | $N_e/N_d$ | $N_e/N_h'$ |
|---|---|---|---|---|---|---|---|---|---|---|---|---|
| 0.052 | 18 | 13 | 5 | 2 | 0.14 | 5.16 | 0.08 | 4.76 | | 1.1 | 64.5 | |
| 0.083 | 7.4 | 44 | 5 | 9 | 1.0 | 8.52 | 1.35 | 7.97 | | 1.1 | 6.3 | |
| 0.053 | 2.4 | 153 | 5 | 9.2 | 0.25 | 7.05 | 1.04 | 4.35 | | 1.6 | 6.8 | |
| 4 | 21 | 13 | 5 | 8.9 | 64 | 14.94 | 30.5 | 57.8 | 16.98 | 0.26 | 0.49 | 0.88 |
| 4 | 14 | 29 | 5 | 10 | 40 | 14.22 | 28.6 | 59.6 | 16.16 | 0.24 | 0.50 | 0.88 |
| 4 | 10 | 46 | 10 | 10.6 | 55 | 25.68 | 55 | 145 | 29.2 | 0.18 | 0.47 | 0.88 |

[a] $\delta$ is the thickness of the diffusion layer in the absence of a chemical reaction.
[b] $N$ is the interfacial flux density of hypochlorite, i.e., three times the rate of anodic chlorate formation: $N_e$, experimental value; $N_d$, calculated for pure diffusion; $N_h$, $N_h'$, calculated for coupling of diffusion with homogeneous chemical reaction, from Eqs. (29) and (31), respectively.

of ClO⁻ and HClO it is assumed that reaction (6) proceeds at the limiting rate. The boundary conditions for the integration of Eq. (26) thus are

$$c_2 = 0 \quad \text{at } y = 0, \qquad c_2 = c_{2c} \quad \text{at } y = \delta$$

where $\delta$ is the thickness of the diffusion layer in the absence of a chemical reaction and $c_{2c}$ is the steady state concentration which builds up in the bulk solution (Figure 5).

One obtains for the concentration profiles of $Cl_2$ and HClO, respectively,

$$c_1 = (j\varepsilon/2FD_1 a) \exp(-ay) \qquad \text{with } a = (k_h/D_1)^{1/2} \tag{27}$$

$$c_2 = (j\varepsilon/2FD_2 a)[1 - \exp(-ay)] + (y/\delta)\{c_{2c} - (j\varepsilon/2FD_2 a)[1 - \exp(-a\delta)]\} \tag{28}$$

where $j$ is the current density, and $\varepsilon$ the current efficiency for chlorate production. Two of the concentration profiles given by Eq. (28) are shown in Figure 6. The broken line represents the profile which one would have if the hydrolysis (3) would take place in the bulk and not in the diffusion layer so that the hypochlorite would be transported to the anode through the whole diffusion path $\delta$ of the usual diffusion layer. The generation of HClO within the

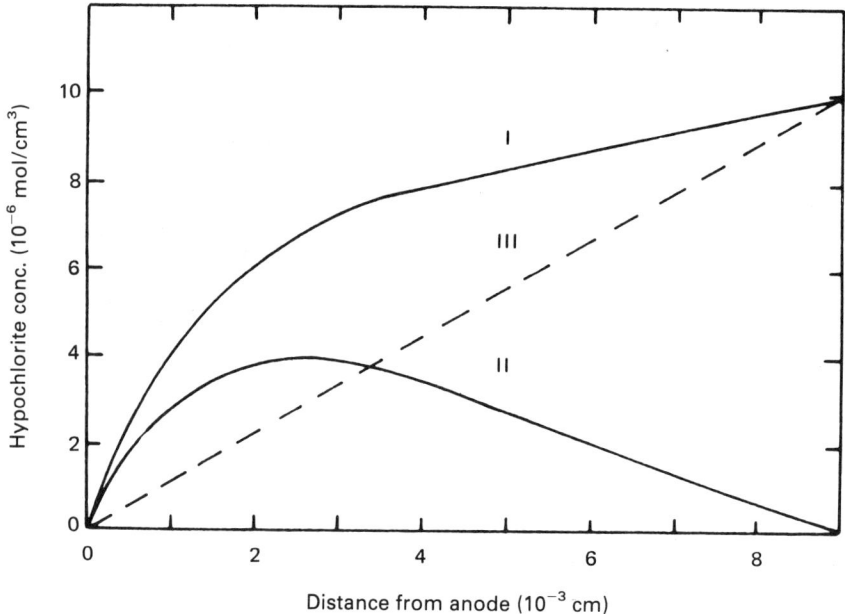

Figure 6. Concentration of hypochlorite in diffusion layer. Curves I and II: calculated from Eq. (28) (diffusion coupled with homogeneous reaction), for $c_{sc} = 10^{-5}$ and 0 (mol/cm³), respectively. Curve III: diffusion-controlled process. The curves are for $j = 20$ mA/cm² and $\delta = 0.09$ mm, which corresponds to experiments of Landolt and Ibl.[24]

# INORGANIC ELECTROSYNTHESIS

diffusion layer makes the interfacial concentration gradient steeper and the rate of hypochlorite oxidation higher. The interfacial flux density $N_h$ of HClO, which is proportional to the rate of anodic chlorate formation, is given by

$$N_h = \frac{j\varepsilon}{2Fa\delta}(1 - a\delta - e^{-a\delta}) - \frac{D_2 c_{2c}}{\delta} \quad (29)$$

Equation (29) can be written as

$$\varepsilon = \frac{1 - D_2 F c_{2c}/j\delta}{3/2 - [1 - \exp(-a\delta)]/2a\delta} \quad (30)$$

The values of $\varepsilon$ calculated from Eq. (30) can be compared with the experimental current efficiencies and the $N_h$ obtained from Eq. (29) can be compared directly to the experimental rates of electrochemical chlorate formation. The latter were measured by Landolt and Ibl[24,25] under well-defined hydrodynamic conditions, i.e., for well-defined values of $\delta$. The agreement was good for dilute NaCl solutions (0.1 $M$) but very poor for concentrated ones (4 $M$) (Table 1). The diffusion coefficients used for the calculation were the effective values, based on concentration gradients, measured by Chao.[90]

Jakšić has modified the treatment of Ibl and Landolt by introducing activities instead of concentrations into Eq. (26). He obtains instead of (29) and (30)[91–94]

$$N'_h = (j\varepsilon/2Fa\delta)[1 - a\delta - \exp(-a\delta)] - D_2 f c_{2c}/\delta \quad (31)$$

$$\varepsilon = \frac{1 - D_2 Ff c_{2c}/j\delta}{3/2 - [1 - \exp(-a\delta)]/2a\delta} \quad (32)$$

where $f$ is the activity coefficient of ClO$^-$. In the comparison of the calculated values with the experimental ones the authors used the aforementioned effective diffusion coefficients, measured by Chao.[90] Inserting a value of 0.1 for the activity coefficient of ClO$^-$ they found an approximate agreement (Table 1) with the measurements of Landolt and Ibl in concentrated NaCl solutions (4 $M$). However, it has been pointed out by these authors[95] that if Eq. (26) is written in terms of ClO$^-$ activities instead of concentrations the diffusion coefficients to be used are different from the effective values based on concentration gradients, measured by Chao. If one uses the latter ones Eq. (29) or (30) [and not the modified equations (31) or (32)] should be applied. Furthermore, Imagawa[62] has deduced from his kinetic measurements that the activity coefficient of ClO$^-$ is close to 1. We may conclude that in this model (which we will call the acid diffusion layer model) the introduction of activities instead of concentrations cannot explain the large discrepancies between calculated and measured values in the case of concentrated solutions (Table 1).

Landolt and Ibl[25] have proposed the following alternative model for concentrated NaCl solutions (which we will call the buffered diffusion layer model). At high $Cl^-$ concentrations the hydrolysis equilibrium [Eq. (3)] is shifted to the left. For NaCl 4 $M$, 20°C, and 20 mA cm$^{-2}$ the $H^+$ ion concentration calculated from the diffusion of $H^+$ ion away from the anode is 0.026 $M$ at the point of the maximum of the HClO profile (Figure 6). The *equilibrium* concentration of HClO at this point is then 20 times smaller than that calculated from the integration of Eq. (26). Therefore, the hydrolysis cannot run to the extent corresponding to this integration, and the $H^+$ concentration in the diffusion layer cannot build up sufficiently to ensure the diffusion of the $H^+$ ions *as such* toward the interior of the solution. They are carried away as HClO. The latter dissociates into $H^+$ and $ClO^-$ when it reaches further inward the region where the pH is maintained at a higher level through the $OH^-$ ions coming from the cathode. The $ClO^-$ thus formed diffuses backward toward the anode (Figure 7). One-third of it reacts at the anode to form chlorate and two-thirds react with the $H^+$ ions generated at or near the anode, thus buffering the diffusion layer. HClO is formed and serves as carrier for the transport of the hydrogen toward the bulk. Let us consider a situation where no chlorate is formed chemically in the bulk (owing to low temperature and high or low pH). If the above model applies, the rate of anodic chlorate formation should then be substantially smaller than the value $N_d$ calculated for the usual diffusion of $ClO^-$ through a diffusion layer of a thickness corresponding to the prevailing hydrodynamic conditions. This is roughly the case for concentrated NaCl

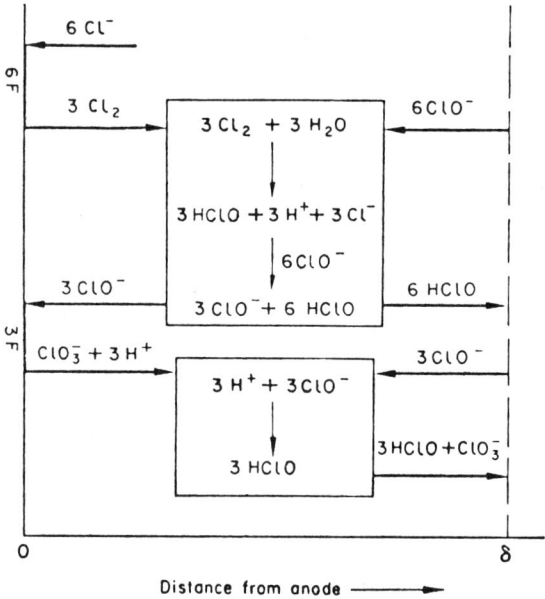

Figure 7. Model representing the reactions proceeding within the diffusion layer in the electrolysis of concentrated NaCl solutions according to Landolt and Ibl.[25]

solutions, as can be seen from Table 1, where the experimental values $N_e$ measured with graphite anodes are compared to the calculated ones. These results were confirmed by experiments in which an inert buffer (carbonate) was used. The ratio of the rate of anodic chlorate formation $N_e$ to the bulk hypochlorite concentration $c_{2c}$ was then several times larger than for pure NaCl solutions because the buffering action of the $ClO^-$ is taken over by the external buffer and all the $ClO^-$ diffusing toward the anode is available there for oxidation to chlorate. Similar results were obtained by Foerster (see Figure 8).

Note that the above model assumes that the $ClO^-$ reacts at the limiting rate, i.e., at zero interfacial concentration. This premise seems to be fulfilled at graphite electrodes but it is questionable whether it is still so for platinum or coated titanium electrodes. Indeed, smaller rates of anodic chlorate formation were measured with platinum electrodes,[96] and some differences in the steady state $ClO^-$ concentrations were observed depending upon whether platinum black or bright platinum was used.[20a] Jakšić et al.[97] found that addition of foreign anions ($CrO_4^{2-}$, $SO_4^{2-}$, $CO_3^{2-}$, $F^-$) substantially decreases the current

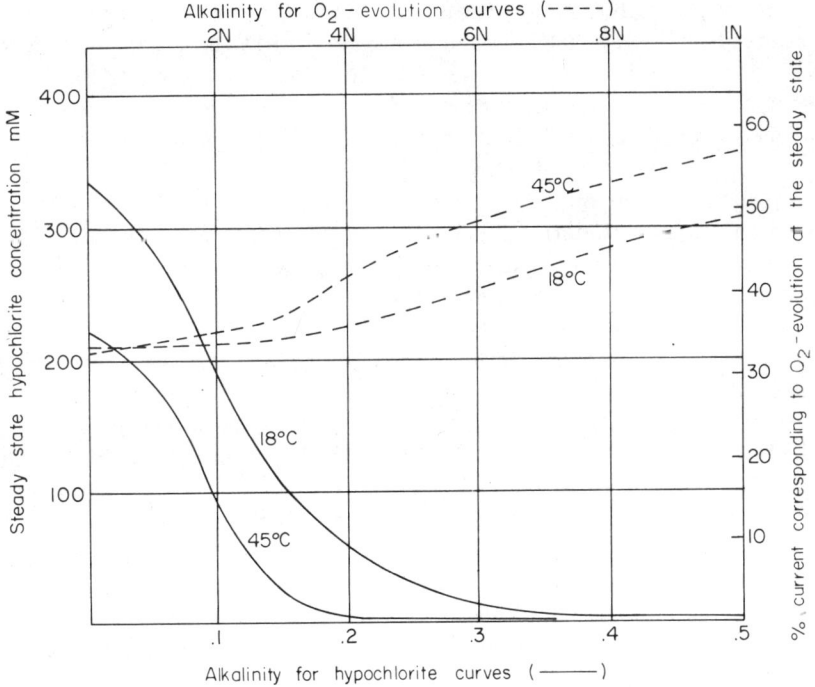

Figure 8. Steady state bulk hypochlorite concentration and current efficiency of $O_2$ evolution as a function of the alkalinity of the solution.[20e]

efficiency, i.e., increases the rate of anodic chlorate formation and the effect increases with increasing concentration of a foreign salt. This can be possibly interpreted by the fact that the potential for $Cl_2$ evolution is lower with the platinum and coated titanium electrodes, so that the $ClO^-$ is no longer oxidized at its limiting rate under normal conditions. The foreign anions tend to restore the situation where the $ClO^-$ is oxidized at the limiting rate.

Note that at the high current densities and high current efficiencies of modern chlorate cells the model of the buffered diffusion layer may also have to be modified for another reason. One may envisage that under these conditions the concentration of dissolved molecular chlorine in the vicinity of the anode rises to such a value that the oxygen bubbles departing from the anode contain substantial quantities of chlorine. Part of the active chlorine is thus transported to the bulk in the form of a gas, i.e., by a mechanism different from the one discussed so far. The chlorine then dissolves from the oxygen–chlorine bubbles into the solution. However, at very high current density this dissolution may no longer be complete and the gas contains chlorine which is lost for the production of chlorate, unless recovered from the exhaust gas.

In the foregoing discussion we have discussed mainly the electrolysis of solutions close to neutrality as used in industrial practice. Of theoretical interest are the experiments made with alkaline electrolytes. Figure 8 shows the steady state $ClO^-$ concentration of Fig. 5 as a function of the $OH^-$ ion concentration.[20e] The $ClO^-$ concentration in the bulk drops soon to zero with increasing alkalinity. However, chlorate formation is still going on together with increasing $O_2$ evolution. Foerster[20f] has interpreted these results by a so-called primary chlorate formation from $ClO^-$ reacting at the anode with electrolytically generated O [Eq. (23)].

However, in the model of the buffered diffusion layer (see above) there is no need to assume reaction (23). The formation of chlorate is explained by the buffering of the diffusion layer by the $OH^-$ ions, which makes unnecessary the build up of $ClO^-$ in the bulk.[25] We may note the high current efficiency for $O_2$ evolution at high alkalinities (Figure 8). This is due to the fact that with increasing $OH^-$ concentration the discharge of the latter competes more and more with the $Cl^-$ discharge. Quite generally it should be pointed out that the potentials of the various possible anode reactions [discharge of $Cl^-$, Eq. (1), oxidation of $ClO^-$ [Eq. (6)], or of $HClO$, Eq. (6'), oxidation of $ClO^-$ through Eq. (23), or discharge of $OH^-$, Eq. (7')] are relatively close. Therefore, the material of the electrode and its preparation, which influence the respective overpotentials, may considerably shift the relative rates of the above electrode reactions and, as a consequence, those of the subsequent processes taking place within the solution. In addition, the other electrolysis parameters (current density, composition of solution, pH, temperature, hydrodynamic conditions), obviously also play an essential role. One may thus obtain a great variety of results, depending on the circumstances.

## 2.4. Discussion of the Factors Relevant for Industrial Chlorate Electrolysis

### 2.4.1. Generalities

The foregoing fundamentals are of importance as a basis for the optimization of chlorate electrolysis. Minimization of the overall cost is quite generally an important aim of electrochemical engineering. Optimization consists in a choice of the operational variables such that the consumption of energy and material (both for the running operation and for the investment) is minimized. The main operational parameters are (1) volume of chemical reactor, (2) temperature, (3) current density, (4) electrolyte flow rate, (5) interelectrode distance, (6) electrode material, (7) pH, and (8) composition of electrolyte. Another important parameter is the hypochlorite concentration. However, this is not an independent variable since its value is determined by the setting of the other aforementioned parameters. The influence of many of these parameters is interconnected (for a discussion see reference 71). A complete quantitative optimization, taking all these interactions into account would be a most complex enterprise. Apparently, in industry and in the literature, only partial optimizations, with respect to some variables taken separately, have been carried out so far. A number of calculations have been reported (see, for instance, References 26, 93, 98, and 111). But in many cases the optimization has been an experimental one because much of the data needed for a computation (which would be the aim of modern electrochemical engineering) are lacking. However, the fundamentals discussed above can provide a guide and it is in that sense that we will discuss in the following the influence of the aforementioned parameters. It should be noted that the conclusions which can be drawn depend upon whether one accepts for the concentrated solutions the model of the acid or that of the buffered diffusion layer. Jakšić[26] has calculated the optimum flow velocity with the first model.

### 2.4.2. General Equations for Current Efficiency

The current efficiency is an important quantity because it much affects the energy consumption. We consider the arrangement of Figure 2, assuming that the chemical chlorate formation in the electrolysis cell and cathodic losses are negligible. We assume further that the electrochemical and chemical reactors can be approximated as perfectly mixed systems with average concentrations of $HClO + ClO^-$ of $c_{2c}$ and $c_{2h}$, respectively. A mass balance then yields the relationship derived earlier [see Eq. (24)], which for Foerster's stoichiometry ($b = \frac{2}{3}$) can be written as

$$\varepsilon = \tfrac{2}{3} + \tfrac{2}{3}F\dot{V}_L I^{-1}(c_{2c} - c_{2h}) \tag{33}$$

On the other hand the number of moles of $ClO^-$ and $HClO$ consumed per unit time in the chemical reactor through the chemical chlorate formation [Eq. (5)] is

$$\dot{V}_L(c_{2c} - c_{2h}) = V_r(dc_{2h}/dt) = V_r f(c_{2h}) = V_r B c_{2h}^3 \tag{34}$$

where $V_r$ is the volume of the chemical reactor. The function $f(c_{2h})$ and the value of $B$ are given by the rate equation (18). Combining Eqs. (33) and (34) one obtains

$$\varepsilon = \tfrac{2}{3} + \tfrac{2}{3}(V_r/I)FBc_{2h}^3 \tag{35}$$

Finally, we can make use of the fact that in concentrated NaCl solutions the anodic formation of chlorate through reaction (6), i.e., the anodic loss through oxygen evolution, is proportional to the average hypochlorite concentration $c_{2c}$ in the electrochemical reactor (see Section 2.3.6 and Figure 4):

$$1 - \varepsilon = AFk^* c_{2c}/I \tag{36}$$

where $k^*$ is a proportionality factor related to the mass transport coefficient† and $A$ is the anode area. Combination of Eqs. (33) and (36) yields

$$\varepsilon = \tfrac{2}{3} + \tfrac{2}{3}\dot{V}_L A^{-1}[(1-\varepsilon)/k^* - F(A/I)c_{2h}] \tag{37}$$

In Eqs. (35) and (37) $V_r/I$ and $A/I$ are the inverse of the current concentration and of the current density, respectively. It is seen that the current efficiency depends on *both* of these quantities. The first dependence is due to the fact that the loss-free (chemical) chlorate formation [Eq. (5)] proceeds in the *volume* of the solution, the second dependence is linked with the fact that the anodic loss reaction [electrochemical chlorate formation through reaction (6)] takes place at the *surface* of the anode.

Combination of Eqs. (35) and (37) eliminates $c_{2h}$ and yields a relationship of the form

$$\varepsilon = f(I/V_r, I/A, \dot{V}_L/A) \tag{38}$$

This equation allows one to calculate the current efficiency from the relevant variables, including volumetric flow rate $\dot{V}_L$ and temperature $T$, provided that one knows (a) the dependence of the rate constant $B$ on $T$ and (b) the dependence of the coefficient $k^*$ of Eq. (36) on $\dot{V}_L$ and $T$. Jakšić[26] has calculated $\varepsilon$ as a function of $\dot{V}_L$ by means of a relationship similar to Eq. (38) and found that $\varepsilon$ goes through a maximum. However, he did not use Eq. (36) to express the relationship between $c_{2c}$ and $\varepsilon$ and assumed instead that the acid diffusion layer model (see Section 2.3.6) is applicable to concentrated as well as to dilute NaCl solutions.

In the following we will refrain from solving Eq. (38) numerically and will instead discuss qualitatively the influence of the main operational parameters of chlorate electrolysis. Indeed, the assumption of a perfectly mixed reactor is

† According to the buffered diffusion layer model, applicable to concentrated NaCl solutions (see Section 2.3.6) $k^*$ is approximately proportional to the mass transfer coefficient.

not well realized in industrial cells (especially not in the electrolytic reactor where considerable concentration differences from inlet to outlet may occur). However, this will hardly affect the qualitative behavior of the system as discussed below.

### 2.4.3. Influence of Electrolyte Volume and of Temperature

The rate coefficient $B$ in Eq. (34) increases with increasing temperature. Therefore, from Eq. (35) it is seen that at constant cell current $I$ an increase of the temperature $T$ or of the volume $V_r$ of the chemical reactor causes an increase of $\varepsilon$, unless $c_{2h}$ would decrease sufficiently to compensate the influence of $B$ or $V_r$. However, a decrease of $c_{2h}$ would either decrease $c_{2c}$ (at constant $c_{2c} - c_{2h}$) or increase $c_{2c} - c_{2h}$ (at constant $c_{2c}$). In the first case, $\varepsilon$ must then increase because of Eq. (36). A mass balance for the electrochemical reactor shows that (at constant $\dot{V}_L$ and $I$) the second case also corresponds to an increase of $\varepsilon$. In reality, therefore, both $c_{2c}$ and $c_{2h}$ will decrease, but the latter will not decrease enough to prevent an increase of the amount of hypochlorite converted to chlorate in the chemical reactor. We may conclude that an increase of temperature or of chemical reactor volume improves the current efficiency. However, there is a limit to this because $c_{2h}$ can obviously not drop below zero. Therefore, $c_{2c}$, and thus also $1 - \varepsilon$ [see Eq. (36)], keeps a finite value. Thus if one increases the reactor volume $V_r$ or the temperature more and more the current efficiency tends toward a maximum. The above conclusions are in agreement with experimental results of Jakšić et al.,[76] who measured $\varepsilon$ as a function of temperature (Figure 9). The current efficiency increased from 85% to 97% when the temperature was raised from 15 to 80°C but the increase leveled in the range from 60 to 80°C.†

The increase of $V_r$ is beneficial for the current efficiency but it also increases the investment. From the viewpoint of cost there are thus two effects which act in opposite directions. Furthermore, we have seen that there is a limit to the increase of $\varepsilon$ with increasing $V_r$. Therefore, the reactor volume $V_r$ has an optimum value at which the overall cost is minimized.

† It is to be noted that the influence of temperature is not as straightforward as that of the reactor volume $V_r$ because an increase of temperature increases the diffusion coefficient and thus the factor $k^*$ in Eq. (36). This increases at constant $c_{2c}$ the loss $1 - \varepsilon$, as has been experimentally found by Hammar and Wranglén.[89] In general, however, the decrease of $c_{2c}$ appears to overcome this effect but it is possible that under certain circumstances the function $\varepsilon = f(T)$ goes through a maximum. In fact, with increasing temperature a decrease in the current efficiency, accompanied by an increasing current lost on $O_2$, was observed by Shlyapnikov et al.[281] It may be noted that the proportionality between $1 - \varepsilon$ and $c_{2c}$ is well established for graphite electrodes but no information seems available about the situation for the case of oxide-coated titanium anodes. The latter substantially decrease the overpotential for chlorine evolution and thus tend to shift the anodic chlorate formation away from a diffusion-controlled process (with $c_2 = 0$ at the interface). On the other hand, industrial titanium anodes are operated at higher current densities than graphite, which would tend to restore the limiting current situation for the anodic chlorate formation. No experimental results seem available which would allow a statement regarding the influence of the coated titanium on the relationship governing the anodic chlorate formation.

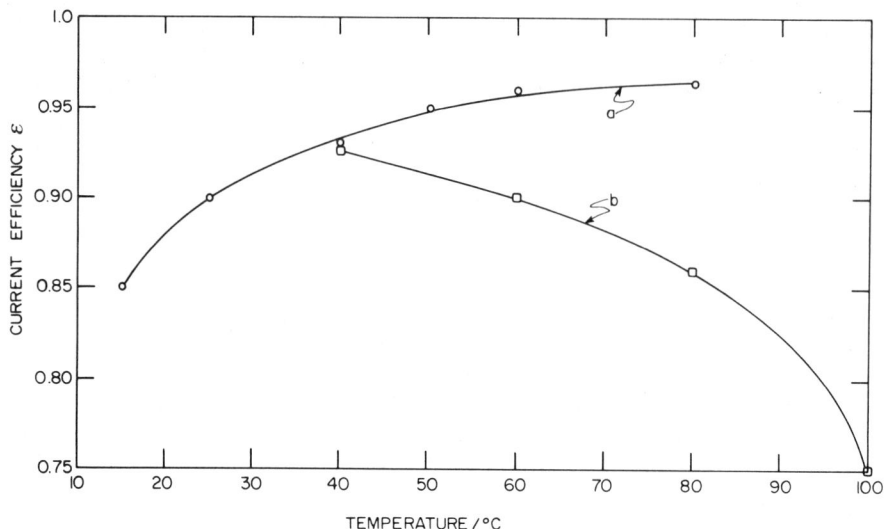

Figure 9. Dependence of current efficiency of chlorate formation on temperature. (a) NaClO$_3$, Pt anode, pH = 6[76]; (b) KClO$_3$, RuO$_2$ anode.[281]

Instead of discussing the influence of $V_r$, as we have done here one could also lump this influence together with that of $I$ and discuss the influence of the current concentration $I/V_r$, which is a quantity often used in the earlier literature.[31] However, this influence is more complex than that of $V_r$ and more difficult to deal with qualitatively. Modern (high-temperature) systems are designed for values of $I/V_r$ of the order of 7–10 A/l at 80°C,[112,119] or even higher (up to 50 A/l), whereas older systems were operated at 3 A/l (35°C).[120]

The optimum value of $V_r$ depends on the temperature. An increase of the latter tends to increase $\varepsilon$ and/or allows one to work with a smaller $V_r$. However, it may cause additional heating cost. Furthermore, there is the problem of the durability of the electrode material. With graphite anodes the temperature could not be raised above some 40°C because otherwise the wear losses become excessive. With Pt or DSA electrodes the temperature limitation is virtually removed. With the system designs of recent years 80°C seems to be the optimum. A higher temperature of 105–115°C has been proposed in a patent by Holmes (Mathieson),[113] and experimental results have been reported by Newberry et al.[49] The system needs no refrigeration of the outgoing stream because cooling is provided by vaporization, which also concentrates the solution and facilitates the separation of the chlorate by crystallization. However, it does not seem that this process has been applied industrially, probably because of corrosion problems. As calculation based on the equations of Maxwell or Bruggeman shows, above 80°C the generation of steam enriched

# INORGANIC ELECTROSYNTHESIS

bubbles increasingly raises the resistance of the solution and causes a disturbing nonuniform current distribution.† With modern cell operation current efficiencies as high as 95% can be achieved.

### 2.4.4. Influence of Current Density

An increase of current density tends to increase the number of moles of hypochlorite to be converted to chlorate in the chemical reactor. At constant $V_r$ and $T$ it thus tends to increase $c_{2h}$. On the other hand, a mass balance for the electrolytic cell shows that an increase of current density tends to increase $c_{2c} - c_{2h}$. Therefore, $c_{2c}$ tends to increase, but the effect on $\varepsilon$ in Eq. (36) is overcompensated by the increase in $I$. Indeed by combining Eqs. (35) and (37) it can be shown that an increase of c.d. tends to increase $\varepsilon$. ‡

However, an increase of current density increases substantially the energy consumption by making the ohmic losses in the solution between the electrodes larger. But the increase of current density has also a beneficial effect (in addition to the increase in $\varepsilon$): it decreases the electrode area needed for a given production and thus the investment cost. Therefore, there is an optimum current density at which the sum of the investment and energy cost is minimum.[114] For the Krebs process[115] (rated at 3 kA m$^{-2}$) with Pt–Ir-clad electrodes, the optimum current density is 6 kA m$^{-2}$ for an energy price of 1 U.S. cent per kWh.† Today's higher energy cost tends to decrease, while the expensive Pt and DSA anodes tend to increase the optimum current density. However, at increasing current density factors other than the decrease in electrode size or the increase of energy consumption may substantially influence the cost. For instance, with graphite electrodes current densities higher than 0.6 kA m$^{-2}$ can hardly be applied because of the electrode wear. With DSA electrodes high current densities are attainable without problems, but other effects arise such as losses through the formation of perchlorate at too positive potentials. Nevertheless, the trend of the last two decades has been toward the use of higher current densities in industrial electrolysis[116,117] (see Table 2).

### 2.4.5. Influence of Flow Velocity

The influence of flow velocity is less straightforward than that of the reactor volume or of the current density. An increase of flow velocity has two effects acting in opposite directions: (a) It accelerates the mass transport and therefore increases $k^*$, so that according to Eq. (36) $\varepsilon$ tends to decrease, and (b) it increases the volumetric flow rate $\dot{V}_L$; at constant $I$ according to Eq. (33)

---
† Unpublished calculation of P. Gallone, University of Genoa, Genoa, Italy.
‡ It may be noted that an increase in current density also increases the rate of gas evolution and thus accelerates the mass transfer. The result is an increase in the rate of the electrochemical chlorate formation and thus a further decrease of $\varepsilon$.

Table 2
Typical Operational Data for Chlorate Cells[129,139,315]

| Anodes | Graphite | Coated titanium |
|---|---|---|
| Cell potential, V | 2.9–3.8 | 2.9–3.3 |
| Current density, $A/m^2$ | 300–600 | 1500–4000 |
| Current, kA | 6–30 | 6–100 |
| Current efficiency | 0.82–0.87 | 0.92–0.95 |
| Temperature, °C | 40–45 | 60–80 |
| pH | 6–7 | 6–6.5 |
| Concentration, g/liter | | |
| NaCl | 310–100 | 310–50 |
| $NaClO_3$ | 0–500 | 0–650 |
| $Na_2Cr_2O_7$ | 1–6 | 1–6 |
| Electricity consumption, kAh/t $NaClO_3$ | 1740–1840 | 1590–1640 |
| Energy consumption, kWh/t $NaClO_3$ | 5000–7000 | 4600–5400 |
| Anode wear, kg/t $NaClO_3$ | 7–18 | $(0.1–0.5) \times 10^{-3}$ |
| Current concentration, A/l | 2–6 | 20–50 |

$c_{2c} - c_{2h}$ thus tends to decrease and $c_{2c}$ does likewise, so that according to Eq. (36) $\varepsilon$ tends to increase. Either the first or the second effect predominates depending on the value of $c_{2c} - c_{2h}$. Under the conditions prevailing in industrial cells the first effect is probably the more important one.

It is to be noted that $\dot{V}_L$ may not be an independent variable. This is the case in most modern cells where the circulation of the solution is not ensured by a mechanical pump but by a gas lift, caused by the electrolytically evolved gas[34,35,139] or by gas introduced from outside.[121] The flow velocity thus increases with increasing current density and with increasing length of the vertical pipe connecting the electrochemical cell with the chemical reactor. This velocity can be calculated from the buoyancy force and the friction losses in the pipes and reactors. Typical flow velocities are 0.2–1 m/sec.[122]

### 2.4.6. Influence of Interelectrode Distance

At constant volumetric flow rate $\dot{V}_L$ a decrease of the interelectrode distance increases the linear velocity $v$ and thus tends to accelerate the loss reaction (6), i.e., it tends to decrease the current efficiency $\varepsilon$. However, a decrease of the interelectrode distance has the great advantage of decreasing the ohmic losses in the solution, i.e., of decreasing the energy consumption. One thus ususally tries to achieve a good current efficiency by adjusting the other parameters (expecially $V_r$ and $T$) and to make the interelectrode distance very small. With graphite anodes this was difficult to realize, but distances of 3–5 mm are common today when metal anodes are applied. Even smaller distances (0.8–1.6 mm) have been claimed.[118] The decrease in electrode distance is limited by several factors: The filling of the interelectrode space with gas bubbles results in an increase of cell resistance after the distance has dropped below an optimum value. The resistance of bubble-filled electrolyte can be calculated from the Bruggeman equation. Another limiting factor is the design problem of how to keep narrow electrode distances from endangering a safe operation of the cell by shortcircuits, especially by solid contaminations of the electrolyte. Finally, the increase in pressure drop by reduced electrode distance has to be considered carefully.

### 2.4.7. Influence of pH

The pH influences the rate of the chemical chlorate formation [Eq. (5)] because it requires the simultaneous presence of HClO and ClO⁻ and because the ratio $c_{HClO}/c_{ClO^-}$ depends on pH owing to the dissociation equation (4). For the reasons indicated in Section 2.3.3 the rate of reaction (5) should be as large as possible. Differentiating Eq. (18) with respect to $a_{H_3O^+}$ and equating to zero $[d(dc_2/dt)/da_{H_3O^+} = 0]$ yields the optimum pH for the maximum rate of dismutation:[68]

$$pH_{opt} = -\log 2K^* \qquad (39)$$

The dependence on the temperature $T$, derived from Eq. (19), may therefore be directly expressed by[70]

$$pH_{opt} = C - 0.0075\, T/°C \qquad (40)$$

The fact that a higher temperature requires a lower pH has currently been accepted in industrial practice.[35,112] However, the value of $C$ is questionable.† Experimental results of Jakšić et al.[71] lead to $C = 6.3$ and hence to optimum pH values of 6.0 and 5.7 for temperatures of 40 and 80°C, respectively. Similar optimum pH values were experimentally obtained by Shlyapnikov and Filippov.[78] Recently, the optimum values pH = 6.7 (40°C) and pH = 6.4 (80°C) were found.[281] It must be pointed out that, owing to Eq. (2) in modern closed-loop electrolyte systems as shown in Figure 2, the pH at the exit of the cell is usually considerably higher than the mean value in the chemical reactor. That is where the optimum pH value under discussion applies. Industrial cells are operated in the range of pH = 6–6.5. The pH is controlled or at least recorded.

### 2.4.8. Influence of Anode Material

There is also a certain influence of the anode material on the operating conditions. Apart from the fact that the material limits the electrolyte temperature (see Table 2), it also affects the admissible chloride concentration. Cells with graphite anodes are operated with chloride concentrations not below

---

† Note that the optimum pH depends upon whether one accepts Eq. (5) or (5′).

100 g/liter, whereas somewhat lower concentrations are acceptable with coated titanium anodes. Ruthenium-coated anodes exhibited a stable behavior if the NaCl concentration was maintained at 50–60 g/liter, but showed a considerable increase in potential if the concentration dropped below 30 g/liter.[131] This finding was confirmed recently.[367] The mechanism and sequence of degradation of the ruthenium-oxide-coated titanium anode in chloride-containing acidic solutions was interpreted as an anodic oxidation of $RuO_2$ and a subsequent chemical decomposition of $RuO_4$ or a formation of soluble chloro-complexes.[367]

It is generally known that the material and its surface condition influence the potential and thus affect the suitability to chloride discharge. The current efficiency of a chlorate cell alternatively equipped with anodes of various materials—$PbO_2$, $MnO_2$, Pt, and $RuO_2$—was studied.[110] It was found that the titanium anodes coated with platinum and ruthenium dioxide exhibited the distinctly superior yields. The four anode materials also showed different behaviors in view of the optimum pH and temperature of the electrolyte.[281]

Selection of anode coating materials is limited by the fact that some of the wear products act as catalytically decomposing on hypochlorite, e.g., iridium.[45]

## 2.5. Industrial Cells

Features common to nearly all industrial cells of the past decade are:

(1) use of titanium anodes with coating of noble metals or of their alloys or oxides,
(2) small and invariable electrode distance,
(3) high velocity of electrolyte flow through the cell, and
(4) separate chemical reactor, integrated into the electrolyte flow system.

Some typical designs will be discussed in the following.

Figure 10 shows the whole electrolyte system with the cell (a) (or electrochemical reactor), the chemical reactor (b) and the cooler (c) to remove the excess heat developed under steady state operation condition. The electrolyte is kept in motion by the gas lift caused by the electrolytically evolved bubbles, mainly consisting of hydrogen. A centrifugal pump as applied in older cells[123] is not required and energy is saved. The unipolar electrodes are formed in sheets. Cathodes are made of carbon steel. A sufficient cathodic polarization is required during shutdown, i.e., at least 15–70 $A/m^2$, depending on the temperature.[124,125]

Another cell with self-circulation of the electrolyte[34] is shown in Figure 11. The titanium anodes are surrounded by steel cathodes which act simultaneously as cooling surfaces and make a separate cooler superfluous.[117]

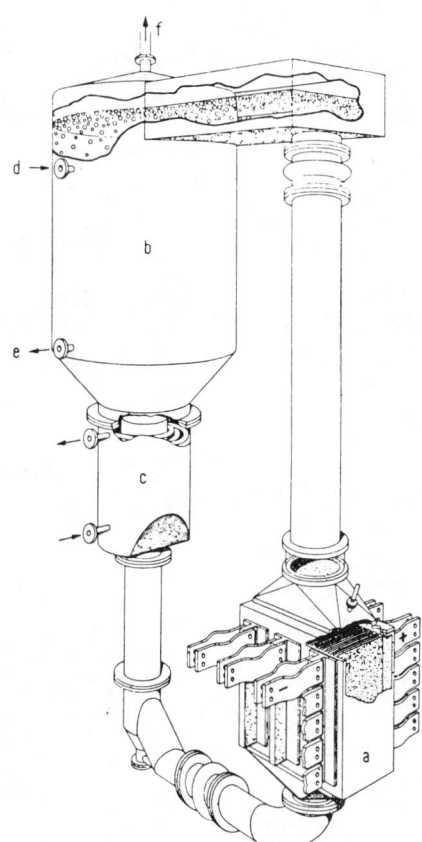

Figure 10. Industrial chlorate cell with natural bubble-induced electrolyte convection (Krebs & Cie, Paris).[139] (a) Cell, (b) chemical reactor, (c) cooler, (d) electrolyte feed, (e) electrolyte outlet, (f) hydrogen outlet.

Unipolar cells directly electrically interconnected without intercell bus bars[112,126,137] and combined with a common chemical reactor (Figure 12) reflect the strenuous efforts to substantially reduce capital investment.[127]

In an even more pronounced way this trend can be recognized in a novel cell[141] which has been tested on the industrial scale and which has no internal cooling coils nor a separate external reactor (Figure 13). The electrolyte volume required for the chemical reaction is incorporated in the cell tank, thus resuming older cell design. However, owing to the use of coated titanium anodes the temperature may be increased to 90°C where the required reaction volume is not large. The electrode distance is small and the current density can be kept at high values. Gas bubbles are removed from the interelectrode space through horizontal slots in the cathode.

Cells with diaphragms were again taken into consideration recently.[128] Chlorine gas and hydroxide are produced separately and reacted outside the cell to avoid the production of hypochlorite inside the cell with consequent anodic oxidation and cathodic reduction. The idea, however, is not new and is based on a proposal of Kershaw (Reference 13, p. 34).

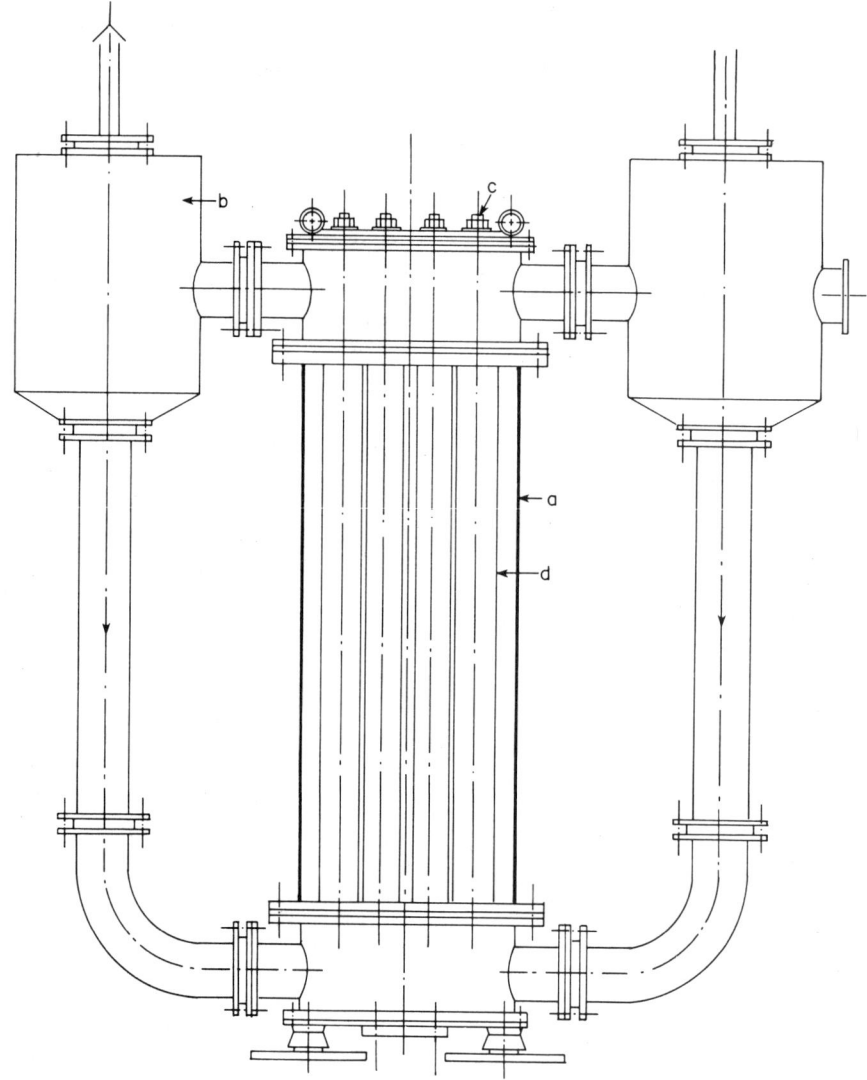

Figure 11. Chlorate cell with bubble-induced electrolyte circulation (Krebskosmo).[117] (a) Cell, (b) chemical reactor, (c) anode, (d) cathode.

The anode substrate material of modern cells is usually titanium. Variations in coating materials, compositions, and preparation methods are immense. Customary coatings consist of platinum, platinates, mixed oxides (e.g., titanium oxide and ruthenium oxide or indium oxide and rhodium oxide) with or without special conductive subcoatings such as tin oxide and antimony oxide.

Although none of the coatings is completely resistant, wear has been reduced to satisfactory values (0.1–0.5 g/t $NaClO_3$). In any case, the coating is thin (about 1 $\mu$m) and the wear only affects the coating, whereas the substrate material remains unattacked. Thus the interelectrode distance virtually does not change in the course of operation, whence the name "dimensionally stable anodes" given for commercially marketing this type of electrode. Small interelectrode distances are possible, allowing a compact cell design and, furthermore, high current densities (Table 2). A great advantage is the low overpotential of the noble-metal-coated anodes for $Cl^-$ discharge (see Chapter 2 on chlorine electrolysis) and the possibility to operate at high temperature. The latter is also favorable for the chemcial chlorate formation as well as for a subsequent chlorate crystallization stage.

Cathodes are currently made of mild steel or low-carbon steel. Dichromate is added to the electrolyte (3–7 g/liter) to avoid the cathodic reduction of hypochlorite[89,125] and—to a smaller extent—of chlorate.[133] Dichromate may be omitted when using chromium-plated steel cathodes[47]

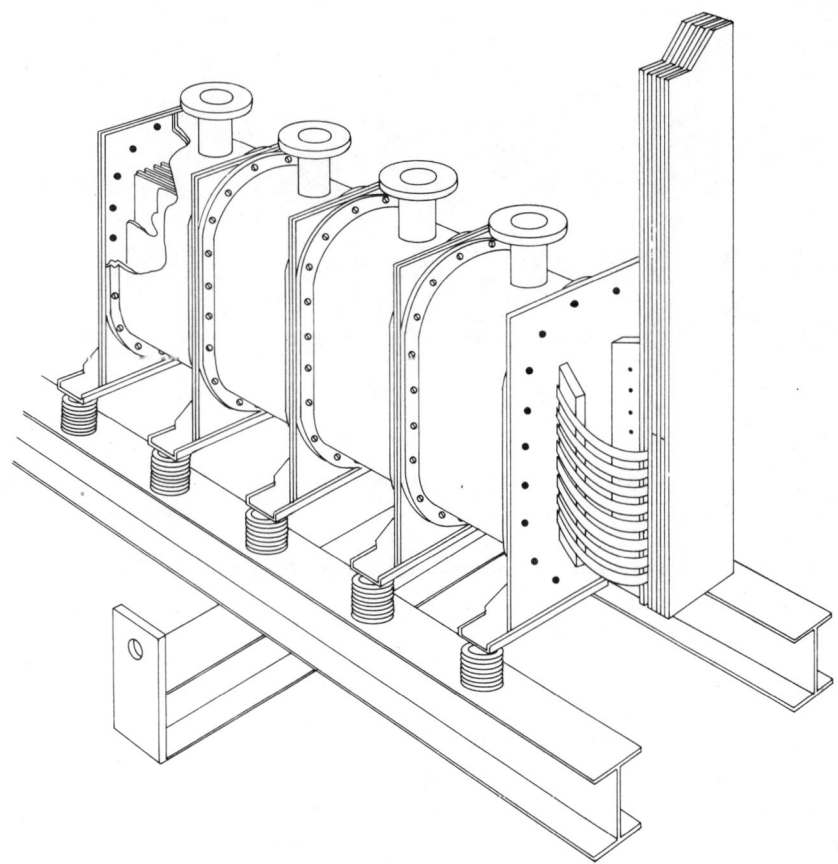

Figure 12. Directly connected unipolar chlorate cells without intercell bus bars (Chemetics).[127]

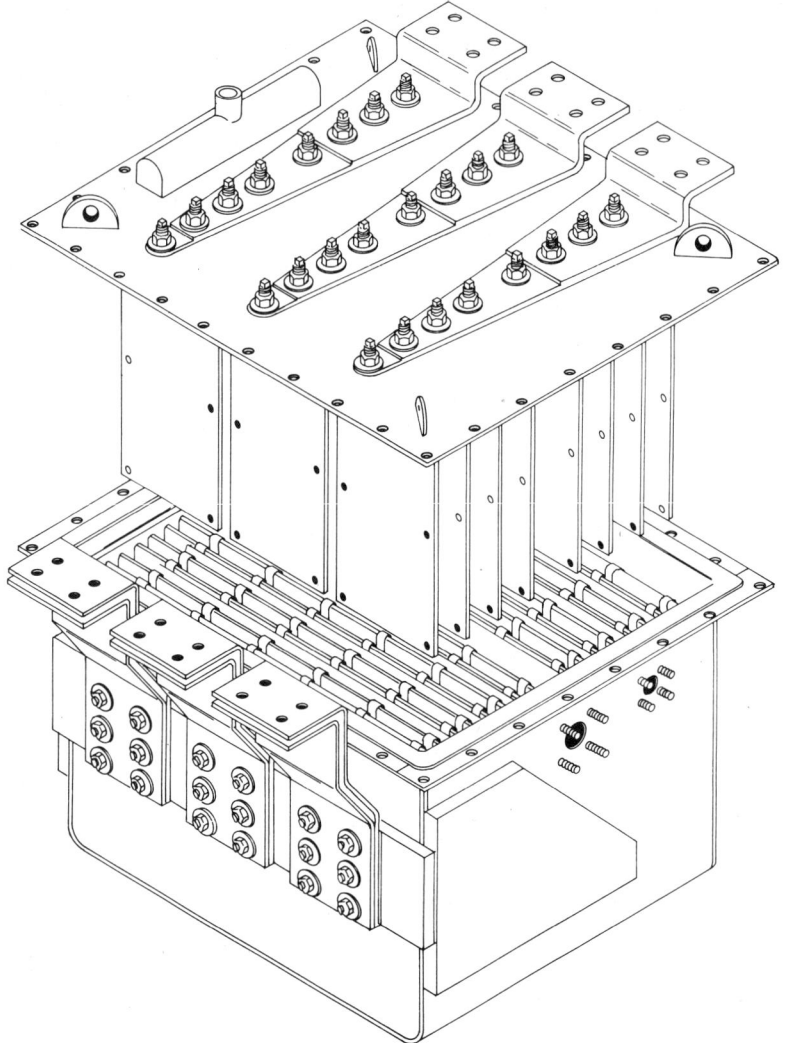

Figure 13. Chlorate cell with integrated chemical reaction volume (Pennwalt).

and chromium cathodes.[134] Negligible cathodic reduction in the absence of dichromate was also found[135] with austenitic steel cathodes, containing Cr, Ni, Mo, and Ti alloys. In any case, cathodes of carbon steel or iron are superior in regard to their lower overpotentials.

Titanium *cathodes* were formerly considered unserviceable[49] owing to the danger of diffusion of hydrogen into the coating substrate and consequent formation of titanium hydride making the cathode brittle. The result is a severe diminution of the coating life time as compared to unipolar anodes. There have been several attempts to prepare substrates to eliminate hydrogen diffusion, but none appear successful.[50] In context with the great technical importance of bipolar electrode arrangements,[140] titanium cathodes were recently investigated and did not exhibit a corrosion risk, especially at high temperatures.[51] Bipolar titanium electrodes are industrially applied[52] in spite of the higher coating loss.

## 3. Hypochlorite

### 3.1. Industrial Significance

Hypochlorite in aqueous solution of high and low concentrations has been used as a bleaching agent since the close of the eighteenth century, above all in the cellulose and textile industry. The bleaching action of the hypochlorous acid formed is based on either chlorination, oxidation, or chlorhydrination, and depends on the pH.[143]

The electrosynthesis of hypochlorite competes with the chemical production by reaction of gaseous chlorine in slight stoichiometric excess with sodium hydroxide. This process is thoroughly treated in Reference 143a. It is superior to the electrosynthesis in view of the high hypochlorite concentration attainable [170–220 g of available chlorine per liter† (av. $Cl_2$/liter) and even more if the residual chloride concentration is extremely low[144]].

Electrosynthesis cells without diaphragms exhibit considerable efficiency losses mainly due to both cathodic reduction and anodic oxidation of hypochlorite. As discussed in Section 2.3, the extent of these losses depends on the hypochlorite bulk concentration. That is why in industrial cells the hypochlorite concentration is limited to 10 g av. $Cl_2$/liter which, however, is sufficient for disinfection and common bleach processes. The value applies to cells operated with feed brine of 2–3% NaCl made from purchased salt. The situation is different if cells are fed with seawater. As salt is free, it is no longer necessary to make stronger hypochlorite solutions to optimize the use of NaCl. The final hypochlorite concentration can be kept low and amounts to only 0.4–2 g av. $Cl_2$/liter. As a consequence, losses are cut down and the total energy cost can be lowered noticeably, nearly reaching the value needed for the chemical route from chlorine and caustic (about 3–3.7 kWh/kg av. $Cl_2$).

Industrial hypochlorite solutions, being not free of catalytically active contaminations, undergo a homogeneously or heterogenously catalyzed decomposition,[145] which is detrimental mainly for solutions generated from seawater. Long time storage and transport over long distances must be avoided. Highly concentrated hypochlorite solutions are only prepared if necessary.

† The "available chlorine" indicates the oxidizing power of an agent as compared to elementary chlorine (Reference 143b).

These remarks outline the practical range of application: Electrosynthesis of hypochlorite is appropriate, where small concentrations are satisfactory and when for small capacities the greater expenditure of chloride and energy can be compensated for by the advantage to dispense with transport and storage of chlorine or of hypochlorite solutions. As the cells for electrosynthesis of hypochlorite can be operated with considerably lower safety risk and little maintenance by unskilled labor, they are most suitable for an on-site production. Many small plants were in operation during the first two decades of this century in laundries, but also in the pulp and textile industry. However, when solid calcium hypochlorite with high content of available chlorine and good stability in storage was introduced on the market the electrosynthesis of hypochlorite solutions was no longer competitive. With the rapidly expanding chlorine–caustic industry great quantities of cheap waste chlorine became available which could be used even at low concentration to prepare bleach liquor by absorption in alkaline solution. Consequently, the electrosynthesis of hypochlorite has been completely out of use in the last few decades.[146,147] The history of the electrosynthesis of hypochlorite would thus have been settled if it were not true that some years ago new fields of industrial application opened up, where the disadvantages of electrosynthesis are effectively compensated for.

Following the trend to replace chlorine by hypochlorite for the disinfection of drinking water and sewage—which can also be observed in municipal plants[148]—electrosynthesis cells of small capacities are now used for the disinfection of swimming pools and in the food industry.

A special advantage is the possiblity of using seawater directly as the electrolyte. Owing to the high amount of extraneous ions favoring catalytic decomposition, the application is restricted to cases where immediate consumption or short time storage of the generated solution is ensured. Thus, electrosynthesis is used to prevent fouling in condensers of ocean-going vessels. Cooling water circuits of coastal power stations are now often cleaned or kept free of fouling by use of hypochlorite. Its direct generation by electrosynthesis has replaced mechanical cleaning procedures. Municipal plants for sewage treatment have been erected on the seaside. Seawater is electrolyzed and the resulting hypochlorite solution is used for disinfection of sewage. The first of these plants was opened on the island of Guernsey in 1966. Input comes from cesspools and domestic sewage; it is freed of hydrogen sulfide gas and mixed with hypochlorite solution made in an electrolytic cell. The stream passes a disintegrator where solids are reduced to below 0.3 mm and enters agitated reactor vessels from where the odorless effluent is discharged to the sea.[149,150] On-site electrolytic production of hypochlorite is also used to deodorize vent air, which is contacted in scrubbers with the hypochlorite solution to oxidize or chlorhydrinize organic compounds.

A fast-growing major use for hypochlorite cells is the disinfection of seawater for injection into oil fields for secondary oil recovery. The largest plant for this purpose, opened in 1978 in Saudi Arabia, produces 5400 kg/day of available chlorine. The disinfected seawater is then filtered, deaerated, and pumped a distance of 100 km inland for injection.[339]

Owing to all these new applications, the electrosynthesis of hypochlorite solutions has been vigorously revived. The facilities initially installed were small and the production figures were modest. An estimate for the capacity of hypochlorite cells both installed and under construction is now 200 tons/day of av. $Cl_2$[339] and the figure is growing fast. The future is considered bright,[151] e.g., because impending legislation in a number of countries would forbid road transport of liquid chlorine, thus making *in situ* hypochlorite plants in power stations, for example, a virtual necessity.

## 3.2. Development of Theory and Technology

Hypochlorite solutions were prepared by electrosynthesis as early as 1801.[152,153] The discovery of hypochlorite is attributed[154] to Balard.[155]

The reactions involved were investigated at the turn of the century in connection with the preparation of chlorate.[156,157] At that time, under application of these results industrial cells were designed and constructed. A good review of the older technology is presented in References 158 and 159.

The anodes were of platinum or graphite, the cathodes of graphite. The arrangement was unipolar or bipolar. A cooler inside or outside the cell was provided for removal of heat. One of the most remarkable cells made use of the buoyancy forces of the cathodically evolved hydrogen to effect a desired self-circulation of the electrolyte through the interelectrode space.[160,320] This principle was much later extensively applied in chlorate synthesis.[34,141] Current density was 1500 $A/m^2$, temperature was 30–40°C. To prevent the cathodic reduction of hypochlorite, organic additives, calcium chloride, or dichromate were used. Although the cells exhibited an amazing maturity for that time and were working safely, the decay of hypochlorite electrosynthesis could not be stopped in view of the mentioned competition with the chemical production.

With the reappearance of hypochlorite electrosynthesis the cell construction was improved mainly by introducing the noble-metal-coated anodes which, as for chlorate manufacture, have the advantage of stable interelectrode distance and of a substantial reduction of capital investment as compared to massive platinum. On the other hand, as a result of increased current density the flow rate in the cell was increased, ensuring good mixing of anode and cathode products.

## 3.3. Reaction Fundamentals

The reactions are essentially the same as those involved in chlorate electrolysis, which have been discussed in Section 2.3. Primarily, chloride ions are discharged and the chlorine thus formed hydrolizes in the solution to HClO and ClO⁻. The hypochlorite and hypochlorous acid are not stable and are consumed in a number of loss reactions. The latter all tend to increase with increasing hypochlorite concentration and they thus limit the concentration that can be attained or maintained over a longer period of time. The main loss reactions are as follows:

(a) Catalytic decomposition by the action of substances such as Co, Ni, and Cu (see Section 2.3.2).

(b) Chemical chlorate formation by reaction (5) (see Section 2.3.4). The rate of this reaction depends on pH and temperature. It can be minimized by working at low temperatures in an adequate pH range.

(c) Cathodic reduction by reaction (9) (see Section 2.3.2). It can be prevented by inserting a diaphragm separating the cathode from the anode or minimized by addition of small amounts of chromic acid to the solution. The latter, however, is practically impossible if—as is the common arrangement in industrial plants—the cell is not integrated into a closed electrolyte circuit. For this reason relatively considerable cathode losses can occur in industrial cells (Figure 14).

(d) Anodic oxidation by reaction (6) (see Section 2.3.6). The mechanism of this reaction depends on the chloride concentration. In on-site plants (which are the most common ones today) this concentration is usually much smaller than in chlorate production. In the electrolysis of seawater it is of the order of 0.5 $M$. For graphite anodes one is then in the transition region between the validity of the model of the acid diffusion layer and that of the buffered diffusion layer.[24,25] In the latter case it is to be expected that the rate of the anodic loss reaction (6) is proportional to the concentration of the active chlorine (HClO + ClO⁻) in the bulk solution. Such a proportionality has been postulated by Heal, Kuhn, and Lartey in their computerized modeling of electrolytic hypochlorite production.[170]

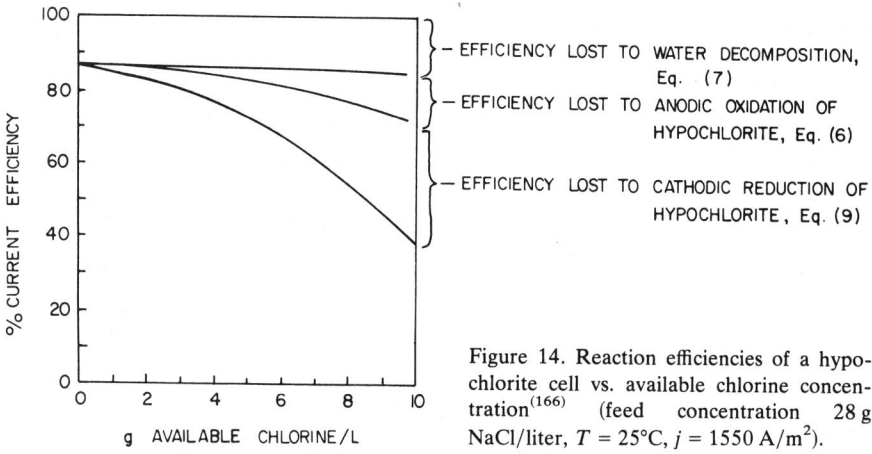

Figure 14. Reaction efficiencies of a hypochlorite cell vs. available chlorine concentration[166] (feed concentration 28 g NaCl/liter, $T = 25°C$, $j = 1550$ A/m²).

After the beginning of electrolysis the concentration of active chlorine increases and eventually a steady state is reached, in which, in the absence of external consumption, the generation of hypochlorite is compensated for by the above-mentioned loss reactions. The steady state value of this concentration depends on various parameters, among others on current density and, in the transition region between the models of the acid and buffered diffusion layer, also on chloride concentration.

An unwanted side reaction is the decomposition of water by reaction (7) (see Section 2.3.1). As in electrosynthesis of chlorate, the reaction competes with the anodic oxidation of hypochlorite. However, in hypochlorite cells the decomposition of water is much more pronounced since it is favored as the chloride concentration decreases and the common electrolyte is a weak brine.

### 3.4. Industrial Cells

For the reasons discussed in Section 3.3, the current efficiency strongly depends on the final concentration of available chlorine (Figure 14). The latter is always small. Solutions produced in cells operated with seawater do not exceed available chlorine concentrations of about 2 g/liter. If the solution is made of pure brine, the energy loss due to reduction and oxidation of hypochlorite on one hand and the salt utilization on the other hand balance economically at an optimum concentration of 8–10 g/liter.[166]

Specific energy consumption is also sensitive to the anode material. Graphite anodes with a lead dioxide coating as used in chlorate manufacturing are also industrially applied in making hypochlorite.[99,169] Lead dioxide anodes were found superior as compared to magnetite, graphite, and platinum coated on titanium with respect to their energy consumption.[151] The material is competitive with coated titanium since the investment cost is lower and lifetime is similar. However, the problem of electrolyte pollution by corrosion products (heavy metal) may in certain cases be a serious objection in regard to environomental protection.

The most commonly used anode material at present is titanium with various coatings. Hypochlorite electrosynthesis from seawater was among the first applications to serve as a test for these new anodes in the early sixties.[161,162] It is known from experience today that platinum loss rises considerably if the electrolyte temperature drops to values common in seawater electrolysis (5°C or below). It was shown,[163] however, that the temperature has only a secondary effect and the real cause may be the influence of the high anode potential at low temperatures when current density is kept constant.

Graphite is no longer used as an anode material in cells going into use at present.

Cathodes are commonly made of nickel alloy or bare titanium. Tentatively, the use of chromium-clad cathodes was claimed[164] to have the benefit of protection from cathodic hypochlorite reduction as experienced in chlorate manufacture.[46,47] Some bipolar cells are equipped with platinum-coated titanium cathodes[165] to counter the fouling problem of the cathodes by slow and regular current reversal.[151] This procedure, however, again raises the question of the possibility of successfully using coated titanium cathodes, which has currently been rejected because of the formation of titanium hydride.

With the revival of electrolytic hypochlorite generation, cell design has been thoroughly modified. The first systems used for disinfection purposes were simple unipolar cells, later followed by bipolar constructions.[165,167] The bipolar electrodes allow a more compact cell design and are somewhat lower in capital investment due to the absence of interelectrode bus bars. However, they always have the problem of current leakages between the cells within a battery and require a very careful design. Small electrode distance is a must in hypochlorite cells because of the poor conductivity of the dilute electrolyte used. But the resulting high liquid velocities enhance mass transfer of hypochlorite to the cathode and contribute to an increase of the cathodic reduction. On the other hand, high electrolyte velocity keeps the hypochlorite concentration low, thus lowering mass transfer. We arrive at a similar optimization problem as discussed for chlorate cells in Section 2.4.

Emphasis in modern industrial cell design—predominantly equipped with vertical electrodes, scarcely with inclined[72] electrodes—is laid on high electrolyte flow for the reasons already mentioned. Furthermore, the temperature of the electrolyte passing through the cell is only slightly increased, thus slowing down chemical chlorate formation. High flow rates are also advantageous in avoiding deposits of magnesium and calcium hydroxides and carbonates on the cathode from seawater, as these solids are taken off in suspended form. Electrolyte velocities of 0.6–1.4 m/sec in the interelectrode space at a current density of 1330 $A/m^2$ [72,164] are claimed to keep the pH value in the cathodic boundary layer at values low enough to prevent disturbing deposits.

Typical operational data of a modern hypochlorite cell are given in Table 3. A cell especially designed for disinfection of sewage is shown in Figure 15. Pure sodium chloride solution or seawater is pumped through a pipe with two wire electrodes in its middle. The hypochlorite solution generated is discharged through a nozzle and mixed with the sewage. Other industrial tubular cells were described by Kuhn.[169]

His analysis[170] of the influence of various parameters involved in hypochlorite formation under technical conditions is expected to form a valuable basis for industrial cell design. The model did not, however, study the effect of cathode surface roughness and texture, or the effect of variable anode: cathode area ratios, which were studied elsewhere.[136,340]

Energy costs for disinfection of municipal wastes by electrolytically generated hypochlorite were reported in a review to amount to 6–10 U.S. cents per kilogram of available chlorine, in agreement with recent information.[171]

Table 3
Typical Operational Data of the Diamond Shamrock Cells for Hypochlorite Generation[166,339]

|  | Brine feed | Seawater feed |
|---|---|---|
| Current density, A/m$^2$ | 1500 | 1500 |
| Current efficiency, % | 40 (single stage) | 90 |
|  | 65 (multistage) |  |
| Temperature, °C | 25 | 5–25 |
| Concentration |  |  |
|   Entrance cell, g NaCl/liter | 30 | 30 |
|   Exit cell, g av. Cl$_2$/liter | 8–10 | 1.0 |
| Energy consumption, kWh/kg av. Cl$_2$ | 5.1–5.3 | 3.3–3.9 |
| Sodium chloride consumption, kg NaCl/kg av. Cl$_2$ | 3–3.5 |  |
| Anode | Coated totanium | Coated titanium |
| Cathode | Bare titanium | Nickel alloy |

Capital investment was reported as 2000–3000 U.S. dollars per kg av. Cl$_2$/hr. Capacities of the cells range between 10 and 1000 kg av. Cl$_2$/day.[171]

Several trends for future cell design are recognizable. Modern ion exchange membrane technology permits the possibility of hydraulically separating the anode and cathode sections and manufacturing of hypochlorite by subsequent reaction of chlorine and hydroxide outside the cell, thus preventing the undesired cathodic reduction of hypochlorite.[151,172] Current efficiency could be raised to 75% in producing solutions of 5–12 wt % NaClO. Yurkevich and Vrevskii used ion-exchange membranes[48,72,173] and showed that the current efficiency can be increased up to 64%. This is considerably higher than for usual undivided cells and evidences an energetic advantage in favor of the ion exchanger in spite of the higher cell potential. Another field of

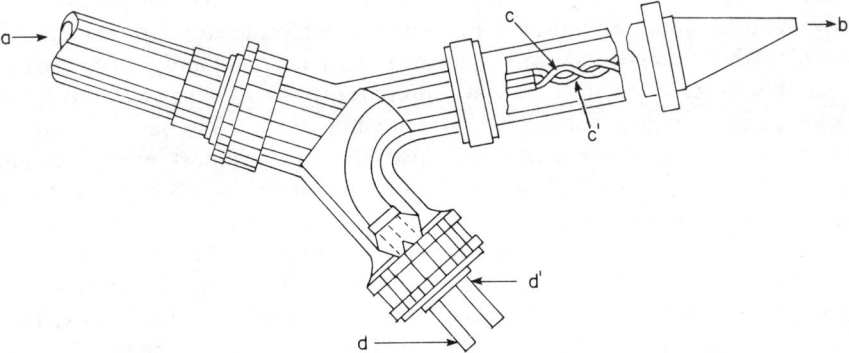

Figure 15. High-velocity hypochlorite cell for disinfection of sewage.[168] (a) Electrolyte feed, (b) electrolyte outlet, (c), (c'), electrodes, (d), (d') contacts.

## 4. Perchloric Acid and Perchlorate

### 4.1. Methods of Preparation

Various methods for manufacturing perchloric acid and perchlorates are known,[175] such as the thermal decomposition of chloric acid to perchloric acid and that of potassium chlorate to potassium perchlorate; the chemical oxidation of chlorates to perchlorates with strong oxidizing agents, such as ozone, sodium persulfate, and lead dioxide; and the reaction of a strong mineral acid with a chlorate to produce perchlorate, chloride, and chlorine dioxide. Electrosynthesis, however, is the only method that has achieved industrial importance. Two variants are industrially applied: (1) electrolysis of hydrochloric or hypochlorous acid to form perchloric acid, which, if needed, can subsequently be chemically converted to perchlorates, and (2) anodic oxidation of chlorate dissolved in aqueous solution. A third variant, a direct electrolysis starting from a chloride solution, without isolation of the chlorate formed intermediately, thus leading to perchlorate in one-stage electrolysis, might gain industrial significance.

Perchloric acid can be prepared by reacting sodium perchlorate with hydrochloric acid,[209,319] or ammonium perchlorate with a mixture of nitric and hydrochloric acid.[317] A direct electrosynthesis of perchloric acid is possible, too, by anodic oxidation (1) of dilute aqueous hydrochloric acid (at low temperature and high current density),[16,177] (2) of hydrogen chloride dissolved in dilute sulfuric acid,[178] or (3) of chlorine dissolved in perchloric acid.[179]

The usual method of manufacturing sodium perchlorate is the anodic oxidation of sodium chlorate, which is used as such or as intermediate for conversion to other perchlorates. Potassium perchlorate can also be made by electrosynthesis directly. However, owing to its smaller solubility as compared to sodium perchlorate it is currently prepared from sodium perchlorate by conversion with potassium chloride and precipitated.[213] The same method is applied to the manufacture of the industrially highly important ammonium perchlorate, prepared by conversion with ammonium chloride or ammonia plus hydrochloric acid.[318]

As chlorate is in any case prepared by electrosynthesis, it is obvious to attempt to manufacture perchlorate by electrolysis of sodium chloride in such a way that the oxidation of chloride and of chlorate occur simultaneously. This procedure, however, results in low current efficiencies, amounting to only 49% at 32°C.[180] Various salts are different in this regard: Lithium chloride (in contrast to sodium or potassium chloride) used as educt for total direct

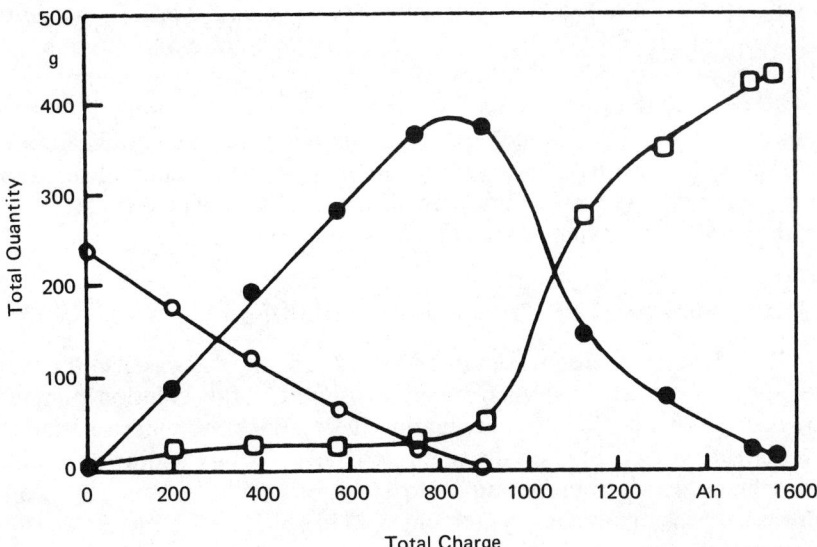

Figure 16. Development of concentrations in electrolysis of chloride directly to perchlorate without isolation of chlorate according to Nagalingam et al.[184] ○, NaCl; ●, NaClO$_3$; □, NaClO$_4$.

electrosynthesis of perchlorate shows high current efficiency, while rubidium chloride cannot be oxidized electrolytically beyond the chlorate stage Reference 175, p. 77. Early attempts to prepare sodium perchlorate in one-stage electrolysis of sodium chloride with or without a diaphragm failed[181,182] as the current efficiency was too low. More recently, an electrolysis of sodium chloride in the presence of fluoride,[183] resulting in an increase of the electrode potential, was successful[184,185] (see Figure 16). Alternatively, an addition of $K_2S_2O_8$ was found to be even more effective.[186]

## 4.2. Industrial Significance

For a long time after the discovery of perchlorate no commerical use developed. Industrial perchlorate manufacture started at the end of the nineteenth century to make explosives. But world production amounted only to 2000–3000 t/year for all perchlorates.[175] During World War I the world production of perchlorates increased up to 50,000 t/year.[187] After the war production dropped to low values, and new explosives were placed into the foreground. The use of ammonium perchlorate as a jet propellant, being the main use today, gave rise to perchlorate production. Ammonium perchlorate used either as a jet propellant or as an explosive is advantageous inasmuch as only gaseous components develop in combustion. Owing to the military use of perchlorates, a reliable estimation of production data is problematic. World production from 1962 to 1964 was estimated to be on the order of magnitude of 50,000–100,000 t/year;[187,188] U.S. production alone amounted to 24,000 t/year in 1974.[1] Within the past few years the importance of perchloric acid has grown considerably with the increased demand for ammonium perchlorate, which can easily be obtained by neutralization with ammonium hydroxide. Lithium perchlorate has been considered superior to both potassium and ammonium perchlorate as a propellant oxidizer,[189] but has the disadvantage that more exhaust products are solids.

### 4.3. Development of Theory and Technology

Perchlorate was discovered in 1816 by Stadion, who was also the first to prepare it by electrosynthesis from a potassium chlorate solution using electrodes of platinum,[190] thus anticipating the features of modern perchlorate manufacture. Later, Stadion obtained perchloric acid by anodic oxidation of hydrochloric acid.[191] Preparation of perchloric acid by electrolysis of dilute chloric acid was carried out by Berzelius[192] in 1835 and by Riche.[193] But all electrochemical preparation methods found little interest until the end of the nineteenth century. In 1890, Carlson was granted a patent,[194,13] for diaphragmless cells for sodium perchlorate synthesis and in 1895 he operated the first commercial electrosynthesis of ammonium and potassium perchlorate.[44] Production was started in France, Germany, Switzerland, and the U.S.A. at the turn of the century.

As occurred for chlorate, theoretical interest in the reaction fundamentals arose only after the product had attained commercial importance. At that time, in 1898, Foerster,[195] Haber and Grinberg,[16] and Winteler[181,196,197] studied the fundamentals methodically. The view was taken that perchlorate is formed by simple anodic oxdiation of chlorate. Based on experimental investigations Oechsli[198] developed the theory of a self-decomposition of chlorate to form perchlorate. Valuable contributions to the perchlorate theory were made by Bennett and Mack,[199] Williams,[200] Knibbs,[60] and more recently by Grotheer and Cook[201] and by de Nora et al.[202]

### 4.4. Reactions and Mechanisms

#### 4.4.1. Oxidation of $ClO_3^-$ to $ClO_4^-$

Chlorate is oxidized anodically to perchlorate according to the overall reaction

$$ClO_3^- + H_2O = ClO_4^- + 2H^+ + 2e, \quad E° = 1.19 \text{ V} \qquad (41)$$

The standard potential of this process is very positive and close to that for the oxidation of water:

$$2H_2O = O_2 + 4H^+ + 4e, \quad E° = 1.228 \text{ V} \qquad (7)$$

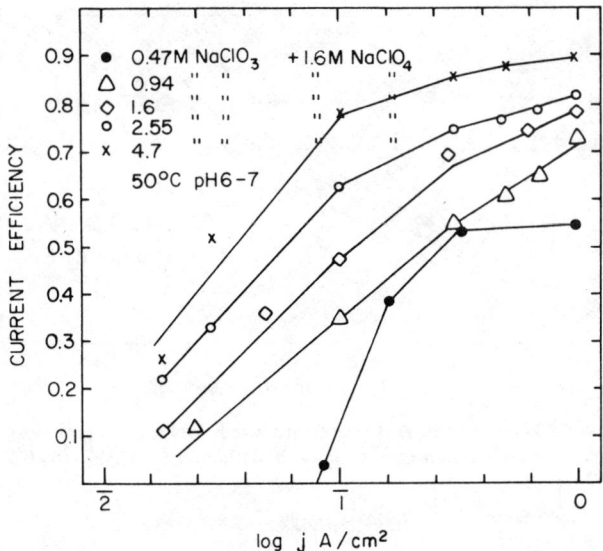

Figure 17. Current efficiency of perchlorate formation vs. current density on smooth platinum at different $NaClO_3$ concentrations according to de Nora et al.[202]

There are thus two strongly competing reactions and the achievement of a high current efficiency in perchlorate production is a problem. According to Nernst's law the pH dependence of the equilibrium potential is the same for the two reactions. Furthermore, with Pt anodes the overpotential depends little on pH in both cases.[187,202] The acidity of the solution is thus not an important variable. Indeed, perchlorate can be produced with a good current efficiency in a fairly broad pH range.[187]

A sensitive parameter is the anode potential, which is determined in practice by the current density (c.d.) imposed to the electrode. The most relevant study in this connection is that of de Nora et al.,[202] who investigated systematically the overpotential and the current efficiency at Pt and $PbO_2$ electrodes. As Figure 17 shows, the current efficiency strongly decreases with decreasing overall c.d., i.e., with decreasing anode potential. This can best be discussed in terms of the partial currents for $O_2$ and $ClO_4^-$ formation, the ratio of which determines the current efficiency for perchlorate production. Figure 18 shows the anode potential as a function of these partial c.d.'s. The increase of partial c.d. with overpotential is much faster for the perchlorate formation than for the $O_2$ evolution, so that a high current efficiency for the production of the former can be achieved only at a high overall c.d. An interesting finding of de Nora et al. is the influence of $ClO_3^-$ concentration. From the comparison of the broken lines of Figure 18 it is seen that the overpotential for $O_2$ evolution substantially increases with increasing $ClO_3^-$ content of the solution, suggesting an increasing adsorption of their ions at the

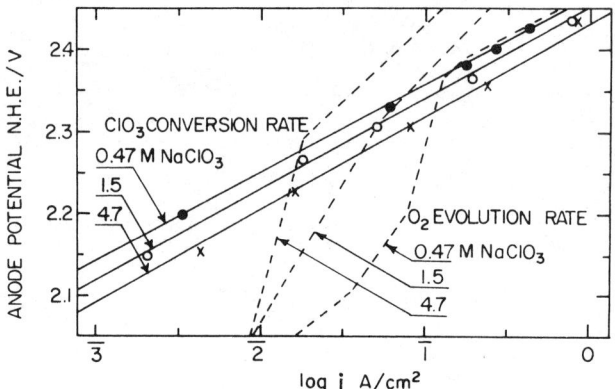

Figure 18. Rates of chlorate conversion to perchlorate and of oxygen evolution expressed in partial current densities $j$ vs. anode potential.[202] Smooth platinum anode, pH = 6–7, $T = 50°C$, 1.6 $M$ $NaClO_4$.

anode. The inhibition of the oxygen formation by $ClO_3^-$ is a lucky feature from the technical viewpoint, which allows high current efficiencies, especially at large $ClO_3^-$ concentrations. Conversion of chlorate is virtually quantitative.

Figures 17 and 18 apply to Pt anodes. Somewhat different results but by and large the same trends (increase of current efficiency with increasing c.d. or $ClO_3^-$ concentration) were found for $PbO_2$. The highest current efficiency (observed at the highest partial c.d. used, namely, 1 $A/cm^2$) was somewhat higher for Pt (90%) than for $PbO_2$ (83%).[202] An addition of NaF causes a further inhibition of the oxygen evolution and current efficiencies of 92–94% can then be reached with $PbO_2$ electrodes according to Schumacher et al.[203] Several authors[202,203] have also studied the influence of another additive, $Na_2Cr_2O_7$. In industrial practice the current efficiencies range from 85% to 97%.[204]

Two main mechanisms have been proposed for the anodic oxidation of $ClO_3^-$:

(a) The $ClO_3^-$ reacts with discharged oxygen resulting from the oxidation of water and chemisorbed at the electrode surface:[199,201]

$$H_2O \rightarrow O + 2H^+ + 2e \qquad (42)$$

$$ClO_3^- + O \rightarrow ClO_4^- \qquad (43)$$

(b) The primary step is the direct discharge of $ClO_3^-$ at the anode[60,198,205,206]

$$ClO_3^- \rightarrow ClO_3 + e \qquad (44)$$

Several follow-up steps are possible and have been envisaged in the literature.

Let us mention two of these subroutes (References 60 and 205, respectively):

$$2ClO_3 \rightarrow O_2Cl-O-O-ClO_2 \xrightarrow{H_2O} ClO_4^- + ClO_3^- + 2H^+ \quad (45)$$

$$ClO_3 \xrightarrow{H_2O} ClO_4^- + 2H^+ + e \quad (46)$$

A recent study and discussion of the mechanism of perchlorate formation is due to Wang et al.[365]

The partial c.d.–potential curves for $O_2$ evolution exhibit marked changes in slope (Figure 18) suggesting a modification of the corresponding reactions and of the relevant adsorption layer. In contrast to this, the log $\eta$–log $j$ plots for $ClO_3^-$ conversion are straight lines over the three decades of c.d. investigated (Figure 18). Their slope is 0.11–0.12 V and the influence of $ClO_3^-$ concentration is $(d \ln j)/(d \ln c) \cong 0.5$, which corresponds to a one-electron step electrode reaction with $\alpha = 0.5$. From these facts and some other findings de Nora et al.[202] concluded that the rate-determining step is the direct discharge of $ClO_3^-$ according to reaction (44), probably followed by subroute (46). The detailed mechanism is still open to question.

### 4.4.2. Direct Oxidation of $Cl_2$

Chlorine dissolved in a perchloric acid solution can be oxidized anodically to $HClO_4$ with no side reaction other than oxygen evolution.[179,326] The reaction has been studied by Krasilova et al.[326] At low potentials only oxygen was formed but at 2.9 V they obtained $HClO_4$ with 60% current efficiency at 6 N $HClO_4$. This is the optimum concentration. At higher concentrations the partial current of $HClO_4$ drops off drastically, owing probably in part to adsorption of perchlorate at the interface. On the other hand, at low concentrations, below 6 N, it is the rate of oxygen evolution which increases strongly when the $HClO_4$ concentration is lowered at constant potential.

In the low potential range, where there is only oxygen formation, the rate of the latter at constant potential decreases with increasing concentration of dissolved $Cl_2$. This suggests an inhibition of oxygen evolution by adsorption of Cl at the anode. This inhibition facilitates the attainment, at increasing c.d., of a potential sufficiently high for $HClO_4$ formation. It appears that adsorption phenomena play an important role in the electrosynthesis of per- compounds, through their influence on the competing $O_2$ evolution or through a more direct action on the formation of the per- compound itself. Further examples of such effects are encountered in 4.4.1 (formation of $ClO_4^-$ from $ClO_3^-$) and 7.1 (synthesis of peroxodisulfate). The kinetics of various anodic processes at high potentials have been studied by Veselovsky et al. and discussed in terms of the structure of the electric double layer.[368]

The direct oxidation of $Cl_2$ to $HClO_4$, which involves the transfer of not less than seven electrons per atom of Cl, is most probably a very complex process. Nothing seems to be known so far about the intermediate steps.

## 4.5. Industrial Cells

### 4.5.1. Perchlorate Cells

Homogeneous reactions are not of importance here since perchlorate is formed anodically only; therefore, perchlorate cells do not show the features that are typical of chlorate cell systems. The problem of a minimized residence time of the electrolyte in the cell which plays an eminent role in chlorate electrosynthesis (owing to the competing reaction routes) are of subordinate significance. Moderate electrolyte flow is satisfactory. Hence, cell design is relatively simple.

The anode material is very important. Only two have attained industrial application. Smooth platinum has been industrially applied as it was for a long time the only material to yield the high overpotential required for all peroxidation processes (as applies to persulfate and perborate, too). The considerable capital investment due to the use of massive platinum induced a search for alternative means. Cladding of smooth platinum foil on copper or silver was tried, but a foil thickness of about 70 $\mu$m is needed to provide satisfactory protection of the nonresistant copper substrate, thus requiring a still uneconomic platinum quantity. To substitute platinum many materials such as tungsten, molybdenum, graphite, magnetite, and manganese dioxide have been considered (Reference 175, p. 82) without satisfying results. Alone, the use of lead dioxide was reported[207,208] to yield a relatively high current efficiency (79% and 58% respectively). This anode material has found a wide application in combination with graphite as the substrate[99] though the current efficiency is lower by some percent as compared to the platinum anode.[186,187] Very recently, a coating of platinates on titanium was reported to be successful for perchlorate manufacture.[45]

Cathodes are made of bronze, carbon steel, CrNi steel, or nickel. As in chlorate synthesis, dichromate is added to the electrolyte solution to minimize cathodic reduction by the mechanism discussed in Section 2.3.2. Dichromates, however, are not admissible if lead dioxide anodes are used because on these electrodes, the dichromate catalyzes the oxygen evolution.[202] For this case, nickel or CrNi steel were found to be the most effective cathode material.[203]

Operational data for the production of sodium perchlorate are shown in Table 4. Further details of the perchlorate process, including cell design, can be found in the reviews of Schumacher[175] and Legendre[187]; the latter also contains interesting economic considerations.

### 4.5.2. Perchloric Acid Cells

A process used commercially for perchloric acid manufacture[209] starts from sodium perchlorate which is reacted with an excess of concentrated hydrochloric acid to form perchloric acid and precipitated sodium chloride. To

Table 4
Typical Operational Data for Perchlorate Cells[1,187]

| | |
|---|---|
| Current, A | 500–12,000 |
| Current density, $A/m^2$ | 1,500–5,000 |
| Cell potential, V | 5–6.5 |
| Current efficiency, % | |
|    Platinum anodes | 90–97 |
|    Lead dioxide anodes | 85 |
| Electrolyte temperature, °C | 35–50 |
| Electrolyte pH | 6–10 |
| Concentration, g/liter | |
|    $Na_2Cr_2O_7$ | 0–5 |
|    Entrance cell | |
|       $NaClO_3$ | 400–700 |
|       $NaClO_4$ | 0–100 |
|    Exit cell | |
|       $NaClO_3$ | 3–50 |
|       $NaClO_4$ | 800–1,100 |

obtain a very pure product, a direct electrosynthetic process for perchloric acid is also industrially applied.[179] Gaseous chlorine is dissolved in chilled perchloric acid and anodically oxydized to form perchloric acid (see Section 4.4.2). The anodes are made of platinum foils fixed by spot welding to a tantalum rod. The cathodes are horizontally slitted silver sheets, placed inside frames of PVC, which are arranged like filter press elements. The interelectrode space is divided by PVC fabric. Operation data are compiled in Table 5. Platinum consumption of 0.025 g/t of 70% $HClO_4$ is remarkably low as compared to perchlorate manufacture (2–7 g/t $NaClO_4$). Platinum dissolved at the anode is partly deposited at the cathode and can be recovered. An anode life time of two years has been reported, but can be increased up to seven years.[211] Current efficiency can be improved by application of even lower temperatures in the range of −5 to −30°C.[212]

Current efficiency obtained is lower than for perchloric acid prepared from chloride via chlorate and perchlorate, but the process directly yields products

Table 5
Operational Cell data in Perchloric Acid Manufacture by Direct Anodic Oxidation of Chlorine[179,210]

| | |
|---|---|
| Current | 5000 A |
| Current density | 2000–2500 $A/m^2$ |
| Cell potential | 4.4 V |
| Current efficiency | 0.60 |
| Electrolyte temperature | −5 to +3°C |
| Chlorine concentration, entrance cell | 3 g/liter |

of high purity, which dispenses with separation stages, thus compensating for the low current efficiency by a lower plant investment and lower labor cost. Further, unusual perchlorates (of magnesium, barium, copper, lead) can be prepared by direct conversion with perchloric acid, not requiring the roundabout route via sodium chlorate.

## 5. Bromate

The industrial significance of the electrosynthesis of bromates is small as compared to the chlorate process: U.S. production in 1968 amounted to only 300 tons.[214] However, the production rate has steadily grown within the past few decades.

In context with the investigation of the electrochemical chlorate formation at the turn of the century, the mechanism of bromate electrosynthesis was studied by Pauli,[215] Sarghel,[216] Müller,[217] Kretzschmar,[218] Bray,[219] and Skrabal,[220,221] whose contributions form the basis of today's bromate theory.

Formation of bromate by the electrochemical route essentially corresponds to chlorate formation. The reactions involved are quite similar but the values of the thermodynamic and kinetic parameters are very different for chlorate, bromate, and iodate (Table 6). This explains the differences observed in the electrosynthesis of bromate and iodate as compared to that of chlorate. The equilibrium

$$Br_2 + Br^- \leftrightarrows Br_3^- \qquad (47)$$

lies much more on the right than in the case of chlorine, where the formation of $Cl_3^-$ is negligible. Further, the rate constants for the hydrolysis of $Br_2$ and quite particularly that for the chemical bromate formation [corresponding to Eqs. (3) and (5), respectively] are much larger than in the case of chlorine. The chemical bromate formation is thus much facilitated as compared to that of chlorate. Finally, according to Kretzschmar[218] the anodic oxidation of $BrO^-$ to $BrO_3^-$ with $O_2$ evolution, corresponding to Eq. (6), virtually does not take place under the usual conditions, whereas the primary anodic oxidation (involving the simultaneous discharge of $OH^-$) analogous to Eq. (23) is much more important that in the case of chlorine.

A detailed study of the reactions occurring in the electrolysis of bromide solutions was made by Kretzschmar.[218] In the electrolysis of neutral NaBr 1 $M$ solutions a behavior similar to that shown by curve 1 of Figure 5 is observed: the bulk hypobromite concentration increases slowly with time and reaches finally a steady state value. It depends more than in the case of chlorine upon whether a bright or a black plantinum electrode is used. The current efficiency of oxygen evolution is about 16% for the first electrode but almost zero for the second one (Figure 19). The amount of oxygen generated could not be brought into correlation with the formation of bromate and was ascribed to the

## Table 6
### Comparison of Physicochemical Data for Cl, Br, I

|  | Cl | Br | I | Reference |
|---|---|---|---|---|
| $X_2 + 2e \rightleftarrows 2X^-_{(aq)}$ |  |  |  |  |
| Electrode standard potential (25°C), V | 1.396 | 1.087 | 0.621 | 327 |
| Solubility of $X_2$ in $H_2O$ (25°C), mol liter$^{-1}$ | 0.092 | 0.21 | 0.0013 | 20, 328 |
| Solubility of $X_2$ in 1 N KX (25°C), mol liter$^{-1}$ | 0.058 | 1.35 | 0.055 | 20, 328 |
| $X_2 + X^- \rightleftarrows X_3^-$ |  |  |  |  |
| $K_{X_3^-} = \dfrac{a_{X_3^-}}{a_{X_2} a_{X^-}}$ | 0.18 (25°C) | 17 (25°C) | 830 (20°C) | 278 |
| Hydrolysis of the halogen (20°C) |  |  |  |  |
| $K_h = \dfrac{a_{H^+} a_{X^-} a_{HXO}}{a_{X_2}}$ | $3.88 \times 10^{-4}$ | $4.45 \times 10^{-9}$ | $4.3 \times 10^{-13}$ | 278 |
| Dissociation of the hypohalogenous acid (25°C) |  |  |  |  |
| $K_d = \dfrac{a_{XO^-} a_{M^+}}{a_{HXO}}$ | $3.2 \times 10^{-8}$ | $2 \times 10^{-9}$ | $2.3 \times 10^{-11}$ | 278 |
| Rate constant of hydrolysis, sec$^{-1}$ 20°C, $[X_2] = 0.1$ M |  |  |  |  |
| $X_2 + H_2O \xrightarrow{k} H^+ + X^- + HXO$ | 11.0 | 110 | 3.0 | 278 |
| Rate constant of halogenate formation $l^2$ mol$^{-2}$ sec$^{-1}$ |  |  |  |  |
| $2HXO + XO^- \rightarrow XO_3^- + 2X^- + 2H^+$ | 0.16 | 18 | — | 20 |
| $-\dfrac{d[XO^-]}{dt} = k[HXO]^2[XO^-]$ |  |  |  | 33 |

discharge of OH$^-$ ions which takes place mainly at the bright platinum because of the higher overvoltage for Br$^-$ discharge as compared to the black platinum. It was concluded that chemical bromate formation predominates. The electrolysis of alkaline solutions also exhibited a characteristic difference as compared to the behavior shown on Figure 8 for chlorine. The steady state concentration of ClO$^-$ drops virtually to zero at high OH$^-$ concentrations, whereas with concentrated Br$^-$ solutions this is not the case for BrO$^-$ even in very alkaline solutions. Anodic oxidation of BrO$^-$ takes place but according to the primary oxidation mechanism [Eq. (23)]. In general, according to Foerster[20b] it is easier to reach high BrO$^-$ concentrations in the solution than ClO$^-$ concentrations. This would be of importance for the production of hypobromite which has, however, no industrial importance in contrast to the

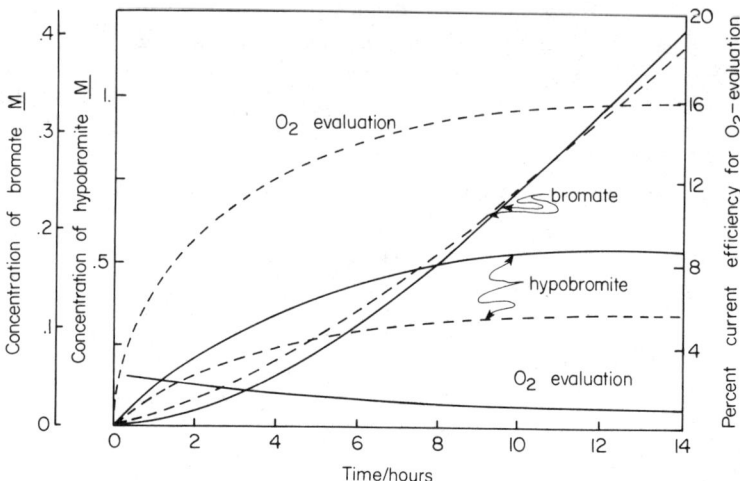

Figure 19. Evolution with time of bulk concentration of hypobromite and of current efficiency for $O_2$ formation in the electrolysis of a 3.5 N NaBr solution[20b] (9–10°C, $j = 27$ mA/cm²). ——, Platinized; ---, bright Pt anode.

manufacturing of hypochlorite. But the high hydrolysis rate and the easy formation of $BrO^-$ is the reason why the diaphragmless electrolysis of bromide solutions has been preferred recently by some authors to the electrolysis of chlorides in the electrosynthesis of propylene oxide through the reaction of electrolytically generated hypohalogenite with propylene and subsequent saponification by the $OH^-$ ions produced at the cathode (see Chapter 4 on Electro-Organic Syntheses). On the other hand, owing to the high rate of chemical bromate formation and to the absence of $BrO^-$ oxidation by the reaction corresponding to Eq. (6) bromate is readily obtained with high current efficiencies by electrolysis of neutral to slightly alkaline bromide solutions of some 1 mol $Br^-$ per liter. According to Kretzschmar[218] the yield is a quantitative one with platinum black anodes. More recently, two groups of authors[222,223] have reported that current efficiencies of 95% (measured in part on the pilot plant scale) can be achieved with $PbO_2$ electrodes.

Owing to the considerably smaller vapor pressure of bromine, losses due to vaporization of bromine are negligible, but liquid bromine can be found in the cell under unfavorable conditions, especially at pH values lower than 8 and unsatisfactory stirring of the electrolyte.[222]

Initially, cells with diaphragms were proposed[224] but modern cell design and operation corresponds largely to the chlorate cells. Essential features are again the addition of dichromate to the electrolyte to prevent a cathodic reduction of hypobromite by formation of a chromium oxide layer on the cathode, and the benefit of elevated temperatures to accelerate bromate formation. However, corrosion by the electrolyte is more severe.

Cathode and anode materials are the same as in chlorate electrosynthesis, i.e., stainless steel or copper as cathodes and graphite, platinum-sheeted copper or platinized titanium as anodes. Recently, glassy carbon anodes have been proposed.[225] Industrial application of lead dioxide anodes was fostered by the work of Sugino[222]; their performance was also studied by the Udupa school[223] and by Dzhafarov and Efendieva.[226] The product was reported to be free of lead,[223] thus not requiring filtration of the electrolyte.

To prepare the electrolyte, bromine is added to a potassium hydroxide or—as the solubility of sodium bromate is higher—to a sodium hydroxide solution, thus providing some chemical bromate formation prior to electrolysis. The use of calcium bromide has been studied by Sarghel[216] but resulted in considerable deposits on the cathode, a problem which is well known from technical chlorate electrosynthesis, too.

A valuable review of electrochemical bromate technology was presented by Radford.[227]

## 6. Iodate and Periodate

In anodic oxidation of solutions of iodide or iodine the final products hypoiodite,[228] iodate, and periodate are possible. In further analogy to the chloride electrolysis as previously discussed, solutions of iodate and iodic acid are oxidizable to form periodate and periodic acid. Industrial productions of iodate and periodate are closely linked. As can be seen from an estimate for the total U.S. production of 25 t for the year 1968,[214] their industrial significance is by far not comparable to the chlorate and perchlorate production.

### 6.1. Iodate

The formation of $I_3^-$ (and also $I_5^-$) is even more pronounced than that of $Br_3^-$. The potential of $I^-$ discharge is much less positive than for $Br^-$ or $Cl^-$ (Table 6). The competing discharge of $OH^-$ (which leads to a strong increase of the current efficiency of oxygen evolution in alkaline Cl solutions) is not observed even in alkaline $I^-$ solutions. The chemical formation of $IO_3^-$ by a reaction analogous to reaction (5) is even faster than that of bromate, and hypoiodite is quite unstable. No hypoiodite builds up in the bulk with neutral or slightly alkaline iodide solutions. Curves similar to those of Figures 5 and 8 are observed only in strongly alkaline solutions.[20c] According to Foerster[20c] iodide solutions of about 0.2 M are well suited for the production of iodate with a high yield. A detailed study of the kinetics of electrolytic iodate formation was made recently by Landsberg et al.[229]

Performance of the anode process is decisively influenced by the anode material. The main one is lead dioxide, which can be used massive, or as a coating on lead, on platinum, or on graphite.

*Table 7*
*Typical Operation Data for an Industrial Potassium Iodate Cell*[234]

| | |
|---|---|
| Cell voltage | 2.2–2.25 V |
| Anodic current density | 225 A/m$^2$ |
| Temperature | 50°C |
| Current efficiency | 94–96% |
| Specific energy consumption | 1.8 kWh/kg KIO$_3$ |
| Anode material | Graphite |
| Cathode material | Nickel |
| Cell case material | Carbon steel |

The use of lead dioxide anodes electrodeposited on graphite in large-scale manufacture of sodium iodate from alkaline iodine solutions was investigated by Venkatachalapathy *et al.*[230] Current efficiencies of about 75% were found, not tending to fall off even approaching complete conversion to iodate. It was further shown that apart from the anodic oxidation, iodate was also formed due to chemical reaction. The wear of the lead dioxide anode was reported[230] as 3 g/kAh. The cathodes were plates of stainless steel covered with asbestos fiber, allowing to dispense with the use of dichromate. The product was found[230] colorless, whereas graphite anodes result in a product coloration.[222,231] Lead dioxide on graphite anodes is industrially applied for the manufacture of iodate.[161]

Nevertheless, graphite has been used in industrial plants and showed virtually no wear after 5 years of operation.[234] Typical operation data are given in Table 7. Smooth platinum anodes, investigated by Foerster and Gyr[232,233] may lead to a sudden increase in electrode potential resulting in oxygen evolution.[234] Production of iodic acid starting from powdered iodine has also been described in the literature.[237]

### 6.2. Periodate

Anodic oxidation of iodate or iodic acid to form periodate or periodic acid has received an attention which appears less due to the industrial relevance at the present time but rather to its possible use in the near future.

A method for preparation of periodate, described by Willard and Ralston,[235] utilizes an anolyte of iodine disolved in hydrochloric acid and a catholyte of nitric acid. Due to the severe corrosiveness of the electrolytes necessitating costly precious metals, the process was considered[236] inadequate for industrial use. Mehltretter and Wise[236] have proposed as electrolyte iodine in aqueous sodium hydroxide for a first-stage preparation of sodium iodate, followed by an anodic oxidation in acid electrolyte to form periodate in the same cell. An alkaline anolyte, however, implies the problem of scaling at the anodes by trisodium paraperiodate if iodine is contained at higher concen-

# INORGANIC ELECTROSYNTHESIS

trations. The effect is reduced by addition of sodium sulfate, which results in an acid reaction at the anode surface. To prevent excessive liberation of iodine the anolyte must therefore be kept alkaline in bulk by periodically adding sodium hydroxide solution.[236]

It is understood that for periodate manufacture lead dioxide anodes play an important role, the use of which being known since decades.[235,238] Iodine dissolved in bromine water was successfully oxidized to periodic acid with a current efficiency of 70–75% at massive lead dioxide anodes. Coated or massive lead dioxide was found to be largely superior to platinum and graphite.[222] A process for periodic acid manufacture starting from iodine has been patented[239] using lead dioxide anodes as well. In all cases, the lead dioxide was either massive or on lead or a platinum substrate.

## 7. Peroxodisulfate

Manufacture of peroxodisulfate (persulfate) consists in the anodic oxidation of sulfuric acid or sulfates. Production figures were large at the time when peroxodisulfate was used as intermediate in the manufacturing of hydrogen peroxide, but has gone down considerably since the electrochemical process of hydrogen peroxide preparation was substituted by a chemical one (see Section 8). Now as before, however, the electrochemical route of peroxodisulfate manufacture is the only one industrially applied.

### 7.1. Reaction Fundamentals

The mechanism of formation† of peroxodisulfate is believed to start from an anodic oxidation of sulfuric acid or sulfates according to the following reactions:

$$2HSO_4^- \rightarrow 2HSO_4 + 2e \tag{48}$$

$$2HSO_4^- \rightarrow 2SO_4^- + 2H^+ + 2e \tag{49}$$

$$2SO_4^{2-} \rightarrow 2SO_4^- + 2e \tag{50}$$

$$2SO_4^- \rightarrow S_2O_8^{2-} \tag{51}$$

$$2HSO_4 \rightarrow S_2O_8^{2-} + 2H^+ \tag{52}$$

to result in the following overall reactions:

$$2HSO_4^- \rightarrow S_2O_8^{2-} + 2H^+ + 2e, \quad E° = 2.123 \text{ V} \tag{53}$$

$$2SO_4^{2-} \rightarrow S_2O_8^{2-} + 2e, \quad E° = 2.010 \text{ V} \tag{54}$$

---

† Several authors (see, for instance, References 357–359) have studied the kinetics and the mechanism of the *reduction* of the peroxodisulfate, which will, however, not be reviewed here.

Alternatively to this mechanism as proposed by Richarz,[309] Foerster[20d] suggested a primary formation of atomic oxygen,

$$H_2O \rightarrow O + 2H^+ + 2e, \qquad E° = 2.421 \text{ V} \tag{42}$$

followed by a reaction with sulfate:

$$2SO_4^{2-} + O + 2H^+ \rightarrow S_2O_8^{2-} + H_2O \tag{55}$$

A third mechanism, proposed by Haber,[310] assumes the intermediate formation of hydrogen peroxide,

$$2H_2O \rightarrow H_2O_2 + 2H^+ + 2e, \qquad E° = 1.776 \text{ V} \tag{56}$$

subsequently reacting with sulfate to form peroxodisulfate,

$$H_2O_2 + 2SO_4^{2-} + 2H^+ \rightarrow S_2O_8^{2-} + 2H_2O \tag{57}$$

From the viewpoint of thermodynamics the Foerster mechanism appears less probable, since the standard potential of reaction (42) is more positive than the one of peroxodisulfate formation. Moreover, it was experimentally confirmed that the oxygen contained in the peroxodisulfate formed originates from the sulfate and not from water.[292,355,356] Objections are also justified regarding the mechanism proposed by Haber. If hydrogen peroxide is formed intermediately, it would be expected to undergo an anodic decomposition according to

$$H_2O_2 \rightarrow O_2 + 2H^+ + 2e, \qquad E° = 0.682 \text{ V} \tag{58}$$

at a much less positive potential. Further, hydrogen peroxide could never be detected in the beginning of an electrolysis[292] and is formed only as a hydrolysis product of the Caro acid. Finally, an increase of the current efficiency would be expected when hydrogen peroxide is added to the solution in case the Haber mechanism applies. However, a decrease was experimentally observed[292] possibly owing to reaction (58).

At present, reactions (48)–(52) are the most generally accepted ones.[341–343,352] However, the view existed that $S_2O_8^{2-}$ ions are formed by discharge of either $HSO_4^-$ ions,[341,342] Eq. (48), or of $SO_4^{2-}$ ions,[343] Eq. (50), but not of both simultaneously. Very recently, Balej and Kadeřávek[344] concluded from experimental findings that most probably both anions participate by their discharge in the formation of peroxodisulfate where their rates depend on the reaction conditions. In the range of optimum concentration for the formation of peroxodisulfate the discharge of $SO_4^{2-}$ ions is likely to be many times higher than that of $HSO_4^-$.

The evolution of oxygen according to

$$H_2O \rightarrow \tfrac{1}{2}O_2 + 2H^+ + 2e, \qquad E° = 1.23 \text{ V} \tag{7}$$

is an independent side reaction.[324] As seen from the standard potentials of reactions (53) and (54), anode potentials of at least 2 V are required to make peroxodisulfate. Nevertheless, the participation of reaction (7) must be kept small. High overpotentials of the oxygen evolving reaction are thus necessary and can be achieved mainly through the material and the surface condition of the anode. The requirements of high overpotential and corrosion resistance are best met by smooth platinum, which is the common industrial anode material.[243,244,324] From their study of the shape of the overpotential–time curves at constant c.d. (establishment of the steady state) and of the overpotential versus c.d. curves Smit and Hoogland[343] concluded that at high anodic potentials the coverage of the surface by oxygen is increasingly replaced by patches of adsorbed $SO_4^{2-}$ on which $S_2O_8^{2-}$ formation proceeds instead of oxygen evolution [a mechanism which is in contrast to earlier interpretations of similar measurements by Hickling and Jones[353] and Rius and Garcia[354]]. In practice, the anodic current density must be large enough not only from the viewpoint of overpotential but also from that of capital investment (high cost of platinum) and is kept in industrial cells in the range 5000–10,000 A/m². Further, oxygen evolution is inhibited by low temperature. The adsorption of sulfate ions or radicals[311,343] makes it understandable that the kinetics of the peroxodisulfate formation depends on the concentration[292] of $SO_4^{2-}$ ions which are present in considerable amounts even in concentrated $H_2SO_4$ solutions. According to recent measurements by Balej and Kadeřávek[344] with sulfuric acid of various concentrations the current efficiency for the formation of peroxocompounds goes through a maximum at 8 M $H_2SO_4$ for current densities of 0.35–1 A/cm². Large sulfuric acid concentration and low temperature (15–25°C) are thus favorable for industrial applications. The anode potential can also be raised by use of appropriate additives to the electrolyte (rhodanides, halogenides, particularly fluoride, cyanides, thiourea), the influence of which was interpreted by blocking active centers which favor recombination of atomic oxygen.[312–314,343] In addition to the anions ($SO_4^{2-}$, $F^-$) the adsorption of cations[343,350] plays a considerable role, in competition with the $SO_4^{2-}$. According to Smit and Hoogland[343] and Balej and Kadeřávek[369] the adsorption of $SO_4^{2-}$ is enhanced in the order $Na^+ < NH_4^+ < K^+ < Cs^+ < Rb^+$.

Instead of the hitherto mostly used and investigated platinum anodes Fukuda et al.[346] recently studied the overpotential behavior of Ti-supported $RuO_2$ anodes at high potentials in mixed $H_2SO_4$–$(NH_4)_2SO_4$ solutions.

Hydrolysis of peroxodisulfuric acid (Caro's acid) in the cell is another problem of importance in practice. Peroxomonosulfuric acid forms even at low temperatures with a noticeable rate,

$$H_2S_2O_8 + H_2O \rightarrow H_2SO_5 + H_2SO_4 \tag{58a}$$

and is anodically oxidized to oxygen and sulfuric acid:

$$H_2SO_5 + H_2O \rightarrow H_2SO_4 + O_2 + 2H^+ + 2e^- \tag{58b}$$

Apart from the product loss, the depolarization effect is detrimental. The hydrolysis is favored as the concentrations of sulfuric acid and of peroxodisulfate and the temperature increase.

Some time ago Smit and Hoogland[343] reinvestigated the formation of Caro's acid and claimed that it takes place at the anode rather than by hydrolysis, but a very recent study by Balej et al.[347,348] confirmed the earlier conclusions,[244] i.e., that Caro's acid is formed and consumed by reactions (58a) and (58b).

The cathodic performance involves no problems. Hydrogen is evolved at the lowest potential possible. The common industrial cathode materials are lead and graphite.

### 7.2. Industrial Cells

A valuable recent review of various industrial cells was published by Thiele and Matschiner[324] (see Table 8). In the original Weissenstein cell of 1905, cooling coils of glass were used. The low current efficiency is obvious in view of an electrolyte residence time of several hours. Large current density at the anode as compared to the cathode was applied in the cell of Teichner and Baum in 1930.[330] An anodic current concentration of more than 200 A/liter made it possible to attain concentrations of 25–30% $H_2S_2O_8$ in half an hour. A post World War II cell[253] used platinum wire anodes wound around a tantalum rod with silver core. Cooling coils of lead were arranged in the cathode compartment only. The tubular diaphragms were of porcelain. The current concentration amounted to 1000 A/liter. Other industrial anodes consisted of platinum foils connected by welding with a tantalum rod bearing a copper core[258,259] (Figure 20). A cell developed in the U.S.S.R. uses a microporous PVC diaphragm and net anodes of platinum. Graphite serves as the cathode material. Two recent East German cells are described in detail[324] and reflect the continuous efforts to lower energy consumption and increase the capacity without increasing the floor space.

The lowered resistivity of modern cells allows one to process sulfate solutions of low conductivity, too. The manufacture of solid peroxodisulfates of ammonium, sodium, and potassium, which have recently gained increas-

Table 8
Characteristics of Peroxodisulfate Cells[324]

|  | Current, A | Cell potential, V | Current efficiency | Energy consumption, kWh/kg $H_2S_2O_8$ |
|---|---|---|---|---|
| Original Weissenstein cell, 1905 | 300 | 6.0 | ≤0.6 | ≥2.76 |
| Weissenstein cell, 1930[330] | 1,000 | 5.7 | 0.70 | 2.25 |
| Degussa cell, 1951[253,331] | 7,000–8,000 | 4.3 | 0.74 | 1.61 |
| U.S.S.R. cell, 1957[332] | 1,000 | 4.3 | 0.72 | 1.65 |
| Eilenburg cell, 1963[324,333] | 12,000 | 3.8 | 0.70 | 1.50 |
| Eilenburg cell, 1972[324,334] | 14,000 | 3.7 | 0.78 | 1.31 |

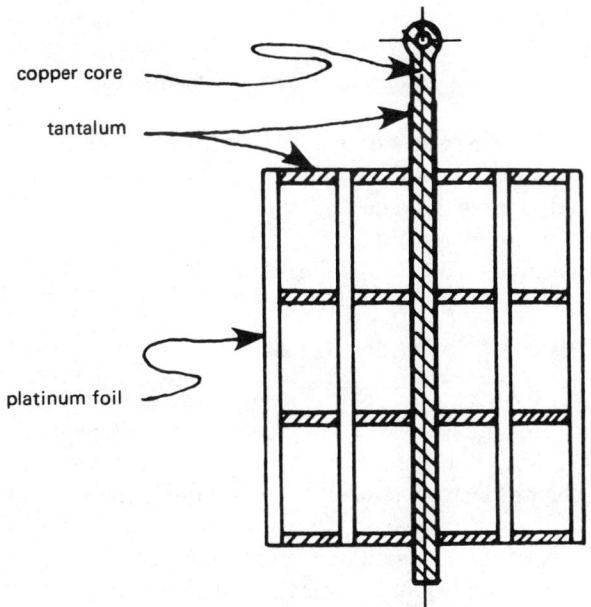

Figure 20. Industrial anode for persulfate manufacture.[258]

ing economical relevance as initiators in polymerizations, as oxidants and bleaching agents, was originally interconnected with the hydrogen peroxide production. The direct electrosynthesis, developed after the decay of the electrochemical hydrogen peroxide preparation, could prevail over the routes starting from hydrogen peroxide. The easily soluble sodium and ammonium peroxodisulfates are prepared in diaphragm cells with current efficiencies of 70–75% at cell potentials of 3.9–4.3 V. Even the less soluble potassium peroxodisulfate can be obtained by direct electrosynthesis if a dilute potassium sulfate solution is used as the starting electrolyte,[335] which could replace the conventional practice of converting solid ammonium peroxodisulfate with caustic lye or with solutions of sulfate or hydrogensulfate of potassium.

Recently, a pulse electrolysis of alkali metal persulfate solutions between a Zr or Zr-based cathode and a platinum anode was reported.[260] Current was periodically interrupted; the duration of the on and off periods was controlled.

Industrial cells are going to be improved partly by application of scientific optimization methods[325] with the target of larger capacities (up to $10^5$ A [324]), partly by use of ion exchange membranes which would allow one to process simultaneously electrolytes of different nature, thus opening up industrial processes with more efficient cathode performances than the evolution of hydrogen.

## 8. Hydrogen Peroxide

Various routes are available for the electrochemical preparation of hydrogen peroxide:
  (1) Anodic oxidation of sulfates or sulfuric acid with formation of peroxodisulfate and its subsequent hydrolysis.
  (2) Direct cathodic reduction of oxygen.
  (3) Indirect reduction of oxygen using an organic redox couple which is cathodically regenerated.

### 8.1 Hydrogen Peroxide by Anodic Oxidation

The conventional electrosynthesis of hydrogen peroxide consists in the anodic oxidation of aqueous sulfuric acid or of ammonium sulfate to the corresponding peroxodisulfates. Perhaps, the process is the most striking example of how a magnificent electrochemical process, after being steadily improved throughout decades and finally operating at enormous capacities, was substituted within a short time by a chemical process. Since in 1953 the chemical autoxidation of alkyle anthraquinone was applied industrially,[336-338] electrosynthesis has undergone a steady decay, and is now restricted to some existing plants, whereas the electrochemical route has been abandoned for new installations (except for very few special cases). Its absolute production data as well as its share in total $H_2O_2$ production drop steadily.[364]

The economic disadvantage as compared to the alkyle anthraquinone process is greater capital investment, mainly owing to the platinum anodes used, and additionally greater energy consumption (10 kWh/kg as compared to less than 3 kWh/kg[240]). Solely for plants with small capacities an economic advantage for the electrosynthesis was calculated some years ago,[241] the economic break-even point being reported to be at 50 tons $H_2O_2$/day. The increase of the energy price within the past years has shifted this point to the disadvantage of the electrochemical route.

Older-type anodes consisted of a silver rod clad with platinum of a thickness to exclude corrosion of the nonresistant silver. As occurred in other electrosynthesis processes, efforts were made to replace the costly anodes by a corrosion-resistant substrate (tantalum, tentatively titanium) coated by only a thin layer of platinum to ensure electrocatalytic activity but not necessarily corrosion protection, thus achieving a considerably smaller capital investment. These inherently successful attempts, however, have not been able to endanger the economic predominance of the anthraquinone route.

Three different processes for electrosynthesis of hydrogen peroxide are available. They all have in common the first electrochemical step of peroxodisulfate preparation. (The reaction fundamentals and technology are outlined in Section 7.) In a further step the peroxodisulfate hydrolyzes to form first Caro's acid and then hydrogen peroxide:

$$H_2S_2O_8 + H_2O \rightarrow H_2SO_5 + H_2SO_4 \quad H_2SO_5 + H_2O \rightarrow H_2SO_4 + H_2O_2 \quad (59)$$

The three formerly highly important electrosyntheses (the Weissenstein or Degussa process, the Löwenstein–Riedel process, and the Pietzsch–Adolph process) differ by the composition of the electrolyte, the process operation, and consumption data. All these processes are thoroughly discussed in readily available monographs[242-244] and handbooks[245,246] and need not be treated here further.

## 8.2. Hydrogen Peroxide by Direct Cathodic Reduction

Since the experiments of Traube[176] in the nineteenth century, it has been known that hydrogen peroxide is formed when oxygen is contacted with a cathode in aqueous acid solution. More than 40 years ago E. Berl[329] and W. G. Berl investigated the cathodic reduction of oxygen using alkaline electrolytes as an alternative to the anodic oxidation of acid solutions which was industrially applied at that time. Oxygen was forced to pass porous electrodes to build up large three-phase interface areas in small electrode volumes (like in fuel cells). Use of activated carbon electrodes[247,316] at current densities of 1500-A/m$^2$ apparent electrode area yielded a current efficiency of virtually 100%.[248] The attempts did not find technical realization because the gas cathodes used did not stand for long (only some hundreds of hours), the electrolyte temperature needed was low (5–10°C) and thus uneconomical, and a suitable separation process of hydrogen peroxide was not available at that time.[240] Using the knowledge gained by modern fuel cell technology the cathodic reduction of oxygen has found new interest and was the object of a renewed study of the technological aspects of the process.[240,249,250]

The reactions wanted are

$$O_2 + H_2O + 2e \rightarrow HO_2^- + OH^- \quad (60)$$

$$HO_2^- + H_2O \rightarrow H_2O_2 + OH^- \quad (61)$$

Cathodic loss reactions include the direct reduction of oxygen to $OH^-$ ions,

$$O_2 + 2H_2O + 4e \rightarrow 4OH^- \quad (62)$$

formation of hydrogen,

$$2H_2O + 2e \rightarrow H_2 + 2OH^- \quad (2)$$

and the reduction of hydrogen peroxide already obtained, occurring noticeably at noble metal cathodes, e.g., platinum,[142]

$$HO_2^- + M \rightarrow MO + OH^- \quad (63)$$

$$MO + H_2O + 2e \rightarrow M + 2OH^- \quad (64)$$

Reactions (63) and (64), and a catalyzed decomposition of hydrogen peroxide, by combination of Eq. (63) with

$$MO \rightarrow M + \tfrac{1}{2}O_2$$

can be largely suppressed by a low steady state concentration of hydrogen peroxide, i.e., by sufficiently high electrolyte flow rate through the cell.

Kastening[250] used porous gas cathodes, the effectivity of which was increased by hydrophobizing. They were made of activated carbon with 30 wt % PTFE powder. The catalytic activity of activated carbon for the decomposition of hydrogen peroxide formed has been previously decreased by heating over 900°C *in vacuo* or in hydrogen atmosphere. A pilot plant cell is shown in Figure 21. The electrodes are separated by a diaphragm or a cation exchanger membrane. Potassium hydroxide solution is an unproblematic electrolyte. The gas used is oxygen with a current efficiency of at least 90% or air with maximum 82%. Almost the same results have been obtained by Balej *et al.*[370] with porous cathodes analogously made of activated carbons previously heated over 900°C with alkali sulfides, or made of appropriate sorts of

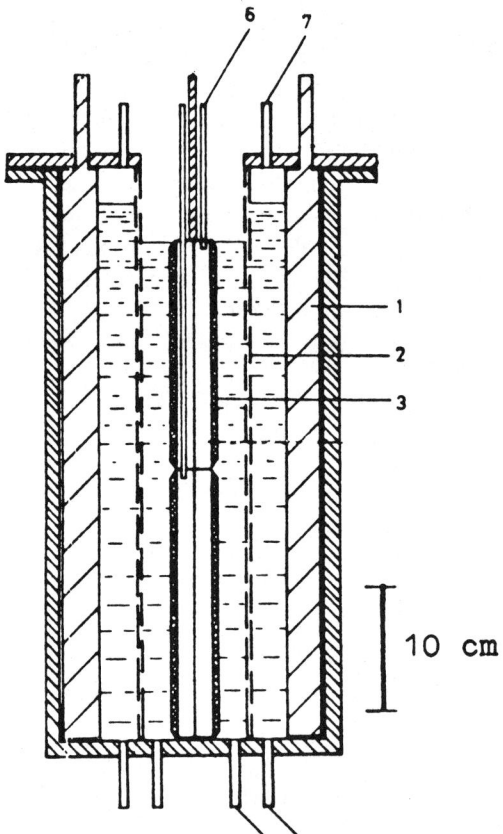

Figure 21. Cell for direct reduction of oxygen to hydrogen peroxide according to Kastening.[250] 1, Anode; 2 diaphragm; 3, cathode; 4, electrolyte feed; 5, electrolyte outlet; 6, gas feed; 7, gas outlet.

carbon black.[251] Recently, promising results have been obtained with a trickle-bed electrochemical reactor.[371]

An alternative electrolyte is neutral or slightly acid aqueous potassium chloride solution turning alkaline during operation. The possible use of various carbonaceous materials as porous gas electrodes was tested by Špalek et al.[251] The greatest current yields were found at low-area-type carbon black which simultaneously had the lowest catalytic effect on the spontaneous decomposition of hydrogen peroxide.

### 8.3. Hydrogen Peroxide by Indirect Reduction

An alternative which seems not to have reached the technical stage consists of an indirect reduction of oxygen bubbled through the cathode compartment of a cell.[321] Reduction proceeds in dissolved form with an organic redox couple, such as the sodium salt of 2,7-anthraquinone disulfonic acid. What makes the process an electrochemical one is the fact that the redox couple ($Q/HQ^-$) is regenerated at the cathode according to

$$Q + H_2O + 2e^- \rightarrow HQ^- + OH^- \qquad (65)$$

The reduced form of the redox couple reacts with oxygen:

$$HQ^- + O_2 \rightarrow Q + HO_2^- \qquad (66)$$

The alkaline solution leaving the cell is reacted with calcium hydroxide, which results in precipitation of calcium peroxide, which in turn is treated with carbon dioxide to yield hydrogen peroxide. The cell, equipped with a cation-exchange membrane and cathodes of graphite or glassy carbon, is reported to operate at a current efficiency of at least 95%.[322]

### 8.4. Development of Technology

The prerequisites for the development of a big industry were fulfilled when in 1905 the first process for electrosynthesis of hydrogen peroxide was patented[252] and industrial manufacture was started in 1909 in Weissenstein (Austria). Fundamental modifications in the cell design and the whole process, mainly related to the problem of hydrolysis of the Caro acid in the cell (instead of in the subsequent process stage), were introduced by Pietzsch and Adolph[254] as early as 1910 and by Löwenstein and Riedel[255] in 1924. All processes were persistently improved up to the 1960s.[253]

The electrochemical processes removed the two important impediments of hydrogen peroxide application, i.e., the formerly low attainable concentration and the rapid decomposition, by introducing a distillation subsequent to the electrosynthesis to get any desired concentration and high product purity. By addition of stabilizers the product quality could be further raised. Production rates could be increased by opening new markets for the improved hydrogen peroxide solutions. The competing chemical route (from barium peroxide with acids) could be substituted nearly completely. During World War II production boomed and was particularly high in Germany and Austria. A broad market was also available after the war in the paper and textile industry.[256,257] Although further substantial process improvements were not expected,[244] the electrochemical processes were perfected (e.g., by addition of some cations which form more easily soluble peroxodisulfates[323] and by cheaper anodes as mentioned before) until the overwhelming competition of the anthraquinone route came up.

The economic competitiveness of the cathodic or indirect reductions as compared to the alkyle anthroquinone process is at present an open question. As long-time current efficiency at technical current densities is satisfactory and energy consumption is comparable to that in the alkyle anthraquinone process,[240] large-scale electrochemical preparation of hydrogen peroxide might perhaps regain industrial relevance.

## 9. Perborate

The fate of industrial electrosynthesis of perborate is closely linked with that of hydrogen peroxide. As well as for hydrogen peroxide, nowadays the demand for perborate is expanding steadily, mainly for use as a component in washing agents (world production amounted to 500,000 t/year in 1974[261]), but electrochemical perborate synthesis has been substituted by chemical routes within the last few years.

Initially, perborate was manufactured industrially using $Na_2O_2$ according to[262]

$$NaBO_2 + Na_2O_2 + CO_2 + 4H_2O \rightarrow NaBO_3 \cdot 4H_2O + Na_2CO_3 \quad (67)$$

the $Na_2O_2$ itself being obtained from electrolytically produced Na.

At the beginning of this century, efforts were made to use a direct electrosynthesis by anodic oxidation of borate which proceeded with unsatisfactory current efficiency, until an addition of sodium carbonate to the electrolyte was proposed[263] and a larger plant went into production in 1920. By anodic oxidation at platinum, peroxocarbonate is generated:

$$2CO_3^{2-} \rightarrow C_2O_6^{2-} + 2e \quad (67a)$$

It hydrolyzes in the aqueous electrolyte. The resulting hydrogen peroxide reacts with sodium metaborate to perborate. The overall reaction starting from borax is

$$Na_2B_4O_7 \cdot 10H_2O + Na_2CO_3 + 10H_2O \xrightarrow{8F} 4NaBO_2(OH)_2 \cdot 3H_2O + CO_2 + 4H_2 \quad (68)$$

Carbon dioxide was bound by $Na_2CO_3$ under formation of $NaHCO_3$. To prevent a precipitation, sodium hydroxide was added to the electrolyte and reconverted the $NaHCO_3$ to $Na_2CO_3$. As an alternative, hydrogen carbonate could be reacted with $Na_2O_2$ and borax to produce in addition perborate[264]:

$$6NaHCO_3 + 4Na_2O_2 + Na_2B_4O_7 \cdot 10H_2O + 3H_2O$$
$$\rightarrow 4NaBO_2(OH)_2 \cdot 3H_2O + 6Na_2CO_3 \qquad (69)$$

Detailed technical process data are given by Machu.[244] The current efficiency is 60%, the energy consumption 3.8 kWh/kg borate at 5 kA/m² anodic c.d.

The process was operated discontinuously and therefore required high labor costs. It is evident that the process could only be operated economically as long as the hydrogen peroxide was itself produced by electrosynthesis. Since hydrogen peroxide is now manufactured cheaper by a chemical route, the existing electrosynthesis plants for perborate have been shut down within the last few years.

The kinetics of the primary anodic oxidation reaction (67a) has been studied by Wiel, Janssen, and Hoogland.[349] In the absence of $H_2O_2$ the standard exchange c.d. is small, less than 1.5 mA/cm². The overall current efficiency for perborate formation decreases with increasing concentration of $H_2O_2$ in the solution.

## 10. Permanganate

### 10.1. Industrial Significance

In the manufacture of potassium permanganate, too, electrochemical routes compete with chemical ones. However, today only the electrochemical production is of importance. Here, the route of anodic oxidation of potassium manganate is predominant as compared to the alternative anodic oxidation of manganese or of its alloys. The following treatment is restricted to the first route. Several variants of the second one are dealt with in References 265 and 266, but it seems that apart from a few exceptions[266b] the direct manganese oxidation is not industrially applied.

Permanganate manufacture is not a big industry. World production of potassium permanganate amounted to about 10,000 t/year before World War II and later expanded to about 30,000 t/year.[266b] Production of other permanganates is comparatively insignificant.

### 10.2. Reaction Fundamentals

Potassium manganate is anodically oxidized to permanganate:

$$MnO_4^{2-} - e \rightarrow MnO_4^- \qquad (70)$$

whereas hydrogen is evolved at the cathode. The kinetics of redox reaction (70) were studied by several authors.[360,361,362] It is a fast reaction. At a Pt rotating disk in a NaOH solution the steady state potential–c.d. curve is reversible.[361] The exchange current densities $j_0$ could be obtained from potentiostatic and galvanostatic transients.[361,360] Extrapolation from the cathodic and anodic

Tafel range yields different values of $j_0$. This was interpreted by the occurrence of two consecutive partial reactions consisting of adsorption–desorption processes of the redox components.[360]

Loss reactions are the cathodic reduction of permanganate

$$MnO_4^- + e \rightarrow MnO_4^{2-} \tag{71}$$

the anodic oxidation of hydroxide, when the manganate concentration has fallen to low values,

$$OH^- - e \rightarrow \tfrac{1}{2}H_2O + \tfrac{1}{4}O_2 \tag{72}$$

and the chemical reduction of permanganate when the hydroxide generated at the cathode has attained high concentrations,

$$MnO_4^- + OH^- \rightarrow MnO_4^{2-} + \tfrac{1}{2}H_2O + \tfrac{1}{4}O_2 \tag{73}$$

Reaction (71) is the major difficulty in cell design and can be circumvented by use of diaphragms. It is to the merit of Klonowski[267] to have shown that cathodic reduction can effectively be cut down by applying a cathodic current density which is substantially higher than the anodic one. In this case, a diaphragm may be omitted without damage to current efficiency.

### 10.3. Development of Technology

Development of industrial cells is characterized by the efforts to suppress the loss reactions (71)–(73). The first industrial cells, started in 1884[268] and at the turn of the century,[269] used a diaphragm to divide the interelectrode space. It was later proposed by Tumanow[270] to install two diaphragms and to introduce the electrolyte in the middle chamber to prevent all the loss reactions. The stratification of a KOH solution over a specifically heavier solution of potassium manganate with the intention to avoid a diaphragm was proposed by Deissler.[271] It is not known, however, whether these proposals were industrially applied. Immediate consideration was paid to the idea of Klonowski[267] to apply a relatively high cathodic current density. Only one year later, Schütz[272] presented the technical design of a diaphragmless cell applying the new concept. The ratio of cathodic and anodic current density is 5. The anode is made of a nickel sheet, the cathodes are of carbon steel sheets. A central agitator kept the electrolyte in permanent motion, preventing the formed permanganate to settle. The cell was operated in a batch mode. The design of Schütz served as basis for the industrial cells of Bitterfeld (Germany).[273] A good review of older processes is given in Reference 265.

### 10.4. Industrial Cells

An interesting cell, developed during World War II by Müller and Suter, provided a heating below and a cooling beside the electrodes to induce a thermal self-convection of the electrolyte and to dispense with the agitator.[265] The flow ensured a mixing sufficient to keep the concentration below saturation. Crystallization at the anodes and subsequent evolution of oxygen according to Eq. (72) was thus prevented. Current efficiency was 80% and power consumption only 0.5 kWh/kg permanganate with a c.d. of 1 and 1.5 kA/m² at the anode and cathode, respectively.[363]

A continuous process industrially applied in Czechoslovakia[266a] uses compressed air instead of an agitator.

The cell of Carus Chemical Company[274,275] features inclined electrodes (Fig. 22). Cathodically evolved hydrogen causes a self-convection of the

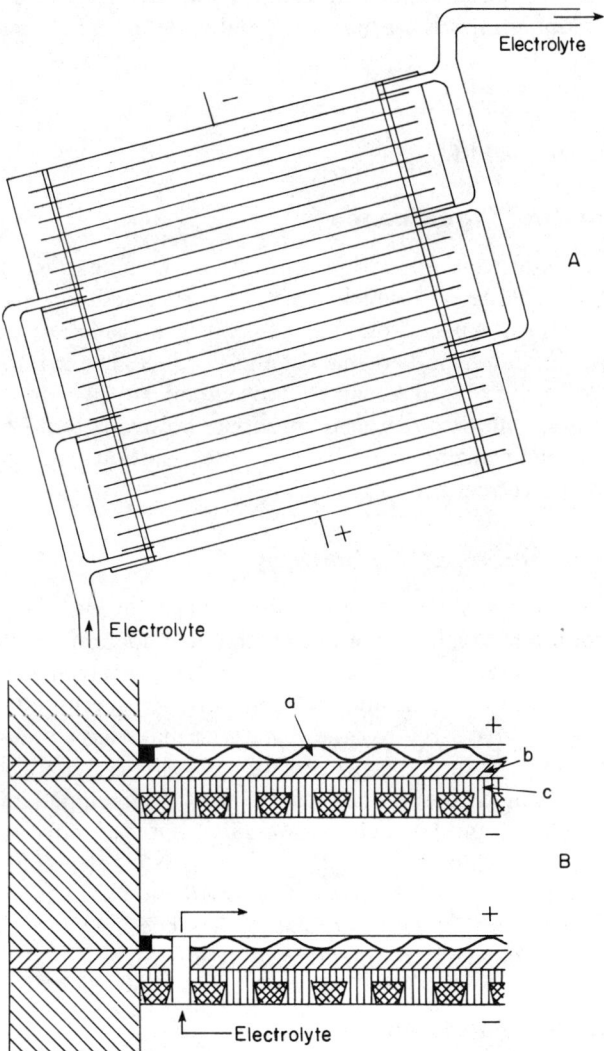

Figure 22. Industrial permanganate cell (Carus).[274] A, Cross section of the cell; B, detail; a, anode consisting in a Monel wire net; b, basic sheet of the bipolar electrode; c, cathode with nonconductive areas.

electrolyte which makes external agitation superfluous. Contrarily to the aforementioned cells, the Carus cell is equipped with bipolar electrodes. The cathodic current density amounts to 6000–15,000 A/m$^2$ and is greater than the anodic current density by the factor 100. The current efficiency is correspondingly high.

Anodes used are or were of nickel,[272,276] copper–nickel alloys such as Monel,[274] stainless steel, or nickel-plated steel.[266] Cathodes are generally made of carbon steel.[276]

Production of potassium permanganate worldwide is in the hands of a few producers. Competition is strong, and reliable detailed information on technology is rare.

## 11. Chromic Acid

### 11.1. Industrial Significance

Several industrial oxidation processes involve the formation of chromium(III) sulfate,[285] which must be processed for economical and ecological reasons. Apart from the possibility of isolating chromium(III)-containing compounds and to market the product, the regeneration of the solution is to be considered. It can be done chemically by oxidation with air in an alkaline medium. Strongly acid solutions, however, require considerable quantities of neutralizing agent. For these solutions an electrochemical regeneration is economically superior.

### 11.2. Development of Technology

In 1893, Häussermann[279] described a process producing chromate from sodium chromite in alkaline solution by anodic oxidation. At the turn of the century it was shown[280,282,283] that chromium(III) salts in an acid solution can be oxidized anodically to chromic acid. Further, the superior effectivity of lead dioxide anodes as compared to other electrode materials was already known. At the same time, a continuous process for electrolytic preparation of chromic acid solutions using chromium(III) wastes[284] was developed and industrially applied in 1902[285] and has been repeatedly improved since then. Different processes were developed in the following years. Up to the present time, commercial interest in the field has not dropped and has included recently the electrolytic reduction of chromic acid in connection with the intensified requirements of active environmental protection.[286,287]

### 11.3. Reaction Fundamentals

The electrolyte subjected to regeneration is usually a strongly acid solution of Chromium(III) sulfate, the chromium ions of which are complex bound, sometimes accompanied by organic acids.

The anodic process at lead dioxide is understood as a reaction of adsorbed complex chromium(III) ion with oxygen to form an intermediate Cr–O complex which is anodically oxidized to chromate[285]:

$$Cr^{3+} + O \rightarrow CrO^{3+} \tag{74}$$

$$CrO^{3+} + 3H_2O + e \rightarrow CrO_4^{2-} + 6H^+ \tag{75}$$

At room temperature the current efficiency is virtually 100% at lead dioxide anodes and current densities of about 100 A/m².[285] Smooth platinum, however, leads predominantly to evolution of oxygen and yields a current efficiency of about 1% only. Experimental investigations[288,289] made evident that the overpotential cannot be decisive. This finding contradicts[290] the assumption of a solely electrochemical mechanism according to

$$Cr^{3+} - 3e \rightarrow Cr^{6+} \tag{76}$$

$$Cr^{6+} + 4H_2O \rightarrow CrO_4^{2-} + 8H^+ \tag{77}$$

On the other hand, a purely chemical mechanism

$$Cr^{3+} + 1.5O + 2.5H_2O \rightarrow CrO_4^{2-} + 5H^+ \tag{78}$$

is also considered unlikely.[285]

Pure chromic acid free of extraneous ions does not undergo a cathodic reduction. In this case, only hydrogen is evolved at the cathode.[291]

## 11.4. Industrial Cells

Design of technical cells is complicated by the fact that the chromic acid obtained by anodic oxidation is in reality not pure and thus easily reduced at the cathode. For this reason, diaphragms have been used industrially since the early days of technical cells. The solution to be regenerated passes through all cathode compartments of the cell battery before it enters the anode compartments where the oxidation takes place. The aim of this electrolyte flow scheme[282,284] is to compensate continuously for the diminution of sulfuric acid in the catholyte by the corresponding increase in the anolyte. Of course, any residual chromic acid contained in the chromium(III) solution coming from the preceding chemical oxidation and subjected to regeneration is completely reduced at the cathode.

Originally, the technical diaphragms which were made of clay, and later of sintered alumina and silica, caused big problems.[282] A modern process[285] operating with diaphragms and based on the design principles as outlined above uses a current density of 300 A/m²,[293] cell voltage of 3–4 V, and an electrolyte temperature of 50–80°C. The current efficiency is near 100% except for the first few anode compartments (where a decomposition of organic components occurs) and for the last few anode compartments [where, owing to the decreasing chromium(III) concentration, the limiting current is attained]. Therefore, it is not economical to completely oxidize the chromium(III).

Energy consumption amounts to 4 kWh/kg $CrO_3$. The diaphragm seems to contribute significantly to the voltage drop.

Cells that are industrially applied to regenerate solutions used for etching butadiene-containing plastics[366] also exhibit the diaphragm problem. The current efficiency is limited to 60–70%, partly owing to the oxidation of organic compounds. The process, however, is considered economical.[345]

As an alternative, various cells without diaphragms were developed: To avoid the diaphragm problems in general as well as to circumvent the disadvantage of complete reduction of chromic acid in the cathode compartment, Le Blanc designed a cell with only a partial separation of anode and cathode compartment.[294] The anolyte was moved slowly so as not to stir the stagnant catholyte. Current efficiency was surprisingly large: 83%. A similar cell was industrially applied.[293,295] Another cell without a diaphragm uses a packed bed anode.[297] Other means to dispense with a diaphragm are the addition of ammonium sulfate and sodium acetate to the electrolyte in considerable concentrations.[298] Their influence was studied by Askenasy and Révai.[301]

Chromium(III) solutions (nearly) free of sulfate may in any case be regenerated in cells without a diaphragm.[299] Using lead anodes and steel cathodes a final concentration of 600 g $CrO_3$/liter was obtained with current efficiencies of 95% and an energy consumption of 2.5 kWh/kg $CrO_3$.[296]

The influence of sulfuric acid, chromic acid, temperature, current density, and anode material on the performance in a continuous regeneration was investigated by McKee and Leo.[300] The benefit of a low anodic current density was confirmed and was later applied industrially.[277]

## 12. Sodium Sulfate Electrolysis

Caustic lye and sulfuric acid may be prepared by electrosynthesis from sodium sulfate solution. The process was applied in Germany[302–304] and had considerable significance for working off sodium sulfate obtained as byproduct in the manufacture of rayon.

The solution was electrolyzed between a graphite anode and a moving mercury cathode. As in alkali chloride electrolysis by the amalgam process, the amalgam formed was subsequently reacted to generate sodium hydroxide and hydrogen. Corrosion problems in the electrolysis cell caused by the anodically formed sulfuric acid lead to replacement of the graphite anodes by a lead–silver alloy. A diaphragm was used to prevent an extensive dilution of the acid by sulfate and the mixing of oxygen with the cathodically evolved hydrogen.

A cell with a double diaphragm is described in Reference 305 where the influence of the process parameters is also discussed.

Owing to the high energy consumption and to the substantial capital investment needed, the electrochemical route is not considered competitive in the present price situation and seems not to be practiced industrially.[306]

However, an application of the electrochemical route for removal of sulfur dioxide from waste gases was reported.[307,308] The sodium pyrosulfite solution resulting in alkaline absorption is reacted to sodium sulfate which is electrochemically processed. The sodium hydroxide and sulfuric acid obtained are recycled. The cell is operated with an ion-selective membrane.

## Acknowledgment

The authors are indebted to Prof. P. Gallone (Milano, Italy) for his valuable comments.

## Auxiliary Notation

| | | | |
|---|---|---|---|
| $a_i$ | activity of species $i$ | $N$ | interfacial flux density of hypochlorite, equal to three times the anodic chlorate formation. $N_d$, $N_e$, $N_h$, $N'_h$ (see Table 1) |
| $a$ | $=(k_h/D_1)^{0.5}$ | | |
| $b$ | number in Eq. (24) | | |
| $B$ | factor in Eq. (34) | | |
| $c$ | concentration | $t$ | time |
| $D$ | diffusion coefficient | $T$ | temperature |
| $f$ | activity coefficient | $y$ | spatial coordinate perpendicular to the electrode surface |
| $F$ | Faraday constant | | |
| $I$ | electrolysis current | $V_r$ | volume of chemical reactor |
| $j$ | current density | $\dot{V}_L$ | liquid flow rate |
| $k$ | rate constant of chemical chlorate formation, Eq. (13) | $\alpha$ | charge transfer coefficient |
| | | $\delta$ | diffusion layer thickness |
| $k^*$ | proportionality factor in Eq. (36) | $\varepsilon$ | current efficiency |
| $k_h$ | rate constant of chlorine hydrolysis, Eq. (3) | | |
| $K$ | dissociation constant of hypochlorous acid, Eq. (14) | **Subscripts** | |
| $K_a$ | dissociation constant of hypochlorous acid, Eq. (16) | c | cell |
| | | h | chemical reactor |
| $K^*$ | dissociation constant of hypochlorous acid, Eq. (17) | 1 | $Cl_2$ |
| | | 2 | total hypochlorite, $HClO + ClO^-$ |

## References

1. W. C. Gardiner, Advances in electrolytic production of the industrial chemicals from 1952 to 1977, *J. Electrochem. Soc.* **125**, 22C–29C (1978).
2. H. V. Casson and G. J. Crane, Technology and economics of on-site sodium chlorate production, *Paper Trade J.*, 65–68 (May 15, 1967).
3. H. V. Casson, G. J. Crane, and G. E. Styan, Technology and economics of on-site sodium chlorate production, *Pulp Paper Mag. Canada*, 39–51 (1968).
4. Anonymous, Sodium chlorate: Bright pulp industry future, *Chem. Eng. (N.Y.)*, 74–76 (September 1967).
5. E. Kesting, Ger. Pat. 831,542 (1948).
6. W. H. Rapson, Recent developments in the manufacture of chlorine dioxide, *Can. J. Chem. Eng.* **36**, 262–266 (1958).
7. W. H. Rapson, Pulp bleaching as of today, *Tappi* **62**(6), 14–17 (1979).

8. G. C. Lyttle, Economics of oxygen applications for the pulp and paper industry, *Paper Trade J.*, 25–27 (Feb. 28, 1979).
9. A. R. Despić, in *Electrochemistry. The Past Thirty and the Next Thirty Years*, H. Bloom and F. Gutmann, eds., Plenum, New York (1977), p. 15.
10. W. von Hisinger and J. Berzelius, Versuche, betreffend die Wirkung der electrischen Säule auf Salze und auf einige von ihren Basen, *Neues Allg. J. Chem.* **1**, 115–149 (1803).
11. H. Kolbe, Beobachtungen über die oxydirende Wirkung des Sauerstoffs, wenn derselbe mit Hülfe einer elektrischen Säule entwickelt wird, *J. Prakt. Chem.* **41**, 137–139 (1847).
12. C. Watt, Brit. Pat. 13,785 (1851).
13. J. B. Kershaw, *Die elektrolytische Chloratindustrie*, Knapp, Halle (Saale) (1905).
14. F. Oettel, Studien über die elektrolytische Bildung von unterchlorigsauren Salzen und chlorsauren Salzen, *Z. Elektrochem.* **1**, 354–361 (1894); **1**, 474–480 (1895).
15. F. Oettel, Zur Elektrolyse von Chlorcalcium-Lösungen, *Z. Elektrochem.* **5**, 1–9 (1898).
16. F. Haber and S. Grinberg, Über die Elektrolyse der Salzsäure, *Z. Anorg. Chem.* **16**, 198–228, 329–361 (1898).
17. H. Wohlwill, Über die Elektrolyse der Alkalichloride, *Z. Elektrochem* **5**, 52–76 (1898).
18. F. Foerster, E. Müller, and F. Jorre, Zur Kenntnis der Vorgänge bei der Elektrolyse der Alkalichlorid-Lösungen, *Z. Elektrochem.* **6**, 11–23 (1899).
19. F. Foerster and F. Jorre, Zur Kenntniss der Beziehungen der unterchlorigsauren Salze zu den chlorsauren Salzen, *J. Prakt. Chem.* **59**, 53–101 (1899).
20. F. Foerster, *Elektrochemie wässeriger Lösungen*, 3rd ed., Barth, Leipzig (1922): (a) p. 672, (b) pp. 780–786, (c) pp. 789–792, (d) p. 842, (e) p. 681, (f) p. 663.
21. V. de Valera, On the theory of electrochemical chlorate formation, *Trans. Faraday Soc.* **49**, 1338–1351 (1953).
22. V. de Valera, The mechanism of chloride electrolysis, *Trans. Faraday Soc.* **52**, 250–260 (1956).
23. V. de Valera, Anodic oxygen evolution in chlorate cells, *Trans. Faraday Soc.* **60**, 1450–1456 (1964).
24. N. Ibl and D. Landolt, On the mechanism of anodic chlorate formation in dilute NaCl solutions, *J. Electrochem. Soc.* **115**, 713–720 (1968).
25. D. Landolt and N. Ibl, On the mechanism of anodic chlorate formation in concentrated NaCl solutions, *Electrochim. Acta* **15**, 1165–1183 (1970).
26. M. M. Jakšić, Mass transfer and optimization of Faradaic yields in a chlorate cell process, *Electrochim. Acta* **21**, 1127–1136 (1976).
27. J. Landin, Swedish Patent 8820 (1897).
28. P. Imhoff, German Patent 110,505 (1898).
29. P. Imhoff, U.S. Patent 627,063 (1898).
30. C. Wagner, The cathodic reduction of anions and the anodic oxidation, *J. Electrochem. Soc.* **101**, 181–185 (1954).
31. E. Müller and P. Koppe, Einfluss der Stromkonzentration auf die elektrolytische Chloratbildung, *Z. Elektrochem.* **17**, 421–430 (1911).
32. A. Schumann-Leclercq, French Patent 772,326 (1933).
33. D. V. Kokoulina and L. I. Krishtalik, The volume reaction forming sodium chlorate in the anolyte of the chlorate electrolyzer, *Sov. Electrochem.* **7**, 325–330 (1971); *Elektrokhim.* **7**, 346–352 (1971).
34. E. Hausmann, West German Patent 957,937 (1957).
35. J. Fleck, Chlorat-Elektrolyse, *Chem. Ing. Tech.* **43**, 173–177 (1971).
36. T. Matsumura, R. Itai, M. Shibuya, and G. Ishi, Electrolytic manufacture of sodium chlorate with magnetite anodes, *Electrochem. Technol.* **6**, 402–404 (1968).
37. M. Janes, Graphite anodes in brine electrolysis. II. Application to chlorate cells, *Trans. Electrochem. Soc.* **92**, 23–44 (1947).
38. N. N. Nechiporenko, P. Kh. Voroshilov, N. V. Sivokon, and V. K. Beidin, Investigation of the anode process in electrolysis of sodium chloride solution, *Zh. Prikl. Khim.* **33**, 1818–1828 (1960); *J. Appl. Chem. USSR* **33**, 1798–1807 (1960).

39. R. Weiner and G. Klein, Lebensdauer von Graphit-Anoden bei der technischen Chlorat-Herstellung, *Chem. Ing. Tech.* **29**, 339–344 (1957).
40. M. M. Jakšić and I. M. Czonka, Improvements in the impregnation of graphite anodes for electrolysis of alkali chlorides, *Electrochem. Technol.* **5**, 473–478 (1967).
41. M. M. Jakšić, The effect of pH on graphite wear in a chlorate cell process, *J. Appl. Electrochem.* **3**, 219–225 (1973).
42. G. Wranglén, B. Sjödin and B. Wallén, A new test method for graphite anodes in alkali chloride electrolysis, *Electrochim. Acta* **7**, 577–587 (1962).
43. T. Nagai and T. Takei, Electrolytic production of chlorate I. Mechanism of chlorate formation, *Denki Kagaku* **24**, 557–561 (1956).
44. B. Carlson, Die elektrochemische Industrie Schwedens, *Z. Elektrochem.* **6**, 471–472 (1900).
45. K. R. Koziol, K. H. Sieberer, H.-C. Rathjen, J. B. Zenk, and E. F. Wenk, Neue Entwicklungen und Möglichkeiten mit beschichteten Titan-Anoden, *Chem. Ing. Tech.* **49**, 292 (1977).
46. G. Grube and A. Burkhardt, Die Verwendung verchromter Kathoden bei der elektrolytischen Darstellung der Chlorate, *Z. Elektrochem.* **30**, 67–72 (1927).
47. T. Nagai and T. Takei, Reduction at cathode and its prevention by chromium plated cathode in chlorate production, *J. Electrochem. Soc. Jpn* **25**, E-79 (1957).
48. B. D. Yurkevich and B. M. Vrevskii, Mechanism of electrolysis of solutions of alkali metal chlorides with ion-exchanger cathodes (in Russian), *Tr. Probl. Lab. Leningr. Inst. Tekst. Logk. Prom.*, 345–349 (1971); cited in *Chem. Abstr.* **78**, 51708 (1973).
49. J. R. Newberry, W. C. Gardiner, A. J. Holmes, and R. F. Fogle, High current density chlorate cell using platinized anodes, *J. Electrochem. Soc.* **116**, 114–118 (1969).
50. J. R. Hodges, private communication, 1978.
51. L. M. Elina, Yu. V. Dobrov, and V. M. Gitneva, Study of titanium as a material for cathodes in chlorate baths, *Zh. Prikl. Khim.* **47**, 1655–1657 (1974); *J. Appl. Chem. USSR* **47**, 1699–1701 (1974).
52. J. R. Hodges, Pennwalt Corp., U.S. Patent 4,075,077 (1978).
53. F. Foerster and E. Müller, Beiträge zur Theorie der Elektrolyse von Alkalichloridlösungen, *Z. Electrochem.* **9**, 171–185, 195–208 (1903).
54. F. Foerster and P. Dolch, Die Umwandlung von Hypochlorit in Chlorat in alkalischer Lösung, *Z. Elektrochem.* **23**, 137–147 (1917).
55. F. Foerster, The electrolysis of hypochlorite solutions, *Trans. Am. Electrochem. Soc.* **46**, 23–50 (1924).
56. M. W. Lister, Decomposition of sodium hypochlorite: the uncatalyzed reaction, *Can. J. Chem.* **34**, 465–478 (1956).
57. M. W. Lister and R. C. Petterson, Oxygen evolution from sodium hypochlorite solutions, *Can. J. Chem.* **40**, 729–733 (1962).
58. M. W. Lister, Decomposition of sodium hypochlorite: the catalyzed reaction, *Can. J. Chem.* **34**, 479–488 (1956).
59. J. M. González Barredo, Demostración de la naturaleza autocatalítica de la descomposición del ión hipocloroso, *An. Fis. Quim.* **37**, 123–157 (1941).
60. N. V. S. Knibbs, Perchlorate formation, *Trans. Faraday Soc.* **16**, 424–433 (1920).
61. N. V. S. Knibbs, Chlorate formation, *Trans. Faraday Soc.* **16**, 415–424 (1920).
62. H. Imagawa, Chemical reactions in the chlorate manufacturing electrolytic cell. (a) Part I: The vapour pressure of hypochlorous acid on its aqueous solution, *J. Electrochem. Soc. Jpn.* **18**, 382–385 (1950). (b) Part II: The vapour pressure of hypochlorous acid on its mixed aqueous solution with sodium chlorate, *J. Electrochem. Soc. Jpn.* **19**, 271–274 (1951). (c) Part III: Reaction of chlorate formation with sodium hypochlorite and free hypochlorous acid, *J. Electrochem. Soc. Jpn.* **20**, 25–28 (1952). (d) Part IV: The effect of sodium chlorate on the reaction of chlorate formation, *J. Electrochem. Soc. Jpn.* **20**, 571–574 (1952). (e) Part V: The role of NaCl in chlorate formation and a new reaction scheme of chlorate formation, *J. Electrochem. Soc. Jpn.* **21**, 520–525 (1953).

63. F. Foerster, Zur Kenntniss des Überganges der unterchlorigsauren Salze in chlorsaure Salze, *J. Prakt. Chem.* **63**, 141–166 (1901).
64. A. Skrabal and A. Berger, Determination of the dissociation constants of hypochlorous acid by the kinetic method, *Monatsh. Chem.* **70**, 168–192 (1937).
65. A. Skrabal, Zur Theorie der Halogenathbildung, *Monatsh. Chem.* **84**, 102–115 (1953).
66. J. d'Ans, H. E. Freund, Über die Chloratbildung aus Hypochlorit, *Z. Elektrochem.* **61**, 10–18 (1957).
67. I. E. Flis and M. K. Bynyaeva, Autoxidation processes in hypochlorite solutions, *Zh. Prikl. Khim.* **30**, 339 (1957); *J. Appl. Chem. USSR* **30**, 359–365 (1957).
68. M. M. Jakšić, B. Ž. Nikolić, I. M. Csonka, and A. B. Djordjević, Effect of neutral salts on conversion of available chlorine to chlorate, *J. Electrochem. Soc.* **116**, 684–687 (1969).
69. J. C. Morris, The acid ionization constant of HOCl from 5 to 35°, *J. Phys. Chem.* **70**, 3798–3805 (1966).
70. H. Vogt, Comments on Jakšić [Reference 71], *J. Electrochem. Soc.* **121**, 1606 (1974).
71. M. M. Jakšić, Mutual effect of current density, pH, temperature, and hydrodynamic factors on current density in the chlorate cell process, *J. Electrochem. Soc.* **121**, 70–79 (1974).
72. M. Ya. Fioshin, *Uspekhi v oblasti elektrosinteza neorganiceskikh soedinenii* (Advances in the electrosynthesis of inorganic compounds), Idsdatelstvo "Khimiya" Moscow (1974).
73. V. A. Shlyapnikov, The mechanism of formation of sodium chlorate, *Sov. Electrochem.* **7**, 1080–1082 (1971); *Electrokhim.* **7**, 1128–1131 (1971).
74. V. A. Shlyapnikov, Theory of electrochemical synthesis of sodium chlorate, *J. Appl. Chem. USSR* **46**, 1076–1082 (1973); Zh. Prikl. Khim. **46**, 1014–1018 (1973).
75. A. Rius and J. Llopis, Sobre la oxidacion anodica de los hipochloritos alcalinos, *An. Fis. Quim.* **41**, 1030–1053 (1945).
75a. A. Rius and J. Llopis, oxidacion anodica de hipocloritos con anodos distintos de platino, *An. Fis. Quim.* **41**, 1282–1293 (1945).
76. M. M. Jakšić, A. R. Despić, I. M. Csonka, and B. Ž. Nikolić, Studies on chlorate cell process. V. Theory and practice of a modified technology for electrolytic chlorate production, *J. Electrochem. Soc.* **116**, 1316–1322 (1969).
77. M. M. Jakšić, A. R. Despić, B. Ž. Nikolić, and S. M. Maksić, Effect of some anions on the chlorate cell process, *Croat. Chem. Acta* **44**, 61–66 (1972).
78. V. A. Shlyapnikov and T. S. Filippov, On the formation of chlorates in the electrochemical method of their production. I, *Electrokhim.* **2**, 1273–1281 (1966); *Sov. Electrochem.* **2**, 1165–1172 (1966).
79. V. A. Shlyapnikov and T. S. Filippov, On the problem of chlorate formation in the electrochemical method of its production. II, *Electrokhim.* **4**, 15–18 (1968); *Sov. Electrochem.* **4**, 20–23 (1968).
80. V. A. Shlyapnikov, The mechanism of sodium chlorate formation, *J. Appl. Chem. USSR* **42**, 2051–2056 (1969); *Zh. Prikl. Khim.* **42**, 2182–2188 (1969).
81. E. I. Yakovleva, K. I. Rozental, and T. S. Filippov, The electrochemical formation mechanism of chlorine–oxygen compounds on a smooth platinum electrode. I, *Zh. Fiz. Khim.* **30**, 937–944 (1956).
82. I. E. Flis and I. M. Vorob'ev, Kinetic investigation of the processes on the platinum electrode in hypochlorite solutions, *Russ. J. Phys. Chem.* **37**, 973–977 (1963).
83. J. S. Mayell and S. H. Langer, Electrochemical kinetics of chloride-ion oxidation at a bright platinum electrode, *Electrochim. Acta* **9**, 1411–1416 (1964).
84. I. Atanasiu and L. Stancu, Electrochemical mechanism of formation of oxygenated chlorine compounds at graphite electrodes, *Bul. inst. politek. Bucuresti* **20**, 61–73 (1958); *Chem. Abstr.* **55**, 1238h (1961).
85. T. S. Filippov and E. I. Yakovleva, Study of the mechanism of electrochemical formation of oxychloride compounds by anodic polarography, Trudy Chetvertogo Soveschchaniya po Elektrokhimii, Moscow (1956), pp. 257–262; *Chem. Abstr.* **54**, 8368a (1960).
86. F. Foerster and E. Müller, Über das Verhalten der unterchlorigen Säure und ihrer Salze bei der Elektrolyse, *Z. Elektrochem.* **8**, 633–638, 665–672 (1902).

87. A. B. Djordjević, B. Ž. Nikolić, I. V. Kadija, A. R. Despić, and M. M. Jakšić, Kinetics and mechanism of electrochemical oxidation of hypochlorite ions, *Electrochim. Acta* **18**, 465–471 (1973).
88. F. Foerster and E. Müller, Zur Kenntnis der Elektrolyse, zumal der Alkalichloride, an platinierten Elektroden, *Z. Elektrochem.* **8**, 515–540 (1902).
89. L. Hammar and G. Wranglén, Cathodic and anodic efficiency losses in chlorate electrolysis, *Electrochim. Acta* **9**, 1–16 (1964).
90. M. S. Chao, The diffusion coefficients of hypochlorite, hypochlorous acid, and chlorine in aqueous media by chronopotentiometry, *J. Electrochem. Soc.* **115**, 1172–1174 (1968).
91. M. M. Jakšić, Individual ionic activities and mass transfer in anodic chlorate formation, *J. Appl. Electrochem.* **3**, 307–314 (1973).
92. A. R. Despić, M. M. Jakšić, and B. Ž. Nikolić, The effect of kinetic and hydrodynamic factors on current efficiency in the chlorate cell process, *J. Appl. Electrochem.* **2**, 337–343 (1972).
93. M. M. Jakšić, B. Ž. Nikolić, and M. D. Spasojević, Die Grundlagen der optimalen Auslegung von Chloratzellen, *Chem. Tech. (Leipzig)* **27**, 158–162, 534–538 (1975).
94. M. M. Jakšić, A. R. Despić, and B. Ž. Nikolić, The latest technological developments in the electrolytic production of chlorates, *Elektrokhim.* **8**, 1573–1584 (1972); *Sov. Electrochem.* **8**, 1533–1542 (1972).
95. N. Ibl and D. Landolt, to be published.
96. D. Landolt and N. Ibl, Anodic chlorate formation on platinized titanium, *J. Appl. Electrochem.* **2**, 201–210 (1972).
97. I. Taniguchi and T. Sekine, Chemical reaction of chlorate formation. *Denki Kagaku* **43**, 715–720 (1975).
98. J. Claus, Evaluation of reactions in chlorate cells by digital computer, paper 256 presented at the Electrochemical Society Meeting, Boston, 1968.
99. Anonymous, New anodes show off for chemical producers, *Chem. Eng. (N.Y.)* **72**, 82–83 (July 19, 1965).
100. K. C. Narasimham and H. V. K. Udupa, Preparation and applications of graphite substrate lead dioxide (GSLD) anode, *J. Electrochem. Soc.* **123**, 1294–1298 (1976).
101. M. W. Lister, The decomposition of hypochlorous acid. *Can. J. Chem.* **30**, 879–889 (1952).
102. V. I. Éberil' and T. S. Filippov, Behavior of graphite anodes under the conditions of electrochemical production of sodium chlorate at different current densities, *Zh. Prikl. Khim.* **40**, 2482–2487 (1967); *J. Appl. Chem. USSR* **40**, 2377–2382 (1967).
103. T. S. Filippov, V. I. Éberil', R. A. Agapova, and G. N. Razygraeva, Behavior of graphite electrodes under the conditions of electrochemical production of sodium chlorate: effects of added sodium chromate and of electrolyte pH, *Zh. Prikl. Khim.* **40**, 2488–2491 (1967); *J. Appl. Chem. USSR* **40**, 2383–2386 (1967).
104. V. I. Éberil' and F. V. Kupovich, Influence of the NaCl concentration on magnitude of the anode potential and wear of the graphite anodes in the electrochemical preparation of chlorates, *Elektrokhim.* **6**, 332–335 (1970); *Sov. Electrochem.* **6**, 324–326 (1970).
105. R. A. Agapova and L. M. Elina, Influence of electrolyte pH on the behaviour of graphite anodes under the conditions of electrochemical production of sodium chlorate, *Zh. Prikl. Khim.* **44**, 1302–1307 (1971); *J. Appl. Chem. USSR* **44**, 1320–1324 (1971).
106. R. A. Agapova and L. M. Elina, Influence of sodium chloride concentration and electrolyte temperature on the stability of graphite anodes under the conditions of electrolytic production of chlorate, *Zh. Prikl. Khim.* **44**, 1514–1518 (1971); *J. Appl. Chem. USSR* **44**, 1536–1539 (1971).
107. V. I. Éberil' and L. M. Elina, Some peculiarities of the behaviour of graphite anodes in the electrolysis of solutions of NaCl and, in particular, in the production of chlorates, *Elektrokhim.* **6**, 782–786 (1970); *Sov. Electrochem.* **6**, 758–762 (1970).
108. V. I. Éberil', D. V. Kokoulina, L. I. Krishtalik, and L. M. Elina, The problem of the reasons for the increased internal wear of the graphite anodes in the electrochemical preparation of chlorate, *Sov. Electrochem.* **5**, 304–307 (1969), *Elektrokhim.* **5**, 336–340 (1969).

109. R. T. Atanasoski, B. Ž. Nikolić, M. M. Jakšić, and A. R. Despić, Platinum–iridium catalyzed titanium anode. I. The properties and use in chlorate electrolysis, *J. Appl. Electrochem.* **5**, 155–158 (1975).
110. V. A. Shlyapnikov, Role of the anode material in electrosynthesis of chlorates, *J. Appl. Chem. USSR* **49**, 86–89 (1976); *Zh. Prikl. Khim.* **49**, 90–94 (1976).
111. T. R. Beck, A contribution to the theory of electrolytic chlorate formation, *J. Electrochem. Soc.* **116**, 1038–1041 (1969).
112. G. D. Westerlund, Canadian Patent 914,610 (1970/1972).
113. A. J. Holmes, U.S. Patent 3,043,757 (1962).
114. N. Ibl and E. Adam, Optimierung in der elektrochemischen Verfahrenstechnik, *Chem. Ing. Techn.* **37**, 573–581 (1965).
115. J. Fleck, West German Auslegeschr. 1,667,574 (1967/1973).
116. R. Bauer, Die Chlorat-Elektrolyse, *Chem. Ing. Tech.* **34**, 376–379 (1962).
117. K. Hass and W. Tromm, Chloralkali-, Chlorwasserstoff- und Wasser-Elektrolyse, *Chem. Ing. Tech.* **40**, 557–564 (1968).
118. J. C. Harke *et al.*, West German Offen. 2,248,552 (1973).
119. V. A. Shlyapnikov and E. I. Adaev, The electrosynthesis of Berthollet's salt, potassium chlorate, *Sov. Electrochem.* **12**, 1001–1003 (1976); *Elektrokhim.* **12**, 1089–1092 (1976).
120. G. O. Westerlund, Canadian Patent 828,147 (1966/1969).
121. V. A. Shlyapnikov, Role of mixing of the electrolyte in electrosynthesis of chlorates, *J. Appl. Chem. USSR* **49**, 371–373 (1976); *Zh. Prikl. Khim.* **49**, 370–372 (1976).
122. British Patent 1,185,507 Krebs & Cie, Paris (1967).
123. P. Remirez, Producing captive sodium chlorate in integrated pump mills, *Chem. Eng. (London)* **74**, 136–138 (14 August 1967).
124. V. I. Ginzburg and M. A. Mel'nikov, Conditions for protecting steel cathodes during electrolytic production of sodium chlorate, *Sov. Chem. Ind.*, 414–416 (1971).
125. V. A. Shlyapnikov and T. S. Filippov, Cathodic reduction during chlorate production, *Elektrokhim.* **5**, 866–868 (1969); *Sov. Electrochem.* **5**, 806–807 (1969).
126. Solvay & Cie, Austrian Patent 292027 (1971).
127. Chemetics Int. Ltd., Sodium chloride, Chlorine dioxide, Information, 1978.
128. British Patent 1,161,678 (1969).
129. P. Wintzer, Entwicklung und Trend der Chlordioxidbleiche mit integrierter Chlorat-Elektrolyse für die Zellstoffindustrie, *Chem. Ing. Tech.* **52**, 392–398 (1980).
130. V. de Nora and J.-W. Kühn-von Burgsdorff, Der Beitrag der dimensionsstabilen Anoden (DSA) zur Chlor-Technologie, *Chem. Ing. Tech.* **47**, 125–128 (1975).
131. L. M. Elina, V. M. Gitneva, V. I. Brystov, and N. M. Shmygul', Use of ruthenium oxide anodes in chlorate electrolysis, *Sov. Electrochem.* **10**, 59–61 (1974); *Elecktrokhim.* **10**, 68–70 (1974).
132. E. Hausmann, E. Kramer, and H. Vogt, Chloratelektrolyse und Herstellung von Chlordioxid, *Chem. Anlagen Verfahren* **1970**(5), 59–62, **1970**(8), 66.
133. I. E. Veselovskaya, E. M. Kuchinskii, and L. V. Morochko, The cathodic reduction of chlorate, *Zh. Prikl. Khim.* **37**, 76–83 (1964); *J. Appl. Chem. USSR* **37**, 85–91 (1964).
134. T. Nagai and T. Takei, Prevention of cathodic reduction with chromium cathode in chlorate production, *J. Electrochem. Soc. Jpn.* **25**, E-108 (1957).
135. V. I. Skripchenko, Ye. P. Drozdetskaya, and K. G. Il'in, Production of sodium chlorate without introducing protective additives, *Sov. Chem. Ind.* 813–815 (1971); *Khim. Prom.* (1971).
136. A. T. Kuhn and H. B. H. Hamzah, The effect of electrode roughness and the ratio of anode to cathode area on the performance of an undivided hypochlorite cell, *Chem. Ing. Tech.* **52**, 762–763 (1980).
137. A. Suter, French Patent 947,230 (1949).
138. M. Antler and C. A. Butler, Degradation mechanisms of platinum- and rhodium-coated titanium anodes in the electrolysis of chloride and chloride–chlorate solutions, *Electrochem. Technol.* **5**, 126–130 (1967).

138a. R. Piontelli, Degradation mechanisms of platinum- and rhodium-coated titanium anodes in the electrolysis of chloride and chloride–chlorate solutions, *Electrochem. Technol.* **5**, 558–559 (1967).
139. H. Vogt, in *Ullmanns Encyklopädie der technischen Chemie*, 4th ed. Vol. 9, Verlag Chemie, Weinheim (1975), pp. 553–565.
140. I. Roušar, Calculation of current density distribution and terminal voltage for bipolar electrolyzers; application to chlorate cells, *J. Electrochem. Soc.* **116**, 676–683 (1969).
141. Pennwalt-Catalytic, information leaflet, 1978.
142. W. Baucke and A. Winsel, Zum $H_2O_2$-Problem in Sauerstoff-Diffusionselektroden, *Electrochim. Acta* **12**, 31–40 (1967).
143. W. H. Sheltmire, in *Chlorine. Its Manufacture, Properties and Uses*, J. S. Sconce, ed., Reinhold, New York (1962): (a) p. 512ff, (b) p. 516.
144. Société Ugine, French Patent 1,352,198 (1962).
145. G. Zenker, Untersuchungen über die katalytische Zersetzung von NaClO-Lösungen, Dissertation, Technical University of Berlin, 1954.
146. H. Walde, *Elektrische Stoffumsetzungen in Chemie und Metallurgie in energiewirtschaftlicher Sicht*, Klepzig, Düsseldorf (1968).
147. E. Heubach, in *Ullmanns Encyklopädie der technischen Chemie*, 4th ed., Vol. 9, Verlag Chemie, Weinheim (1975), p. 544.
148. Anonymous, Chlorine may on the way out, *Chem. Week* (30 December 1967).
149. Anonymous, British to demonstrate electrolytic treatment of sewage in coastal area, *Chem. Eng. (N.Y.)* **72**, 9–10 (4 January 1965).
150. Anonymous, Electrolyzed seawater sterilizes sewage wastes, *Chem. Eng. (London)* **73**, 98 (9 May 1966).
151. A. T. Kuhn and R. B. Lartey, Electrolytic generation "In-Situ" of sodium hypochlorite, *Chem. Ing. Tech.* **47**, 129–135 (1975).
152. C. W. Böckmann, Versuch und Beobachtungen über die Wirkungen der galvanischen Electricität durch Volta's Säule, *Ann. Phys. (Leipzig)* **8**, 137–162 (1801).
153. J. K. P. Grimm, Einige Versuche mit Volta's Säule; dass Electricität die thierische Ausdünstung vermehrt; ist Wasser ein Nichtleiter der Wärme? *Ann. Phys. (Leipzig)* **7**, 348–362 (1801).
154. J. R. Partington, *A History of Chemistry*, Vol. 4, Macmillan, London (1964), p. 97.
155. A. J. Balard, Recherches sur la nature des combinaisons décolorantes du chlore, *Ann. Chim. Phys.* **57**, 225–304 (1834).
156. P. H. Prausnitz, Studien über die elektrolytische Herstellung von Natriumhypochlorit, *Z. Elektrochem.* **18** 1025–1080 (1912).
157. E. Abel, *Hypochlorite und elektrische Bleiche. Theoretischer Teil*, Knapp, Halle (Saale) (1900).
158. W. A. Müller, in *Ullmanns Encyklopädie der technischen Chemie*, 3rd ed., Vol. 5, Urban und Schwarzenberg, Munich (1954), pp. 501–508.
159. V. Engelhardt, *Hypochlorite und elektrische Bleiche. Technisch-konstruktiver Teil*, Knapp, Halle (Saale) (1900).
160. M. Haas, F. Oettel, German Patent 114,739 (1901).
161. A. T. Kuhn and P. M. Wright, in *Industrial Electrochemical Processes*, Elsevier, Amsterdam (1971), p. 546.
162. G. Wranglén, Anodes for cathodic protection in sea-water, *Ind. Tekn.* **90**, 75–79 (1968).
163. C. Marshall and J. P. Millington, Loss of platinum from platinized titanium in hypochlorite cells at low electrolyte temperatures, *J. Appl. Chem.* **19**, 298–301 (1969).
164. West German Patent 1,956,156 (1971).
165. A. F. Adamson, B. G. Lever, and W. F. Stones, The production of hypochlorite by direct electrolysis of sea water: Electrode materials and design of cells for the process, *J. Appl. Chem.* **13**, 483–495 (1963).
166. J. E. Bennett, Non-diaphragm electrolytic hypochlorite generators, *Chem. Eng. Progr.* **70**(12), 60–63 (1974).

167. J. B. Cotton and A. C. Wood, Titanium in electrochemical processes, *Trans. Inst. Chem. Engrs.* **41**, 354–359 (1963).
168. Engelhard, U.S. Patent 3,544,442 (1970).
169. A. Kuhn, On-site hypochlorite generation, *Proc.* **21**(3), 6–7, (4), 10–12 (1975).
170. G. R. Heal, A. T. Kuhn, and R. B. Lartey, A parametric study and computer-based simulation of an undivided sodium hypochlorite electrolyzer, *J. Electrochem. Soc.* **124**, 1690–1697 (1977).
171. Anonymous, Available chlorine in situ, information by Diamond Shamrock Corp., Lurgi, 1977.
172. H. B. H. Cooper, U.S. Patent 3,390,065 (1958).
173. B. D. Yurkevich and B. M. Vrevskii, Electrolytic preparation of hypochlorites using ion-exchange cathodes (in Russian), *Izv. Vyssh. Ucheb. Zaved. Khim. Khim. Tekhnol.* **13**, 1493–1495 (1970); cited in *Chem. Abstr.* **74**, 49003 (1971).
174. V. I. Monasyrskii, Prospects of using highly mineralized thermal waters in the electrolytic preparation of disinfection reagents in water purification installations, *Izuch. Ispol' z. Glubin. Tepla Zemli,* 247–249 (1973); cited in *Chem. Abstr.* **80**, 74204 (1974).
175. J. C. Schumacher, *Perchlorates. Their Properties, Manufacture and Uses,* Reinhold, New York (1960).
176. M. Traube, Über die elektrolytische Entstehung des Wasserstoffhyperoxyds an der Kathode, *Sitzungsber. Kgl. Preuss. Akad. Wiss. Berlin,* 1041–1050 (1887).
177. H. M. Goodwin and E. C. Walker, The electrolytic oxidation of hydrochloric acid to perchloric acid, *Trans. Amer. Electrochem. Soc.* **40**, 157–166 (1922).
178. F. C. Mathers, Electrolytic oxidation of hydrochloric acid to perchloric acid, *Proc. Indiana Acad. Sci.* **63**, 138–139 (1953).
179. W. Müller and P. Jönck, Herstellung von Perchlorsäure durch anodische Oxydation von Chlor, *Chem. Ing. Tech.* **35**, 78–80 (1963).
180. Y. Kato, K. Sugino, K. Koizumi, and S. Kitahara, Sodium perchlorate production with a pure lead peroxide anode, *Electrotech. J. (Jpn.)* **5**, 45–48 (1941).
181. F. Winteler, Ueber die Bildung von Perchloraten der Alkalien und alkalischen Erden durch Elektrolyse, *Chem. Ztg.* **22**, 89–90 (1898).
182. K. Sugino, Preparation, properties, and application of the lead peroxide electrode manufactured by a new method, *Bull. Chem. Soc. Jpn.* **23**, 115 (1950).
183. S. Kitahara and T. Ohsuga, The preparation of electrodes of lead peroxide and their applications. Making flat and compact anodes of lead peroxide and their use in electrolytic production of sodium perchlorate, *J. Electrochem. Assoc. Jpn.* **10**, 409–413 (1942).
184. M. Nagalingam, P. Govinda Rao, C. J. Raju, K. C. Narasimham, S. Sampath, and H. V. K. Udupa, Direkte Oxidation von Natriumchlorid zu Natriumperchlorat, *Chem. Ing. Tech.* **41**, 1301–1303 (1969).
185. H. V. K. Udupa, K. C. Narasimham, *et al.*, Large-scale preparation of perchlorates directly from sodium chloride, *J. Appl. Electrochem.* **1**, 207–212 (1971).
186. J. C. Grigger, H. C. Miller, and F. D. Loomis, Lead dioxide anode for commercial use, *J. Electrochem. Soc.* **105**, 100–102 (1958).
187. A. Legendre, Herstellung von Perchloraten durch Elektrolyse, *Chem. Ing. Tech.* **34**, 379–387 (1962).
188. J. C. Schumacher and R. D. Stewart in: Kirk-Othmer, Encyclopedia of chemical technology. 2nd edn., Vol. 5, New York: Wiley (1964), p. 62.
189. J. E. Reynolds and T. W. Clapper, The manufacture of perchlorates, *Chem. Eng. Progr.* **57**(11), 138–143 (1961); **57**(12), 94–97 (1961).
190. F. von Stadion, Von den Verbindungen der Chlorine mit dem Sauerstoff, *Ann. Phys. (Leipzig)* **52**, 219 (1816).
191. F. von Stadion, Sur les combinaisons du chlore avec l'oxigène, *Ann. Chim. Phys.* **8**, 406–414 (1818).
192. J. J. Berzelius, *Lehrbuch der Chemie,* Vol. 2, 3rd ed., Arnoldi, Dresden (1835), p. 77.
193. A. Riche, Recherches sur l'action du courant electrique sur le chlore, le brome, l'iode en présence de l'eau, *C.R. Acad. Sci. Paris* **46**, 348–358 (1858).

194. O. Carlson Swedish Patent 3,614 (1892).
195. F. Foerster, Über die Darstellung der Überchlorsäure und ihrer Salze, Z. Elektrochem. **4**, 386–388 (1898).
196. F. Winteler, Studien über die Elektrolyse der Chloralkalien, Z. Elektrochem. **5**, 49–51, 217–221 (1898).
197. F. Winteler, Über die Bildung von überchlorsauren Salzen durch Elektrolyse, Z. Elektrochem. **7**, 635–642 (1901).
198. W. Oechsli, Über die elektrolytische Chloratbildung, Z. Elektrochem. **9**, 807–828 (1903).
199. C. W. Bennett and E. L. Mack, Electrolytic formation of perchlorate, Trans. Am. Electrochem. Soc. **29**, 323–346 (1916).
200. J. G. Williams, The electrolytic formation of perchlorate from chlorate, Trans. Faraday Soc. **15**, Part 3, 134–137 (1920).
201. M. P. Grotheer and E. H. Cook, Mechanism of electrolytic perchlorate production, Electrochem. Technol. **6**, 221–224 (1968).
202. O. de Nora, P. Gallone, C. Traini, and G. Meneghini, On the mechanism of anodic chlorate oxidation, J. Electrochem. Soc. **116**, 146–151 (1969).
203. J. C. Schumacher, D. R. Stern, and P. R. Graham, Electrolytic production of sodium perchlorate using lead dioxide anodes, J. Electrochem. Soc. **105**, 151–155 (1958).
204. E. Hausmann and E. Kramer, Verarbeitung von Chlorat zu Perchlorat und Chlordioxid, Chem. Ing. Tech. **43**, 170–173 (1971).
205. K. C. Narasimham, S. Sundararajan, and H. V. K. Udupa, Lead dioxide anode in the preparation of perchlorates, J. Electrochem. Soc. **108**, 798–805 (1961).
206. K. Sugino and S. Aoyagi, Studies on the mechanism of the electrolytic formation of perchlorate, J. Electrochem. Soc. **103**, 166–171 (1956).
207. G. Angel and H. Mellquist, Versuche, um einen Ersatz für das Platin als Anodenmaterial bei elektrolytischen Oxydationsprozessen zu finden, Z. Elektrochem. **40**, 702–707 (1934).
208. Y. Kato and K. Koizumi, A new process for the lead peroxide anode, J. Electrochem. Assoc. Jpn. **2**, 309–312 (1934).
209. J. C. Pernert, U.S. Patent 2,392,861 (1946).
210. P. Jönck, in *Ullmanns Encyklopädie der technischen Chemie*, 3rd ed., Suppl. Vol., Urban und Schwarzenberg, Munich (1970), pp. 445–446.
211. P. Jönck, private communication.
212. A. A. Rakov and I. V. Shimonis, U.S.S.R. Patent 512, 677 (1976); cited in Chem. Abstr. **86**, 98,099 (1977).
213. E. Blau and R. Weingand, Notizen über die Erzeugung von Kaliumperchlorat, Z. Elektrochem. **27**, 1–10 (1921).
214. C. L. Mantell, Cell design for bromates, iodates, and periodates, J. Electrochem. Soc. **115**, 91c (1968).
215. H. Pauli, Beiträge zur Elektrolyse der Alkali-Bromide und -Fluoride, Z. Elektrochem. **3**, 474–478 (1897).
216. J. Sarghel, Über die Elektrolyse der Bromide der Erdalkalien, Z. Elektrochem. **6**, 149–158, 173–188 (1899).
217. E. Müller, Über ein elektrolytisches Verfahren zur Gewinung der chlor-, brom- und jodsauren Salze der Alkalien, Z. Elektrochem. **5**, 469–473 (1899).
218. H. Kretzschmar, Über die Einwirkung von Brom auf Alkali und über die Elektrolyse der Bromalkalien, Z. Elektrochem. **10**, 789–817 (1904).
219. W. C. Bray, The hydrolysis of iodine and of bromine, J. Am. Chem. Soc. **32**, 932–938 (1910); **33**, 1485–1487 (1911).
220. A. Skrabal, Die Halogenbleichlaugen-Reaktionen, Z. Elektrochem. **40**, 232–246 (1934).
221. A. Skrabal, Über die Stärke der unterhalogenigen Säuren, Z. Elektrochem. **48**, 314–327 (1942).
222. T. Osuga and K. Sugino, Electrolytic production of bromates, J. Electrochem. Soc. **104**, 448 (1957).
223. S. Sundararajan, K. C. Narasimham, and H. V. K. Udupa, Electrolytic preparation of bromates, Chem. Process Eng. **43**, 438–441, 447 (1962).

224. W. Vaubel, Ueber ein neues Verfahren zur elektrolytischen Darstellung von Chloraten, Bromaten, Jodaten, sowie Hypochloriten, *Chem. Ztg.* **22**, 331 (1898).
225. E. Eisner and U. Eisner, German Patent 2,540,926 (1977); *Chem. Abstr.* **86**, 130, 041a (1977).
226. E. A. Dzhafarov and Sh. M. Efendieva, Electrosynthesis of bromates on a lead dioxide anode (in Russian), *Azerb. Khim. Zh.* 1967(5), 166–169; *Chem. Abstr.* **69**, 24005 (1968).
227. P. J. M. Radford, in *Bromine and its Compounds*, (Z. E. Jolles, ed., Benn, London (1966), pp. 164–173.
228. W. Geissler, R. Nitzsche, and R. Landsberg, Über die elektrochemische Oxydation von Jodid und Jod zum Hypojodit an Graphitelektroden, *Electrochim. Acta* **11**, 389–400 (1966).
229. R. Landsberg, R. Nitzsche, and W. Geissler, Über die elektrochemische Oxydation von Jodid zum Jodat an Graphitelektroden, *Electrochim. Acta* **11**, 495–506 (1966).
230. M. S. Venkatachalapathy, S. Krishnan, M. Ramachandran, and H. V. K. Udupa, Electrochemical preparation of sodium iodate from iodine using graphite substrate lead dioxide anode, *Electrochem. Technol.* **5**, 399–404 (1967).
231. E. A. Dzhafarov, Sh. M. Efendieva, F. G. Bairamov, and A. M. Musaev, Electrosynthesis of iodates at Pb dioxide anodes, *Azerb. Khim. Zh.* 1966 (2), 125–129 (Russ.) *Chem. Abstr.* **65**, 11763b (1966).
232. F. Foerster and K. Gyr, Über die Einwirkung von Jod auf Alkalien, *Z. Elektrochem.* **9**, 1–10 (1903).
233. F. Foerster and K. Gyr, Zur Kenntnis der Elektrolyse von Jodkalium-Lösungen, *Z. Elektrochem.* **9**, 215–226 (1903).
234. J. C. Schumacher, Electrolytic production of potassium iodate, *Chem. Eng. Progr.* **56**(5), 83–84 (1960).
235. H. H. Willard and R. R. Ralston, The electrolytic oxidation of iodine and of iodic acid, *Trans. Electrochem. Soc.* **62**, 239–254 (1932).
236. C. L. Mehltretter and C. S. Wise, An electrolytic process for making sodium metaperiodate, *Ind. Eng. Chem.* **51**, 511–514 (1959).
237. E. Torigai and E. Ishii, *Bull. Osaka Ind. Res. Inst.* **7**, 195 (1956).
238. E. Müller and O. Friedberger, Die Darstellung der freien Überjodsäure durch Elektrolyse. *Ber. Dt. Chem. Ges.* **35**, 2652–2659 (1902).
239. Sh. Sh. Khidirov, D. P. Semichenko, and V. I. Lyubushkin, U.S.S.R. Patent 217, 384.
240. B. Kastening, Synthese von Wasserstoffperoxid, paper read at the 13th Tutzing Symposium of Dechema, 1976.
241. A. H. H. Schmidt, Technisch-elektrochemische Herstellung von Wasserstoffperoxyd, *Chem. Ing. Tech.* **37**, 832–834 (1965).
242. R. Powell, *Hydrogen Peroxide Manufacture*, Noyes, Park Ridge, N.J. (1968).
243. W. C. Schumb, C. N. Satterfield, and R. L. Wentworth, *Hydrogen Peroxide*, Reinhold, New York (1955).
244. W. Machu, Das Wasserstoffperoxyd und seine Perverbindungen, Springer, Vienna (1951).
245. F. Beer, in *Chemische Technologie*, Vol. 1, K. Winnacker and L. Küchler, eds., Hanser, Munich (1970), pp. 525–529.
246. J. Müller in: *Ullmanns Encyklopädie der technischen Chemie*, vol. 13, 3rd ed., Urban und Schwarzenberg: Munich (1962), p. 212–227.
247. E. Berl and H. Burkhardt, Über die Herstellung von aktiven Kohlen, *Z. ang. Chemie* **43**, 330–333 (1930).
248. W. G. Berl, A reversible oxygen electrode, *Trans. Electrochem. Soc.* **83**, 253–270 (1944).
249. D. H. Grangaard, U.S. Patent 3,462,351 (1969), 3,507,769 (1970), 3,592,749 (1971).
250. B. Kastening and W. Faul, Production of hydrogen peroxide by cathodic reduction of oxygen, *Ger. Chem. Eng.* **1**, 183–190 (1978). Herstellung von Wasserstoffperoxid durch kathodische Reduktion von Sauerstoff, *Chemie-Ing.-Technik* **49**, 911 (1977).
251. O. Špalek, J. Balej, and K. Balogh, Preparation of hydrogen peroxide by cathodic reduction of oxygen in porous electrodes made of different carbonaceous materials, *Collect. Czech. Chem. Commun.* **42**, 952–959 (1977).
252. G. Teichner, German Patent 217,539 (1905).

253. J. Müller, Technisch-elektrochemische Herstellung von Wasserstoffperoxyd, *Chem. Ing. Tech.* **35**, 389–392 (1963).
254. A. Pietzsch, German Patents 243,366; 241,702; 256,148.
255. L. Löwenstein and J. D. Riedel, German Patent 510,064.
256. H. Schröter, Neue Erkenntnisse auf dem Gebiet der Holzschliffbleiche, *Papier* 21 (1967), 760.
257. F. L. Fennell and N. J. Stalter, Hydrogen peroxide for bleaching kraft pulp, *Tappi* **51**(1), 62A–66A (1968).
258. J.-W. Kühn-von Burgsdorff, Platinierte Sondermetall-Elektroden in der Elektrochemie, *Chem. Ztg.* **88**, 597–601 (1964).
259. J.-W. Kühn-von Burgsdorff, Der Einsatz aktivierter Metallanoden in elektrochemischen Prozessen, *Chem. Ing. Tech.* **49**, 294–298 (1977).
260. J. Terraz and J. Malafosse, German Offen. 2,528,204 (1976); cited in *Chem. Abstr.* **85**, 38,723 (1976).
261. F. Beer, G. Düsing, and H. Pistor, Wasserstoffperoxid und Peroxoverbindungen in der Anorganischen Chemie *Chem. Ztg.* **99**, 120–125 (1975).
262. Degussa, German Patent 218,569 (1905).
263. K. Arndt, German Patent 297,233 (1912).
264. H. Pistor, in *Chemische Technologie*, K. Winnacker and L. Küchler, eds., Vol. 1, 3rd ed., Hanser, Munich (1970), pp. 574–575.
265. G. Schaufler, in *Ullmanns Encyklopädie der technischen Chemie*, 3rd ed., Vol. 12, Urban und Schwarzenberg, Munich (1960), pp. 231–233.
266. H. Marcy, in *Anorganisch-technische Verfahren*, F. Matthes and G. Wehner, eds., VEB Deutscher Verlag für Grundstoffindustrie, Leipzig (1964), (a) p. 750, (b) p. 762.
267. S. Klonowski, Über die Manganatschmelze und die Überführung von Kaliummanganat in Kaliumpermanganat auf elektrolytischen Wege, Dissertation, TH Karlsruhe, 1910.
268. Schering, German Patent 28,782 (1884).
269. Salzbergwerk Neustassfurt, German Patent 101,718 (1899).
270. Tumanow, U.S.S.R. Patent 51,390 (1936).
271. Deissler, German Patent 105,008 (1898).
272. E. Schütz, Das Kaliumpermanganat, *Z. Angew. Chem.* **24**, 1628–1631 (1911).
273. B.I.O.S. Final Report No. 964, Item No. 22; B.I.O.S. Final Report No. 1577, Item No. 22 (1947).
274. Carus Chemical Comp., U.S. Patent 2,908,620 (1957).
275. W. L. Faith, D. B. Keyes, and R. L. Clark, *Industrial Chemicals*, Wiley, New York (1965).
276. Anonymous, New British plant boosts potassium permanganate output, *Brit. Chem. Eng.* **9**, 383 (1964).
277. Montecatini, Italian Patent 292,502 (1932).
278. M. Eigen and K. Kustin, The kinetics of halogen hydrolysis, *J. Am. Chem. Soc.* **84**, 1355–1361 (1962).
279. C. Häussermann, Beiträge zur Technologie der Alkalidichromate, *Dinglers Polytechn. J.* **287**, 161–162 (1893); *Z. Angew. Chem.* **6**, 363 (1893).
280. F. Regelsberger, Über Regeneration von Chromsäure aus chromoxydhaltigen Materialen, *Z. Angew. Chem.* **12**, 1123–1128 (1899).
281. V. A. Shlyapnikov, S. I. Statkevich, E. I. Adaev, and L. V. Perevozchikova, Influence of electrolyte pH and temperature on the electrosynthesis of Berthollet's salt, *Zh. Prikl. Khim.* **49**, 2247–2252 (1976); *J. Appl. Chem. USSR* **49**, 2257–2260 (1976).
282. M. Le Blanc, Die elektrolytische Regeneration von Chromsäure und die Herstellung säurebeständiger Diaphragmen, *Z. Elektrochem.* **7**, 290–295 (1900).
283. M. Le Blanc, Die Darstellung des Chroms und seiner Verbindungen mit Hilfe des elektrischen Stroms, Knapp, Halle (Saale) (1902).
284. Farbwerke Höschst, German Patent 103,860 (1898).
285. M. Käppel, Elektrolytische Regeneration von Chromsäure, *Chem. Ing. Tech.* **35**, 386–389 (1963).

286. G. Schulze, Die elektrochemische Reduktion chromsäurehaltiger Abwässer, *Galvanotechnik* **58**, 475–480 (1967).
287. N. Ibl and A. M. Frei, Untersuchungen eines neuen Weges zur Entgiftung chromathaltiger Abwässer durch elektrolytische Reduktion, *Galvanotech. Oberflächenschutz* **5**, 117 (1964).
288. E. Müller and M. Soller, Die Rolle des Bleisuperoxyds als Anode bei der elektrolytischen Oxydation des Chromsulfates zu Chromsäure, *Z. Elektrochem.* **11**, 863–872 (1905).
289. R. F. J. Gross and A. Hickling, The anodic oxidation of chromic salts to chromates, *J. Chem. Soc.* 325 (1937).
290. F. Regelsberger, Einfluss des Elektrodenmaterials auf den Reaktionsverlauf bei Elektrolysen, *Z. Elektrochem.* **6**, 308 (1899).
291. M. Käppel and H. Gerischer, Zum Mechanismus der elektrolytischen Chromatabscheidung durch Reduktion von Chromsäure, *Z. Elektrochem.* **64**, 235–244 (1960).
292. J. Balej, H. Matschiner, and W. Thiele, Zum Mechanismus der anodischen Bildung von Peroxodisulfaten, *Chem. Techn. (Leipzig)* **30**, 578–581 (1978).
293. J. Billiter, *Die technische Elektrolyse der Nichteisenmetalle*, Springer, Vienna (1954), p. 105.
294. M. Le Blanc, German Patent 182,287 (1905).
295. Verein für chemische und metallurgische Produktion zu Aussig, Austrian Patent 34,562 (1908).
296. P. Dilthey, in *Chemische Technologie*, K. Winnacker and E. Weingärtner, eds., Vol. 2, Hanser, Munich (1950), p. 484.
297. German Patent 251,694 (1912).
298. G. Adolph and A. Pietzsch, German Patent 199,248 (1906); British patent 9636 (1907); U.S. Patent 895,930 (1907).
299. M. J. Udy, U.S. Patent 1,739,107 (1929).
300. R. H. McKee and S. T. Leo, A continuous process for electrolytic regeneration of chromic acid, *J. Ind. Eng. Chem.* **12**, 16–26 (1920).
301. P. Askenasy and A. Révai, Beiträge zur Kenntnis der elektrolytischen Regenerierung von Chromsäure aus Lösungen von Chromsulfat, *Z. Elektrochem.* **19**, 344–362 (1913).
302. FIAT Final Report No. 429 (1945).
303. FIAT Final Report No. 831 (1946).
304. C. Hampel, *Encylopaedia of Electrochemistry*, Reinhold, New York (1964), p. 1065.
305. W. W. Stender and I. J. Seerak, Electrolysis of aqueous solutions of alkali sulfates, *Trans. Electrochem. Soc.* **68**, 493–520 (1935).
306. C. Jackson and A. T. Kuhn, in *Industrial Electrochemical Processes*, Elsevier, Amsterdam (1971), p. 517.
307. Anonymous, Stack gas scrubber uses electrochemical cell, *Europ. Chem. News*, 30 (February 12, 1971).
308. A. T. Kuhn, Electrochemical methods for $SO_2$ flue gas treatment, *J. Appl. Electrochem.* **1**, 41–44 (1971).
309. F. Richarz, Zur Kenntniss der Entstehungsweise von Wasserstoffsuperoxyd an der Anode bei der Electrolyse verdünnter Schwefelsäure, *Ann. Physik Chem.* **31**, 912–924 (1887).
310. F. Haber, Über die Autoxydation und ihren Zusammenhang mit der Theorie der Ionen und der galvanischen Elemente, *Z. Elektrochem.* **7**, 441–448 (1901).
311. J. W. Schultze, Die Adsorption von Wasser an Platinelektroden in wässrigen Elektrolyten, *Ber. Bunsenges. Phys. Chem.* **73**, 483–492 (1969).
312. A. Rius Miro and J. Ocon Garcia, Modificacion del poder oxidante de un anodo durante su trabajo, *An. Fis. Quim.* **40**, 861–885 (1944).
313. J. C. Schumacher and D. R. Stern, Large-scale continuous production of ammonium perchlorate, *Chem. Eng. Progr.* **53**(9), 428–432 (1957).
314. Yu. M. Tyurin, G. F. Volodin, L. A. Smirnova, and Yu. V. Battalova, Effect of solution composition on the properties of oxide films produced on platinum anodes at high positive potentials, *Elektrokhim.* **9**, 532–536 (1973); *Sov. Electrochem.* **9**, 512–516 (1973).
315. P. Wintzer, private communication.
316. E. Berl and H. Burkhardt, Über die Herstellung von aktiven Kohlen, *Z. Angew Chem.* **43**, 330–333 (1930).

317. H. H. Willard and G. Frederick Smith, The perchlorates of the alkali and alkaline earth metals and ammonium. Their solubility in water and other solvents, *J. Amer. Chem. Soc.* **45**, 286–297 (1923).
318. J. C. Schumacher and D. R. Stern, *Chem. Eng. Progr.* **53**(9), 428–432 (1957).
319. T. W. Richards and H. H. Willard, Further investigation concerning the atomic weights of silver, lithium and chlorine, *J. Am. Chem. Soc.*, 15ff (1910).
320. F. Oettel, Die elektrischen Bleichapparate "System Haas und Oettel," *Z. Elektrochem.* **7**, 315–320 (1901).
321. B. Kastening and H. Schmitz, West German Patent 2,453,739 (1974).
322. B. Kastening and H. Schmitz, Indirect electrosynthesis of hydrogen peroxide, Symposium on the Engineering Aspects of Electrochemical Synthesis, Dubrovnik, 1975.
323. H. Schmidt, German Patent 941,543 (1942).
324. W. Thiele and H. Matschiner, Wasserstoffperoxid und Peroxodischwefelsäure, *Chem. Tech. (Leipzig)* **29**, 148–154, 682 (1977).
325. M. Schleiff, W. Thiele, and H. Matschiner, Optimierung von Elektrolysezellen, dargestellt am Beispiel eines Elektrolyseurs für Peroxodischwefelsäure, *Chem. Tech. (Leipzig)* **29**, 679–682 (1977).
326. T. Ya. Krasilova, É. V. Kasatkin, and V. I. Veselovskii, Effect of perchloric acid and molecular chlorine concentration on formation of $HClO_4$ and $O_2$ during oxidation, *Elektrokhim.* **6**, 356–358 (1970); *Sov. Electrochem.* **6**, 349–351 (1970).
327. A. J. Bard, ed., *Encyclopedia of Electrochemistry of the Elements*, Vol. 1, Marcel Dekker, New York (1978).
328. M. S. Sherrill and E. F. Izard, The solubility of chlorine in aqueous solutions of chlorides and the free energy of trichloride ion, *J. Am. Chem. Soc.* **53**, 1667–1674 (1931).
329. E. Berl, A new cathodic process for the production of $H_2O_2$, *Trans. Electrochem. Soc.* **76**, 359–369 (1939).
330. G. Teichner and G. Baum, German Patent 567,542 (1930).
331. J. Müller, German Patent 975,825 (1951).
332. C. A. Adzemjan *et al.*, U.S.S.R. Patent 167,492 (1957).
333. W. Thiele, East Ger. Patent 27,961 (1963).
334. W. Thiele, East Ger. Patent 99,548 (1972).
335. W. Thiele, K. Wildner, and H. Matschiner, Peroxodisulfate in kristallisierter Form, *Chem. Tech. (Leipzig)* **31**, 198 201 (1979).
336. Anonymous, Peroxide by non-electrolytic processes, *Chem. Eng.* **60**(10), 108, 112 (1953).
337. P. W. Sherwood, Wasserstoffperoxyd über petrochemische Rohstoffe, *Chem. Ing. Tech.* **32**, 459–461 (1960).
338. O. von Schickh, Herstellung von Peroxyden durch Autoxydation, *Chem. Ing. Tech.* **32**, 462 (1960).
339. J. E. Bennett, private communication, 1979.
340. R. B. Lartey, PhD thesis, Salford, 1977.
341. E. A. Efimov and N. A. Izgaryshev, Kinetic study of the sulfuric acid electrolytic oxidation, *Zh. Fiz. Khim.* **31**, 1141–1149 (1957).
342. E. V. Kasatkin and A. A. Rakov, Kinetics and mechanism of low temperature electrochemical oxidation at high anode potentials, *Electrochim. Acta* **10**, 131–140 (1965).
343. W. Smit and J. G. Hoogland, The mechanism of the anodic formation of the peroxodisulphate ion on platinum, *Electrochim. Acta* **16**, 1–18, 821–831, 961–979, 981–993 (1971).
344. J. Balej and M. Kadeřávek, Influence of sulphuric acid concentration on current yield of peroxodisulphate, *Collect. Czech. Chem. Commun.* **44**, 1510–1520 (1979).
345. R. D. Apfelbach, Regeneration von Beizen Zur Kunststoffgalvanisierung, *Galvanotechnik* **70**, 144–148 (1979).
346. K.-I. Fukuda, C. Iwakura, and H. Tamura, Anodic processes on a titanium-supported ruthenium dioxide electrode at high potentials in a mixture of sulfuric acid and ammonium sulfate, *Electrochim. Acta* **23**, 613–618 (1978).
347. J. Balej, Effect of Caro's acid in electrolytic preparation of peroxodisulphates, *Collect. Czech. Chem. Commun.*, in press.

348. J. Balej, M. Thumorá, and M. Kadeřávek, Mechanism of formation of peroxomonosulphuric acid during electrolytic preparation of peroxodisulphates, *Collect. Czech. Chem. Commun.*, **45**, 3254 (1980).
349. P. M. v.d. Wiel, L. J. J. Janssen, and J. G. Hoogland, The electrolysis of a carbonate–borate solution with a platinum anode, *Electrochim. Acta* **16**, 1217–1234 (1971).
350. A. Frumkin, Adsorption des cations à des potentiels anodiques, *Electrochim. Acta* **5**, 265–290 (1961).
351. P. Gallone, Trattato di ingegneria elettrochimica, Tamburini, Milano (1973).
352. Yun-Tsao Ts'u and T'ien-Yin Mi, Mechanism for the anodic formation of the persulfate ion, *Doklady Akad. Nauk S.S.S.R.* **125**, 1069–1072 (1959).
353. A. Hickling and A. O. Jones, Anodic formation of persulphate using pulsed current, *Trans. Faraday Soc.* **62**, 494–502 (1966).
354. A. Rius Miro and J. Ocon Garcia, Accion de las adiciones en la oxidacion anodica del acido sulfurico y sus sales, *An. Fis. Quim.* **40**, 886–896 (1944).
355. A. N. Frumkin, R. I. Kaganovich, M. A. Gerovich, and V. I. Vasil'ev, The mechanism of the anodic persulfate formation, *Doklady Akad. Nauk S.S.S.R.* **102**, 981–983 (1955).
356. A. I. Brodskii, I. F. Franchuk, and V. A. Lunenok-Burmakina, Isotope method of investigation of the electrolytic formation and hydrolysis of persulfate, *Doklady Akad. Nauk S.S.S.R.* **115**, 934–937 (1957).
357. A. N. Frumkin, O. A. Petry, and N. V. Nikolaeva-Fedorovich, On the determination of the value of the charge of the reacting particle and of the constant $\alpha$ from the dependence of the rate of electro-reduction on the potential and concentration of the solution, *Electrochim. Acta* **8**, 177–192 (1963).
358. R. Memming, Mechanism of the electrochemical reduction of persulfates and hydrogen peroxide, *J. Electrochem. Soc.* **116**, 785–790 (1969).
359. N. V. Nikolaeva-Fedorovich, B. B. Damaskin, and O. A. Petrii, Effect of surface-active organic substances on the electroreduction of anions, *Coll. Czech. Chem. Commun.* **25**, 2982–2992 (1960).
360. W. J. Plieth, Kinetik der Manganat/Permanganat-Redoxelektrode an Platin und Gold, *Ber. Bunsenges.* **74**, 1042 (1970).
361. H. Schurig and K. E. Heusler, Anwendung der rotierenden Scheiben-Ring-Elektrode zur Untersuchung der Reduktion von Permanganat in alkalischen Lösungen, *Fresenius Z. Analyt. Chem.* **224**, 45–62 (1967).
362. R. Thiele and R. Landsberg, Zum Mechanismus der Manganat-VI-Permanganat-Redoxelektrode, *Z. phys. Chem.* **236**, 261–270 (1967).
363. A. Schmidt, *Angewandte Elektrochemie*, Verlag Chemie, Weinheim (1976).
364. W. M. Weigert, *Wasserstoffperoxid und seine Derivate*, Hüthig, Heidelberg (1978).
365. P. C. Wang, Y. C. Chu, and F. Y. Cheng, Mechanism of anodic formation of perchlorate ions on platinum electrodes. *Chun-Kuo K'o Hsueh Yuan Ying Yung Hua Hsueh Yen Chiu Chi K'an* **1966** (16). 18–24: cited in *Chem. Abstr.* **67**, 17232–17234 (1967).
366. H. Narcus, Plating on Plastics: practical plant operation and trouble shooting, *Plating* **55**, 816–820 (1968).
367. F. Hine, M. Yasuda, T. Noda, T. Yoshida, and J. Okuda, Electrochemical behavior of the oxide-coated metal anodes, *J. Electrochem. Soc.* **126**, 1439–1445 (1979).
368. V. I. Veselovsky, E. V. Kasatkin, A. A. Yakovleva, and A. A. Rakov, Structure of the double layer and kinetics of anodic processes at high potentials, *Electrochim. Acta* **17**, 2095–2101 (1972).
369. J. Balej and M. Kadeřávek, Effect of various cations on the initial rate of formation of peroxodisulphates, *Coll. Czech. Chem. Commun.*, **45**, 2272–2282 (1980).
370. J. Balej, K. Balogh, and O. Špalek, Possibility of producing hydrogen peroxide by cathodic reduction of oxygen. Influence of pretreatment of active carbon on the properties of porous carbon electrodes for preparing hydrogen peroxide by cathodic reduction of oxygen, *Chem. Zvesti* **30**, 384, 611 (1976).
371. C. Oloman and A. P. Watkinson, Hydrogen peroxide production in trickle-bed electrochemical reactors, *J. Appl. Electrochem.* **9**, 117–123 (1979).

# 4
# Electro-Organic Syntheses

## KLAUS KÖSTER and HARTMUT WENDT

## 1. Introduction

### 1.1. Historical Survey

The history of organic electrochemical synthesis goes back to the earlier days of electrochemistry. As early as 1834 Faraday described the anodic conversion of acetate anions to $CO_2$.[1]

Some years later, in 1849, Kolbe[2] discovered that by electrolysis of valeric acid in aqueous solution the hydrocarbon "isobutyl" ($n$-octane) was produced:

$$2RCOO^- - 2e \rightarrow R-R + 2CO_2 \qquad (1)$$

It is this established Kolbe synthesis that has recently regained interest and is used to produce the dimethylester of sebacinic acid[3] on a large scale. For 75 years after Kolbe's discovery, various different types of electro-organic synthetic reactions were discovered, and until the mid-twenties, electro-organic synthesis was considered a promising tool for industrial application.

By that time, many of the present-day electro-organic synthetic reaction types had been discovered. Most of the work concentrated on the preparative aspects of electrochemical conversion of functional groups ($-NO_2$ to $-NO$, $-NHOH$ and $NH_2$; CO to $-CHOH$; C=C to CH—CH; C=N to HC—$NH_2$, etc.). Also, cathodic and anodic coupling reactions (e.g., $2R_1R_2CO$ to

---

**KLAUS KÖSTER and HARTMUT WENDT** • Institute of Chemical Technology, Technical University of Darmstadt, Darmstadt, West Germany.

pinacoles), which are so important today, were already quite well known by 1930.

The compilations of Swann,[4-7] who from 1936 to 1952 published four reviews of the progress in electro-organic chemistry in the *Journal of the Electrochemical Society*, are testimony to the scientific and industrial activities of this time. Recently this work was continued by Alkire.[98]

Fichter's[8] book (1942) summarized the results of an epoch of intensive preparative work. However, for many years electro-organic synthesis became a routine preparative method in the laboratories of organic chemists, and no electro-organic industrial process developed. The main reason for this was that until 1940 the kinetics and detailed mechanisms of organic electrochemical reactions were not easily understood. In the late 1940s fast electronic equipment was developed, and an increasing interest in the kinetic aspect of electrode reactions led to remarkable progress in the theory of general electrode kinetics.

By 1960, one could speak of a completely new understanding of electrochemical phenomena. Nearly all important contributions to electrochemical kinetics until 1960 have been discussed by Vetter.[9] Furthermore, the electrochemical double layer and its kinetic consequences as well as the phenomenon of electrosorption were well known in the 1950s and 1960s.[9-13] However, ten years passed before the newly gained knowledge was transferred and applied to the field of electro-organic synthesis.

In 1965 Baizer developed the electrochemical synthesis of adipodinitrile [Eq. (2)] into an economically attractive process.[14] Baizer's success and the electrosynthesis of tetraethyl lead[15] [Eq. (3)] created many new efforts in university and industrial laboratories:

$$2CH_2=CH-CN + 2e + 2H^+ \rightarrow NC-CH_2CH_2CH_2CH_2-CN \quad (2)$$

$$4CH_3CH_2MgCl - 4e + Pb \rightarrow Pb(CH_2CH_3)_4 + 4MgCl^+ \quad (3)$$

Numerous communications and patents, approximately 1180 between 1968 and 1974, have been published on electro-organic synthesis. Also, books appeared between the years 1971–1976.[16-22]

## 1.2. Electro-organic Synthesis and Electrochemical Engineering

The application of electro-organic synthesis for industrial processes resulted in new discoveries in electrochemical engineering. The new electro-organic synthetic processes were based primarily on nonaqueous solutions with low electrical conductivities. This was reason enough to completely reconsider the problem of the construction principles for electrochemical cells.

Due to the demand for low current densities the concept of using an electrode either in the form of a packed bed or a fluidized bed was invented by

Goodrich[23] and Fleischmann.[24] To treat low-conductivity electrolytes often used in electro-organic synthesis, the capillary-gap cell was invented by Beck.[25] This cell uses anode to cathode distances on the order of 1 mm and smaller.

### 1.3. Mediated Electrochemical Synthesis

The technique of reprocessing inorganic redox couples acting as electron transfer agents to oxidize or reduce organic substances gained new interest and was employed in many cases on a large scale. However, little has been published in this field. Methods of "indirect" or "mediated" electrochemical synthesis are promising because they avoid pollution of the environment by recycling the respective redox agent.

### 1.4. Future Development

Electrosorption and its influence on charge transfer and subsequent chemical reaction is still (with respect to organoelectrochemistry) understood only qualitatively. The industrial application of electro-organic synthesis is not widespread. However, for small processes, especially in the pharmaceutical industries, some successful processes are expected. Also, there is good reason to believe that electro-organic synthesis will become more important in this field.

### 1.5. Scope of this Chapter

The purpose of this chapter is not to give a complete survey of all the work done in electro-organic synthesis since the appearance of Baizer's book[17] and Beck's treatise.[18] Rather, it is desired to give a comprehensive and concise survey of the different reaction types together with some HMO (Hückel molecular orbital)-theory-based theoretical background. More important reaction types will be explained in selected examples. Additionally, important published examples of electro-organic synthetic processes that have gained technical or economical relevance are presented.

## 2. Direct Electrochemical Oxidation and Reduction of Organic Molecules

### 2.1. Introduction and Survey

The anodic oxidation and cathodic reduction of any substance at an electrode is characterized by two physical entities: the number of the electrons that are exchanged between the molecule and the electrode and the electrode potential at which electron transfer is observed. The electrode potential is

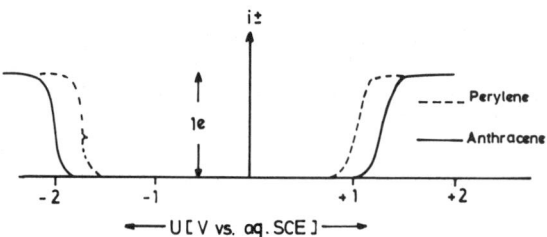

Figure 1. Reversible polarographic 1e reduction and oxidation waves of anthracene and pyrylene[26,27] (solvent: acetonitrile).

usually measured against the potential of a suitable reference electrode, which may be different for different solvents.

In Figure 1, for example, the cathodic and anodic polarograms (cathode: dropping Hg; anode: rotating Pt) of anthracene and pyrylene in the solvent acetonitrile are measured against a calomel electrode.† The hydrocarbons concerned are reduced and oxidized in one-electron waves. From the slopes of the current–voltage curves [i.e., $\log[(i_L - i)/i_L]$ vs. $E$: $(60 \text{ mV})^{-1}$], it is evident that the electrochemical oxidations and reductions are reversible 1e steps.

The half-wave potentials of such reversible steps have a thermodynamic meaning and approximately equal the normal potential of the respective redox couple. In Table 1 are collected the cathodic and anodic half-wave potentials of some other aromatic hydrocarbons which can be oxidized and reduced in acetonitrile[26,27] reversibly. Unsaturated hydrocarbons, which are less easily oxidized or reduced, show, because of fast consecutive reactions which follow their reduction and oxidation, irreversible polarographic 2e waves. Their half-wave potential has only a limited thermodynamic meaning and is partially kinetically controlled.

## 2.2. Molecular Orbital (MO) Representation and Energy Demand for Anodic Radical Cation and Cathodic Radical Anion Formation from Unsaturated Hydrocarbon Molecules[26,27]

Reversible 1e oxidation and reduction reactions which can be observed for some unsaturated hydrocarbons are to be understood as being due to the

---

† Contrary to a very common opinion it is not necessary for electrochemical measurements in nonaqueous solution to use a reference electrode within the same medium. Instead, a common aqueous reference, e.g., aq. sat. calomel, may be used. Provided that the measuring time is long enough, and the bridge connection between aqueous and nonaqueous solution is arranged so that a stationary diffusion potential exists at the electrolyte boundaries, this diffusion potential is well defined and reproducible and is an unknown but definite part of the measured potential. Therefore, in the following text—unless otherwise mentioned—all potentials refer to the aqueous saturated calomel electrode (sat. cal.).

## Table 1
### Cathodic and Anodic Half-Wave Potentials for a Series of Unsaturated Aromatic Hydrocarbons[26,27]

| Hydrocarbon | Cathodic $E_{1/2}$, V vs. sat. cal. | Anodic $E_{1/2}$, V vs sat. cal. |
|---|---|---|
| Benzene | — | +2.79 |
| Naphthalene | −2.50 | +2.19 |
| Phenanthrene | −2.49 | +1.84 |
| Anthracene | −1.97 | +1.37 |
| Perylene | −1.66 | +1.06 |
| Styrene | −2.30 | +1.70 |
| Diphenylbutadiene | −1.90 | +1.14 |

reversible generation of solvated radical cations and radical anions from hydrocarbon molecules [Eqs. (4) and (5)]:

$$A - e \rightarrow A^+ \quad \text{(oxidation)} \tag{4}$$

$$A + e \rightarrow A^- \quad \text{(reduction)} \tag{5}$$

Figure 2a presents the simplified HMO energy scheme for the polyene $H(CH=CH)_n H$.

This energy scheme results from Hückel treatment of linear $\pi$-electron systems. Although oversimplified, it enables reasonable and semiquantitative prediction of the energy changes involved in redox reactions and the estimation of respective anodic and cathodic half-wave potentials of conjugated $\pi$ systems of polyenes or other unsaturated conjugated hydrocarbons.

The HMO energy scheme is characterized by a number, $n/2$, of bonding molecular orbitals possessing MO energies which lie negative to the energy, $\alpha$. The quantity $\alpha$ is negative and represents the binding energy of a $p$ electron in the $p$ orbital of an $sp^2$-hybridized isolated C atom. Additionally, there exist $n/2$

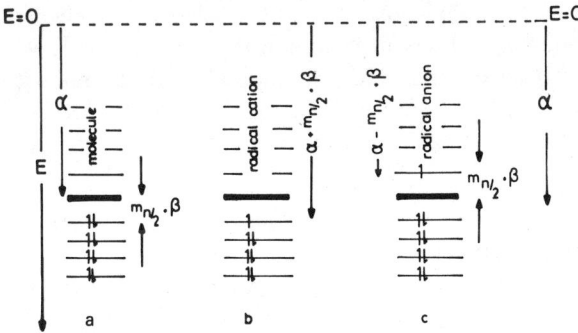

Figure 2. Hückel $\pi$-electron–MO-energy scheme of an arbitrary unsaturated conjugated hydrocarbon (e.g., octatetraene). (a) Molecule, (b) radical cation, (c) radical anion.

antibonding MOs, the energies of which are arranged above the value of $\alpha$. The MO energies of bonding and antibonding molecular orbitals are grouped symmetrically with respect to the energy reference value, $\alpha$. For the neutral polyene molecule all bonding MOs are occupied by electron pairs. Figure 2b shows that in the HMO picture the generation of a radical cation by a $1e$ oxidation of the polyene is to be described as the removal of one electron out of the highest occupied molecular orbital (HOMO). For this process, the binding energy $(\alpha + \beta \cdot m_{n/2})$, which equals the ionization energy, is necessary [Eq. (6)]:

$$\text{polyene molecule} \rightarrow e + \text{polyene radical cation} \qquad (6)$$

$$\Delta E_1^+ \text{ (vacuum)} = I, \qquad \Delta E_1^+ \approx -(\alpha + \beta \cdot m_{n/2}) \qquad (6a)$$

For the generation of a radical cation as well as of a radical anion from a neutral molecule *in vacuo*, first of all the ion self-energy $(e^2/8\pi\varepsilon_0 r)$ has to be supplied. Therefore, it seems advisable to break up the ionization energy into two parts:

(a) The electrical self-energy of the ion $\Delta E_{ion}$, which is given for a spherical molecule in Eq. (7).
(b) The extra Coulomb-interaction energy $\alpha'$ of a $p$ electron which interacts with the atomic core of an $sp^2$-hybridized C atom being in a lone $p$ orbital.

For a spherical molecule:

$$E_{ion} = \frac{e_0^2}{8\pi\varepsilon_0 r} \qquad \text{(for } r = 2 \text{ Å: } E_{ion} \approx 7.2 \text{ eV)} \qquad (7)$$

Thus, for the ionization energy of the molecule one obtains

$$\Delta E_1^+(\text{vacuum}) = \Delta E_{ion} - (\alpha' + \beta \cdot m_{n/2}) \qquad (8)$$

With the aid of a Born–Haber cycle, the amount of energy necessary to generate a radical cation from a polyene molecule not in a vacuum but in an electrolytic solution (dielectric constant $\varepsilon$) at an electrode can be derived.

Thus, one has to take into account that the electrode potential, which is measured versus a suitable reference electrode, determines the energy necessary for the anodic oxidation of a depolarizer molecule. The important differences between ionization in solution and ionization in a vacuum are as follows:

(a) It is a solvated, not a bare, radical cation which is formed from a solvated molecule.
(b) The electron is not removed to infinity but (with respect to the fact that the electrode potential is measured against a reference, e.g., the calomel) is transferred to the reference electrode where it is used to carry out a reduction (e.g., $HgCl + e \rightarrow Hg + Cl^-$).

## ELECTRO-ORGANIC SYNTHESES

This process is described in Eq. (9) and the energy terms are summed up in Eq. (10):

(polyene molecule)$_{solvated}$ + (HgCl)$_{reference}$ →

$$\text{(polyene radical cation)}_{solv} + (\text{Hg} + \text{Cl}^-_{solv})_{ref} \quad (9)$$

$$\Delta E_2^+(\text{solvent}) = -[\alpha' + \beta \cdot m_{n/2}] + \frac{e_0^2}{8\pi\varepsilon_0 r_{ion}}$$

$$+ \frac{1}{N_A}[\Delta G°(\text{radical})_{solv} - \Delta G°(\text{molecule})_{solv}] \quad (10)$$

$$+ \frac{1}{N_A}[\Delta G°(\text{Hg}) + \Delta G°(\text{Cl}^-)_{solv} - \Delta G°(\text{HgCl})]_{ref}$$

$\Delta E_2^+(\text{solvent})$ is a physical entity which is measured by the normal potential of the redox couple: (polyene molecule/polyene radical cation) with respect to a chosen reference potential which might be well approximated by the anodic half-wave potential, $E_{1/2}^{ox}$:

$$\Delta E_2^+(\text{solvent}) \approx E_{1/2}^{ox} \cdot e \quad (11)$$

$\Delta G°(\text{rad. cat.})_{solv}$ can be expressed according to the Born model [Eq. (12)] which uses the macroscopic dielectric constant, $\varepsilon$:

$$\Delta G°(\text{rad. cat.})_{solv} = -(e_0^2/8\pi\varepsilon_0 r)(1 - 1/\varepsilon) \quad (12)$$

Equations (7) and (12) introduced into Eq. (10) yield

$$\Delta E_2^+(\text{solvent}) = -[\alpha' + \beta \cdot m_{n/2}] + \frac{e_0^2}{8\pi\varepsilon_0 \varepsilon r_{ion}} - \frac{1}{N_A}[\Delta G°(\text{molecule})]_{solv}$$

$$+ \frac{1}{N_A}[\Delta G°(\text{Hg}) + \Delta G°(\text{Cl}^-)_{solv} - \Delta G°(\text{HgCl})]_{ref} \quad (13)$$

$$E_{1/2}^{ox} = \Delta E_2^+/e_0 \quad (14)$$

[If $\Delta E_2^+(\text{solvent})$ is expressed in eV, then its numerical value equals the value of $E_{1/2}^{ox}$.] For the generation of a radical anion in the same electrolyte solution, the energy, and the corresponding cathodic half-wave potential, is obtained in a similar way, i.e.,

$$\Delta E_2^-(\text{solvent}) = +[\alpha' - \beta \cdot m_{n/2}] + \frac{e_0^2}{8\pi\varepsilon_0 \varepsilon r_{ion}} - \frac{1}{N_A}[\Delta G°(\text{molecule})_{solv}]$$

$$+ \frac{1}{N_A}[\Delta G°(\text{HgCl}) - \Delta G°(\text{Hg}) - \Delta G(\text{Cl}^-)_{solv}]_{ref} \quad (15)$$

$$E_{1/2}^{red} = -\Delta E_2^-/e_0 \quad (16)$$

Comparison of Eqs. (13) and (14) with (15) and (16) gives an explanation for the observed correlated changes of anodic and cathodic half-wave potentials, which can be summarized as follows:

For a series of related, unsaturated hydrocarbon molecules for which the different terms in Eqs. (13) and (15), especially $\alpha'$, $r_{ion}$, and $\Delta G°$ (molecule)$_{solv}$, can be assumed to be nearly constant, the shifts of $E_{1/2}^{red}$ and $E_{1/2}^{ox}$ are due to changes in $m_{n/2}$ (see Figure 2). Members of such series which are more difficult to oxidize than others are also more difficult to reduce and vice versa. Thus the anodic and cathodic half-wave potentials of comparable groups of unsaturated hydrocarbons are grouped symmetrically around a common midpoint potential which, according to a proposal of Parker,[26,27] might be used as a reference potential for the comparison of the voltage series for different solvents (for acetonitrile this midpoint potential turns out to be approximately 0.36 V vs. sat. cal.). The difference of the anodic and cathodic half-wave potential is directly proportional to $2(m_{n/2} \cdot \beta)$, the energy difference of LUMO (lowest unoccupied molecule orbital) and HOMO. This difference is smaller the more extended the conjugated systems of unsaturated hydrocarbons. Highly conjugated molecules can be converted into their respective radical cations and radical anions much more easily than can the less conjugated systems.

### 2.3. Hückel Molecular Orbital (HMO) Representation of the Electrochemical Oxidation and Reduction of Double Bonds between Carbon and Heteroatoms

Upon exchange of a carbon atom, involved in a conjugate $\pi$-electron system by a heteroatom, i.e., by an atom which has an electronegativity that is different from that of carbon, the energetic symmetry with respect to oxidation and reduction vanishes. This is shown in Figure 3, where the MO energy scheme for a $>$C$=$C$<$ bond is compared with a double bond between carbon and a heteroatom, which is more electronegative than carbon (i.e., $>$C$=$N or $>$C$=$O). Since the energy of a $p$ electron in a $p$ orbital of $sp^2$-hybridized O or N atoms is more negative than in a $p$ orbital of carbon, the MO energy of the bonding MO formed between the hetero- and the carbon atom is significantly more negative than that of a $>$C$=$C$<$ bond.

The MO energy of the antibonding MO is then less positive with respect to the antibonding energy of a $>$C$=$C$<$ bond. So, the $\pi$ system of such double

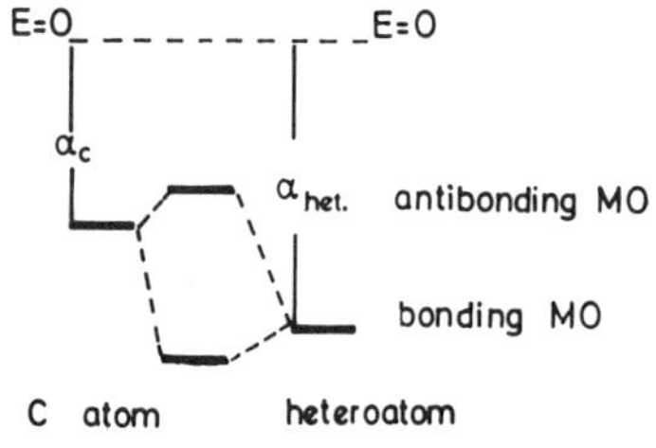

Figure 3. Schematic MO energy presentation for a carbon–heteroatom double bond.

## Table 2
### Cathodic Half-Wave Potentials of Vinyl, Imide, and Carbonyl Compounds

| Compound | $E_{1/2}$, V vs. sat. cal. |
|---|---|
| Methyl-styrene | −2.3 |
| Benzaldehyde-oxime | −1.8 |
| Acetophenone | −1.5 |
| Propylene | −2.5 |
| Acetone-imide | −1.44 |
| Acetone | −2.1 |

bonds is more difficult to oxidize and easier to reduce than the carbon–carbon double bond. Table 2 presents the cathodic half-wave potentials for analogous compounds carrying the $>C=CR_1R_2$, $>C=NR$, and $>C=O$ group and shows clearly that the vinyl compound can be reduced only at relatively high cathodic potentials. Also, the reduction half-wave potentials for the imide and the carbonyl groups are less negative. Oppositely, the anodic oxidation of these compounds cannot be observed. The ease of reduction of C=N and C=O double bonds is even more enhanced if the heteroatoms are protonated and the functional groups bear a positive charge which additionally favors the insertion of an electron in the LUMO by increasing further the effective electronegativity of the heteroatom.

### 2.4. Reactivity of Radical Cations and Radical Anions Generated from Unsaturated Hydrocarbon Molecules

The ion self-energy demand for the generation of radical cations and anions *in vacuo* as expressed by Eq. (7) is approximately identical in magnitude and sign for radical anions and cations, respectively. However, the formation of a radical cation means a loss of binding energy $-(\alpha + m_{n/2} \cdot \beta)$ which is also approximately gained by the formation of a radical anion. This explains the very different magnitudes of the ionization potentials (6–8 eV) and electron affinities of unsaturated hydrocarbons (0.1–0.6 eV). This does not mean, however, that radical anions generated from unsaturated hydrocarbons in solutions are stable, or that the respective radical cations are very unstable. Actually, both are unstable and in most cases both radical anions and radical cations generated from the same hydrocarbon do react readily. The reason for this is that the relative stability of a species is determined by both the amount of energy needed for its generation, and the free energy with respect to possible products (which determines its thermodynamic instability and, thus, its reactivity). This is comparable for radical cations and radical anions. This conclusion is drawn from observed short lifetimes of both radical ions in a solvent like purified acetonitrile, which is a stable and chemically inert solvent possessing low acidity and relatively low basicity ($pK_a \approx 20$).

From Eqs. (13) and (15), it is clear that the higher the value of $m_{n/2}$ and the more negative and more positive the cathodic and anodic half-wave potentials, the higher are the reactivities of the radical ions formed at these potentials.

HMO theory is a tool which, beyond the estimation of free energies of generation, allows predictions to be made concerning the reactive sites of the radical ions. Thus one can predict *a priori* the C atom at which the subsequent reaction will occur.

In Table 3, the charge densities, $q^2$, for LUMO and HOMO (identical in alternate unsaturated hydrocarbons) are given for butadiene, hexatriene, for naphthalene, anthracene and for styrene.[28]

The charge densities of HOMO and LUMO determine the following for radical cations: positive spin densities and charge densities; and for radical anions: negative charge densities and spin densities. It should be stressed that in the frame of HMO theory, there is no spatial separation of spin and positive charge density for radical cations. The data of Table 3 show that charge

Table 3
Charge Densities of LUMO and HOMO of Some Unsaturated Hydrocarbons[28]

| Hydrocarbon | Structure | Charge densities ($q^2$) | |
|---|---|---|---|
| Butadiene | C=C−C=C  (1 2 3 4) | $q^2(1) = 0.362$ $q^2(2) = 0.138$ $q^2(3) = 0.138$ $q^2(4) = 0.362$ | |
| Hexatriene | C=C−C=C−C=C  (1 2 3 4 5 6) | $q^2(1) = 0.271$ $q^2(2) = 0.058$ $q^2(3) = 0.174$ | $q^2(4) = 0.174$ $q^2(5) = 0.058$ $q^2(6) = 0.271$ |
| Naphthalene | (fused bicyclic, positions 1–10) | $q^2(1, 4, 5, 8) = 0.1806$ $q^2(2, 3, 7, 6) = 0.0692$ $q^2(9, 10) = 0$ | |
| Anthracene | (fused tricyclic, positions 1–10) | $q^2(1, 4, 5, 8) = 0.0967$ $q^2(2, 3, 6, 7) = 0.0484$ $q^2(9, 10) = 0.193$ $q^2(\text{C angular}) = 0.0093$ | |
| Styrene | C$_6$H$_5$−CH=CH$_2$ (ring 1–6; 7, 8 vinyl) | $q^2(8) = 0.354$ $q^2(7) = 0.155$ $q^2(1) = 0.112$ | $q^2(6, 2) = 0.094$ $q^2(5, 3) = 0.017$ $q^2(4) = 0.155$ |

densities are remarkably different for different C atoms in these compounds. For instance, in the polyenes butadiene and hextriene, the C atoms at the end of the chain carry highest charge densities, and the terminal sites of the radical ions are the most reactive.

For naphthalene, the $\alpha$ position will be the most reactive in the radical ion; in anthracene, the reactivity of the radical ions is concentrated on the 9- and 10-carbon atom. Radical ions produced from styrene will be most reactive at the terminal (or $\beta$) carbon atom of the vinyl group. Further, radical ions from 1,4-diphenylbutadiene are most and (equally) reactive at the carbon atoms attached to the phenyl rings (the benzylic C atom) of the butadiene chain.

## 2.5. Typical Chemical Reactions of Radical Cations and Radical Anions Generated Electrochemically from Unsaturated Hydrocarbons

Due to the very high quantum mechanical improbability of a two-electron charge transfer between an electrode and a depolarizer molecule[29] and because a second charge transfer would need a remarkably higher energy than the first charge transfer step, in most cases conversion of any neutral molecules at electrodes will, at first, generate reversibly radical ions.

The frequently observed polarographic irreversible two-electron waves are generally due to a rapid chemical reaction following the initial one-electron charge transfer. This consecutive reaction usually produces a species which can undergo a further very rapid charge transfer at the applied electrode potential leading to an irreversible overall $2e$ oxidation or reduction (i.e., $d \log [(n_L - i)/i]/dU \geq (120 \text{ mV})^{-1}$):

$$\text{molecule} \xrightarrow{\pm e} (\text{rad. ion}) \pm \xrightarrow[\text{fast}]{\text{reaction}} \text{intermediate} \xrightarrow[\text{fast}]{\pm e} \text{product} \quad (17)$$

It is the aim of this chapter to discuss the features of typical chemical reaction steps for radical cations and anions, respectively, and to show the far-reaching analogy for the chemcial reactions of electrochemically produced radical cations and radical anions.

Radical cations being unstable species possess, even *in vacuo* (e.g., in a mass spectrometric experiment where they cannot interact with other molecules) relatively short lifetimes. An isolated radical cation is very often stabilized by intramolecular rearrangement reactions, proton dissociation, fragmentation, etc. In solution, however, the electrochemically generated unstable radical ions are in direct contact with surrounding solvent molecules and other species dissolved in the electrolyte, e.g., electrolyte ions, molecules of the unsaturated hydrocarbon from which the radical ion was just generated, etc. Thus they have unhindered access to a great multitude of reactants, and, from this point of view, the consecutive chemical reactions which follow the radical ion generation are not expected to proceed *a priori* very selectively.

Indeed, this is the experience in the performance of electrochemical oxidation and reduction of organic substances. Therefore the general assumption that electrochemical reactions proceed more smoothly and more selectively than reactions that are performed by simple thermal activation has to be modified. Only the first charge transfer determination by a precise potential is a rather selective step. The following chemical reactions, however, are manifold and have to be made selective through a detailed knowledge of, or at least an intuitive understanding of, the kinetics of the different competitive consecutive reactions.

### 2.5.1. Chemical Reactions of Radical Cations

The chemical reactivity of radical cations (comparable to that of radical anions) is determined by both: the existence of an unpaired electron, i.e., its radical character, and, the existence of the positive charge, i.e., the ionic character of this species.

As radical ions possess an unpaired spin, one should expect that they undergo typical radical reactions, i.e., H abstraction, radical addition to unsaturated hydrocarbons, radical disproportionation (giving the respective hydrocarbon di-cation and the intact, unsaturated hydrocarbon) and radical-radical dimerization [Figure 4a, parts (1), (2), (3)]. Due to its positive charge the radical cation is also expected to behave as an electrophile, adding to any nucleophile available in its neighborhood, e.g., the basic part of the solvent molecules (e.g., —OH in $H_2O$, —OMe in HOMe, —C$\equiv$N group in acetonitrile, etc.), basic anions of the electrolyte (e.g., $OH^-$, $OR^-$, halide, or pseudohalide anions, etc.) [Figure 4a, part (4)].

Radical cations may even add to unsaturated hydrocarbon molecules which, in their electronic $\pi$ systems, offer a reactant of some nucleophilicity [Figure 4a, part (5)]. In addition, radical cations can react further by proton dissociation from a vicinal carbon atom [Figure 4a, part (6)] to form an allyl radical. It must be stressed that the six different reaction paths given in the schematic presentation do not proceed with equal, or even comparable, probability.

Radical cations which are more easily produced (i.e., at lower oxidation potentials) and hence are of greater stability, preferably react according to the typical radical reactions [parts (2) and (3) of Figure 4a]. The more reactive radical cations, however, behave much more like carbo-cations and add to all available nucleophiles. They either undergo coupling reactions with C–C coupling, i.e., they add to basic unsaturated hydrocarbons [part (5) of Figure 4a], or they react with the solvent or other (mostly anionic) nucleophiles present in the electrolyte. Proton dissociation [Figure 4a, part (6)] is observed if, in the course of this reaction, relatively stable allyl radical groups can be formed, i.e., if a secondary or tertiary C atom is available vicinal to the oxidized vinyl group to form a relatively stable allyl radical.[30]

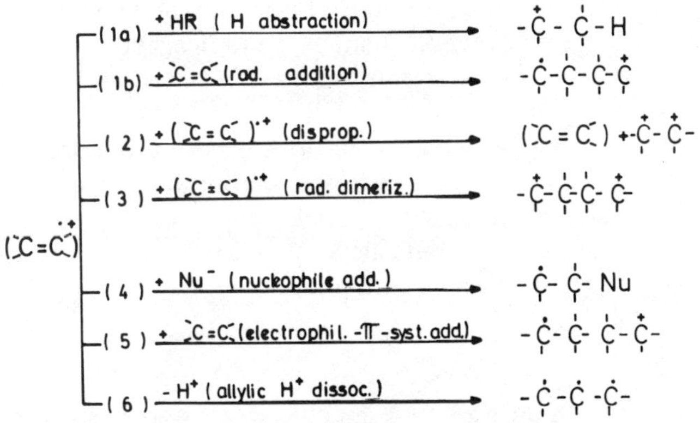

## ELECTRO-ORGANIC SYNTHESES

$$(\overset{..}{C}=\overset{.}{C})^{\cdot+}$$

- (1a) $\xrightarrow{+HR\ (H\ abstraction)}$ $-\overset{+}{C}-\overset{|}{C}-H$
- (1b) $\xrightarrow{+C=C\ (rad.\ addition)}$ $-\overset{+}{C}-\overset{|}{C}-\overset{|}{C}-\overset{\cdot}{C}$
- (2) $\xrightarrow{+(C=C)^{\cdot+}\ (disprop.)}$ $(C=C) + \overset{+}{-C}-\overset{+}{C}-$
- (3) $\xrightarrow{+(C=C)^{\cdot+}\ (rad.\ dimeriz.)}$ $-\overset{+}{C}-\overset{|}{C}-\overset{|}{C}-\overset{+}{C}-$
- (4) $\xrightarrow{+Nu^-\ (nucleophile\ add.)}$ $-\overset{.}{C}-\overset{|}{C}-Nu$
- (5) $\xrightarrow{+C=C(electrophil.\ -\pi-syst.add.)}$ $-\overset{.}{C}-\overset{|}{C}-\overset{|}{C}-\overset{+}{C}-$
- (6) $\xrightarrow{-H^+\ (allylic\ H^+\ dissoc.)}$ $-\overset{.}{C}-\overset{|}{C}-\overset{.}{C}-$

Figure 4a. Alternative reaction routes for radical cations obtained from unsaturated hydrocarbons: (1a)–(3), radical reactions; (4)–(6), cationic reactions.

The hydrogen-abstraction reaction [Figure 4a, part (1)] obviously does not play an important part in the different consecutive reactions of radical cations. Rearrangement reactions similarly are only rarely found for anodically formed radical cations. Irrespective of whether the radical cations react as radicals [Figure 4a, reactions (2) and (3)] or as cations [Figure 4a, reactions (4), (5), and (6)], the intermediate product obtained is an unstable and still remarkably reactive species: for reaction (2) it is a di-carbo-cation with the two positive charges distributed either: (1) on a more or less extended conjugate system, or (2) only on two geminal C atoms (if a vinyl group was oxidized). For reaction path (3), it is an intermediate with two carbo-cations which are separated by two C atoms.

For reaction paths (4)–(6) (carbo-cation reactions) the obtained species are unstable as well. The cationic species obtained by paths (1) and (2) eventually react with nucleophiles. These nucleophiles may be solvent molecules, or added anions, or (less frequently) some basic unsaturated hydrocarbon molecule, i.e., still intact unoxidized depolarizer molecules [Eq. (18)]. The reaction of the carbo-cations with an anionic nucleophile terminates

$$-\overset{+}{\underset{|}{C}}- + Nu^- \rightarrow -\overset{|}{\underset{|}{C}}-Nu \tag{18}$$

the total reaction chain, which starts with the anodic formation of a (relatively unreactive) radical cation and passes via the two chemical steps to a stable end product (ecc sequency).†

Since the "radicalic" reactions are typical of the more stable radical cations with longer lifetimes (i.e., about $10^{-3}$ sec) these reactions occur

---

† Here and throughout this chapter, in expressions such as ecc sequency and epe sequency, e stands for electron, c stands for cation, and p stands for proton.

frequently at a larger distance from the electrode in homogeneous solutions. Oppositely, in the further reaction chain of the radicals obtained by reactions (4)–(6) (Figure 4a) a second charge transfer step is generally involved. The reason for this is twofold:

(a) The C radicals produced by reactions (4)–(6) are more accessible to anodic oxidation than the unsaturated hydrocarbon from which the reaction started.[31,32] So, under the potential conditions necessary to perform the primary oxidation step, these radicals will be oxidized immediately at the anode [Eq. (19)] to the respective carbo-cation:

$$-\overset{\cdot}{\underset{|}{C}}- - e \rightarrow -\overset{+}{\underset{|}{C}} \quad (19)$$

(b) It was pointed out that the more reactive radical cations preferably follow the "carbo-cation" reactions. Thus, it is to be understood that rates of these reactions are very fast. Quite often they are so fast indeed that, to our knowledge,[33,34] the reacting species has little time to escape from the electrode surface into the bulk of the solution. This means the radicals produced by reaction path (3)–(6) are either generated at the electrode surface or at such short distance from the electrode surface that they can diffuse back to the anode quickly. Since they are either still in physical contact or at least close to the anode, they are further oxidized immediately, or very shortly after their appearance, to carbo-cations which eventually react with nucleophiles.

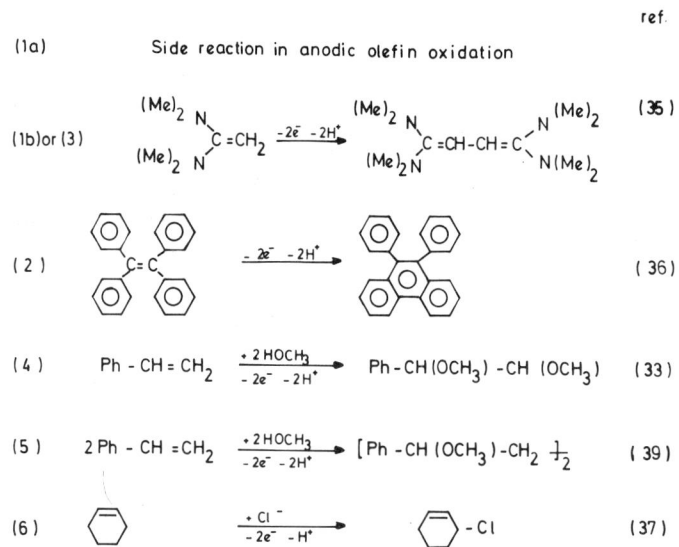

Figure 4b. Some examples for reaction products of radical cations of vinyl compounds obtained by different consecutive reactions (numbering of reaction routes corresponds to numbering in Figure 4a).[34–39]

# ELECTRO-ORGANIC SYNTHESES

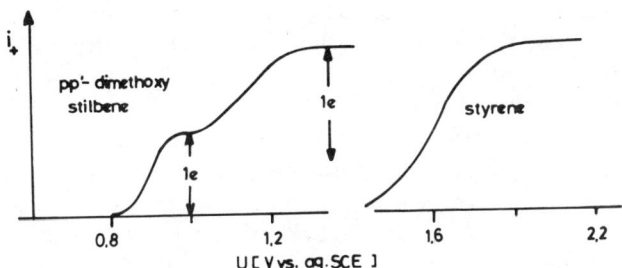

Figure 5. Anodic polarogram of an easily oxidizable unsaturated hydrocarbon showing a reversible $1e$ oxidation wave at low potentials ($pp'$-dimethoxystilbene) and of a less easily oxidizable unsaturated hydrocarbon showing an irreversible $2e$ oxidation wave (styrene).[34]

The total reaction chain for the more reactive radical cations, therefore, is typically an ecec sequency. An electrode-kinetic investigation method which is as simple as is the anodic voltammetry at rotating solid electrodes reveals this principle: Figure 5 compares the anodic voltammogram in acetonitrile of a more easily oxidizable vinyl compound ($pp'$-dimethoxy-stilbene) with a half-wave potential of ($E_{1/2} = 0.8$ V vs. sat. cal.) with the polarogram of a vinyl (styrene) compound which is remarkably less accessible to anodic oxidation ($E_{1/2} = 1.6$ V vs. sat. cal.). The polarogram of $pp'$-dimethoxy-stilbene is characterized by a reversible $1e$-oxidation wave and a second, irreversible $1e$-oxidation wave. At the potential range which is determined by the first wave, a relatively stable and long-living radical cation is formed. At the potential range of the second wave, two electrons are transferred in an ee sequency (one after another but the second immediately following the first charge transfer). Here a much more reactive carbo-di-cation is formed which reacts in two steps with available nucleophiles (eecc).

The polarogram of styrene, however, shows only one two-electron wave, which is determined by a very fast ecec sequency. The first chemical step (c) can be shown to be the reaction with the solvent acetonitrile, since after appropriate work-up procedure, anodic acetamidation is found:

$$(C_6H_5CH-CH_2)^{\cdot+} + CH_3CN \rightarrow C_6H_5-\dot{C}H-CH_2(\overset{+}{N}C-CH_3) \quad (20)$$

$$C_6H_5\dot{C}H-CH_2\overset{+}{N}C-CH_3 - e + CH_3CN \rightarrow$$
$$C_6H_5CH(\overset{+}{N}C-CH_3)-CH_2(\overset{+}{N}CCH_3) \quad (21)$$

$$\overset{\diagdown}{\underset{\diagup}{C}}-\overset{+}{N}C-CH_3 + H_2O \rightarrow \overset{\diagdown}{\underset{\diagup}{C}}-NHCOCH_3 + H^+ \quad (22)$$

An appropriate borderline between less reactive radical cations which follow the radical reaction sequence and more reactive radical cations which react more like carbo-cations can be drawn according to the half-wave potentials necessary for their generation. This borderline value is around 1.2 to 1.3 V vs. sat. cal.[34] In Figure 4b, one example for each of the reaction paths (2)–(6) is collected from recent literature.[33-39] This part of the figure shows that, owing to an overall two-electron transfer per molecule of the final product, generally doubly substituted (or doubly unsaturated) compounds are produced. The substituents are alkoxy groups in cases where the anodic oxidation is performed in alcoholic solutions, or they are acetamide groups in cases where the solvent is acetonitrile.

One obtains products that are subject to further chemical substituent-exchange reactions and a multiple of other chemical conversions. Thus the products of anodic conversion of unsaturated hydrocarbons are valuable substances. It should be pointed out that the anodic C–C coupling reactions in particular are of great practical interest and potential importance in preparative organic chemistry. It must, however, be emphasized that in many cases preparative work is hindered by undesired polymerization reactions proceeding at the electrode surface itself. Thus, very often, soon after beginning an electro-organic synthesis which is based on the anodic conversion of unsaturated hydrocarbons, the current density decreases; i.e., the synthesis slows down significantly since the anode becomes electrically insulated by a dense and closed polymer layer.

### 2.5.1.1. Anodic Side-Chain Substitution as Typical Reaction Path for Radical Cations of Alkyl Aromates

The pronounced stability of benzyl radicals and benzyl radical cations vs. phenyl radical and phenyl radical cations gives reason for special pathways of the anodic reactions of alkyl aromatics.

Very often radical cations of alkyl aromates can be stabilized by proton dissociation from the benzylic carbon atom:

$$Ar-CH_2-R - e \rightarrow (Ar-CH_2-R)^{\cdot +} \quad (23)$$

$$(Ar-CH_2-R)^{\cdot +} - H^+ \rightarrow (Ar-\dot{C}H)-R \quad (24)$$

In a subsequent fast charge transfer, a benzyl carbo-cation is generated which reacts with any nucleophile available. Thus, in methanol methoxylation and in acetonitrile, acetamidation at the benzylic C atom is observed:

$$(Ar-\dot{C}H-R) - e + {}^-OMe \rightarrow Ar-CH(OMe)-R \quad (25)$$

$$(Ar-\dot{C}H-R) - e + CH_3CN \xrightarrow{+H_2O} Ar-CH(NHCOCH_3)-R + H^+ \quad (26)$$

### 2.5.1.2. Anodic Aromatic and Allylic Substitution

A special and very interesting reaction path starting from the initiating step of the anodic generation of radical cation enables the anodic substitution of aromatic compounds or substitution at the allyl position on an isolated $>C=C<$ bond.

The anodic oxidation of naphthalene in acetic acid was shown to proceed via primary formation of the naphthalene radical cation adding acetate anion and forming an allylic radical which, immediately after its formation, undergoes a second $1e$-oxidation step since it is oxidized more easily than naphthalene. The produced allylic cation then becomes stabilized by deprotonation, i.e., it forms $\alpha$-acetyl-naphthalene as the final and stable oxidation product:

$$C_{10}H_8 - e \rightarrow (C_{10}H_8)^{\cdot+} \tag{27a}$$

$$C_{10}H_8^{\cdot+} + {}^-OAc \rightarrow \begin{array}{c}\text{H} \quad \text{OAc}\end{array} \tag{27b}$$

$$\begin{array}{c}\text{H} \quad \text{OAc}\end{array} \xrightarrow{-e-H^+} C_{10}H_7OAc \tag{27c}$$

In a similar manner, an anodic allylic substitution can be observed for the anodic oxidation of olefins. For instance, the anodic conversion of cyclohexene not only yields products which are obtained by the anodic oxidation of the $>C=C<$ bond, but also yields products which are anodically substituted at the C atoms vicinal to the $>C=C<$ bond:[37]

$$R_2C=CR-CHR_2 \xrightarrow{-e} [R_2C=CR-CHR_2]^{\cdot+} \xrightarrow{-H^+} [R_2\dot{C}-\dot{C}R-\dot{C}R_2] \tag{28a}$$

$$R_2\dot{C}-\dot{C}R-\dot{C}R_2 \xrightarrow{-e+Nu^-} R_2C=CR-C(Nu)R_2 \tag{28b}$$

### 2.5.1.3. Radical Cations from Aliphatic and Aromatic Amines and Their Typical Reactions[40]

#### 2.5.1.3.1. Radical Cations from Alkyl Amines and Their Reactions.

Alkyl amines are relatively easy to oxidize, i.e., they undergo anodic oxidation at $+1$ V to $+1.3$ V vs. sat. cal. Although detailed kinetic studies for the oxidation and the relevant consecutive reactions of alkyl amines are not available, some mechanistic details can be deduced from the most important products obtained.

Obviously, the primarily generated N radical cation stabilizes by proton dissociation from the carbon atom to which the amino group is attached:

$$CH_3-NHR - e \rightarrow (CH_3-NHR)^{\cdot+} \quad (29a)$$

$$(CH_3-NHR)^{\cdot+} - H^+ \rightarrow (CH_2=\dot{N}HR) \quad (29b)$$

Subsequent second charge transfer to form the iminium cation accompanied by proton dissociation from the [Eqs. (30a) and (30b)] nitrogen atom leads to the stable imide. The imide usually hydrolyzes in the working-up procedure to form carbonyl compounds:

$$(CH_2=\dot{N}HR) \xrightarrow{-e} (CH_2=NHR)^+ \quad (30a)$$

$$(CH_2=NHR)^+ - H^+ \rightarrow CH_2=NR \xrightarrow{+H_2O} H_2CO + NH_2R \quad (30b)$$

### 2.5.1.3.2. Radical Cations from Aryl Amines and Their Reactions.

With aromatic amines the primary charge transfer generates a radical cation, the unpaired spin and positive charge of which may distribute over the aromatic system. Thus for phenyl amine radical cations the N atom and the ortho and para C atoms represent reactive sites. The result is N–N, N–C, and C–C coupling reactions in which a second aryl amine molecule is involved:

$$ArNH_2 - e \rightarrow (ArNH_2)^{\cdot+} \quad (31a)$$

$$(ArNH_2)^{\cdot+} + ArNH_2 \xrightarrow{-3e-4H^+} Ar-N=N-Ar \quad (31b)$$

$$(ArNH_2)^{\cdot+} + ArNH_2 \xrightarrow{-e-2H^+} Ar-NH-Ar-NH_2 \quad (31c)$$

$$(ArNH_2)^{\cdot+} + ArNH_2 \xrightarrow{-e-2H^+} H_2N-Ar-Ar-NH_2 \quad (31d)$$

### 2.5.1.4. Radical Cations from Aliphatic and Aromatic Hydroxy Compounds[41]

#### 2.5.1.4.1. Radical Cations from Alcohols.
Few details are known of the anodic oxidation kinetics for aliphatic alcohols leading eventually to carbonyl compounds or carboxylic acids (or their esters). The reason for this is seen in the very fast consecutive reactions following the first charge transfer. Proton loss and subsequent second oxidation to the respective carbo-cation are likely to be the next steps.

#### 2.5.1.4.2. Radical Cations from Phenols.
Contrary to radical cations formed by anodic oxidation of alcohols, the radical cation generated from phenols is remarkably more stable. In these species unpaired spin and positive charge does not concentrate solely on the hydroxyl group; rather, it spreads over the aromatic system. Consequently, as in aromatic amines, anodic O–O

coupling (which is the exception), O–C coupling, and C–C coupling are observed. Often, an addition of nucleophiles from the solvent–electrolyte system leads to anodic substitution in the ortho or para position of the aromatic.

The most important reaction being observed in electrolytes with sufficient water content is the oxidation of phenols to quinones. This overall reaction is initiated by nucleophilic attack of hydroxyl ions or of water on activated C atoms ($o$ or $p$) of the phenol radical cation.

### 2.5.1.5. Radical Cations from Organic Sulfides[42]

#### 2.5.1.5.1. Radical Cations from Mercaptans and Dialkyl Disulfides.
Due to the relatively low anodic half-wave potentials of mercaptans, the radical cations of these compounds are more stable than radical cations of alcohols. The most important consecutive reaction following the first anodic charge transfer is dissociation of the sulfide proton and subsequent radical–radical dimerization by which dialkyl disulfides are formed:

$$\text{HSR} - e(\text{HSR})^{\cdot+} \xrightarrow{-H^+} \cdot S-R \xrightarrow{\text{dimer.}} R-S-S-R \qquad (32)$$

Thus cysteine is oxidized to cystene reversibly and thiourea dimerizes anodically to dithiourea.

In presence of strong nucleophiles, especially in the presence of water, the typical anodic reaction of disulfides is essentially a reaction in which the valence shell of sulfur is extended. Thus disulfides are oxidized to the corresponding disulfoxides. Several disulfides, on anodic oxidation, undergo anodic cleavage of the sulfur–sulfur bond forming sulfinic and, finally, sulfonic acids:

$$R-S-S-R + 2H_2O \xrightarrow{-4e-4H^+} R-SO-SO-R \qquad (33a)$$

$$R-SO-SO-R + 4H_2O \xrightarrow{-6e-6H^+} 2RSO_3H \qquad (33b)$$

#### 2.5.1.5.2. Radical Cations from Dialkyl Sulfides.
The chemical consecutive reactions following the anodic oxidation of a dialkyl sulfide to its radical cation are determined by the enhanced C–H acidity of the carbon atoms linked to sulfur (which in the unoxidized compound is remarkably high). Thus proton dissociation and formation of a cation by a second charge transfer step are observed. The cation may finally combine with any nucleophile, e.g., may couple with the sulfur atom of another dialkyl sulfide molecule. Thus anodic self-coupling with the formation of sulfonium ions is observed:

$$CH_3-S-CH_3 \xrightarrow{-e} CH_3-S^{\cdot+}-CH_3 \xrightarrow{-H^+} CH_3-\dot{S}=CH_2 \qquad (34)$$

$$CH_3-\dot{S}=CH_2 - e + (CH_3)_2S \rightarrow CH_3-S-CH_2-S^+(CH_3)_2 \qquad (35)$$

In presence of high water concentrations, the primarily formed radical cation adds $OH^-$ and finally dialkyl sulfoxides $(alk)_2SO$ are formed (ecec).

*2.5.1.5.3. Radical Cations from Diaryl Sulfides.*[43] Diaryl sulfides show relatively low half-wave potentials (between +1 and 1.6 V vs. sat. cal.) in acetonitrile. In the primarily formed radical cation (as with N-aryl and O-aryl compounds), spin and positive charge spreads over the aromatic $\pi$ systems of the aryl rests attached to the sulfur atom. Therefore, the radical cations of diaryl sulfides are relatively stable and show little reactivity, respectively. They add only strong nucleophiles (e.g., $OH^-$ from water) to form sulfoxides eventually. It can be shown that, in nonaqueous media, the S di-cation is formed by disproportionation of the radical cation:

$$(Ar)_2S - e \rightarrow (Ar)_2S^{\cdot +} \tag{36}$$

$$2(Ar)_2S^{\cdot +} \rightleftarrows (Ar)_2S^{2+} + (Ar)_2S \tag{37}$$

The S di-cation, however, is much more reactive than the radical cation and may add to different weak nucleophiles—even to aromatics. A typical second consecutive step, then, would be the formation of an anodic dimer [Eq. (38)]—possessing the sulfonium cation function—or the coupling with an added aromatic molecule to form a triaryl sulfonium cation:

$$(Ar)_2S^{2+} + (Ar)_2S \rightarrow (Ar)_2-\overset{+}{S}-Ar-SAr + H^+ \tag{38}$$

$$(Ar)_2S^{2+} + Ar' \rightarrow (Ar)_2Ar'S^+ + H^+ \tag{39}$$

### 2.5.1.6. Radical Cations from Alkyl and Aryl Halides[44]

The ease of the anodic oxidation of alkyl and aryl halides increases from the compounds of the lighter towards the heavier halogens. Iodides (alkyl and aryl) are remarkably easy to oxidize, e.g., $E_{1/2}$ (methyliodide) = 2.2 V; $E_{1/2}$ (phenyliodide) = 2 V vs. sat. cal.[44] Consecutive reactions and respective reaction products following the first anodic $1e$ charge transfer have only been investigated for iodides.

Radical cations derived from alkyl iodides decompose by C–I bond breaking, yielding a carbo-cation and iodine (which is further oxidized at potentials necessary to oxidize the alkyl iodide). The carbo-cation generated by the decomposition of the primary radical cations then undergoes the usual chemical reactions. The most important reaction is the addition of nucleophiles. Thus, if the oxidation is performed in acetonitrile, then acetamidation is observed:

$$alk-I - e \rightarrow (alk-I)^{\cdot +} \rightarrow alk^+ + I \tag{40}$$

$$alk^+ + CH_3CN \xrightarrow[-H^+]{+H_2O} alk-NH-CO-CH_3 \tag{41}$$

With aryl iodides, obviously, the C—I bond of the radical cation is remarkably stronger; furthermore, the positive charge of the radical cation seems to be concentrated mainly at the heteroatom and is not distributed over the aromatic system as evidenced in aryl bromides and aryl chlorides.[45] Thus, an anodic attack of the iodobenzene radical cation on other aromatic systems is observed with the final reaction product being a diaryl iodonium cation:

$$ArI - e \rightarrow ArI^{\cdot+} \xrightarrow[-H^+ -e]{+Ar'} Ar-\overset{+}{I}-Ar \qquad (42)$$

This coupling reaction is very similar to that which yields sulfonium cations by anodic oxidation of diaryl sulfides. However, because of the higher anodic potential necessary to oxidize aryl iodides, their radical cations are more reactive and couple easily (at the anode surface) to other aromatic compounds, whereas radical cations from diaryl sulfides do not couple, but disproportionate, to form the more reactive S di-cation, which may couple similarly to another aryl compound, although in homogeneous solution.

### 2.5.2. Chemical Reactions of Radical Anions

#### 2.5.2.1. Radical Anions from Unsaturated Hydrocarbons

Since the only successful cathodic and technical large-scale electro-organic synthesis, the Baizer process [Eq. (2)], is based on the initial cathodic generation of radical anions, the question of the kinetic behavior of radical anions seems of special importance. Generally, one can state that the reactive behavior of radical anions obtained by cathodic reduction of unsaturated hydrocarbons is determined by the ionic character as well as by the radical character of this species. Figure 6 shows the typical reactions which a radical anion may undergo, demonstrating in a far-reaching analogy to the reaction scheme of radical cations (cf. Figure 4a).

Typical radical reaction paths as shown in Figure 6 are H abstraction (1), radicalic reaction with the electrode metal (1a) to form organometal

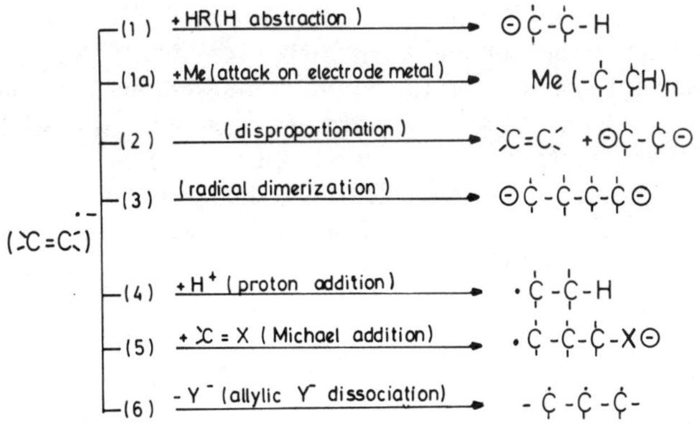

Figure 6. Alternative reaction routes for radical anions obtained from unsaturated hydrocarbons (e.g., vinyl compounds): (1)–(3), radical reactions; (4)–(6), anionic reactions.

compounds and radical anion disproportionation (2) to di-carbanions and the original hydrocarbon. A further radical reaction is radical dimerization. The radical attack at the electrode metal leading to the formation of metal organic compounds is a special feature of cathodically formed radical anions and radicals. This is observed scarcely in anodic reactions since oxidations are performed usually at noble metal anodes or even metal oxide anodes which are not attacked by radicals. Furthermore, organometallic compounds would not be at all stable at more anodic potentials but would be oxidized and decomposed. The typical anionic reactions of radical anions are mainly determined by the basicity fo hydrocarbon radical anions which is—although less than the basicity of true carbanions—still remarkably high (i.e., the p$K$ of the conjugate acids of C radical anions is around 20).

Protonation of the radical anion [reaction (3), Figure 6] is very frequently observed, even in solvents of low acidity, and leads to the formation of a radical. Michael additions of the radical anion to C=C or C=X [reaction (4), Figure 6] represent one possible route to cathodic C–C coupling products.

For radical cations, proton dissociation from carbon atoms vicinal to oxidized vinyl groups is observed. The analog reaction of radical anions is to be seen in the dissociation of anions of electronegative heteroatoms, or negatively charged groups, from C atoms vicinal to the conjugate $\pi$ system constituting the radical anion. As has been discussed in Section 2.4, the free energy of formation of a radical anion is less than the free energy of formation of a radical cation from the same unsaturated hydrocarbon. Thus one should expect a radical anion to be less reactive than the respective radical cation.

By analogy to the relative importance of "radical" and "cationic" behavior of radical cations in relation to their reactivity, one should expect that, for radical anions, the radical reactions (1) to (3) of Figure 6 should be more important than the "anionic" reactions.

This indeed seems to hold with one exception: the proton addition to radical anions to form the respective radical which is a typical ionic reaction is the most important of all reactions (1)–(6) in Figure 6. This is so, because on one hand, proton donors cannot be eliminated completely from any practical electrolyte–solvent system. (It is to be pointed out that even acetonitrile with p$K \approx 20$ is to be assumed a relatively strong proton donor for radical anions!) On the other hand, performance of cathodic electro-organic synthesis without any proton donor is not at all desirable: lack of any proton donor means that a cathodically started reaction sequence cannot be terminated by proton addition to a carbanion and would thus lead to undesired electropolymerization.[46] So, radical anions apart from reaction (3) in Figure 6 [rather the reactions (1), (1a), and (2)] are of synthetic importance rather than anionic reactions (4) and (5).

The proton addition reaction

$$(\text{rad. anion})^{\cdot -} + HA \rightarrow (\text{rad. anion} - H)^{\cdot} + A^- \quad (43a)$$

is a very fast diffusion-controlled reaction, provided HA is a stronger acid than is the conjugate acid of the radical anion.

Under practical electrolysis conditions where proton donors are generally available at the cathode, the protonation of the radical anion proceeds so fast that the radical anion species has no chance to diffuse away from the electrode. Only in thoroughly purified and dried aprotic solvents such as tetrahydrofuran or dioxane can the protonation be slowed down so that the cathodic polarograms of more easily reducible hydrocarbons are characterized by two separate $1e$ waves. In protic solvents, however, owing to very fast protonation with subsequent second charge transfer (epe), irreversible $2e$ waves for the reductions of many unsaturated hydrocarbons are observed.

It was shown[31,32] that carbon radicals are very easily oxidized and reduced. Thus, in a fast subsequent reduction step, the protonated radical anions are expected to be reduced further to a carbanion (epe sequency):

$$(\text{rad. anion} - \text{H})^{\cdot} + e \rightarrow (\text{anion})^{-} \tag{43b}$$

This carbanion may (1) be protonated again [Eq. (44)] which will stop the reaction sequency,

$$(\text{anion})^{-} + \text{HA} \rightarrow \text{hydrocarbon} + \text{A}^{-} \tag{44}$$

or (2) it will add in a Michael addition to a molecule with an activated double bond [e.g., olefins, imides, etc.; see Eq. (45)] as the analog of the typical anodic coupling reaction, namely, the addition of a carbonium to a double bond. This addition reaction leads to cathodic C–C coupling and extension of the carbon skeleton of the primary unsaturated hydrocarbon:

$$\text{R}^{-} + \underset{/}{\overset{\backslash}{\text{C}}}=\text{X} \rightarrow (\text{R}-\overset{|}{\underset{|}{\text{C}}}-\text{X})^{-} \tag{45}$$

The anion will eventually be protonated [Eq. (46)], resulting in a stable cathodic coupling product,

$$(\text{R}-\text{C}-\text{X})^{-} + \text{HA} \rightarrow \text{R}-\overset{|}{\underset{|}{\text{C}}}-\text{XH} + \text{A}^{-} \tag{46}$$

In the absence of proton donors another Michael addition may occur as the second step of an anionic polymerization leading to cathodic oligomers or polymers.[46] For the Baizer process these reactions are well established by electrode kinetic and preparative results of Baizer's and others.

For acrylonitrile under practical synthesis conditions, Beck[47] had shown the first (rate-determining charge transfer) to be first order both with respect to acrylonitrile and water, which is added as proton donor. This can be understood in terms of a concerted electron–proton transfer:

$$\text{CH}_2=\text{CH}-\text{CN} + e + \text{H}^+ \rightarrow (\cdot\text{CH}_2-\text{CH}_2-\text{CN}) \xrightarrow{+e} {}^{-}\text{CH}_2-\text{CH}_2-\text{CN} \tag{47}$$

The proton adds to the $\alpha$-carbon atom which, because of the inductive effect of the nitrilo group, should carry the lowest negative charge density in the radical cation. It is the adsorptive interaction of the radical anion and the carbanion, respectively, with the cathode metal surface which favors protonation at the less basic carbon atom.

The propionitrile anion would be protonated immediately [Eq. (50)] to propionitrile if this protonation could not be delayed:

$$(^-CH_2-CH_2-CN) + HA \rightarrow CH_3-CH_2-CN + A^- \qquad (48)$$

The use of McKee salts with quaternary alkyl ammonium cations leads to a drastic change of the chemical composition of the electrolyte layer adjacent to the electrode surface.[48] The surface-active lyophylic cations are strongly adsorbed at the cathode displacing the water and other proton donors to an extent that the proton activity at the electrode still allows for the concerted protonation charge transfer reaction [Eq. (47)]. However, the rate of the further proton addition [Eq. (48)] is diminished to the point that the Michael addition [Eq. (49)], the key step for the hydrodimerization of acrylonitrile, can compete successfully:

$$NC-CH_2-CH_2^- + CH_2=CH-CN \rightarrow$$
$$(NC-CH_2-CH_2-CH_2-CH-CN)^- \qquad (49)$$

Indeed, the Michael addition can be made to proceed with 95% current yield and 90% mass yield. (The lower mass than current yield is due to a certain amount of polymers being formed by side reaction, e.g., anionic polymerization.) Diffusing some angstroms away from the electrode, the dimeric carbanion will be protonated readily to the desired end product, adipodinitrile.

The rate of the undesired side reaction [Figure 6, (1a)] (the consumption of the cathode material by attack of radical anions) is mainly determined by the nature of the cathode metal. The less noble and the more organophilic this metal, the greater will be the rate of its consumption by radical attack.

Whereas mercury and lead cathodes are nearly completely stable during the reduction of acrylonitrile, tin electrodes are readily consumed by the formation of tetrapropionitrile tin:

$$4CH_2=CH-CN + 4e + 4H^+ + Sn \rightarrow Sn(CH_2-CH_2-CN)_4 \qquad (50)$$

To avoid any production of the relatively unstable and toxic metal alkyls, the use of carbon cathodes is advisable since highly purified carbon or graphite shows hydrogen overpotentials comparable to those of Hg, Pb, and Cd and offers comparably good conditions to perform cathodic reactions.

Summarizing the reaction scheme of Figure 6, it must be stressed that cathodic reduction of unsaturated hydrocarbons presents reactions of some synthetic value, especially in the C–C coupling reactions (2) and (5) of Figure 6 and the Michael addition which follows the reductive protonation with subsequent reduction of the radical to a carbanion. Dimerization of primary radical anions due to radical radical dimerization could only be identified positively in cases where the radical cation was stabilized by ion pair formation with a suitable cation.[49]

### 2.5.2.2. Reactions of Radical Anions Obtained by Cathodic Reduction of Carbon Heteroatom Double Bonds

The most important difference between radical anions obtained from unsaturated hydrocarbons and from systems containing C=X (X representing an heteroatom of greater electronegativity than carbon) is that the basicity of these radical anions is even higher than that of the hydrocarbon radical anions. Therefore the most important reaction is protonation at the heteroatom:

$$(>C=X)^{\cdot -} + H^+ \rightarrow -\overset{\cdot}{\underset{|}{C}}-XH \qquad (X: NR, OS) \qquad (51)$$

and the consecutive reactions of the generated C radical will then determine the composition of the cathodic product.[50]

Additionally, it is well known that the cathodic reduction of carbonyl groups (>C=O) and imide groups (>C=NR) is catalyzed by protons or proton donors. This observation is to be understood in terms of a protonation preceding the cathodic charge transfer:

$$(>C=X) + HA \rightarrow (>C=XH)^+ + A^- \qquad (52a)$$

$$(>C=X)H^+ + e \rightarrow (>\overset{\cdot}{C}-XH) \xrightarrow[+e+H^+ +>C=X]{\text{dim.}} HX-\overset{|}{\underset{|}{C}}-\overset{|}{\underset{|}{C}}-XH \qquad (52b)$$

Reduction of the protonated species yields a radical in which the more electronegative heteroatom is protonated. The carbon atom, however, carries the unpaired electron. Therefore, in the hydrocoupling reactions of carbonyl and imide compounds only the carbon atoms which carry these groups are involved, whereas the heteroatoms are accepting the transferred proton.

Radical–radical pairing with accompanying C–C coupling as well as further reduction with subsequent protonation are the most important reactions observed in C=X reduction.

### 2.5.2.3. Reactions of Cathodically Generated Carbon Radical Anions and Carbanions with Inorganic Electrophiles (Except Protons)[51–54]

Electrogenerated carbon radical anions and carbanions may react with any electrophile present in the solvent–electrolyte system provided the respective electrophile itself is inert against cathodic reduction at the applied working potential. From the preparative point of view, the most interesting inorganic electrophiles are carbon dioxide, sulfur dioxide, and sulfur trioxide, the sulfur oxides being very stable at cathodic potentials.

Thus, by cathodic reduction of aromatic hydrocarbons as well as of polyenes in aprotic solvents (DMF, THF), in presence of carbon dioxide the dianions of dicarboxylic acids are produced as shown in Eqs. (53a) and (53b) for the reduction of butadiene and anthracene, respectively:

$$CH_2=CH-CH=CH_2 + 2e + 2CO_2 \rightarrow {}^-O_2C-CH_2-C=C-CH_2-CO_2^- \tag{53a}$$

$$C_6H_6(CH)_2C_6H_6 + 2e + 2CO_2 \rightarrow C_6H_6(CHCO_2^-)_2C_6H_6 \tag{53b}$$

Baizer[55,56] reported on a systematic study of the cathodic carboxylation of vinyl compounds (activated towards cathodic reduction by electron-donating groups). Cathodic carboxylation is performed by reduction of unsaturated compounds in aprotic solvents saturated with $CO_2$ or even working under enhanced $CO_2$ pressure. Similarly, the cathodic production of disulfonate anions is performed by electrolysis in $SO_3$-saturated electrolytes.

### 2.5.2.4. N-Radical Anions from Nitro and Nitroso Compounds and Their Consecutive Reactions[57]

Aromatic and aliphatic nitro compounds can be easily reduced at the cathode (aromatic more easily than aliphatic).

The first charge transfer usually produces a radical anion of moderate stability. Since oxygen is more electronegative than nitrogen, in radical anions obtained from nitro compounds, negative charge and spin density are more concentrated at the oxygen atom than on nitrogen.

Proton addition is the most important consecutive reaction following the first charge transfer. This is followed again by a second charge transfer. This second charge transfer may either proceed at the electrode or in the bulk of the solution by homogenous charge transfers from another radical anion and produces a nitroso compound by reductive cleavage of the N—O bond. The nitroso compound is not usually isolated because it is more easily reduced than the starting nitro compound [Eq. (55)]. However, its condensation product with the hydroxylamine compound produced by further reduction of the nitroso compound [Eq. (56)] may be obtained [Eq. (57)]. The primary $1e$ reduction of nitro- and nitroso- groups is acid catalyzed. In many cases, the protonation may be the first reaction step preceding the charge transfer in acidic solutions:

$$R-NO_2 + e \rightarrow (RNO_2)^{\cdot -} \xrightarrow{+H^+} (RNO_2H)^{\cdot} \tag{54a}$$

$$(RNO_2H)^{\cdot} + e \rightarrow RNO + OH^- \tag{54b}$$

The reduction of the nitroso compound, usually leads to the formation of the respective hydroxylamine compound via a proton-catalyzed reaction, epep or pepe [Eq. (55)]. Hydroxylamines are only reduced further to the respective amine at higher cathodic potentials and lower pH values:

$$RNO + 2H^+ + 2e \rightarrow RNHOH \qquad (55)$$

$$RNHOH + H^+ \rightarrow (RNHOH_2)^+ \xrightarrow[+H^+]{+2e} RNH_2 + H_2O \qquad (56)$$

Hydroxylamines couple with nitroso compounds to azoxy compounds:

$$R-NO + H(OH)N-R \rightarrow R-NO=N-R + H_2O \qquad (57)$$

## 2.6. Electrochemical Generation of C Radicals and Their Reactions

### 2.6.1. Cathodic Generation of C Radicals by Cathodic Cleavage of Carbon Heteroatom Bonds

#### 2.6.1.1. Cathodic Cleavage of Carbon–Halogen Bonds

The cathodic electrochemistry of carbon–halogen bonds has been summarized recently by Rifi[58] and Casanova and Eberson.[19] The ease of reduction of carbon–halogen bonds depends on the nature of the halogen as well as on the nature of the carbon skeleton attached to the halide-bearing carbon atom. The experimental data demonstrate that the reduction half-wave potentials increases remarkably from iodides to chlorides.

This parallels the increase in bond strength of the respective carbon–halogen bonds. The carbon–fluorine bond is so stable that the reduction potential (of the usually inactive carbon–fluorine bonds) is expected to be at such negative cathodic potentials which no electrolyte–solvent system could withstand. The cathodic attack at C–halogen bonds results in a cleavage of this bond, producing the respective halide-anion and a carbon radical [Eq. (58)]. Inductive destabilization of the C–halogen bond in compounds such as benzylchloride leads to a decrease of $E_{1/2}$:

$$R-hal + e \rightarrow R^{\cdot} + hal^- \qquad (58)$$

The inductive stabilization of the produced radical on one hand and the inductive destabilization of the C–halogen bond by a second halogen attached either geminal or vicinal (i.e., to the halogen bearing C atom) explains the low half-wave potentials of vicinal dihalides or aromatic ortho and para dihalides, or compounds like bromoform, chloroform, carbon tetrachloride, and tetrabromide, respectively.

The most important consecutive reactions following reaction (58) is radical–radical dimerization [Eq. (59)] and radical attack on the used electrode material with formation of metal–organic compounds (Eq. (60)). Further reduction of the produced radical to carbanion with

$$2R^{\cdot} \rightarrow R-R \qquad (59)$$

$$n \cdot R^{\cdot} + Me \rightarrow Me(R)_n \qquad (60)$$

typical subsequent carbanion reactions may also occur. The most important consecutive reaction is reductive protonation, presenting the way to electrohydrogenation of organic halides:

$$R \cdot X + 2e + HA \rightarrow RH + X^- + A^- \qquad (61)$$

The cathodic reduction of vicinal dihalides under aprotic conditions leads to the formation of a $>C=C<$ between the two carbon atoms involved [Eq. (62)] in a sort of intramolecular radical–radical pairing:

$$X-\underset{|}{\overset{|}{C}}-\underset{|}{\overset{|}{C}}-X + 2e \rightarrow \phantom{X}>C=C< + 2X^- \qquad (62)$$

Rifi and others demonstrated that the intramolecular cathodic halide elimination of $\alpha$-$\omega$ dihalides opens the way to cyclic compounds with a small number of carbon atoms. So 1,3-dihalides form cyclopropanes and 1,4-dihalides, cyclobutanes:

$$CH_2X-(CH_2)_2-CH_2X + 2e \rightarrow \overline{CH_2-(CH_2)_2-CH_2} + 2X^- \qquad (63)$$

### 2.6.1.2. Cathodic Reduction of Onium Compounds[59]

Onium cations can generally be reduced at mercury electrodes where cathodic $H_2$ evolution does not seriously interfere. For ammonium cations, reduction half-wave potentials of 2–3 V (vs. sat. cal.) are observed; for phosphonium and arsonium cations, 1.7–2.3 V; and for sulfonium cations, half-wave potentials of 1–2 V are observed.

The cathodic reduction of onium cations, e.g., tetraalkyl or tetraaryl ammonium or trialkyl or triaryl sulfonium cations, in the presence of proton donors (e.g., $H_2O$ or carboxylic acids) yields the alkyl or aryl hydrocarbon and the corresponding trialkyl (triaryl) amine or dialkyl (diaryl) sulfide:

$$(X(R)_{n+1})^+ + 2e + H^+ \rightarrow XR_n + HR \qquad (64)$$

In aprotic media, radical dimerization of the hydrocarbon radicals can be observed, which is evidence that the next chemical step following charge transfer is the homolytic cleavage of the heteroatom–carbon bond:

$$(XR_{n+1})^+ + e \rightarrow (XR_{n+1})^{\cdot} \rightarrow XR_n + R\cdot \qquad (65)$$

Further consecutive reactions are the same as those discussed for radicals generated by the cathodic cleavage of carbon–halogen bonds. The cathodic reduction of triphenylsulfonium cation at mercury cathodes offers an interesting path to form diphenylmercury. With high concentrations of the triphenylsulfonium cation (which is strongly adsorbed like other alkyl- or aryl-onium cations at the electrode surface) the formation of diphenylmercury is observed at the expense of benzene, which is the expected reduction product:

$$2(C_6H_5)_3S^+ + Hg + 2e \rightarrow (C_6H_5)_2S + Hg(C_6H_5)_2 \qquad (66)$$

## 2.6.2. Anodic Generation of Radicals

### 2.6.2.1. Anodic Oxidation of Carbanions or Metal–Alkyl or Metal–Aryl Compounds

The anodic oxidation of carbanions is the simplest method of generating C radicals at the electrode–electrolyte interface, to study their reactions, and to make use of radical reactions for synthetic purposes. The only difficulty in this sort of radical generation is to be seen in the preparation of the carbanions. Being strong nucleophiles, they are very reactive species which are stable only in aprotic media. By using compounds with enhanced carbon acidity (e.g., diesters of malonic acid or alkyl-nitro and dinitro compounds which possess $pK$ values of 11 or smaller) the C anions can be prepared more easily. So, the Na salt of the dimethyl ester of malonic acid $[CH_3O(CO)-CH^{(-)}-(CO)OCH_3]$ can be obtained by neutralizing the ester with sodium-methanolate in methanolic solutions.[60,61] The generation of C anions of less acidic hydrocarbons like acetylenes (e.g., phenylacetylene) is only possible in strictly aprotic solvent, i.e., in anhydrous THF.

Anodic polarograms of carbanions from stronger C acids show $1e$ waves at potentials around +1 V vs. sat. cal. In most cases the reaction product of the anodic oxidation of carbanions is the dimer formed by radical–radical dimerization of the primarily formed radical. As a minor by-product, a hydrocarbon is formed by H abstraction from the solvent molecules. The $1e$ oxidation, dimerization, and the H abstraction are written in Eqs. (67a–67c) for the phenylacetylide anion[62]:

$$C_6H_5C\equiv C^- - e \rightarrow C_6H_5C\equiv C\cdot \qquad (67a)$$

$$2C_6H_5C\equiv C\cdot \rightarrow (C_6H_5-C\equiv C)_2 \qquad (67b)$$

$$C_6H_5C\equiv C\cdot \rightarrow C_6H_4C\equiv CH \qquad (67c)$$
(H abstraction)

The dimerization reaction is diffusion controlled and needs no thermal activation. Thus, the anodic dimerization (67b) is very often a rather selective reaction. The anodic oxidation of metal alkyls and aryls as well as that of alkyl or aryl Grignard compounds proceeds in a way very similar to the oxidation of carbo-anions since alkyl and aryl rests in Grignard compounds are understood to be crypto C anions. The Nalco process (Section 4.2.4) uses the anodic Grignard reagent oxidation.

### 2.6.2.2. Anodic Oxidations of Nitrogen Anions (Amide Anions)[63]

Similar to the anodic oxidation of C anions, the anodic oxidation of N anions proceeds in a $1e$ step to the respective N radical which either reacts by hydrogen abstraction or dimerizes to form hydrazine derivatives.

Thus, in THF, alkylamide anions form $NN^-$-dialkylamine-hydrazines which are deprotonated and oxidized to azoalkyl:

$$^-NH-alk - e \rightarrow \cdot NH-alk \qquad (68a)$$

$$2\cdot NH-alk \rightarrow (alk-NH)_2 \qquad (68b)$$

$$(alk-NH)_2 - 2H^+ - 2e \rightarrow alk-N=N-alk \qquad (68c)$$

Alkylamines can be greatly acidified by acylation.

So, for carboxylacylamines and sulfoacylamines, p$K$ values of 10–12 are observed. These compounds are acidic enough to be converted into their conjugate base even in protic solvents like methanol and water. The anodic oxidation of carboxyl amide anions, however, does not lead to N–N coupling products, since O radicals are formed by their oxidation rather than N radicals. O radicals react preferably by hydrogen abstraction. By acylating the amine with sulfuric acid, however, a sulfoacylamide is formed, the anion of which can be oxidized to a nitrogen radical. So, the anodic oxidation of $N,N'$-dialkylsulfamides in alkaline methanolic or aqueous solution finally yields azoalkyls. This azo- compound formation is explained by thermal decomposition of the unstable cyclic intermediate, $N,N'$-dialkylthiadiazridine-1,1-dioxide:

$$O_2S(NH-alk)_2 - 2e - 2H^+ \rightarrow O_2S\begin{array}{c}N-alk\\|\\N-alk\end{array} \rightarrow SO_2 + alk-N=N-alk \qquad (68d)$$

### 2.6.2.3. Radical Formation by Anodic Oxidation of Carboxylate Anions[64]

The anodic $1e$ oxidation of anions of carboxylic acids to carboxylate radicals produces very short-lived intermediates which decompose nearly immediately by decarboxylation:

$$RCOO^- - e \rightarrow RCOO\cdot \rightarrow R\cdot + CO_2 \qquad (69a)$$

The anodic potential necessary to oxidize carboxylate anions (more than +2 V vs. sat. cal.) lies well above the oxidation potential of alkyl radicals. Nevertheless, performing the oxidation at platinum, rhodium, or glassy carbon electrodes produces mainly the dimer of the alkyl radical:

$$2R\cdot \xrightarrow[GC]{Pt\ metals} R-R \qquad (69b)$$

It is the fact that metals of the platinum group adsorb radicals only very weakly[95] which forces the radicals to react according to homogenous kinetics in solution. This favors radical dimerization after decarboxylation and prohibits further oxidation of radicals to carbo-cations. At graphitic carbon anodes, however, the reaction proceeds according to the expected route; i.e., one obtains products which are typical for the chemical consecutive reactions of carbonium because at C anodes the adsorbed C radicals undergo further oxidation to carbo-cations immediately after decarboxylation.

The anodic decarboxylation of carboxylate anions with the formation of hydrocarbons by radical–radical dimerization was one of the first reaction types of organoelectrochemistry discovered by Kolbe (see Section 1.1). Today, it is a synthesis principle which is used even in a large-scale semitechnical process (see Section 4.2.3). Recently, Schäfer[65,66] showed the remarkable versatility using mixed Kolbe synthesis for the elegant synthesis of some pheromones, which can be synthesized otherwise only with some effort.

### 2.6.2.4. Performance of Organoelectrosynthetic Reactions Induced by Radical-Addition Reactions to Unsaturated Hydrocarbons[67–69]

The anodic oxidation of halide anions ($Cl^-$, $Br^-$, $I^-$), pseudohalide anions (like $N_3^-$, $SCN^-$), nitrite anions ($NO_2^-$), alkoxy- anions ($AlkO^-$), or carbanions ($R_1R_2R_3C^-$) produces the respective radical at the electrode surface in the first step. In the presence of appropriate olefins (like styrene or butadiene), radical addition to these acceptors may occur. This reaction may then start a reaction sequence which might well be of synthetic interest, with respect to the structure of the obtained end products as well as with respect to the obtained mass and current yields. It is, as in the Kolbe synthesis, the use of platinum or other platinum electrodes which favor typical radical consecutive reactions instead of further oxidation of the radicals to carbonium.

Thus, for instance, if azide anions are oxidized (the same holds for alkoxy anions like methoxylate, and for appropriate carbanions) in methanol or acetic acid at platinum anodes in the presence of styrene, one obtains a diazido monomer and a diazido dimer compound:

$$N_3^- - e \rightarrow \dot{N}_3 \xrightarrow{C_6H_5-CH=CH_2} C_6H_5-\dot{C}H-CH_2N_3 \xrightarrow[(Pt)]{+\dot{N}_3}$$

$$C_6H_5-CHN_3-CH_2N_3 \quad (70a)$$

$$2 C_6H_5-\dot{C}H-CH_2N_3 \xrightarrow[(Pt)]{} (C_6H_5-CH(CH_2N_3))_2 \quad (70b)$$

It was Schäfer[68,69] who fully explored the applicability of this reaction type and showed that these reactions proceed very often with remarkably high selectivity.

Performing these oxidations at carbon anodes, however, implies that, after the first radical-addition step, the other typical radical reactions are no longer favored. Instead, the C radical is oxidized to the respective carbonium [Eq. (70c)]. The carbonium prefers solvolysis, i.e., undergoes acetamidation in acetonitrile, acetylation in acetic acid, or methoxylation if the reaction is performed in methanol. Thus at carbon electrodes typical products are monomers with two different substituents; the first is the added radical group, the second the nucleophilic part of the solvent:

$$C_6H_5-CH-CH_2-N_3 - e + HOCH_3 \rightarrow C_6H_5-CHOCH_3-CH_2N_3 + H^+ \quad (70c)$$

A comparably useful C–C coupling reaction at carbon anodes is observed if Kolbe electrolysis is performed in the presence of appropriate vinyl and polyene compounds which are not oxidized at the applied potential but are able to react as acceptors for the radicals produced according to Eq. (70d).

In these cases the Kolbe radical adds to the unsaturated hydrocarbon, producing a radical which stabilizes by combination with another Kolbe radical [Eq. (70d)] or by radical–radical dimerization [Eq. (70e)]:

$$R\cdot + \,\,{>}C{=}C{<} \rightarrow R-\overset{|}{\underset{|}{C}}-\overset{|}{\underset{|}{C}}\cdot \xrightarrow{+R\cdot} R-\overset{|}{\underset{|}{C}}-\overset{|}{\underset{|}{C}}-R \quad (70d)$$

$$2R-\overset{|}{\underset{|}{C}}-\overset{|}{\underset{|}{C}}\cdot \rightarrow R-\overset{|}{\underset{|}{C}}-\overset{|}{\underset{|}{C}}-\overset{|}{\underset{|}{C}}-\overset{|}{\underset{|}{C}}-R \quad (70e)$$

## 2.6.2.5. Electrochemical Preparation and Conversion of Heterocycles[70–72]

The electrochemistry of heterocycles is a very interesting part of organo-electrochemistry where, among other authors, Lund, Cauquis, and Tabacovic have been especially successful. The electrochemical preparation of heterocycles is not based on special reaction principles. The electrosynthesis of heterocycles simply makes use of the described reaction schemes for achieving intramolecular ring closure. The cathodic ring closure by reductive C–C bond formation in 1,1'-ethylene-bis-(3-carbamidopyridinium) dibromide is seen below[73]:

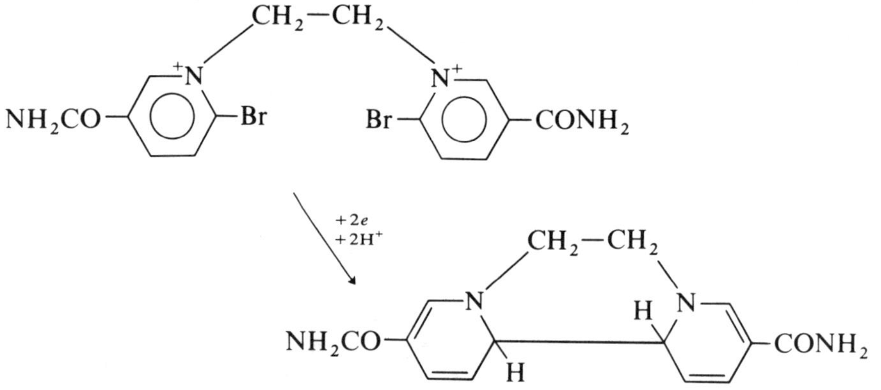

(71)

# ELECTRO-ORGANIC SYNTHESES

The anodic ring closure plays an important part in the electrochemical synthesis of alkaloids.[21] The electrochemical oxidation coupling of coclaurine to different bisbenzylisoquinoline alkaloids makes use of the well-known anodic O–C coupling reaction of phenols[21] (see Section 2.5.1.3.2).

However, in many of the usual ring closure reactions, neither radicals nor radical ions are involved. Instead, it may be electrochemical preparation of a nucleophilic group in the appropriate neighborhood of an electrophilic group which may give rise to a conventional condensation or any other reaction which brings about the desired ring closure. Thus, by partial reduction of $o,o'$-dinitrobiphenyl, $o$-hydroxylamino-$o'$-nitro-biphenyl is produced, which undergoes rapid intramolecular condensation with N–N coupling to the azoxy-compound which is readily reduced to 5,6-phenanthroline:

$$\text{(NO}_2\text{)(NO}_2\text{)-biphenyl} + 6e + 6H^+ - 2H_2O \rightarrow \text{(HNOH)(NO)-biphenyl} \quad (72)$$

$$\text{(HNOH)(NO)-biphenyl} - 2H_2O \xrightarrow[+2H^+]{+2e} \text{N–N azoxy} \quad (73)$$

Table 4 shows six examples for such ring closure synthesis which are based essentially on classical condensation reactions between the electrogenerated nucleophilic hydroxylamine groups and electrophilic groups like $-COOH$, $-CO-R$, $-\overset{|}{C}(R)NO_2$, and $-\overset{|}{C}(OR)_2$.

## 2.7. Summary of Reaction Types for the Direct Electrochemical Conversion of Organic Substances

### 2.7.1. Anodic and Cathodic Coupling Reaction

#### 2.7.1.1. Carbon–Carbon Bond Formation

Starting from unsaturated hydrocarbons (especially starting from vinyl compounds), C–C coupling reactions can be performed by anodic oxidation as well as by cathodic reduction according to Eqs. (74a) and (74b) (see Section 2.5):

$$2 \overset{\diagdown}{\underset{\diagup}{C}}=\overset{\diagup}{\underset{\diagdown}{C}} - 2e + 2Nu^- \rightarrow Nu-\overset{|}{\underset{|}{C}}-\overset{|}{\underset{|}{C}}-\overset{|}{\underset{|}{C}}-\overset{|}{\underset{|}{C}}-Nu \quad (74a)$$

$$2 \overset{\diagdown}{\underset{\diagup}{C}}=\overset{\diagup}{\underset{\diagdown}{C}} + 2e + 2H^+ \rightarrow H-\overset{|}{\underset{|}{C}}-\overset{|}{\underset{|}{C}}-\overset{|}{\underset{|}{C}}-\overset{|}{\underset{|}{C}}-H \quad (74b)$$

## Table 4
### Some Typical Cathodic[70,71] and Anodic[68,72] Ring Closure Reactions Leading to Formation of Heterocycles

Cathodic reduction of olefins and arenes in the presence of carbon dioxide leads to $CO_2$ addition and formation of dicarboxylic acids. Coupling of olefins by a radical–radical dimerization reaction may be initiated in an alternate fashion by anodic generation of organic or inorganic radicals, $X\cdot$, followed by their addition to the unsaturated hydrocarbon (especially at Pt anodes):

$$\mathrm{\underset{}{\Large >}C=C\underset{}{\Large <} + X\cdot \rightarrow X-\overset{|}{\underset{|}{C}}-\overset{|}{\underset{|}{C}}\cdot \xrightarrow{\text{dim.}} X-\overset{|}{\underset{|}{C}}-\overset{|}{\underset{|}{C}}-\overset{|}{\underset{|}{C}}-\overset{|}{\underset{|}{C}}-X} \qquad (74c)$$

Aromatic hydrocarbons may undergo similar anodic or cathodic C—C coupling reactions as vinyl compounds and polyenes.

The anodic oxidation of aromatic hydroxy- compounds and amines will yield C–C coupling as well as O–C and N–C coupling, respectively, and N–N coupling. The anodic decarboxylation of carboxylate anions at Pt anodes is

# ELECTRO-ORGANIC SYNTHESES

another route of C–C coupling (Kolbe synthesis). Anodically generated C radicals which are obtained from carbanions react in the same way as Kolbe radicals.

### 2.7.1.2. N–N Bond Formation

Electrochemically induced homogeneous condensation of nitro- and hydroxylamino groups to form the azoxy- link is an important and easy-to-handle method of forming N–N bonds. This method is supplemented by the possibility of oxidizing N anions, either in aprotic solvents or N anions of acylated amines in protic solvents, to the respective amino radicals which will combine to form hydrazine or azocompounds.

### 2.7.1.3. C–Heteroatom Bond Formation

Anodic addition reactions to unsaturated hydrocarbons, aromatic and allylic substitution, and anodic side-chain substitution of alkyl arenes open the way to form C–halogen, C–oxygen, and C–nitrogen bonds anodically. The first is induced by anodic oxidation of halide anions (especially $Cl^-$). The second is obtained by anodic electrolysis in alcohols as solvents, leading to alkoxylation.

The C–N bond formation may be initiated by oxidation of the pseudohalide $N_3^-$ in the presence of the less easily oxidizable unsaturated hydrocarbons, or by anodic oxidation of unsaturated hydrocarbons in acetonitrile, leading to acetamidation. Heteroatom–carbon bond formation is possible, too, by the anodic oxidation of aryl iodides and diaryl and dialkyl sulfides. The anodic oxidation of these compounds in the absence of water will yield diaryl iodonium and triaryl sulfonium or trialkyl sulfonium compounds, respectively.

By cathodic reduction of olefins and arenes in presence of $SO_3$ (and sometimes $SO_2$ in cases where the olefin is easier to reduce than is sulfur dioxide), disulfonic and disulfinic acids are obtained.

### 2.7.1.4. Formation of S–S Bonds and S–O Bonds

S–S bonds are formed anodically very easily and in some cases in a reversible reaction by the anodic oxidation of mercaptans, especially with exclusion of water. Anodic oxidation of mercaptans, of disulfides, and of dialkyl and diaryl sulfides in aqueous solutions or electrolyte–solvent systems, which contain water, yields sulfoxides ($>S=O$) and sulfones ($>SO_2$).

## 2.8. Influence of Electrosorption for Product Composition and Yields of Electro-organic Synthetic Processes[96]

### 2.8.1. General Remarks

Due to the remarkably high reactivity of the reaction intermediates like radicals and radical ions, organosynthetic reactions often do not proceed very selectively. The knowledge of mechanisms and rates for the different reaction paths will enable the preparative electrochemist to select electrolysis conditions in a way to perform the most selective synthesis with high yields for only one product.

For reactive intermediates of moderate stability the relevant consecutive reactions often proceed away from the electrode, at least in the range of the diffusion layer or even in the inner bulk of the solution. For such cases optimization means matching concentration conditions in the electrolyte in order to fasten only one of the competing reactions with respect to the others. Very reactive intermediates, however, cannot escape from the electrode surface into the solution and will react directly at the electrode surface. In this case one has to change the reaction conditions at the electrode surface to match them with the demands of the desired reaction.

Surface concentration ratios of competing reactions—since they are governed by electrosorption—in general are different and, in special cases, are extremely different from the concentration conditions in the bulk of the electrolyte. The latter is especially true if a certain species is preferentially adsorbed.

This will lead to a marked increase in concentration ratios in favor of this species. If the surface concentration of a species is depleted by electrochemical consumption, however, this may result even in zero surface concentration, if the electrochemical conversion is performed under diffusion-limited current conditions. So, mass transport hindrance may result in a remarkable decrease of the surface concentration of the depolarizer or a consumed reactant. At present, little is known about kinetic data of reactions which proceed at the electrode surface.

However, for the cathodic production of optically active compounds from chiral compounds carrying the electroactive carbonyl function, more details are known. Another example concerning the kinetics of the anodic oxidation of vinyl compounds treats the problem of competing surface reactions semiquantitatively. Both examples are reported here in order to give an idea of the most important kinetic principles of electrochemical surface reactions.

### 2.8.2. Cathodic Production of Optically Active Compounds by Reduction of Carbonyl Functions in the Presence of Electrosorbed Optically Active Cations[74,75]

Grimshaw detected[76] that the addition of optically active cations induces an optical activity (although small) in the alcohol obtained by cathodic reduction of carbonyl compounds. Horner and co-workers and a number of other scientists investigated the optical induction of optically active cations. It was Kariv[77] and co-workers who finally succeeded in reducing 2-acetyl-pyridine at mercury cathodes in the presence of strychnine with an optical yield of about 50%.

From tensammetric measurements,[78] it is well established that the optically active cations are strongly adsorbed at the mercury cathode. There they form a nearly closed layer which is likely to be highly ordered. This ordered, closed layer forms a sort of chiral matrix which forces the radical anions and radicals, respectively, to adjust themselves sterically and to react in a well-defined way which favors one of the stereoisomers in particular. Cathodic dimerization, which is known to compete with reduction of carbonyl compounds to the carbinol, is known to proceed, not at the cathode surface, but in front of it in the electrolyte. Here the ordered matrix is not available; consequently, the dimer shows no optical activity at all.

Although the principles of cathodic induction are understood, the knowledge is still so restricted that generalizations are not yet possible. It should be stressed that the cathodic inductive power of different optically active cations is very different for different cations and for different substrates.

### 2.8.3. Electrosorption and Mass Transfer as Dominating Factors for the Product Composition of the Anodic Oxidation of Styrene[34]

As Schäfer had shown[30] the anodic oxidations of vinyl compounds (among them styrene) in methanol yields dimeric and monomeric products:

$$C_6H_5CH=CH_2 - 2e + 2HOCH_3 \rightarrow C_6H_5-CH(OCH_3)-CH_2(OCH_3) + 2H^+ \tag{75a}$$

$$2C_6H_5CH=CH_2 - 2e + 2HOCH_3 \rightarrow (C_6H_5-CH(OCH_3)-CH_2)_2 + 2H^+ \tag{75b}$$

In order to optimize the anodic synthesis with respect to either the monomer or the dimer (which are both synthetically useful products), it is necessary to clarify the reaction mechanism, especially for the dimer formation. The most important question is whether the key step for the anodic dimerization consists in the electrophilic addition of the primarily formed radical cation to an intact olefin molecule,

$$\ce{>C=C<} - e \rightarrow (\ce{>C=C<})^{\cdot+} \xrightarrow{+\text{olefin}} -\overset{+}{C}-\overset{|}{C}-\overset{|}{C}-\overset{\cdot}{C}- \quad \text{(dimer)} \tag{75c}$$

or whether two radical cations dimerize,

$$2(\ce{>C=C<})^{\cdot+} \rightarrow -\overset{+}{C}-\overset{|}{C}-\overset{|}{C}-\overset{+}{C}- \xrightarrow[-2H^+]{+2HOR} RO-\overset{|}{C}-\overset{|}{C}-\overset{|}{C}-\overset{|}{C}-OR \quad \text{dimer} \tag{75d}$$

The influence of the current density on the rates of the two different coupling reactions can be predicted to be opposite. In case the dimerization is a second-order radical dimerization, the reaction rate will be increased with respect to the first-order solvolysis rate [Eq. (75e)], which eventually leads to monomer formation by high stationary concentrations of the radical cations:

$$(\diagdown C=C\diagup)^{\cdot+} + HOR \xrightarrow{-H^+} -\overset{|}{\underset{|}{\dot{C}}}-\overset{|}{\underset{|}{C}}-OR \xrightarrow[+HOR]{-H^+-e} RO-\overset{|}{\underset{|}{C}}-\overset{|}{\underset{|}{C}}-OR \quad (75e)$$
$$\text{(monomer)}$$

A quantitative treatment predicts a linear dependence of the dimer/monomer ratio on the stationary radical cation surface concentration.

Since the stationary radical cation surface concentration increases with the rate of its formation (i.e., the current density), it is obvious that, for the radical dimerization mechanism (75d), the dimer yield should increase significantly with increase in current density. For the anodic styrene and butadiene oxidations, the opposite result is observed: By approaching current density 0, the maximal dimerization yield is obtained; whereas, under diffusion limited current conditions, i.e., applying the highest possible current density, the dimer yield decays to zero. This clearly proves that the radical cation addition mechanism [Eq. (75e)] is operative for the anodic styrene dimerization. With this knowledge the styrene oxidation can be optimized with respect to the formation of the dimer (usually expected to be the most interesting product) and with respect to the formation of monomer (which, being a $\alpha$-$\beta$-disubstituted compound may also be of interest).

In conclusion, the reactions of the radical cations of styrene and other less easily oxidizable hydrocarbons are so fast that they proceed at the place of their formation, e.g., at the anode surface. That they do react at the anode surface and not in the outer Helmholtz plane can be shown by comparing the anodic half-wave potentials for styrene at different anode materials. The half-wave potentials of the styrene oxidation is strongly dependent on the anode material used. This reflects the different olefin surface concentrations of different anode materials but identical bulk-phase concentrations of the olefin.

Therefore, the conditions for the coupling reaction to overcome the solvolysis depend on the anode material. Under otherwise identical electrolysis conditions, the styrene oxidation obtains nearly 100% current yield for the dimer at carbon anodes and only 20%, 10%, and 5% at magnetite, lead dioxide, and platinum electrodes, respectively. This shows that the surface olefin concentrations are approximately 5, 10 and 20 times less on $Fe_3O_4$, $PbO_2$, and Pt, respectively, than on carbon anodes. Evidently the carbon anode offers optimal conditions for the anodic styrene dimerization because of the pronounced tendency of carbon to adsorb olefins.

## 2.9. Solvent Electrolyte Systems and Electrolyte Materials Used in Electro-organic Synthesis[79]

Processing organic substances in electrolytic reactions based on the use of aqueous solvents often demands the technique of emulsion electrolysis.[99] Many of the electrolysis reactions in which organic substrates are converted at the anode or cathode have to be performed in nonaqueous solvents or solvents which have only a very limited water content. For cathodic processes, basic solvents like N-methyl-formamide or dimethylformamide are advised. Ethers like dioxane, tetrahydrofurane, and diglyme can similarly be used, but these solvents possess lower dielectric constants and lower solubility for electrolytes. Electrolyte solutions in ethers show remarkably low degrees of dissociations and accordingly low electric conductivity.

Ethers offer the advantage of high volatility, i.e., can be removed easily in the working-up procedures by distillation. Dimethylsulfoxide (DMSO) possesses good solubilities for inorganic electrolytes and good electric conductivities for their solutions due to high solvation energies for cations and high dielectric constant. However, DMSO is disadvantageous with respect to the working-up of the electrolysis products because of its high boiling point.

Alcohols, especially methanol, are solvents which offer a good compromise concerning moderate proticity and basicity, respectively, electrolyte solubility, and volatility. They can be used in many cases for both cathodic reactions and anodic reactions.

Acetonitrile, a solvent of remarkably low basicity and acidity as well as chemical and electrochemical inertness, is especially suitable for anodic synthesis. It possesses moderate solubility for organic and inorganic electrolytes and is volatile enough to be removed easily in the usual working-up procedure. The choice of the electrolytes used for electro-organic synthesis depends mainly on the stability of the cationic and anionic components at the working potential of the process involved.

For cathodic processes very stable organic cations which are not proton donors are often chosen because alkali cations may be already deposited cathodically at the applied potentials, and proton donor cations (e.g., $[NH-R_3]^+$) are reduced by $H_2$ evolution. Therefore, tetraalkylammonium cations are frequently used.

Anions such as $ClO_4^-$ (which may be hazardous in the final working-up), $BF_4^-$, $PF_6^-$, and some selected organic anions (especially sulfonates like benzene sulfonate and tolyl sulfonate) are often used. For anodic processes, alkali-metal salts, especially $Li^+$ and $Na^+$ salts of the aforementioned anions, are recommended since they are cheap, possess good solubility (especially Li salts), are stable under anodic conditions (which does not hold for tetraalkylammonium cations), and can be separated in the working-up procedure easily by extraction with water. Electrode materials for cathodic processes are Hg, Cd, Sn, Zn, Pb, and highly purified carbon or graphite all of which possess high $H_2$ overpotentials.

For anodic processes, platinum, other platinum metals, gold, carbon, and glassy carbon are used. Less inert oxidic materials like $PbO_2$ cannot be used in nonaqueous media. They are thermodynamically unstable in the absence of water and are reduced or dissolved more or less readily in contact with organic substrates, even at high anodic potentials by chemical reduction.

## 3. Mediated Electrochemical Conversion of Organic Substrates

### 3.1. General Description of Mediated Electro-organic Synthesis

Oxidation and reduction processes in preparative organic chemistry are usually executed by using homogeneous redox reactions with conventional oxidants or reductants rather than by using the electrode as electron donor or acceptor, respectively. Such conventional redox reactions in many cases, however, can be supplemented by an electrolysis cell which allows the regeneration of the used redox system. This sort of coupled process is generally called "mediated" or "indirect" electro-organic synthesis. Very little is published about the details of such processes. Some are, however, working, and especially in the pharmaceutical industries one may expect that mediated anodic or cathodic processes will spread owing to their advantages with concern to effluent pollution.

### 3.2. Anodic Mediator System

Usual oxidants are bichromate, permanganate, periodate, or $Tl^{3+}$ as typical oxidants, transferring more than one electron and $Fe(CN)_6^{3-}$, $Fe^{3+}$, $Co^{3+}$, etc., as $1e$ oxidants.[80] Most of them are suitable for anodic regeneration.

There are at least two larger technical-scale processes which use bichromate as oxidant with anodic regeneration. Farbwerke Heochst has long used recycled bichromate[81] to bleach paraffin waxes. The conversion of L-sorbose to L-ascorbic-acid is performed at different pharmaceutical firms by means of continuously regenerated bichromate. The anodic oxidation of toluene to benzaldehyde mediated by $Co^{3+}/Co^{2+}$ and $Mn^{3+}/Mn^{2+}$ in strongly acidic aqueous $HClO_4$ has been reported recently.[82,97] The anodic oxidation of benzene at $PbO_2$ anodes to $p$-benzoquinone, which had previously been developed at the pilot-plant stage (see Section 4.2.2), is a mediated anodic oxidation too. The chemical oxidation of benzene by the lead-dioxide surface, and not the anodic electron transfer from benzene to the anode, very likely governs the rate of the reaction. The anodic process consists only in regeneration of $PbO_2$ from $PbSO_4$.

## 3.3. Cathodic Mediator System and Typical Mediated Homogeneous and Heterogeneous Cathodic Conversions[83-85]

The reduction of organic compounds, especially of nitro compounds, with electrochemically regenerated alkali-metal amalgams is one of the most popular examples of mediated cathodic synthesis.

Apart from alkali amalgams as heterogeneous reduction mediators, many homogeneous reducing agents are available which open a multitude of interesting synthetic possibilities. However, they have not yet been investigated for their suitability for induced cathodic organoelectrosynthesis. In this context the redox systems $Eu^{2+}/Eu^{3+}(E° = -0.43\ V)$, $Cr^{2+}/Cr^{3+}(E° = -0.41\ V)$, $Ti^{2+}/Ti^{3+}(E° = -0.37\ V)$, $V^{2+}/V^{3+}(E° = -0.255\ V)$, and $Sn^{2+}/Sn^{4+}(E° = +0.15\ V)$ (all vs. NHE) should be mentioned. Of these, only the use of $Sn^{2+}/Sn^{4+}$ as a mediator in the reduction of nitro groups to amines has been reported. $Eu^{2+}$, $Cr^{2+}$, $Ti^{2+}$, and $V^{2+}$ can be used for reductive cleavage of carbon–halogen bonds, selective hydrogenation of $>C=C<$ and $>C\equiv C<$, the reduction of epoxides to olefins, and the reduction of other products like ketones or $\alpha$-hydroxyketones.

The selective reduction of nitro compounds to hydroxylamino compounds is also possible. It is feasible to reduce carbonyl functions with $Ti^{2+}$, $V^{2+}$, and $Cr^{2+}$ and to produce hydrodimers (pinacols) and monomeric reduction products (alcohols). Last, but not least, the so-called electrocatalytic hydrogeneration of organic compounds is to be mentioned. This sort of cathodic synthesis is based on the catalytic activation of electrogenerated hydrogen. Here again, not the cathodic generation of hydrogen, but its catalyzed chemical reaction with the substrate governs the rate of the synthesis. Solvated electrons can be produced by cathodic electron ejection into appropriate solvents. These solvated electrons react in a sort of mediated reduction with different electron acceptors, e.g., aromates or benzene (electrochemical Birch reaction).

# 4. Semitechnical and Technical Electrochemical Organic Synthesis

## 4.1. General Remarks

Many older reports in the scientific and technical literature about organoelectrosynthesis processes are concerned only to a very limited extent with the working-up procedure and all other supplementary equipment and apparatus which, together with the electrolytic cell, make up a production process. They deal mainly with the layout of the cell construction and the operation variables of the electrolysis. However, recent publications show the growing conscience for the fact that the electrolysis cell is only one—although an important—part of a complete process.

## 4.2. Anodic Processes

### 4.2.1. Propylene Oxide Production with Membrane-Equipped Chlorine Electrolysis[86]

Simmrock reported on a new version of the synthesis of propylene epichlorohydrine and the consecutive pH-controlled conversion of the chlorohydrine to propylene epoxide. The special novelty of this process is a membrane cell with cation-exchange membranes based on sulfonated polytetrafluoroethylene (Nafion®). These membranes allow the anolyte (to which the propylene is added) to remain weakly acidic. In the anodic compartment there proceeds only the epichlorhydrine formation,

$$CH_3-CH_1=CH_2 + Cl_2 + H_2O \rightarrow CH_3-CH(OH)-CH_2Cl + HCl \quad (76)$$

and uncontrolled base-catalyzed hydrolysis, yielding the undesired propane-1,2-diol, can be avoided.

The desired dehydrochlorination

$$CH_3-CHOH-CH_2Cl \xrightarrow{OH^-} CH_3-CH\underset{O}{-}CH_2 + HCl \quad (77)$$

can be performed selectively only if a strict pH control prohibits that the pH increased beyond a critical value around 12. In the conventional Kellog process this was achieved by the addition of lime suspensions. In Simmrock's process the caustic solution produced at the cathode is added in portions to the different stages of a special dehydrochlorination column. The effluent of this reactor is simply sodium chloride solution, which is recycled into the divided electrolysis cell. The proposed process avoids the production of $CaCl_2$, an undesired co-product of the Kellog process, and this offers a remarkable economical advantage for the new process version. Figure 7 is a schematic flow sheet of the process proposed by Simmrock.

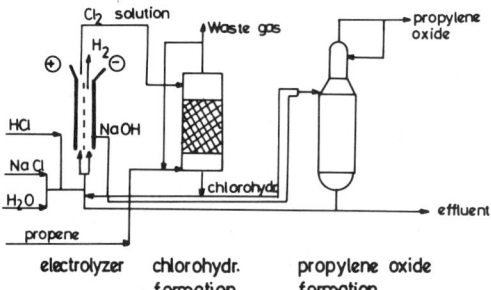

Figure 7. Flow sheet for propylene oxide production.[86]

## 4.2.2. Benzoquinone Synthesis from Benzene

In the patent literature the proposal has existed since 1800 to oxidize benzene in aqueous sulfuric acid solutions to benzoquinone at $PbO_2$ anodes:

$$C_6H_6 - 6e + 2H_2O - 6H^+ \to C_6H_4O_2 \qquad (78)$$

Only recently, two papers[87,88] were published which report on the details of a technical verification of this process and its economic implications.

In both cases benzene emulsions are processed in aqueous sulfuric acid solutions. The special mass transport problems for this sort of electrolyte are tackled by the use of special electrode structures (electrodes equipped with equally spaced ledges). A careful matching of applied current densities to mass transfer velocities of benzene to the anode and to the heterogeneous oxidation rate of benzene by $PbO_2$ is essential for high current and mass yields as well as the economic feasibility of the process.

## 4.2.3. Kolbe Synthesis of Sebacinic Acid

By 1891 Brown and Walker had reported on the Kolbe coupling of semiesters of dicarboxylic acids to diesters of dicarboxylic acids[89]:

$$RO(CO)-(CH_2)_n-COO^- - 2e - 2CO_2 \to (RO-(CO)-(CH_2)_n)_2 \qquad (79)$$

Later, Fioshin[90] described a continuous process which used an undivided cell with steel cathodes and massive Pt anodes. The semiester is dissolved in methanol and is only partly neutralized. Beck investigated the question of whether a capillary gap cell could be used for this reaction. This is advantageous since low electric conductivities of the electrolyte demands a remarkable decrease of the electrode distances in order to minimize Ohmic voltage drops across the cell.[91]

## 4.2.4. Lead Tetraethyl by Anodic Oxidation of Alkyl Magnesium Chloride at Lead Anodes (Nalco Process)

Since 1964, the Nalco Company has been operating a plant which produces lead tetraethyl and tetramethyl, respectively, in an anodic process.[92] The first chemical step of the process consists of the large-scale preparation of Grignard reagent from magnesium turnings and ethyl or methyl chloride, respectively, in etheral solution. The produced Grignard reagent supplies sufficient electric conductivity for the etheral solutions (THF or tetraethyleneglycol-dimethylether) so that no further addition of supporting electrolyte is necessary.

The electrolytic cell is made of a tube-and-shell heat exchanger. The heat exchanger tubes are used as cathodes which are separated from the anode by an insulating perforated polymer sleeve covering the inner side of the tubes. The

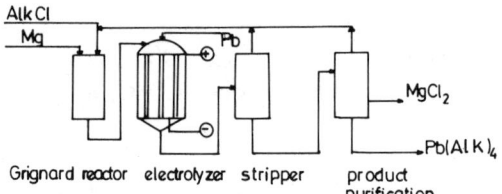

Figure 8. Flow sheet of the Nalco process (lead tetraalkyl production).[92]

tubes contain lead-shot which forms the sacrificial anode and which is continuously refilled through a feeder on top of the tubes. The etheral solution is pumped through the tube which is cooled from the outside to a working temperature of 40–50°C.

The anodic reaction directly yields lead tetraalkyl:

$$4(\text{alkMgCl}) + \text{Pb} - 4e \rightarrow 4\text{MgCl}^+ + \text{Pb(alk)}_4 \qquad (80)$$

At the cathode, magnesium is deposited and reacts with excess alkylchloride to regain Grignard reagent. Figure 8 presents the flow sheet of the Nalco process.

### 4.3. Cathodic Processes

#### 4.3.1. Baizer Synthesis of Adipodinitrile from Acrylonitrile[93]

Figure 9 gives the flow sheet of the Baizer–Monsanto process. The mechanistic details of the Baizer hydrodimerization of acrylonitrile have been described in Section 2.5.2. The technical cell construction uses the filter press arrangement with hollow bipolar lead electrodes (with internal cooling). The divided cells are equipped with a cation-exchange membrane, separating the aqueous anolyte (diluted aq. $H_2SO_4$) from the catholyte (i.e., aqueous solution of McKee salts saturated with acrylonitrile). The product is separated from the catholyte by extraction with acrylonitrile and the resaturated catholyte is recycled into the cathode compartment.

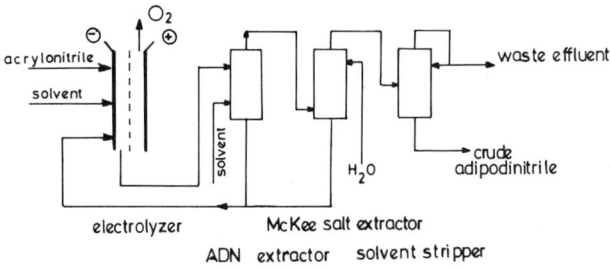

Figure 9. Flow sheet of the Monsanto process.[93]

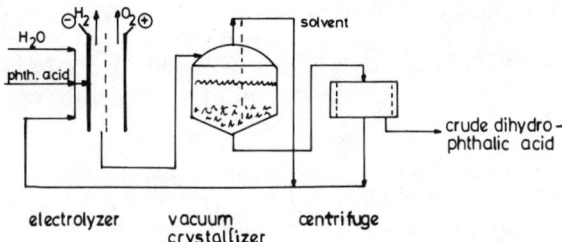

Figure 10. Flow sheet for the cathodic reduction of phthalic acid to dihydrophthalic acid.[94]

### 4.3.2. Dihydrophthalic Acid

Nohe[94] reported of another cathodic process which has been developed up to a technical scale. 1,2-dihydrophthalic acid is obtained by cathodic reduction of phthalic acid. The reduction is performed in a mixed electrolyte containing approximately 25 wt % $H_2O$, 5 wt % $H_2SO_4$, 15 wt % phthalic acid, and 55 wt % of an organic solvent, e.g., THF, dioxane, or methylglycol at lead cathodes. Aqueous $H_2SO_4$ solution is used as the anolyte. The electrolysis is performed in divided cells equipped with Nafion membranes. The cells use hollow bipolar steel electrodes through which a water stream is passed to cool the electrolysis cell. The two sides of these hollow electrodes are plated with lead so that a bipolar lead electrode is formed (cf. Section 4.3.1).

Figure 10 shows the flow sheet of the process operated by BASF, since 1974, producing about 100 tons of dihydrophthalic acid per year.

### 4.3.3. Concluding Remarks

There are few examples of successful working electro-organic processes on an industrial scale, and the mass and value of products being produced by these processes is rather limited. Nonetheless, the practical experience and the theoretical knowledge which has been gained in this field during the last 15 years promises that further progress in industrial electro-organic synthesis is very likely to occur in the future.

## References

1. M. Faraday, *Poggendorfs Ann. Phys. Chem.* **33**, 438 (1834).
2. H. Kolbe, Untersuchungen uber die Electrolyse organischer Verbindungen, *Liebigs Ann. Chem.* **69**, 257–294 (1849).
3. Al. Kamneva, M. Y. Fioshin, L. J. Kazukova, Sh. M. Itenberg, and A. Yu. Ershov, Synthesis of dimethyl-sebacate by anodic condensation, *Khim. Prom.*, 263 (1963).
4. S. Swann, Jr., Electro-Organic Chemical Preparations, *Trans. Electrochem. Soc.* **69**, 287–342 (1936).

5. S. Swann, Jr., Electro-organic chemical preparations II, *Trans. Electrochem. Soc.* **77**, 459–499 (1940).
6. S. Swann, Jr., Electro-organic chemical preparations III, *Trans. Electrochem. Soc.* **88**, 103–120 (1945).
7. S. Swann, Jr., Electro-organic chemical preparations IV, *Trans. Electrochem. Soc.* **95**, 219–226 (1952).
8. F. Fichter, *Organische Elektrochemie*, Steinkopff, Leipzig/Dresden (1942).
9. K. J. Vetter, *Elektrochemische Kinetik*, Springer, Berlin (1961).
10. R. Parsons, in *Advances in Electrochemistry and Electrochemical Engineering*, Vol. 1, P. Delahay and C. W. Tobias, eds., Interscience, New York (1961), pp. 1–64.
11. E. Gileadi, *Electrosorption*, Plenum Press, New York (1967).
12. B. E. Conway, *Electrode Processes*, Ronald, New York (1965).
13. P. Delahay, *Double Layer and Electrode Kinetics*, Interscience, New York (1966).
14. M. M. Baizer, Electrolytic reductive coupling I, acrylonitrile, *J. Electrochem. Soc.* **111**, 215–222 (1964).
15. E. Guccione, Electrolysis: New route to alkyl lead compounds, *Chem. Eng.* **72**(13), 102 (1965).
16. L. Eberson and H. Schafer, Organic chemistry, in *Topics in Current Chemistry*, Vol. 21, Springer, Berlin (1971).
17. M. M. Baizer, ed., *Organic Electrochemistry*, Marcel Dekker, New York (1973).
18. F. Beck, *Elektro-Organische Chemie*, Verlag Chemie, Weinheim (1973).
19. J. Casanova and L. Eberson, in *The Chemistry of the Carbon-Halogen Bond*, S. Patai, ed., Part 2, Interscience, New York (1973), pp. 979–1047.
20. H. L. Lehmkuhl, Preparative scope of organometallic chemistry, *Synthesis*, 377–396 (1973).
21. J. M. Bobbit, Electro-oxidation and isoquinoline-alkaloid biosynthesis, *Heterocycles* **1**, 181–221 (1973).
22. L. Eberson and K. Nyberg, Synthetic Use of Anodic Substitution Reactions, *Tetrahedron* **32**, 2185–2206 (1976).
23. J. R. Backhurst, J. M. Coulson, F. Goodrich, and R. E. Plimley, Preliminary investigations of fluidized-bed electrodes, *J. Electrochem. Soc.* **116**, 1600 (1969).
24. M. Fleischmann, J. W. Oldfield, and C. L. K. Tennakon, Electrochemical Bipolar Particulate Cells, in Abstracts of the Symposium on Electrochemical Engineering, Newcastle-upon-Tyne, Vol. 3 (1971), pp. 53–69.
25. F. Beck and H. Guthke, Entwicklung neuerer Zellen für elektroorganische Synthesen, *Chem. Ing. Tech.* **41**, 943–950 (1969).
26. V. D. Parker, On the Problem of Assigning Values of Energy Changes of Electrode Reactions, *J. Am. Chem. Soc.* **96**, 5656–5659 (1974).
27. V. D. Parker, Energetics of electrode reactions II: The relationship between redox potentials, ionization potentials, electron affinities and solvation energies of aromatic hydrocarbons, *J. Am. Chem. Soc.* **98**, 98–103 (1976).
28. E. Heilbronner and Bock, *Das HMO-Modell und Sein Anwendung*, Band I, II, III, Verlag Chemie, Weinheim.
29. J. O'M. Bockris and A. K. N. Reddy, *Modern Electrochemistry*, Plenum/Rosetta Edition, New York (1973), p. 955.
30. H. Schafer, Oxidative Addition von Anionen an Olefine und Oxidative Dimerisierung von Olefinen, *Chem. Ing. Tech.* **42**, 164–170.
31. I. Lillie, G. Beck, and A. Henglein, Pulsradiolyse und Polarographie: Halbstufenpotentiale fur die Oxidation und Reduktion von kurzlebigen organischen Radikalen an der Hg Elektrode, *Ber. Bunsenges Phys. Chem.* **75**, 458–465 (1971).
32. J. K. Kochi, Oxidation–reduction reactions of free radicals, in *Free Radicals*, J. K. Kochi, ed., Wiley–Interscience, New York (1973).
33. M. Katz, Oe. Saygin, and H. Wendt, Process variables in electro-organic synthesis II, the direct anodic oxidation of butadiene leading to the production of bimethoxylated $C_4$, $C_8$ and $C_{12}$ olefins, *Electrochem. Acta.* **19**, 193–200 (1974).

34. V. Plzak, H. Schneider, and H. Wendt, Process variables in electro-organic synthesis III, electrosorption and mass-transfer as dominating factors in the anodic styrene oxidation, *Ber. Bunsenges Phys. Chem.* **78**, 1373–1379 (1974).
35. J. M. Fritsche and H. Weingarten, Electrolytic oxidation of organics I, oxidative coupling of vinylidenebisdimethylamine to 1.1.4.4 dimethylaminobutadiene, *J. Am. Chem. Soc.* **90**, 793–795 (1968).
36. E. Steckhan, Spectrochemical study of olefins I, *Electrochem. Acta.* **22**, 395–399 (1977).
37. G. Faifa, M. Fleischmann, and D. Pletcher, Anodic oxidation of cyclohexene–chloridione mixtures in acetonitrile, *J. Electroanal. Chem.* **25**, 455–459 (1970).
38. H. Schafer and E. Steckhan, Anodische Dimerisierung und Funktionalisierung von Olefinen, *Chem. Ing. Tech.* **44**, 186–187 (1972).
39. H. Schafer, Oxidative Dimerisierung von Olefinen, *Angew. Chem.* **81**, 532 (1969).
40. V. D. Parker, Anodic oxidation of amines, in *Organic Electrochemistry*, M. M. Baizer, ed., Marcel Dekker, New York (1973), pp. 509–530.
41. V. D. Parker, Anodic oxidation of oxygen-containing compounds, in *Organic Electrochemistry*, M. M. Baizer, ed., Marcel Dekker, New York (1973), pp. 531–550.
42. V. D. Parker, Anodic oxidation of sulfur-containing compounds, in *Organic Electrochemistry* M. M. Baizer, ed., Marcel Dekker, New York (1973), pp. 551–562.
43. H. Hoffelner, S. Yorgiyadi, and H. Wendt, Anodic phenyl–onium-cation formation I, tris-anisylfulfonium cation, *J. Electroanal. Chem.* **66**, 138–142 (1975).
44. L. L. Miller and A. K. Hoffman, The electrochemical formation of carbonium and jodonium ions from alkyl and aryliodides, *J. Am. Chem. Soc.* **89**, 593–597 (1967).
45. H. Hoffelner, H. W. Lorch, and H. Wendt, Anodic phenyl-onium cation formation II, reaction mechanism and optimization for the anodic formation of diphenyliodonium cations, *J. Electroanal. Chem.* **66**, 183–194 (1975).
46. F. Beck and H. Leitner, Elektrochemische Initiierte Polymerisation von Arylnitril, *Angew. Makromol. Chem.* **2**, 51–63 (1968).
47. F. Beck, Die Rolle des Elektrolyten bei der kathodischen Dimerisierung des Acrylnitrils, *Ber. Bunsenges. Phys. Chem.* **72**, 379–388 (1968).
48. J. E. Gillet, Ueber den Mechanismus det elektrochemischen Reduktion aktivierter Olefine, *Chem. Ing. Tech.* **40**, 573–575 (1968).
49. J. P. Petrowitch and M. M. Baizer, Electrolytic reductive coupling XIX, Effect of counterions in cathodic reductions of 1,2 diactivated olefins, *J. Electrochem. Soc.* **118**, 447–450 (1971).
50. L. G. Feoktistov and H. Lund, Saturated carbonyl compounds and derivatives, in *Organic Electrochemistry*, M. M. Baizer, ed., Marcel Dekker, New York (1973) pp. 347–398.
51. J. W. Loveland, Electrolytic Production of Acylic Carboxylic Acids from Hydrocarbons, U.S. Patent 3,032,489, *Ca* **57**, 4470 (1962).
52. S. Wawzonek and D. Wearing, Polarographic studies in acetonitrile and dimethylformamide IV (stability of anion-free radicals), *J. Am. Chem. Soc.* **81**, 2067–2069 (1959).
53. S. Bank and D. A. Noyd, Reactions of aromatic radical anions III, reaction with sulfur dioxide (1), *Tetrahedron Lett.*, 1314–1415 (1969).
54. F. Beck, *Electro-Organische Chemie*, Verlag Chemie, Weinheim (1973), p. 253.
55. D. A. Tysee and M. M. Baizer, Electrocarboxylation I, mono- and dicarboxylation of activated olefins, *Org. Chem.* **39**, 2819–2823 (1974).
56. D. A. Tysee and M. M. Baizer, Electrocarboxylation. II. Electrocarboxylation dimerization and cyclization, *J. Org. Chem.* **39**, 2823–2828 (1974).
57. J. Lund, Cathodic reduction of nitro-compounds, in *Organic Electrochemistry*, M. M. Baizer, ed., Marcel Dekker, New York (1973), 315–346.
58. M. R. Rifi, Electrochemical reduction of organic halides, in *Organic Electrochemistry*, M. M. Baizer, ed., Marcel Dekker, New York (1973), pp. 279–314.
59. L. Horner, Onium compounds, in *Organic Electrochemistry*, M. M. Baizer, ed., Marcel Dekker, New York (1973), 429–446.

60. R. Brettle and J. G. Parkin, Anodic oxidations II, the electrolysis of dialkyl sodiumalonates, *J. Chem. Soc. B*, 1352–1355 (1967).
61. N. Nyberg, Oxidative coupling, in *Organic Electrochemistry*, M. M. Baizer, ed., Marcel Dekker, New York (1973), pp. 705–730.
62. R. Bauer and H. Wendt, Anodic formation of diacetylenes, *J. Electroanal. Chem.* **80**, 395–399 (1977).
63. Z. Ali, R. Bauer, W. Schön, and H. Wendt, Anodic oxidation of N anions III. Anodic oxidation of N anions of diacyclamides, *J. Appl. Electrochemistry* **10**, 97–107 (1980).
64. L. Eberson, Carboxylic acids, in *Organic Electrochemistry*, M. M. Baizer, ed., Marcel Dekker, New York (1973), pp. 469–508.
65. J. Knolle and H. J. Schafer, Synthesis of brevicomine by Kolbe-electrolysis, *Angew. Chem. Int. Ed. Engl.* **14**, 785 (1975).
66. W. Seidel, J. Knolle, and H. J. Schafer, Syntheses von Z-7-Dodecenylacetat (LoopLure) durch Kolbe-Electrolyse, *Chem. Ber.* **110**, 3544–3552 (1977).
67. T. Inoue and S. Tsutsumi, Electrochemical synthesis III, the homolytic methoxylation of some arylated olefins by the anodic oxidation of methanol, *Bull. Chem. Soc. Jpn.* **38**, 661–666 (1965).
68. H. Schafer and A. Alazrak, Oxidative Addition von Natrium-malon-sauredimethylester und acetylacetonat an Olefine, *Angew. Chem.* **80**, 485–486 (1968).
69. H. Schafer, Oxidative Addition des Azidions an Olefine, Ein einfacher Zugang zu Diaminen, *Angew. Chem.* **82**, 134 (1970).
70. H. Lund, Electrolysis of Heterocyclic Compounds, *Organic Electrochemistry*, M. M. Baizer, ed., Marcel Dekker, New York (1973), pp. 563–620.
71. H. Lund, Electrolysis of N. Heterocyclic Compounds, *Advances in Heterocyclic Chemistry*, Vol. 12, A. R. Katritzky and A. J. Boulton, eds., Academic, New York (1970), pp. 213–316.
72. D. Koch and H. Schafer, Einstufige Pyrol Synthese durch anodische Dimerisierung von Enaminketonen odor Estern, *Angew. Chem.* **85**, 264–265 (1973).
73. D. J. Clemens, A. K. Garrison, and A. L. Underwood, Electrochemical reduction of 1,1'-ethylene-bis-(3-carbamide-pyridiumbromide), *J. Org. Chem.* **34**, 1867–1871 (1969).
74. L. Eberson and H. Horner, Stereochemistry of Organic Electrode Processes, *Organic Electrochemistry*, M. M. Baizer, ed., Marcel Dekker, New York (1973), pp. 869–900.
75. A. J. Fry, *Stereochemistry of Electrochemical Reductions in Topics in Current Chemistry*, Vol. 34, Springer, Berlin/New York (1972), pp. 1–48.
76. R. N. Gourley, J. Grimshaw, and P. G. Millar, Electrochemical reduction in the presence of tertiary amines: An asymmetric synthesis of 3,4-dihydro-4-methylcoumarin, *Chem. Commun.*, 1278–1279 (1967).
77. J. Kopilov, E. Kariv, and L. L. Miller, Asymmetric, cathodic reductions of acetylpyridines, *J. Am. Chem. Soc.* **99**, 3450–3454 (1977).
78. W. J. M. van Tilborg, Phase-sensitive detection of adsorption phenomena at mercury electrode, *J. Royal Netherlands Chem. Soc.* **96**, 213–230 (1977).
79. R. N. Adams, *Electrochemistry at Solid Electrodes*, Marcel Dekker, New York (1969), p. 19.
80. R. C. Augustine, ed., *Oxidation Techniques and Applications in Organic Chemistry*, Marcel Dekker, New York (1969).
81. M. Kappel, Elektrolytische Regeneration von Chromsaure, *Chem. Ing. Tech.* **35**, 386–389 (1963).
82. N. Ibl and H. W. Lorch, private communication.
83. W. Funke, Reduktion Organischer Verbindungen mit Na-triumamalgam, *Chem. Ing. Tech.* **35**, 336–340 (1963).
84. H. C. Rance and J. M. Caulson, Electrolytic preparation of *p*-aminophenol, *Electrochem. Acta* **14**, 283–292 (1969).
85. J. R. Hanson and E. Premuzici, Die Reduktion Organischer Verbindungen mit CrII-Salzen, *Angew. Chem.* **80**, 271–276 (1968).

86. K. H. Simmrock, Die Herstellungsverfahren fur Propylen Oxid und ihre Elektrochemischen, *Alternative Chem. Ing. Tech.* **48**, 1085–1076 (1976).
87. M. Fremery, H. Hover, and G. Schwarzlose, Electrochemische Benzol-Oxidation, ein Nebenproduktfreier Weg zu Hydrochinon, *Chem. Ing. Tech.* **46**, 635–637 (1974).
88. J. P. Millington and J. Trotenau, A pilot plant for the electrolytic production of *p*-benzoquinone and hydroquinone, in *Abstracts of the Spring Meeting of the Electrochemical Society, Washington*, 1976, Electrochemical Society, Princeton, New Jersey (1976), p. 700.
89. A. C. Brown and J. Walker, Elektrolytische Synthesen Zweibasiger Sauren, *Liebigs Ann. Chem.* **261**, 107–128 (1891); 41–47 (1893).
90. M. Ya. Fioshin and A. I. Kamneva, *Chem. Ind.* (in Russian), 159 (1960); 263 (1963).
91. F. Beck, Kolbesynthese von Sebacinsaureestern in der Kapillarspaltzelle, *Electrochem. Acta* **18**, 359–368 (1973).
92. D. Danley, Industrial electro-organic electrochemistry, in *Organic Electrochemistry*, M. M. Baizer, ed., Marcel Dekker, New York (1973), pp. 907–943, esp. 939–943.
93. D. Danley, Industrial electro-organic electrochemistry, in *Organic Electrochemistry*, M. M. Baizer, ed., Marcel Dekker, New York (1973), pp. 936–939.
94. H. Nohe, Probleme der elektrochemischen Hydrierung aufgezeigt am Beispiel der Hydrierung von Phthalsaure zu Dihydrophthalsaure, *Chem. Ing. Tech.* **46**, 594–602 (1974).
95. V. Plzak and H. Wendt, Charge transfer and consecutive kinetics of azide anion oxidation at platinum, glassy carbon and carbon anodes in acetonitrile. An analogy to Kolbe synthesis, *Ber. Bunsenges. Phys. Chem.* **83**, 481–486 (1979).
96. K. Köster, P. Riemenschneider, and H. Wendt, Influence of electrosorption on kinetics and selectivity of electroorganic synthetic reactions, *Isr. J. Chem.* **18**, 141–151 (1979).
97. M. S. Venkatachalapathy, R. Ramaswamy, and H. V. K. Udupa, Electrically regenerated manganic sulphate for the oxidation of aromatic hydrocarbons I. Oxidation of toluene to benzaldehyde, *Bull. Acad. Pol. Sci., Ser. Sci. Chim.* **8**, 361–368 (1960).
98. S. Swann, Jr., and R. Alkire, Bibliography of electro-organic syntheses (1801–1975), Port City Press Inc., Baltimore Md. (1980).
99. H. Feess and H. Wendt, Performance of electrolysis with two-phase electrolyte, *J. Chem. Tech. Biotechnol.* **30**, 297–312 (1980).

# 5
# Electrometallurgy of Aluminum

**WARREN E. HAUPIN and WILLIAM B. FRANK**

## 1. Hall–Heroult Cell for Alumina Reduction

A typical modern aluminum reduction cell (or pot) consists of a rectangular steel shell, 9–12 m long by 3–4 m wide, and 1–1.2 m high. It is lined with refractory thermal insulation that surrounds an inner lining of either carbon blocks or monolithic carbon baked in place. Few materials other than carbon can withstand the combined corrosive action of a fluoride electrolyte (or bath) and molten aluminum. The thermal insulation is adjusted to provide sufficient heat loss to freeze a protective coating of electrolyte, known as ledge, on the inner walls, but not on the bottom, which must remain substantially bare for electrical contact with the molten aluminum cathode. A crust of frozen electrolyte and alumina covers the electrolyte. Electric current enters the cell either through prebaked carbon anodes (Figure 1) or through a continuous self-baking Soderberg anode (Figure 2) and flows through 3–6 cm of electrolyte, forming $CO_2$ at the anode and aluminum at the cathode. Steel collector bars joined to the carbon lining at the bottom conduct electrical current from the cell. Today's cells range from 50- to 250-kA current capacity.

Prebaked anodes are molded from petroleum coke and coal tar pitch binder into blocks typically 70 cm wide by 125 cm long and 50 cm high, and baked to 1000–1200°C. These materials are used because of their high purity. The more noble impurities such as iron and silicon deposit in the aluminum,

---

**WARREN E. HAUPIN and WILLIAM B. FRANK** • Aluminum Company of America, Alcoa Laboratories, Alcoa Center, Pennsylvania 15069.

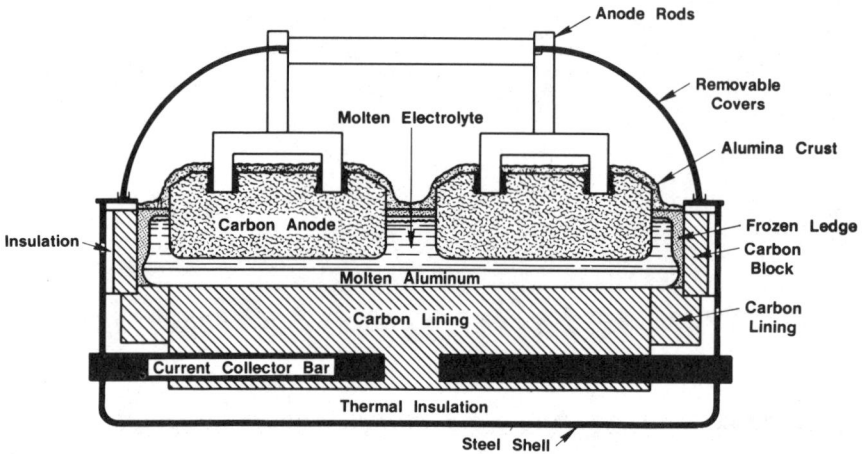

Figure 1. Aluminum electrolysis cell with prebaked anodes. [By permission of *Kirk–Othmer Encyclopedia of Chemical Technology*, Vol. 2, 3rd ed., Wiley, New York (1978).]

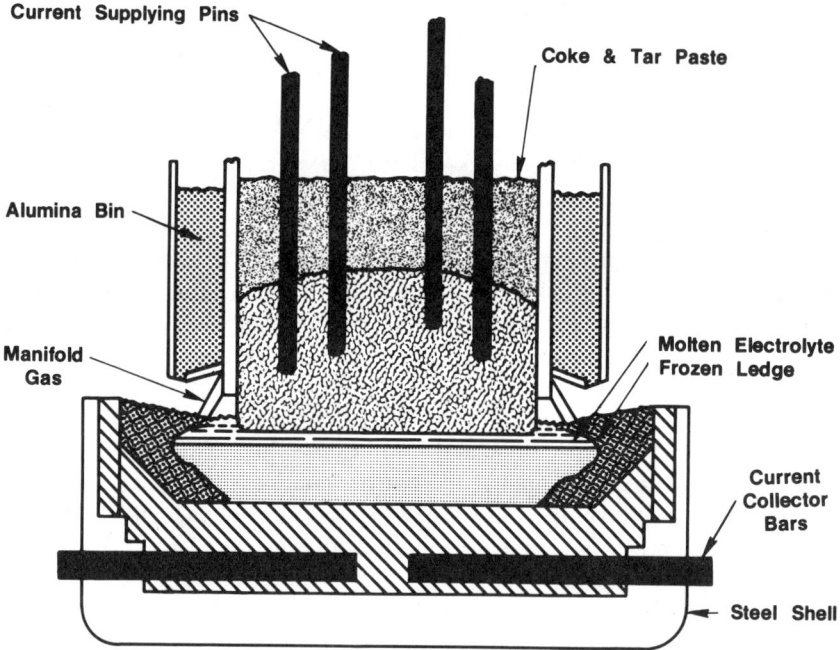

Figure 2. Aluminum electrolysis cell with Soderberg anode. [By permission of *Kirk–Othmer Encyclopedia of Chemical Technology*, Vol. 2, 3rd ed., Wiley, New York (1978).]

while less noble impurities such as calcium and magnesium tend to accumulate as fluorides in the bath. Steel stubs are keyed into the anode with cast iron to support the anodes in the electrolyte and conduct electric current into the anodes. Electrical resistivity of prebaked anodes ranges from 0.005–0.006 $\Omega$ cm. Anode current density ranges from 0.6–1.3 A/cm$^2$.

Soderberg anodes are formed continuously from a paste of petroleum coke and coal tar pitch. This mixture is added to the top of a rectangular steel casing, typically 6–8 m long by 2 m wide and 1 m high. While passing through the casing, the paste bakes forming carbon to replace the carbon being consumed at the bottom. The baked portion extends past the casing and into the molten electrolyte. Electric current enters the anode through vertical or sloping steel pins or spikes. Periodically the lowest spikes are reset to a higher level. The resistivity of Soderberg anodes is about 30% higher than prebake anodes, while the current density is lower, ranging from 0.6–0.9 A/cm$^2$. Although Soderberg anodes save the capital and labor required to form and bake the prebaked anodes, Soderberg anode cells are being phased out because of the greater difficulty in collecting and disposing of baking fumes.

Alumina is added in appropriate increments, either manually or by automatic feeders. Molten aluminum is removed from the cells, generally daily, by siphoning into a crucible. The aluminum normally is 99.6% to 99.9% pure. Typical impurities are Fe, Si, Ti, V, and Mn, largely from the anode but also from impurities in the alumina.

## 2. Electrolyte

### 2.1. Composition and Physical Properties

The electrolyte used to produce aluminum is essentially a solution of aluminum oxide or alumina ($Al_2O_3$) dissolved in fused cryolite ($Na_3AlF_6$). Pure cryolite melts at about 1010°C; the cryolite–alumina eutectic at 960°C contains about 10.5 mass percent alumina (Figure 3). In industrial operation the alumina content of the electrolyte ranges from 2 to 6 wt %. Calcium fluoride ($CaF_2$) is always present in the electrolyte, arising from low levels of calcium oxide impurity in the alumina. It attains a steady state concentration of 4–8%, at which level calcium is removed in the aluminum product and emitted in the off gases at a rate essentially equal to its introduction.

Aluminum fluoride is usually added to the electrolyte in excess of the cryolite composition. The relative amounts of sodium fluoride (NaF) and aluminum fluoride ($AlF_3$) in the electrolyte are expressed as the *cryolite ratio* (CR), the molar ratio of NaF to $AlF_3$, or the *bath ratio*, the mass ratio of NaF to $AlF_3$. Because the molecular weight of $AlF_3$ is almost exactly twice that of NaF, the cryolite ratio is double the value of bath ratio. Addition of aluminum

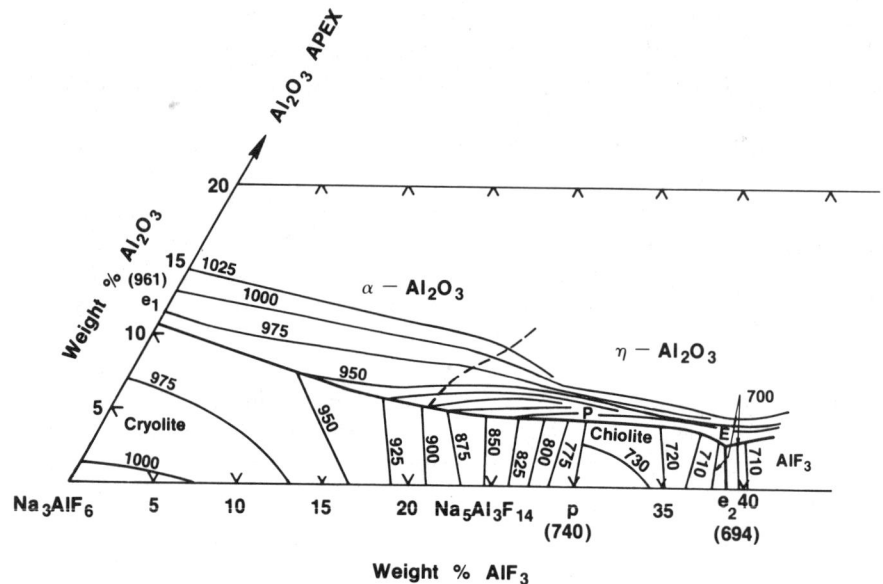

Figure 3. The system $Na_3AlF_6$–$AlF_3$–$Al_2O_3$. P, ternary peritectic point (28.3% $AlF_3$–4.4% $Al_2O_3$–67.3% $Na_3AlF_6$, 723°C); E, ternary eutectic point (37.3% $AlF_3$–3.2% $Al_2O_3$–59.5% $Na_3AlF_6$, 684°C); p, binary peritectic point; e, $e_1$, binary eutectic points. From Foster.[79]

fluoride improves current efficiency. Industrial cells operate at cryolite ratios between 2 and 3. Operating temperature is usually in the range 940–980°C.

Other additives are sometimes used to further reduce the freezing point and alter the physicochemical properties of the electrolyte. Additive compounds must be more stable cathodically than aluminum fluoride to avoid contamination of the aluminum produced by the cation of the additive. Addition of virtually any compound to molten cryolite reduces the equilibrium solubility of alumina and its rate of solution. Because rapid dissolution and high activity of alumina in the electrolyte are desirable, the additive is usually controlled to less than 10 wt % as fluoride.

The density of liquid aluminum at 950–1000°C is 2.3 g cm$^{-3}$. Experience has shown that the density of the molten electrolyte should be less than 2.1 g cm$^{-3}$ to maintain a good separation between the metal and fused salt phases. Additions which decrease the density of cryolite–alumina melts are beneficial.

The specific electrical conductivity of the electrolyte is important in the energy efficiency of the process. Based on a correlation derived by Choudhuri[1] most measurements found in the literature fit the relationship

$$\kappa = \exp[2.0156 - 0.0207(\%\,Al_2O_3) - 0.005(\%\,CaF_2) - 0.0166(\%\,MgF_2) \\ + 0.0178(\%\,LiF) + 0.0063(\%\,NaCl) + 0.217(CR) - 2068.4/T] \quad (1)$$

where $\kappa$ is in $\Omega^{-1}\,\mathrm{cm}^{-1}$ and $T$ is in degrees Kelvin. Improved electrical conductivity can be achieved by addition of lithium compounds. Most other additives reduce conductivity. Lithium compounds have been investigated as electrolyte additives more than any other material. However, the cost–benefits aspects for lithium additives have not been convincing enough to result in widespread use.

Other physical properties of the electrolyte important in cell operation are vapor pressure, viscosity, and surface tension. The vapor pressure of the electrolyte and its reactivity with water vapor are important with respect to fluoride emissions. Grjotheim et al.[2] and Rolin[3] provide comprehensive reviews of the physicochemical properties of the electrolyte and their implications.

## 2.2. Ionic Constitution of the Electrolyte

The structure and properties of fused salts can be described in terms of ionic species. From measurements of the electrical conductivity of solid and liquid cryolite,[4] analysis of the conductivity of $NaF-AlF_3$ melts[5,6] and from transference number measurements of cryolite–alumina melts,[7,8] it has been established that in cryolite-rich melts, ionic conductivity results almost exclusively from the migration of sodium ions toward the cathode. The contribution of simple fluoride anions is small and there is no evidence for the existence of trivalent or any aluminum cations. Thus the primary ionization of cryolite on melting can be represented as

$$Na_3AlF_6 \rightleftharpoons 3Na^+ + AlF_6^{3-} \qquad (2)$$

There is general agreement that the hexafluoroaluminate ion undergoes partial dissociation in the melt,

$$AlF_6^{3-} \rightarrow AlF_4^- + 2F^- \qquad (3)$$

$$K = (a_{AlF_4^-})(a_{F^-}^2)/a_{AlF_6^{3-}} = 4\alpha^3/[(1-\alpha)(1+2\alpha)^2] \qquad (4)$$

The mechanism and the degree of dissociation, $\alpha$, of the hexafluoroaluminate anion have been developed from cryoscopy,[9–12] density studies,[6,13–16] Raman spectroscopy,[17,18] and viscosity.[19] Values for the dissociation constant, $K$, and its temperature dependence developed from these studies vary considerably. The authors feel that the models for melt constitution developed by Paucirova et al.[15,16] and Sato et al.[6] best represent the ionic structure of $NaF-AlF_3$ melts near the composition $3NaF \cdot AlF_3$. The degree of dissociation, $\alpha$, of hexafluoroaluminate ion at the melting point of cryolite is about 0.3. The major ions present in fused cryolite are $Na^+$, $AlF_6^{3-}$, $F^-$, and $AlF_4^-$. Below a cryolite ratio of 3, lowering the ratio decreases $AlF_6^{3-}$ and $F^-$ and increases $AlF_4^-$ concentrations.

## 2.3. Dissolution of Alumina

The fused alkali metal hexafluoroaluminates are rather unique solvents for aluminum oxide; few other melts exhibit sufficient stability and dissolve enough alumina to permit electrolytic deposition of aluminum. The maximum solubility for alumina in the $NaF-AlF_3$ system occurs at or near the composition $3NaF \cdot AlF_3$.

When alumina is dissolved in fused cryolite the physicochemical properties of the melt are changed appreciably.[2,3] The density, vapor pressure, electrical conductivity, and surface tension decrease while the viscosity increases with alumina content. The trends and extent of these variations suggest strong chemical interaction between the solvent and solute, rather than a "sugar-in-water" type of solution. The variations in physical properties have been studied by many researchers who have developed a wide variety of models to describe the ionic species formed in the solution of alumina. The nature of the oxygen-containing ions has also been studied by cryoscopy,[20] calorimetry,[21] Raman spectroscopy,[22] and vapor pressure studies.[23] There is evidence that more than one oxygen-containing anion is present in the electrolyte. Complex, rather bulky oxyfluoroaluminates, e.g., $Al_xO_yF_z^{(3x-2y-z)}$, are more probable structures than simpler aluminates such as $AlO_2^-$, $AlO^+$, $Al_2O_4^{2-}$, and $AlO_3^{3-}$. There is probably a single oxygen atom per complex ion, particularly at low concentration of alumina. At infinite dilution, three foreign particles are created for each $Al_2O_3$ that dissolves. Gilbert et al.[22] report that the presence of species such as $AlOF_2^-$, $AlOF_3^{2-}$, and $AlOF_5^{4-}$ is not consistent with their results by Raman spectroscopy; only those species with bridging Al–O–Al bonds should be considered.

Reasonable dissolution mechanisms are, for low alumina contents,

$$4AlF_6^{3-} + Al_2O_3 \rightarrow 3Al_2OF_6^{2-} + 6F^- \qquad (5)$$

and for more concentrated solutions,

$$AlF_6^{3-} + Al_2O_3 \rightarrow 3AlOF_2^- \qquad (6)$$

## 3. Electrode Reactions

### 3.1. Cathode Reaction

As pointed out previously, no Al cations are present in the melt. Instead, aluminum is bound in various anionic complexes. The fact that $Na^+$ ion is the principal current carrier[7] has led some investigators to the assumption that sodium is the primary product discharged at the cathode, and aluminum results from a secondary reaction. The difference between the decomposition potentials for producing aluminum, reaction (7), and producing sodium, reaction (8), can be calculated from Eq. (9). The reversible emf for formation of sodium at

1 atm pressure is about 0.24 V less favorable than for the formation of liquid aluminum over the range of compositions and temperatures used industrially.

$$Al_2O_3(\text{soln}) + \tfrac{3}{2}C(s) \rightarrow 2Al(l) + \tfrac{3}{2}CO_2(g) \tag{7}$$

$$6NaF(\text{soln}) + Al_2O_3(\text{soln}) + \tfrac{3}{2}C(s) \rightarrow 2AlF_3(\text{soln}) + \tfrac{3}{2}CO_2(g) + 6Na(g) \tag{8}$$

$$\Delta E = (\Delta G^\circ_{(7)} - \Delta G^\circ_{(8)})/6F - (RT/6F)\ln(a^2_{AlF_3}/a^6_{NaF})(a^6_{Na}) \tag{9}$$

where $\Delta G^\circ_{(7)}$ and $\Delta G^\circ_{(8)}$ are the standard free energy changes for reactions (7) and (8). Activities of $AlF_3$ and $NaF$ are from Kvande.[23] The activity of Na was calculated from the vapor pressure of pure Na at temperature $T$. This difference, $\Delta E$, decreases with increasing temperature and increasing cryolite ratio. Since the cathodic overvoltage for normal current densities and electrolyte compositions is always appreciably less than 0.24 V, the cathodic product is mainly aluminum with sodium present at very low activity.

Early investigators studying potential decay curves found plateaus suggesting three-electron, two-electron, and one-electron processes: $Al^{3+} + 3e^- \rightarrow Al$, $Al^{3+} + 2e^- \rightarrow Al^+$, $Al^+ + e^- \rightarrow Al$, $Na^+ + e^- \rightarrow Na$, and $2Na^+ + e^- \rightarrow Na_2^+$.

This early work did not rule out discharge of $Na^+$ to Na at low activity, followed immediately by reaction with $AlF_4^-$ or $AlF_6^{3-}$ to form Al, NaF, and $F^-$. If the second reaction is very fast, it could lower the sodium activity sufficiently to discharge $Na^+$ at the $Al^{3+}$ potential and look like a three-electron transfer. However, one would expect that this would lead to a lower exchange current than was observed. Thonstad and Rolseth[24] measured the charge transfer process by a double-pulse technique and found the exchange current density to be 12 A/cm$^2$ for $n = 3$ (36 A/cm$^2$ for $n = 1$).

Bowman[25] using cyclic voltammetry, stationary electrode polarography, differential pulse polarography, and chronopotentiometry at tungsten and glassy carbon electrodes, recently found evidence only for a reversible three-electron transfer process. There was no evidence for a chemical reaction following the electron transfer process, ruling out discharge of $Na^+$ at low activity followed by chemical reaction to form Al. Bowman also saw no evidence for a chemical reaction preceding electron transfer, which would be present if $Al^{3+}$ discharge were preceded by dissociation: $AlF_6^{3-} \rightleftharpoons Al^{3+} + 6F^-$ and/or $AlF_4^- \rightleftharpoons Al^{3+} + 4F^-$.

The potentials observed by Bowman at platinum electrodes were very similar to those that Kubik et al.[26] attributed to the formation of $Al^{2+}$, $Al^+$, Na, and $Na_2^+$. Bowman offered additional X-ray fluorescence evidence indicating that these electrochemical signals resulted from oxidation of a series of Al–Pt alloys.

Three-electron transfer processes involving oxyfluoroaluminate anions are possible. However, the most probable cathode reactions are

$$AlF_4^- + 3e^- \rightarrow Al + 4F^- \tag{10}$$

and

$$AlF_6^{3-} + 3e^- \rightarrow Al + 6F^- \tag{11}$$

Reaction (10) would predominate at low cryolite ratios (<2), while reaction (11) would predominate at high cryolite ratios (>3).[18] Total cathodic overvoltage fits a Tafel relationship over the range of commercial interest. Thonstad[27] explains that this is fortuitous and should not be taken as an indication of charge transfer or reaction overvoltage. Laboratory measurements by Thonstad and Rolseth[24] without stirring gave Tafel plots for electrolytes of industrial interest with coefficient "$a$" ranging from 0.19 to 0.21 and "$b$" ranging from 0.23 to 0.25. Stirring reduced the overvoltage 40–50%, suggesting concentration overvoltage. Thonstad[24,27] calculated the change in electrolyte composition at the metal-to-electrolyte interface, caused by (1) the inward migration of $Na^+$ ions, (2) the inward diffusion of $3Na^+ \cdot AlF_6^{3-}$ and $Na^+ \cdot AlF_4^-$, (3) the discharge of $AlF_6^{3-}$ and $AlF_4^-$ to $F^-$ ions and Al at the interface, and (4) the resultant outward diffusion of $Na^+F^-$. To demonstrate that this enrichment of NaF at the cathode interface explained the cathode overvoltage, Thonstad compared and found good agreement between their measured overvoltages and the emf between two aluminum half-cells, one containing electrolyte of the bulk composition and the other containing electrolyte corresponding to the calculated interfacial composition.

Further confirmation that a high concentration of sodium fluoride develops at the cathode interface comes from the observation that, when operating at low cryolite ratio (where the increase in liquidus temperature with increasing sodium fluoride concentration is large), a crust forms over the cathode when current is increased and melts when the current is lowered.

Cathodic overvoltage in industrial cells lies between the laboratory value for unstirred cells and that for vigorously stirred cells. Thonstad and Rolseth[24] found cathodic overvoltage in industrial cells to be about 0.1 V. This is in good agreement with our data,[28] which can be represented by

$$\eta = [RT(1.375 - 0.125CR)/(1.5F)] \ln (i/0.257) \tag{12}$$

where CR is the cryolite ratio (mole ratio $NaF/AlF_3$) and $i$ is the cathodic current density ($A/cm^2$).

### 3.2. Anode Process

The anode process may be even more complicated than the cathode process. Thermodynamically, oxygen depositing onto carbon at the cell operating temperature should equilibrate to essentially all CO with very little $CO_2$. However, based upon the volume of gas produced or net carbon consumption, the primary anode product is essentially all $CO_2$. Anodic formation of $CO_2$ requires four electrons per molecule, while formation of CO requires only two. The volume of anode gas produced per faraday gives the primary $CO/CO_2$ ratio formed at the anode. A subsequent reduction of $CO_2$ to CO by aluminum or sodium dissolved in the electrolyte will not change the gas volume. Measurements, however, can be confused by the presence of carbon in the melt that is not part of the active anode, electronic conduction through the melt, or by reduction of $CO_2$ by metal all the way to carbon. Carbon dust of several origins is found in industrial cells. When carbon reacts with $CO_2$, it increases the volume of cell gas the same as though CO were the primary product. Permeation of cell gas through the anode has a similar effect.

Ginsberg and Wrigge[29] overcame these problems by use of a cell in which the anode compartment was isolated by an oxygen ion permeable diaphragm of calcium oxide stabilized zirconia. Using graphite anodes, almost pure $CO_2$ was obtained except at very low current densities ($<0.06$ A/cm$^2$).

The primary anode reaction can also be calculated from net carbon consumption if air burning is eliminated. Investigators using this technique nearly always find slightly higher carbon consumption than should result from quantitative formation of $CO_2$, because nearly all carbon anodes dust during electrolysis. When a correction is made for dusting[30,31] or when special nondusting carbon is used,[32] carbon consumption very closely agrees with $CO_2$ being the primary anode product. It can be concluded that $CO_2$ is the primary anode product at normal current densities.

It is surprising that the anode gas composition is displaced so far from thermodynamic equilibrium. Revazyan[33] believes that $CO_2$ cannot react with the anode to form CO because, being positively charged, there are no valence electrons available at the surface. Thonstad and Hove[34] point out that reaction kinetics cause the discharged oxygen to form $CO_2$ rather than thermodynamically favored CO. They assume that the initial oxygen-containing ions are transported through the double layer and discharged with comparatively little overvoltage. This oxygen is chemisorbed on the carbon surface as proposed by Blyholder and Eyring,[35] for ordinary combustion (Figure 4a). A carbon site occupied by oxygen is not available for further oxygen discharge—in effect the site is electrically insulated. As sites become occupied, it takes

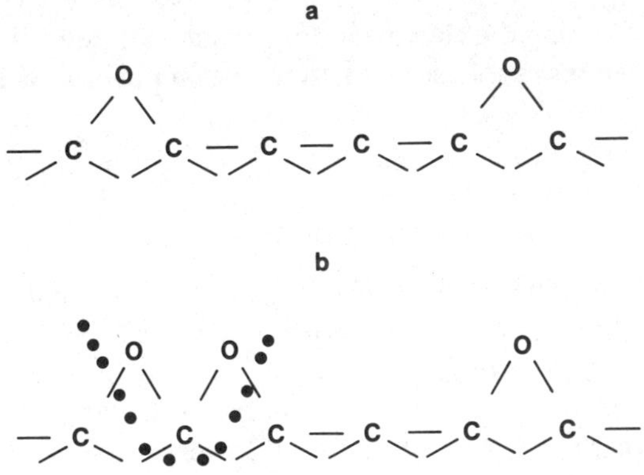

Figure 4. Schematic representation for formation of $CO_2$ on carbon surface. (a) Initial oxygen deposits on most active sites first forming $C_2O$, a stable surface compound. (b) As surface becomes covered $C_3O_2$ is formed, requiring additional energy (overvoltage). $C_3O_2$ cleaves easily, as indicated by dotted line, forming $CO_2$ and new surface.

additional energy (overvoltage) to deposit oxygen at increasingly less favorable sites. The surface $C_2O$ compound is quite stable and breaks down to CO only at a very slow rate:

$$C_2O(\text{surface}) \rightarrow CO(\text{ads}) + C(\text{surface}) \qquad (13)$$

Any CO formed will desorb from the surface at a moderate rate but reaction (13) provides new sites far too slowly[35] to meet the oxygen deposition rate at commercial current densities. Hence, adjacent sites start to become occupied as shown in Figure 4b, raising the overvoltage still higher. The $C_3O_2$ surface compound, formed by two adjacent occupied sites, decomposes by reaction (14) at a rapid rate[35]:

$$C_3O_2 \rightarrow CO_2(\text{ads}) + 2C(\text{surface}) \qquad (14)$$

The rate of CO formation is so slow that at commercial current densities, the anode product is nearly 100% $CO_2$. As discussed later the Tafel slopes indicate a reaction order of one-half. Blyholder and Eyring[35] point out that when oxygen reacts both in pores and on the surface, a half-order reaction results. The high anode overvoltage would force oxygen to be deposited in the pores (less favorable sites), thus accounting for the half-order chemical reaction observed. Based on present knowledge, the following anode reactions are suggested: At low alumina concentrations and high cryolite ratios,

$$2Al_2OF_6^{2-} + 2AlF_6^{3-} + C \rightarrow 6AlF_4^- + CO_2 + 4e^- \qquad (15)$$

At low alumina concentration and low cryolite ratios,

$$2Al_2OF_6^{2-} + 4F^- + C \rightarrow 4AlF_4^- + CO_2 + 4e^- \qquad (16)$$

At high alumina concentration and high cryolite ratios,

$$2AlOF_2^- + 2AlF_6^{3-} + C \rightarrow 4AlF_4^- + CO_2 + 4e^- \qquad (17)$$

At high alumina concentration and low cryolite ratios,

$$2AlOF_2^- + 4F^- + C \rightarrow 2AlF_4^- + CO_2 + 4e^- \qquad (18)$$

Adding the corresponding anode and cathode reactions plus the corresponding alumina solution reaction gives an overall cell reaction,

$$2Al_2O_3 + 3C \rightarrow 4Al + 3CO_2 \qquad (19)$$

### 3.2.1. Anode Overvoltage

The anode overvoltage can be broken down into at least four components. First, there is a component that fits a Tafel relationship. This generally indicates charge transfer control but can also represent reaction control. Most observations favor the latter because (1) addition of combustion catalysts to the carbon ($Na_2CO_3$, $Fe_2O_3$) lowered overvoltage, while inhibitors ($H_3BO_3$)

increased overvoltage [Thonstad and Hove[34]]; (2) impedance measurements indicate reaction control [Drossbach et al.,[36,37] Thonstad[38]]; (3) the potential rise and decay are slow when a current pulse is applied; and (4) the primary anode product is $CO_2$ rather than the thermodynamically favored CO.

The considerable spread in anode overvoltage data obtained by various authors probably results from the considerable variation in reactivity of various carbons. Data representative of industrial practice fit the relationship proposed by Vetter[39] for reaction overvoltage:

$$\eta = RT/[p(n/\nu)F] \ln i/i° \qquad (20)$$

where the slow step involves $n/\nu = 2$ electrons per oxygen molecule, the reaction order $p$ ranges from 0.4 to 0.6, varying with carbon reactivity and porosity. The reaction limiting current density $i°$ ranges from 0.0039 to 0.0085 A/cm$^2$ as $Al_2O_3$ concentration, $c$, varies from 2 to 8 wt %. Vetter's relationship $d(\log i°)/d(\log c) = p$ leads to a reaction order of 0.56, well within the range observed from the Tafel slopes.

The next two components of overvoltage result from the increase in effective resistivity of the electrolyte caused by gas bubbles. Since this is an ohmic effect, it can be argued that it should not be called an overvoltage. It must be included, however, in analyzing the cell voltage and is a nonequilibrium effect. Gas bubbles on the anode, in addition to increasing electrical resistance, also increase true overvoltage by raising the local current density in areas not covered by bubbles. The bubble effect can be measured by superimposing alternating currents of various frequencies on the direct current and observing the change in ac resistance with frequency and dc current level. The ac frequency effect is complex. At high frequencies the bubbles are bypassed, behaving like small capacitors. At low frequencies there is polarization of the alternating current, causing wave shape distortion and polarization resistance, but the full effect of the bubbles is seen. Fourier analysis of either pulse, step changes, or white noise also can be used.

De La Rue and Tobias[40] have shown that the effect of bubbles in the bath can be calculated:

$$\kappa_e = \kappa(1-\varepsilon)^{1.5} \qquad (21)$$

where $\kappa_e$ is the effective conductivity including bubbles, $\kappa$ is the gas-free conductivity, and $\varepsilon$ is the gas fraction.

Smaller, more slowly rising gas bubbles form with increasing alumina concentration. The resulting increase in gas fraction can be determined from the increase in the apparent volume of the electrolyte when current is flowing.

Gas bubbles adhering to the anode have an even greater effect. Landau[41] has derived an expression to calculate the increase in resistance caused by bubbles on the electrode:

$$R = R_0\{1 + 0.5(a/L)^3/[1-(A_b/A_t)^{1.5}]\} \qquad (22)$$

where $R_0$ is the bubble-free resistance, $a$ is the average bubble radius, $L$ is the interelectrode gap, $A_t$ is the total electrode area, and $A_b$ is the total bubble cross sectional area.

The effect of bubbles has not in the past received sufficient attention. Haupin[42] has shown that the presence of bubbles contributes between 0.09 and 0.35 V.

The fourth component is concentration, or diffusion, overvoltage at the anode. It can be correlated by

$$\eta = RT/2F \ln[i_c/(i_c - i)] \qquad (23)$$

where $i_c$ is the critical, or concentration-limiting current density. Data from Thonstad et al.[43] for 0–6% $Al_2O_3$, using a 6-mm-diameter graphite electrode, can be represented

$$i_c = [0.2 + 1.37(\text{wt \% } Al_2O_3)][1 + 0.0022(T - 1293)] \qquad (24)$$

The temperature coefficient is from his earlier measurements.[44] The relationship by Piontelli et al.[45] seems to represent better industrial cells, probably because it accounts for the anode size effect:

$$i_c = \phi[5.5 + 0.018(T - 1323)]A^{-0.1}[(\% Al_2O_3)^{0.5} - 0.4] \qquad (25)$$

where $\phi$ is a shape factor equal to 1 for a horizontal planar anode and is equal to 1.4 for a hemispherical anode, and $A$ is area in $cm^2$.

The critical current can be calculated from first principles:

$$i_c = (\eta/\nu)F \cdot D(C/\delta) \qquad (26)$$

where $\eta/\nu$ is two electrons per oxygen molecule (slow step), $D$ is the diffusion coefficient of the oxygen-containing complex, $C$ is the bulk concentration of oxygen in the reacting species, and $\delta$ is the thickness of the diffusion layer.

However, $\delta$ is exceedingly difficult to measure or calculate. It is easier to measure $i_c$ as the current density required to produce an anode effect.

### 3.2.2. Anode Effect

An anode effect is a phenomenon common to many fused salt electrolytic processes. Its physical manifestation is the growth of larger and larger bubbles on the anode as the contact angle of the bubble diminishes. Finally, a few large bubbles coalesce forming a single large bubble covering most if not the entire surface of the anode. With a constant potential source the current falls to a low value. With a constant current source, as is used industrially, the applied potential rises to about 30 V and the current penetrates the gas film by a multitude of small electric arcs.

In aluminum electrolysis, the electrical manifestation is typical of concentration overvoltage. The chemical manifestation is a change in anode gas composition from largely $CO_2$ to largely CO with significant quantities (3–25%) of $CF_4$ and minor amounts of $C_2F_6$. Many investigators[43,46,47] find that fluorine production precedes the anode effect producing a carbon–fluorine surface compound that causes dewetting of the anode. However, electrolytes containing large amounts of LiF can produce several tenths percent $CF_4$ in the anode gas without producing an anode effect. A few investigators believe that the anode effect is caused by volatilization of the electrolyte.[47] Some investigators report a loss of current efficiency during anode effect. Others find no effect on current efficiency. Haupin[28] has found an initial increase in current efficiency as the cell went on the anode effect, followed by lowered current efficiency as the temperature rose owing to the high voltage. All are in agreement that power efficiency suffers.

While there is no general agreement among various investigators, the following mechanism is proposed as plausible. As alumina is depleted, overvoltage increases. The increase in surface tension of the bath with decreasing alumina concentration coupled with the electrocapillary effect, resulting from the higher anode overvoltage, reduces wetting of the anode and the bubble contact angle decreases. At about a 1.2 V[48] anode overvoltage, sufficient activity of fluorine is produced to result in fluorine bonding to the carbon. Fluorocarbon compounds have low surface energy and promote dewetting and growth of large bubbles. When a few very large bubbles are present it is easy for them to coalesce, producing the anode effect. Once the cell is on anode effect, the production of significant quantities of $CF_4$ further promotes dewetting, and even if the alumina concentration is restored, the cell will remain on the anode effect until the current is interrupted to allow the bubble to collapse.

### 3.2.3. Inert Anodes

Beginning with the original patent of Charles Martin Hall, the possibility of a nonconsumable anode has intrigued investigators. An inert anode would eliminate carbon consumption, save the labor of changing anodes, and allow energy-saving changes in cell design such as a bipolar configuration. Material for such an anode must resist attack by oxygen and the molten fluoride electrolyte, must have reasonably high electrical conductivity, and this conductivity must be electronic rather than ionic. It must be mechanically strong and resistant to thermal shock.

The equilibrium (decomposition) potential for an inert anode with oxygen evolution will be slightly more than 1 V greater than for a carbon anode. However, overvoltage is about 0.5 V less on an inert anode than a carbon anode, making the counterelectromotive force (cemf) of the cell only about 0.5 V higher than the conventional cell. Because frequent changing of consumable anodes is eliminated, the higher cemf can probably be compensated for by maintaining an ideal electrode shape and decreasing the anode–

cathode spacing. Judging from recent patents,[49-56] there is much activity in the field with a considerable degree of success.

## 4. Current Efficiency

Cathode current efficiency of industrial Hall–Heroult cells ranges from 80% to 95%. The principal loss mechanism is recombination of anodic and cathodic products. This involves several steps taking place at different locations in the electrolyte as depicted in Figure 5. At zone A, the metal-to-electrolyte interface, aluminum reacts introducing reduced species and possibly dispersed metal into the melt. Across zone B, there is diffusion of reactants, reduced species, and other products through the boundary layer at the interface. Zone C is a region of convective transport of reduced species away from the metal–electrolyte interface. Zone D is a region of convective transport of carbon dioxide gas away from the anode. Some solution of $CO_2$ into the electrolyte also occurs in this region. Zone E is a region of mixing and chemical reaction between reduced species and the $CO_2$. Following chemical reaction, there must be dissolution and transport of the product metal oxide and CO away from the reaction site.

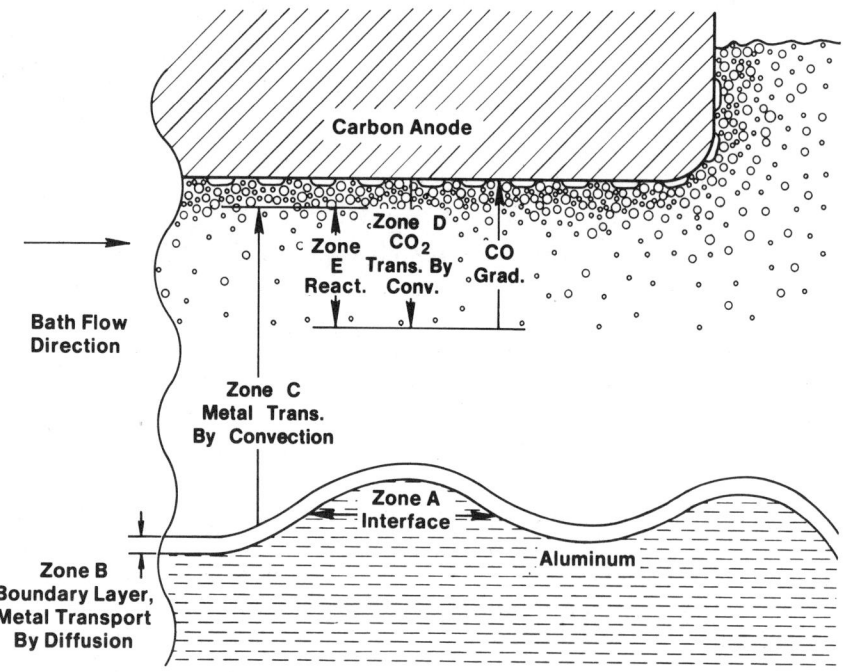

Figure 5. Mechanism of current efficiency loss.

While there is general agreement that the electrolyte contains reduced species, there is controversy over whether the reduced species is dissolved metallic sodium, monovalent aluminum, subvalent sodium, or a colloidal suspension of aluminum metal. Reaction (27) will produce an activity of sodium in the melt, while reaction (28) will produce an activity of aluminum monofluoride:

$$Al(l) + 3NaF(soln) = 3Na(soln) + AlF_3(soln) \tag{27}$$

$$2Al(l) + AlF_3(soln) = 3AlF(soln) \tag{28}$$

Quantitative measures of the reduced species in solution have been made by quenching a sample of electrolyte, grinding it in a dry box, and adding it to water[57] or dry alcohol[58] and measuring the hydrogen evolved. Following this first hydrogen evolution with water or alcohol, the investigators observed a much smaller second evolution by reacting the residue with acid. They concluded that the first evolution represented sodium, its concentration in the electrolyte increasing with increasing cryolite ratio. The second evolution was assumed to be aluminum. It decreased with increasing cryolite ratio. Similar results were obtained by Vetyukov and Vinokurov[59] by extracting the sodium with bromobutane followed by treatment of the samples with HCl.

A more accurate method of determining quantitatively the dissolved reductant in the melt involves electrolytic oxidation. Yoshida and Dewing[60] first saturated the melt with aluminum, then transferred it to a separate compartment where oxygen was evolved on a platinum anode. The quantity of reduced species (reported as aluminum) was determined from the amount of oxygen consumed. They obtained the relationship

$$\text{wt \% Al in solution} = -0.2877 + 0.0134(CR) + 0.0003(°C)$$
$$-0.0019(\% CaF_2) - 0.0017(\% Li_3AlF_6)$$
$$-0.0029(\% NaCl) \tag{29}$$

Thonstad and Rolseth[61] measured the reduced species concentration by determining the limiting current on a platinum anode. The anode reaction is

$$M(soln) \rightarrow M^{n+} + n(e^-) \tag{30}$$

which gives the relationship

$$I_{\lim} = nFK_DC \tag{31}$$

where $K_D$ is the mass transfer coefficient and $C$ the concentration of reduced species in the melt. This method has the advantage of speed and ease of measurement over the Yoshida and Dewing method but requires knowledge of the mass transfer coefficient, which changes with stirring and convection.

When aluminum is added to the melt directly or by electrolysis, foglike streamers come from the metal and soon render the melt opaque. This has been taken by some as evidence for a colloidal dispersion of aluminum in the melt. Hydrogen bubbles also rise from the aluminum if the melt is not completely anhydrous. Gerlach et al.[62] and Choudhuri[63] observed aluminum droplets 1–15 μm in diameter in quenched electrolyte samples and concluded that these particles were too large to have been formed by disproportionation of AlF during cooling; hence aluminum droplets probably existed in the melt. On the other hand, adding aluminum to molten cryolite lowers the melting point,[9,64] indicating that at least some true solution is formed. Reactions (27) and (28) produce activities of Na and AlF in the electrolyte which are manifest as vapor pressures of these species over the electrolyte. Kvande[23] has measured the vapor pressures of Na and AlF over NaF–AlF$_3$ melts in contact with Al (Figure 6). The reducing power of the electrolyte (reactivity with $CO_2$) can be assumed proportional to $p_{Na} + 2p_{AlF}$.

Gaseous $Na_2F$ has been postulated but its existence has not been established, nor are thermodynamic data available for it.

Droplets of aluminum dispersed in molten sodium chloride will not react with oxidizing gas bubbles unless either the gas is somewhat soluble in the melt

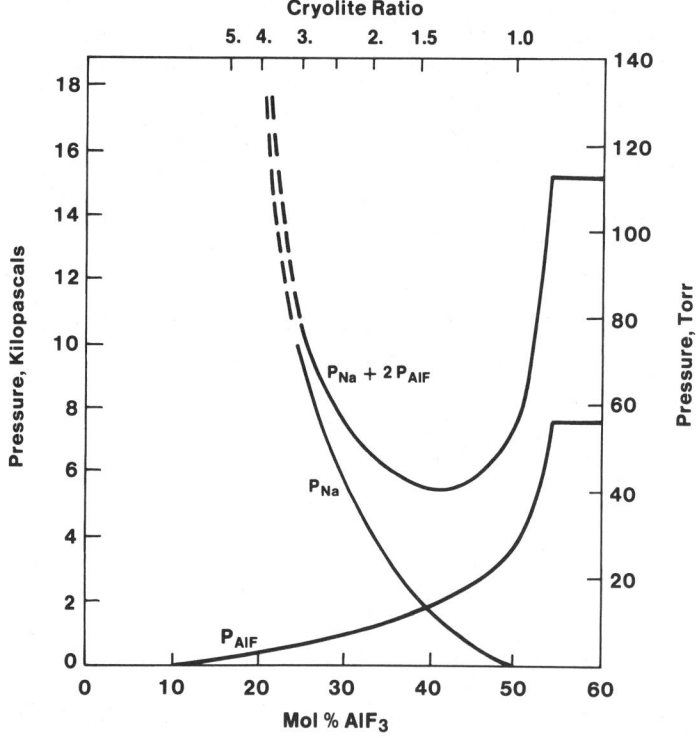

Figure 6. Equilibrium vapor pressure of Na and AlF over the electrolyte with Al at 1278 K from Kvande.[23] The reducing power of the electrolyte is taken proportional to $P_{Na} + 2P_{AlF}$ and is an important factor in loss of current efficiency.

or the metal somewhat soluble in the melt. It appears that the gas bubble and the aluminum droplet are kept separated by interfacial tension. If additions such as aluminum chloride are made to the melt to promote solubility of the aluminum, some reaction does occur. Or if a gas such as chlorine, having some solubility in the melt, is used some reaction will occur. By analogy, oxidation of dispersed droplets of aluminum will occur in Hall–Heroult cells only to the extent that $CO_2$ is soluble in the melt. Bratland et al.[65] have measured the solubility of $CO_2$ in the electrolyte and found it to be an order of magnitude smaller than the solubility of reduced species.

## 4.1. Factors Controlling Reoxidation Rate

In summary, the major loss mechanism appears to be dissolved metal reacting with $CO_2$ bubbles as depicted in Figure 5. The rate-controlling step, however, is not clear. Chemical reactions appear to be fast compared to mass transport. Under certain operating conditions, the rate of diffusion of dissolved metal through the boundary layer at the metal–electrolyte interface, zone B of Figure 5, appears to be rate controlling, but under other operating conditions the rate of convective transport of metal to the reaction zone, zone C, appears to be rate controlling. Figure 5 is merely schematic and not intended to represent the true relative dimensions of the layers and zones. Indeed, the reaction zone is much thinner and much nearer the anode surface than shown. There is evidence that the current efficiency of industrial cells is dependent upon the anode–cathode separation. Where mass transport is rate controlling, improvement in current efficiency can be achieved through modifications of cell design as well as through changes in bath chemistry.

## 4.2. Other Losses in Current Efficiency

Current efficiency correlations based on reoxidation of reduced species by $CO_2$ generally leave 1–2% loss in efficiency unexplained. There are several other mechanisms for loss of faradaic efficiency. Aluminum carbide formed by reaction of metal with the carbon lining represents a loss. New cell linings absorb sodium with reaction (27) maintaining an equilibrium activity of Na. The lining saturates early in the cell's life, but until this occurs, current efficiency is low. When a metal dissolves in a molten salt, it usually imparts electronic conductivity to the melt, thereby lowering current efficiency in electrolysis. Borisoglebsky et al.[66] found 1.5% electronic conductance in aluminum-saturated melts of industrial composition. These investigators suggested that a corresponding reduction in current efficiency would be expected. On the other hand, Dewing and Yoshida[67] found *no* electronic conductance, with a possible error of 1.5%.

Elements such as phosphorus, vanadium, and iron, which can be reduced at the cathode yet not alloy completely with the aluminum, may either escape in the anode gases, e.g. iron, or be reoxidized by $CO_2$ at the anode, e.g., phosphorus. In either case, current efficiency is lowered.

## 5. Energy Considerations

Power consumption depends upon cell voltage and current efficiency, CE, according to the relationship kWh/kg = $298.1(E_{cell})/\%CE$. Minimizing cell voltage and maximizing current efficiency are very important. Current efficiency was discussed in the last section. The total applied potential (cell voltage) is the sum of electrochemical and ohmic voltage drops as shown in Figure 7. The equilibrium potential, $E_3$, can be calculated from thermodynamic data:

$$\alpha\text{-}Al_2O_3(\text{soln}) + \tfrac{3}{2}C(c) \rightarrow 2Al(l) + \tfrac{3}{2}CO_2(g) \tag{32}$$

$$E_3 = (-\Delta G°_{(32)}/6F) - (RT/6F)\ln[(a_{Al}a_{CO_2}^{1.5})/a_{Al_2O_3}a_C^{1.5})] \tag{33}$$

where $\Delta G°_{(32)}$ is the free energy change for reaction (32). The activities of Al, C, and $CO_2$ can be assumed to be essentially unity and the activity of alumina, $a_{Al_2O_3} = (c_{Al_2O_3}/c_{Al_2O_3,\text{sat}})^{2.77}$, where $c_{Al_2O_3,\text{sat}}$ is alumina concentration at saturation. The exponent 2.77 is from Figure 2 of Rolin's treatise.[68] This exponent is assumed to be independent of temperature and bath additives. For an electrolyte at 970°C, with the alumina content 65% of saturation, $E_3 = -1.207$ V. Alternately, one can assume that alumina is decomposed by reaction (34).

$$\alpha\text{-}Al_2O_3(\text{soln}) \rightarrow 2Al(l) + \tfrac{3}{2}O_2(g) \tag{34}$$

requiring electrical energy $E_1 = -2.233$ V for the above conditions.

$$E_1 = [-\Delta G°_{(34)}/6F] + [RT/6F]\ln a_{Al_2O_3} \tag{35}$$

In both cases, the alumina is assumed to be $\alpha$, the phase in equilibrium with the bath, although the feed to the cell is largely $\gamma$-alumina. The oxygen deposited on the anode is depolarized by the carbon reaction

$$O_2(g) + C(c) \rightarrow CO_2(g) \tag{36}$$

Adding $E_2 = +1.026$ V gives the equilibrium potential $E_3 = -1.207$ V.

$$E_2 = -\Delta G°_{(36)}/4F \tag{37}$$

Adding to the equilibrium potential are several overvoltages described previously. The anode reaction overvoltage, $E_4$, is normally about $-0.5$ V referred to a $CO_2$ electrode. The anode concentration overvoltage, $E_5$, is insignificant above 2.5 wt% alumina but rises rapidly below 2% alumina, leading to the anode effect. Finally, there is the cathodic overvoltage, $E_6$, generally about $-0.1$ V in industrial cells. These voltages combined give $E_7$,

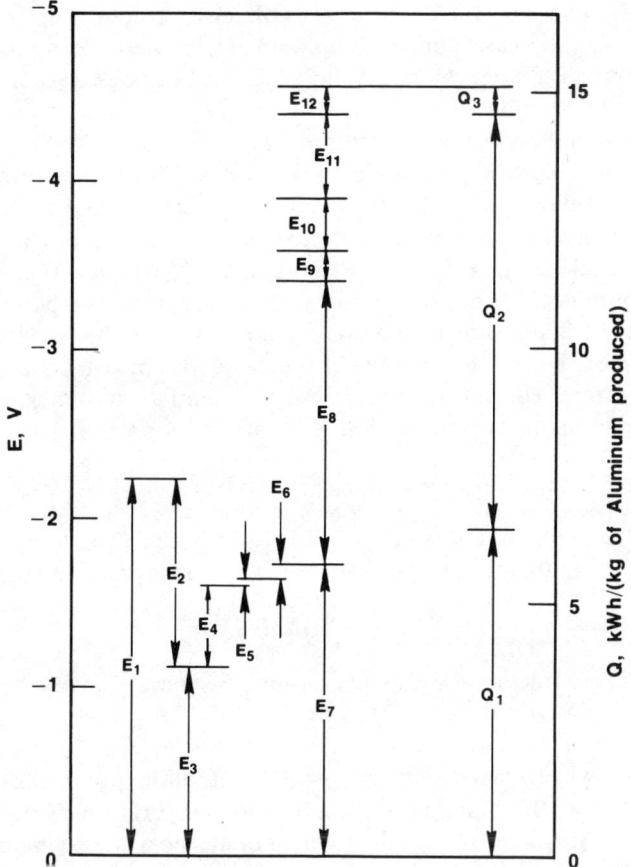

Figure 7. Energy requirements of the Hall–Heroult cell. $E_1$, decomposition of alumina; $E_2$, depolarization by carbon; $E_3$, equilibrium potential; $E_4$, anode reaction overvoltage; $E_5$, anodic concentration overvoltage; $E_6$, cathodic overvoltage; $E_7$, cemf of cell; $E_8$, electrolyte voltage drop; $E_9$, electrolyte bubble voltage; $E_{10}$, anode voltage drop; $E_{11}$, cathode voltage drop; $E_{12}$, external voltage drop; $Q_1$, enthalpy to produce aluminum + CO and $CO_2$; $Q_2$, cell heat losses; $Q_3$, bus heat loss.

the cemf of the cell, typically about $-1.8$ V. Linear extrapolation of cell volts versus current over the normal operating range to zero current as applied for process control gives a value of about $-1.65$ V.

The remainder of the applied potential overcomes ohmic losses. The hypothetical, bubble-free potential drop through the electrolyte is given by

$$E_8 = IL/\kappa A \tag{38}$$

where $I$ is current, $L$ the anode-to-cathode spacing (3–5 cm), $\kappa$ the electrical conductivity of the melt (2.0–2.4 $\Omega^{-1}$ cm$^{-1}$), and $A$ the effective bath area.

Typically, $E_8$ is about $-1.6$ V. The actual voltage drop across the interelectrode gap is increased by gas bubbles an amount, $E_9$, as described in the previous section. This will add typically $-0.15$ to $-0.2$ V. The anode ohmic voltage, $E_{10}$, represents the potential drop through the anode, its connection stubs and the contact between them. It ranges from $-0.25$ to $-0.30$ V with prebake anodes and $-0.45$ to $-0.55$ V with Soderberg anodes. The cathode voltage, $E_{11}$, includes the voltage loss in the lining, collector bars, and the contact between the two. It typically ranges from $-0.45$ to $-0.55$ V. There is also a voltage loss, $E_{12}$, in external electrical conductors ranging from $-0.1$ to $-0.2$ V. This does not contribute heat to the cell, but nevertheless represents a power loss.

The right-hand side of Figure 7 shows where the electrical energy represented by the voltages on the left is expended. The two scales correspond at 90% current efficiency. Reduction of alumina consumes carbon and produces aluminum, carbon dioxide, and carbon monoxide:

$$Al_2O_3(T_A) + (3 - X)C(T_A) \rightarrow 2Al(T_E) + XCO_2(T_E) + (3 - 2X)CO(T_E) \quad (39)$$

where $T_A$ is ambient temperature and $T_E$ is electrolyte temperature. $Q_1$ represents the enthalpy for reaction (39). The corresponding voltage $Q_{1E}$ is

$$Q_{1E} = (CE)\Delta H_{(39)}/6F \quad (40)$$

where CE is the fractional cathodic current efficiency.

Assuming a typical $CO/CO_2$ ratio of 0.265 (CE = 0.90) the energy, $Q_1$, required to reduce alumina ($\gamma$-$Al_2O_3$) with carbon at 970°C is 6.419 kWh/kg of aluminum, which corresponds to 1.940 V. This enthalpy includes the energy required to heat the alumina and carbon to temperature. Note that $Q_1$ is greater than $E_7$. A small part of the joule or ohmic heat is used in the reduction of alumina. $Q_2$ represents the heat that must be dissipated from the cell for it to operate isothermally. In this particular example, 8.12 kWh/kg aluminum must be dissipated. Heat generated external to the cell is represented by $Q_3 = 0.53$ kWh/kg aluminum. Dividing the productive energy $Q_1$ by the total electrical energy input $(Q_1 + Q_2 + Q_3)$ gives an electrical power efficiency of 42.6%, which is typical of a modern Hall–Heroult cell.

### Heat Balance

A proper heat balance is critical to the design of an aluminum smelting cell. Having obtained $Q_2$, the amount of heat available for heat losses, the cell may be designed. Since the anodes with their electrical connections and the cathode with its electrical connections both conduct heat from the cell and generate joule heat internally, either a finite element or finite difference solution of the combined electrical and heat flow is recommended.[69] In order to protect the cavity walls from erosion, wall insulation must be adjusted to extract an amount of heat that will maintain a ledge of frozen electrolyte of the desired thickness. These calculations have been described by Haupin[70] and Ikeuchi and Arita.[71]

## 6. Electromagnetic Effects

Magnetic fields are generated in the cell by electric current flow within the cell and external to the cell. Electric current flowing through the molten aluminum and the electrolyte interacts with magnetic fields present in these fluids to generate forces that cause movements of these fluids. These forces and flows can produce static pile-up of aluminum and waves. Proper location of conductors, magnetic shielding, and use of other magnetic improvement devices[72] require calculation of the electrical and magnetic fields, the resultant forces, and the movements of aluminum that these forces produce.[73,74] Undesirable movements of the molten aluminum and electrolyte are aggravated strongly by improper electrical current balance in the anodes and cathode structure.

## 7. Producing High-Purity Aluminum

Electrolytic refining is often used for the few applications requiring higher-purity aluminum than is produced by the conventional Hall–Heroult process just described. Also, impurities tend to increase in recycled metal. As the percent recycled aluminum increases some of the metal may require electrolytic refining. Hoopes in 1901, with the assistance of Frary and Edwards, developed a three-layer electrolytic refining process.[75] Aluminum is electrolytically transported from an anodic bottom layer of molten impure aluminum through an intermediate electrolyte layer to a cathode of pure aluminum floating on top. Copper is added to the bottom layer to increase its density. Aluminum and elements less noble than aluminum are preferentially extracted (oxidized) at the anode. Since the activity of aluminum in the electrolyte is maintained high, elements less noble than aluminum will not be reduced at the cathode but slowly accumulate in the electrolyte. The composition of the electrolyte is selected to have a density less than the alloy layer but greater than pure aluminum. Electrical connection is made to the alloy through carbon or graphite blocks. Graphite is used for the cathode connection. Hoopes used carbon walls to resist attack by a fluoride electrolyte. Short circuiting between the anode and cathode layers was prevented by maintaining a frozen ledge of electrolyte on the walls. The cell operated at 950–1000°C and produced 99.98% pure aluminum.

In 1932, The Pechiney Company in France developed an electrolyte consisting of chlorides and fluorides (Table 1) having a 720°C liquidus. The lower operating temperature (750°C) and less aggressive electrolyte allowed

Table 1
Electrolytic Purification Processes[a]

| Characteristic | Hoopes | Pechiney | AIAG Neuhausen |
|---|---|---|---|
| Cathode layer | | | |
|   Al purity | 99.98% | 99.99 + % | 99.99 + % |
|   Density, g/cm$^3$ | 2.29 | 2.30 | 2.30 |
| Electrolyte | | | |
|   Composition, % | | | |
|     NaF | 25–30 | 17 | 18 |
|     AlF$_3$ | 30–38 | 23 | 48 |
|     CaF$_2$ | | | 16 |
|     BaF$_2$ | 30–38 | | 18 |
|     BaCl$_2$ | | 60 | |
|     Al$_2$O$_3$ | 0.5–7.0 | | |
|   Density, g/cm$^3$ | 2.5 | 2.7 | 2.6 |
|   Resistivity, Ω cm | 0.3 | 0.75–0.85 | 1.1 |
| Anode layer | | | |
|   Composition, % | | | |
|     Al | 75 | 67 | 70 |
|     Cu | 25 | 33 | 30 |
|   Density, g/cm$^3$ | 2.8 | 3.14 | 3.05 |
| Operating characteristic | | | |
|   For a cell of amperage | 20,000 | 25,000 | 14,000 |
|   Volts | 5–7 | 6.9 | 5.5–5.3 |
|   Current density, A/cm$^2$ | 0.95 | 0.40 | 0.36 |
|   Current efficiency, % | 90–98 | 96–98 | 92–95 |
|   Operating temperature, °C | 950–1000 | 750 | 740 |

[a] References 75 and 76.

use of electrically nonconductive magnesia brick for the cell walls. This cell produced aluminum as high as 99.995% pure. In 1937 Aluminium-Industrie Aktiengesellschaft at Neuhausen patented a low-melting (720°C), all-fluoride electrolyte that also allowed use of a magnesia brick lining.

The segregation sump pictured in Figure 8 is an important design feature of modern electrolytic purification cells. Operating at least 30°C cooler than the main cell, it serves as the charging port for the cell. As impurities concentrate in the bottom layer, saturation is reached and crystals preferentially form in the cooler sump where they can be removed. This feature greatly extends the operating life of the cell and makes scrap recovery practical.

## 8. Electrolysis of Aluminum Chloride

Pure molten aluminum chloride does not ionize sufficiently to permit electrolysis but it can be electrolyzed from an alkali metal chloride melt to give

# ELECTROMETALLURGY OF ALUMINUM

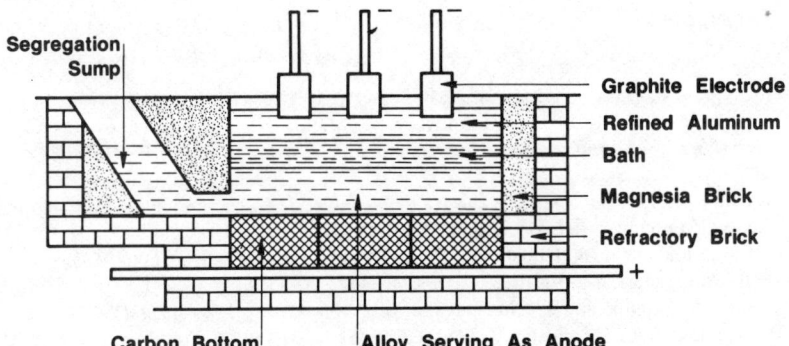

Figure 8. Cell for electrolytic refining of aluminum. [By permission of *Kirk–Othmer Encyclopedia of Chemical Technology*, Vol. 2, 3rd ed., Wiley, New York (1975).]

chlorine and either solid aluminum or molten aluminum, depending upon the operating temperature. The idea is not new. In 1854 Bunsen[77] and Deville[78] produced aluminum by electrolyzing aluminum chloride in sodium chloride. Numerous problems, however, prevented commercialization of the process until 1973 when Alcoa announced construction of a plant to produce aluminum based upon aluminum chloride electrolysis. Because the anode is not consumed, the chloride process permits design of a highly energy-efficient cell with bipolar electrodes. The cathode reaction is

$$AlCl_4^- + 3e^- \rightarrow Al + 4Cl^- \qquad (41)$$

while the anode reaction is

$$2Cl^- \rightarrow Cl_2 + 2e^- \qquad (42)$$

The reversible decomposition potential is about 1.95 V. The anodic overvoltage is negligible and there is slightly under 0.1 V cathodic overvoltage. Gas bubbles on the anode and in the electrolyte significantly increase the electrical resistance of the cell. Careful design of the anode configuration and adequate bath circulation is required to minimize this effect. The cell consumes about 9 kWh/kg of aluminum or one-third less electrical energy than the best Hall–Heroult cells.

## Acknowledgments

The authors wish to thank Dr. Allen Russell, Dr. Rudolf Keller, and Dr. Jomar Thonstad for their many helpful suggestions.

## References

1. G. Choudhuri, *J. Electrochem. Soc.* **120**, 381 (1973).
2. K. Grjotheim, C. Krohn, M. Malinovsky, K. Matiasovsky, and J. Thonstad, *Aluminium Electrolysis, The Chemistry of the Hall–Héroult Process*, Aluminium Verlag GmbH, Düsseldorf (1977).
3. M. Rolin, *La Physicochemie des Bains D'Electrolyse de L'Aluminium*, Institut National Des Sciences Appliquees de Lyon, Villeurbanne, France (1975).
4. G. J. Landon and A. R. Ubbelohde, *Proc. R. Soc. (London)* **240A**, 160 (1957).
5. W. B. Frank and L. M. Foster, *J. Phys. Chem.* **64**, 310 (1960).
6. V. Sato, A. Kojima, and T. Ejima, *J. Jpn. Inst. Met.* **41**(12), 1249–1256 (1977).
7. W. B. Frank and L. M. Foster, *J. Phys. Chem.* **61**, 1531 (1957).
8. A. Tuol and M. Rolin, *Electrochim. Acta* **17**, 2277–2291 (1972).
9. K. Grjotheim, *Contribution to the Theory of Aluminium Electrolysis*, Kgl. Norske Videnskabers Selskabs Skrifter, Nr 5, F. Bruns Bokhandel, Trondheim, Norway (1956).
10. J. Brynestad, K. Grjotheim, and S. Urnes, *Met. Ital.* **52**, 495 (1960).
11. M. Rolin, *Bull. Soc. Chim. France*, 671, 677, 681 (1960).
12. M. Rolin and M. Rey, *Bull. Soc. Chim. France*, 2785, 2791, 2794 (1966).
13. W. B. Frank and L. M. Foster, *J. Phys. Chem.* **64**, 95 (1960).
14. K. Matiasovsky, M. Paucirova, and M. Malinovsky, *Rev. Roum. Chim.* **17**, 801–808 (1972).
15. M. Paucirova, M. Malinovsky, and K. Matiasovsky, *Rev. Roum. Chim.* **17**, 807–817 (1972).
16. K. Matiasovsky, M. Paucirova, and M. Malinovsky, *Collect. Czech. Chem. Comm.* **37**(6), 1913–1916 (1972).
17. B. Gilbert, G. Mamantov, and G. M. Begun, *Inorg. Nucl. Chem. Lett.* **10**, 1123 (1974).
18. B. Gilbert, G. Mamantov, and G. M. Begun, *J. Chem. Phys.* **62**, 950 (1975).
19. W. Brockner, K. Torklep, and H. Øye, *Ber. Bunsenges Phys. Chem.* **83**, 12–19 (1979).
20. K. Ito and E. Nakamura, Structural Entities of Electrolytes and Electrode Reactions in Aluminum Electrolysis, *Sumitomo Light Met. Tech. J.* **17**(1-2), 61–82 (1976).
21. J. L. Holm, Thermodynamic properties of molten cryolite and other fluoride mixtures, dissertation, The University of Trondheim, Norway, 1971.
22. B. Gilbert, G. Mamantov, and G. M. Begun, *Inorg. Nucl. Chem. Lett.* **12**, 415 (1976).
23. H. Kvande, Thermodynamics of the system $NaF-AlF_3-Al_2O_3-Al$ studied by vapour pressure measurements, The University of Trondheim, Norway, 1979.
24. J. Thonstad and S. Rolseth, Proceedings of 3rd ICSOBA Conference September 1973, p. 657, Nice.
25. K. A. Bowman, Ph.D. dissertation, University of Tennessee, Knoxville, March 1977.
26. C. Kubik, K. Matiasovsky, M. Malinovsky, and J. Zeman, *Electrochim. Acta* **9**, 1521 (1964).
27. J. Thonstad, Model of diffusion layer for cathodic deposition of aluminum, in *Molten Salt Electrolysis in Metal Production*, Institute of Mining and Metallurgy, London (1977), pp. 1–6.
28. Unpublished results, Alcoa Laboratories.
29. H. Ginsberg and H. C. Wrigge, *Metall* **26**, 997 (1972).
30. D. Schlain, C. Kenahan, and J. H. Swift, U.S. Bureau of Mines Report No. 6265, 1963.
31. W. E. Haupin, *J. Electrochem. Soc.* **120**(1), 85 (1973).
32. K. Grjotheim, M. Malinovsky, K. Matiasovsky, A. Silny, and J. Thonstad, *Can. Met. Quart.* **11**, 295 (1972).
33. R. R. Revazyan, *Izv. Vyssh. Ucheb. Zav. Tsvet. Met.* **3**(1), 101 (1960).
34. J. Thonstad and E. Hove, *Can. J. Chem.* **42**, 1542 (1964).
35. G. Blyholder and H. Eyring, *J. Phys. Chem.* **61**, 682 (1957).
36. P. Drossbach, T. Hashino, P. Krahl, and W. Pfeiffer, *Chem. Ing. Tech.* **33**, 84 (1961).
37. P. Drossbach and T. Hashino, *J. Electrochem. Soc. Jpn.* **33**, 229 (1965).
38. J. Thonstad, *Electrochim. Acta* **15**, 1581 (1970).

39. K. Vetter, *Electrochemical Kinetics*, Academic, New York (1967).
40. R. E. De La Rue and C. W. Tobias, *J. Electrochem. Soc.* **106**, 827 (1959).
41. U. Landau, Case-Western Reserve University, private communication.
42. W. E. Haupin, *J. Met.* **23**(10), 46 (1971).
43. J. Thonstad, F. Nordmo, and J. K. Rødseth, *Electrochim. Acta* **19**, 761 (1974).
44. J. Thonstad, *Electrochim. Acta* **12**, 1219 (1967).
45. R. Piontelli, B. Mazza, and P. Pedeferri, *Metallurgia Ital.* **57**(2), 1, 51 (1965).
46. N. Watanabe, Anode effect in molten fluoride systems, in *Extended Abstracts of the 30th Meeting, International Society of Electrochemistry*, Trondheim, Norway, 1979, p. 102.
47. A. J. Calandra, C. M. Ferro, and C. E. Castellano, *Electrochim. Acta* **25**, 201 (1980).
48. C. Brunet and P. Mergoult, Electrode effects in cryolitic baths, in *Extended Abstracts of the 30th Meeting, International Society of Electrochemistry*, Trondheim, Norway, 1979, p. 167.
49. B. Marincek, U.S. Patent 3,562,135 (1971).
50. B. Marincek, U.S. Patent 3,692,645 (1972).
51. K. Yamada *et al.*, German Patent application 2,547,168 (1976).
52. K. Yamada *et al.*, British Patent 1,461,155 (1977).
53. H. Klein, U.S. Patent 3,718,550 (1973).
54. H. Alder, U.S. Patent 3,960,678 (1976).
55. H. Alder, U.S. Patent 3,930,967 (1976).
56. H. Alder, U.S. Patent 4,057,480 (1977).
57. W. E. Haupin, *J. Electrochem. Soc.* **107**, 232 (1960).
58. J. Thonstad, *Can. J. Chem.* **43**, 3429 (1965).
59. M. M. Vetyukov and V. B. Vinokurov, *Physical Chemistry and Electrochemistry of Molten Salts and Slags*, Naukova Dumka, Kiev (1969), p. 367.
60. K. Yoshida and E. W. Dewing, *Met. Trans.* **3**, 1817 (1972).
61. J. Thonstad and S. Rolseth, in *Proceedings of 4th ICSOBA Conference*, Vol. 2, October 1978, p. 437.
62. J. Gerlach, W. Schmidt, and H. Schmitt, *Erzmetall* **20**, 111 (1967).
63. K. B. Choudhuri, Dissertation, Technische Universitat Berlin, 1968.
64. J. Gerlach, U. Hennig, and R. Roedel, *Metallography* **29**, 267 (1975); R. Roedel, Dissertation, Technische Universitat Berlin, 1972.
65. D. Bratland, K. Grjotheim, C. Krohn, and K. Motzfeldt, *J. Met.* **19**, 13 (1967).
66. Yu. V. Borisoglebsky, M. M. Vetyukov, and S. Abuzeyd, *Tsvet. Met.* **1978**(7), 41–43.
67. E. W. Dewing and K. Yoshida, *Can. Met. Quart.* **15**(4), 299 (1976).
68. M. Rolin, Le Procede Heroult Etude De L'Electrolyse, Institut National Des Sciences Appliquees De Lyon, Villeurbanne, France, 1977.
69. P. C. Kohnke and J. A. Swanson, Thermo/electric finite elements, in ICCAD Conference on Numerical Methods in Electrical and Magnetic Field Problems, Santa Margarita Ligure, Italy.
70. W. E. Haupin, *J. Met.* **42**(7), 41–44 (1971).
71. H. Ikeuchi and Y. Arita, *Electrochim. Acta* **25**, 201 (1980).
72. R. F. Robl, *Light Met.* **1**, 97 (1976).
73. R. F. Robl, *Light Met.* **1**, 179 (1975).
74. P. Entner, J.-M. Blanc, and W. Schmidt-Hatting, Observations, measurement and calculations concerning magnetic effects in aluminum electrolysis cells, in *Proceedings of the 30th Meeting, International Society of Electrochemistry*, Trondheim, Norway, 1979.
75. T. G. Pearson, The Chemical Background of the Aluminium Industry, Royal Institute of Chemistry Monograph, No. 3, 1955.
76. L. Evans and W. B. C. Perrycoste, B.I.O.S. Final Report No. 1757, Item No. 21, Vereinigte Aluminium Werke, Erftwerk, Grevenbroich, March and July 1946.
77. R. W. Bunsen, *Poggendorf. Anal.* **97**, 648 (1854).
78. H. Sainte-Claire DeVille, *Ann. Chim. Phys.* **43**, 27 (1854).
79. P. A. Foster, Jr., *J. Am. Cer. Soc.* **58**(7-8), 288 (1975).

# 6
# Electrolytic Refining and Winning of Metals

## V. A. ETTEL and B. V. TILAK

## 1. Introduction

In electrolytic refining, the plates of crude metal are anodically dissolved in a suitable electrolyte, while "pure" metal is deposited on the adjacent cathodes. The electrorefining process was introduced about a hundred years ago to produce a substitute to fire-refined copper. Today, practically all of the world's copper production (~8,000,000 metric tons/year) is electrorefined, constituting by far the largest electrolytic refining industry. Much smaller, but also important, are the electrolytic refining industries producing lead, nickel, silver, and other minor metals.

The soluble anodes in nickel refining can also be cast from nickel matte containing ~20% S. This electrolytic process, although technically very similar to refining with a metallic anode, is not a refining process in the true sense and is sometimes called electrowinning with a soluble anode.[1]

A process closely related to electrorefining is electrowinning with insoluble anodes. In this process, the metal is dissolved chemically, e.g., by leaching calcined ore, etc. The pure metal is then "electrowon" using insoluble anodes, e.g., lead anodes, producing oxygen:

$$2H_2O \rightarrow O_2 + 4H^+ + 4e$$

---

**V. A. ETTEL** • J. Roy Gordon Research Laboratory, INCO Metals Company, Sheridan Park, Mississauga, Ontario L5K 129, Canada.   **B. V. TILAK** • Hooker Chemical Corporation, Research Center, Long Road. Grand Island, New York 14072.

Currently some 70% of the world's total zinc production (~5,000,000 metric tons/year) is made by electrowinning with insoluble anodes, and this percentage is increasing.

Although the anodic reactions are different, most other aspects of refining and electrowinning processes using aqueous electrolytes are similar and will be discussed in the next section, with prime emphasis on the technological details. Due to their specific nature, the processes employing fused salt electrolytes, such as aluminum electrowinning, are not included in this review.

## 2. Electrochemical Principles of Electrorefining and Electrowinning

### 2.1. Electrochemical Selectivity

In an ideal electrorefining process, the metal would dissolve "nearly reversibly" on the anode and plate "nearly reversibly" at the cathode. The impurities which are more electropositive (noble) than the electrorefined metal will not dissolve at this potential and will remain in the anode slime. The impurities which are more electronegative (common) than the refined metal will dissolve anodically but will not deposit at the cathode, hence the accumulation in the electrolyte.

As could be expected, this simple picture reflects reality reasonably well only with the metals which exhibit low dissolution and deposition overpotentials (Ag, Cu, Pb) and with impurities whose standard potentials are sufficiently different from that of the refined metal.

#### 2.1.1. Anodic Process

The selectivity of the anodic process is often helped by the refractory nature of some impurity-containing compounds present in the anode metal (Ag, Se, and Te are present in copper anodes as quite refractory selenides and tellurides of silver and copper). In spite of this, a very small amount of silver still tends to dissolve from copper anodes and plate onto cathodes. This partial dissolution is suppressed by maintaining ~30 mg/liter of $Cl^-$ ion in the copper refining electrolyte, so that $Ag^+$ concentration in the electrolyte will not exceed the value given by the solubility product of AgCl.

The "common" impurities which normally dissolve from the anode can also be present in the anodes in a poorly soluble form. (Nickel is often present in copper anodes as the refractory form of NiO, thus appearing at least partially in the anode slime.) In sulfate-ion-containing electrolytes lead is insolubilized as $PbSO_4$. One component which, contrary to the simple theory, is always found as a major component of anode slimes is the refined metal itself.

The typical behavior of anode impurities in various refining processes is summarized in Table 1. It can be seen that in Au, Ag, Cu, and Pb refining

Table 1
Disposition of Impurities in Some Refining Processes

| Process | PM[a] | Au | Ag | Se | Te | S | Cu | Sb | As | Bi | Ni | Co | Fe | Pb | Zn |
|---|---|---|---|---|---|---|---|---|---|---|---|---|---|---|---|
| Au refining | E[c] | — | S | — | — | — | E | — | — | — | — | — | — | S | — |
| Ag refining | Se[e] | S | — | — | S | — | E | — | — | S | — | — | — | sE[d] | — |
| Cu refining | S | S | S | S | S | S | — | S | sE | sE | sE | — | E | S | E |
| Ni refining | S | S | S | — | — | S | sE | — | E | — | — | E | E | Se | — |
| Pb refining | S | S | S | S | S | — | S | Se | S | S | E | E | E | — | E |

[a] PM = platinum group metals.
[b] S = slime.
[c] E = electrolyte.
[d] sE = mostly electrolyte.
[e] Se = mostly slime.

processes, the impurities more "noble" than the refined metal remain undissolved in the anode slimes. Consequently, they cannot plate and contaminate the cathodes. In nickel refining, on the other hand, some noble impurities partially dissolve in the electrolyte and will plate on the cathode if it is immersed in the same electrolyte. This is the main reason why nickel refining cells have diaphragms separating the catholyte from the anolyte.

In electrowinning processes with insoluble anodes, the possibility of eliminating noble impurities by selective dissolution is usually lost. This disadvantage can, however, be overcome by cementation using a powder of the metal to be refined, whereby the more noble impurities are precipitated from the "pregnant" electrowinning electrolyte, without adding any foreign ions. Cementation with zinc powder is used in all zinc electrowinning operations to remove Cu, Co, Ni, As, Sb, Cd, Ge, and etc. Because of its efficiency, cementation is also used to offset the poor selectivity of nickel anode dissolution with respect to copper. Anodic dissolution of nickel usually results in anolyte containing 0.5–1 g/liter Cu, which is easily reduced to 1 mg/liter by cementation with active nickel powder.

### 2.1.2. Cathodic Processes

The selectivity of the cathodic process with respect to the soluble (less noble) impurities in electrorefining and in electrowinning is better with "reversible" metals (Ag, Cu, Pb, Zn) than with nickel. Electrolytically refined copper usually contains <5 ppm Ni and <1 ppm As and Bi, although these impurities are present in most copper refining electrolytes in relatively high concentrations (~20 g/liter Ni, ~1 g/liter As, ~0.1 g/liter Bi). Similarly, high-purity silver can be produced from electrolytes containing ~1 g/liter Cu and Pd, and high-purity zinc is electrowon from electrolytes containing

~1 g/liter Mn. The least favorable situation arises with nickel because many soluble anode metal impurities (Cu, Fe, Co, Zn, Pb, As) readily co-deposit with nickel.

The most dramatic exception from the simple deposition selectivity rule is the behavior of zinc in nickel refining or electrowinning. It co-deposits readily into nickel, although the potential of the polarized nickel cathode is not negative enough to permit the deposition of zinc metal. The explanation probably lies in the reduced thermodynamic activity of Zn when dispersed in the matrix of nickel.[2]

Poor selectivity of the cathodic process with respect to some impurities can also result from excessive cathode polarization. Although arsenic does not usually co-deposit on copper cathodes, it could, if concentration polarization becomes high because of low copper concentration in the electrolyte or because of high current density.

The most frequent cause of cathode contamination, however, is a simple mechanical occlusion of anode slime (especially "floating slime") and electrolyte when the cathode is rough and porous. It is believed that most of the impurities found in copper cathodes have been incorporated by this mechanism. It is certainly the reason why electrowon copper often contains >20 ppm of Pb.

## 2.2. Addition Agents

Metals with high plating overpotential, i.e., nickel, can be deposited in compact, microcrystalline form without any additives because the high overpotential facilitates nucleation. Metals with low plating overpotential, such as, Ag, Pb, and to some extent Cu, tend to produce coarse, crystalline deposits with roughness increasing with increased thickness. This is due to the relatively difficult nucleation process at low polarization which causes the new metal to grow preferentially on the few existing nuclei. Such type of deposition is sometimes acceptable, as in the Thum cell for the refining of silver, where loose crystals of the electrodeposited metal are collected on a horizontal cathode. In most refining cells with vertical electrodes, it is important, however, that the deposits grow thick without becoming rough, dendritic, and porous. This is achieved by addition of various inhibitors for the cathodic reaction to the electrolyte. These inhibitors, usually called addition or leveling agents, are adsorbed on the cathode, thus increasing the overpotential of the cathodic reaction and, therefore, the ease of nucleation. A preferential adsorption on the most active growth sites also helps to inhibit the growth of large crystallites which have the tendency to become fast-growing dendrites.

### 2.2.1. Copper Refining

In copper refining the most important addition agent is animal glue which, in some form or the other, is used by most refineries. At typical addition rates

(30–100 g/ton of cathode weight), it increases cathode polarization by ~30–50 mV. However, glue is unstable in the strongly acidic and hot copper-refining electrolyte and undergoes gradual hydrolysis, which eventually destroys its beneficial effects. According to Andersen et al.[3] the activity of glue in the refining electrolyte is completely destroyed by heating at 95°C for 40 min. It is also lost by incorporation into the deposit and by adsorption onto anode slimes.

Another addition agent used by all copper refineries is chloride ion, which is maintained at a concentration between 10 and 40 mg/liter in the electrolyte. It also increases cathode polarization,[4] but above ~40 mg/liter, the deposit quality rapidly deteriorates,[5] probably due to the precipitation of CuCl.

Most copper refineries use at least one more addition agent selected from the following group: (1) thiourea, (2) lignin-related additives (Goulac, Bindarene or Lignone, Orzane A), (3) Avitone (sulfonated petroleum product), (4) safranin (dye), and (5) casein.

Thiourea is perhaps the most widely used addition agent from this group. It increases polarization through formation of a sulfide film on the cathode and results in incorporation of ~10 ppm S into the copper cathodes.

### 2.2.2. Lead Refining

The traditional addition agents used in lead refining in hydrofluosilicic acid electrolyte (Betts process) are lignin sulfonate and glue. In one refinery, the glue has recently been replaced with a natural extract "aloes" which affected a dramatic improvement in the quality of the deposits. It is known that $o$-quinones enhance the cathodic polarization of lead in this electrolyte, and it is therefore believed that a complex between $o$-quinone (present in aloes) and lead is causing the effect.[6]

### 2.2.3. Electrowinning

Addition agents used in electrowinning zinc include sodium silicate, gum arabic, glue, cresylic acid, and a soya bean extract. Copper electrowinning operations often use still another natural colloid—derivatives of guar gum. (See Bockris and Razumney,[7] for a discussion on the role of inhibitors during metal deposition.)

## 2.3. Mass Transport in Refining and Electrowinning Cells†

### 2.3.1. Natural Convection in Refining Cells

In refining cells with vertical electrodes, the transport of metal ions from the anode surface into the bulk electrolyte and from there onto the cathode surface is provided by a combination of diffusion and natural convection. That

† See Chapters 1 and 3 of Vol. 3 for a more comprehensive treatment of mass transport phenomena.

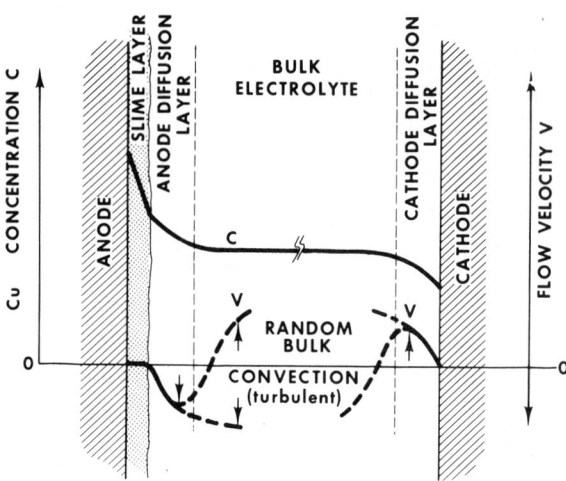

Figure 1. Natural convection and concentration profiles in copper refining.

is, the concentration gradient near the electrode causes the electrolyte (near the electrode) to have a "surface" concentration of the metal salt different from the bulk. Consequently, this electrolyte will also have a different specific gravity than the bulk, i.e., it will be more dense in the diffusion layer near the anode and less dense near the cathode, relative to the bulk electrolyte. The specific gravity differences induce a slow flow within these diffusion layers— downward flow at the anode and upward flow at the cathode. The diffusion gradient profiles have the usual shape (see Figure 1), and the thickness of these diffusion layers is determined by the flow velocities in these convective layers.

At low current densities and/or at low concentration of electrolyte, the density gradient is small and the convective flow is slow and laminar. At conditions prevailing in the refining cells, the density gradient is, however, so large that the convective flow is turbulent.[8]

The flow of depleted electrolyte on the cathode tends to segregate into parallel, vertical microstreams separated by a distance of ~0.5–2 mm. This improves the access of fresh electrolyte to the electrode surface between these microstreams, resulting in a vertical region with efficient mass transfer relative to the vertical regions which are sources of the microstreams. The periodically changing mass transfer efficiency across the cathode face can cause a nonuniform growth rate such that the cathode becomes striated.[9]

Mass transfer by natural convection with turbulent flow in the diffusion layer can be described by a semiempirical correlation of dimensionless numbers[8]:

$$Sh = 0.31 \, (Sc \, Gr)^{0.28}$$

where

$Sh = jh/nF\,\Delta C D = h/\delta_N$

$Sc = \nu/D$

$Gr = h^3 g\,\Delta\rho/\nu^2 \rho_B = h^3 g\alpha\,\Delta C/\nu^2$

Sh = Sherwood number (a measure of the mass transfer rate)

Sc = Schmidt number (describes the diffusion process)

Gr = Grashof number (describes the process of natural convection)

$j$ = current density

$h$ = electrode height

$\Delta C$ = concentration difference

$\delta_N$ = thickness of Nernst diffusion layer

$\nu$ = kinematic viscosity

$D$ = diffusion coefficient

$g$ = acceleration of gravity

$\rho_B$ = density of bulk electrolyte

$\Delta\rho$ = difference of electrolyte density between the bulk and the layer adjacent to the electrode

$\alpha$ = densification coefficient

From this expression, it can be seen that the mass transfer rate is proportional to $\Delta C^{1.28}$, which means that if the Nernst diffusion layer $\delta_N$ is defined by

$$\delta_N = \frac{nFD\,\Delta C}{j}$$

it is not independent of $\Delta C$, i.e., of current density, as originally proposed by Nernst. Increasing current density, therefore, causes the convective flow velocity to increase and the diffusion layer $\delta_N$ to become thinner.

Under normal operating conditions, for example, in copper refining, the thickness of the diffusion layer is ~0.3 mm.[10] This implies that the theoretical limiting current density is two to three times greater than the usual operating current density of 200–250 A/m².

Mass transfer near the anode is also driven by natural convection, but the flow in the diffusion layer is directed downwards. The mass transfer near the anode is, however, impaired by the layer of slime which is usually present on the surface of dissolving anodes. Since convection is not possible inside a porous layer, the mass transport across this layer is by diffusion only. The steady state thickness of this layer and its porosity depend on the composition of the anode metal, and substantial variation of these factors has been reported with copper refining anodes.[11] The concentration gradients across the natural convection layer and across the slime layer can become so high that the electrolyte inside the slime layer becomes supersaturated and crystallizes. This phenomenon is known in copper refining as anode passivity. To reduce the incidence of anode passivation, the electrolyte should be kept as hot as possible, and the concentration of electrolyte components ($CuSO_4$, $NiSO_4$, and $H_2SO_4$) should be as low as practical. Cu concentrations of ~40 g/liter (used by most refineries) is a compromise to give satisfactory mass transfer at both electrodes. Anodes with high content of silver are particularly prone to passivation.

The critical role of the slime layer on the anode mass transfer process is illustrated by theoretical calculations predicting that, in the absence of the slime layer, copper refining anodes should not passivate below 1400 A/m$^2$.[12] This conclusion is also supported by an experimental finding that slime-forming copper anodes are passivated at current densities lower (by a factor of 3) than anodes made of pure copper.[11]

The natural convective flows in copper-refining cells "pump" the depleted electrolyte up the electrodes to the surface of the cell and the enriched electrolyte down the anodes to the bottom. This not only leads to electrolyte stratification but also to an intensive turbulence in the electrolyte between the electrodes, as has been demonstrated using tracer techniques.[13]

The artificial electrolyte circulation used in refining cells does not appreciably enhance the mass transfer caused by natural convection, even at very high circulation rates. It can be calculated from the theory of mass transfer in forced flow systems that the electrolyte flow velocity between electrodes in a copper refining cell would have to be $\gg$3 cm/sec to increase the mass transfer rates near the electrodes. By comparison, the fastest electrolyte circulation velocity used in the copper-refining industry is only 0.26 cm/sec (at Onahama #3 tankhouse). Electrolyte circulation is, however, necessary to maintain a steady supply of addition agents to control electrolyte temperature and to prevent stratification.

### 2.3.2. Enhanced Convection in Electrowinning Cells

In electrowinning cells without diaphragms (electrowinning of zinc, copper, or cobalt), mass transfer near the upper sections of the cathode is enhanced by evolution of oxygen ~2–3 times compared to the bottom parts of the cathodes where pure natural convection prevails.[10]

In zinc electrowinning, ~10% of the current is wasted due to the unavoidable reduction of H$^+$ ions. The growing H$_2$ bubbles slowly displace electrolyte from the diffusion layer. When the bubble separates from the cathode surface, the void is quickly filled with fresh electrolyte from the bulk which amounts to an efficient enhancement of the mass transfer. Using available data on mass transfer enhancement of cathodic processes in the region of hydrogen evolution,[14,15] one can estimate the thickness of the cathodic diffusion layer in zinc electrowinning to be ~0.1 mm. This is a factor of 3 less than would be obtained with pure natural convection and explains why compact zinc can be electrowon at much higher current densities than Cu or Ni.

## 3. Technological Principles of Refining and Electrowinning

The "two-dimensional" character of these processes and the relatively slow mass transfer rate present rather unique technological problems, not encountered in other process industries. For example, an average copper refinery producing 500 tons of copper per day needs ~0.2 km$^2$ of total electrode area. This corresponds to about 50,000 anodes and 50,000 cathodes suspended in about 1500 tanks occupying a total floor area of about 6000 m$^2$.

A similar tankhouse will be the heart of any electrorefining and electrowinning operation, regardless of the metal produced. The unique technical problems involved result from the fact that for each of the many cells provisions must be made for (1) continuous supply of fresh electrolyte, (2) continuous withdrawal of the "spent" electrolyte, (3) supply of electrical energy to cathodes and anodes, (4) periodic supply of "new" cathodes, and (5) periodic removal of "finished" cathodes. The electrorefining cells must also have (6) periodic supply of soluble anodes, (7) periodic removal of spent anodes (anode scrap), and (8) periodic removal of anode slime.

The technological principles used to achieve these requirements in various refining and electrowinning operations are the same (or similar) and will therefore be described in this section functionally, rather than by individual industry.

### 3.1. Soluble Anodes

Soluble anodes cast from the impure metal are usually 80–100 cm wide, 90–110 cm long, and 3–6 cm thick, depending on the anode life cycle and operating current density. Some typical anode life cycles are presented in Table 2.

Most often the anodes are cast with suspending lugs (see Figure 2A), although other shapes are also used, particularly in the nickel industry.

A typical anode-casting machine is a rotating table with ~24 horizontal molds placed around its perimeter. The wheel slowly revolves, bringing the molds through a spray of mold wash, allowing them to be filled with molten metal, and then passing through a water-sprayed cooling section. The solidified

Table 2
Some Typical Anode Life Cycles

| Metal | Anode life, days | Anode thickness, cm | Number of cathodes from one anode |
|---|---|---|---|
| Copper | 28 | 4–5 | 2 |
| Nickel | 30–32 | 4.5–5.5 | 3–8 |
| Lead | 8–14 | 3–4 | 2 |

anodes are partially lifted in the mold by lifting pins located in the bottom of the mold and subsequently removed by a pickup device.

Manual control of mold filling with metal results in anodes of varying weight. This is overcome by the newly developed automatic mold-filling systems which fill the molds with a preweighed amount of metal. These systems were developed by the copper refineries of Mitsubishi, Outokumpu, and others. The Outokumpu system (see Figures 3 and 4) has become particularly popular since it has the capability of filling two molds at the same time, thus giving a high casting rate (100 tons/hr).[16] This corresponds to 200 anodes per hour. The molds for casting copper anodes are made from anode copper and usually last ~800 casting cycles. Molds for casting nickel anodes are also made of copper and are equipped with intensive water-cooling capability. Nickel sulfide anodes are formed in cast iron molds.

Cast anodes often have imperfections on the upper set surface, i.e., blisters, fins, and edge rims, which can cause shorting with the cathodes.

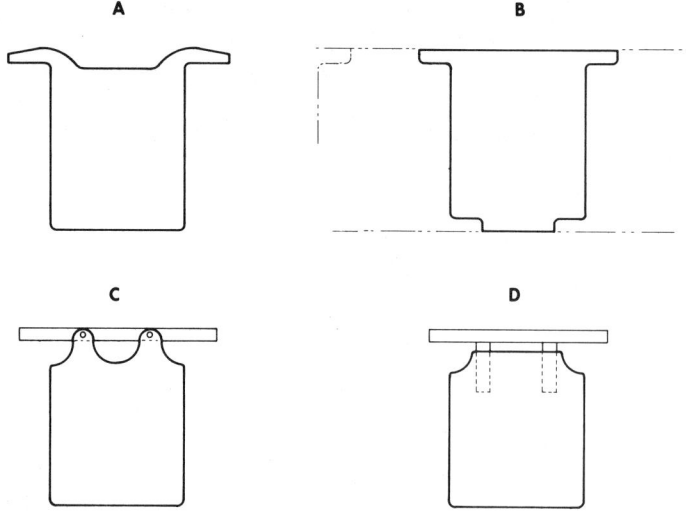

Figure 2. Shapes of soluble anodes. (A) Classical anode (Cu, Ni, Pb), (B) continuously cast anode (Onahama Cu anode), (C) bolted cross bar anode (Ni), (D) cast-in crossbar anode (Ni sulfide).

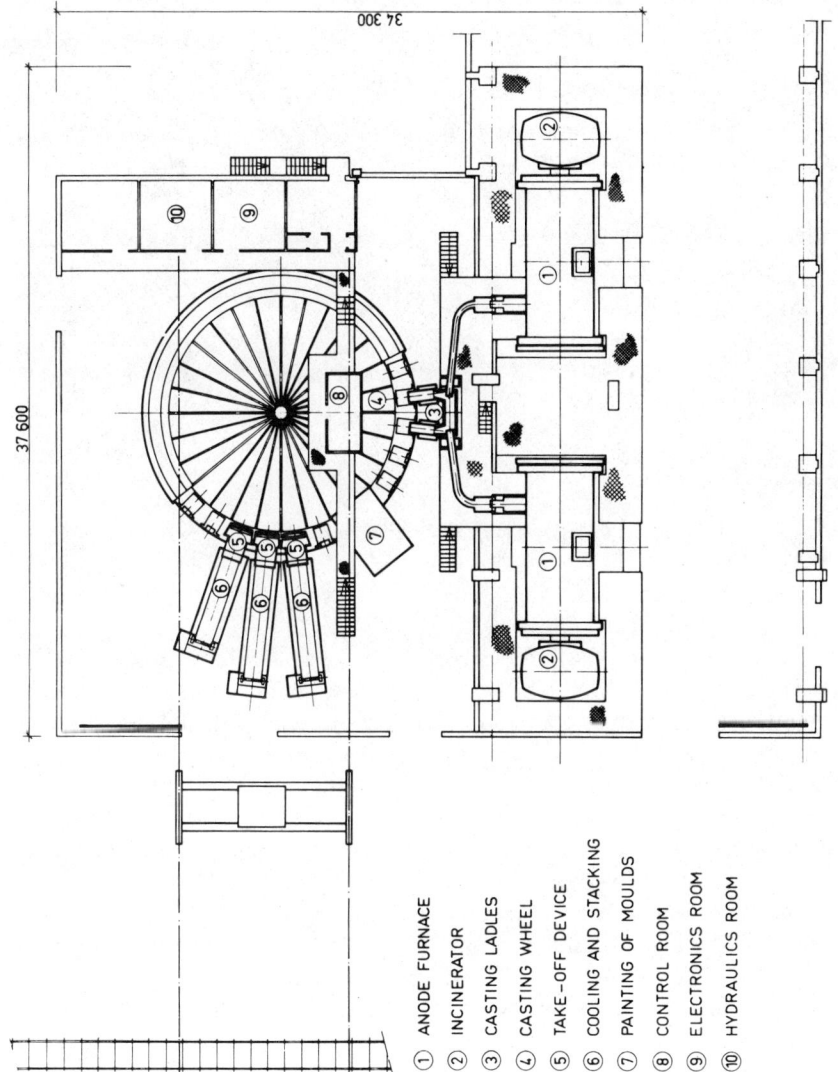

Figure 3. Outokumpu automated anode casting system (with permission from Outokumpu Oy).

Figure 4. View of the metal metering ladles at Outokumpu Oy (with permission from Outokumpu Oy).

Therefore, the cast anodes are usually inspected and large imperfections manually corrected by chipping. Several Japanese refineries use large presses to flatten the set surface of copper anodes. The press also straightens the lugs and flattens the contact area. Other refineries use an automatic milling machine to produce clean reproducible contact areas on the lugs.

The newest development in anode fabrication is the use of the Hazelett continuous strip casting machine. This approach was pioneered by the BCR refinery in Britain and perfected by the Japanese refinery Onahama.[17] The Onahama continuous casting system involves casting a strip ~106 cm wide and 1.3 cm thick, which is subsequently sheared into individual anodes (see Figures 2 and 5) of uniform thickness and with smooth surfaces.

In the course of the refining process, the anodes cannot be dissolved completely, and between 10% and 30% of the anode weight remains at the end of the cycle as anode scrap. The scrap is recycled to the anode furnace.

### 3.2. Insoluble Anodes for Electrowinning

Anodes for electrowinning metals from sulfate electrolytes are usually made of various lead alloys. Zinc-electrowinning anodes are made from a lead–silver alloy usually containing ~0.5% Ag. This alloy is chosen since it promotes formation of adherent $MnO_2$ coating on the anode from $Mn^{2+}$ ions added to the electrolyte and thus allows the production of zinc cathodes with only 20–40 ppm lead.

Anodes used in electrowinning copper are usually made from a lead–antimony alloy (~5% Sb). Although lead ions will not co-deposit with copper at the cathodic potential at which copper is deposited, electrowon copper cathodes are usually contaminated with 10–50 ppm Pb which is incorporated into the deposits as $PbSO_4$ particles which have been released from the corroding anodes. Anode corrosion in copper electrowinning electrolytes is reduced significantly by adding cobalt to the electrolyte. This lowers the operating potential of the oxygen producing anode[18] by ~50 to 100 mV, depending on the concentration of cobalt. Low anode corrosion rates are also obtained with a Pb–Ca alloy containing ~0.05% Ca,[19] which is finding increased use in copper electrowinning operations.

Entrapment of the anode produced $PbSO_4$ slime can also be suppressed when smooth cathodes are produced. Under these conditions it is possible to produce copper cathodes with <5 ppm Pb.[18] Nickel electrowinning anodes are often made from pure lead.[20]

Metal silicide anodes (Chilex anodes) were used in the past for electrowinning copper, and similar anodes are still used in some cobalt electrowinning operations. The low conductivity and brittleness have prevented wider use of these anodes.

Coated titanium anodes used extensively in the chlorine industry (e.g., DSA®, Electrode Corp.) have not yet been employed widely in electrowinning

Figure 5. Hazelett caster for continuous anode casting (with permission from Onahama Smelting and Refining Co., Ltd.).

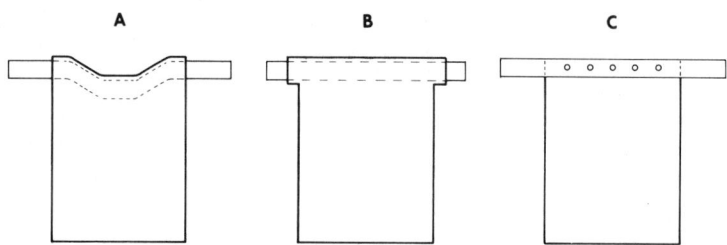

Figure 6. Typical shapes of insoluble anodes. (A) Cu electrowinning anode, (B) Zn electrowinning anode, (C) Ni electrowinning anode.

operations, mainly because of the higher cost of the oxygen-evolving DSA variety. Small quantities of cobalt and nickel are produced by electrowinning from the chloride electrolyte, where the anode material is graphite or the chlorine-evolving DSA.

Typical shapes of electrowinning anodes are shown in Figure 6.

### 3.3. Cathodes

Two types of cathodes are used for electrodeposition of metals: (1) reusable cathode blanks, and (2) sacrificial starter sheets.

### 3.3.1. Reusable Cathode Blanks

Reusable blanks are used mainly in the zinc and cobalt electrowinning industry. Zinc electrowinning blanks are rigid sheets of pure Al permanently attached (by welding) to aluminum suspension bars and masked on the edges with plastic strips. The "rabbit ears" (see Figure 7A) are used for blank handling. Zinc is electrodeposited on the blanks for several days. The blanks

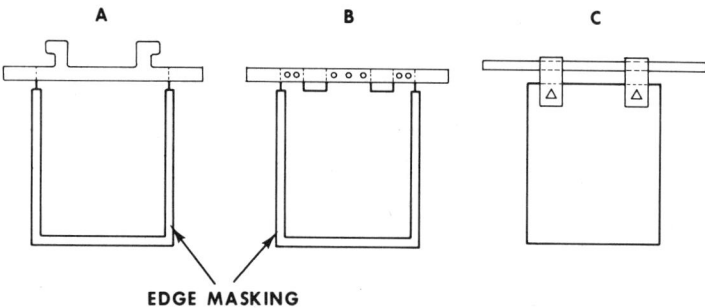

Figure 7. Cathodes for electrodeposition of metals. (A) Reusable Al Blank for Zn Electrowinning, (B) Ti mother blank for preparing copper starter sheets, (C) Cu refining starter sheet with inserted suspension bar.

are then removed from the cell and transported to manual stripping stands or to automatic stripping machines where zinc deposits are removed. The stripped blanks are sometimes cleaned by brushing before placement in the cells. There are several types of stripping machines used in the zinc industry.[21] Most of these are based on knife edge devices traveling down along the blank surface (or, conversely, the blank traveling up while the knife edges are stationary). In some instances, the deposits are first loosened at the top by an automatic hammering device.

Cobalt electrowinning blanks are made of stainless steel, and the brittle cobalt deposits are removed by crushing with hammers. Reusable titanium blanks are also used in some small-scale copper refining and electrowinning operations.

A unique type of cathode blank is used by INCO to produce buttons of sulfur-depolarized nickel (SD rounds), used in the plating industry. The surface of these stainless steel blanks is masked with a proprietary dielectric coating, except for a pattern of small circular areas. Metal buttons deposited on these areas are periodically separated by an automatic mechanical device, and the stripped blanks are returned to the refining cells.

### 3.3.2. Starter Sheets

Starter sheets are used in most other electrowinning and refining industries. In copper refining, the starter sheets are prepared by deposition for ~24 hr on copper, stainless steel, or titanium blanks (see Figure 7B) in a separate section of the tankhouse called the stripper section. To be able to subsequently grow good copper cathodes, the starter sheets must be smooth and soft to permit straightening. The stripper sections usually have separate electrolyte circuits to permit independent control of electrolyte purity and addition agent feed rate. Stripper sections are often operated at a current density lower than the rest of the refinery (commercial sections). The blanks are either masked on the edges with cemented plastic strips or have a groove along the edges where the deposit may be easily broken. Starter sheets are stripped from the blanks manually or by various automatic stripping machines developed over the last ten years by Japanese refineries and by Outokumpu (see Figure 8). To permit easy stripping of starter sheets, the stripped blanks are wetted with a solution of soap, with a water emulsion of mineral oil, or with a proprietary parting agent developed by Noranda.[22]

Stacks of stripped sheets are then processed through a starting sheet machine, where the sheets are straightened, trimmed (not practiced in all refineries), suspension loops (with preinserted suspension bars) attached by punching, and the sheets embossed with a shallow pattern to give them mechanical rigidity. Because of the fear of slime sedimentation on the horizontal parts of the embossed pattern, some refineries prefer a rolled vertical pattern, while other refineries avoid both. Some refineries have

1. Mother blank
2. Starting sheet
3. Stripping conveyor
4. Stripping blade
5. Stripping jaw
6. Stripping arm
7. Guide beam
8. Pile straightener
9. Adjustable hydraulic table
10. Pallet
11. Control desk

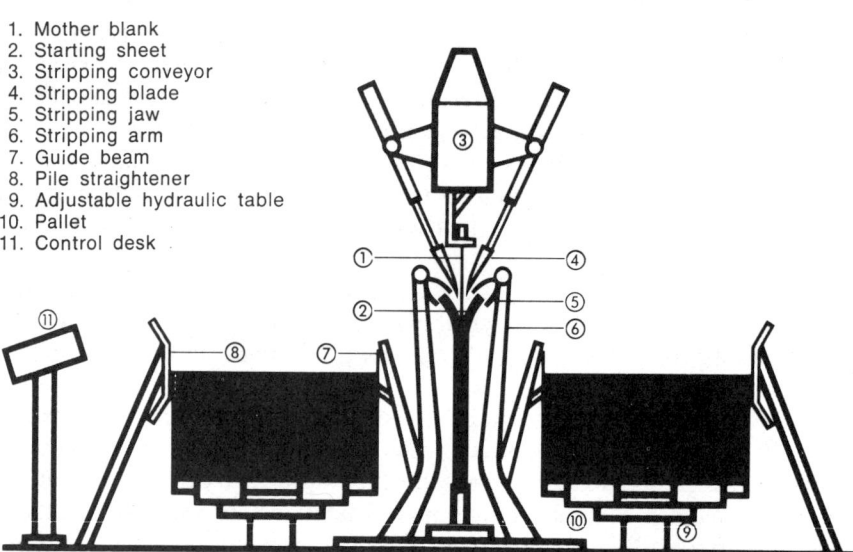

Figure 8. Outokumpu starter sheet stripping machine (with permission from Outokumpu Oy).

developed their own starting sheet machines, but the majority probably use the commercially available "Wennberg" machine from Sweden, shown in Figure 9.

To keep the cathodes straight,[23] the Phelps Dodge refinery removes two-day-old cathodes from the commercial cells and straightens them on a special mobile press.

Figure 9. Wennberg starter sheet preparation machine (with permission from INCO Metals Company).

Identical starter sheets are also used in copper electrowinning tankhouses. In fact, many smaller copper electrowinning operations purchase their starter sheets from the local refineries.

Nickel refining and electrowinning starter sheets are grown on masked stainless steel blanks, usually for two days. Starter sheets used in refining lead are conveniently prepared by continuous casting, using a water-cooled drum. The endless sheet of lead ~1 mm thick is automatically cut into individual sheets which are further processed into finished cathodes with inserted suspension bars.[24]

Dimensions of refining and electrowinning cathodes are always somewhat larger than anode dimensions so that current density at cathode edges is not excessive. In most newer tankhouses, the cathodes are 90–100 cm wide and 95–100 cm long, regardless of the metal produced or the type of operation. Some older installations still employ smaller cathodes (~70 cm wide, ~90 cm long). The latest trend, however, is towards still larger cathodes to reduce operating costs. The "jumbo" cathodes used in some zinc electrowinning[21] and lead refining operations[24] have a total surface area ~2.5 m² each.

### 3.4. Electrolytic Cells

Electrolytic cells are rectangular tanks, usually made of concrete and lined with lead or another suitable inert material. Lead or rigid PVC is the preferred lining material in copper refineries, while some copper and zinc electrowinning operations use premolded drop-in paraliners[25] made of thermoplastic materials (e.g., PVC), manufactured by the Barber-Webb Co. in California. Preformed FRP liners are used in nickel refining.

Individual tanks are arranged side-by-side in units (banks) of 10–20 tanks (see Figures 10 and 12). Only nickel refining and electrowinning tanks are usually built as side-by-side twin tanks in order to facilitate distribution of

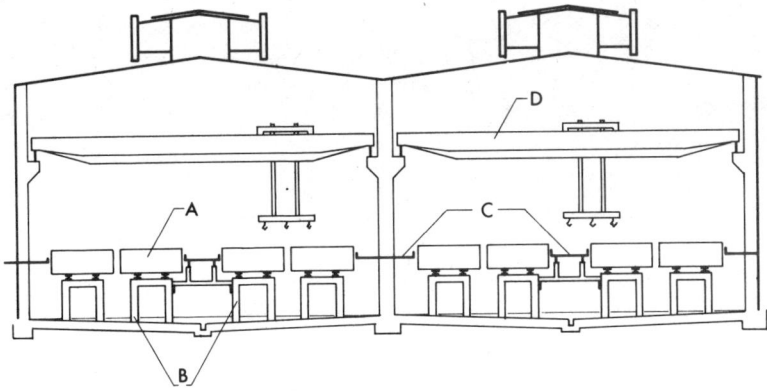

Figure 10. Tank arrangement in a copper refinery. A, tanks; B, support pillars; C, floor; D, crane.

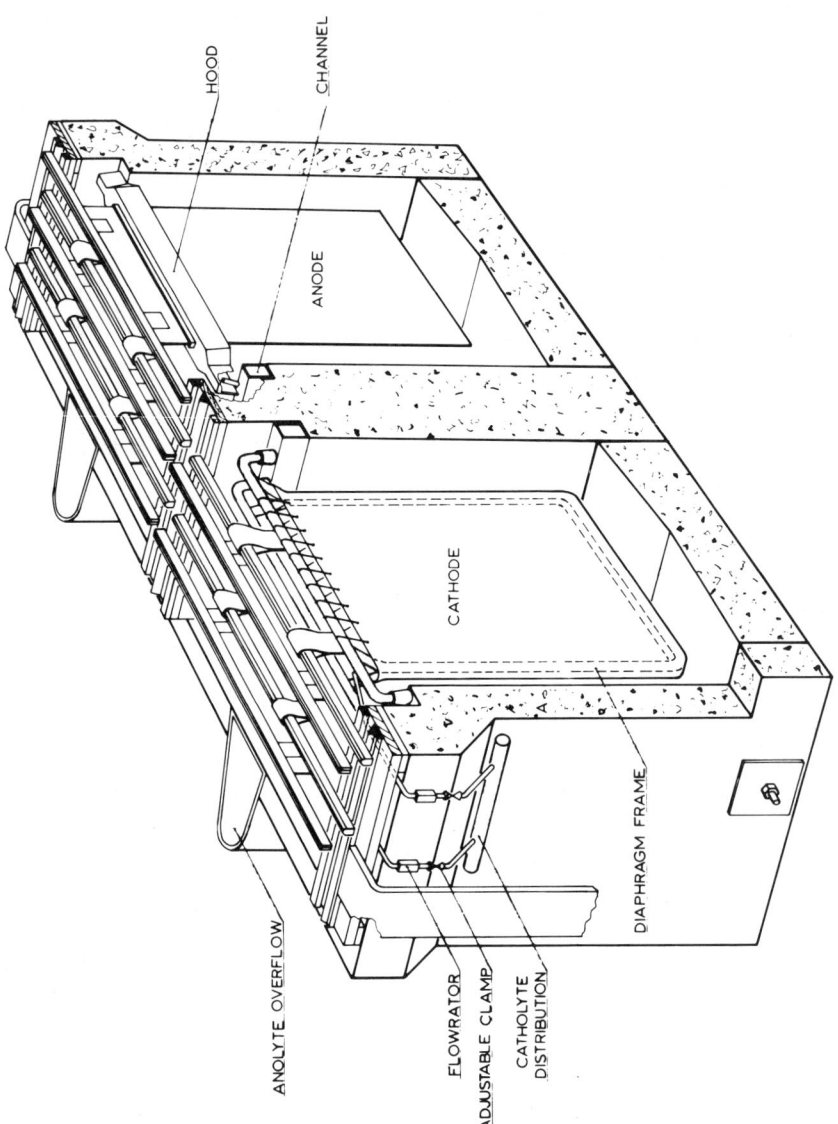

Figure 11. Modified Hybinette cell used in nickel electrowinning. The Hybinette cell used in Ni refining is quite similar (with permission from Outokumpu Oy).

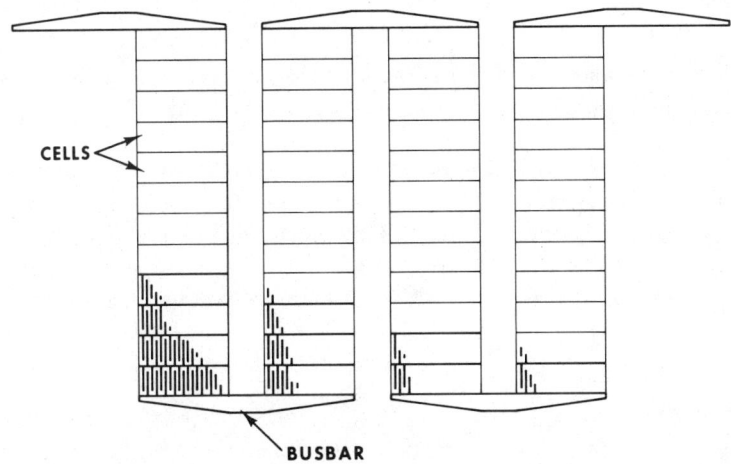

Figure 12. Electrical connection of tanks in series. Electrodes in one cell are connected in parallel (multiple system).

electrolyte into individual cathode compartments (see Figure 11). Tank internal dimensions are such that a 5–10-cm clearance exists between the cathodes and the vertical walls of the tank. The clearance between the cathodes and the cell bottom is much larger (25 cm or more), to allow sufficient room for settling and accumulation of anode slime. Tank length varies, depending on the number of anodes in one tank (20–50) and their spacing (9–15 cm). Nickel refining and electrowinning tanks are fitted with removable cathode frames with attached diaphragms (see Figure 11). Tanks are usually built on pillars, and the "basement" is used for electrolyte circulation and slime collection systems.

A new concept in tank design was used in #3 tankhouse of the Onahama refinery (Mitsubishi concern). The bank of adjacent tanks is replaced with one large tank accommodating 20 rows of 50 anodes each. Pillar-supported current distribution bars are used instead of the intermediate walls, thus permitting lateral electrolyte circulation (flow parallel with electrode faces). These "swimming pool" tanks measure $4 \times 28$ m and are built of PVC-lined steel. This design helped to reduce the capital outlay for tankhouse construction by ~30% compared to a conventional design.[17]

A disadvantage of this design is a small loss of current efficiency due to stray currents, as the individual rows (cells) within the large tank are connected electrically in series.

## 3.5. Electrical Circuits

In modern installations, the direct current is produced with silicon-diode- or thyristor-equipped rectifiers.[26] The latter is a solid state equivalent of thyratron and is used when periodic current reversal is needed.

### 3.5.1. Busbar Systems

Copper busbars carry the current to the cells which are the ends of the electrical circuit. To minimize the length of busbar, the cells in one bank are electrically connected in series, and the neighboring banks are connected in a zig-zag fashion as shown in Figure 12. The zig-zag connection of banks makes it possible to short-circuit and bypass one section of a tankhouse. Large tankhouses often have several independent electrical circuits. This is especially important in tankhouses with high cell voltage, for example, zinc electrowinning tankhouses, in order to limit the maximum voltage against the ground.

### 3.5.2. Contact Systems

Various contact systems are used to connect the cells in series (see Figure 13). However, the previously popular Whitehead system has been virtually abandoned because it complicates electrode handling. The Walker system (connecting the anodes of one cell with the cathodes of the adjoining cell through an intermediate triangular contact bar) is the most widely used one at present. Some refineries use a modified Walker contact system with a flat or twin triangular contact bar which further facilitates electrode handling. This type of contact was further improved at the Japanese refinery Tamano[27] by incorporating a contact wetting device into the twin contact bar. Wet contacts remain clean and give a low contact resistance. Consequently, a very uniform distribution of current between individual electrodes is achieved. Improving anode contacts by contact area milling was described in Section 3.1.

### 3.5.3. Current Density Distribution

Ideally, all cathodes should operate with the same and uniform current density. In practice, however, neither is fully achieved. Because of contact differences and irregular interelectrode gap, the individual cathodes, which are electrically connected in parallel, never draw the same current. These differences are particularly large in copper refining where cathodes operating at 50% higher current than the average value are not uncommon.[28]

Since neither of the electrodes is ideally flat nor vertical, there is also an appreciable variation of local current density over one cathode surface. Both effects are undesirable, as cathode areas operating with excessive current density will develop roughness, porosity, and shorts.

The means to reduce this problem include high-quality anodes and cathodes, uniform spacing, good "plumbing" (vertically), clean contacts, and (to an appreciable degree) also a suitable electrolyte composition. High electrolyte "throwing power" as a result of high electrolyte conductivity combined with sufficiently high polarization (in most cases, achieved by addition agents) will tend to reduce the nonuniform distribution of current density.

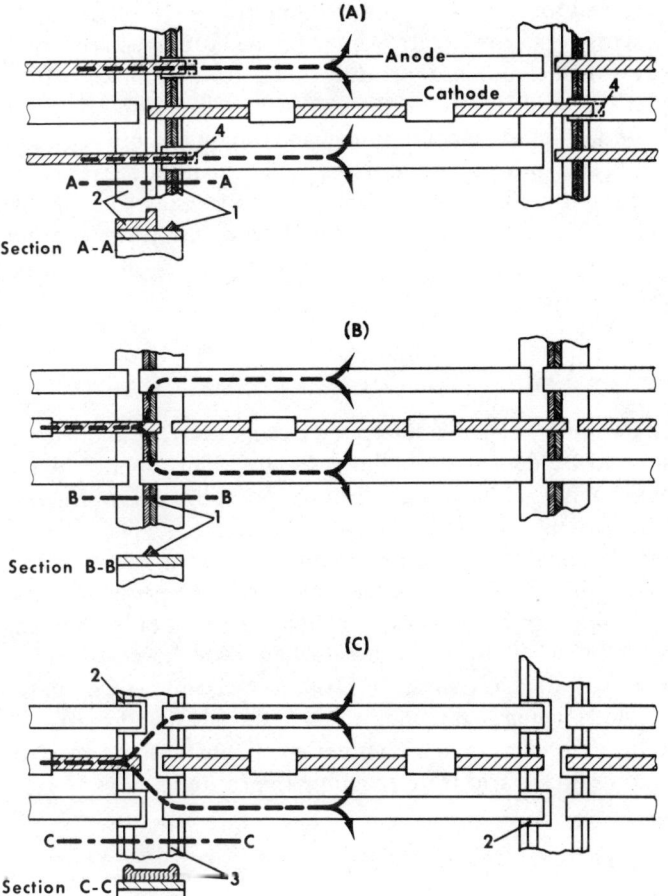

Figure 13. Electrical contact systems. (A) Whitehead systems with Baltimore Groove, (B) Walker system, (C) modified Walker system.

### 3.6. Electrolyte Circuits

#### 3.6.1. Refining

In refining cells the purpose of electrolyte circulation is to provide a steady supply of addition agents, to maintain a suitable cell temperature, to correct electrolyte composition by removing the soluble impurities from the cell, and to adjust the concentrations of the main electrolyte components. In copper refining, the first two purposes are achieved by circulating electrolyte through the cells through steam heat exchangers and addition agent feed tanks at a rate of 20–40 liter/min per cell. A very small portion of this recirculation stream

(~0.1%) is treated for removal of soluble impurities (Ni, As, and Sb) and for removal of excess copper by electrowinning. The latter step is needed because the anode current efficiency is, depending on the oxygen content of the anodes, slightly higher than 100%, due to the chemical dissolution of cuprous oxide.

Soluble impurities in nickel refining can co-deposit onto the cathodes and must, therefore, be removed before they reach the cathode. This is achieved in the Hybinette cell (see Figure 11), in which cathodes are placed in diaphragm-equipped boxes. The impure anolyte is withdrawn from the common anode compartment, purified (to be described in Section 4.3), and returned to the cathode boxes.

### 3.6.2. Electrowinning

In electrowinning cells the electrolyte circulation system must also supply the metal ions to be deposited at the cathodes. The "spent" electrolyte is, therefore, sent to the chemical replenishing operation and returned as "pregnant" electrolyte.

In electrowinning zinc, a large amount of heat is released in the cells because of the high cell voltage. To prevent excessive hydrogen evolution, the electrolyte must be kept below 40°C. The large amount of heat is removed by recirculating cell electrolyte at a fast rate (e.g., ~60 liter/min/cell) through large air-cooled towers. Only about 10% of this electrolyte recirculation flow is sent through the leaching plant, where it is replenished and purified.

In nickel electrowinning, the Hybinette cell must also be used to prevent anode-liberated sulfuric acid from reaching the cathode.

### 3.6.3. Circulation within a Cell

The circulation within the conventional cells is usually directed from one end of the cell to the overflow at the opposite end. One notable exception is the lateral circulation used in the Onahama #3 tankhouse.

## 3.7. Slime Handling

In copper and nickel refining cells the bulk of anode slimes separate from the dissolving anodes and accumulate at the cell bottom. Periodically, the slimes are discharged through a bottom opening, via a system of launders, into the slime-collecting sump. Additional anode slimes are also collected in the anode scrap (residue) washing operation. A new concept in slime removal involves cleaning cell bottoms by a moving submerged "vacuum cleaner" type of nozzle.[17]

In lead refining and nickel refining with a sulfide anode, the slimes remain attached to the anode and are recovered from the "spent" anodes. This operation can be facilitated using a slime removal machine with water spray and rotating brushes.[24]

Small amounts of anode-derived sludge (e.g., $PbSO_4$) can also accumulate in electrowinning cells. The anode sludge produced in Zn electrowinning cells contains mostly $MnO_2$ which precipitates from the $Mn^{2+}$ ion-containing electrolyte on the anode.

### 3.8. Material Handling

Efficient material handling[29] is of utmost importance to refining and electrowinning operations because of the large number of electrodes that are involved. For example, in a large copper refinery producing ~500 tons of Cu per day, about 2000 new anodes, 2000 spent anodes, 4000 starter sheets, and 4000 finished cathodes must be handled every day, usually in a single shift.

In older refineries, rail cars and bridge cranes provide the basic transportation, moving anodes from the anode casting wheel into the cells. Many new refineries prefer forklift trucks to rail cars. Loading anodes and cathodes into the cells is often facilitated by automatic electrode spacing machines. In some refineries, anodes and cathodes are also interspaced together to permit loading the cell in a single crane trip. A similar machine may also be used to disassemble the finished cathodes from the anode scrap.

An unusual extension of the automatic electrode handling system is used in the #3 tankhouse of the Onahama refinery. The preassembled and prespaced cell load of anodes and cathodes is placed on a driverless transfer rail car which automatically moves along the central aisle of the tankhouse to the position of the bridge crane. The crane operator then only needs to make a short trip sideways to the empty cell and back with the finished stack (finished cathodes and anode scrap).

The automatic transfer rail car principle is also used in the Vielle Montagne zinc electrowinning plant in Belgium.[21] In the majority of modern zinc electrowinning tankhouses, however, the cathodes are removed from the cells with a crane and placed on a monorail conveyor which transports the cathodes to the stripping machine and back.

### 3.9. Electrolyte Mist in Electrowinning

Oxygen evolving from insoluble anodes generates a fine mist of electrolyte, thus making the tankhouse atmosphere unhealthy, unless appropriate precautions are taken. Many local health authorities are limiting the content of mist to ~1 mg of major electrolyte component (e.g., $H_2SO_4$) per 1 $m^3$ of air above the cells.

The simplest and most efficient method of mist elimination involves addition of a frothing agent (e.g., Dowfax 2AO) to the electrolyte.

Unfortunately, this method, which is quite popular in some copper electrowinning operations, cannot be used in solvent extraction–electrowinning plants because the frothing agent causes decomposition of the solvent extraction reagent LIX64N. Many electrowinning operations use several layers of plastic balls to suppress the mist.[30] An oil layer on the electrolyte is much less efficient and not as practical but is also used.

Mist elimination in nickel electrowinning is achieved by withdrawing anode oxygen with the help of small hoods attached to every anode (see Figure 11). The hoods are vented into a common channel which removes the mist containing oxygen from the cells. The efficiency of these hoods increases by placing loose bags around the anodes.

The additives used in zinc electrowinning also serve to limit frothing, but there is still too much mist generated above the cells to meet the statutory limit. Many new electrowinning plants are eliminating the acid mist by the recently developed forced horizontal ventilation system.[31]

## 4. Operating Practices

### 4.1. Copper Refining

Most of the copper refineries existing today (list of the major refineries is given in Table 3) are designed in a very similar way (see Figure 14 for a typical

Table 3
Major Copper Refineries[a]

| Country and plant | Location of plant | Annual capacity at end of 1975, metric tons |
|---|---|---|
| United States | | |
| The Anaconda Company | Great Falls, Montana | 230,000 |
| ASARCO Incorporated | Amarillo, Texas | 380,000 |
| | Perth Amboy, New Jersey | 150,000 |
| | Tacoma, Washington | 140,000 |
| Kennecott Copper | Garfield, Utah | 170,000 |
| Kennecott Refining Corporation | Anne Arundel County, Maryland | 250,000 |
| Magma Copper Company | San Manuel, Arizona | 180,000 |
| Phelps Dodge Refining Corporation | El Paso, Texas | 380,000 |
| United States Metals Refining Co. | Carteret, New Jersey (subsidiary of AMAX, Inc.) | 160,000 |
| Canada | | |
| Canadian Copper Refiners | Montreal, East, Quebec | 440,000 |
| International Nickel Co. of Canada, Ltd. | Copper Cliff, Ontario | 190,000 |

Table 3 (continued)

| Country and plant | Location of plant | Annual capacity at end of 1975, metric tons |
|---|---|---|
| **Chile** | | |
| Compania De Cobre Chuquicamata, S.A. | Chuquicamata | 290,000 |
| Empress Nacional De Mineria (ENAMI) | Las Ventanas | 112,000 |
| Sociedad, Minera El Teniente S.A. | Caletones | 130,000 |
| **Belgium** | | |
| Metallurgie Hoboken–Overpelt, S.A. | Olen | 330,000 |
| **Germany, Federal Republic** | | |
| Norddeutsche Affinerie | Hamburg | 240,000 |
| **Spain** | | |
| Rio Tinto Patino, S.A. | Huelva | 105,000 |
| **United Kingdom** | | |
| B.I.C.C. Metals Limited (formerly BCR Ltd.) | Prescot and Windes, Lancashire | 165,000 |
| **Zaire** | | |
| La Generale Des Carrieres Et Des Mines | Likasi–Shituru | 250,000 |
| | Luilu, near Kolwezi | 180,000 |
| **Zambia** | | |
| Roan Consolidated Mines Limite Limited Ndola Copper Refinery Division | Ndola | 135,000 |
| Mufulira Division | Mufulira | 245,000 |
| Nchanga Consolidated Copper Mines, Ltd. | Chingola | 150,000 |
| | Kitwe | 225,000 |
| **Japan** | | |
| Mitsubishi Metal Mining Co., Ltd. | Naoshima, Kagawa-Ken | 163,200 |
| Nippon Mining Co., Ltd. | Hitachi, Ibaragi-Ken | 192,000 |
| | Saganoseki, Oita-Ken | 168,000 |
| Onahama Smelting & Refining Co., Ltd. | Onahama, Fukushima-Ken | 234,000 |
| Sumitomo Metal Mining Co., Ltd. | Besshi, Ehime-Ken | 180,000 |
| **Australia** | | |
| Copper Refineries Pty., Ltd. | Townsville, Queensland | 155,000 |

[a] Refineries with an annual capacity of >100,000 tons at the end of 1975. Copper refineries in U.S.S.R. and other satellite countries are omitted owing to lack of accurate data. Source: American Bureau of Metal Statistics, Inc.[36]

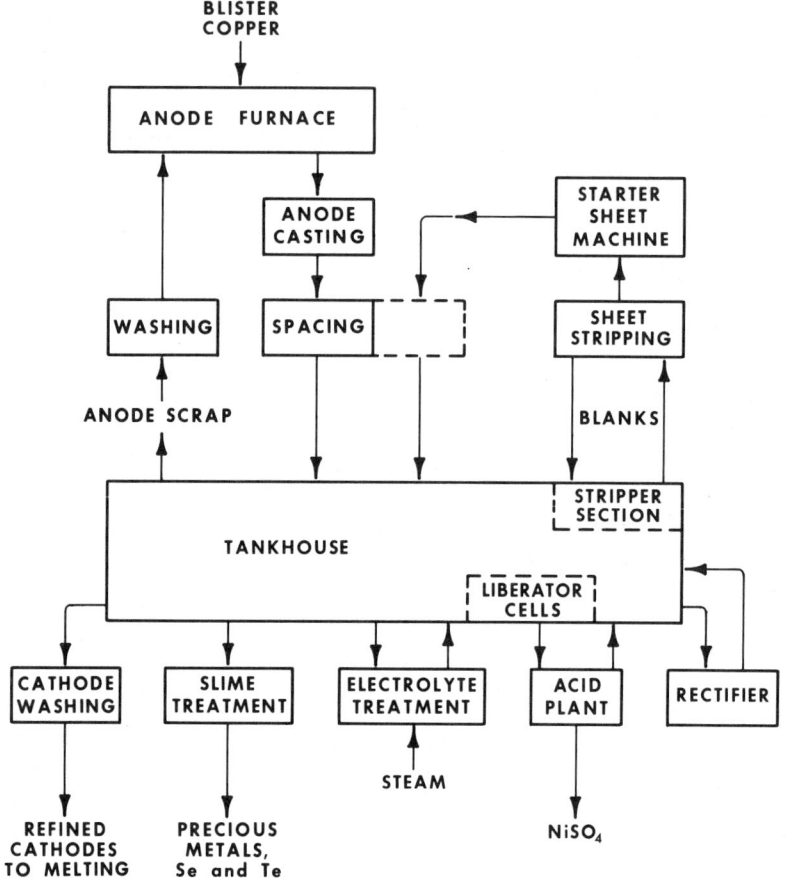

Figure 14. Flow sheet of copper refining operation.

flow sheet used in copper refining and Figure 15 for a view of a copper refining tankhouse) and employ similar operating practices. The chief differences appear in the degree of automation, which has been rapidly increasing over the last 15 years.

### 4.1.1. Anodes

In a modern, but conventional copper refinery, the anodes (weighing 300–380 kg) are formed on a casting wheel equipped with an automatic metal weighing mechanism which maintains constant anode weight to within ±2%. The anodes are automatically lifted, transported to the tankhouse by rail car or forklift truck, inspected, and put on a spacing machine. The prespaced anodes (usually 10 cm center-to-center) are removed with a crane and placed into an empty cell.

Figure 15. View of a copper refining tankhouse (with permission of INCO Metals Company).

### 4.1.2. Cathodes

Starter sheets are grown on copper or titanium blanks for ~24 hr in a "stripper" section of the tankhouse. The 24-hr blanks (with deposits) are placed on an endless conveyor and stripped manually or automatically. The stripped sheets are transported to the starter sheet machine where they are trimmed, flattened, equipped with suspension loops and a crossbar, and spaced. The finished starter sheets are transported to the cell and lowered between the anodes. A final spacing adjustment is done manually.

### 4.1.3. Electrolysis

The electrolysis is carried out at ~200 A/m$^2$ for about 14 days, after which time the grown cathodes are removed and replaced with a new set of starter sheets. After another 14 days, both anodes and cathodes are removed, the electrolyte is discharged, and the slimes from the cell bottom are removed.

During the deposition cycle, some cathodes tend to develop electrical short circuits with the anodes. These "shorts" develop most frequently (1) during the first few days of every anode cycle, since parts of some cathodes are unavoidably too close to the anodes, and (2) during the last few days of every cathode cycle when the fully grown cathodes tend to become rough and nodular. The shorts may be detected manually by gaussmeters, by heat-sensitive paint, by cell voltage monitoring, or by infrared scanning[32] which produces a "photographic" tankhouse picture with "hot" (shorted) electrodes appearing whiter than others. The detected short circuits should be corrected very quickly in order to maximize current efficiency and minimize production losses. Finished cathodes and anode scrap are separated, washed, weighed, and transported to the cathode and anode furnaces, respectively. Anode scrap is usually ~15% of the original anode weight.

### 4.1.4. Electrolyte

The electrolyte is recirculated through the cells at a rate of ~40 liter/min, reheated to ~60°C, and replenished with addition agents. The concentration of copper and nickel in the electrolyte containing 150–200 g/liter $H_2SO_4$ is maintained at ~40 g/liter Cu and ~15 g/liter Ni by withdrawing a small bleed stream of electrolyte. The bleed stream is first decopperized, usually in three stages. The first stage often consists of "Pyne-Green" tanks with soluble anodes in which the depleted electrolyte which is transported to the cell surface by natural convection is skimmed off, while most of the electrolyte is removed from the cell bottom. The partially depleted electrolyte (~20 g/liter Cu) is then passed through two stages of liberator cells with insoluble anodes. The second liberator cell removes copper to ≪1 g/liter and also induces codeposition of any arsenic present. Because of the danger of arsine evolution, these cells are usually located outside the tankhouse and well vented. Copper produced in these cells is very impure and powdery and must be recycled. Some refineries are now removing arsenic from the electrolyte by solvent extraction. The decopperized electrolyte is concentrated to crystallize nickel sulfate and other impurities, while the reclaimed, strong sulfuric acid (black acid) is returned to the tankhouse. Some refineries also employ dialysis to separate nickel sulfate from the acid. The decopperization is needed to compensate for the "chemical" dissolution of copper due to the $Cu_2O$ content of the anodes and due to the leaching of copper contained in the anode slime during processing.

Refining electrolyte can become saturated with sparingly soluble arsenates of Sb and/or Bi. Under certain conditions, these compounds precipitate as a "floating slime" which does not settle and tends to contaminate the cathodes.[33] This situation may be remedied by either increasing the electrolyte bleed rate, by increasing the concentration of As(III) as in Bolidens process,[33] or by reducing the concentrations of Bi, Sb, and As by adsorption on stannic acid as in the Norddeutsche process.[33]

### 4.1.5. Operating Costs

The cost of copper refining has three major components:

(1) labor,
(2) tankhouse capital-related costs (depreciation and interest), and

(3) cost of copper inventory (interest on the value of copper locked-in the process).

The costs of electrical energy and steam are, at present prices, several times smaller[34] than the above three components. The typical electrical energy requirement is 0.25 kWh/kg and its components are given in Table 4.

### 4.1.6. New Developments

There have been two major new developments in electrolytic refining aimed at reducing the cost of the conventional process. The first approach has been to use a substantially higher current density ($\sim 300$ A/m$^2$) to increase the plating rate and to reduce the second two major cost components. This is made possible by periodically reversing the current direction, e.g., 10 sec every 200 sec. During the current reversal, some deposited copper is redissolved while the nucleating nodules are presumably dissolved preferentially through an electropolishing effect. This process was patented by a Bulgarian copper refinery[35] and is presently being used by several copper refineries (Brixlegg in Austria, IMI in Britain, Tamano in Japan, and, most recently, Rio Tinto in Spain). Several other refineries have installed the system (rectifier) but have not adopted the process on a full production scale. The use of periodic current reversal (PCR) increases the consumption of electricity by $\sim 40\%$ but is still economical at the present energy prices (e.g., 2¢/kWhr).

One attractive feature of the PCR process is that it permits increasing the capacity of an older tankhouse without excessive capital investment. It is not certain, however, if high-quality cathodes can be produced with the PCR process under all circumstances (e.g., with very impure anodes).

Another new cost-saving approach in copper refining technology was developed by the Onahama refinery in Japan. The use of a continuously cast thin anode resulted in reducing copper inventory by $\sim 50\%$ and permitted placing $\sim 20\%$ more electrodes into the cell (higher production rate per cell), while providing $\sim 20\%$ greater and more uniform gaps between the electrodes.

Table 4
Typical Requirement of Electrical Energy in Copper Refining

| | |
|---|---|
| Current density | 210 A/m$^2$ |
| Current efficiency | 97% |
| Cell voltage components | |
|   Cathode polarization | $\sim 80$ mV |
|   Anode polarization | $\sim 30$ mV |
|   Ohmic drop in electrolyte | $\sim 100$ mV |
|   Ohmic drop in cell hardware | $\sim 70$ mV |
| Average cell voltage | 280 mV |
| Electrical energy requirement | $\sim 0.25$ kWh/kg |

This last feature reduced the short-circuiting problem so dramatically that cell inspection work could be completely abolished. Unlike the PCR process, the energy requirement in the Onahama process is not increased. The only disadvantages of the system are the higher anode-casting cost and a greater amount of anode scrap that needs to be remelted (~30% of the anode weight).

One of the early technological developments in copper refining was the Nicholls series system in which copper anodes in a single cell are connected electrically in series on a bipolar principle (one side anodic and the opposite side cathodic). There are no starter sheets in this approach, but at the end of each cycle, the anode scrap and the deposited copper must be mechanically separated from each other by peeling. Although this process was abandoned many years ago, there are some recent attempts to revive and modernize it using Hazelett-type anodes.

### 4.2. Lead Refining

Lead bullion has been refined since 1903 by the Betts process, which is unique in producing refined lead with a very low bismuth content, retaining all the impurities in the anode slimes. Betts' process is practiced by major producers in Canada, Peru, U.S.A., Sardinia, Japan, and U.S.S.R., with a combined annual capacity of about 450,000 tons, representing about 12% of the world's production of lead.[37]

#### 4.2.1. Anode Preparation

The feed material for producing lead anodes is impure lead bullion containing Cu, Sn, Sb, and Bi. After decopperizing, Sn is removed along with some As, Sb, and Bi by treatment with caustic soda. The anodes typically analyze[24] 98% Pb, 0.5% Bi, 0.01% As, 1% Sb, 0.02% of Cu and Sn.

#### 4.2.2. Electrorefining

Employing lead cathode starting sheets, electrolysis is carried out in aqueous electrolyte containing ~90 g/liter of free $H_2SiF_6$ and ~70 g/liter Pb as lead fluosilicate at ~30–40°C and a current density of ~160–200 A/m². Animal bone glue and lignin sulfonate or "aloes" are also added to obtain smooth Pb deposits. The cell voltage increases from ~0.35 to 0.65 V owing to building up of adherent anode slimes. Pb is deposited at a current efficiency of ~95%, which corresponds to an energy requirement of ~0.13 kWh/kg. The electrolyte enriched in Sb and Bi may be purified by electrolysis at low current density.[24]

#### 4.2.3. Sulfamic Acid Process

The sulfamic acid process[38] involves electrolysis using solutions containing 80 g/liter Pb and 100 g/liter free sulfamic acid at a current density of

100–120 A/m² and a cell voltage of 0.5–0.65 V, with a current efficiency of ~95%. The main disadvantage of this process is the high cost of sulfamic acid.

### 4.3. Nickel Refining

Nickel is electrolytically refined by the metal anode process and the sulfide anode process. The metal anode process is practiced by the International Nickel Company of Canada (INCO) at the Port Colborne refinery in Canada,[1] the Falconbridge Nikkelwerk Aktieselskap at Kristiansand in Norway,[39] the Sumitomo nickel refinery at Niihama in Japan, at Larymna Nickel in Greece,[40] at the Norilsk in Severonickel, and Yuzhural Nickel refineries in the U.S.S.R.,[41] whereas the sulfide anode technology is employed by INCO at the Thompson nickel refinery in Canada, by Sumitomo at Niihama, and Shimura Kako refinery at Tokyo in Japan.

An important feature of these operations is the degree of purification or refining which depends on the ultimate use of nickel. Thus the presence of Sb, As, Bi, Cu, Pb, Zn, Sn, S, and P in nickel is totally unacceptable for manufacturing nickel alloys. Cobalt in nickel is undesirable for nuclear applications because of the formation of Co-60, which is radioactive (half-life ~5 years) and for some special alloys, since cobalt is ferromagnetic.

Nickel is also refined by a nonelectrolytic, carbonyl process,[1] which yields nickel pellets (or powder) of exceptionally high purity.

#### 4.3.1. Metal Anode Preparation

The feed material for preparing the metal anodes at INCO's Port Colborne nickel refinery[1] is NiO produced by roasting nickel matte. This oxide is reduced with petroleum coke to metallic nickel at ~1540°C and cast into anodes which typically analyze 94% Ni, 1% Co, 4% Cu, and small amounts of other impurities.

A different method of treating the sulfide ores is practiced in the Falconbridge process.[1] In this practice, Cu (usually associated with nickel) is separated by roasting the converter matte to oxide and acid-leaching the calcine. Smelting the leached calcine produces anode metal typically analyzing 76% Ni, 1.5% Co, 17% Cu, and other impurities.

Larymna nickel prepares[40] metal anodes from ferronickel containing ~90% Ni.

#### 4.3.2. Sulfide Anode Preparation

The starting material for preparing the anodes at the Thompson nickel refinery in Manitoba, Canada, is converter nickel matte containing 20% S and 2.6% Cu. This is heated to ~1100°C and cast as anodes. The sulfide anodes, prone to cracking (unlike the metal anodes), require slow cooling and

*Table 5*
*Typical Requirement of Electrical Energy in Nickel Refining with Metal Anode*

| | |
|---|---|
| Current density | 200 A/m$^2$ |
| Current efficiency | 96% |
| Cell voltage components | |
|   Cathode polarization | 0.25 V |
|   Anode polarization | 0.30 V |
|   Ohmic drop in electrolyte | 0.90 V |
|   Ohmic drop in diaphragm | 0.15 V |
|   Ohmic drop in cell hardware | 0.30 V |
| Average cell voltage | 1.90 V |
| Electrical energy requirement | 1.9 kWh/kg |

careful handling and typically analyze 76% Ni, 0.5% Co, 2.6% Cu, 0.5% Fe, and 20% S.

### 4.3.3. Electrorefining with Metal Anodes

The cathode starting sheets are prepared by stripping nickel deposited on stainless steel blanks, and stamping the sheets with a shallow pattern to impart rigidity and prevent warping during the growth of nickel cathodes to full size. Electrodeposition is carried out in Hybinette divided cells (see Section 3.4 for details) at a current density of ~200 A/m$^2$ from the purified electrolyte typically containing 60 g/liter Ni$^{2+}$, 95 g/liter SO$_4^{2-}$, 35 g/liter Na$^+$, 55 g/liter Cl$^-$, and 10 g/liter H$_3$BO$_3$ (pH 3 to 4) at ~60°C.

The average cell voltage is ~1.9 V. The distribution of the electrical energy requirement of the metal anode process is presented in Table 5.

INCO metal anodes weigh ~220 kg and produce three nine-day crops of cathodes leaving anode scrap which is ~20% of the original anode weight.

Anode slimes are remelted and cast into secondary anodes enriched in precious metals. The secondary anode slimes are processed to recover the precious metals.

### 4.3.4. Electrorefining with Sulfide Anodes

This process is, in most aspects, similar to the metal anode process.[1,42] However, sulfide anodes must be placed into loose bags to collect the voluminous sulfur-containing sludge. The anode produces only two crops of cathodes and more scrap is produced (~30%). The anode sludge is melted and processed, by pressure filtration, to pure elemental sulfur and a filter cake containing precious metals.

### Table 6
### Typical Requirement of Electrical Energy in Nickel Refining with Sulfide Anode

| | |
|---|---|
| Current density | 200 A/m$^2$ |
| Current efficiency (cathode) | 96% |
| Cell voltage components | |
|     Reversible cell voltage | 0.35 V |
|     Cathode polarization | 1.50 V |
|     Anode polarization | 0.25 V |
|     Ohmic drop in electrolyte | 1.10 V |
|     Ohmic drop in diaphragm | 0.15 V |
|     Ohmic drop in cell hardware | 0.35 V |
| Average cell voltage | 3.7 V |
| Electrical energy requirement | 3.5 kWh/kg |

Although similar current density to that in the metal anode process is used, the cell voltage is much higher, mainly because of the potential drop across the layer of porous sludge on the anode. The electrical energy requirement of the sulfide anode process is ~3.5 kWh/kg; its components are listed in Table 6.[34]

While the main anode reaction is

$$Ni_3S_2 \rightarrow 3Ni^{2+} + 2S + 6e \quad (E° = 0.1 \text{ V})$$

a small part of electrical current is also used to oxidize sulfur and to produce oxygen. Consequently, the anodic current efficiency is somewhat smaller than the cathodic current efficiency, or, in other words, more nickel is plated out on the cathodes than was dissolved on the anodes. The electrolyte must, therefore, be continuously replenished, e.g., by leaching ground anode scrap.[1]

### 4.3.5. Electrolyte Purification

The purification schemes adopted by various refineries vary somewhat, but, except for the presence or absence of the nickel-replenishing operation, they do not depend on whether the metal or the sulfide anode process is used.

Copper is usually removed by cementation on nickel powder in two countercurrent stages. The rate of the cementation can be enhanced by adding a small amount of sodium thiosulfate. The Thompson Refinery uses $H_2S$ for precipitating copper. When $H_2S$ is used, most of the Pb and As present will also be removed.

Iron is removed by oxidative hydrolysis using $Cl_2$ and $NiCO_3$ as a base. When small amounts of As are present, they will be efficiently removed during iron hydrolysis as insoluble ferric arsenate.

Cobalt is also removed by oxidative hydrolysis as $Co(OH)_3$, but higher pH and oxidation potential must be used in this case. Any lead present will also precipitate during this operation as $PbO_2$. Cobalt and iron are sometimes precipitated in a single step and separated from each other by redissolution and reprocessing of the filter cake. Chlorine required for oxidation is sometimes produced in one section of the refinery using insoluble anodes.[39] Nickel carbonate used in Fe and Co precipitation is produced by precipitating a portion of the electrolyte (usually washings, etc.) with soda ash.

Lead can also be removed with $BaCO_3$, with coprecipitates $PbSO_4$ into the $BaSO_4$ lattice. This process can be used when cobalt is not being removed from the electrolyte.

A flow sheet of the INCO Thompson nickel refinery operations in Figure 16 is shown as an example of the nickel refining process.

### 4.3.6. Production of Sulfur-Depolarized (SD) Nickel

Nickel produced by the processes described above is the regular electrolytic nickel containing ~10 ppm of sulfur and is predominantly used in metallurgical applications. However, for electroplating purposes, instead of regular nickel, SD nickel containing ~200 ppm of sulfur is often used as the anode material, since the latter dissolves readily in nickel plating bath at lower potentials than pure nickel.

SD Ni is produced by adding $SO_2$ or another sulfur-incorporating compounds to the purified electrolyte, which is then fed to the cathode boxes.

## 4.4. Silver Refining

The feed material for silver refining is the anode slimes from Cu, Ni, Pb, and Zn refining operations and also concentrates from desilverization of lead by the Parkes process. Typically, these materials contain 10–20% Ag, 0.1–0.5% Au, and various concentrations of other metals such as Se, Te, Pt, Pd, Cu, Ni, etc.

### 4.4.1. Anode Preparation

The slime-processing practices vary from refinery to refinery. One flow sheet used by many copper refineries involves leaching raw slimes in aerated electrolyte to remove copper, roasting to volatilize Se and Te, and finally smelting and refining in a Doré furnace to remove $SiO_2$, Pb, Sb, and additional Cu. The "Doré" anodes typically contain ~95% Ag—the balance being Cu, Au, and Pt group metals and the finished anodes are usually about $20 \times 40 \times 2$ cm in size.[43,44]

### 4.4.2. Electrorefining

Electrolysis is carried out in two types of cells—Thum (also called Thum-Balbach cells) or Moebius cells. The Thum cells consist of shallow tanks

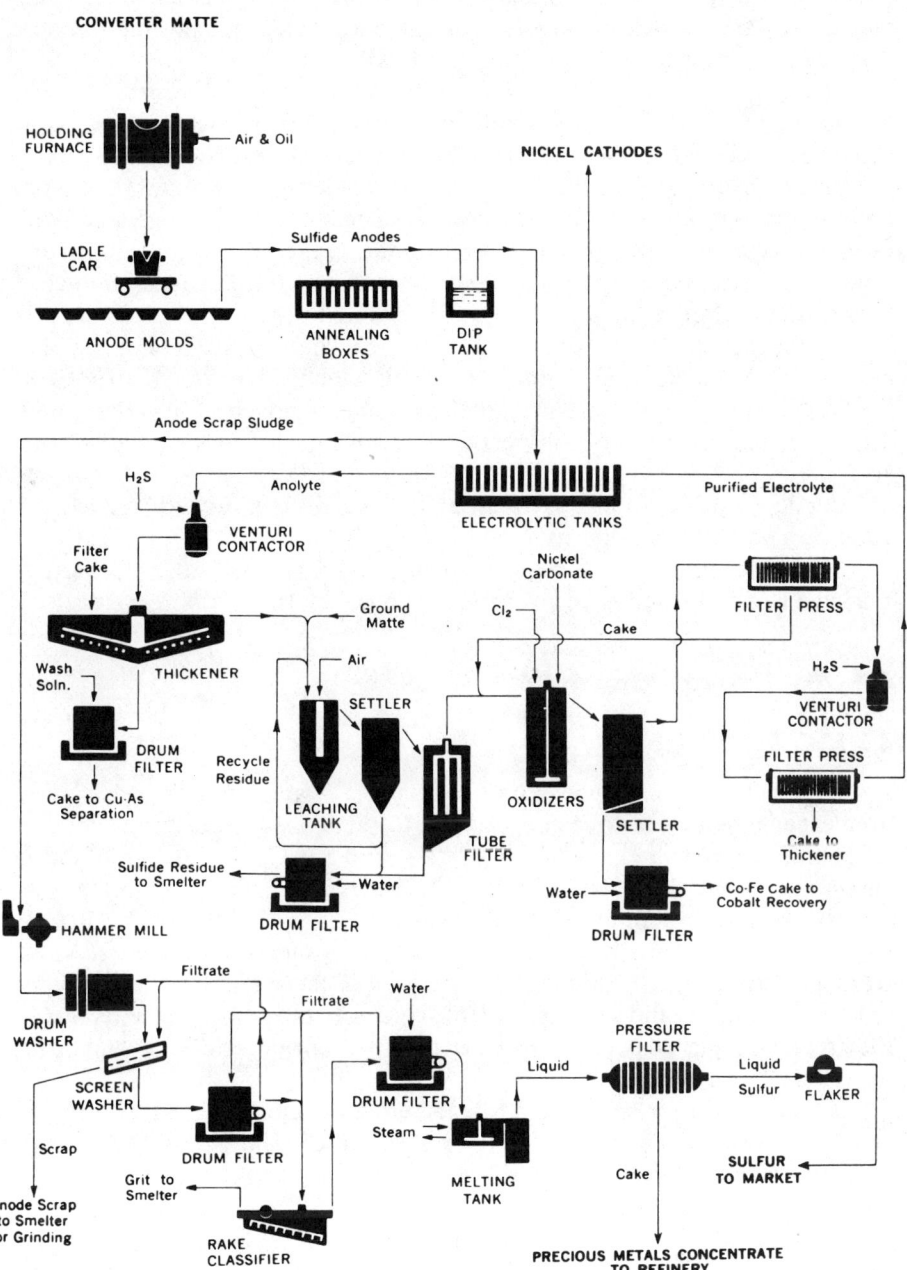

Figure 16. Flow sheet of INCO's nickel refinery at Thompson (with permission of INCO Metals Company).

equipped with graphite or stainless steel bottoms acting as cathodes. The anodes are arranged horizontally above the cathodes, resting on wooden frames with canvas bottoms to retain the anode slime.

In the Moebius cells, the Doré anodes and the stainless steel cathodes are arranged vertically as in copper refining. Each anode is encased in a polypropylene or canvas bag which collects the gold mud as the anode dissolves. Ag is deposited from an acidic nitrate bath containing 150 g/liter Ag (pH 1–1.5) at a current density of ~300 A/m$^2$ and a temperature of ~35°C. Electrolyte mixing is accomplished by an oscillating cathode scraper which also brushes the silver crystals off the cathode plates. The crystals are collected in a removable basket with a filter cloth bottom, which is situated beneath the electrode.[44]

In the course of the refining process, Cu and Pd accumulate in the electrolyte. When the copper content in the electrolyte exceeds ~10 g/liter, or the Pd content exceeds ~1 g/liter, the solution is removed, desilverized, and discarded, and new electrolyte is made by dissolving scrap anodes in hot nitric acid.

Ag deposited in a loose crystalline form is collected, melted, and cast as 1000-oz bars of 99.99% purity.

The slimes in the anode bag containing Au, Pt, Pd, and other metals, are collected and processed to recover the precious metals.

### 4.5. Refining of Other Metals

#### 4.5.1. Gold Refining

Gold is recovered by the Wohlwill process,[43] wherein the anode slimes from silver refining are collected, purified to remove Ag, and cast as anodes containing approximately 94% Au, 5% Ag, and 1% Cu, Pt, Pb, and Pd and measuring about $20 \times 10 \times 1$ cm.[44] Electrolysis is carried out in acidic chloride solutions containing 150–200 g/liter Au and 140 g/liter HCl at ~70°C and at a current density of 1200 A/m$^2$, using thin gold ribbon cathodes. During electrolysis, the cathodes are lifted out of the solution periodically to remove the nodules and thereby prevent shorting between cathode and anode. Electrolyte is replaced periodically when the total concentration of Pt and Pd in the electrolyte exceeds ~75 g/liter.

The electrolyte from the gold refining process contains Pt, Pd, etc. These metals are recovered by precipitation as $(NH_4)_2 PtCl_6$ and $(NH_4)_2 PdCl_6$, and reduction to sponge metal.

#### 4.5.2. Tin Refining

Crude tin from smelting ore concentrates and from slags is processed by fire-refining and electrolytic refining techniques to produce high-grade tin metal. Electrolytic refining plants use either an acidic or alkaline bath. The electrolytic tin refining plant at Texas City employs a sulfate bath[45] containing 60 g/liter $H_2SO_4$ and 25 g/liter Sn. 2.4 g/liter Cresylic acid, 0.01 g/liter Aloin, and ~48 g/liter phenol sulfonic acid are also added to prevent nodular growth of Sn during deposition. Electrolysis is carried out at a current density of ~95 $A/m^2$ at ~40°C (cell voltage: 0.3 V; current efficiency: 85%; energy consumption: 0.22 kWh/kg) using tin starter sheets prepared by pouring molten tin on steel plates.[45]

The slimes are quite adherent to the anode and are periodically removed by scrubbing with revolving wire brushes.

The alkaline bath contains sodium or potassium stannate and free NaOH and operates without addition agents at ~80°C.[46] Tin dissolves in this medium as stannate only from anodes coated with a film of hydrated oxide ($SnO_2 \cdot 2H_2O$), which is formed either by operation at high current densities or by slow insertion of the anode into the cell. Ease of operation and the capability of using low-grade anodes are claimed as the advantages of electrorefining from alkaline electrolytes.

Sb, Bi, and In are also produced by electrorefining crude anodes, and the reader is referred to Mantell[43] for a brief description of these processes.

### 4.6. Zinc Electrowinning

#### 4.6.1. Roasting

The feeds to zinc electrowinning plants are zinc sulfide (sphalerite) concentrates usually containing ~50% Zn and ~10% Fe. The sulfide concentrates are converted into acid-soluble oxides by roasting at ~930°C in fluid bed roasters. Two types of roasters are mainly used in the industry: (1) slinger-fed turbulent layer roaster (Lurgi), and (2) slurry-fed Dorr–Oliver roaster.

The $SO_2$-containing roaster gas passes through a waste heat boiler, cyclone, and dust precipitator, all collecting the produced calcine. Part of the calcine is also collected directly from the roaster bed overflow. The dust-free (calcine) roaster gas is fed into a standard sulfuric acid plant, often equipped with various end-gas scrubbing systems to meet the local $SO_2$ emission standards. The calcine produced is cooled, sometimes ground and stored in bins before forwarding to the leaching plants.

#### 4.6.2. Leaching

The leaching is usually done in two distinct phases. In the first phase, the easily soluble zinc oxide is dissolved in spent electrolyte under mild conditions, that is, moderate temperature and a neutral pH of about 5. This operation is sometimes preceded with wet grinding of the concentrate in the spent electrolyte. The neutral leaching leaves most of the iron undissolved, since it is present in the form of a poorly soluble zinc ferrite.

The ferrite-containing residue is releached under more severe conditions (up to 200 g/liter $H_2SO_4$ and 95°C), sometimes in several stages (cold, hot, and superhot) depending on the iron content of the calcine. This dissolves most of the residual zinc and iron and leaves a residue containing mostly lead and silver.

### 4.6.3. Electrolyte Purification

Two different processes are employed to precipitate and reject iron from the acid leach liquor. In the first process, ferric iron is precipitated as ammonium, sodium, or other jarosite [$NaFe_3(SO_4)_2(OH)_6$, for example] which is a highly crystalline material. To this end, zinc calcine is added to reduce the concentration of free acid, $MnO_2$ is added to keep iron oxidized, and $Na_2SO_4$, for example, is added to supply the jarosite-stabilizing cation. Jarosite, which precipitates between pH 1 and 1.5, is removed from the process and the liquor, which still contains a small amount of iron (1–2 g/liter), is returned to the neutral leach. The jarosite process is patented and is presently used by a great majority of the zinc plants (see Table 7 for a list of major electrolytic zinc producers). A typical leaching–jarosite precipitation flow sheet[47] is in Figure 17.

An alternative iron rejection process[21] developed in Belgium is based on precipitating iron as geothite ($\alpha$-FeOOH). This is achieved by first reducing the ferric iron, in the hot leach liquor with zinc sulfide concentrate, to the ferrous state, neutralizing the liquor with calcine to pH ~2 and aerating at 90–95°C for 5–7 hr.

The neutral leach liquor is purified by cementation on zinc powder, usually in two stages. The impurities that must be removed are Cu, Cd, and Pb, which would contaminate the product, and As, Sn, Se, Co, Ni, Sb, and Ge, which catalyze the hydrogen discharge reaction and can completely prevent the deposition of zinc.[48–50] The last two impurities are especially harmful and must be reduced to <0.1 mg/liter level.

In one of the purification schemes, Cu and Cd are precipitated first at low temperature. The filtrate is then heated to 90°C and Co (along with Ge) is precipitated on zinc dust in the presence of antimony. Other purification schemes employ the addition of $As_2O_3$ and Cu to catalyze the cementation of cobalt or $\alpha$-nitroso-$\beta$-naphthol to precipitate cobalt. The cementation cake is usually reprocessed to recover metallic cadmium, which is a valuable by-product of most zinc plants.

The purified, neutral liquor containing ~170 g/liter Zn is bled into the stream of electrolyte, which is recirculated between the cells and the cooling towers.

### 4.6.4. Electrowinning

The typical composition of electrolyte fed to the electrowinning cells is 55–70 g/liter Zn and 150–200 g/liter $H_2SO_4$; the spent electrolyte has the zinc

Table 7
Major Electrolytic Zinc Plants[a]

| Country and plant | Location of plant | Annual capacity at end of 1975, metric tons |
|---|---|---|
| **United States** | | |
| ASARCO Incorporated | Corpus Christi, Texas | 100,000 |
| The Bunker Hill Company | Silver King, Idaho | 95,000 |
| **Canada** | | |
| Canadian Electrolytic Zinc Limited | Valleyfield, Quebec | 205,000 |
| Cominco Ltd. | Trail, British Colombia | 250,000 |
| Texasgulf Canada Ltd., Div. Texasgulf, Inc. | Timmins, Ontario | 110,000 |
| **Mexico** | | |
| Met-Mex Penoles, S.A. | Torreon, Coahuila | 95,000 |
| **Europe** | | |
| Soc. Anon. de la Vielle Montagne | Balen, Belgium | 170,000 |
| | Viviez, Aveyron, France | 100,000 |
| Outokumpu Oy | Kokkola, Finland | 150,000 |
| Preussag-Weser-Zink GmbH | Nordenham, West Germany | 130,000 |
| Ruhr-Zink GmbH | Dattein, West Germany | 140,000 |
| Soc. Min e Met. di Petrusoia | Crotone, Italy | 90,000 |
| A.M. & S. Europe-Billiton Elzet Campine | Budel, Netherland | 165,000 |
| Det Norske Zinkkompani, A.S. | Eitrheim, near Odda, Norway | 85,000 |
| Asturiana de Zinc S.A. | San Juan de Nieva, Spain | 120,000 |
| **Asia** | | |
| Akita Smelting Co., Ltd. | Akita-ken, Japan | 155,000 |
| Mitsubishi Metal Mining Co. Ltd. (Akita Refinery) | Akita-shi, Akita-ken, Japan | 105,000 |
| Mitsui Mining & Smelting (Hikoshima) | Shimonoseki-shi, Yamaguchi-ken, Japan | 85,000 |
| Toho Zinc Co., Ltd. (Annaka Refinery) | Annaka-shi, Gunma-ken, Japan | 140,000 |
| **Australia** | | |
| Electrolytic Zinc Co. of Australia | Risdon, Tasmania | 210,000 |

[a] Listing only plants with capacity 80,000 tons or more. Source: American Bureau of Metal Statistics, Inc.[36]

concentration, 5–10 g/liter less. The electrowinning is done at 35–38°C to limit $H_2$ evolution. The cathode operates at a potential $-0.9$ V with respect to the standard hydrogen electrode. This explains why the electrowinning process is so sensitive to the impurities catalyzing the $H_2$ discharge reaction. Contaminating the zinc electrolyte with barely detectable concentrations of such impurities can even result in negative current efficiencies. That is, instead of

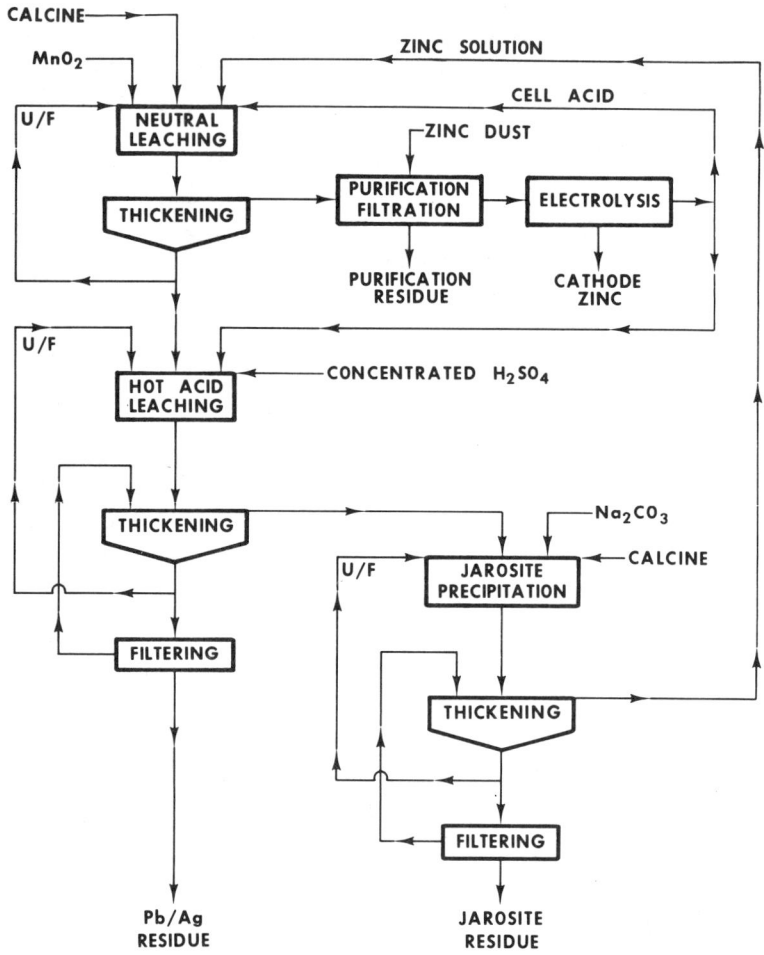

Figure 17. Typical leaching–jarosite precipitation flow sheet.

plating, the already formed deposit is dissolving, in spite of the current passing through the cell.

Current densities used in individual plants vary between 400 and 800 A/m². Some plants reduce the current density during the peak energy demand hours to values as low as 300 A/m².

Zinc is deposited on aluminum blanks usually for 48 hr. At the end of the cycle, $\frac{1}{3}$ to $\frac{1}{2}$ of the cathodes (e.g., every other) are removed from the cell and placed onto the transport mechanism that will deliver them to the stripping machine. At the same time, the finished cathodes are replaced with stripped blanks. The usual cathode spacing is 7.5–9 cm. Anodes are lead containing 0.5–1% Ag, often equipped with attached porcelain or plastic spacers to

Table 8
Typical Requirement of Electrical Energy in Zinc Electrowinning

| | |
|---|---|
| Current density | 570 A/m$^2$ |
| Current efficiency | 90% |
| Cell voltage components | |
|   Reversible cell voltage | 2.00 V |
|   Cathode polarization | 0.15 V |
|   Anode polarization | 0.60 V |
|   Ohmic drop in electrolyte | 0.50 V |
|   Ohmic drop in cell hardware | 0.25 V |
| Cell voltage | 3.5 V |
| Electrical energy requirement | 3.3 kWh/kg |

maintain a uniform distance from the cathodes. Anodes are periodically cleaned to remove the scale of $MnO_2$ depositing on their surfaces.

The major trends in the zinc cellroom practices over the last 20 years include increasing mechanization and increasing the cathode size from ~1.2 m$^2$ of total immersed area to ~1.8 m$^2$ in most of the newer plants, some even using "jumbo" cathodes with 2.6 m$^2$ of the immersed area. Lower current densities (400–600 A/m$^2$) and longer deposition times (48 hr instead of 24 hr) are also typical features of the new plants, designed to reduce cathode handling.

The energy requirement in a zinc electrowinning cell operating at 570 A/m$^2$ with a current efficiency of 90% will be ~3.3 kWh/kg. The breakdown of typical energy requirement is presented in Table 8.

### 4.7. Copper Electrowinning

More than 600,000 tons of copper is produced worldwide by electrowinning by major producers (see Table 9). The method of preparing the electrolyte solution entering into the copper electrowinning cells varies considerably, depending on the ore type or the source to be leached. These various practices may be classified into four broad categories, described below.

### 4.7.1. Leaching–Electrowinning Process

This technology is extensively practiced (~10% of the world's production)[51] with oxide ores and generally involves leaching with sulfuric acid generated during copper electrowinning. In the case of sulfide ores, leaching is promoted by maintaining a high ferric content in the electrolyte, which is regenerated during electrolysis.

The procedure practiced with silicate ores at the Ray Mines Division of Kennecott Copper Corporation[52] consists in vat leaching the deslimed ore

## Table 9
### Copper Electrowinning Operations

| Operations | Estimated capacity, tons/year | Process used |
|---|---|---|
| Anamax<br>Twin Buttes, Arizona | 33,000 | leach, SX, EW[a] |
| Anaconda<br>Anaconda, Montana | 30,000 | Ammonia leach, SX, EW |
| Cities Service<br>Miami, Arizona | 5,000 | Leach, SX, EW |
| Cyprus mines<br>Bagdad, Arizona | 6,500 | Leach, SX, EW |
| Ranchers<br>Miami, Arizona | 5,000 | Leach, SX, EW |
| Cyprus Mines<br>Johnson Camp, Arizona | 5,000 | Leach, SX, EW |
| Kennecott<br>Ray Mines, Arizona | 22,000 | Vat leach, EW |
| Inspiration<br>Inspiration, Arizona | 15,000 | Leach, EW |
| Hecla Mining<br>Lakeshore, Arizona | 30,000 | Roast, leach, EW |
| INCO Metals<br>Copper Cliff, Ontario | 15,000 | Pressure leach, EW |
| Chile Exploration<br>Chuquicamata, Chile | 130,000 | Leach, EW |
| Duisburger Kupferhütte<br>Duisburg, West Germany | 10,000 | Leach, EW |
| Gecomines<br>Shituru, Zaire | 125,000 | Roast, leach, EW |
| Gecomines<br>Luilu, Zaire | 90,000 | Roast, leach, EW |
| Mufulira<br>Chambishi, Zambia | 20,000 | Roast, leach, EW |
| Anglo American<br>Nchanga, Zambia | 110,000 | Leach, EW |
| Anglo American<br>Nchanga, Zambia | 100,000 | Leach, SX, EW |

[a] SX and EW refer to solvent extraction and electrowinning, respectively.

with spent electrowinning electrolyte. Fe and Al buildup in the electrolyte is prevented by using a portion of the spent electrolyte to leach the slimes, recovering copper from this liquor by cementation on shredded iron, and discarding the final liquor containing Fe and Al. The cement copper is redissolved in fresh $H_2SO_4$ and blended with vat leach solutions for feed to the electrowinning cells.

### 4.7.2. Roasting–Leaching–Electrowinning (RLE) Process

Hecla Mining Company[53] and Chambishi RLE Plant[54] in Zambia have adopted this process, dictated by the economics in treating ore bodies consisting of both oxidized and sulfide ores that could be mined concurrently. In the Chambishi operations, the copper concentrates are fluid bed roasted to produce $SO_4^{2-}$ containing calcine. The resulting calcine is leached with a portion of the spent electrolyte from the copper electrowinning cells, while the other portion is used for leaching oxide ores. This solution is purified to remove Fe prior to electrowinning. The advantage of the RLE process is that the $H_2SO_4$ losses are compensated by the sulfate ion content in the calcine.

### 4.7.3. Solvent Extraction–Electrowinning Process

The leaching–cementation–refining operations generally involve a large consumption of scrap iron and acid. These disadvantages are largely overcome in the solvent extraction route (see Figure 18) in which copper is selectively extracted from acid solutions (containing 0–5 g/liter $H_2SO_4$) by LIX-64® dissolved in kerosene. This reagent is highly selective for copper in the presence of other metals. The copper-loaded organic is stripped of its copper content by contacting it with 100–150 g/liter $H_2SO_4$ to produce high-grade purified copper-containing electrolyte, from which copper is electrowon directly. The reagent and the solvent are recovered and recycled.

This technology developed by General Mills was originally applied by the Bluebird Mine of Ranchers Development Corporation[55] and is practiced, among others, by the Nchanga Tailings leach plant in Zambia.[56]

### 4.7.4. Leaching Copper-Containing Residues and Precipitates

There are several copper electrowinning plants which process copper-containing residues (e.g., from nickel carbonyl refining process) and

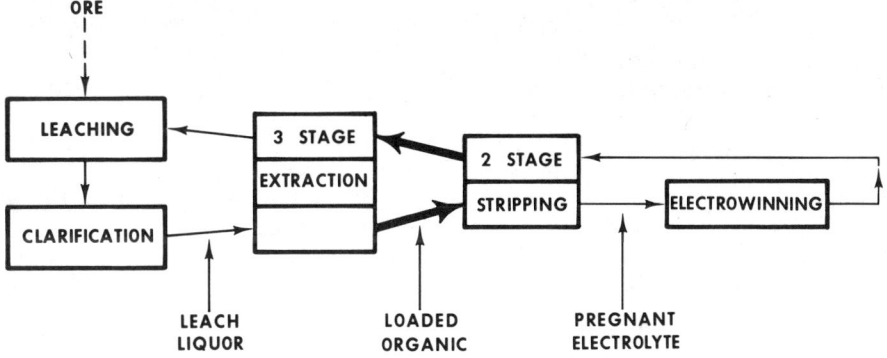

Figure 18. Leaching–solvent extraction–electrowinning flow sheet.

precipitates. As an example, the process used by Duisburger Kupferhütte is outlined here.[57]

The feed to this plant consists of pyrite roasting residues (cinders) produced in the $H_2SO_4$ manufacturing plants. In order to convert pyrite cinders into a Fe ore suitable for pig iron production, the cinders are subjected to a chloridizing–sulfating roast and are vat leached with water. The chloride-sulfate leach liquor containing small concentrations of Cu, Zn, Co, Cd, and precious metals is reduced with cement Cu, where most Cu is precipitated as cuprous chloride. Cuprous oxide, formed by reaction of the cuprous chloride intermediate with lime at elevated temperatures, forms the feed material for the leaching–electrowinning operation. Since the cuprous oxide contains a number of elements detrimental to the copper quality, the leach liquor is subjected to a two-stage purification prior to electrowinning.

The leaching–purification procedure consists of leaching the feed with the spent electrolyte, during which half of the copper in cuprous oxide is converted to metallic Cu which is solubilized by oxidative leaching. The purification operation involves (1) extended oxidation of the above suspension (when Fe and other impurities are precipitated out), and (2) reduction of the solution with an acid feed suspension containing finely divided copper, during which most of the chloride is removed as CuCl along with Se. (Zn, Mg, Al, Ni, and Co from the solution are removed by bleeding off part of the electrolyte.) The filtered solution is fed to the electrolytic cell.

### 4.7.5. Selenium Removal from Electrowinning Electrolyte

Copper electrolytes prepared by direct leaching of sulfide ores or residues often contain small concentrations of Se and Te. Se and Te codeposit into copper readily and must, therefore be removed from the pregnant electrolyte. The Se and Te content in the electrolyte can be lowered to ~1 ppm by contacting leach liquors with $SO_2$ at 90°C to reduce and precipitate Se, followed by aging and pressure filtration,[58] or by fluid bed cementation on granulated copper.[59]

### 4.7.6. Electrowinning Conditions

Electrolysis is carried out from solutions containing 25–60 g/liter Cu, 50–180 g/liter $H_2SO_4$, and 5–10 g/liter Fe at 50–60°C and at a current density of 200–300 A/$m^2$, using Pb–6% Sb or Pb–0.05% Ca anodes and copper starter sheet cathodes. Typical cell voltage is ~2 V; the breakdown of electrical energy requirement is presented in Table 10.[34]

Copper is deposited at a current efficiency of 80–95%, depending on the amount of Fe present in the solution,[60] and usually contains ~10–50 ppm Pb. Some carefully designed electrowinning operations are, however, producing copper equal in purity to the refined product.[58]

Table 10
Typical Requirement of Electrical Energy in Copper Electrowinning

| | |
|---|---|
| Current density | 300 A/m$^2$ |
| Current efficiency | 85% |
| Cell voltage components | |
|   Reversible cell voltage | 0.90 V |
|   Cathode polarization | 0.05 V |
|   Anode polarization | 0.60 V |
|   Ohmic drop in electrolyte | 0.40 V |
|   Ohmic drop in cell hardware | 0.50 V |
| Cell voltage | 2.0 V |
| Electrical energy requirement | ~2.0 kWh/kg |

During electrolysis, addition agents (e.g., Jaguar plus) are added to obtain smooth Cu deposits and surfactants (e.g., Dowfax) to suppress the acid mist (see also Sections 2.2 and 2.3). One of the unusual features observed with solvent extraction electrolytes is the phenomenon of "organic burn"[61]—a dark, chocolatelike material deposited mainly at the top edge of the cathode. This phenomenon is avoided by removing the finely dispersed organic droplets from the feed electrolyte by flotation or other techniques.

### 4.7.7. Some Novel Approaches

In an attempt to eliminate the need for smelting sulfide ores, several chloride-based leaching–electrowinning processes were developed. A small commercial plant using one-electron electrowinning from a cuprous-ion-containing electrolyte has been built by Duval Corporation in Arizona.[62] The energy consumption of this process is ~1 kWh/kg, but the product requires further refinement.

### 4.8. Nickel Electrowinning

This technology, which is practiced at the Outokumpu nickel refinery in Finland[20] and two refineries in Africa,[63] involves atmospheric pressure leaching of high-grade nickel matte containing ~60% Ni, 30% Cu, and 7% S and electrowinning the purified leach solution in Hybinette-type electrolytic cells using insoluble anodes. Nickel is deposited at the cathode, and oxygen evolution takes place at the anode. The acid generated from the anodic reaction is used by recirculating the anolyte to the leach circuit.

In the Outokumpu operations, summed up in the flow sheet in Figure 19, the granulated and ground low sulfur matte is leached with the anolyte from the copper electrowinning cells in the presence of air and $BaCO_3$. During this

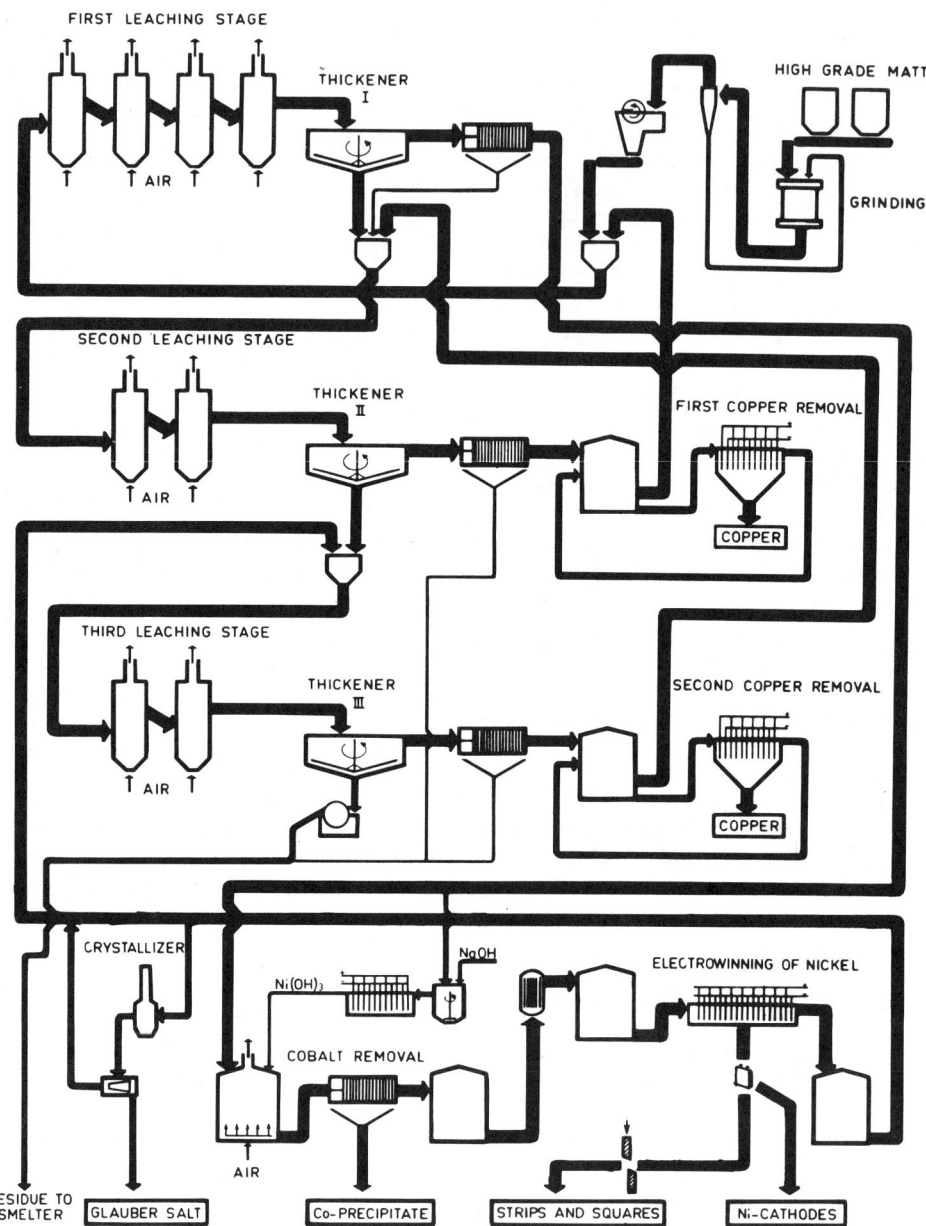

Figure 19. Leaching–Ni–electrowinning flow sheet used by Outokumpu Oy (with permission from Outokumpu Oy).

process, copper is first leached and subsequently cemented out in exchange for the dissolution of more nickel, according to the following reaction scheme:

$$2Cu + O_2 + 2H_2SO_4 \rightarrow 2Cu^{2+} + 2SO_4^{2-} + 2H_2O$$

$$Cu^{2+} + Ni \rightarrow Cu + Ni^{2+}$$

The second reaction is promoted in the last leaching stage by restricting the oxygen flow until the pH attains a value of ~6. At this point, most Pb has been precipitated with $BaSO_4$ and most Fe hydrolyzed as $Fe(OH)_3$ so that Co is the only major impurity remaining in the solution.

### 4.8.1. Electrolyte Purification

Cobalt is removed from filtered pregnant liquor in two stages as $Co(OH)_3$, according to the reaction scheme

$$Ni(OH)_3 + Co^{2+} + 2OH^- \rightarrow Ni(OH)_2 + Co(OH)_3$$

In the first stage, the electrolyte is contacted with partially reacted $Ni(OH)_3$, where ~50% of the cobalt in the solution is precipitated. The solution, after filtration, is fed to the second stage and heated to ~85°C before reacting with fresh $Ni(OH)_3$. The purified solution is filtered and used as the feed electrolyte to the electrowinning tanks.

The $Ni(OH)_3$ required for cobalt removal is prepared electrolytically by oxidizing $Ni(OH)_2$ in electrolytic tanks using iron cathodes and nickel anodes.

### 4.8.2. Electrolysis Conditions

The purified electrolyte, at a temperature of 65°C and pH ~3.5, is electrolyzed in Hybinette cells using nickel starter sheets and lead anodes at a current density of ~200 A/m². Under these conditions, nickel is deposited at a current efficiency of ~94% and a cell voltage of ~3.7 V, the electrical energy requirement being ~3.7 kWh/kg. The finished cathodes analyze 99.95% Ni, 0.05% Co, 0.001% Cu, 0.0005% Fe, 0.0006% Pb, 0.0001% As, 0.002% S, in addition to 16 ppm of $H_2$ and ~40 ppm of $O_2$.

### 4.8.3. New Developments

Electrowinning nickel from chloride electrolyte had briefly been practiced around the year 1900 (Hoepfner process) without commercial success, mainly due to the lack of suitable materials of construction. This problem has now been overcome, and there is a renewed interest in nickel (and cobalt) electrowinning from chloride electrolyte. Advantages of chloride electrolyte include high conductivity, thus permitting the use of higher current densities without excessive energy demand. Furthermore, in chloride solutions, the anodic reaction (i.e., $Cl_2$ evolution) can be carried out at low overpotentials using the DSA® electrodes. Other advantages are high salt solubility and easier Co–Ni separation which can be accomplished by solvent extraction or by ion exchange. The "chloride nickel electrowinning route" is already being pursued actively by several nickel producers.[64,65]

### 4.9. Cobalt Electrowinning

About 60% of world cobalt production is produced at Shituru and Luilu electrowinning plants in Zaire[51] and the Rhokana electrowinning plant in Zambia. The source for cobalt electrowinning is usually the spent electrolyte from the copper electrowinning circuit, which contains ~40 g/liter Cu and 12 g/liter Co at Shituru, and 32 g/liter Cu and 40 g/liter Co at Luilu. These two plants use different techniques and are described below.[66,67]

#### 4.9.1. Shituru Operations

The feed liquor is subjected to a partial electrolytic removal of Cu, followed by a series of pH-controlled precipitations using lime to remove the remaining copper and iron. The last traces of Cu are removed by cementation with cobalt granules.

The pH of this copper-free cobalt electrolyte is subsequently raised with lime to ~8, to precipitate cobalt as $Co(OH)_2$. This is filtered and reacidified to a pH of ~7 to redissolve a portion of $Co(OH)_2$. The resulting pulp is electrolyzed in air-agitated cells using stainless steel cathodes and Pb–Sb–Ag anodes at a current density of ~400 A/m$^2$ and a temperature of ~50°C. Cobalt, deposited at a cell voltage of ~5 V with a purity of 93–95% (energy requirement: 6.5 kWh/kg of metal), is removed from the cathodes by hammering and fire refined to remove Zn. The final product is ~99% pure.

#### 4.9.2. Luilu Operations

The spent electrolyte from the copper electrowinning circuit is first subjected to Fe removal at pH = 3.3 with lime and copper hydroxide. Copper is then removed from the electrolyte in two stages at pH ~5 and ~6.2, followed by NaHS + $H_2S$ treatment, where Ni, Zn, and the remaining traces of Cu are removed. After filtration, cobalt is precipitated as $Co(OH)_2$ and leached with the spent electrolyte from cobalt electrowinning cells to a pH of ~6.5. The clear liquor analyzing ~45 g/liter Co is electrolyzed at 65°C and at a current density of 500 A/m$^2$ to produce high-purity cobalt containing 99.95% Co. About 300–400 ppm Ni, 10–30 ppm Zn, 30 ppm Fe, 20 ppm S, 8 ppm Cu, 50 ppm C, 2 ppm $H_2$, and 1–2 ppm Pb are also present in the cobalt metal produced at Luilu.

### 4.9.3. New Developments

Electrowinning of cobalt from $CoCl_2$ electrolyte is becoming increasingly popular, for the same reasons discussed in the nickel electrowinning section. Several small operations of this type are in production already, but no technical details have been published as yet.

## 4.10. Electrowinning of Other Metals

Cd, Cr, Mn, Ga, Tl, Te, and In are also produced in a limited quantity by electrowinning leach solutions. The annual production capacity of these metals is only in the range of 500 kg to 5000 tons (except Mn), at few locations throughout the world, and a brief description of these electrowinning operations is presented below.

### 4.10.1. Cadmium

About 6000 tons/year of cadmium is produced[51] electrolytically in the U.S.A., U.S.S.R., and Japan from the Cu–Cd–Zn residue from the electrolytic zinc plants. This feed material is treated with the spent electrolyte from Zn electrowinning cells followed by Cu removal using zinc dust. During a subsequent stage of Zn dust treatment, cadmium is precipitated, and the resulting spongy Cd is allowed to oxidize in air for several weeks (to promote dissolution) prior to leaching with the spent electrolyte from the cadmium electrowinning cells. Electrolysis is carried out from acidic sulfate solutions containing 100–200 g/liter Cd and 60 g/liter Zn at a current density of $\sim 80$ A/m$^2$ (cell voltage: 2.5–2.7 V; current efficiency: 90%; energy consumption: 1.5 kWh/kg), using Pb–Ag alloy anodes and Al cathodes. To prevent "treeing" during electrolysis, addition agents such as glue[68] or flocculating agents (e.g., polyacrylamide[69]) are added and bath temperature is maintained at $\sim 20$ to 30°C using cooling coils in the cells.[68]

### 4.10.2. Chromium

According to the U.S. Bureau of Mines,[70] Union Carbide Corporation Metals Division and Shieldalloy Corporation were the only manufacturers of electrolytic chromium metal in the U.S. in 1965 (annual production probably <5000 tons).

The starting material for chromium electrowinning is ferrochromium, which is crushed and leached with the spent anolyte, chrome alum mother liquor, and $H_2SO_4$. After removing suspended solids and iron sulfate, chrome alum is precipitated from the solution and redissolved to prepare the pregnant electrolyte for electrowinning.

The feed solution containing 130 g/liter $Cr^{3+}$ and 3 g/liter $H_2SO_4$ is passed to the cathode compartment where it is mixed with the circulating mixture containing ~65 g/liter $Cr^{3+}$ and 1 g/liter $H_2SO_4$, resulting in a catholyte analyzing ~12 g/liter $Cr^{3+}$ and 13 g/liter $Cr^{2+}$ (pH ~2.1–2.4). The amount of divalent chromium in the solution, which reduces the current efficiency for Cr deposition, is controlled by regulating the amount of circulating catholyte. The anolyte typically analyzed 13 g/liter $Cr^{6+}$ and 2 g/liter $Cr^{3+}$ and is separated from the catholyte by a diaphragm. Electrolysis is carried out at ~700 $A/m^2$ at ~55°C using Pb–Ag alloy anodes and 316 stainless steel cathodes, over a period of 72 hr. Under these conditions, 99.8% pure Cr is deposited at a current efficiency of 45% (cell voltage: 4.2 V; energy consumption: 18 kWh/kg), and the brittle chromium deposits are removed from the cathode blanks by hand stripping.[71]

### 4.10.3. Manganese

About 34,000 tons of manganese are electrowon annually in the free world—the major manufacturers (producing more than 10,000 tons/year) being Foote Mineral Company at New Johnsonville, Tennessee, and Union Carbide Corporation at Marietta, Ohio.[72]

Since manganese is the most electronegative metal electrodeposited from aqueous solutions, the electrolytes must be purified from all the more electropositive elements and pH controlled to suppress the hydrogen evolution reaction.

The feed solution is prepared by leaching the ores with $(NH_4)_2SO_4$ + $H_2SO_4$ solutions, followed by removal of Fe and Al at pH ~6.5, when Mo, As, and silica are also removed. Cu, Zn, Ni, and Co are removed by sulfide precipitation and Mg by crystallization of Mn–Mg–$NH_4$ sulfate at low temperatures. Electrowinning is carried out from the purified sulfate[73] solutions containing 30–40 g/liter Mn, 125–150 g/liter $(NH_4)_2SO_4$, and 0.1 g/liter $SO_2$ (also 0.008–0.016 g/liter glue) at pH = 7.2 and at a current density of 400 to 600 $A/m^2$ (cell voltage: 5.1 V; energy consumption: 8–9 kWh/kg; current efficiency: ~60%), in divided cells using Pb–1% Ag anodes and Hastelloy or stainless steel or Ti cathode blanks. Anodes are bagged to prevent the $MnO_2$ sludge (formed at the anode) from contaminating the catholyte.

Mn deposited at the cathode is chipped off the blanks by flexing or striking the cathode with a rubber mallet. (See Schlain,[74] for a more complete description of the flow sheet.)

### 4.10.4. Gallium

Gallium is produced in limited quantities throughout the world by, among others, Aluminum Company of America, Eagle-Picher Company in the U.S.A., Johnson Matthey and company in the U.K., Spolana plant in Kaznejov (Czechoslovakia), and the Nippon Light Metal Company in Japan. Ga present in aluminum minerals, Zn and Pb ores, and coal is processed by three different routes.

In France, Ga from Bayer process liquor, present only at concentrations <0.2 g/liter, is electrolyzed at a mercury cathode, and the resulting dilute amalgam is redissolved to prepare a more concentrated solution for further electrolysis.[75]

The U.S. practice involves caustic leaching of concentrates containing ~0.45% $Ga_2O_3$ and electrolysis in stainless steel vessels at 90°C at a cathodic current density of ~6000 $A/m^2$ (cell voltage 7–9 V). Ga deposited at a low current efficiency is collected and refined electrolytically to produce high-grade metal.

Johnson Matthey Company[76] treats the residual solution after Ga recovery from flue dust of producer gas plants. The solution is treated with Al scrap to cement heavy metals and Ga is extracted with isopropyl ether. After removing As as sulfide and Fe as hydroxide, the alkaline solution is electrolyzed at Pt electrodes for an initial period, and then the gallium pool is used for further electrolysis. The resulting Na–Ga amalgam is further purified.

### 4.10.5. Thallium

The annual production of Tl is only of the order of a few tons and involves treatment of flue dusts which are sulfate-leached and purified from heavy metals.[76] The saturated thallous sulfate solution is electrolyzed at ~30°C using Pt (Ni or stainless steel may also be used) cathodes and Pt anodes. (Under alkaline conditions, current efficiency decreases.)

It may be noted that the preferred alternative route for Tl recovery is cementation on Zn or Al.

### 4.10.6. Tellurium

Electrolytic tellurium is produced by Canadian Copper Refiners in Montreal.[77] Tellurium is electrowon from a sodium tellurite electrolyte using stainless steel cathodes and anodes at ~70 $A/m^2$ and 45°C. The spent electrolyte is replenished with $TeO_2$ to give ~250 g/liter Te and pH >14. Tellurium deposits are removed after three days by knocking off and breaking.

## References

1. J. R. Boldt, Jr., *The Winning of Nickel*, Longmans Canada Ltd., Toronto (1967).
2. M. J. Nicol and H. I. Philip, Underpotential deposition and its relation to the anomalous deposition of metals in alloys, *J. Electroanal. Chem.* **70**, 233 (1976).
3. T. N. Andersen, R. D. Budd, and R. W. Strachan, A rapid electrochemical method for measuring the concentration of active glue in copper refinery electrolyte which contains thiourea, *Met. Trans. B* **7B**, 333 (1976).
4. V. A. Mukhin and A. I. Levin, The Effect of the Chloride Ion on the Quality of Electrolytic Copper, *Tsvetn. Metal.* **37**(11), 36 (1964).

5. C. W. Dichrodt and J. H. Schloen, Electrolytic copper refining, in *Copper—Science and Technology of the Metal, Its Alloys, and its Compounds*, A. Butts, ed., Hafner, New York (1959), Chap. 8, p. 165.
6. C. J. Krauss, Cathode deposit control in lead electrorefining, *J. Met.* **28**(11), 4 (1976).
7. J. O'M. Bockris and G. A. Razumney, *Fundamental Aspects of Electrocrystallization*, Plenum, New York (1967).
8. M. G. Fouad and N. Ibl, Natural convection mass transfer at vertical electrodes under turbulent flow conditions, *Electrochim. Acta* **3**, 233 (1960).
9. U. Landau and J. Osterwald, Formation of grooves in copper cathodes in electrorefining, *Erzmetall* **29**(3), 103 (1976).
10. V. A. Ettel and A. S. Gendron, The role of mass transfer in designing electrowinning cells, *Chem. Ind. (London)*, No. 9, 376 (May 3, 1975).
11. S. Abe, B. W. Burrows, and V. A. Ettel, Anode passivation in copper refining, paper presented at the Annual AIME Meeting, Atlanta, March, 1977.
12. N. Ibl, Optimization of copper refining, *Electrochim. Acta* **22**, 465 (1977).
13. J. R. Rawling and L. D. Costello, Mixing characteristics of a copper refinery tankhouse cell, *J. Met.* **21**(5), 49 (1969).
14. M. G. Fouad and G. H. Sedahmed, Effect of gas evolution on the rate of mass transfer at vertical electrodes, *Electrochim. Acta* **17**, 665 (1972).
15. N. Ibl, R. Kind, and E. Adam, Mass transfer at electrodes with gas stirring, *An. Quim.* **71**, 1008 (1975).
16. A. Kapanen, R. Rantanen, and T. Mäntymäki, Recent trends of mechanization in copper refining at Outokumpu Oy., in *Extractive Metallurgy of Copper*, Vol. 1, J. C. Yannopoulos and J. C. Agarwal, eds., AIME, New York (1976), Chap. 8, p. 554.
17. H. Ikeda and Y. Matsubara, No. 3 tankhouse at the Onahama smelter and refinery, *Extractive Metallurgy of Copper*, Vol. 1, J. C. Yannopoulos and J. C. Agarwal, eds., AIME, New York (1976), Chap. 30, p. 588.
18. A. S. Gendron, V. A. Ettel, and S. Abe, Effect of cobalt added to electrolyte on corrosion rate of lead–antimony anodes in copper electrowinning, *Can. Metall. Q.* **14**(1), 59 (1975).
19. T. N. Andersen, D. L. Adamson, and K. J. Richards, Corrosion of lead anodes in copper electrowinning, *Met. Trans.* **5**, 1345 (1974).
20. Outokumpu Nickel Production, Outokumpu News, Outokumpu Oy., Helsinki, No. 2, 1967.
21. G. M. Meisel, New generation zinc plants, design features, and effect on costs, *J. Met.* **26**(8), 25 (1974).
22. E. M. Elkin, Electrodeposition of strippable metal coatings and compositions and articles useful therefore, U.S. Patent 3,523,873 (1970).
23. B. H. Spoon, Cathode for electrolytic refining of metal, such as copper, U.S. Patent 3,661,756 (1972).
24. E. Nomura, M. Aramaki, and Y. Nishimura, Electrolytic lead refining with large cell and its mechanization at Takehara refinery, paper presented at 103rd AIME Meeting, 1974.
25. Copper Electrowinning Plant for Union-Corp., *S.A. Min. Eng. J.*, 157 (July 1969).
26. F. T. Bennell, Thyristors or Copper Refining, *Chem. Ind. (London)*, No. 4, 141 (February 19, 1977).
27. T. Kitamura, T. Kawakita, Y. Sakoh, and K. Sasaki, Design, construction, and operation of periodic reverse current process at Tamano, in *Extractive Metallurgy of Copper*, Vol. 1, J. C. Yannopoulos and J. C. Agarwal, eds., AIME, New York (1976), Chap. 26, p. 525.
28. P. J. Mackey, P. Tarassoff, and G. Lemelin, Current distribution and resistances in copper refinery tankhouse cells, *Inst. Min. Metall. Trans. Sec. C*, **84**, C42 (1975).
29. R. A. Bengtsson, Polish refinery designed to streamline handling of copper anodes and cathodes, *Eng. Min. J.* 92 (February 1976).
30. Cheaper electrolytic ore extraction, *Min. Mag.* **118**(6), 397 (1968).
31. J. A. Davis, The unidirectional flow ventilation system, *Heat. Piping Air Cond.* **49**(3), 63 (1977).

32. J. Smith, Infrared scanning spots tankhouse problems, *Eng. Min. J.* **175**(1), 96 (1974).
33. T. B. Braun, J. R. Rawling, and K. J. Richards, Factors affecting the quality of electrorefined cathode copper, *Extractive Metallurgy of Copper*, Vol. 1, J. C. Yannopoulos and J. C. Agarwal, eds., AIME, New York (1976), Chap. 25, p. 511.
34. V. A. Ettel, Energy requirements in electrolytic winning and refining of metals, *CIM Bull.* **70**(783), 179 (1977).
35. D. A. Petrov, L. T. Lachev, and J. D. Popov, Method for electrolytic refining of copper at high current densities (300–700 A/m$^2$), Bulgarian Patent 10,188 (1963).
36. Non-ferrous metal data, American Bureau of Metal Statistics Inc. (1976).
37. M. Barak, The technological significance of the electrochemistry of lead, *Chem. Ind. (London)*, No. 20, 871 (October 16, 1976).
38. E. R. Freni, Electrolytic lead refining in Sardinia, *J. Met.* **17**(11), 1206 (1965).
39. F. R. Archibald, The Kristiansand nickel refinery, *J. Met.* **14**(9), 648 (1962).
40. L. P. Nicolaidis, Larymna nickel, *Mining Mag.* **125**(3), 200 (1971).
41. L. V. Volkov, The present state of electrolysis in the nickel industry and prospects for its improvement, *Tsvetn. Metall. (Non-Ferrous Metals)* **7**, 30 (1975).
42. W. W. Spence and W. R. Cook, The Thompson Refinery, *Trans. Can. Inst. Min. Metall.* **67**, 257 (1964).
43. C. L. Mantell, *Electrochemical Engineering*, McGraw-Hill, New York (1960).
44. A. H. Leigh, Precious metals refining practice, *International Symposium on Hydrometallurgy*, D. J. I. Evans and R. S. Shoemaker, eds., American Institute of Mining, Metallurgical, and Petroleum Engineers, Inc., New York (1973), Chap. 5.
45. T. S. Mackey, The electrolytic tin refining plant at Texas City, Texas, *J. Met.* **21**(6), 32 (1969).
46. R. M. MacIntosh, Tin and tin alloys—metallurgy, in *Kirk–Othmer Encyclopedia of Chemical Technology*, Vol. 20, H. Mark, ed., Interscience Publishers, New York (1969), p. 278.
47. F. S. Gaunce, G. M. Freeman, J. E. Dulson, C. E. Paden, E. G. Sharp, E. R. Hamilton, P. Salmon, and D. J. Kemp, The electrolytic zinc plant, *CIM Bull.* **67**(745), 116 (May 1974).
48. C. J. VanNiekerk and D. R. Allen, The electrolytic extraction of zinc at the zinc corporation of South Africa Limited, *J. S. Afr. Inst. Min. Met.*, 146 (February 1977).
49. L. Mager, Processes for zinc electrolysis, *Chem. Ing. Tech.* **45**, 158 (1973).
50. E. VanDen Neste, Metallurgie Hoboken-Overpelt's zinc electrowinning plant, *CIM Bull.* **70**(784), 173 (August 1977).
51. D. Gilroy, *Industrial Electrochemical Processes*, A. T. Kuhn, ed., Elsevier, Amsterdam (1971).
52. D. L. Simpson, B. H. Ensign, and K. F. Marquardson, The design of the process and facilities for the recovery of copper from silicate ores at Ray Mines Division, Kennecott Copper Corporation, paper presented at TMS Operating Metallurgy Conference, December, 1967.
53. W. A. Griffith, H. E. Day, T. S. Jordan, and V. C. Nyman, Development of the roast–leach–electrowin process for Lakeshore, *J. Met.* **27**(2), 17 (1975).
54. L. R. Verney, J. E. Harper, and P. N. Vernon, Development and operation of the Chambishi process for the roasting, leaching, and electrowinning of copper, *Electrometallurgy*, T. A. Henrie and D. H. Baker, eds., AIME, New York (1968), p. 272.
55. K. R. Rawling, Commercial solvent extraction plant recovers copper from leach liquors, *World Min.*, 34 (December 1969).
56. J. A. Holmes, L. N. Stewart, A. D. Denchar, and J. D. Parker, Design, construction, and commissioning of the Nchanga Tailings leach plant, in *Extractive Metallurgy of Copper*, Vol. II, J. C. Yannopoulos and J. C. Agarwal, eds., AIME, New York (1976), Chap. 46, p. 907.
57. H. Kudelka, R. Dobbener, and N. L. Piret, Copper electrowinning at Duisburger Kupferhuette, *CIM Bull.* **70**(784), 186 (1977).
58. A. S. Gendron, R. R. Matthews, and W. C. Wilson, Production of high quality electrorefined and electrowon copper at INCO's Copper Cliff copper refinery, *CIM Bull.* **70**(784), 166 (1977).
59. P. Charles and P. Hannaert, Fluid bed cementation of selenium contained in a copper electrolyte, *Copper Metallurgy*, R. P. Ehrlich, ed., AIME, New York (1970), p. 240.

60. W. W. Harvey, Material balance and current efficiency in electrowinning, paper presented at the 67th Annual AIChE Meeting, December, 1974.
61. W. R. Hopkins, G. Eggett, and J. B. Schuffham, Electrowinning of copper from solvent extraction electrolytes—problems and possibilities, in *International Symposium on Hydrometallurgy*, D. I. J. Evans and R. S. Shoemaker, eds., American Institute of Mining, Metallurgical, and Petroleum Engineers, Inc., New York (1972), Chap. 7.
62. Hydrometallurgy: New processes move to commercialization, *Eng. Min. J.* **177**(6), 244 (1976).
63. H. T. Brown and P. G. Mason, Electrowinning of nickel at the Bindura Smelting and Refining Company, *J. S. Afr. Inst. Min. Met.*, 143 (February 1977).
64. L. R. Hougen, R. Parkinson, J. Saetre, and G. Vanweet, Operating experiences with a pilot plant for the electrowinning of nickel from all-chloride electrolyte, *CIM Bull.* **70**(782), 136 (1977).
65. P. Mardine, S. Gratien, and L. Allais, Electrolytic cell, U.S. Patent 3,959,111 (1976).
66. R. J. M. Wyllie, Why Gecomines is a world leader in copper and cobalt hydrometallurgy, *World Min.* 42 (September 1970).
67. M. A. Bouchat and J. J. Saquet, Electrolytic cobalt recovery in Katanga, in *Extractive Metallurgy of Copper, Nickel, and Cobalt*, P. Queneau, ed., Interscience, New York (1961).
68. C. A. Hampel, Zinc and Cadmium Electrowinning, in *The Encyclopedia of Electrochemistry*, Reinhold, New York (1964), p. 1180.
69. R. J. Hopkins and J. C. Nixon, Minor metals, in *The Australian Mining, Metallurgical, and Mineral Industry*, J. T. Woodcock, ed., The Iron and Steel Institute, The Institute of Metals, Joint Library, London (1965), Chap. 10, p. 229.
70. Mineral facts and problems, *U.S. Bur. Mines Bull.* **630** (1965).
71. F. E. Bacon, Chromium electrowinning, in *The Encyclopedia of Electrochemistry*, Reinhold, New York (1964), p. 198.
72. A. G. Thomson, Manganese, *Min. Ann. Rev.*, 69 (May 1968).
73. M. Harris, D. M. Meyer, and K. Auerswald, The production of electrolytic manganese in South Africa, *J. S. Afr. Inst. Min. Met.*, 137 (February 1977).
74. D. Schlain, Preparation of primary purified metal by electrowinning and electrorefining from aqueous and non-aqueous electrolytes, Vol. 1, Part 2, *Techniques of Material Preparation and Handling*, Techniques of Metal Research, R. F. Bunshah, ed., Interscience, New York (1968), Chap. 2, p. 493.
75. L. K. Hudson, Gallium as a by-product of alumina manufacture, *J. Met.* **17**(9), 948 (1965).
76. W. Ryan, *Non-Ferrous Extraction Metallurgy in the United Kingdom*, Institute of Mining and Metals, London (1968).
77. G. Bridgestock, E. M. Elkin, and S. S. Forbes, Operations at Canadian Copper Refiners Limited, *Can. Min. Met. Bull.*, 773 (October 1960).

# 7
# Electroplating

***CHRISTOPH J. RAUB***

## 1. Introduction

In one of the earliest textbooks for the electroplater (*Der Galvanotechniker*, M. Zapfe, Leipzig, 1911) one can read: "in electroplating, theory and practical applications are closely interconnected. Theory possesses such an overwhelming importance as rarely in an artistic trade or if one prefers in a trade art." This touch of an artistic trade was, for a long time, connected with practical electroplating, and only in the last 50 years has it moved to a modern industrial production process, which is applied to a wide range of products.

## 2. Present-Day Technology of Electroplating with Metals

Electrodeposited layers today serve different purposes:

(1) to ensure corrosion protection to materials which otherwise would be harmed,
(2) to provide decorative surface layers for metals or nonmetals,
(3) to give materials certain technological properties which they otherwise would not possess and which do not change over a long period of time and

---

**CHRISTOPH J. RAUB** • Forschungsinstitut für Edelmetalle und Metallchemie, D7070 Schwäbisch Gmünd, West Germany.

(4) to facilitate savings of expensive metals and reduce materials costs by using thin layers which are just thick enough to last for the expected lifetime of the whole part.

For many applications one layer has to fulfill more than one of these requirements, a typical example being electrodeposited gold for modern electrical connectors.[1-5] Firstly, gold deposits have to prevent base materials, usually copper- and nickel-base alloys, from forming high-resistance tarnishing or corrosion layers. Secondly, they have to provide a surface with low electrical contact resistance and superb wear properties which do not change under corrosive conditions, even at longer times and which, in addition, maintain good welding and soft soldering properties. Thirdly, instead of using gold alloys for producing solid connector pins, thin layers of gold of about 2-$\mu$m thickness are deposited. Only these enable mass production of many millions of pieces at a reasonable price.

Another example is chromium layers in the automobile industry. They have to protect the underlying steel material against corrosion, fulfill a decorative purpose, provide a highly wear-resistant surface in piston rings and on shafts, and do all this on mass-produced articles at thicknesses between about 0.1 and 100 $\mu$m (depending on the application).

The various applications of silver show how electrodeposits can fulfill technical demands due to their chemical and physical properties. In the beginning of this century, silver plating of cutlery or tableware made of brass or German silver replaced, to a high extent, traditionally handcrafted products, since the decorative effect was identical, yet the price quite different. However, it is not only the aesthetic appearance which makes silver a choice material for cutlery; it is also its high resistance against aggressive organic acids in food.

In addition, the bactericidic action of Ag ions ensures that silver-coated articles are in general sterile. For technical purposes the properties of high reflectivity, low resistivity, and high heat conductivity of electrodeposited silver layers are used for constructing mirrors, HF waveguides, and medium to heavy-duty electrical contacts. The superb gliding properties and good heat conductivity are the reasons for the extensive use of silver layers for heavy-duty engine bearings or pivots of which extreme reliability is demanded. Again, massive silver or its alloy would be too expensive and, furthermore, would not exhibit the mechanical properties of, e.g., a steel bearing shell coated with silver.

These examples already show a trend which has occurred more frequently in the last years, namely, the change of interest in electrodeposits for functional, rather than decorative, applications.

In the beginning of electroplating, fairly little was known about properties of deposits. Extensive investigations have provided us now with new insights into the mechanism of deposition from practical electrolytes and into the properties of the deposits in order to ensure production quality and reproducibility. This is especially true for gold, chromium, and nickel layers. The decorative effect has changed little since the first layers were produced, but their technical properties have changed substantially. There are Au–Co alloys for connectors, Au–Cu–Cd layers for slip rings, hard chromium for use as a wear-resistant coating, black chromium for solar collectors, and microcracked and multilayered chromium and nickel for better corrosion protection. Another well-known use is tin electroplates on thin steel for the food-processing industry which today has replaced traditional tin galvanizing and requires about 6 tons/year of tin.[6]

In general, most of the electrodeposits should protect a surface from which the attack by chemical corrosion or mechanical wear starts. Plastics, for example, can be provided with highly scratch- and wear-resistant surfaces by these coatings, which makes the material (in its technical use) comparable to a piece made from metal.

## 3. Electrolytically Produced Metallic and Nonmetallic Layers

Besides protective coatings made by applications of an outside voltage to cathodically polarized articles in an electrolyte, the term "electroplating" encompasses the catalytic or electroless deposition of metals and alloys and the electrolytic production of nonmetallic, mostly oxidic layers. These layers, produced mostly anodically, not only ensure corrosion protection, but are now used, because of their dielectric properties, for capacitors, e.g., $Al_2O_3$, $Nb_2O_5$, and $Ta_2O_5$. The anodization of aluminum and its alloys has become one of the most important branches of electroplating in recent years.

Anodizing produces corrosion-resistant and protective layers which can be colored for architectural applications in a wide range of shades. Hard anodized layers improve the wear resistance of aluminum to a high extent. Furthermore, electrolytically generated $Al_2O_3$ layers are highly absorbent and can be impregnated with a variety of materials to form the base for gluing metal to metal, as in the aircraft industry.[7–9]

The electroless deposition of, e.g., copper, silver, gold, nickel–phosphorus, and nickel–boron alloys on other metals or nonmetals can be achieved either electrochemically or by addition of chemical reductants to an electrolyte. These methods have found a wide range of technical interest, especially in the coating of nonmetallics, e.g., ceramics, plastics, and in the printed circuit industry.[10] Electroplating shops often do metal coloring also, which is a kind of controlled corrosion or tarnishing of metallic surfaces which have often been previously electroplated. Other fields, considered to be part of electroplating even if they work without application of outside voltage, are the production of protective coatings, e.g., of the chromate, phosphate, or oxalate type.

## 4. Electroforming

Electroplating is used to produce thin layers on surfaces of various materials or to generate quite heavy articles by a process called "electroforming." Electroformed gold vessels were among the first articles to be produced in the nineteenth century by the newly discovered gold electroplating processes.[11] Today heavy, self-supporting objects are electroformed which are not in sufficient demand for large-scale mass production. Previously, the process had been used almost exclusively for artistic purposes, e.g., to produce copies of works of art or to manufacture printing plates.

In recent years it has been applied more often to generate pieces of elaborate design for technical applications which are too difficult to manufacture otherwise by turning, milling, etc. Metals to be used for this purpose are, out of technical and price considerations, mostly copper and nickel, e.g., for rocket nozzles with built-in cooling windings, for HF waveguides, and for tools and forms for the plastics industry. It should be mentioned, too, that quite a high amount of copper foil used in the printed circuit industry is produced electrolytically on a continuing basis.

## 5. Solutions Used in Electroplating

Electrolytes for electroplating are mostly aqueous solutions. Organic electrolytes have been developed for deposition of various metals, but have until now not found a larger-scale technical application. In the 1960s anhydrous electrolytes for deposition of aluminum gained some interest in the electronics and precision parts industry.[12] Electrolysis of molten salts was one of the earliest methods to be used for reactive metal production. Today it is applied commercially to produce ductile layers of platinum from cyanides[13,14] and tantalum from halogenide-containing melts.

Yet these electrolytes have certain drawbacks which prevent their general use. They react with air and moisture, and electrolysis has to run at a deposition temperature of at least several hundred degrees. It is not always easy to deposit smooth, coherent layers from salt melts, since the metals often tend to crystallize as coarse dendrites or crystalline powders which cannot be used for coating purposes. Therefore this deposition is generally handled only in specialized shops or divisions.

Aqueous electrolytes used for practical electroplating limit the choice of metals to be deposited often to those which have deposition potentials nobler than the one for hydrogen. Others can be discharged only because of the high hydrogen overvoltage, but in most cases hydrogen is codeposited. Metals which cannot be deposited from aqueous electrolytes are, e.g., the light metals: beryllium, magnesium, silicon, and aluminum, which are of great interest as oxide-forming protective layers and the refractory elements: titanium, zirconium, hafnium, vanadium, niobium, tantalum, molybdenum, and tungsten. Experiments have been made to deposit some of these elements as alloys, but the deposits normally have a high concentration of built-in oxides or other salts; thus deposits are difficult to get in a dense, coherent form.

In addition to electrolytic processes discussed, dense layers of metals and alloys may be obtained by mechanical cladding, flame spraying, and evaporating or sputtering in a vacuum. For low-melting metals, e.g., aluminum, zinc, and tin, methods have been developed to produce layers by dipping the base metal into the molten metal. The disadvantage of this method is that, due to the fairly high temperature, the coating reacts to form brittle, intermetallic layers which may cause problems in later stages of manufacturing, but which, on the other hand, may be advantageous, such as those of $FeSn_2$ on tin. For some refractory elements and compounds, processes which take advantage of the thermal decomposition of metal salt vapors with or without addition of reducing gases (e.g., hydrogen, carbohydrides, or carbon monoxide) have been developed and are in a production stage for tungsten, molybdenum, tantalum, and iridium deposits.

## 6. Alloy Plating

Despite inherent limitations, electrolytic deposition distinguishes itself from other methods by the variety of the properties of the deposits which can be adjusted in selecting the deposition parameters such as electrolyte composition, temperature, current density, etc. By varying only the current density, we can, for example, change the tensile strength of Au–Co or Au–Ni alloys by nearly a factor of 2. An increase in deposition temperature of the same gold electrolyte from 20 to 50°C reduces internal tensile stresses by a factor of 10 (Figure 1).

The different electrochemical properties of the metals make it impossible to deposit certain technically interesting alloys. There is at present no process for the electrolytic production of stainless steel films or Pd–Ag alloys for electrical contacts. On the other hand, alloy deposits which cannot be produced metallurgically with the same properties have found many special and potential applications. Among the electrodeposited alloy systems that have attained commercial importance are brass, bronze,[15] nickel–cobalt, lead–tin, tin–zinc, nickel–tin, copper–tin–zinc, and a wide variety of gold alloys.[16]

An interesting example is presented by electrodeposited nickel–tin alloys.[17,18] Prepared in the usual metallurgical way by melting, casting, etc., alloys with 65% tin and 35% nickel have attained little technical interest. Electrodeposited alloys of the same composition are of a crystal structure, not found in the equilibrium state at room temperature, and form hard and

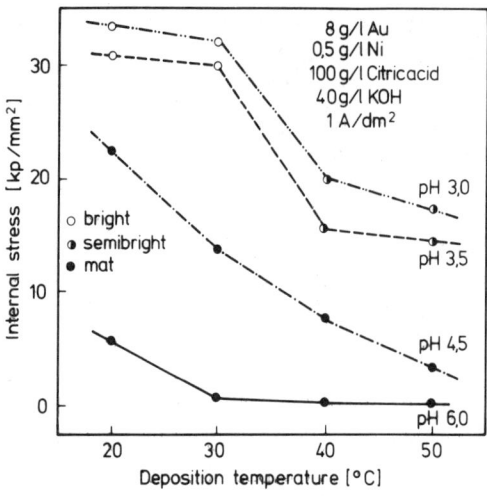

Figure 1. Stress of gold–nickel deposits vs. electrolyte temperature for various pH values.

wear-resistant surface layers which are significant in contact applications. Gold–cobalt alloys deposit in metastable solid solution.[19] Heat treatment causes phase segregation and hardening of the deposits.

Another, indirect way to produce alloys by electroplating is by heat treatment of separately deposited sandwich structures, forming alloys between the various deposits and/or the base metals. The process has been developed to the extent where it is possible to continuously produce homogeneous brass coatings on a steel strip about 3 ft wide by alternate electrodeposition of copper and zinc.[20] Heat treatment is also used to increase homogeneity of Au–Cu–Cd layers and improve their wear properties. However, only by very careful process control is it possible to obtain uniform layers in this way.

Void formation results from the Kirkendall effect. Decomposition of included foreign matter is possible,[4] as is the occurrence of brittle, intermetallic phases between the deposited layers themselves and/or the base metal. An intermetallic phase is desirable for better corrosion behavior of tin plate. Therefore, after deposition the tin layer is heated above its melting point (flow melting) in order to form a layer of $FeSn_2$, which is corrosion resistant to attack by acids from food. In another application, flow melting of electrodeposits has proved very advantageous.

Electrodeposited tin films in printed circuit boards tend to develop tin whisker growth, which may cause harmful short circuits.[21] By flow melting these layers the whisker growth can be reduced or avoided completely.

Furthermore, *in situ*, solder can be produced, completely covering a complicated shape, e.g., if the thicknesses of silver and copper sandwich layers are adjusted in such a way that they correspond to a eutectic composition of Ag-70, Cu-30, a heat treatment of the coated part at temperatures above 780°C will generate the liquid solder and produce a joint even in small crevices, where solder would not ordinarily flow. Certainly these liquid–solid reactions have to be carefully controlled, since they proceed much faster compared to solid–solid diffusion, a process which is used to generate layers of a nickel–cadmium intermetallic compound on turbine blades by diffusing electrodeposited layers of nickel and cadmium, alternately. The melting point of the compound is higher than that of cadmium, and it possesses high resistance to combustion gases.

Electroplating processes differ from electrowinning or refining methods by the fact that the latter attempt to deposit the metal as cleanly as possible— the cleaner the better. The structure of the deposits is of little importance compared to purity. Electrodeposited 24 carat gold results from cyanide-based electrolytes in a shiny smooth surface, while gold refined in a chloride electrolyte of a purity of better than 99.999 is obtained as a coarse, spongy, dendritic structure adhering to the cathode. Therefore, in technical language the expression "electrolytic metal" is always connected with the understanding that it describes a metal of high purity. This is not true for electrodeposits.

For these the deposition form and the properties of the layers are most important; purity is generally a secondary consideration. Therefore, different electrolytes are used for electrowinning and electroplating. The former ones in general are rather clean metal–salt solutions, while electroplating electrolytes contain an additional variety of organic or inorganic substances to influence the deposition form in the desired way. Therefore, the electroplated layers are generally fairly impure, containing various nonmetallic inorganic or organic impurities.

Bright gold–cobalt deposits from acid cyanide electrolytes, e.g., possess up to several hundred ppm of hydrogen, oxygen, nitrogen, and carbon[2] (Figure 2). Bright deposits are often rather impure. They incorporate molecular

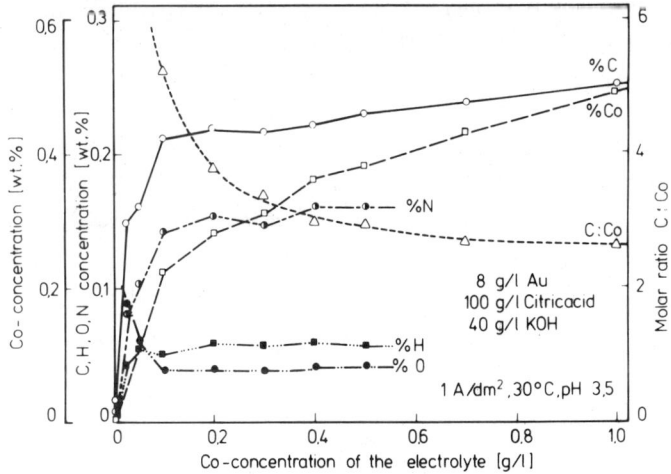

Figure 2. Cobalt, carbon, and gas contents, C:Co molar ratio of deposits vs. cobalt concentration of an Au–Co electrolyte.

inclusions of organic substances which cause large changes in the properties of the metal. Only in the lower concentration range is this change proportional to the impurity concentration; at higher levels, no clear relationship exists.

## 7. Properties of Deposits

By molecular inclusion the hardness of copper can increase to more than 300, of silver to 200,[16] and of iron to nearly 800 kp/mm$^2$ (Figure 3). These hard iron deposits equal chromium in their wear properties and are used for combustion engine parts.[22] Higher inclusion rates also have a strong influence on the electrical resistivity. Due to the presence of insulating films or filaments, lattice defects, etc., in the electroplated metal, its resistivity may be increased by a factor of 10$^3$ compared to that of the pure metal. Chemical properties change as well. Sulfur-containing bright nickel deposits corrode more easily than semibright or matte layers free of sulfur. In order to take advantage of the special properties of bright nickel coatings, new ways and methods had to be discovered to further improve corrosion resistance. This is done by using double nickel layers, the lower one being a sulfur-free, matte nickel on top of which a sulfur-containing bright layer is deposited.

Organic inclusions also account for differences in density between electrodeposited fine gold (24 carat) and hard gold–cobalt (23 carat) deposits. An alloy with 0.118% cobalt was found to have a density of 16.6, compared to 19.3 g/cm$^3$ for the pure metal.[23]

Heat treatment strongly influences the properties of materials with included nonmetallics. Hard gold layers with too high a carbon content cannot be soft soldered properly and lose their good wear resistance. Silver with included 3.5 wt % citrate shows a decrease of resistivity from 10$^3$ to 4 $\mu\Omega$ cm, after being annealed for a short time at 110°C. An interesting behavior is

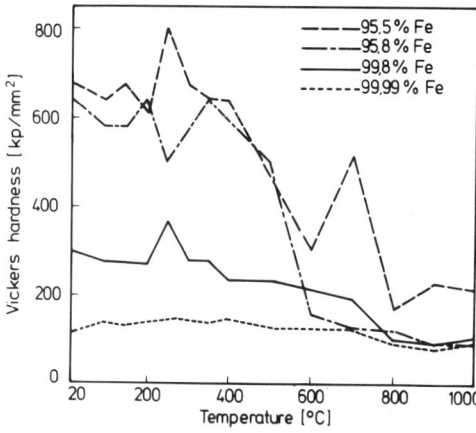

Figure 3. Hardness of various Fe deposits vs. annealing temperature.

exhibited by a 95.5 wt % iron sample containing citrate. After a decrease in hardness due to decomposition of citrate at lower temperatures at 700°C, a hardness peak is observed, which is caused by carbon diffusing into the nickel layer.[22] The hardness of electrodeposited chromium, depending on the deposition parameters, is mostly due to codeposition of chromium oxide and chromium hydride and is not a property of the metal itself.

Deposits with included substances generally have a characteristic lamellar microstructure which may be coarse or fine. They sometimes show typical fiber texture, depending on the substrate and the amount of additions, but also can be completely disordered. The X-ray Debye–Scherrer patterns are indistinct and may not show up at all at high diffraction angles. With high amounts of included material, they may approach an amorphous state, the famous example being amorphous antimony. X-ray studies showed that the inclusion of foreign matter in the metal lattice causes numerous defects and only small, coherent lattice ranges remain. Because of the close relationship between alloy and single-metal deposits with inclusions of nonmetallic substances, it is sometimes very difficult to discern the influence of an alloying metal from that of an inclusion.

Electrodeposited metals incorporated with nonmetallic substances are comparable to extremely age-hardened, supersaturated solid solutions. A difference exists in alloys age hardened by diffusion and agglomeration; after lengthy heat treatment, the normal properties of the alloy are regained. In electrodeposits the influence of nonmetallic impurities cannot be removed since nonmetallic compounds do not diffuse in the lattice. Inclusions often decompose by gas evolution, and the remaining metal shows gas porosity and blisters; therefore, the properties of the annealed state are not completely attained. Small age-hardening effects (as observed in the tensile strength of gold deposits) are caused by the greater thermal expansion of incorporated, nonmetallic impurities. Age hardening in differently prepared gold alloys is seen in Figure 4. Similar effects can be observed in silver with a very low inclusion concentration.

Bright deposits are obtained primarily by organic additions agents, the so-called "brighteners," which were discovered rather empirically. Their great number and variety makes it difficult to establish some kind of scientific order.[24] Brightness of the surface is caused by the adsorption of brighteners or their decomposition products on the cathodic surface. The adsorption is irreversible so that a partial covering of the cathode is formed, and incorporation of the adsorbed substances in the crystallizing metal will occur. Depending on the type and quantity of the molecule incorporated, addition agents or their decomposition products influence the electrolytic crystallization of the deposited metal.

It is typical of bright deposits that the position of the crystal faces at the surface of the base metal influences the deposit to a very small thickness (1–2 $\mu$m), and the typical structure of the bright metal begins to grow independently

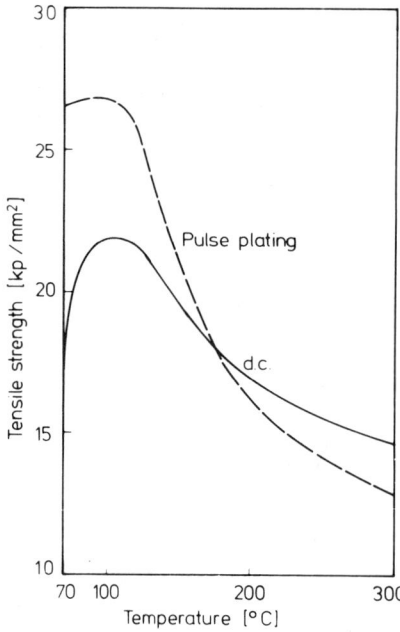

Figure 4. Tensile strength of gold deposits vs. annealing temperature.

of the structure of the basis surface. Growth may occur disorderedly or with a characteristic texture. In matte deposits the structure of the base surface can still be observed up to very high thicknesses. The typical brighteners are also typical crystallization inhibitors by which the cathodic potential is changed to more negative values.

On the other hand, compounds which contain cathodically reducible S compounds can cause a depolarization of metal deposition, as in nickel. Addition agents which serve as brighteners mostly polarize the deposition of a metal more than that of hydrogen. Furthermore, they are often cathodically reduced. The current efficiency for bright deposits is thus lower than 100%.

## 8. Inhibitors

The sensitivity of metals to inhibitors varies. Metals of low sensitivity are, e.g., lead and tin; of medium sensitivity, silver and copper; and of high sensitivity, palladium and nickel. As a rule one can state that inhibitor sensitivity is proportional to the melting point of the metal.

It has been shown in many investigations that bright deposits have adsorbed surface layers of two types: one consists of a netlike loose structure, which is difficult to observe. The included brightener or its decomposition product are generally found (in this case) near the bordering planes of crystal

Figure 5. Lamellar structure of a bright nickel deposit (magnification 100×). Reproduced at 76%.

columns. The other type of layer is relatively dense and can be detected as a separate amorphous film on the cathode surface.

These layers, which can be penetrated by ions, are the reason for a lamellar structure, as is best known for bright nickel (Figure 5). Technical electrolytes contain several brighteners which make systematic investigations rather tedious. Sometimes impure products are used, and it is difficult to establish the influence of the various components since decomposition, polymerization, and condensation processes may occur.

## 9. Throwing Power

Of great practical importance for electroplating are "macro- and microthrowing power."[25,26] The macrothrowing power is determined by the primary current distribution or the local electrical field strength. The electrochemical properties of the systems change this primary current distribution to the secondary one, depending upon the various polarization factors. In general, polarization increases with rising current density. The current density on protruding areas is therefore significantly reduced, more than in recesses. Polarization thus acts as an equalizer which is more pronounced the higher the polarization. Good macrothrowing properties of electrolytes can be expected if deposition occurs with high concentration polarization.

Macrothrowing power of the electrolytes is determined experimentally in small electroplating cells of special geometrical shape (Haring–Blum and Hull cells). Acid electrolytes which contain high concentrations of dischargeable metal ions in a hydrated form have a worse macrothrowing power, compared to alkaline solutions in which dischargeable complexes are present only in small concentrations.

Of similar importance is the "microthrowing power," a term used for surface profiles on which the primary current distribution has zero influence. In this case at identical current densities, only processes responsible for the transport of dischargeable ions through the cathodic diffusion layer are important for metal distribution. Microthrowing is therefore confined to surface dimensions in which effective thickness differences within the cathodic diffusion layer exist. In electrolytes with high concentrations of dischargeable ions, the difference in metal ion concentration within the diffusion layer is relatively small, and the concentration of dischargeable ions is nearly uniform over the whole cathode. Therefore concentration polarization is low and microthrowing power good. Only at high current densities may critical thickness differences of the cathodic layer be found.

In electrolytes with low concentrations of dischargeable complexes, the depletion of the diffusion layer is pronounced. Local thickness differences become very effective because of the high concentration polarization. At places of low dischargeable metal concentration, the polarization increases and current density (or growth rate) drops. A typical example is cyanide–copper electrolytes. Copper is normally discharged via the complex $[Cu(CN)_2]^-$. During copper deposition the concentration of this complex in the diffusion layer is reduced rather quickly, but that of $CN^-$ increases. This higher $CN^-$ concentration certainly does not influence the $[Cu(CN)_2]^-$ migration into the implied diffusion layer, but it disfavors the formation of new $[Cu(CN)_2]^-$ according to the equation

$$[Cu(CN)_3]^{2-} \rightleftarrows [Cu(CN)_2]^- + CN^-$$

and increases concentration polarization.[25]

From the earlier discussions it is realized that there is a change in metal distribution going from macro- to microthrowing. In electrolytes with little concentration polarization, macrothrowing is bad but microthrowing power good. Solutions with high macrothrowing power have low microthrowing properties. All factors which tend to increase migration of dischargeable complexes through the diffusion layer such as stirring, increase in temperature, improve microthrowing, but reduce macrothrowing properties of the electrolyte. In silver electrolytes, microthrowing power strongly depends upon the $CN^-$ concentration. At low $CN^-$ concentrations, microthrowing power is good; at high ones, bad. This is because at low $CN^-$ concentration, the discharge occurs from the AgCN molecule or directly from hydrated $Ag^+$ (formed from $AgCN \rightleftarrows Ag^+ + CN^-$) and at higher ones via the $[Ag(CN)_2]^-$ complex.

These considerations have to be taken into account for the proper usage of electrolytes. For profiles with sharp corners, the thickness distribution will be better for deposits from a cyanide electrolyte. If the profiles are shaped in such a way as to prevent good convection, their throwing power is likely to be worse compared to acidic ones. By strong stirring, this may be improved since the thickness of the diffusion layer is reduced.

## 10. Leveling

Another effect dependent upon the diffusion layer is "leveling," which describes the smoothing of a surface by an electrodeposit.[25] It is caused by organic "levelers" which are added to the electrolyte in small concentrations and are always incorporated in the deposit, the inclusion dropping with current density. During deposition in the cathodic layer, the concentration of these levelers drops. Local thickness differences in the diffusion layer effect a different supply of levelers. On peaks diffusion layer is thin, and supply of levelers faster than in crevices where there is a thicker diffusion layer. The higher concentration of leveler on these peaks increases polarization and decreases current density, shifting the areas of higher current density towards the deepenings. Deposition rate there becomes faster, and the surface profile is smoothened.

Another kind of leveling is geometric leveling. This happens in narrow notches if, in electrolytes of good microthrowing power with high growth rate in the tip of the notch, the layers growing in field direction from both sides of the notch meet in the center after some time, filling it up. Leveling is very important since it reduces the pretreatment of grinding and polishing.

## 11. Further Work on Alloys

If not organic compounds, but other elements of the periodic system, are incorporated in various concentrations, we enter the field of alloy deposition, which has been mentioned earlier. A. Brenner has published a two-volume book, thoroughly discussing alloy deposition (see "Suggested Reading" at the end of this chapter).

Hydrogen is the most common alloying element, as stated in 1949 by Blum and Hogabom. On zinc and cadmium surfaces the overvoltage of hydrogen is very high. Hydrogen evolution occurs after the limiting current density is exceeded. The deposits themselves do not contain any hydrogen in supersaturated solution, and the hydrogen escapes in molecular form. Hydrogen found within the deposits is formed, e.g., by the reactions[27]

$$Zn + H_2O \rightleftarrows ZnO + H_2$$

or

$$Zn + Zn(OH)_2 \rightleftarrows 2ZnO + H_2.$$

A certain amount of $H_2$ may be trapped in voids as molecular hydrogen, as is also assumed for gold.[28]

Very high supersaturation of hydrogen is observed in elements which can be deposited only simultaneously with hydrogen, e.g., iron, cobalt, nickel, and palladium.[29] For nickel, the hydrogen of the deposits is rated low at low pH values. At high pH values (in chloride electrolytes over pH 5 and in sulfate solutions over pH 6.3), the hydrogen content increases suddenly. The explanation is the different hydrolysis of nickel salts.

In chloride solutions the hydrolysis, followed by adsorption and inclusion of basic nickel salts, occurs at lower pH values. The strong lattice defects caused by these inclusions lead to a very high hydrogen enrichment. These strains can also be found if a second metal is codeposited; the only problem is to distinguish between the influence of the metal and its incorporated salts, as is the case in nickel–zinc alloys. In a Watt's electrolyte containing nickel and zinc, the deposition of zinc is favored due to the strong polarization of nickel deposition and the equally strong depolarization of zinc by zinc hydrolysis products.

This preferred deposition of the less noble metal is called "abnormal electrodeposition." In electrodeposited, face-centered, nickel–zinc solid solutions, the hydrogen content increases strongly with zinc concentrations until the saturation limit of zinc is reached[16] (Figure 6). At higher concentrations, it drops down to zero at about 80% Zn.

It is further possible to stabilize unstable hydrides by electrolysis. By proper electrolysis conditions, e.g., hcp CrH can be produced.[30] The existence of a face-centered crystal Cr hydride was established, yet its composition could not be determined. The supersaturated hydrogen solutions decompose in vacuum (CrH) and even in air (Ni–H, Pd–H).

By alloy electroplating not only intermetallic compounds (not occurring at equilibrium conditions), e.g., Sn–Ni, are produced, but greatly extended solid solutions as well. This may be achieved in the Cu–Pb system by specific inhibition. Normally copper, the nobler metal, would be deposited exclusively from Cu–Pb electrolytes. This can be avoided by specifically inhibiting the

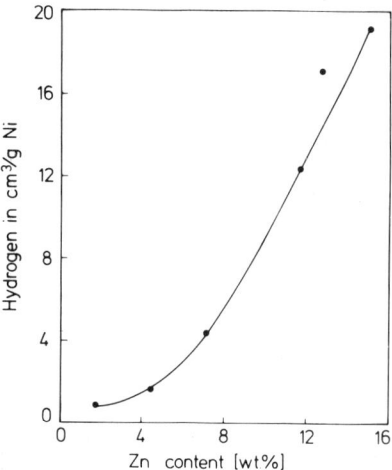

Figure 6. Hydrogen concentration of Ni-Zn deposits.

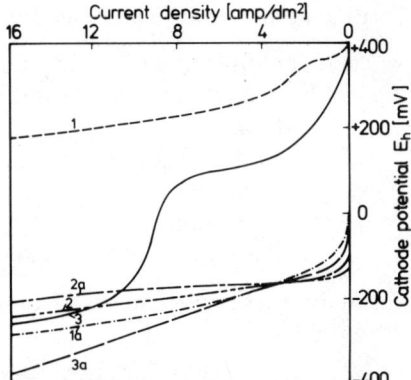

Figure 7. Current density–potential curves for various Cu, Pb, and Cu+Pb perchlorate electrolytes. 1, 60 g/liter thiourea; 1a, +1 g/liter thiourea; 2, 185 g/liter Pb; 2a, +1 g/liter thiourea; 3, 60 g/liter Cu+185 g/liter Pb; 3a, +1 g/liter thiourea.

copper discharge by adding thiourea to a perchloric acid–copper–lead solution.[31,32] Normally the deposition potential of copper is about 0.4 V more noble than that of lead. If only 1 g/liter of thiourea is added, the potential curve of copper is shifted by about 0.45 V to more negative values, the lead potential remaining unchanged (Figure 7). From this, thiourea electrolytes containing Cu-Pb solid solution deposits with up to 12 wt % Pb are produced as fcc solid solutions. A similar extension of the solid solution range by electrodeposition can be obtained in Ag–Bi, Ag–Pb, and Cu–Sn alloys.

## 12. Effect of Complexing

The cathodic discharge of mass particles can be strongly controlled by complexing the metals in solution. This complexing is quite different for the various metals. It also varies depending upon the concentration of the complexing anion, e.g., of $CN^-$. The stability of the predominant complex controls the formation of the lower coordinated dischargeable complex, and thus the discharge potential of a metal. Therefore it is possible to achieve similar discharge potentials of metals to deposit as alloys or reverse potentials by just altering the concentration of the complexing compound. For cyanide $\alpha$-brass plating, the zinc content of the deposits increases with rising $CN^-$ concentration of the electrolyte, because the stability of the copper cyanide complex $[Cu(CN)_3]^{2-}$ or even $[Cu(CN)_4]^{3-}$ is much higher. The discharge of copper in a cyanide bath is controlled by the $[Cu(CN)_2]^-$ complex. Its concentration depends upon the stability of the dominant, more highly coordinated complexes. Addition of NaOH favors codeposition of zinc too, because it facilitates the formation of dischargeable $Zn(OH)_2$ according to the equation

$$[Zn(CN)_4]^{2-} + 4OH^- \rightleftharpoons [Zn(OH)_4]^{2-} + 4CN^-$$

Three general rules can be stated for alloy deposition under normal conditions:

(1) The discharge of the most noble metal is favored, causing deposits to be richer in this metal than would be expected from their concentration.
(2) All variations in electrolysis conditions which increase polarization shift the composition to the side of the less noble metal.
(3) All variations in electrolysis conditions which decrease polarization increase the concentration of the more noble metal in the alloy.

Certainly, these rules are valid only at current densities below the limiting current density.

A further variation in alloy composition can be achieved by codeposition of solid particles producing "dispersion-strengthened" alloys. This is a very modern development for the production of wear-resistant layers (Ni + $Al_2O_3$)[33] or for superconducting cables (Cu + $Nb_3Ge$, Al).[34] It has been used to produce grinding wheels, with diamonds embedded either in electrodeposited nickel or bronze.

## 13. Summary

This short survey can only cover some of the problems in electroplating. Today more and more, the artistic or alchemistic approach is replaced by a detailed knowledge of the basic principles. We are no longer looking at electrodeposits because of their brightness only, but because they are highly estimated elements used in nearly every product from food cans coated with tin to tiny gold-plated connector pins in a computer.

## Suggested Reading

Blum, W., and G. B. Hogaboom, *Principles of Electroplating and Electroforming*, McGraw-Hill, New York (1949).

Brenner, A., *Electrodeposition of Alloys*, New York, Academic, New York (1963).

Drossbach, P., *Grundriss der algemeinen technischen Electrochemie*, Borntraeger, Berlin-Nikolassee (1952).

Fischer, H., *Elektrolytische Abscheidung und Elektrokristallisation von Metallen*, Springer, Berlin (1954).

Kardos, O., and D. G. Foulke, Applications of mass transfer theory, electrodeposition on small scale profiles, in *Advances in Electrochemistry and Electrochemical Engineering*, Vol. 2, Interscience, New York.

H. W. Dettner, and J. Elze, *Handbuch der Galvanotechnik*, Hanser, Munich (1966).

E. Raub and K. Müller, *Fundamentals of Metal Deposition*, Elsevier, Amsterdam (1967).

Milazzo, G., *Elektrochemie. Theoretische Grundlagen und Anwendungen*, Springer, Vienna (1952); first English edition (translated from the Italian manuscript) appeared as *Electrochemistry, Theoretical Principles and Practical Applications*, Elsevier, Amsterdam (1963).

Vagramyan, A. T., and Z. A. Soloveva, *Technology of Electrodeposition*, Izdatel, Akad. Nauk SSSR. Inst. Fiz. Khim. Moscow (1960).
Volmer, M., *Das elektrolytische Kristallwachstum*, Hermann and Cie, Paris (1934).

## References

1. Ch. J. Raub, *Z. Werkstofftech.* **4**, 359 (1973).
2. A. Knödler, *Galvanotechnik* **68**, 3 (1977).
3. R. G. Baker and T. A. Palumbo, *Plating* **58**, 79 (1971).
4. Ch. J. Raub, A. Knödler, and I. Lendvay, *Plating* **63**, 35 (1976).
5. F. H. Reid and W. Goldie, *Gold Plating Technology*, Electrochemical Publications, Ayr, Scotland (1974).
6. Proceedings of the Technical Symposium, September 13, 1974, Bad Neuenahr, Rasselstein, A. G., ed., Neuwied, pp. 5–45.
7. S. Wernick and R. Pinner, *The Surface Treatment and Finishing of Aluminum and its Alloys*, R. Draper Ltd., Teddington (1959).
8. T. Biestek and J. Weber, *Conversion Coatings*, Portcullis, Redhill, England (1976).
9. Ch. J. Raub, in Proceedings of the 24th International Congress of IUPAC, Hamburg, September 2–8, 1973.
10. F. W. Mirth and H. Speckhardt, *Galvanotechnik* **65**, 105 (1974).
11. L. B. Hunt, *Gold Bull.* **8**, 16 (1973).
12. R. Dötzer, *Galvanotechnik* **67**, 967 (1976).
13. A. Weser, *Metalloberfläche* **29**, 581 (1975).
14. R. N. Rhoda, *Plating* **49**, 69 (1962).
15. L. Missel, *Plating*, 36 (October 1978).
16. E. Raub, *Plating* **63**(1), 29 (1976); **63**(2), 30.
17. M. Antler, M. Feder, C. F. Hornig, and J. Bohland, *Plating*, 30 (1976).
18. J. H. Thomas III, *J. Electrochem. Soc.* **124**, 677 (1977).
19. Ch. J. Raub, H. R. Khan, and J. Lendvay, *Gold Bull.* **9**, 123 (1976).
20. G. Falkenhagen, in *Herstellung von kaltgewalztem Band*, Part II, Stahleisen, Düsseldorf (1970), p. 192.
21. A. Politycki and H. P. Kehren, *Z. Metallkde.* **59**, 309 (1968).
22. E. Raub, Mitt., Forschungsgesellsch., Blechverarbeitung und Oberflächenbehandlung, (1951), p. 279.
23. F. H. Reid, *Metalloberfläche* **30**, 453 (1976).
24. H. Fischer, *Metalloberfläche* **1**, 28 (1947).
25. E. Raub and K. Müller, *Fundamentals of Metal Deposition*, Elsevier, Amsterdam (1967), p. 201.
26. O. Kardos, *Plating* **6**, 129, 229 (1974).
27. E. Raub and A. Knödler, *Trans. First Met. Finish.* **38**, 13 (1961).
28. G. J. Clark, C. W. White, D. D. Alfred, B. R. Appleton, and C. W. Magee, *Nucl. Instr. Methods* **149**, 9 (1978).
29. H. D. Hedrich and Ch. J. Raub, *Metalloberfläche* **31**, 512 (1977).
30. A. Knödler, *Metalloberfläche* **17**, 161 (1963).
31. E. Raub, G. Delhoust, and A. Disam, *Metalloberfläche* **22**, 75 (1968).
32. Ph. Javet, N. Ibl, and H. E. Hintermann, *Electrochim. Acta* **12**, 557 (1967).
33. F. K. Sautter, *J. Electrochem. Soc.* **110**, 557 (1963).
34. H. R. Khan and Ch. J. Raub, *J. Less Common Met.* **43**, 49 (1975).

# 8
# Electrochemical Machining

## JAMES P. HOARE and MITCHELL A. LABODA

### 1. Introduction

To fashion a block of metal according to a given pattern, one uses a tool of harder material to scrape or gouge out the softer material of the workpiece. All mechanical machining methods are based on some form of this operation. As the physical properties of new materials, developed to meet the requirements of a continuously advancing technology, reached higher degrees of strength and heat resistance, the demands placed on the properties of the tool materials to machine these "super alloys" (particularly in the aerospace industries) became practically impossible. Consequently, other methods of metal removal have been sought in more recent times.

Among the so-called nontraditional or unconventional methods of metal removal, one may list the following: ultrasonic machining (USM), which is performed by a cutting tool oscillating at high frequencies (about 20,000 Hz) in an abrasive slurry[1-4]; electrical discharge machining (EDM), where metal is removed from a workpiece electrode by a series of discrete electrical discharges across a gap filled with a dielectric fluid between the tool electrode and the workpiece[5-10]; electron beam machining (EBM), which is based on the local vaporization of material by a focused beam of high-energy electrons[8,11,12]; laser beam machining (LBM), which is used to drill holes in various materials

---

*JAMES P. HOARE and MITCHELL A. LABODA* • Electrochemistry Department, Research Laboratories, General Motors Corporation, General Motors Technical Center, Warren, Michigan 48090.

by focusing a high power density laser beam on the workpiece[8,13]; plasma arc machining (PAM), where the molten material of the workpiece is liquified by electron bombardment and the convective heating of a plasma (generated by heating a volume of gas by an electric arc) is blown away by the high velocity gas stream[13]; and ion beam machining (IBM), which is based on the sputtering of material from the workpiece under bombardment of a focused beam of ions.[13-15]

Another group of unconventional machining processes is based on the atom-by-atom removal of metal by controlled anodic corrosion processes. This group includes electropolishing, jet etching, electrolytic or electrochemical grinding (ECG), and electrochemical machining (ECM). It is with this group of machining processes that this chapter will be concerned. At this point, a brief digression into the basic principles of corrosion processes should be helpful.

## 2. Corrosion Process Fundamentals

### 2.1. Local Cell Corrosion

When a metal sample is placed in contact with an electrolyte (in many cases a water solution of an ionic salt such as NaCl or an acid), an atom leaves the metal lattice by giving up an electron to the metal surface and enters the liquid phase as an ion. At the same time, a reduction process (gain of electrons) takes place in the corroding system, providing a sink for the electrons accumulating on the metal surface. The metal dissolution process can continue until the metal phase is consumed.

Since the metal surface sites are microscopically heterogeneous, it is possible that the oxidation or metal dissolution process can take place at active sites such as peaks, while the reduction process (usually reduction of $O_2$ to $H_2O$ or $OH^-$ ions or the evolution of $H_2$ from $H^+$ ions) can occur at protected sites such as valleys on the same metal surface. Such a metal dissolution process is known as local cell corrosion[16] in which the local cell current is carried electronically in the metal by migration of electrons from the active to the protected sites and electrolytically by the migration of ions in the electrolyte phase between the two sites. Because the metal phase exhibits an equipotential surface, the oxidation and reduction processes must proceed at the same potential, which is known as the mixed potential.[17] As a result, either the anodic (oxidation) or the cathodic (reduction) process or both must be polarized from the respective equilibrium potentials to this mixed potential value by the local cell current.

In the absence of an externally applied voltage, all wet corrosion takes place by a local cell mechanism. When the anodic reaction occurs on only a few sites compared to a large number of cathodic sites under certain conditions,[18-22] the anodic local current density becomes very large, causing a highly destructive, localized corrosion condition known as pitting. Such a condition is to be avoided since perforation of a metal wall may occur in a relatively short time. However, corrosive agents are used by metallographers for selective attack on certain crystal faces so that the grains appear under different degrees of shading when viewed through a microscope. When the grain boundaries are selectively attacked, they appear as dark lines.

By choosing the proper electrolyte, a metal sample may dissolve uniformly over the surface by general corrosion. As pointed out by Evans,[23] the attack of the metal begins at susceptible sites (such as lattice defects, corners, or edges). When such active material is removed, the attack shifts to some other, more active site. In this way, the anodic attack wanders over the entire metal surface, resulting in general corrosion. General corrosion is the basis of the process known as chemical milling, chemical solution machining, or pattern etch machining[24-30] in which the desired cutting pattern is applied to the metal surface as a mask of chemically resistant film so that only the unprotected regions of the workpiece surface are corroded when the part is exposed to the etchant. For more precise work, the pattern is reproduced photographically.[26] After the workpiece surface is covered by a film of a light-sensitive, chemically resistant material, known as a photoresist, the photographed mask is applied to the metal surface, and the part is exposed to light. By removing chemically the unexposed portions of the photoresist, the workpiece is exposed to the etchant and the required pattern in the metal is obtained. Since the metal removal rate depends on the kinetics of the local cell reactions, the only control of the etching process is through the control of temperature or concentration of the etchant. Consequently, for a given temperature and concentration, the depth of the cut is determined solely by the length of time for which the part is exposed to the corrosive medium. Applications of this process in microelectronics have been employed in the fabrication of printed circuits[29] and integrated circuits.[27]

## 2.2. Bimetallic Corrosion

Because the rate of local cell corrosion is controlled by the kinetics of the anodic or cathodic reaction or both, the metal dissolution process is relatively slow. If the local cell is under cathodic control,[16] which is the most common case (Reference 18, p. 876), the rate of corrosion may be increased by placing the corroding metal in electrical contact with a metal on which the cathodic reaction can proceed more easily. Such a process is known as bimetallic corrosion[31,32] and is important in the architectural design and construction industries.[33-36]

## 2.3. Anodic Corrosion

To increase the corrosion rate of a given metal still further, one may connect a dc, external source of potential, such as a battery or rectifier, to the corroding metal so that it is made anodic with respect to a physically separate counterelectrode. Only the anodic process takes place on the corroding metal under these conditions, and the electrolyte provides a means of maintaining the anodic and cathodic processes physically separate. The rate of metal dissolution can be raised or lowered by increasing or decreasing, respectively, the applied potential. Such a process is called anodic corrosion.

It is not possible in most cases to obtain an unlimited increase in the metal removal rate with increasing potential because most metals in a given electrolyte become passive at a critical anodic potential value. An idealized potentiostatic polarization curve for a given anode in a given electrolyte is sketched in Figure 1. In this case, the metal is inert below a threshold potential (denoted A in Figure 1), and the corrosion rate is zero (region of cathodic protection). At potentials more noble than A, the current increases with increasing potential, and in the absence of passive film formation, the corrosion rate would increase continuously along the curve toward point C. However, the initiation of a passive film usually begins at the critical point, B, after which the current falls off due to a blockage of the active sites by the protective film. In the passive region, P, the corrosion rate has fallen to very low values since the surface is covered by the protective film. With further increases in the potential, a point, T, is reached where either the protective film breaks down, exposing the metal surface to further attack, or a new electrochemical process (such as the evolution of oxygen) takes place on the filmed surface. In either case, the current increases again with increasing potential in the transpassive region. Film breakdown can occur by the interaction of anions adsorbed from the solution with the protective film on the anode surface.[37-39] Clearly the formation and breakdown of anodic films depends strongly on the electrolyte composition.[39]

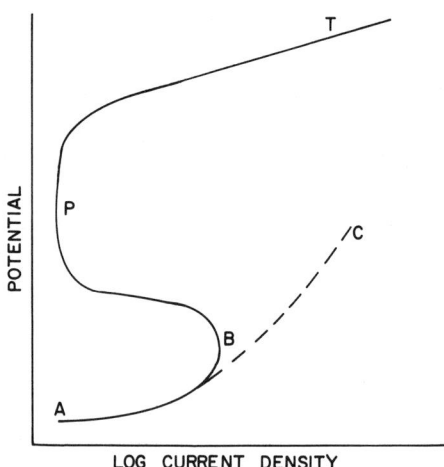

Figure 1. Sketch of idealized, steady-state, potentiostatic polarization curve obtained on a given anode in a given electrolyte.

The development of machining processes based on anodic corrosion mechanisms, will be discussed next.

## 3. Electropolishing

In 1930, Jacquet[40] reported that certain metals could be polished by anodizing them at sufficiently high current densities in certain acids or mixtures of acids. He referred to this process as electrolytic polishing, but in more recent times, the contracted term, *electropolishing*, is more commonly used. A complete history of the development of electropolishing from a laboratory curiosity to an important industrial process has been traced by Jacquet.[41] A partial listing of some electropolishing baths for several metals is recorded in Table 1. More detailed lists can be found in the literature.[44,47,54–56]

In general, the part to be polished is made anodic by connecting it to the positive side of a dc power source (battery or rectifier). The cell is filled with the proper polishing electrolyte. The cathode may be constructed from an insoluble conductor, such as stainless steel, graphite, lead, or platinum. Its area is usually about 100 or 1000 times that of the anodic workpiece. The distance between anode and cathode ranges between 2.5 and 15 mm.[54,57] The applied current density must be sufficiently high enough to produce and maintain a thick anodic film on the metal surface. To obtain a polished surface, between 0.025 and 0.05 mm of metal must be removed. Consequently, electropolishing may be used to size an oversized part or to deburr a conventionally machined piece.[56] In addition, electropolishing increases the corrosion resistance of the machined surface.[47,58] Although electropolishing yields a stress-free, thermally unchanged, polished surface, it is a slow metal-removing process compared to ECG and ECM.

Electropolishing solutions may be classified in two general categories. Those electrolytes having high electrical conductivity comprise the first group and are based on solutions of $H_3PO_4$, $H_2SO_4$, and $CrO_3$ with or without certain organic addition agents, such as glycerol, alcohols, and ethers. To obtain the required polishing current density, low applied voltage is required, generally between 1 and 25 V. Water is often added to the mixture of acids to increase the conductivity of the solution.

In the second category of electropolishing solutions are those having low conductivity and requiring the application of high applied voltages in the range of 50–220 V to obtain the required current densities. These electrolytes are based on mixtures of $HClO_4$ with organic reagents such as acetic acid and acetic anhydride. Due to the explosive nature of $HClO_4$,[59] most industrial processes employ electrolytes based on $H_3PO_4$.

There is a critical current density below which the anodic film is not formed and polishing is not obtained. Above the critical value, the surface becomes more lustrous at the higher current densities, and the polishing time is less.

## Table 1
### Electrolytes for Electropolishing Certain Metals

| Metal | Electrolyte | Current density, mA/cm$^2$ | Temperature, °C | Polishing time, min | Reference |
|---|---|---|---|---|---|
| Aluminum | 20% $HClO_4$ + 80% acetic anhydride (by weight) | 200–400 | 38 max | 15 | 42 |
| Copper | 63% $H_3PO_4$ + 37% $H_2O$ (by weight) | 20–50 | 22 | 5 | 43 |
|  | 65% $H_3PO_4$ + 15% $H_2SO_4$ + 6% $CrO_3$ + 14% $H_2O$ (by weight) | 200–250 | 22–25 | 2–5 | 44 |
| Lead | 30% $HClO_4$ + 70% glacial acetic acid (by volume) | 1300–1600 | 22 | 5 | 45 |
| Nickel | 73% $H_2SO_4$ + 27% $H_2O$ (by weight) | 1000–3000 | 22 | 2 | 46 |
|  | 65% $H_3PO_4$ + 15% $H_2SO_4$ + 6% $CrO_3$ + 14% $H_2O$ (by weight) | 200–250 | 22–25 | 2–5 | 44 |
| Tin | 20% $HClO_4$ + 80% acetic anhydride (by volume) | 600–1000 | 32 | 8–10 | 47 |
| Tungsten | 10% $NaNO_3$ + 90% $H_2O$ (by weight) | 200–400 | 22 | 20–30 | 45 |
|  | $Na_3PO_4$ (160 g/l) | 600 | 38–49 | 10 | 45 |
| Zinc | 25% KOH + 75% $H_2O$ (by weight) | 1000 | 22 | 15 | 48 |
|  | 17% $CrO_3$ + 83% $H_2O$ (by weight) | 2000 | — | — | 49 |
| Brass | 16% $CrO_3$ + 84% $H_2O$ (by weight) | 2000 | — | — | 49 |
|  | 70% $H_3PO_4$ + 30% $H_2O$ (by weight) | 500 | 26 | 5 | 45 |
| Steel (carbon) | 48% $H_3PO_4$ + 40% $H_2SO_4$ + 12% $H_2O$ (by weight) | 3000 | 35–49 | 10 | 50 |
|  | 10% $HClO_4$ + 90% glacial acetic acid (by weight) | 1500–2500 | 26 | 0.5–2 | 51 |
| Steel (stainless) | 30% $H_3PO_4$ + 60% $H_2SO_4$ + 10% $H_2O$ (by weight) | 1800 | 49 | 2 | 52 |
|  | 42% $H_3PO_4$ + 45% glycerol + 13% $H_2O$ (by weight) | 100–600 | 93–149 | 8–15 | 53 |

Operation with the first class of electrolytes may be carried out at temperatures above room temperature, but that with the second class should be performed at or below room temperature.

## Theory

The current density vs. applied voltage curve obtained on Ni in 9 $N$ HCl solution is shown in Figure 2. Similar observations have been made on Cu in $H_3PO_4$,[57,61,62] on Ni in $H_2SO_4$,[63] and on steel in $H_3PO_4$.[64,65] The curve consists of three parts. In the low potential region, the system is inert until the current rises rapidly to a maximum value, at which time the surface is etched. Beyond the maximum, the current falls to a virtually steady value, independent of the applied voltage in this second region. Here, a thick anodic film is formed on the metal surface, and under the film, the surface takes on a mirror-bright polish. The third section of the curve occurs at high anodic potentials where the current rises once more since current can be consumed in the evolution of $Cl_2$. In the case of Cu in $H_3PO_4$, the evolution of $O_2$ accounts for the current increase at high applied voltage.[66]

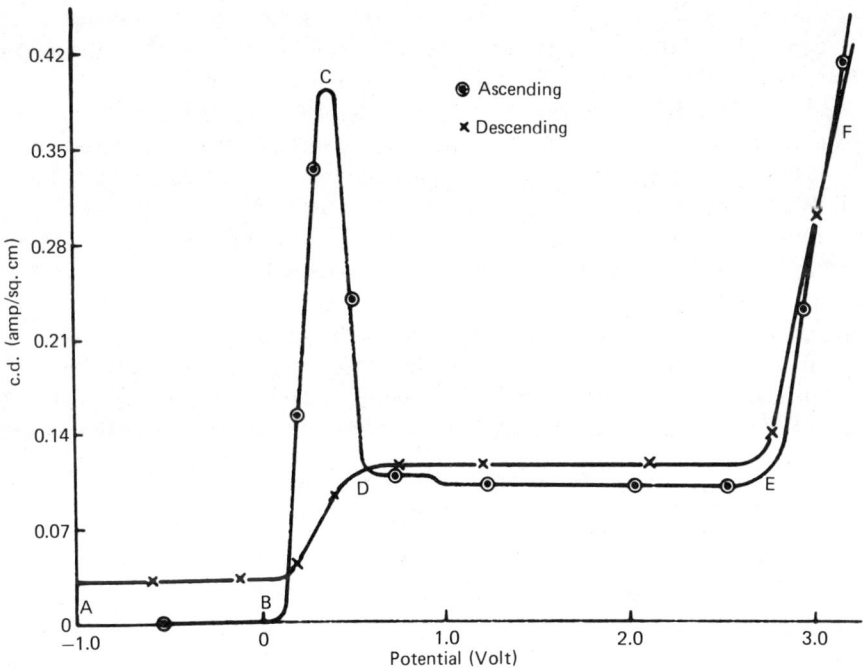

Figure 2. Applied voltage–current-density curve obtained on a Ni anode in 9 $N$ HCl.[60] (With permission of The Electrochemical Society, Inc.)

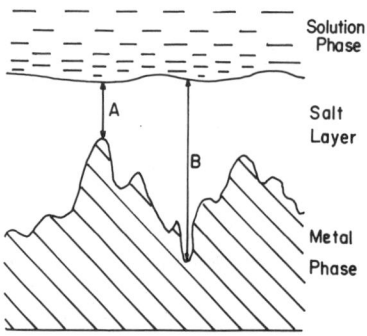

Figure 3. Sketch of smoothing mechanism of a metal anode during electropolishing.

Any theory of electropolishing must take into account a macroscopic effect (a leveling process which wears down the peaks) and a microscopic effect (a brightening process which eliminates etch patterns).[67]

Because electropolishing takes place in the limiting current region, Jacquet[43,68] concluded from studies of Cu in $H_3PO_4$ solutions that the metal surface must be covered with a salt layer, the thickness of which is greater than the surface roughness of the metal, as diagramed in Figure 3. Accordingly, since the distance, A, from a peak to the boundary of the salt film is shorter than the distance, B, from a valley site, the ir drop along path A should be less than along path B. In this way, a higher rate of metal dissolution occurs at the peaks than in the valleys, thus producing a leveling of the surface.

This film resistance model proposed by Jacquet was considered by Elmore to be incomplete since he was convinced that diffusion plays an essential role in the leveling process. According to Elmore, the metal dissolves until a salt film builds up on the anode surface. A limiting current is set up because the rate of the anodic corrosion of the metal is limited by the diffusion of metal ions through the salt layer to the bulk of solution. Since the diffusion path at A is shorter than that at B in Figure 3, the rate of metal removal will be greater at peaks than in valleys, producing a leveling of the surface.

If metal ions leave the anode surface predominantly by diffusion, the current density, $j$, is proportional to the concentration gradient, $\partial c/\partial x$. For a plane anode of area $A$ and located at $x = 0$, the current density is given as

$$j = I/A = FD(\partial c/\partial x)_{x=0} \tag{1}$$

where $I$ is the current, $F$ the Faraday constant, and $D$ the diffusion coefficient for the metal ions. After current has passed for a critical time period (transition time, $t_0$), the concentration of metal ions at $x = 0$ reaches the solubility limit, $c_m$, when the salt film is formed. At time periods greater than $t_0$, the current is limited by the concentration gradient according to Eq. (1).

To analyze the growth of the diffusion layer, the diffusion equation

$$D(\partial^2 c/\partial x^2) = \partial c/\partial t \tag{2}$$

must be solved. At $t = 0$, assume that the concentration of metal ions in solution is zero, and let $j = j^0$. By using standard methods,[61] a solution to Eq. (2) is

$$c(x = 0, t) = [2j^0/F(\pi D)^{1/2}]t^{1/2} \tag{3}$$

When $t = t_0$, $c(x = 0, t) = c_m$, then

$$j^0 t_0^{1/2} = \tfrac{1}{2} c_m F(\pi D)^{1/2} = \text{const} \tag{4}$$

Elmore[61] recorded the proper data obtained on Cu anodes in $H_3PO_4$ electrolyte. For a given acid concentration, the product of $j^0$ and $t_0$ was a constant value. The value of the product varied with the acid concentration because $c_m$ is a function of the acid concentration. These data support the conclusion that the electropolishing process in the limiting current region proceeds by a diffusion-controlled mechanism. The stability of the product of the current density and square root of the transition time has been confirmed in other investigations.[69,70]

From viscosity studies of the electropolishing of Cu in $H_3PO_4$ solutions, Edwards[57] concluded that the polishing action did not depend on reaching a critical $Cu^{2+}$ ion concentration at the anode surface. On the contrary, it was concluded that the limiting step was the diffusion of $PO_4^{3-}$ ions through the salt film to react with $Cu^{2+}$ at the anode surface. This diffusion of a so-called acceptor species can account for the smoothing or leveling action of the process.

Wagner,[71] using this acceptor concept, derived an expression to account for the experimental observations made during the electropolishing of Cu in $H_3PO_4$. He considered a surface with a sine wave profile with wavelength $\lambda$ and amplitude $b_n$. If the diffusing acceptor species complexes rapidly with the dissolving metal ions, then the concentration of acceptors at the metal surface should be zero. Let $y$ be the distance from the average surface plane; then

$$c = 0 \quad \text{and} \quad y = b_n \sin(2\pi x/\lambda) \tag{5}$$

Assuming that $\lambda \ll \delta$ and $b_n \ll \delta$, where $\delta$ is the effective thickness of the hydrodynamic boundary layer, Fick's second law, Eq. (2), applies. At steady state, $\partial c/\partial t = 0$.

A particular solution to Eq. (2) is

$$c(x, y) = B[y - b_n \sin(2\pi x/\lambda) e^{-2\pi y/\lambda}] \tag{6}$$

By differentiating Eq. (6) with respect to $y$, substituting $y = b_n \sin(2\pi x/\lambda)$, expanding the exponential function, and neglecting terms involving higher powers of $(b_n/\lambda)$, the concentration gradient at the surface is

$$\left(\frac{\partial c}{\partial y}\right)_{y = b_n \sin(2\pi x/\lambda)} = \left(\frac{\partial c}{\partial y}\right)_{\text{avg}} \left(1 + \frac{2\pi b_n}{a} \sin\frac{2\pi x}{\lambda}\right) \tag{7}$$

only if $b_n \ll \lambda$ and $\lambda \ll \delta$.

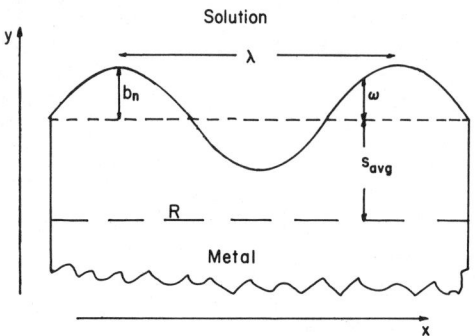

Figure 4. Sine-wave profile of anode surface.

The next consideration pertains to the change in shape of the surface profile with time, $t$. It is convenient to consider a reference plane, $R$, inside of the metal phase as shown in Figure 4 so that $s_{avg}$ is the distance of the average surface plane $R$. Letting $w$ be the distance of a point on the metal surface from the average surface plane, then the distance, $s$, of a point on the surface from $R$ is

$$s = s_{avg} + w \tag{8}$$

and

$$w = b_n \sin(2\pi x/\lambda) \tag{9}$$

Since the decrease in $s$ and $s_{avg}$ per unit time is proportional to the local and average metal dissolution rate, which, in turn, is proportional to the local and average concentration gradient, one may write in accordance with Eq. (7)

$$-ds/dt = c[1 + 2\pi b_n/\lambda \sin(2\pi x/\lambda)] \tag{10}$$

where

$$c = -ds_{avg}/dt \tag{11}$$

By substituting Eq. (8) into Eq. (10), one obtains

$$-dw/dt = c(2\pi b_n/\lambda)\sin(2\pi x/\lambda) \tag{12}$$

Dividing Eq. (12) by Eq. (11) yields

$$dw/ds_{avg} = (2\pi b_n/\lambda)\sin(2\pi x/\lambda) \tag{13}$$

The decrease in the distance $s_{avg}$ is equal to the displacement, $u$, of the average surface plane from its position at $t = 0$. Then,

$$-ds_{avg} = du \tag{14}$$

Substitution of Eqs. (9) and (14) into Eq. (13) gives

$$-db_n/du = 2\pi b_n/\lambda \tag{15}$$

Integrating gives

$$b_n = b_n^0 \, e^{-2\pi u/\lambda} \qquad (16)$$

where $b_n^0$ is the amplitude at $t = 0$, and $b_n$ that at time $t$. As $\lambda$ decreases, $b_n$ becomes less, showing that the amplitude of the short sine waves will decrease more rapidly than that of the longer waves.

For a sawtooth or square wave profile, the surface may be described by a Fourier series. During electropolishing, the shorter wavelengths will disappear until the profile approaches that associated with the longest wavelength of the original pattern. Edwards[57,72] electropolished grooved Cu disks and found that, as the average surface plane receded from its original position, the surface profile approached a pure sine wave pattern as predicted by Wagner's model.

Recently, Zamin and co-workers[73] attempted to verify Wagner's model by studying the electropolishing of Al in a $H_3PO_4$-based electrolyte. From Eq. (16), the loss of metal per unit area, $\Delta m/A$, is

$$u = \Delta m/A = (a/2\pi) \ln (b_n^0/b_n) \qquad (17)$$

where $b_n$ and $b_n^0$ have the same definition as in Eq. (16) and the charge, $Q$, involved is

$$Q = it = (F/W_E)(a/2\pi) \ln (b_n^0/b_n) \qquad (18)$$

where $F$ is Faraday's constant and $W_E$ is the equivalent weight of Al. When, however, they determined the ratio of $Q_{cal}$ to $Q_{obs}$ from weight-loss and profile measurements, it was not possible to distinguish between the Elmore and the Wagner models from the data obtained.

Edwards[57] and Hickling and Higgins[74] electropolished laminated anodes composed of sheets insulated from one another. Alternate sheets were connected in parallel with one set recessed from the other. When the end of this composite electrode was placed in solution, it approximated a surface with periodic peaks and valleys. Current density measurements showed that the current at the peaks was higher than that in the valleys. Nicholas and Tegart[75] postulated from these reports[57,74] that the current density (metal removal rate) at any point on the anode surface exceeds the average current density by an amount proportional to the curvature of the surface at that point. For the rate of change of the distance, $s$, of a point on the surface to the reference plane, $R$, in the anode, they write

$$-\partial s/\partial t = c - \alpha (\partial^2 s/\partial t^2) \qquad (19)$$

where $\alpha$ is a constant with the dimensions of a diffusion coefficient. Assuming a sine wave profile, a solution was obtained of the form

$$s = s_1 - c_t + b_n^0 \sin (2\pi x/a) \, e^{-4\pi^2 \alpha t/a^2} \qquad (20)$$

from which a linear relation between $\Delta m/A$ and $\ln (b_n^0/b_n)$ can be derived. Since Edwards' data[57,75] give a linear plot of $\Delta m/A$ as a function of

($\ln b_n^0/b_n$), the electropolishing process can be described by a model which does not require the knowledge of the diffusing species.

Such diffusion-controlled mechanisms can account for the leveling action but not the microscopic or brightening action of the electropolishing process. The brightening of the surface results from the suppression of selective metal dissolution of preferred crystal planes which causes etch patterns and a matte surface.

To account for the brightening action, Edwards[57,72] suggested that the diffusing acceptor species (either $PO_4^{-3}$ ion or $H_2O$) complexed with the metal ions liberated at the metal surface. He reasoned that the complexing process did not affect the crystal structure of the surface, and hence, a brightening effect would be expected. It is difficult to see how the homogeneous complexing process can prevent the heterogeneous selective attack of the surface.

Hoar[63,76] and later Higgins[60] observed that the current–voltage curves obtained on metals electropolished in various acid electrolytes (e.g., Figure 2) were very similar to the anodic polarization curves of Figure 1. The initial peak in the electropolishing curve suggested to Hoar that a compact oxide film was formed on the metal surface. According to this model, a metal ion can leave the anode surface only at sites below a pore in the oxide film. Since there is a random distribution of pores in the film, there is a random dissolution of metal at the anode surface. As a result, preferential attack of certain crystal faces does not occur, etch patterns are not generated, and a brightening of the metal surface is observed. This dual layer accounts for the experimental observations. Leveling is obtained from the diffusion-controlled passage of metal ions through a relatively thick salt film to the bulk solution, and brightening is obtained from the random metal dissolution through the pores of a thin oxide film lying beneath the salt film on the anode surface.

Hoar and Rothwell[77] studied the effect of solution flow on the electropolishing characteristics as well as on the shape of the polarization curves. As the flow rate is increased, the limiting current region of the polarization curve is shifted to higher current values, as shown in Figure 5. Similar results have been obtained by others.[60,66,70] Such behavior is indicative of a diffusion-controlled process. It has also been noted that brightening of the surface occurred in only part of the limiting current region in agreement with the reports of Powers.[78]

The dual film theory can account for these observations. In general, the curves of Figure 5 show that a peak in the current is obtained before the limiting current region appears. This peak signals the formation of a salt film, and in the initial part of the limiting current region electropolishing is not observed. Only smoothing is observed here. With increasing potential, a second peak (or dip) appears which is associated with the formation of a complete, uniform, porous, thin oxide film under the salt film. At this point, brightening of the surface begins as designated by the triangles in Figure 5. If the flow rate is increased to even higher flow rates, such as 20 cm/sec, the first peak disappears, the dip is

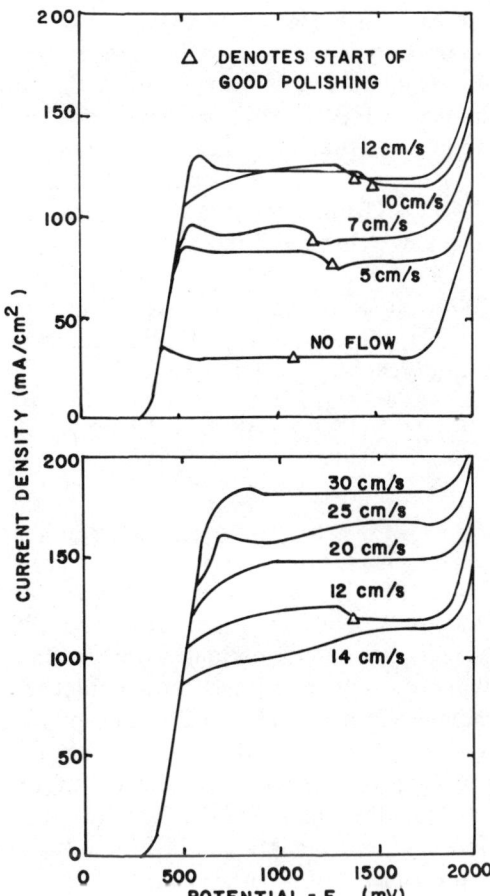

Figure 5. The effect of solution flow rate[77] on the polarization curves obtained on Cu anodes in 6 $M$ orthophosphoric acid at $18 \pm 1°C$. (With permission of Pergamon Press.)

absent, and the surface is not electropolished. At such high flow rates, the salt film is swept away so rapidly that the uniform oxide film cannot be formed. Similar results have been observed[79] during the electrochemical machining of steel. When the oxide film is formed from a salt film (steel in $NaClO_4$ solutions), brightening of the surface is dependent on the rate of solution flow; but if it is formed by direct oxidation of the surface (steel in $NaClO_3$ solutions), brightening is independent of flow rate.

As a demonstration that an oxide film is formed during electropolishing, Hoar and Farthing[62] designed a cell such that small drops of Hg could be dropped from a microburette onto the Cu anode at any location during the electropolishing of Cu in $H_3PO_4$ solutions. In those regions where etching of the surface takes place, the Hg drops wetted the surface; but in the limiting current region where brightening occurs, the Hg beaded and did not wet the surface. Since Hg will wet only an oxide film-free surface, electropolishing occurs only when the thin oxide layer exists under the salt film.

A number of investigations of the nature of the salt film formed on Cu anodes in $H_3PO_4$ solutions have been reported.[66,80-83] Although reduction studies of Allen[81] indicate that the salt layer is composed of $Cu_3(PO_4)_2$, Balasev and Nikitin[82] detected the presence of both $CuH(PO_4)_2$ and $Cu_3(PO_4)_2$ from determinations of the composition of the electrolyte at various stages of polishing. The electron diffraction studies by Williams and Barrett (83) show the film to be $Cu_3(PO_4)_2$. Although $CuH(PO_4)_2$ is formed in the first 200 sec of polishing, it is converted afterwards to $Cu_3(PO_4)_2$.

Copper dissolves in $H_3PO_4$ according to Laforgue-Kantzer[84] by the overall reaction involving water,

$$Cu + H_3PO_4 + H_2O \rightarrow (CuOH)H_2PO_4 + 2H^+ + 2e \quad (21)$$

or according to Krichmar and Galushko,[85]

$$Cu + 2H_3PO_4 \rightarrow Cu(H_2PO_4)_2 + 2H^+ + 2e \quad (22)$$

To test the Wagner model,[71] Kojima and Tobias[70] made a galvanostatic study of the electropolishing of Cu in $H_3PO_4$ as a function of time; in addition weight-loss determinations were conducted on the system. According to Wagner,[71] there should be a depletion of either $H_3PO_4$ or $H_2O$ at the metal–solution interface if Eq. (21) applies or a depletion of $H_3PO_4$ if Eq. (22) is valid. From derived kinetic equations relating the current density to the transference numbers of the concerned species and the concentration gradient and from the solution of the diffusion equation [Eq. (2)], they showed that the data assuming depletion of $H_3PO_4$ at the anode surface did not fit the derived expressions. For example, plots of $jt^{1/2}$ vs. $H_3PO_4$ concentration were not straight lines. However, a plot of $jt^{1/2}$ vs. critical copper phosphate solubility did yield straight line plots.

It appears that the diffusion of metal ions through a salt film is a more favorable mechanism for the smoothing action than the depletion of acid molecules at the anode surface.

The results of ellipsometric studies of Cu anodes in concentrated $H_3PO_4$ solutions confirmed the conclusion from polarization studies[60,70,77,86] that a thin oxide film exists under a thick salt layer at the anode surface being electropolished. From an analysis of the ellipsometric parameters, Novak and co-workers[87] estimated that a thin metal oxide film about 40 Å thick was formed on the anode surface beneath a thick viscous salt layer between 2000 and 3500 Å thick. The salt layer was not composed of a colloidal suspension of precipitate particles. It is known[57] that increasing the viscosity of the electrolyte improves the electropolishing action, and glycerine is used in a number of formulations.[47] The addition of certain surface-active agents improves the properties of various electropolishing baths.[88-90]

At the present time, it seems that the dual anodic film theory proposed by Hoar[67,77] to account for the electropolishing of metals in strong acid solutions is the preferred model. There is evidence[77,91] that brightening can be obtained in the absence of an oxide film if the salt film is compact enough since effectively a random pore system can be set up. Comprehensive reviews of the electropolishing process exist.[44,67,92]

## 4. Jet Etching

Since electropolishing is a general corrosion process, the anodic corrosion must be localized if one wishes to drill small holes through or etch fine lines on a metal sample. Uhlir[93] described an apparatus in which the electrical current flowing between the cathode and anode is directed down a nonconducting tube made of glass or plastic as shown in Figure 6. The end of the glass tube is drawn down to a fine tip, and the electrolyte is passed from a reservoir to the submerged anode through the tip which is positioned exactly over the point on the workpiece where the hole is to be drilled. A Pt wire sealed in the reservoir through the wall of the container serves as the cathode. When a direct current is passed between the cathode and the anode, most of the current follows a path to sites on the metal surface directly beneath the tip because all other paths are longer and have a higher electrical resistance. Consequently, metal is removed only from the region of the anode directly under the glass tip.

Since the tip of the glass tube behaved electrochemically the same as if the tip were replaced by a metal cathode, Uhlir referred to this tip as a "virtual cathode." After a cavity has been formed in the workpiece surface, a hole of uniform bore may be obtained by advancing the virtual cathode into the anode. A wide variety of shapes can be generated by relative motion between the virtual cathode and the workpiece anode such as a series of equally spaced shallow grooves 2 $\mu$m wide in the metal surface. There is a lower limit to the size of holes drilled or the width of grooves cut in a metal anode determined by the fact that as the tip diameter is made smaller, the ohmic resistance of the virtual cathode increases to the point where the current density falls and the drilling rate is reduced to unsatisfactory, low values. For a given tip diameter, the current density (metal removal rate) increases with applied voltage until a critical value is reached where the rate suddenly falls to zero. At this point, the $I^2R$ heating of the circuit has caused the solution temperature to reach the boiling point.

In the jet etching process described by Tilley and Williams,[94] a jet of solution was forced by compressed air through a virtual cathode against the anodic workpiece a few millimeters away and enclosed in a splash container. Metal remover occurs directly under the jet because of the large ohmic resistance of other current paths. A major advantage of this forced flow system over that described by Uhlir is the elimination of uneven etching due to the

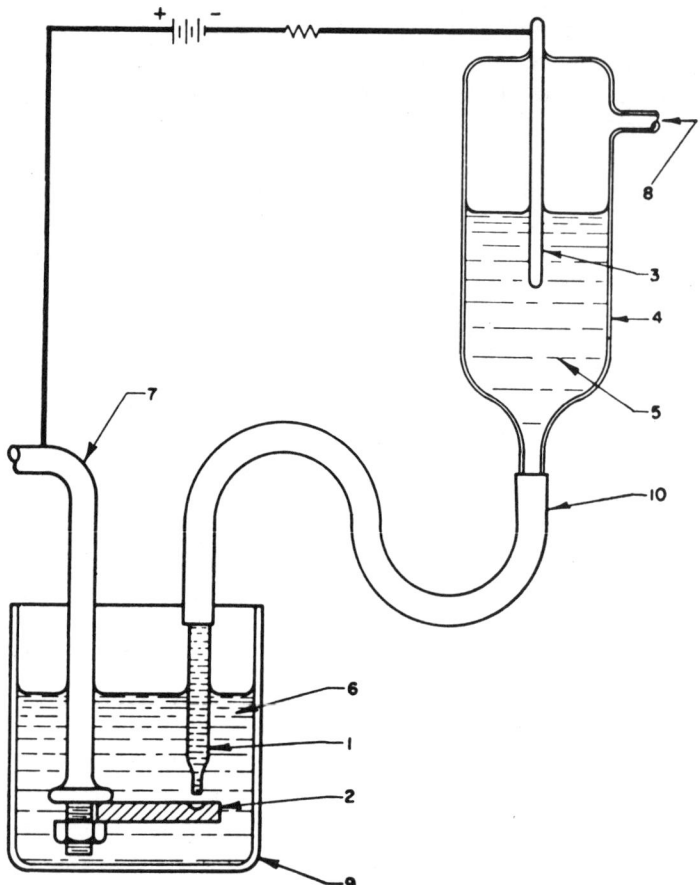

Figure 6. Apparatus for virtual cathode[93] micromachining; 1, glass tip (virtual cathode); 2, work; 3, actual cathode; 4, cathode compartment; 5 and 6, electrolyte; 7, support; 8, apply air pressure; 9, battery jar; and 10, flexible tubing. (With permission of the American Institute of Physics.)

presence of gas bubbles and the accumulation of corrosion products at the anode surface. The diameter of a jet-etched hole is about twice the diameter of the jet.[95] A listing of some of electrolytes used for jet etching various metals is found in Table 2.

The jet etching process was employed by Kelly and Nutting[100] to make thin uniform foils of iron and steel as transmission samples for electron microscope analysis. Since the surface layers of the metal are greatly modified by mechanical machining methods, such samples cannot be obtained by conventional machining methods. The jet of electrolyte was directed through a copper tube fitted with a glass tip (1-mm ID) perpendicularly against the metal sample about 2 mm from the glass tip. By means of cardioid concentric cams in

## Table 2
### Electrolytes for the Jet Etching of Various Metals

| Metal | Electrolyte | Reference |
|---|---|---|
| Aluminum | 2 $M$ KNO$_3$ | 91 |
|  | 2.6 $M$–0.5 $M$ KCl | 91 |
|  | 10% HNO$_3$ | 96 |
| Copper | 2 $M$ KNO$_3$ | 91 |
|  | 2.6 $M$–0.5 $M$ KCl | 91 |
|  | 50% H$_3$PO$_4$ | 95 |
| $p$-Germanium | 0.075–0.3% NaOH | 97 |
|  | 0.1 $M$ NaClO$_4$ + 0.02 $M$ NaOH | 98 |
|  | 0.2 $N$ H$_2$SO$_4$ | 98 |
| $n$-Germanium | 0.075–0.3% NaOH + light | 97 |
|  | 0.1 $N$ In$_2$(SO$_4$)$_3$ + light | 94 |
| Gold | 10% HCl | 99 |
| Iron | 2 $M$ KNO$_3$ | 91 |
|  | 2.6–0.5 $M$ KCl | 91 |
|  | 1 part 10% Cu(NH$_3$)$_4$Cl$_2$ to 3 parts HCl (conc.) | 95 |
|  | 1 part saturated solution CuCl$_2$ to 3 parts HCl (conc.) | 100 |
| $p$-Silicon | 1 $M$ NH$_4$F + 0.2 $M$ NaF | 98 |
|  | 0.2 $M$ NaF + 0.0025 HF | 98 |
| $n$-Silicon | 0.2 $M$ NaF + 0.0025 HF + light | 101 |
| Steel (type 1020) | 2 $M$ KNO$_3$ | 91 |
|  | 2.6–0.5 $M$ KCl | 91 |
| Stainless steel (type 304) | 2 $M$ KNO$_3$ | 91 |
|  | 2.6–0.5 $M$ KCl | 91 |
| Titanium | 10% NaOH | 111 |
| Tungsten | 10% NaOH | 102 |

the $x$- and $y$-directional drive mechanisms, the jet of electrolyte was made to scan the surface horizontally back and forth 3 times/sec and vertically 6 min/cycle. With an applied voltage between 40 and 50 V (to obtain an applied current of 2 A) the metal removal rate in a CuCl$_2$–HCl electrolyte is 10 $\mu$m/min. Under these conditions, the metal removal rate is fairly constant at about 300 $\mu$m/A hr so it is possible to estimate when the desired thickness could be reached from a knowledge of the original thickness of the sample and the applied current. After jet etching a sample to 50 or 100 $\mu$m, final thinning to 20 $\mu$m was obtained with electropolishing in an acetic acid–perchloric acid mixture. Samples of iron and steel for analysis were produced rapidly with smooth surfaces and no modification of the metal structure because no cold working of or thermal damage to the metal surface was incurred.[100,103] Similar results were obtained[99] for the production of gold foils in HCl electrolytes.

Another method for thinning metal samples for electron microscope observations has been reported by Bulckaen and Paganini.[104,105] The electrodes are in the form of circular disks. The rotating circular anode rests on a jet of electrolyte produced by pumping the fluid through a hole in the center of the horizontal circular cathode. To assure uniform metal removal, the Reynolds number must be kept low enough to remain in the laminar flow region. Use of the floating anode, permits a leveling of the surface independent of the radius. Metal removal rates from 2 $\mu$m/min to 20 or 30 $\mu$m/min were obtained on mild steel samples in $HNO_3$ electrolytes. Some difficulty was experienced in controlling the metal removal at the center of the anode disk.

Theler[96] has reported that Al samples may be thinned by directing a spray of 10% $HNO_3$ against each side of the sample at a rate of 1 l/min by a pump. Platinum wires extending into the spray serve as the cathode with the Al as the anode. When a voltage of 60 V is applied across the electrode gap ($\sim$1 cm), a current of about 1 A flows. Final thinning is obtained by electropolishing.

Tensile samples for stress–strain measurements prepared by jet etching give more reliable and reproducible results since jet etching does not give rise to cold working and thermal damage.[95,102] The test sample is rotated[95] at 2000 rpm in a jet of solution flowing by gravity from a reservoir through a copper tube fitted with a glass tip. The jet is moved back and forth along the anode by means of a cam rotating at 75 rpm in the drive system. An iron rod (3-mm OD) was reduced to 1.5-mm OD over a length of 7.5 mm in 15 min using a jet 0.8 mm in diameter and an applied voltage of 70 V. The highly polished cut had a uniform diameter within ±0.01 mm.

In an immiscible-layer, electrolytic cell, Bilello[102] shaped tungsten rods for tensile samples. The cell contained three liquid layers with the top and bottom ones as insulating layers which isolate the conducting layer in between. The top layer was ether, the bottom $CCl_4$, and the middle a 10% aqueous solution of NaOH. A stainless steel ring cathode surrounding the vertically held tungsten rod was held in the middle layer. To maintain uniform metal removal, the rod is rotated at 10 rpm when current was passed between the cathodic ring and the anodic rod.

A jet etching technique[93,106] can be used to remove parts of tool bits and taps broken off in the workpiece. Important advantages include the ability of the process to remove the broken tool parts regardless of metal hardness, threading, or positioning of the metal in the broken part without any mechanical or thermal damage to the workpiece.

Semiconducting materials can be micromachined by jet etching techniques.[94,97,98,101,107,108] When a $p$-type Ge or Si sample is machined by jet etching, the surface is smooth; however, when the $n$-type is machined, the surface is badly pitted and very little material is removed. Polarization curves have been obtained[109,110] on Ge anodes of both types in 0.1 N $H_2SO_4$ electrolyte in the dark. As noted in Figure 7, the curves are the same in the low-current-dnesity region for $p$- and $n$-Ge anodes. Above a certain current density, the overvoltage on $n$-Ge rises rapidly in a limiting current region, whereas the overvoltage on $p$-Ge remains relatively low (even over two orders of magnitude increase in the current density.)

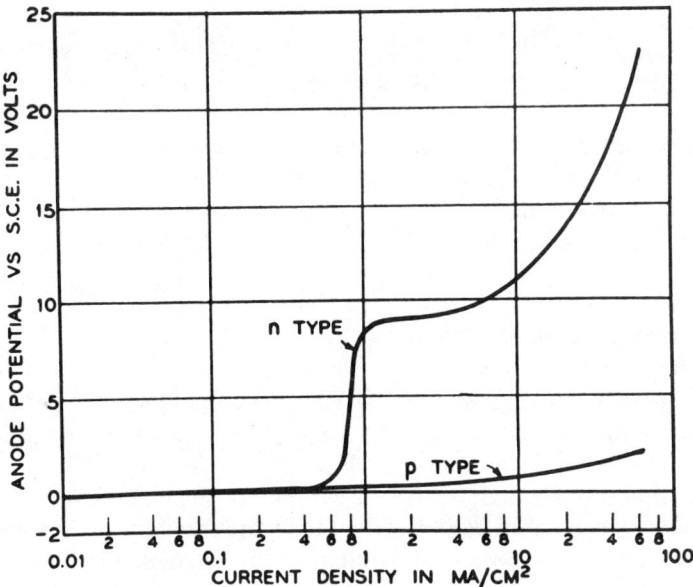

Figure 7. Typical polarization curves obtained[109] on $n$- and $p$-type Ge anodes in 0.1 N $H_2SO_4$ saturated with $GeO_2$ at 25°C in the dark. (With permission of The Electrochemical Society, Inc.)

This behavior has been explained[97,101,107] in terms of minority carriers since the anodic current takes place by hole conduction. In $p$-type Ge or Si, sufficient current can be carried by holes to obtain acceptable jet etching[98] since metal removal is uniform over the entire exposed surface.

With $n$-Ge or $n$-Si, the hole concentration is very small at room temperature, and even at 100°C the hole concentration is not increased enough to support an operating jet etching current.[101] Consequently, as seen in Figure 7, a saturation current is quickly reached at an $n$-Ge anode because the holes are depleted at low current densities. As the potential is increased, a breakdown of the surface barrier in $n$-Ge can occur, resulting in pitting of the surface because the breakdown does not take place evenly over the surface. Such an undesirable condition can be avoided by the generation of electron–hole pairs through the optical excitation of the electrons in the valence band with strong illumination of the semiconducting anode.[97,101] By shining an intense beam of light on the anode surface,[94,97,101,107] $n$-Ge and $n$-Si can be jet etched successfully.

A modification of the jet etching process has been used[97,111,112] to obtain very thin wafers (0.0013 cm thick) of Ge which are free of mechanical surface damage and need little if any surface polishing. The operation has been referred to as the electrolytic sawing or slicing.

In this method, the electrolyte is pumped through a hole in the cathode holder over a taut tungsten wire (0.0068-cm OD) serving as the cathode. The

Ge sample is mounted in an anode holder which is moved by a precision gear train and motor[112] as the cathodic wire cuts through the bar of Ge. The flowing electrolyte sweeps out the corrosion products in the cut in the anode and the electrolyte acts as a coolant for the electrode process. A similar arrangement was used by Metzger[111] to cut stress-free slices from Au, Ti, and Cu bars. As a cathode, a taut piano wire was drawn back and forth by a lever and pulley mechanism with the electrolyte pumped over the wire. Thin wafers of the metals were obtained as the anode was moved toward the wire.

## 5. Electrochemical Grinding

As a metal-removing process industrially competitive with conventional machining methods requiring a narrow gap distance, the electropolishing process encounters a number of serious problems. A major difficulty is in the low corrosion rates obtained (0.1–3 A/cm$^2$ corresponding to 0.0002 to 0.006 cm$^3$ of iron per minute). Since the anode and cathode are held stationary, the distance between the electrodes, initially short, widens as metal is removed until the current falls to such a low value, due to the increasing $IR$ drop across the solution gap, that the anodic corrosion process is virtually stopped. Also, the accumulation of corrosion products in the gap space may bring the metal dissolution process to a halt. Finally, oxide films may be formed on the anode surface in certain electrolytes; blocking the active sites on the metal surface, and preventing further metal removal. These operating deficiencies must be eliminated from any electrolytic machining process before commercial success may be realized.

The first attempt to overcome these difficulties in a successful electrolytic machining operation was the application of electrochemical principles to a mechanical process known as electrolytically assisted grinding, electrolytic grinding,[113] or in recent times, electrochemical grinding (ECG).[114] As described by Keeleric[115,116] in the first published account of ECG, an electrically conducting wheel, mounted on an insulated drive shaft over which an electrolyte is applied, serves as the cathode during the machining of an anodic workpiece, as sketched in Figure 8. The original patents for ECG were granted to Keeleric[117] between 1944 and 1945.

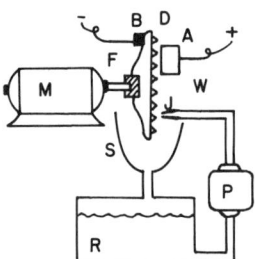

Figure 8. Essential components of an electrolytic grinder: A, workpiece; B, brush; D, imbedded diamond bort; F, insulated bushing; J, electrolyte jet; M, motor; P, pump; R, reservoir; S, splash basin; W, electrically conducting grinding wheel.

The details of the conversion of a mechanical grinder to an ECG machine were published by Metzger and Keeleric,[116] and descriptions of typical ECG machines are found in the literature.[116,118–120] An account of the ECG operation is found in DeBarr and Oliver (Reference 121, p. 35). Although the Soviet workers are recognized[122] for their work with EDM, considerable activity in the field of ECG has been described by Ryabinov[123] dating from a report by Gusev[124] in 1952.

In general, the conducting wheel of the ECG grinder is made of phosphor bronze[119,125] or copper,[126,127] and diamond bort or powder is embedded in the edge or the face of the wheel at a concentration of about 4.4 carats/cm$^3$. The diamonds may be pressed mechanically into the surface of the copper wheel and anchored in place by plating a layer of nickel on the diamond-studded surface.[126] In another method,[128] the diamonds are mounted in the electroformed Ni facing of the wheel, and a layer of either Ni or Cu is plated on this surface. In both cases, the diamonds are exposed to a depth of 12–64 $\mu$m by etching the plated metal away anodically in an electropolishing bath. Chemical etchants may also be used to dress the diamond-studded surface.

The role played by the diamond abrasive in ECG is not mechanical cutting of the workpiece but the maintenance of a constant gap between the cathode wheel and the anode workpiece. Consequently, only a light pressure, sufficient to hold the wheel against the work, is required. Pahlitzsch and Visser[129] studied the effect of the concentration of diamonds in the grinding wheel surface on the metal removal rate of Co-cemented tungsten carbides in NaNO$_3$ solutions. At low (1.5 carats/cm$^3$) and at high (4.5 carats/cm$^3$) concentrations, the metal removal rate was about the same, but at about 3.0 carats/cm$^3$, a maximum in the rate was obtained.

This behavior can be explained as follows. During the ECG process, the presence of insulating anodic films or corrosion products, such as the carbide skeleton of the cemented carbide on the surface of the anode, can inhibit the metal removal rate. As well as maintaining a constant gap between the wheel and work piece, diamonds also scrape away any surface films or nonmetallic components of an alloy which remain after anodic dissolution of the metallic components.[113,116,119,126] If the diamond concentration is too low, the carbides and anodic films are not removed sufficiently, and a reduced metal removal is obtained. When the diamond concentration is high, the void space in the gap between anode and cathode through which the electrolyte must flow is seriously diminished, and the metal dissolution rate is reduced because of the restricted flow of electrolyte through the gap. The resulting high resistance across the gap lowers the flow of electric current between anode and cathode. A judicious balance between these opposing effects results in higher metal removal rates at intermediate diamond concentrations.

By running the ECG grinder with and without the applied voltage across the anode and cathode, it is possible to determine the ratio of metal removed electrochemically to metal removed mechanically. For the grinding of metals and alloys such as steel, the electrochemical dissolution of the anode represents between 95% and 99% of the metal removed.[114,119,130] This percentage falls to values ranging between 60% and 90% for the grinding of cemented carbides[120,129,131,132] because the weakened carbide skeleton remaining after the electrolytic dissolution of the metal components is removed by mechanical grinding.[133]

Other grit materials, such as alumina and silicon carbide, have been used for the grinding of steels[113,119,134,135] with performance equivalent to that of diamonds. However, for the grinding of cemented carbides, diamonds give the best performance.[129,136]

To maintain the narrow gap distance (from 10 to 60 $\mu$m), the constant applied pressure to the grinding wheel is maintained by spring loading,[115] compressed air,[113] applied weights,[119,125] or by a servo-drive system.[137] The effect of pressure applied to the wheel on the metal removal rate has been investigated.[119,129,131] In general, the rate of metal removal increases linearly with current density at a given applied potential. For a given current density, the rate increases with the applied wheel pressure, but the percentage of electrochemical metal removal decreases. Thus the efficiency of the ECG process decreases with applied pressure to the wheel. Pahlitzsch and Visser[129] have reported an empirical relationship obtained from their cemented carbide data relating the rate of metal removal, $W$, to the current density, $i$, and the applied pressure, $p$:

$$W = W_0(i/i_0)^m (p/p_0)^n \tag{23}$$

where $m$ and $n$ are experimentally determined exponents and the subscript zero refers to initial conditions. They found that $m$ had a value of 0.55 for all materials but $n$ varied between 0.25 and 0.54 since the pressure dependence is sensitive to the concentration of tungsten carbide in the cemented carbide material.

If too much pressure is applied to the wheel when grinding metals or alloys such as steel, the diamond grit will cut into the metal surface. Besides reducing the efficiency of the operation, this condition results in the narrowing of the gap; and a point can be reached where deleterious sparking[137a] and impairment of the workpiece surface can occur. To avoid this undesirable circumstance, most machines are protected by a spark-detecting device. Comstock[138] described an electronic detector of the high-frequency noise generated by sparks. The signal from the detector is used to fire a thyratron which reduces the voltage.

Using a beam-type strain gage dynamometer, Cole[119] was able to measure the force exerted on the workpiece during the ECG grinding of steel. If the force exerted by the rotating wheel is $F_m$, then $F_m$ must equal $F_h$, the horizontal force due to the applied pressure against the wheel. A vertical force, $F_v$, is exerted on the workpiece by the rotating wheel and is related to $F_m$ by the mechanical friction relationship,

$$F_v = f^0 F_m \tag{24}$$

From the experimental data, Cole determined that 0.10 is the value of the coefficient of friction, $f^0$.

In this investigation, it was found that $F_v$ decreased as the current density, $j$, was increased, and Cole suggested that enough hydrogen can be generated in the gap at the cathode surface to exert a force, $F_g$, on the workpiece. Then the balance-force equation would be

$$F_h = F_g + F_m \tag{25}$$

and from Eq. (24),

$$F_v = f^0(F_h - F_g) \tag{26}$$

Since the evolution of hydrogen at the wheel cathode surface, and hence $F_g$, is dependent on $j$, this model can account for the dependence of $F_v$ on $j$. By using numerous assumptions and approximations, Cole estimated values for the hydrogen pressure in the grinding wheel interface for given values of $j$ and derived a relationship between $F_v$ and the gas pressure. Good agreement between the calculated and experimental values of $F_v$ was obtained. Others[129,131] have observed an increase in the required applied pressure as indicated by Eq. (25) for a given increase in the current density.

With only minimum applied pressure to the wheel required and with nearly 100% of the metal dissolved electrochemically, the tool wear is greatly reduced for the grinding of fully hardened steels.[139,140] According to Schwartz,[141] a diamond grinding wheel should last an estimated 46 times longer when used in ECG than an equivalent wheel used for mechanical grinding. Cole[119] found virtually no wear of the diamond wheel during a year of grinding steel samples. Electrochemical grinding rates, which are independent of the hardness of the metal, are of the order of 1.6 cm$^3$/min and may reach values nearly twice the rates of mechanical grinding.[115,116,119,138,142]

The ability to remove metal rapidly without inducing stresses in the workpiece metal even in relatively deep cuts[137,143,144] gives ECG a great advantage over conventional grinding methods. Consequently, such delicate structures as stainless steel honeycombs or metal felts can be machined by ECG without burring, curling, warping, or thermal damage.[114,118,126,127,145] Mechanical grinding always leaves scratches, grooves, or tool registry on the machined surface of the metal, but ECG leaves the ground surface smooth and free of registry marks[115,116,144,146] as profilometer measurements of surface roughness range between 0.25 and 1 $\mu$m. When the current density is increased to very high values, it has been noted[120,129] that the surface roughness is increased. Strong electric fields are generated by high applied voltages required to operate in such a current range, and particles of metal may be pulled out of the surface in such fields. If sparking occurs poor surface finish results.

Initially, NaCl solutions were used[116] as the electrolyte, but such solutions were found too corrosive to materials from which the ECG grinders are made. To avoid these corrosion problems, solutions of $NaSiO_4$ (water-glass) were used,[139,147] but grinding[139] was slow in this solution and noxious fumes were given off. From the standpoint of high conductivity with minimum corrosiveness, solutions of $NaNO_3$ between 5% and 10% have been found[113,115,119,139] to be superior to NaCl as an electrolyte for ECG. Although noxious ammonium fumes are generated by the reduction of $NO_3^-$ ion at the cathode, necessitating frequent additions of $NaNO_3$, the overall advantages of the $NaNO_3$ electrolyte are clear. Improvement can be made by adding rust inhibitors, such as $Na_2CO_3$, $Na_3PO_4$, or $NaNO_2$,[114,119,120,129,131,133,144,146] to the nitrate electrolyte. Because the presence of suspended corrosion product particles can lower the conductivity of the electrolyte[148] and thus lower the metal removal rate,[129,131] Pahlitzsch[120] added complexing agents to the electrolyte to keep the corrosion products in solution. $Na_3PO_4$ or Rochelle salts were used for grinding W and Ti, $NH_4NO_3$ for Co, $H_3BO_3$ for ferrous alloys, and $NaNO_3$ or $Na_2CO_3$ for tungsten carbides.

To obtain useful electrochemical rates of metal removal, very high current densities, 12–40 $A/cm^2$ [113,115,126] (fed from the rectifier to the wheel through slip rings or brushes), are required, necessitating a very narrow gap distance (about 0.003 cm) between the anode and cathode. The electrolyte is pumped into the gap by directing a jet of solution against the surface of the wheel[118,119,128] at flow rates of about 1 l/min.[119] The wheel is rotated at the speeds of mechanical grinding (25–30 m/sec) to carry sufficiently large amounts of electrolyte through the gap to remove resistive heat, to sweep out corrosion products, and to provide fresh conducting electrolyte for maintaining the required high anodic corrosion rate. For a given applied voltage, the current density (grinding rate) does not continuously increase with increasing wheel speed because a point is reached where centrifugal forces prevent further quantities of solution to be carried into the gap.[120] If the applied voltage is too high or the rate of wheel rotation too low, sparking can cause localized melting of the surface with the resulting erosion of the wheel, surface damage to the workpiece, and even graphitization of the diamonds.[133]

Electrolyte sputtering and generation of noxious mist can be controlled by enclosing the entire grinding assembly in a plastic box[125] so that the electrolyte can drain down the walls of the box to a solution reservoir. A system is described by Williams[118] in which the wheel is enclosed in a ring of bristles attached to a vacuum source which removes the mist.

Electrochemical discharge grinding, ECDM,[147a] is a modification of ECG. In this process, either ac or pulsed dc is used. Unlike other ECG techniques, which require an electrolytic gap space between the workpiece and the wheel, ECDM must have contact between the two. The wheel can be made of plain graphite without any abrasive spacing particles as no mechanical grinding is involved. Thus ECDM is operated exactly as an ECG device; however, the former requires a sparking condition at the machining interface to expose a relatively untouched surface for continuing electrolytic action.

This is, then, a process combination of ECG and EDM employing an inorganic salt electrolyte rather than a dielectric fluid. While application of this method is not widespread, several pieces of equipment are available.[147a] Such composite machining can in some cases increase the production rate as well as hold close dimensions.

## 6. Principles of Electrochemical Machining

As with mechanical grinding, the shape of the diamond wheel determines the shape of the cut made by ECG in the workpiece, and one is limited in the intricacy of the pattern that can be machined. In the discussion section following a presentation made by Cole[119] at the ASME meeting in November of 1960, Van der Horst proposed that the wheel be eliminated and the diamond-studded cathode be held rigidly against the workpiece. By increasing the solution flow through the gap, the same results could be realized without the commutation problems to the wheel. Cole replied that a pure electrochemical machining machine had been developed by the Anocut Engineering Company of Chicago, and the announcement of its existence in the literature[149,150] was of the first ECM machine offered commercially. Although a British patent was issued to Gusev in 1930[151] for the electrolytic treatment of metals in which all the elements for ECM were described, and although Burgess described the possibilities of using anodic processes for shaping metals in a pamphlet distributed at the 1941 meeting of The Electrochemical Society in Chicago,[128] commercial development of such anodic processes began with the Anocut announcement[118,149] in 1959.

### 6.1. The Gap and Feed Rate

For the ECM process to be commercially satisfactory, metal removal rates of the order of 16.4 cm$^3$/min (1.0 in.$^3$/min) must be realized. Such a cutting rate requires a current density of 160 A/cm$^2$ (1000 A/in.$^2$) to pass across the solution gap between the anode workpiece and the cathode tool. To prevent intolerable $IR$ losses in the gap space at these high current densities, the distance between the electrodes must be maintained at a very low value, between 0.005 cm (0.002 in.) and 0.13 cm (0.05 in.).

Since current is carried across the gap by the ions of the electrolyte, it is imperative that a solution of high conductivity be used such as a 5–25% by weight, aqueous solution of NaCl.[137,149,152–161] Certain acid solutions, such as HCl and $H_2SO_4$, have been used as electrolytes for ECM,[137,161–165] but besides being very corrosive to the machine parts, metal may be plated out on the cathode which would change the shape of the tool and hence the workpiece. As a result, the most practical electrolytes are solutions of neutral salts, such as NaCl, KCl, $CaCl_2$, and $NaNO_3$.[137,152,166] Initially, solutions of NaCl were the most universally employed electrolyte[127,167] (also, Reference 168, p. 63). There are some metals, such as tungsten and molybdenum, which can be machined only in alkaline electrolytes[169,170] because highly protective films are formed on W or Mo anodes in neutral solutions.[171]

During the ECM process, metal removal from the anode produces an increase in the gap distance which causes a decrease in the current; and if the situation is not corrected, the current will fall to a point where machining stops. Consequently, either the anodic workpiece or the cathodic tool is moved by a mechanical-feed drive system to maintain the desired gap distance. If this feed rate is too high for a given metal removal rate (current density), a short circuit between the anode and cathode is possible, and catastrophic damage to the tool due to spark erosion, thermal melting, welding, and metal tearing are possible results. To prevent this costly damage, most ECM machines are protected by sensing devices which detect increasing rates of current or transients in the applied voltage when the gap closes beyond a certain specified safe distance, thus blocking the current. By using silicon-controlled rectifier (SCR) circuits (Reference 121, p. 196; Reference 173, p. 39), it is possible to shut down the current in 10 $\mu$sec. Another cause of sparking may occur in the presence of inclusions in the workpiece metal which cause protrusions in the anode surface due to raised unmachined regions.

Within most values of feed rate and current density, the system rapidly reaches an equilibrium or steady state gap distance due to the balance of the opposing processes of feed rate which tends to narrow the gap and of current density which tends to widen the gap[152,156,174–176] (also Reference 177, p. 36).

Consider the electrolyte to flow at a steady rate through the gap between two plane parallel electrodes as sketched in Figure 9a with a gap distance equal to $h_0$ at zero time, $t_0$. A constant feed rate, $f$, of the cathode is applied with respect to the fixed anode. With the application of a constant dc voltage, $V$, across the cathode tool and the anode workpiece, it is assumed that current flows by the anodic oxidation of the workpiece atoms (of atomic weight, $a$, and atomic number, $z$). It is further assumed that the flow of current results in the evolution of hydrogen at the tool surface by the cathodic reduction of $H_2O$ to $H_2$. Further assumptions are (1) the anodic current efficiency is 100% metal removal, (2) the solution flow is high enough to neglect any gas bubble effects, and (3) overvoltage effects are negligible so that Ohm's law applies.

After time $t$ has elapsed, the system is described by Figure 9b, and the gap distance is

$$h = h_0 + y - ft, \qquad (27)$$

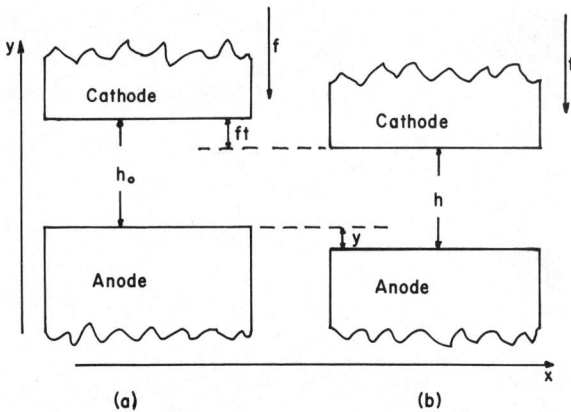

Figure 9. Die-sinking model showing a fixed anode and a moving cathode (feed rate, $f$) at a distance, $h_0$, apart initially (a) and after time, $t$, (b) for equilibrium-gap discussion.

where $h$ is the gap distance at time $t$, $h_0$ is the gap distance at time $t_0$, the anode has receded in time, $t$, because of metal removal. If $A$ is the exposed area of the anode, the current density, $j = I/A$, according to Ohm's law, is

$$j = V\kappa/h \tag{28}$$

where $\kappa$ is the specific conductivity of the electrolyte and $V$ is the applied voltage across the gap. From Faraday's law of electrolysis, the amount of anode material dissolved per second is

$$W = jW_E/F \tag{29}$$

where $W_E$ is the equivalent weight ($W_E = a/z$) of the workpiece metal and $F$ is the Faraday constant (96,500 C). By referring to Figure 9, the amount of workpiece metal removed per unit area per second is

$$W = (dy/dt)\rho_m \tag{30}$$

where $\rho_m$ is the density of the anode material. Combining Eqs. (28), (29), and (30),

$$\frac{dy}{dt} = \frac{V\kappa W_E}{F\rho_m h} = \frac{B}{h} \tag{31}$$

by combining all the constants together in $B$.

By differentiating Eq. (27) with respect to time,

$$\frac{dh}{dt} = \frac{dy}{dt} - f \tag{32}$$

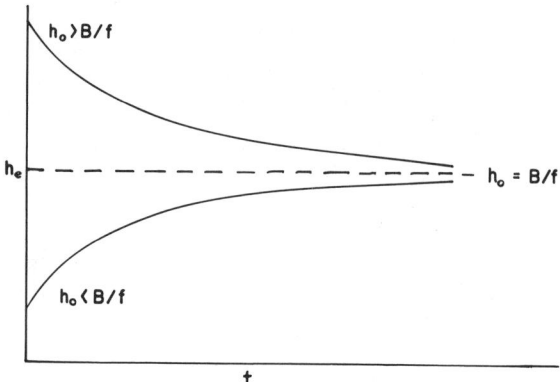

Figure 10. A plot of the attainment of the equilibrium gap, $h_e$, from the initial value, $h_0$, as a function of time.

and substituting Eq. (31) in Eq. (32), one obtains

$$\frac{dh}{dt} = \frac{B}{h} - f \tag{33}$$

Upon integration of Eq. (33),

$$t = \frac{1}{f}(h_0 - h) + \frac{B}{f^2} \ln \frac{h_0 - B/f}{n - B/f} \tag{34}$$

The equilibrium gap distance, $h_e$, is reached when $t$ is constant, and from Eq. (34) this is true when $h_0 = h = B/f$. Consequently,

$$h_e = \frac{B}{f} = \frac{V\kappa W_E}{\rho_m F f} \tag{35}$$

and is the asymptotic value of $h$ since $t$ is a logarithmic function of $h$. When $h_0 > B/f$, $h$ decreases rapidly towards $h_e$; and when $h_0 < B/f$, $h$ increases rapidly towards $h_e$ as shown in Figure 10.

As $h$ approaches $h_e$, $dh/dt$ approaches 0 from Eqs. (33) and (35), and the rate of metal removal, $dy/dt$, approaches the feed rate, $f$, according to Eq. (32). For practical purposes, the steady state gap distance is reached very quickly after the machining operation begins.[156,174]

The quantity $B$ may be considered the machining constant. McGeough and Rasmussen[175] have arranged the parameters included in Eq. (33) in a nondimensional form.

### 6.2. Temperature Effects

If the electrolyte were permitted to remain stagnant in the gap between the tool and workpiece, the $i^2R$ heating due to the passage of these high ECM currents passing through the thin layer of electrolyte would boil the solution away, and the machining process would cease. In addition, the accumulation of corrosion products in the gap space would increase the resistance across the gap to a point where metal dissolution would stop.[178] To avoid this the electrolyte is pumped through the gap with high-pressure pumps delivering between 95 and 950 l/min (25 and 250 gal/min) at 690–2760 kPa (100–400 psi).[149,154,179–181]

The manner in which the temperature changes along the length of the anode–cathode gap is of considerable interest since the conductivity of the electrolyte is a function of temperature.[152,156,183–186] If large variations in the conductivity occur along the gap, uneven metal removal results.

The quantity of heat energy, $q$, required to raise the temperature of a mass, $m$, of electrolyte from $T_1$ to $T_2$ is

$$q = mC_p \Delta T \tag{36}$$

where $C_p$ is the specific heat of the electrolyte. The mass of electrolyte flowing through the gap per unit time is

$$m = F_s \rho_s \tag{37}$$

where $F_s$ is the rate of electrolyte flow in unit volume per unit time and $\rho_s$ is electrolyte density. If we assume that all electrical energy lost in the passage of current across the gap is converted to heat, then

$$j^2 R = V^2/R = Jq \tag{38}$$

where $R$ is the resistance of the electrolyte and $J$ is the mechanical equivalent of heat, 4.187 J/cal. From Eq. (35),

$$f = \frac{V \kappa W_F}{\rho_m F h_e} \tag{39}$$

and the expression for the resistance of the electrolyte

$$R = h_e/\kappa A \tag{40}$$

one obtains by combining Eqs. (36) and (40)

$$f = \frac{W_E}{\rho_m F} \left( \frac{J \kappa F_s \rho_s C_p \Delta T}{h_e A} \right)^{1/2} \tag{41}$$

from which

$$\Delta T = \left( \frac{f \rho_m F}{W_E} \right)^2 \left( \frac{h_e A}{J \kappa F_s \rho_s C_p} \right) \tag{42}$$

Snavely and Cross[182] calculated the rise in temperature from the entrance to exit of the gap for drilling a small hole (0.041-cm ID) in tungsten as 8°C. In an ECM machine, Hopenfeld and Cole[187] found experimentally that the solution temperature rise through the gap, depending on the applied current density and flow rate, varied between 1 and 7.5°C, which corresponds to a variation in conductivity of 1 N NaCl electrolyte of up to 15%. Kubeth and Heitmann[155] measured a $\Delta T$ across the gap as high as 45°C in specified experimental conditions which could cause a change of about 100% in the conductivity of a 10% by weight NaCl solution.

From Eq. (42), it may be seen that $\Delta T$ can be reduced by increasing the solution flow rate, $F_s$, or by increasing the conductivity[186,188] of the electrolyte (usually by increasing the concentration). To avoid uneven metal removal at hot spots, the $\Delta T$ should be kept below 5°C.[153] Heat exchangers are used in the electrolyte flow line of most ECM units to control the temperature of the intake solution to within $\pm 1$°C.[154,179]

In the absence of interfering conditions, such as the presence of particulate matter or of gas bubbles in the electrolyte, the temperature variation along the gap may be calculated by expressing the rate of solution flow, $F_s$, in terms of the velocity, $v_s$, of the solution in unit length per unit time:

$$F_s = v_s b h_e \qquad (43)$$

where $b$ is the width of the exposed anode in the gap. From Eqs. (38) and (40), knowing that the area of the exposed anode is the product of $b$ and its length, $L$,

$$q = V^2 \kappa b L / J h_e \qquad (44)$$

and from Eqs. (36), (37), and (43),

$$q = v_s \rho_s C_p b h_e \, \Delta T \qquad (45)$$

Eliminating $q$ gives

$$\Delta T = \frac{V^2 \kappa}{J v_s \rho_s C_p h_e^2} L \qquad (46)$$

or the temperature, $T_x$, at any point, $L_x$, along the gap is

$$T_x = T_0 + \frac{V^2 \kappa}{J v_s \rho_s C_p h_e^2} L_x \qquad (47)$$

where $T_0$ is the entrance temperature of the electrolyte.

By eliminating $V$ between Eqs. (47) and (39), one obtains

$$T_x = T_0 + \left(\frac{F^2 \rho_m^2}{J W_E^2 C_p \rho_s}\right)\left(\frac{f^2}{v_s \kappa}\right) L_x \qquad (48)$$

Under these conditions, the temperature rises linearly along the gap in the direction of solution flow provided $\Delta T$ is small enough to neglect the change in $\kappa$ with $T$.[189]

In cases where $\Delta T > 5$°C, significant increases in $\kappa$ along the length of the gap[153,183,186,189] may occur. As a result, the current density and hence the metal removal rate will be greater at the exit than at the entrance of the gap, causing an undesirable tapering of the gap, sometimes called[189] the wedge effect. By increasing the solution flow rate, the wedge effect is greatly diminished.[189]

### 6.3. Pressure Effects

Although the $\Delta T$ along the gap can be reduced by increasing the flow rate, a point may be reached where the hydraulic pressure required to achieve the solution flow produces such a steep pressure gradient along the gap that cavitation of the liquid phase occurs.[137,152,156,184,190,191] When the local pressure falls below the vapor pressure of the liquid, bubbles of steam may form in the electrolyte.[156,192,193] Cavitation of the liquid phase causes fluctuations in the conductivity of the electrolyte and nonuniform current density along the gap. Such an event produces striations, grooves, and a general roughening of the machined surface of the workpiece.

Kawafune and co-workers[176] applied a quick-drying ink to the electrode surface, and the flow patterns were observed through a clear plastic plate clamped against the side of the gap. As bubbles of steam were formed when cavitation was present, a white area was observed in the pattern. For a given applied pressure from the pump, they found that cavitation effects were seen to be reduced with an increase in the back pressure at the gap exit, this too in agreement with the findings of Opitz.[190] Consequently, it is possible by proper design of the cathode tool, cell fixturing, and proper control of the ECM parameters to remove cavitation effects from the ECM operation.

### 6.4. Hydrogen Bubble Effects

It was noted[177,185] that theoretical values of $f$ calculated from Eq. (39) or (41) cannot be achieved experimentally. In fact, the maximum feed rate attained is only about half of the calculated value. Cuthbertson and Turner[189] observed that the electrolyte flow was restricted when the voltage was applied across the gap, particularly in cases of low solution flow and high current density. Flow restrictions result from the cathodic evolution of hydrogen at the tool surface.

The cathode tool is made of a metal, such as copper or bronze, which is a good electrical conductor and which is easily machined using conventional procedure. With the application of voltage across the gap, the reduction process at the cathode is the evolution of hydrogen from water,

$$2H_2O + 2e \rightarrow H_2 + 2OH^- \qquad (49)$$

From a study of the effect of gas evolution on the current density in electrolyzers, Tobias[194] has shown that the presence of gas bubbles in the electrolyte produces a decrease in the solution conductivity and a nonuniform current distribution. The presence of such a situation in the ECM gap would have a serious detrimental effect on uniform metal removal.

A number of investigations have been reported[156,160,187,191,192,195,196] in which the gap has been observed optically through transparent windows built into the system. Along the cathode, a two-phase region composed of gas bubbles disbursed throughout the solution phase exists[197] and is known as the "bubble layer."[156,187]

In Figure 11, the influence of flow rate on the bubble layer is observed at a copper cathode in 2 $N$ KCl solution when a current density of 100 A/cm$^2$ is flowing. At low flow rates (400 cm/sec), the average thickness of the bubble layer increases along the length of the gap from the entrance to the exit although there are local variations in thickness. Individual bubbles can be seen, and their average diameter is 35 $\mu$m.[195,198] As the flow rate increases, the bubble diameter decreases so that at flow rates greater than 800 cm/sec the bubble diameter falls below 20 $\mu$m, the limit of optical resolution, and individual bubbles can no longer be distinguished. At a flow rate of 2500 cm/sec, the bubble layer is reduced to the extent that a homogeneous solution phase is attained in the gap space.

The effect of current density on the bubble layer at a constant solution flow rate of 100 cm/sec is recorded in Figure 12. As the current density increases, the average thickness of the bubble layer increases as well as the average diameter of the bubbles (increases from 50 $\mu$m at 5 A/cm$^2$ to 99 $\mu$m at 50 A/cm$^2$). To reduce the effect of the bubble layer at higher current densities requires higher solution flow rates.

It is not possible under ECM conditions to obtain a homogeneous solution from the redissolution of initially formed gas bubbles. Consider a constant flux of mass to the bubble per unit area. Under these conditions, an expression[195] for the decrease in volume of the bubble, $V_b$, is

$$dV_b/dt = -4\pi(RT/P)(3/4\pi)^{2/3} k_m C_{sat} V_b^{2/3} \tag{50}$$

where $C_{sat}$ is the saturation concentration of gas in the bulk solution and $k_m$ is the mass transfer coefficient in centimeters per second. Upon integration and solving for $t_{1/2}$, the time for the initial volume, $V_0$, of the bubble to be reduced to 1/2,

$$t_{1/2} = \frac{2.38 V_0^{1/3}}{4\pi(RT/P)(3/4\pi)^{2/3} kC_{sat}} \tag{51}$$

Landolt et al.[195] estimated $t_{1/2} \sim 1$ sec for 1 atm and 25°C. This value is at least 4 orders of magnitude greater than the residence time ($3 \times 10^{-4}$ sec) in the gap of a bubble moving at 1000 cm/sec.

The effect of the gas bubbles on the ohmic resistance would be reflected in variations in the applied voltage of a controlled current process. A simple theoretical analysis of this very complex bubble layer system is not possible at this time[195] although an attempt has been made by Hopenfeld and Cole[187]

# ELECTROCHEMICAL MACHINING

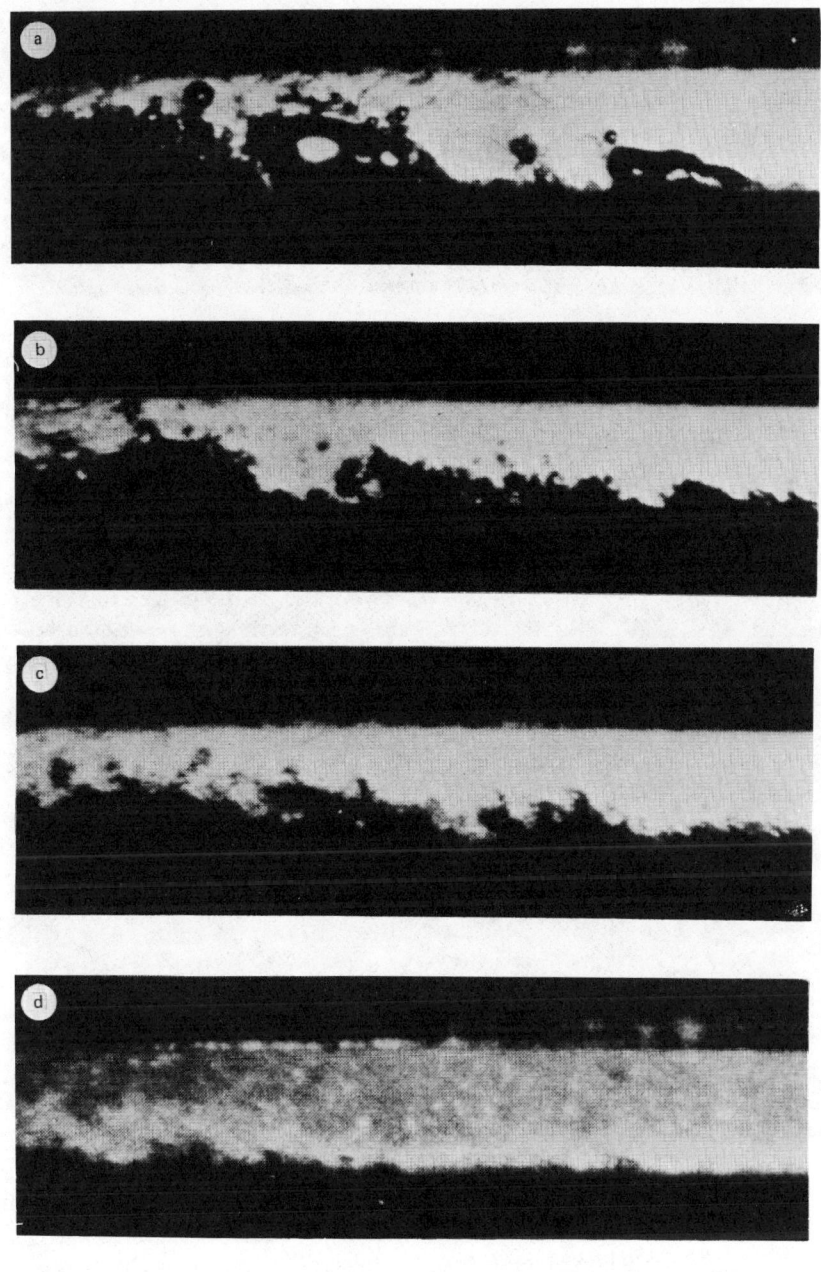

Figure 11. Effect of solution flow rate[195] on gas evolution at a Cu cathode facing up in 2 $N$ KCl at a current density of 100 A/cm$^2$ and flow rates: a, 400 cm/sec; b, 600 cm/sec; c, 1000 cm/sec; d, 2500 cm/sec. (With permission of The Electrochemical Society, Inc.)

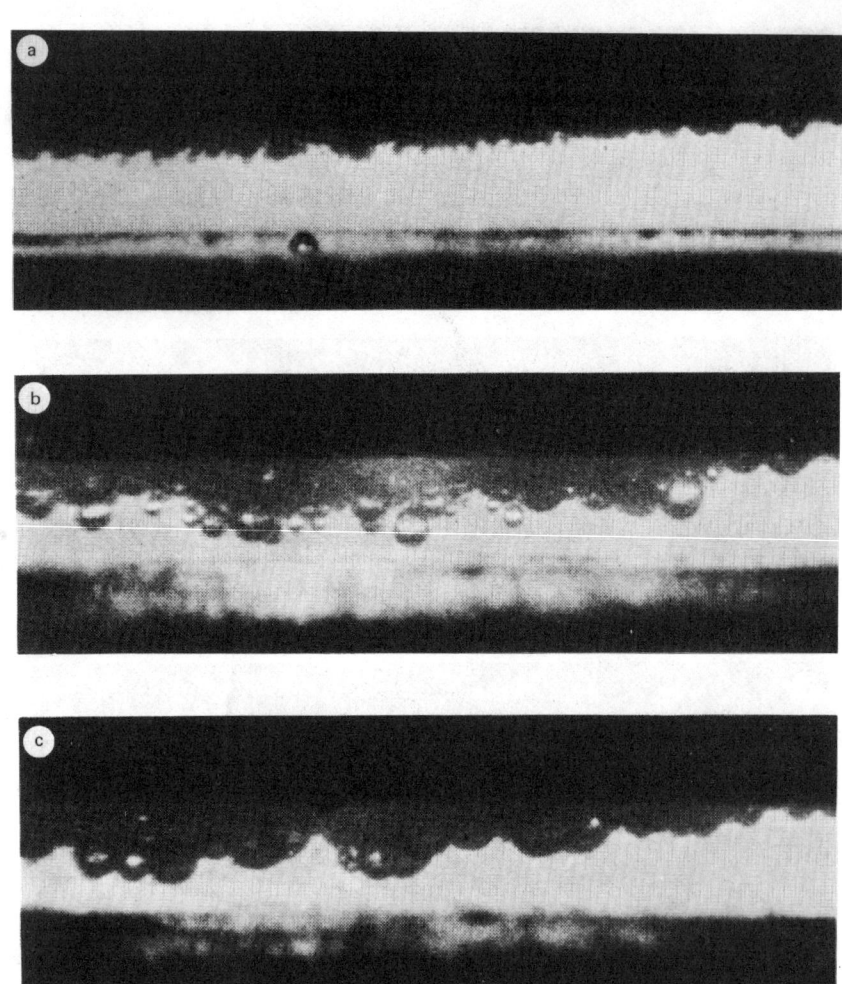

Figure 12. Effect of current density[195] on gas evolution on a Cu cathode facing down in 2 $N$ KCl at a solution flow rate of 100 cm/sec and current densities: a, 5 A/cm$^2$; b, 20 A/cm$^2$; c, 50 A/cm$^2$. (With permission of The Electrochemical Society, Inc.)

using certain simplifying assumptions. At high flow rates (>1500 cm/sec), the effect of the bubble layer on applied voltage was found to be negligible.[195]

Since the bubble layer increases along the length of the gap, the increase in resistance due to this bubble layer compensates[187,196] for the increase in temperature. However, the gas evolution does not entirely compensate for the $\Delta T$ during machining.[188,192,195] To obtain uniform metal removal, one must employ high solution flow rates to minimize conductivity variations due to temperature and bubble layer effects. At high flow rates, such effects as the gas blanketing of electrodes[199] are not observed.[195]

Large amounts of hydrogen evolved during ECM are a safety hazard if allowed to accumulate in the upper regions of the shop. Usually extraction of the hydrogen is accomplished[173,177,199a] by mixing with air and venting to the atmosphere by a fan. Since the lower explosive limit for hydrogen is 4 parts to 100 parts air, the concentration should be kept at 0.4% gaining a safety factor of 10.

## 6.5. Effects of Corrosion Products

Consider the dissolution of iron at a mild steel anode in an NaCl electrolyte. By the transfer of two electrons to the anode metal, iron atoms are oxidized to ferrous ions[200] which pass into the solution phase. Since a dimer of the electron has not been observed, the electron transfer most likely occurs by a minimum of two steps. According to Bockris and co-workers,[201] iron dissolution takes place through an FeOH intermediate either by the discharge of $OH^-$ ions,

$$Fe + OH^- \rightarrow FeOH + e \qquad (52)$$

or by the discharge of water molecules,

$$Fe + H_2O \rightarrow FeOH + H^+ + e \qquad (53)$$

The FeOH radical is discharged,

$$FeOH \rightarrow FeOH^+ + e \qquad (54)$$

and ferrous ions are liberated by the decomposition of $FeOH^+$,

$$FeOH^+ \rightarrow Fe^{2+} + OH^- \qquad (55)$$

In the neutral ECM electrolyte, the $Fe^{2+}$ ions are precipitated as a greenish sludge of $Fe(OH)_2$ which is eventually oxidized by the dissolved oxygen in solution to the red hydrated ferric oxide. In $NaNO_3$ electrolytes, the $Fe^{2+}$ ions are oxidized immediately to $Fe^{3+}$ ions, and a red sludge of $Fe(OH)_3$ appears instead of the dark green $Fe(OH)_2$ sludge.

If the corrosion products are permitted to accumulate in the gap space, the metal removal process will cease. Therefore it is imperative that a solution flow rate high enough to flush out the corrosion products be employed.[137,162,178,202-204] As long as the hydrated oxides remain in a uniform colloidal suspension, good surface finishes with uniform metal removal will be obtained. It is possible to obtain suspensions of insoluble hydroxides in high enough concentrations to increase the viscosity of the electrolyte (Reference 177, p. 25) with a resultant change in metal removal rate.

Another source of concern is the contamination of the electrolyte with particulate matter. These particles may be bits of carbon eroded from high-carbon steels, nonmetallic inclusions, small metallic grains that drop out of the corroding nonuniform alloy, or lumps of coagulated hydroxides. De la Rue and Tobias[148] studied the effective conductivity, $\kappa_e$, of the suspensions of glass beads in electrolytes. Over a range of glass bead sizes from 49 to 6400 $\mu$m in diameter and between bead volume fractions, $f_v$, between 0.05 and 0.4, $\kappa_e$ was found to decrease with $f_v$ according to the Bruggemann[205] equation,

$$\kappa_e/\kappa_c = (1 - f_v)^{3/2} \qquad (56)$$

where $\kappa_c$ is the conductivity of the continuous medium. For nonspherical particles such as sand or small polystyrene cylinders, Eq. (56) predicts values smaller than those experimentally obtained ones.

Consequently, it is highly desirable to remove these particles from the circulating electrolyte. The most common method is through the use of filters[154,173,177,179] in the flow line of the ECM system. Filter presses containing diatomaceous earth have been described (Reference 177, p. 25), but these filters become frequently clogged with sludge and require extensive cleaning. Under such conditions, the solution flow rate is severely restricted. By using single weaves of stainless steel or monel wire or coarse ceramic disks, efficient filtering of the electrolyte is obtained without the need for frequent cleaning (Reference 173, p. 128).

Ito and Shikata[157] have employed centrifuges to remove the sludge, but such a procedure is expensive and requires careful engineering.[167,173,177] Where considerable space is available, settling tanks may be used, but the process is only about 50% efficient even at low flow rates[173,177] Flotation and frothing methods of particle removal have been considered.[137,173,177] Ion-exchange membranes have been used to filter ECM electrolytes with varying degrees of success.

### 6.6. Metal Removal Rate Considerations

The driving force for the electrochemical reaction at a given electrode (oxidation at the anode or reduction at the cathode) is the voltage drop or potential difference across the metal–solution interface. Each electrode reaction has a specific threshold potential below which the reaction will not proceed. A listing of these threshold driving forces is found in a table of redox potentials.[206] The electromotive series of elements is related to redox potentials under standard conditions of temperature, pressure, and unit activity of the reactants. Additional driving force above the redox potential is required because most electrode reactions occur with an activation energy along the reaction path. This excess driving force is called electrochemical polarization, $E_p$, or under special conditions overvoltage or overpotential, $\eta$. The rate of reaction or the current density is an exponential function of the polarization,[207,208] $E_p$:

$$j = j_0(e^{-A'E_p} - e^{B'E_p}) \qquad (57)$$

where $j_0$, $A'$, and $B'$ are constants for a given temperature and electrode reaction. Mass transfer processes (diffusion of reactants to and products away from the electrode–solution interface) can contribute to the polarization, but at the high flow rates in ECM operations, such contributions are negligible, as noted from studies with rotating disk[209] and hemisphere[210,211] electrodes.

In solutions of NaCl or NaNO$_3$, the cathodic reaction is not the reduction of Na$^+$ ion to Na metal at the electrode surface (although current is carried through the bulk of solution by Na$^+$ ions) because Na$^+$ ion reduction occurs at a higher cathodic potential than the reduction of H$_2$O molecules to H$_2$ [Eq. (49)]. Similarly at the anode, the oxidation of metal atoms to metal ions occurs with less driving force (lower anodic potential) than the evolution of O$_2$ from water or that of Cl$_2$ from Cl$^-$ ion unless the anode surface becomes filmed with hydrated oxides, about which more will be discussed later. Under ECM conditions, it is possible to reduce NO$_3^-$ ion to NH$_3$, hydroxylamine, and nitrite.[195,212]

Since the voltage drop across the narrow anode–cathode gap includes the solution $IR$ drop and the polarization terms

$$\Delta E_{gap} = IR_{soln} + (E_p)_{an} + (E_p)_{cath} \tag{58}$$

the applied voltage must be high enough to overcome these voltage drops and supply the high currents demanded by the ECM operation. In practical ECM operation, the applied voltage ranges between 5 and 25 V.[125,149,154,213–215] Since low-voltage, regulated dc power sources are required for ECM, it is not difficult to obtain rectifiers capable of delivering up to 20,000 A.[182] Units of even greater capacity are possible.[167] Using a 20,000-A rectifier, Myerley[179] notes that steel stock may be removed at a rate of 32.7 cm$^3$/min (120 in.$^3$/hr). The first announcement in 1959[149] noted rates between 4.9 and 16.4 cm$^3$/min (18 and 60 in.$^3$/hr).

Typically, metal is removed at current densities between 4.8 and 240 A/cm$^2$ (30 and 1500 A/in.$^2$)[127,153,167,214–216] and as high as 800 A/cm$^2$ (5000 A/in.$^2$)[177,217,218] with feed rates between 0.05 and 1.3 cm/min (0.02 and 0.5 in./min)[182,216,218–220] As the work area to be machined is increased, the feed rate must be lowered for a given applied voltage because the overall amount of metal removed per unit time is directly proportional to the current.[216]

Determination of the shape of the cathode necessary to produce a particular anode configuration is the central problem of the ECM process. A number of attempts with varying degrees of success have been made[156,162,174,183,187,196,221–227] to formulate a mathematical interpretation of the ECM process which would permit prediction of the shape of the cathode required to produce the desired anode configuration. Krylov and co-workers[228] have given an excellent review of the Russian investigations of ECM and the problems encountered in formulating the desired mathematical

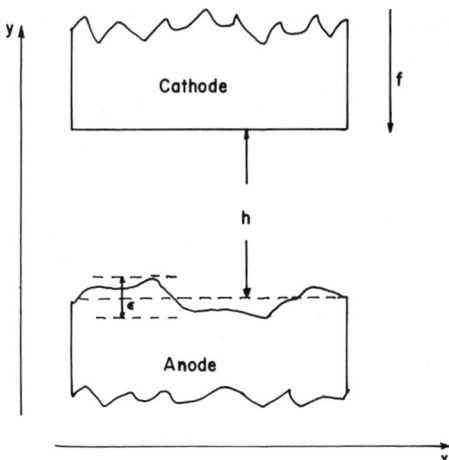

Figure 13. Die-sinking model of fixed anode and moving cathode (feed rate, $f$) for discussion of smoothing process at the anode surface.

model. Nomograms have been constructed[184,224] to aid in the design of an ECM operation.

A suggested mathematical theory of ECM is that formulated by Fitz-Gerald and McGeough.[226] To find the potential, $\varphi$, at any point in the electrolyte in the gap space, Laplace's equation,

$$\nabla^2 \varphi = 0 \tag{59}$$

must be solved. The first problem to be considered is the approach of a smooth plane cathode towards an anode with irregular surface in a smoothing operation, as sketched in Figure 13. At a feed rate of $f$, the cathode moves toward the anode with the average gap distance, $h$.

Irregularities in the anode surface may be described by a Fourier series with time-dependent coefficients. Let the cathode be given by

$$y = 0 \tag{60}$$

and the anode by

$$y = h + \varepsilon_0 \sum_1^\infty a_n \sin \frac{n\pi x}{\lambda} \tag{61}$$

where $2\lambda$ is the fundamental wavelength of the irregularities and $\varepsilon_0$ is the initial value of the maximum amplitude, $\varepsilon$, of the irregularities. It is assumed that $\varepsilon \ll h$. A solution for the Laplace equation in two dimensions is sought with the boundary conditions

$$\varphi = 0 \quad \text{on} \quad y = 0 \tag{62}$$

$$\varphi = V \quad \text{on} \quad y = h + \varepsilon_0 \sum_1^\infty a_n \sin \frac{n\pi x}{\lambda} \tag{63}$$

# ELECTROCHEMICAL MACHINING

and with the solution expanded to a first order in the small parameter $\varepsilon_0/h$,

$$\varphi = \varphi_0 + \frac{\varepsilon_0}{h}\varphi_1 \tag{64}$$

where $V$ is the applied voltage, $\varphi_0$ is the potential between two plane electrodes at a distance, $h$, and $\varphi_1$ is a first-order perturbation to the potential.

Then, $\varphi_0$ satisfies the Laplace equations for the boundary conditions

$$\varphi_0 = 0 \quad \text{on} \quad y = 0 \tag{65}$$

$$\varphi_0 = V \quad \text{on} \quad y = h \tag{66}$$

The solution is

$$\varphi_0 = Vy/h \tag{67}$$

To satisfy the Laplace equation, $\varphi_1$ is chosen with the boundary conditions

$$\varphi_1 = 0 \quad \text{on} \quad y = 0 \tag{68}$$

$$\varphi_0 + \frac{\varepsilon_0}{h}\varphi_1 = V \quad \text{on} \quad y = h + \varepsilon_0 \sum_1^\infty a_n \sin\frac{n\pi x}{\lambda} \tag{69}$$

But on $y = h + \varepsilon_0 \sum_1^\infty a_n \sin(n\pi x/\lambda)$, $\varphi_0 = V + (\varepsilon_0 V/h)\sum_1^\infty a_n \sin(n\pi x/\lambda)$, the condition of Eq. (69) becomes

$$\varphi_1 = -V \sum_1^\infty a_n \sin\frac{n\pi x}{\lambda} \tag{70}$$

The solution is

$$\varphi_1 = \sum_1^\infty A_n \sin\frac{n\pi x}{\lambda} \sinh\frac{n\pi y}{\lambda} \tag{71}$$

From the boundary condition, Eq. (70),

$$A_n = -\frac{a_n V}{\sinh(n\pi h/\lambda)} \tag{72}$$

to a first-order approximation in $\varepsilon_0/h$. From Eqs. (64), (67), (70), and (72), the expression for the potential is

$$\varphi = \frac{Vy}{h} - \frac{V\varepsilon_0}{h}\sum_1^\infty a_n \left(\sin\frac{n\pi x}{\lambda}\right)\frac{\sinh(n\pi y/\lambda)}{\sinh(n\pi h/\lambda)} \tag{73}$$

Let the reduction rate of the anode surface be $dy_a/dt$ in the vertical direction and $-\partial\varphi/\partial y$ be the field in this direction. If the anode surface is described by a vector function $r_a(x,y)$, Ohm's law [Eq. (28)] is written as $i = -\sigma\nabla\varphi$, and Faraday's law is given by Eq. (29), one obtains

$$\frac{dr_a}{dt} = M\nabla\varphi_a \tag{74}$$

where $M = W_e\kappa/E\rho_m$ and the vertical component is

$$\frac{dy_a}{dt} = M\frac{\partial \varphi_a}{\partial y} \tag{75}$$

since the reduction rate of each Fourier component of each of the irregularities is of interest. From Eqs. (61), (73), and (75), one obtains

$$\frac{dh}{dt} + \varepsilon_0 \sum_1^\infty \frac{da_n}{dt} \sin\frac{n\pi x}{\lambda} = \frac{MV}{h} - \frac{MV\pi\varepsilon_0}{h\lambda}\sum_1^\infty na_n\left(\sin\frac{n\pi x}{\lambda}\right)\frac{\cosh(n\pi y/\lambda)}{\sinh(n\pi h/\lambda)} \tag{76}$$

Since the $\sin n\pi x/\lambda$ is linearly independent, the equality of Eq. (76) must hold individually for all the coefficients. By including the effect on $dh/dt$ of the forward motion of the cathode, one obtains

$$\frac{dh}{dt} = \frac{MV}{h} - f \tag{77}$$

which agrees with Eq. (33) ($B = MV$) for the simple theory for two plane parallel electrodes at a distance $h$. The solution of Eq. (77) is Eq. (34).

Also from Eq. (76),

$$\frac{da_n}{dt} = -\frac{MV\pi}{h\lambda}na_n \coth\frac{n\pi h}{\lambda} \tag{78}$$

using the approximation that the arguments of the sinh and cosh function are interchangeable between $n\pi h/\lambda$ and $n\pi y_a/\lambda$ assuming $\varepsilon \ll h$. For very small wavelengths as compared to $h$, it may be assumed that $\coth n\pi h/\lambda \sim 1$, and Eq. (78) becomes

$$\frac{da_n}{dt} = -\frac{MV\pi}{h\lambda}na_n \tag{79}$$

It has been shown[226] that this is a good approximation. When $h$ is constant, Eq. (79) integrates to

$$a_n(t) = a_n(0)\, e^{-MVn\pi t/h\lambda} \tag{80}$$

From Eq. (80), it is noted that the higher harmonics are reduced more rapidly and the shape of the arbitrary irregularity approaches a sine wave with the fundamental wavelength. A similar result was found [Eq. (16)] from an analysis of the electropolishing process. When $h$ is not constant, one obtains, from Eqs. (79) and (77) by integration,

$$a_n(h) = a_n(h_0)\left(\frac{MV - h_0 f}{MV - hf}\right)^{-MVn\pi/\lambda f} \tag{81}$$

which exhibits the same behavior as Eq. (80).

In simple ohmic theory, the electric field lines are considered parallel and the rate of reduction of irregularities may be determined from the length of the solution path across the gap. However, at a point on the anode where the surface curvature is large and convex, the field is greater than at a plane surface. Since short wavelengths have a greater concentration of field at their crests than longer wavelengths, more rapid smoothing is expected in the case of the short wavelengths because of the greater difference in the field between the peak and the trough of the wave. This is just the behavior exhibited by Eq. (80) or (81).

When the wavelength is large compared to $h$ where $\coth n\pi h/\lambda \sim \lambda/n\pi h$, Eq. (78) reduces to

$$\frac{da_n}{dt} = -\frac{MV}{h^2} a_n \qquad (82)$$

which can be shown[226] to lead to the simple ohmic case as expected. When assuming the field lines to be parallel, the ohmic results are obtained.

If one considers the irregularities on the anode surface to be purely sinusoidal, then

$$y = h + \varepsilon \sin kx \qquad (82a)$$

$k$ is the wave number, $n\pi h/\lambda$. In this case, Eq. (78) becomes

$$\frac{d\varepsilon}{dt} = -\frac{MVk}{h} \varepsilon \coth kh \qquad (83)$$

For three-dimensional irregularities, one employs a similar treatment as above with

$$y_a = h + \varepsilon \sin kx \sin lz \qquad (84)$$

where $l$ is the wave number of irregularities in the $z$ direction. By the same procedure described above

$$\varphi = \frac{Vy}{h} - \frac{V\varepsilon \sin kx \sin lz \sinh[(k^2+l^2)^{1/2}y]}{h \sinh[(k^2+l^2)^{1/2}h]} \qquad (85)$$

from which equations for $h$ and $\varepsilon$ are obtained:

$$\frac{dh}{dt} = \frac{MV}{h} - f \qquad (86)$$

$$\frac{d\varepsilon}{dt} = -\frac{MV(k^2+l^2)^{1/2}}{h} \coth[(k^2+l^2)^{1/2}h]\varepsilon \qquad (87)$$

This behavior is similar to the two-dimensional case except that smoothing is more rapid. One might expect this result since the field concentration at "hills" is greater than at waves.

The effect of overvoltage is introduced into the theory by modifying the boundary conditions, Eqs. (62) and (63). Only activation overvoltage, $\eta$, obeying the Tafel equation,[207,208]

$$\eta = a' + b' \log j \qquad (88)$$

is considered.

The potential there is a function of the current density:

$$\varphi = f(j) \quad \text{on} \quad y = 0 \qquad (89)$$

$$\varphi = V - g(j) \quad \text{on} \quad y = h + \varepsilon_0 \sum_1^\infty a_n(t) \sin n\pi x/\lambda \qquad (90)$$

Only the first two terms of a Taylor series expansion of $f(j)$ and $g(j)$ are retained:

$$\varphi = f(\bar{j}) + \left(\frac{\partial f}{\partial j}\right)_{j=\bar{j}} (j - \bar{j})$$

$$= \alpha + \beta(j - \bar{j}) \qquad (91)$$

where $\bar{j}$ is the average current density, $\alpha$, the average cathodic overvoltage, is $f(\bar{j})$, and $\beta$ is $(\partial f/\partial j)_{j=\bar{j}}$.

$$\varphi = V - \left[g(\bar{j}) + \left(\frac{\partial g}{\partial j}\right)_{j=\bar{j}} (j - \bar{j})\right]$$

$$= V - [\gamma + \zeta(j - \bar{j})] \qquad (92)$$

where $\gamma$, the average anodic overvoltage, is $g(\bar{j})$ and $\zeta$ is $(\partial g/\partial j)_{j=\bar{j}}$. Because of these overvoltages, the potential in Eq. (73) will be modified to

$$\varphi = A + \frac{V - A - B}{h}\left\{y - \varepsilon_0 \sum_1^\infty \frac{\sin(n\pi x/\lambda)}{\sinh(n\pi h/\lambda)}\left[(a_n + c_n)\sinh\frac{n\pi y}{\lambda}\right.\right.$$

$$\left.\left. + b_n \sinh\frac{n\pi(h-y)}{\lambda}\right]\right\} \qquad (93)$$

where $A$, $B$, $b_n$, and $c_n$ are evaluated by applying the boundary conditions as before. The behavior of the resulting functions has a complex dependence on the configuration parameters ($h$ and $\lambda$) and on the overvoltage parameters ($\kappa$, $\beta$, and $\zeta$). For certain approximations, it is shown[226] that overvoltage effects cause an increase in the machining time and an increase in the depth of the cut necessary to reach the required degree of smoothness.

Consider now the production of a required shape on an initially flat anode. This shape is thought to be a distribution of irregularities; supposing the cathode shape to be given by the Fourier integral,

$$y_c = \varepsilon_0 \int_{-\infty}^{\infty} b(k) e^{jkx} dk \qquad (94)$$

which represents a small deviation from $y_c = 0$. In other words, $|y_c|/h \gg 1$. Then after a given time, $t$, the anode shape is given by

$$y_a = h + \varepsilon_0 \int_{-\infty}^{\infty} a(k, t) e^{jkx} dk \tag{95}$$

Since the irregularities on both the anode and cathode will influence the potential across the gap, the potential is written as a first-order perturbation to the potential between two parallel plane electrodes as before, Eq. (64). In this case, $\varphi_0$ satisfies boundary conditions $\varphi_0 = 0$ on $y = 0$ and $\varphi_0 = V$ at $y = h$. Then $\varphi_1$ satisfies

$$\varphi_0 + \frac{\varepsilon_0}{h}\varphi_1 = 0 \quad \text{at} \quad y = y_c \tag{96}$$

$$\varphi_0 + \frac{\varepsilon_0}{h}\varphi_1 = V \quad \text{at} \quad y = y_a \tag{97}$$

Now, $\varphi_0$ is given by Eq. (67). If $\varphi_1$ is written in the form

$$\varphi_1 = \int_{-\infty}^{\infty} \Phi_1(k, y) e^{jkx} dk \tag{98}$$

then $\Phi_1$ satisfies the transformed equation

$$\frac{\partial^2 \Phi_1}{\partial y^2} - k^2 \Phi_1 = 0 \tag{99}$$

and the boundary conditions of Eqs. (96) and (97) become

$$\Phi_1 = -b(k) \quad \text{on} \quad y = 0 \tag{100}$$

$$\Phi_1 = -a(k, t) \quad \text{on} \quad y = h \tag{101}$$

to a first-order approximation in $\varepsilon_0/h$. The general solution of Eq. (99) is

$$\Phi_1 = A(k) e^{\pm k[y+B(k)]} \tag{102}$$

After applying the boundary conditions, Eqs. (100) and (101),

$$\Phi_1 = -\frac{a(k, t) \sinh ky + b(k) \sinh k(h - y)}{\sinh kh} \tag{103}$$

The expression for the potential $\varphi$ becomes

$$\varphi = \frac{Vy}{h} - V\frac{\varepsilon_0}{h} \int_{-\infty}^{\infty} e^{jkx} \frac{a \sinh ky + b \sinh k(h - y)}{\sinh kh} dy \tag{104}$$

From Ohm's law and Faraday's law, Eq. (75) is obtained and using the usual procedure, equations for the rate of change of $h$ [Eq. (77)] and of $a$,

$$\frac{\partial a}{\partial t} = -\frac{MVk}{h}(a \coth kh - b \operatorname{cosech} kh) \tag{105}$$

are obtained. For a constant gap distance $h$, the solution of Eq. (105) is

$$a(k, t) = b(k) \operatorname{sech} kh [1 - e^{-(MVkt/h)\coth kh}] \tag{106}$$

assuming $a(k, 0) = 0$, the initially plane anode. From Eq. (106), the amplitude, $a(k)$ of any frequency component on the anode approaches a limiting value,

$$a(k) \to b(k) \operatorname{sech} kh \tag{107}$$

as $t \to \infty$. Components with large $|k|$ values will have little effect on the profile produced on the anode. Thus the range of possible shapes which may be included in this theory is limited.

If the anode shape is specified, what will be the shape of the required cathode? Suppose the anode shape is specified to be

$$y_a = h + \varepsilon_0 \int_{-\infty}^{\infty} c(k) e^{ikx} dk \tag{108}$$

where $c(k)$ is known. If the required cathode profile is

$$y_c = \varepsilon_0 \int_{-\infty}^{\infty} b(k) e^{ikx} dk \tag{109}$$

the equilibrium shape produced by such a cathode according to the method, Eqs. (94)–(107), is

$$y_a = h + \varepsilon_0 \int_{-\infty}^{\infty} b(k) \operatorname{sech} kh\, e^{ikx} dk \tag{110}$$

Therefore, the condition placed on $b(k)$ is

$$b(k) = c(k) \cosh kh \tag{111}$$

Here, once again, one is limited to small amplitudes of the irregularities on the cathode.

Fitz-Gerald and McGeough[226] also considered the effects of overvoltage, and the procedure involving a modification of previous boundary conditions similar to those in discussion above [Eq. (89) *et seq.*] shows that the presence of overvoltage at the electrodes increases the machining time to reach the required shape.

This theory has been applied[226] to a deburring and to a cavity-forming operation; but in all cases, a serious limitation to this theory is the fact that it applies only where the irregularities in the cathode are at least an order of magnitude smaller than the gap distance. Under these conditions, a blurring or washing out of the predicted pattern machined in the anode from a given cathode shape can occur.

## 6.7. Effects of Stray Currents

During penetration of the anodic workpiece by the moving cathodic tool, metal should be removed at those sites on the anode directly opposite the exposed face of the tool since the sides of the tool are covered by an electrically insulating layer,[229] such as a film of an epoxy resin. Removal of metal from other sites is unwanted and distorts the geometry of the finished product. In a hole-cutting operation, the sides of the hole become tapered instead of straight because of stray currents which pass along the longer solution paths, causing metal to be removed from the sides of the cut. The longer a cathode remains in the vicinity of a given region the more metal will be removed from that region. As a result the amount of taper can be reduced by increasing the feed rate of the tool. There are, however, mechanical as well as electrochemical limitations to the rate at which the tool can be fed[230]; consequently, a considerable overcut is always obtained with NaCl-based electrolytes.

Another contribution to the magnitude of the overcut arises from variations in the applied voltage. As a result, a well-regulated power source is necessary for proper ECM operation. It has been estimated by Haggerty[220] that the voltage must be held within 2% variation to maintain a control of dimensions to 25 $\mu$m (0.001 in.). Control of the feed rate must also be held to the same degree to obtain this same tolerance in dimensions.

Overcut effects may be reduced by contouring the cathode and by varying the size and shape of the lip on the cutting face of the tool.[190,216,229] Although allowances for this overcut can be made by undersizing the tool, such hit-and-miss methods may be satisfactory for the gross removal or hogging out of metal but are not acceptable for sizing and finishing operations.

To avoid these arbitrary measures, attempts have been made[162,168,184,196,217,231–234] to predict the amount of overcut from a mathematical description of the ECM process so that a more accurate design of the cathode can be made. Computer programs have been written,[162,184,196,217] but in general, the results have been disappointing, since the agreement between the predicted and the actual anode profiles is not good. For the ECM of steel in NaCl electrolytes, the theoretical curve underestimates the experimental data by as much as 50%.[196]

Since analytical solutions to the ECM problem have offered little assistance to the practicing machinist, empirical methods are widely used in daily shop operation in the shop. One of these methods is the so-called "cosine method,"[234a,234b] which is based on steady state machining conditions but does not take into account solution flow and overvoltage effects.

At any point on the anode surface, consider the angle $\theta$ made between the tool feed direction and a line normal to the anode surface. Then the equilibrium gap distance between cathode surface parallel to the anode surface at that point is $h_e/\cos \theta$.[121,261] Consider the corresponding points $A(x, y)$ on the anode surface and $B(x_1, y_1)$ on the cathode surface at a gap distance $h_e/\cos \theta$. Then,

$y - y_1 = AB \cos \theta = h_e$ and $x_1 - x = AB \sin \theta = h_e \tan \theta = h_e(dy/dx) = h_e y'$. If the general expression for the anode surface is

$$y = a + bx + cx^2 \tag{111a}$$

then

$$y_1 + h_e = a + bx_1 + cx_1^2 - bh_e y' - 2cx_1 h_e y' + ch_e^2(y')^2 \tag{111b}$$

Differentiating Eq. (111a),

$$y' = b + 2cx \tag{111c}$$

Combining Eq. (111c) with $x_1 - x = h_e y'$,

$$y' = (b + 2cx_1)/(1 + 2ch_e) \tag{111d}$$

and substituting into Eq. (111b) gives

$$y_1 = a + bx_1 + cx_1^2 + h_e - h_e(b + 2cx_1)^2/(1 + 2ch_e) \tag{111e}$$

where terms in $h_e^2$ are neglected. The first four terms describe the anode surface displaced through a distance $h_e$ and the fifth gives the correction to the curvature of the anode or the overcut.

Using a similar type of analysis, Moore and co-workers[177] found that the overall radial overcut, $\Omega$, is

$$\Omega = (2h_e L + \Omega_c^2)^{1/2} \tag{111f}$$

where $\Omega_c$ is the overcut at a point on the anode opposite the corner of the cathode tool and $L$ is the land width of the tool. They estimated $\Omega_c$ to be $1.7h_e$ so that

$$\Omega = (2h_e L + 2.9h_e^2)^{1/2} \tag{111g}$$

Good agreement between the overcut predicted from Eq. (111g) and the experimental value was obtained in plung-cut holes using NaCl electrolytes.

Another empirical method is based on electrolytic tank analogs. In this method, a shallow tray made of transparent plastic is filled with an electrolyte (e.g., $CuSO_4$ solution). Thin metal (e.g., Cu) strips which represent the anode and cathode surfaces may be clamped in position in the tray. A dc voltage between 1 and 3 V is applied across the metal strips, and the potential between a fixed point on the anode surface and a movable probe in the solution is measured with a high input impedance voltmeter (Reference 261, p. 221).

To determine the position of the strips in the tray, the fixed shape of the cathode surface is drawn along with the probable shape of the anode surface on a piece of graph paper. Then, knowing the conductivity of the solution, the ratios of the potential differences along lines normal to the anode and a fixed point on the anode surface can be calculated.

After placing the graph paper under the tray, the metal strips are clamped in the indicated positions. The potential ratios are determined with the probe and compared to the calculated ones. If the calculated and experimental values of these ratios do not agree, the shape of the anode strip is changed to compensate. Under such conditions, the anode shape may be used to estimate the overcut.[223]

In a variation of this method, the electrolytic tray may be replaced by conducting paper on which the electrode shapes are drawn with conducting paint (References 121, p. 171; 261, p. 224).

Such simulation techniques are only a first approximation to the actual process[234c] since they do not take into account the effects of solution flow rate and overvoltage effects.

## 6.8. Addition Agent Studies

Because NaCl solutions are very corrosive, the addition of corrosion inhibitors, such as $NaNO_3$, $NaNO_2$, and amines,[177] to the electrolyte has been made.[157,189,235–238] In certain instances, complexing or sequestering agents such as citric acid[173] have been added to NaCl solutions to keep the corrosion products in solution,[238a] thereby achieving better surface finishes. Organic additives have been used[239] with NaCl solutions to decrease the viscosity; but improvements in pumping efficiency amounted to only a few per cent, and the large molecules suffered mechanical degradation in the circulating pump.

The most extensive effort in the use of addition agents is the search for a reagent which when added to the economically advantageous NaCl electrolyte will reduce stray cutting (overcut) to a point where precision ECM can be realized. The results of both organic and inorganic reagents added to NaCl electrolytes have been reported for potassium fluoride,[240] tripotassium phosphate,[241] monoethanolamine,[242] triethanolamine,[242] ethylene glycol, glycerine, imidazole,[243] 2-aminoethanol,[242] 2-mercaptobenzothiazole,[244] and sodium styrene sulfonate.[245] In all cases, films were formed on the anode surface which adhered strongly enough to improve the overcut characteristics of the NaCl electrolyte. Although these addition agents decreased the effects of the stray cutting in NaCl solutions, some reagents produced a roughening of the surface and all drastically altered the cost of the electrolyte. The average cost of a 10% solution of NaCl may be about 0.5 ¢/liter, whereas the addition agent can raise the cost of the electrolyte to values between $0.25 and $2.00/liter. This cost differential is much too great for the advantages gained.

Although Boden and Evans[246–248] have reported that the additions of carbonate, chromate, phosphate, and ferricyanide ions to NaCl solutions have reduced the stray cutting of nickel samples significantly, it has been our experience[249] that the ECM of steels and nickel-based alloys in many electrolytes containing high concentrations of $Cl^-$ ions will produce a certain amount of stray cutting in spite of the presence of passivating anions.

## 6.9. Surface Finish Considerations

One of the attractive features of the ECM process is the elimination of the finishing operations which are often required when a part is machined by conventional methods. Since metal is removed atom by atom by anodic corrosion in ECM, there are no burrs, scratches, or grooves to be removed, and as reported in the literature,[123,149,153,157,160,216,250] Profilometer readings of metal surface finishes range from 0.25 to 1 $\mu$m (10 to 40 $\mu$in.) for carbon steels machined in NaCl solutions. Highly reflective surfaces can be obtained on stainless steels machined in $Cl^-$ ion-based electrolytes.[171,229]

Since the quality of the surface finish of the machined part is strongly dependent on any parameter which affects the current distribution, it is imperative that a uniform flow of electrolyte be maintained through the gap by the proper design of the cell fixture.[160,195,198,251] At the high flow rates required in ECM, the Reynolds number is so high that the desired turbulent flow conditions prevail. Ito and co-workers[160] observed that the surface roughness decreased with increasing feed rate and decreasing electrolyte concentration. If eddies develop in the flow, some solution will remain in the gap space longer than the rest. With the variation of the current density caused by the concentration or depletion of ions in the eddies, undesirable eddy patterns can be reproduced in the workpiece surface.[173] Such problems are avoided by the proper design of the fixture.

The electrolyte flow through the pump must be as smooth as possible since a pulsating flow of electrolyte may produce a poor surface finish.[173] For this reason, multistage centrifugal pumps are used in most commercial ECM machines.

On the surface of some metals, such as stainless steels or aluminum, a protective film is formed which is broken down unevenly in the presence of $Cl^-$ ions. The resultant pitting and uneven metal removal cause excessive roughening of the surface.[252] Poor machining results were also obtained for Cu and brass in NaCl.

Coarse-grained materials such as cast iron are difficult to machine by ECM because the matrix material is etched away, leaving behind grains of carbon-rich material. Extremely rough surface finishes are obtained.[160,190,216,253,254] From a study of the ECM in NaCl solutions of a number of steels containing 0.13% to 1.5% carbon, the rate of the anodic dissolution of the steel decreased with increasing carbon content.[253] The rate, $r$, was found to be an exponential function of carbon content, $C$, of the steel:

$$r = a'' - b'' e^{dC} \tag{112}$$

where $a''$, $b''$, and $d$ are constants. The roughness of the surface increased with increasing carbon content of the steel.[254] Good surface finishes with ECM should be expected only with homogeneous materials.

It was first reported by Gurklis[127,255] and later confirmed[239,256–258a] that the fatigue strength of metals machined by ECM was lower than that of those machined conventionally. The fatigue strength of a given metal is a surface phenomenon.[259] Mechanical machining of a metal surface generates a stressed layer which increases the fatigue strength,[255] but ECM does not generate such layers, rather it produces intergranular surface attack[255-260] with subsequent decrease in fatigue strength. Most of the fatigue strength can be returned by restoring the stressed layer through vapor blasting or shot peening the machined workpiece. Therefore, if the machined part will be subjected to variable stresses, one must take caution in the use of ECM.

Several books have been written on the subject of ECM. The first, by DeBarr and Oliver,[172] gives a general account of ECM. For a fairly detailed account of the practice of ECM, including design considerations and many procedural hints, Wilson's book[173] may be consulted. More recently, McGeough[261] has presented a comprehensive account of the theoretical and mathematical treatments of the ECM processes.

## 7. Electrochemical Machining Operations

ECM is a very versatile process[182,216,220,262,263]; a number of the ECM operations will be described in this section followed by a brief description of several types of ECM machines.

### 7.1. External Shaping

With development of high-temperature, high-strength alloys spurred on by the advent of jet aircraft, it became increasingly difficult to mass produce turbine blades and parts for the jet engines. Consequently, one of the first industrial applications of ECM was the machining of turbine blades[8,153,262-267] (see Figure 14). With the external shaping operation, the tool is designed to fit the desired contours of the finished workpiece. Both sides of the blade can be machined at the same time by moving two cathodes, one on either side of the fixed anode[262] as shown in Figure 15a. Faust and Snavely pointed out[262] that jet engine turbine blades which required from 1 to 2 hr to machine by conventional methods were machined by ECM to tolerances of 0.008 cm (0.003 in.) in 5-10 min.

In the absence of applied pressure to the workpiece, thin foils can be machined as easily as heavy sections without concern for warp, bow, smear, or twist.[262,267] Since mechanical stresses are not induced in the metal lattice, since induced thermal modification of the metallurgical structure is not generated and since burring is absent, the surface finish is acceptable (0.5-1.0 $\mu$m; 20-40 $\mu$in.) in many cases without further treatment. There is no tool wear because there is no physical contact between the tool and the workpiece; if damage to the tool from sparking or short circuits is prevented, reproduction of identical parts may be obtained indefinitely with good accuracy.

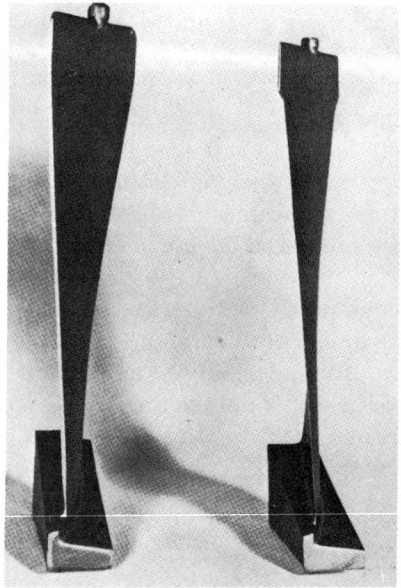

Figure 14. Turbine blade before (left, in as-forged condition) and after (right, finished to dimensions) ECM.[262] (With permission of *Iron Age*.)

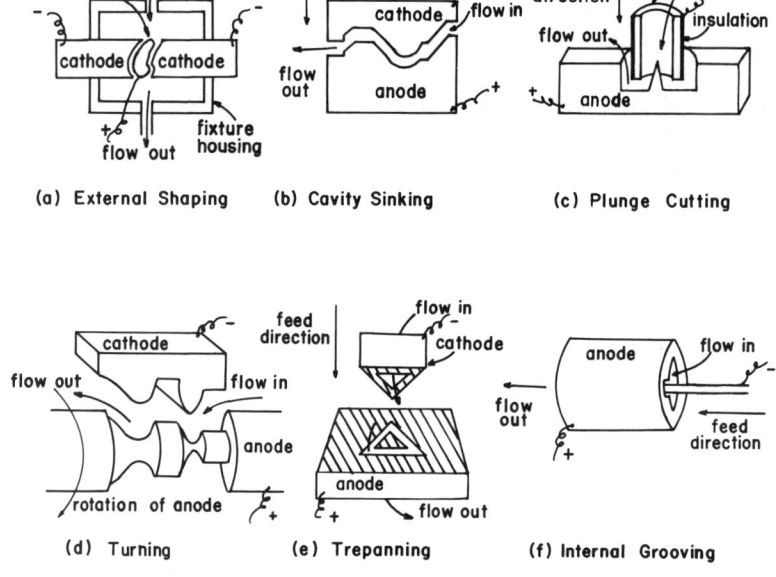

Figure 15. Sketches of various ECM operations.

Although the cathode is usually made of copper, cathodes made of copper–tungsten bronze, stainless steel, brass, and titanium may be used.[216] The only electrochemical process occurring at the cathode is hydrogen evolution in neutral salt solutions, and because hydrogen is liberated only at the cathode, hydrogen embrittlement of the machined anode does not take place.[127,149] To obtain acceptable results, care must be taken in the design of the tool cathode and the cell fixture to provide uniform flow of the electrolyte through the gap. Although the tooling costs may initially be high, the ECM process may be used repeatedly.[230]

In general, many of the remarks made here concerning the external shaping operation are also applicable to the other ECM operations to be discussed.

## 7.2. Die Sinking

The cathode tool is fashioned in the mirror image of the shape of the cavity to be machined in a block of metal in the die-sinking operation, also known as cavity sinking.[152,156] The electrolyte is pumped between the anode and cathode, as shown in Figure 15b. As the face of the tool approaches the workpiece surface, metal is removed opposite the high spots first. As the tool penetrates into the workpiece, the metal opposite the low spots is dissolved until a uniform gap is developed between all points on the anode and cathode surfaces (References 121, p. 22; 262).

Very complex cavities, requiring numerous operations by conventional machining, can be obtained with the die-sinking operation with acceptable finishes in just one quick operation with ECM. For example, the cavity for a pipe elbow die shown in Figure 16 was machined in fully hardened steel according to Haggerty[220] to a depth of 2.83 cm (1.125 in.) in 9.5 min with a finish of 0.65 $\mu$m (25 $\mu$in.) using NaCl electrolyte. Since these complex

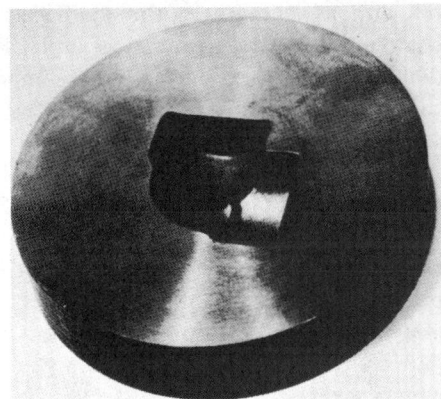

Figure 16. Pipe elbow die cavity.[220] (With permission of the Society of Automotive Engineers, Inc.)

shapes can be machined in steel in the fully hardened state, warping or other deformations of the part during the heat-treating processes required in conventional machining methods are avoided in this ECM operation.[162,179,182,220,268,269]

## 7.3. Plunge Cutting

The cathode in the plunge-cutting operation, also known as hole cutting, is a hollow pipe through which the electrolyte is pumped as sketched in Figure 15c. Solution flow may also be reversed compared to that shown in Figure 15c.[216] The outside walls of the cathode are electrically insulated with a coating of a nonconducting material (e.g., epoxy or polyurethane) so that metal will be removed from the bottom of the hole and not from the sides. In spite of this protection, there is always a certain amount of stray cutting with NaCl electrolytes, which produces tapering of the holes.[162,217] It is for this reason that a valid mathematical model for the ECM process is needed[162,217,226] so that the amount of overcut produced by stray cutting may be predicted. Once this is known, the design of the cathode may be modified so that the desired pattern in the workpiece may be obtained. Unfortunately, the mathematical description required to reach these goals has not been achieved to date.

It is also possible to drill blind holes with the plunge cut operation. Haggerty[220] mentions that a rectangular blind hole, $2.5 \times 3.8 \times 3.8$ cm ($1 \times 1.5 \times 1.5$ in.) was machined by ECM in a steel cylinder with one operation in 6 min. Three separate operations (drilling, milling before heat treatment, and grinding after hardening) requiring 43 min. were necessary to machine the same hole by conventional methods. Depending on the size and complexity of the hole, metal removal rate for plunge cutting varies from 0.15 to 1.5 cm/min. (0.05 to 0.5 in./min.).[220] Holes of many shapes, such as square, elliptic, hexagonal, fluted, etc., may be machined through the hardest alloys by using cathodes of proper design. In general, the metal removal rate falls off with a decrease in the diameter of the hole and an increase in the complexity of the shape of the hole. For the drilling of long holes of small diameter, the cathode should be made of a more rigid material such as tungsten bronze instead of copper so that wobble or swinging of the tool will be avoided.

In some cases, a small slug of metal may be trepanned out in the center of the plunge cut hole as shown in Figure 15c. At the end of the plunge cut, there is danger of a short circuit occurring should the slug hinge and fall against the cathode (Reference 177, p. 9). To avoid costly damage to the cathode, a false or backup workpiece is attached to the bottom of the anode with a screw in the metal to make up the slug. Another advantage in using a backup plate is the fact that loss of solution pressure is prevented when the hole penetrates the workpiece.

## 7.4. Turning

As in the mechanical counterpart, the workpiece anode in the turning operation is held in a rotating chuck. The face of the cathode is contoured to produce the desired shape in the anode as it approaches the rotating workpiece. The electrolyte may be pumped through the gap space as diagrammed in Figure 15d, or it may be passed through a slot in the cathode.[177] If a cathode is used which presents a very narrow exposed area to the anode, current densities as high as 800 A/cm$^2$ (5000 A/in.$^2$) can be attained.

## 7.5. Trepanning

If a large number of identical parts of complex shape are required, they may be cut from metal sheets or slabs by the trepanning operation pictured in Figure 15e in one operation. By using this method good finishes are obtained without burrs.

## 7.6. Internal Grooving

When a thin cathode, which is electrically insulated except for a properly designed, exposed area or "window"—is passed down the inside wall of a hollow cylinder, a slot or groove may be machined in the workpiece wall, as sketched in Figure 15f. If the cylinder is rotated during the internal grooving, a helical groove similar to the rifling in a gun barrel will be obtained.

## 7.7. Wire Cutting

To remove a large volume of metal in a roughing out or hogging process without the use of excessively high currents or large amounts of electrical energy, the wire-cutting operation presented in Figure 17 is an ideal application of ECM. Here, the cathode is either in the form of a wire over which electrolyte is sprayed or a thin metal tube through which the electrolyte is pumped. The electrolyte exits the tube through small holes in the cutting face of the cathode. In general, feed rates are low because the cross-sectional area of the wire or tube is so small that the current-carrying capacity of the cathode is limited.[220] Wire cutting is also useful in cutting off tube ends or in cutting slices from a bar of metal because clean, smooth cuts are obtained without burring or smearing of the metal and without warping.

## 7.8. Deburring

The main difference between the deburring operation and other ECM operations is the fact that no relative motion between the anode and cathode occurs because the tool and workpiece are held fixed with a proper gap space

Figure 17. Demonstration of wire-cutting operation.[220] (With permission of the Society of Automotive Engineers, Inc.)

between them. Consider the wall of a pipe through which a hole has been drilled mechanically, and the internal perimeter of the hole is burred. An insulated metal rod smaller than the hole (by an amount equal to the gap space) is clamped in position with the exposed end of the cathode opposite the burred region of the anode, as shown in the cross-sectional view in Figure 18. The electrolyte is pumped through the gap with the splash shield in place.

Usually the gap space in a deburring machine is wider and the solution flow rate is lower than in other ECM operations because deburring is determined by the kind of burr to be removed. It is difficult to characterize a burr since it is an

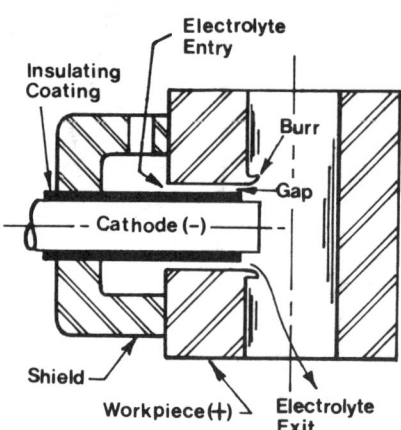

Figure 18. Diagram of a typical deburring machine. (With permission of Chem-Form Division, KMS Industries, Inc.)

# ELECTROCHEMICAL MACHINING

unwanted, undesigned by-product of a manufacturing process. Consequently, the deburring device must be capable of removing the largest burr on the piece as well as any small burrs without damage to the part by excessive metal encountered on the same work.

## 7.9. ECM Machines

The ECM machine includes, besides the electrochemical cell or fixture, the electrolyte circulating pump, flow line filters and heat exchangers, reservoir tanks, electrode drive train, rectifier, electrolyte piping, electrical cables, control panel and components, electronic protection device, hydrogen exhaust fan, and a housing rugged enough to maintain rigidity under the high hydraulic pressures generated. The fixture is composed of the cell housing, the cathode tool, the anode workpiece, and the general design features of the fixture housing, such as baffles, slots, inlet and outlet parts, etc., required for uniform electrolyte flow.

To avoid tangling, the electrolyte hoses and electrical cables may be built into the frame of the machine with permanent connections to the tool platen and the worktable. For multiple fixturing, the electrolyte connections may be

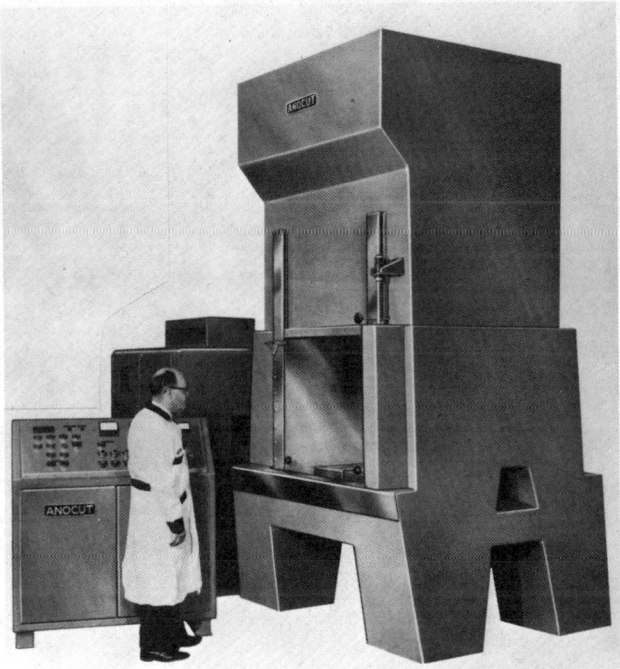

Figure 19. Vertical, 40,000-A, A-frame ECM machine. (With permission of Anocut Engineering Company.)

connected to a manifold. To minimize corrosion, it is good practice to connect the machine to the negative side of the dc power source, with the anodic worktable electrically insulated from the rest of the machine. Most machines have the alignment of the tool platen and the worktable fixed with alignment key slots to be used for rapid tool location. This arrangement reduces the possibility of operator error and extended tooling setup times.

A very rugged type of ECM machine is the vertical A-frame unit shown in Figure 19 which is used for high precision work (less than 25 $\mu$m tolerance) as high electrolyte flow rates (requiring high hydrostatic pressures) are necessary for work requiring small tolerance. Although the ram and control systems are mounted above the work enclosure, which reduces the chances of damage due to electrolyte leakage and which makes it easy to keep the unit clean, there is limited access to the work enclosure, and the handling of heavy or awkwardly shaped parts becomes difficult.

It is also possible to mount the drive systems under the work enclosure in the vertical underdrive A-frame machine shown in Figure 20. In this case, the machine size is greatly reduced although it is still rigid. Electrolyte leakage can cause serious corrosion problems if the machine is not properly drained.

Figure 20. Vertical, underdrive, A-frame ECM machine. (With permission of Anocut Engineering Company.)

Figure 21. C-frame ECM machine. (With permission of Anocut Engineering Company.)

To obtain greater access to the work enclosure, the vertical C-frame ECM machine was designed to be opened on three sides, as demonstrated in Figure 21. The precision obtained with this machine is limited because of reduced rigidity. The rigidity can be increased by using support bars between the machine head and base.

The greatest accessibility to the work enclosure is achieved using the horizontal frame machine pictured in Figure 22. Such a machine can accommodate very heavy tooling and parts by handling the components with overhead lifting equipment. However, this type of machine is not suitable for high precision work.

Figure 22. Horizontal ECM machine. (With permission of Anocut Engineering Company.)

Wilson[173] has summarized in depth both the operational and the economic advantages and disadvantages of the various types of ECM machines currently on the market.

## 8. The Electrochemistry of Electrochemical Machining

In much of the early investigations of ECM, solutions of NaCl were used as the electrolyte because such solutions were inexpensive and possessed high conductivity. With NaCl-based electrolytes, large stray currents are present. Thus large overcuts producing rounded corners and tapered holes are obtained. Large overcuts could be tolerated if they were uniform and predictable; compensating corrections in the design of the tool could then be made to overcome any lack of control of dimensions or geometry.

In practice, however, it appears that the uniformity of the cut decreases as the overcut becomes larger. Consequently, the rate of metal removal due to stray currents depends on the electrochemical machining parameters in a complex manner. Mathematical formulations to predict overcut have been largely unsuccessful[162,184,196,217] and have proven useless in the design of tooling for the ECM production of complex shapes. As a result, the process of tool development is a tedious, empirical, and expensive task to successfully produce the required geometry.

An important goal is the reduction of the presence of stray cutting in the ECM operation. Attempts to modify the NaCl electrolyte through addition agents failed primarily because the electrolyte was considered only as another ohmic part of the ECM system, and the electrochemistry of this anodic corrosion process was ignored. It has been our experience that all NaCl-based electrolytes support large stray currents which produce nonuniform overcuts in ferrous metals and alloys.

### 8.1. New Electrolyte Studies

When attempts were made to apply ECM machines (using NaCl electrolytes) to common operations in the automotive industry, several difficulties arose. For most applications considered, dimensions could not be held, and tolerances could not be met, workpiece integrity was destroyed because of unwanted metal removal due to stray currents. The ECM operation could not be made compatible with other manufacturing operations. Therefore, at General Motors Research Laboratories (GMR), the electrochemistry of the ECM process was investigated in an effort to find a completely new type of ECM electrolyte which would provide its own inherent protection of metal surface sites against unwanted metal removal.

From an intensive investigation of the ECM properties of several hundred electrolytes and combinations of electrolytes, it was reported[270,271] that

# ELECTROCHEMICAL MACHINING

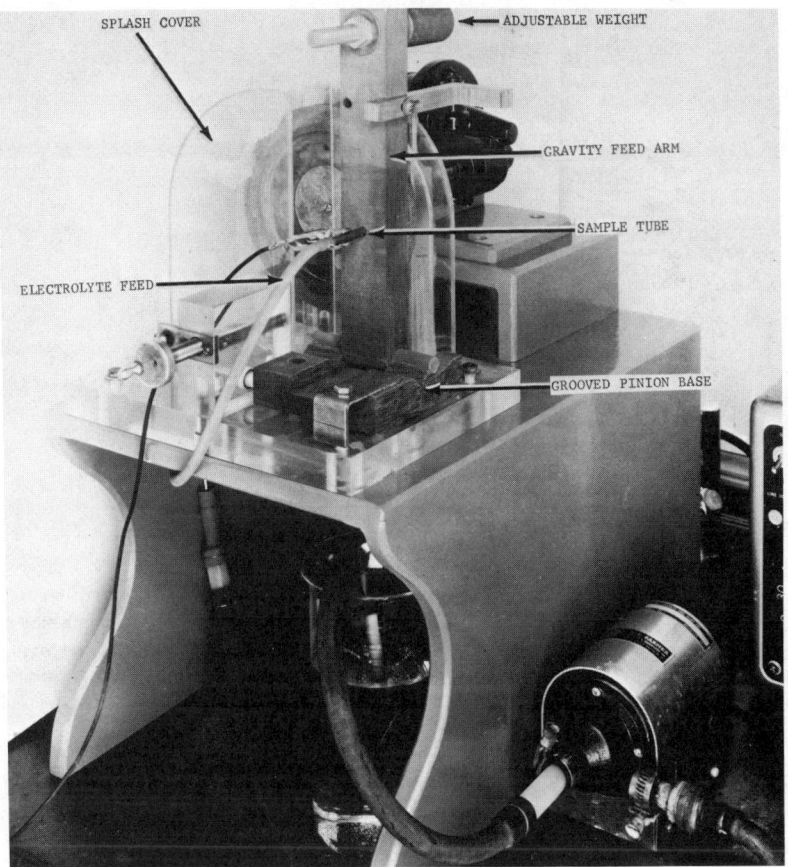

Figure 23. An electrochemical-grinding, laboratory test rig.[168] (With permission of The Electrochemical Society, Inc.)

solutions of $NaClO_3$ are a superior ECM electrolyte for the desired control over workpiece dimensions and geometry at high metal removal rates, leaving excellent surface finishes. In the studies at GMR, small laboratory ECM test rigs were employed, including an electrolytic grinder, Figure 23,[270] a through-hole machine, Figure 24, a plunge cutter, Figure 25,[273] and a closed cell tube end machine, Figure 26.[274,275]

To demonstrate the type of dimensional control obtained with the $NaClO_3$ electrolyte, Figure 27 shows notches cut in fully hardened 5160H steel bars using the test ECG grinder. The bottom sample was cut using NaCl solution, whereas the top sample was cut with $NaClO_3$ solution. In Figure 28, the Proficorder traces recorded on the inside surface of a bushing made of a Carpenter low-alloy, nickel–chromium steel hardened to Rockwell 60C are reproduced before (Figure 28A) and after (Figure 28B) the hole had been sized

Figure 24. A through-hole ECM fixture for sizing and finishing bushings.[270] (With permission of The Electrochemical Society, Inc.)

in the through-hole machine using a $NaClO_3$ electrolyte. Values of the surface finish were reported[270] to range between 0.05 and 0.13 $\mu$m (2 and 5 $\mu$in.). Tolerances of the order of ±5 $\mu$m (±0.0002 in.) were reported. Both the taper and out-of-roundness were below 2.5 $\mu$m (0.0001 in.).

Up to the solubility limit for NaCl in water, the metal removal rate of steel in $NaClO_3$ solutions is about the same as that in NaCl solutions of the same concentration. Above this value, the metal removal rate increases in $NaClO_3$ electrolytes with increasing concentrations up to about 600 g/liter where the rate levels off at twice the maximum rate obtained with NaCl solutions. At these high rates of metal removal, the machined surface still retains a mirror-bright surface.

Subsequently, the excellent ECM properties of $NaClO_3$ electrolytes have been confirmed by other investigators for the machining of steel[276-281] and other metals, such as Co,[282] Ti,[283] and Ni.[284]

## 8.2. ECM Precautions with Oxidizing Salt Electrolytes

Although the use of sodium chlorate electrolyte has taken on considerable commercial importance, certain disadvantages to its use are clear. Much care should be exercised in working with $NaClO_3$.[284a,284b] When in solution, or

# ELECTROCHEMICAL MACHINING 459

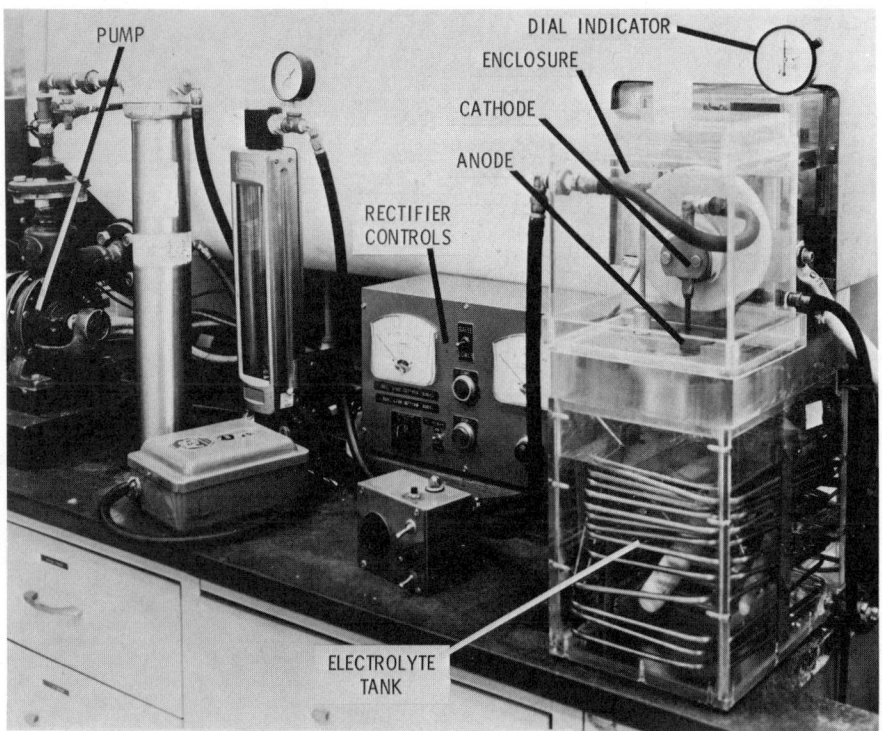

Figure 25. A laboratory plunge-cut ECM machine.[273] (With permission of The Electrochemical Society, Inc.)

when kept wet in an atmosphere of 65–75% relative humidity, the $NaClO_3$ electrolyte does not present a problem. However, when dry $NaClO_3$ comes in contact with relatively finely divided organic materials, such as paper, cloth, or wood, the problems of combustion arise. The combustion danger is severe since it is equivalent to a temporary oxygen-fed fire which lasts until all chlorate is dissipated.

As a result, the electrolyte has been used under closely controlled conditions. Special protective clothing is always worn in the work area. Immediate washing of the clothing used in the work area is necessary to be certain that no chlorate is carried away. Other precautionary measures include (1) mounting of the ECM equipment on raised steel trestle floors, (2) a properly sloped concrete underfloor surrounded by a damming enclosure, (3) sufficiently scrubbed air removal, (4) hot water wash-down hoses, (5) personnel jump tanks, and (6) restricted personnel access to the work area.

The precautions deemed necessary with the use of sodium chlorate electrolyte are no more stringent than for many industrially used chemicals; however, it would be desirable to operate an ECM system without such

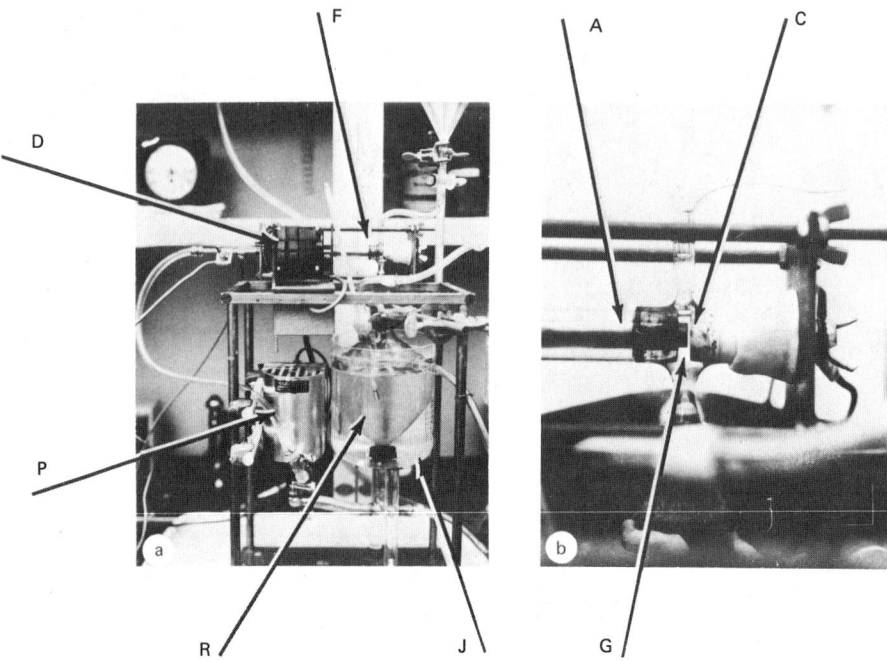

Figure 26. A closed-cell, tube-end, ECM test rig[274] overall view (a) and cell fixture (b): A, tube anode; C, fixed cathode; D, anode drive mechanism; F, cell fixture, G, gap; J, water jacket; P, pump; R, electrolyte reservoir. (With permission of the National Association of Corrosion Engineers.)

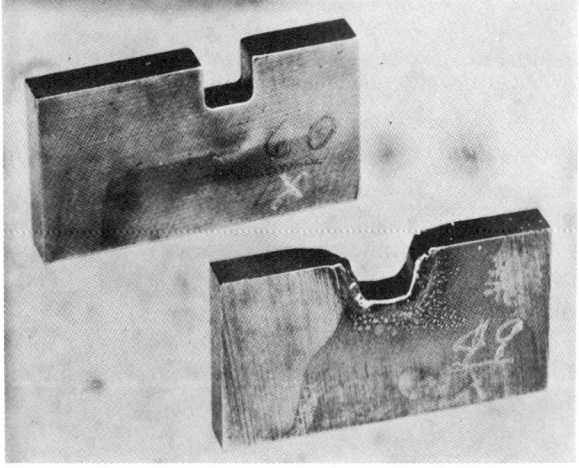

Figure 27. Notches cut[270] in 5160H steel with electrochemical grinder in $NaClO_3$ (top) and NaCl (bottom) electrolyte. (With permission of The Electrochemical Society, Inc.)

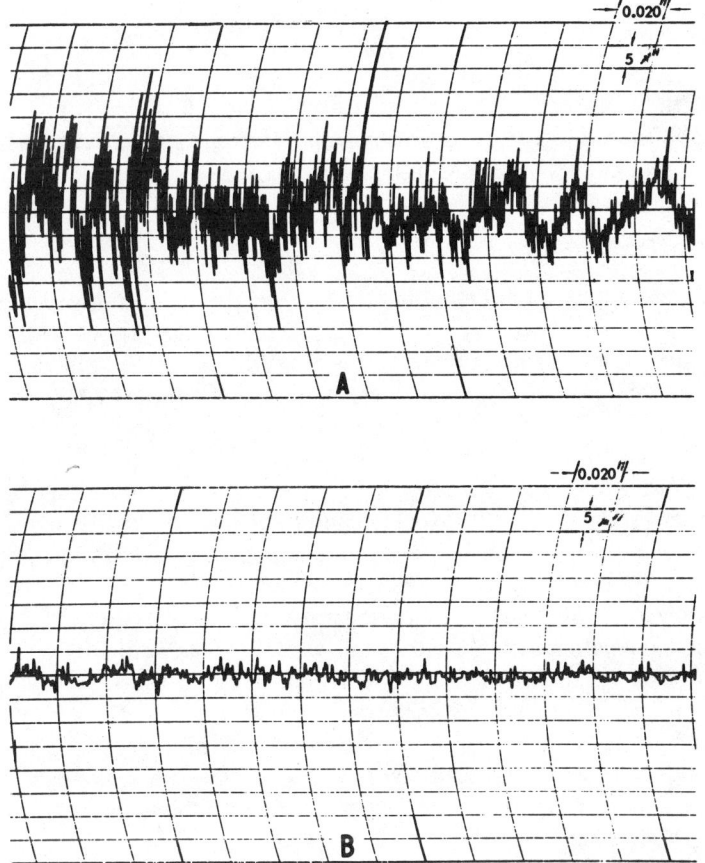

Figure 28. Proficorder traces obtained[270] on the surface of a 5160H steel sample before (A) and after (B) electrochemical machining in NaClO$_3$. (With permission of The Electrochemical Society, Inc.)

constraints. Darling and LaBoda[284c] point out that it is desirable to be able to simply add something to the electrolyte which would reduce the possibility of fire. They described various additions to sodium chlorate electrolyte which reduce the likelihood of combustion and which are at the same time compatible with the unique electrochemical machining characteristics displayed by its use. For relative humidities between 59% and 72%, the additive compounds include lithium bromide, potassium iodide, magnesium perchlorate, lithium citrate, and potassium acetate. When the relative humidity is as low as 35%, lithium bromide and potassium acetate are most effective.

The compatibility referred to above is a subjective matter, since in practice, the addition of any of the above materials did not take precedence to good housekeeping care in production usage of sodium chlorate ECM solutions.

Another safety matter not often considered is the status of sodium nitrate salts and their storage. Such salts are also classified as oxidizers,[284d] as is sodium chlorate, and can form explosive mixtures. Care should be taken in storing large amounts for ECM usage.

### 8.3. Polarization Studies

Since the machining of steel in $NaClO_3$ electrolytes produced such good dimensional control, whereas that in NaCl electrolytes gave poor results due to the presence of large stray currents, it is possible that such behavior may be traced to passivation effects of the highly oxidizing chlorate ions. To understand the passivation behavior of iron and steel in various electrolytes in the absence of the complicating ECM parameters (high electrolyte flow rate, electrode feed rate, etc.), steady state, potentiostatic, anodic-polarization curves were obtained[209,285,286] for iron and steel microelectrodes (small wires) in $O_2$-stirred solutions (the ECM process is exposed to the atmosphere) in conventional polarization cells. The amount of anodic film on the electrode surface associated with each point on the polarization curve was determined by a cathodic, galvanostatic, film-stripping technique using a high-rate (high-current density) pulsing circuit.[287] Only an apparent surface area can be determined because large area changes occur during anodic formation and cathodic stripping of the anodic films. The amount of film present is recorded in terms of the charge, $Q$, associated with the transition time obtained from the trace of the cathodic pulse.[287] Afterwards, these data were correlated with actual ECM measurements.

A typical set of data are presented in Figure 29 with the polarization curves in Figure 29a and the film-stripping curves in Figure 29b for mild steel anodes in $NaClO_3$, NaCl, and $Na_2Cr_2O_7$ solutions. Consider first the curve for a steel anode in $NaClO_3$ solution of 250 g/liter. At potentials below 0.4 V vs. the saturated calomel reference electrode (SCE), iron is active and goes into solution as $Fe^{2+}$ ions, which are rapidly oxidized to $Fe^{3+}$ ions.[288,289] In this active region, the current increases with potential, and no trace of an anodic film can be detected. Above 0.4 V, the current falls because a protective anodic film builds up on the electrode surface and inhibits the anodic dissolution of iron. Then, between 0.8 and 1.4 V, the current is reduced to very low values in the passive region where the anodic film has reached its maximum thickness. Finally, in the transpassive region above 1.4 V, the current once more increases to very large values as the anodic film thickness is diminished. The transpassive current may arise from the destruction of the protective film, allowing the anodic dissolution of metal to take place once more. Another contribution to this current may come from a new anodic process, such as the evolution of

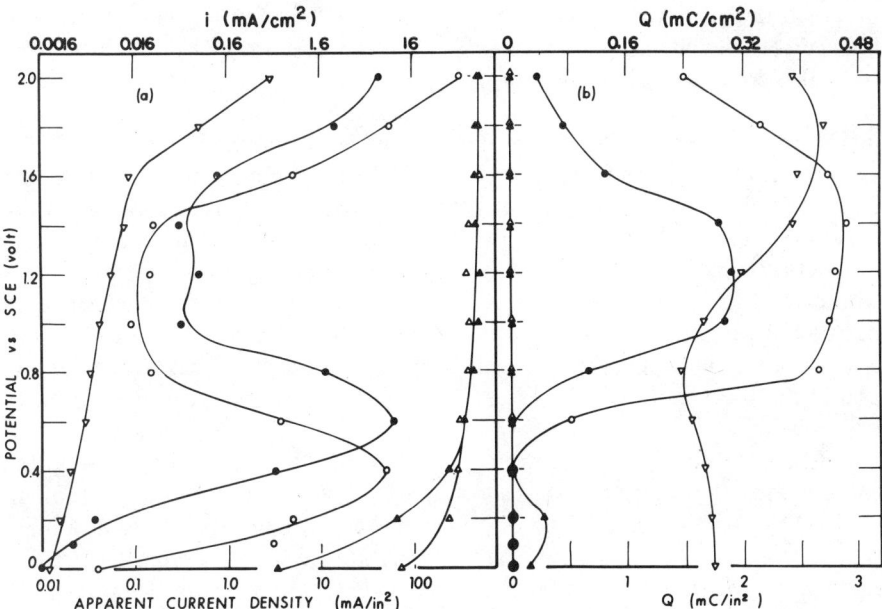

Figure 29. (a) Steady-state, anodic polarization curves obtained[168] on mild steel microelectrodes in solutions of NaCl (△), NaClO$_3$ (○), and Na$_2$Cr$_2$O$_7$ (▽); filled symbols, 50 g/liter; open symbols, 250 g/liter. (b) Anodic oxide film thickness in terms of the amount of charge, Q, determined from constant current stripping pulses as a function of potential; notation same as (a). (By permission of The Electrochemical Society, Inc.)

oxygen occurring on the surface of a conducting film or from a change in the chemical composition of the film.

Consequently, mass balance studies were made in the transpassive region in a closed cell system to determine the current efficiency for both metal dissolution and O$_2$ evolution. For mild steel anodes in NaClO$_3$ solutions, it was found[212] that both the evolution of oxygen and the anodic dissolution of iron contribute to the transpassive current; however, the distribution of current between the two electrode processes is a function of the current density. As the current density increases, the O$_2$ evolution current decreases, and the metal dissolution current increases. For example, in 4.5 M NaClO$_3$ solution, the current efficiency for O$_2$ evolution was 57.9% at 12 A/cm$^2$ and fell to 24.8% at 47 A/cm$^2$, whereas that for iron removal rose from 41.7% at 12 A/cm$^2$ to 78.6% at 47 A/cm$^2$ (see footnote a in Table 5). In a high-velocity flow cell, it was found[290] that metal dissolution of pure iron anodes in 4 M NaClO$_3$ accounted for virtually 100% of the transpassive current above 10 A/cm$^2$.

By comparing the results of the polarization studies with those of the actual ECM behavior obtained on laboratory test rigs, the ECM process was first recorded as occurring in the transpassive region of potentials where very high current densities can be reached.[285] Other investigators[198,228,279,291–295] have also come to this conclusion.

The good control of dimensions and geometry obtained with the machining of iron and steel in $NaClO_3$ solutions has been explained in terms of the protective anodic films formed in this system. To demonstrate this proposal, consider the plunge cut operation sketched in Figure 15c and the polarization curve in Figure 29. For a given applied voltage, $V$, the potential, $E$, controlling the metal removal process at the anode follows the relationship $E = V - IR$, where $R$ is the resistance of the solution path over which the current, $I$, is passed between the anode and cathode. At the high values of $V$ in ECM, it is assumed that the contribution of activation overvoltage to $E$ is relatively small, and at the high solution flow rates, the diffusion overvoltage contribution is kept low.

At anodic sites on the workpiece directly opposite the cutting face of the cathode tool, the solution path is short, the $IR$ term is small, $E$ is large enough to lie in the transpassive range of potentials, and metal is removed at a high rate. Along the sides of the cut where the workpiece sites are remote from the cutting face of the tool, the solution path is long, the $IR$ term is large, $E$ is reduced to values lying in the passive region of potentials, and a protective film is formed on the anode surface with little or no metal removal. In this way, then, significant tapering of the plunge cut hole is prevented as the cathode tool passes through the workpiece.

If the concentration of chlorate is lowered, the polarization curve is shifted toward higher potentials as shown by the curve in Figure 29 for 50 g/liter $NaClO_3$ solution. In this case, the maximum film thickness is less. One would expect for a given applied voltage that the metal removal rate would be lower at the lower $NaClO_3$ concentration. This was observed (Reference 168, p. 227) to be the case. For example, at 10 V applied voltage, mild steel was removed at 0.0127 cm/min (0.005 in./min) in 50 g/liter $NaClO_3$, but 0.178 cm/min (0.070 in./min) in 250 g/liter; at 15 V, 0.117 cm/min (0.046 in./min) in 50 g/liter, but 0.242 cm/min (0.096 in./min) in 250 g/liter. The current efficiency studies at a current density of 47 A/cm$^2$ also showed[212] that the efficiency for metal removal fell from 78.6% to 63.4% and that for $O_2$ evolution rose from 24.8% to 37.5% when the concentration of $NaClO_3$ was decreased from 4.5 $M$ to 2.0 $M$. This behavior was also observed by Kanda.[281]

Figure 29 shows data obtained during the potentiostatic polarization of mild steel anodes in NaCl solutions. It is seen that protective anodic films were not found and the effects of concentration are negligible over most of the potential range studied (Reference 168, p. 227). Although evidence for the presence of salt films at steel anodes in NaCl solutions exists,[186,286] these films are not protective. Consequently, high metal removal rates can take place at low values of $E$, and hence, at sites on the anode remote from the cutting face of the tool. For this reason, large stray currents present in NaCl solutions produce the large tapering of holes or rounding of corners as noted in Figure 27, resulting in the loss of dimensions and geometry. As expected from the polarization curves, the current efficiency for metal removal from mild steel tube ends in both 4.5 $M$ and 2.0 $M$ NaCl is about 98%.[212] Others[121,278] have observed that iron is removed by ECM in NaCl solutions at nearly 100% efficiency.

The lack of dependency of current density on NaCl concentration is pointed out by the following data[168]: at 15 V, 0.19 cm/min (0.075 in./min) in 50 g/liter NaCl and 0.178 cm/min (0.070 in./min) in 250 g/liter.

## 8.4. Transpassive Dissolution of Anodic Films

How a protective film formed on the anode at a given potential may be dissolved away at a higher potential in the transpassive region is of interest. This behavior appears[67] to be associated with the type of anions present in solution. A small anion, such as the $Cl^-$ ion, may penetrate the pores of the protective film, causing breakdown of the film,[37] since it forms soluble complexes with iron (see e.g., Reference 297). But Hoar[296,298] has pointed out that the $ClO_4^-$ is as good as, or in some cases better than, $Cl^-$ as an aggressive agent for the breakdown of iron oxide films. It is difficult to imagine how such a large ion as the $ClO_4^-$ ion can be such an efficient agent for film dissolution by a pore-penetrating mechanism.

To account for the $ClO_4^-$ ion behavior, Hoar[37] suggested the mechanical mechanism of film dissolution. In this scheme, the number of anions adsorbed on the oxide surface increases with increasing potential until a point is reached where the repulsive forces of the adsorbed anions rupture the film at weak spots caused by dislocations, inclusions, etc. On the new surfaces produced by the rupture, more anions can be adsorbed, causing a further rupture of the film and exposure of the underlying metal to attack. Concluding that a truly protective film is a good conductor, Hoar[298] reasoned that passive films on a metal anode cannot be composed solely of metal oxides but must be "contaminated" with anions and water molecules occluded in the oxide layer. Under these conditions, the ionic conductivity of the film is increased; and metal ions may pass through the protective layer.

With these ideas in mind, an ion-exchange model of the anode film was proposed[38,39] to account for the transpassive dissolution of the film and the role played by anions in film formation and removal. According to this model, metal surface sites on the anode may be oxidized by $ClO_3^-$ ions at potentials corresponding to the active–passive transition to form nuclei from which the oxide film may grow.[299-301] It was concluded[38] that the anodic film takes on the form of an open, three-dimensional matrix in which anions and water molecules can be adsorbed on either the internal or external surfaces of the matrix. The elements or building blocks of the matrix are believed[38] to be hydrated iron oxide units with the stoichiometry, but not necessarily the properties, of a bulk oxide. These hydrated oxide units are bound together in the three-dimensional structure through hydrogen bonding with water molecules. From the results of electron diffraction, electron probe, and vacuum fusion studies of the surfaces of steel anodes machined by ECM in NaClO$_3$ solutions,[209] it was further concluded[79] that the structural units of the matrix of the protective film are hydrated Fe$_2$O$_3$ or FeOOH units.

In the passive region, the electrode surface is covered by the three-dimensional matrix of FeOOH units, and initially, the surfaces of the matrix are covered by adsorbed water molecules. As the potential rises to more noble values, the number of ClO$_3^-$ ions adsorbed by displacing the water molecules increases with increasing potential.[302] The low current observed in the passive region is accounted for by the few ferrous ions which can diffuse through the iron oxide matrix. Eventually, a threshold potential is reached above which the concentration of adsorbed anions is large enough to permit the adsorbed ClO$_3^-$ ions to exchange with the oxide ions of the matrix. If it may be assumed[79] that four water molecules are associated with each FeOOH unit, it is estimated that after three ClO$_3^-$ ions had been adsorbed

$$FeOOH \cdot 4H_2O + 3ClO_3^- \rightarrow FeOOH \cdot H_2O \cdot 3ClO_3^- + 3H_2O$$

ion exchange

$$FeOOH \cdot H_2O \cdot 3ClO_3^- \rightarrow Fe(ClO_3)_3 \cdot 3OH^-$$

occurs at the threshold potential, producing soluble salts which dissolve away.

It is expected[302] that the concentration of replaced oxide species, Fe(ClO$_3$)$_3$·3OH$^-$, is greatest at the solution side of the adsorbed film. This situation would account for the observed[285,286] (Figure 29b) thinning of the anodic film in the transpassive region (Figure 29b) with increasing potential by steel anodes in NaClO$_3$ solution. Although one might expect a potential where the anodic film is completely removed from the metal surface to be reached, this is not the case since the presence of a large concentration of ClO$_3^-$ ions with high oxidizing power maintains the surface sites in an oxidized state. At a given high-potential equivalent to ECM values, a thin, porous, uniform oxide film of steady state thickness is formed on the anode surface. A balance between the tendency of the oxidizing ClO$_3^-$ ions to build up the film and that of the ion-exchange reaction to break down the film regulates the film thicknesses. There is evidence[303] that the thin film in the transpassive region does not have the same structure as the thick film in the passive region as might be expected.

There is experimental evidence for the presence of this thin, uniform film formed in the transpassive region. From electron diffraction studies, thin films of $\gamma$-Fe$_2$O$_3$ have been detected[209] on steel parts machined by ECM in NaClO$_3$

solutions but not those machined in NaCl solutions. The surface of a steel part machined in NaClO$_3$ solutions remains mirror bright for years during exposure to the laboratory atmosphere,[281,290] whereas the surface of a steel part machined in NaCl solution begins to rust within a week under the same exposure conditions.

For a given potential, it is likely[302] that more anions will be adsorbed in the oxide matrix of the film with increasing ClO$_3^-$ ion concentration. In this case, the critical concentration of adsorbed ClO$_3^-$ ions for the initiation of the ion-exchange process should be reached at less noble potentials in more highly concentrated solutions. Such a lowering of the threshold potential would appear as a lowering of the transpassive region toward less noble potentials as the concentration of NaClO$_3$ is increased. This behavior is observed in the polarization curves of Figure 29a.

Because the protective film possesses a certain amount of electrical conductivity,[298] a significant portion of the anodic current is consumed in O$_2$ evolution at the beginning of the transpassive region. As the potential is raised toward more noble values, the film dissolves away, and the current efficiency for metal removal increases, in agreement with experimental observations.[168] Since protective anodic films are not formed on steel anodes in NaCl solution,[304] the current efficiency for metal removal in this case is nearly 100%.[212]

## 8.5. Surface Brightening

The mirror-bright surface obtained on steel machined in NaClO$_3$ solutions can be interpreted in terms of the anodic films present on the metal surface in the transpassive region. When a metal anode dissolves at a high rate in any electrolyte, the concentration of metal ions in the solution layers closest to the anode surface reaches such high values that highly porous salt films precipitate out on the metal surface. Under these conditions, the metal dissolution process becomes diffusion controlled, and a leveling of the surface irregularities takes place, as represented in Figure 3. These conditions prevail in the transpassive region even for the machining of steel in NaCl solutions, thus accounting for the usually good surface finishes obtained on parts machined by ECM.

To account for the mirror-bright, polished surface obtained on steel in NaClO$_3$ solutions, the electropolishing mechanism proposed by Hoar and co-workers[62,63,76] is employed. In the far transpassive region where ECM takes place, the steady state, thin, uniform, porous film formed on steel anodes in NaClO$_3$ solution is made as fast as it is removed so that the film follows the receding surface of the workpiece during ECM operation. With the continuous regeneration of the oxide film, a completely random pore distribution results, producing the brilliant electropolished surface. This thin, uniform oxide film is not formed on steel in NaCl solutions, and only a matte surface is obtained in this case.

### 8.6. Solution Flow Effects

When the protective film is formed by the direct oxidation of surface sites by the anions present, the shape of the polarization curve is not affected by the solution flow rate, as shown by studies with a rotating disk.[209,304] In the transpassive region, the thin, uniform oxide film is always present on the anode surface in $NaClO_3$, so it is expected that the reflectivity of the machined surface would not be dependent on the electrolyte flow rate.

However, another way in which a protective film may be formed on the anode is through a salt film. An example of this behavior is the anodic polarization of steel in sodium perchlorate solutions. The steady state, potentiostatic polarization curve for mild steel anodes in $4\,M$ $NaClO_4$ solution is given in Figure 30a; The film stripping curve is given in Figure 30b. These curves are very similar to those of steel anodes in $NaClO_3$ solution except that the active–passive transition is shifted to more noble potentials with increased stirring,[304,305] indicating that the passive film is formed from a salt layer, and no anodic film is detected in the transpassive region.

Because the $ClO_4^-$ ion is such a symmetrical and nonpolarizable ion,[297,306] the oxidizing power of $ClO_4^-$ ion is low. In Figure 30, the active–

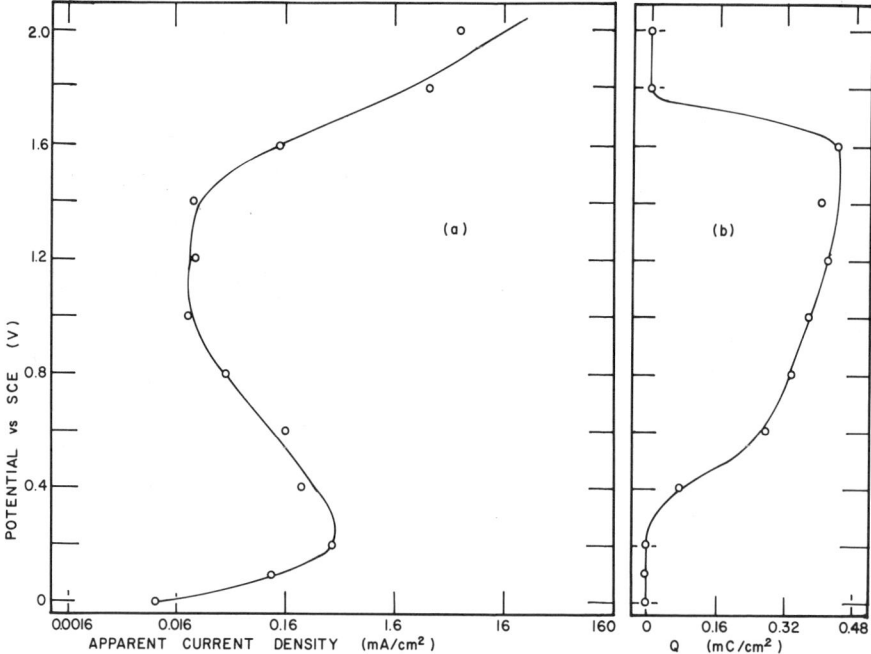

Figure 30. (a) Steady-state, anodic polarization curve obtained[79] on a mild steel microelectrode in $4\,M$ $NaClO_4$; (b) a plot of the quantity of charge associated with the anodic film formed on the metal surface at each point on the polarization curve. (With permission of the National Association of Corrosion Engineers.)

passive transition begins as a layer of iron perchlorate begins to precipitate onto the anode surface. As the potential is increased, the salt layer is oxidized to a protective oxide film by a process similar to

$$Fe(ClO_4)_2 \cdot 2H_2O \rightarrow FeOOH + 2ClO_4^- + 3H^+ + e$$

In this way is the FeOOH matrix generated. By stirring the electrolyte at a higher rate, the metal dissolution products are swept away from the vicinity of the anode more rapidly. Thus a higher potential must be applied to reach the saturation limit of the solution layers next to the anode surface. Dependency of the active–passive transition on stirring or electrode rotation rate is thus accounted for.

When the threshold potential for the ion-exchange reaction is reached, the adsorbed $ClO_4^-$ ions exchange with the oxide ions of the protective film matrix to form soluble salts which dissolve away, leaving the metal surface exposed to attack once more. The presence of an oxide film on the steel anode surface in $NaClO_4$ solutions cannot be detected in the transpassive region (Figure 30b) because of the low oxidizing power of the $ClO_4^-$ ion. Instead, only salt films are observed in the transpassive region.

Under certain hydrodynamic and electrochemical conditions, a salt film at a dissolving anode may be uniform, porous, and compact enough to produce an electropolishing of the metal surface.[307] If these conditions prevail in the transpassive region of anodes in $NaClO_4$ solution, it is predicted that the surface of steel parts machined in $NaClO_4$ electrolytes will be bright, but the quality of the brightness will depend on the solution flow rate. Also, good dimensional control should be obtained under conditions favorable to the formation of anodic films (high voltage, low flow rate) because of the relatively sharp passive–transpassive transition in the polarization curve of Figure 30a.

To put this phenomenological model of ECM to the test, fully hardened steel (5160H) bushings were machined for 30 sec at applied potentials ranging from 15 to 30 V in the through-hole ECM test rig (Figure 24) with an Elkonite (Cu-W) rod cathode extending halfway through the bushing.[79] The electrolyte was pumped from the bottom of the cell fixture through the gap space and out of the top of the cell to the reservoir at a high flow rate (17 liter/min) and a low rate (8 liter/min). Afterwards, the bushings were cut in half longitudinally to assess the quality of the machined surface. The results are presented for $4M$ $NaClO_3$ solution in Figure 31a at 17 liter/min and in Figure 31b at 8 liter/min, and for $4M$ $NaClO_4$ solution in Figure 32a at 17 liter/min and in Figure 32b at 8 liter/min. The bottom part of the open halves of the machined bushings in Figures 31 and 32 is called the high current density (hcd) region because this surface of the bushing is directly opposite the cathode. The top part where metal removal is unwanted is called the low current density (lcd) region.

In Figure 31, the high quality of the bright, hcd surface is independent of both the applied potential and the solution flow rate in $NaClO_3$ solutions. The

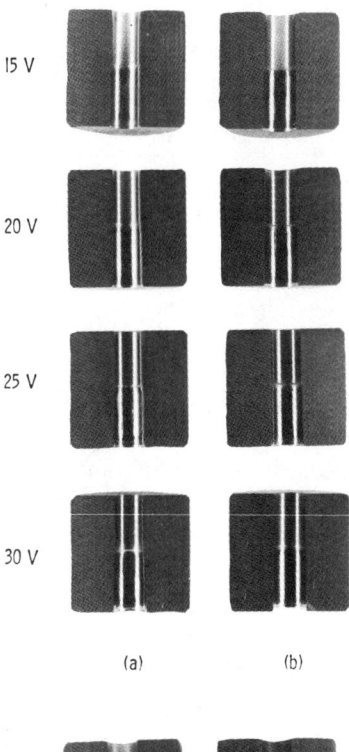

Figure 31. The open halves of fully hardened steel bushings machined[79] in 4 M NaClO$_3$ at solution flow rates of (a) 17 liter/min and (b) 8 liter/min. (With permission of the National Association of Corrosion Engineers.)

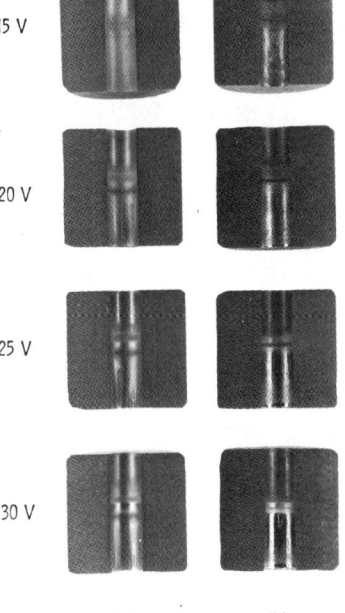

Figure 32. The open halves of fully hardened steel bushings machined[79] in 4 M NaClO$_4$ at solution flow rates of (a) 17 liter/min and (b) 8 liter/min. (With permission of the National Association of Corrosion Engineers.)

sharp demarcation between the hcd and lcd regions points up the excellent control of dimensions obtained with the ECM of steel in $NaClO_3$ solutions.

On the other hand, the quality of the machined surface in $NaClO_4$ solution in Figure 32 is strongly dependent on the applied voltage and the solution flow rate. At low voltages and high flow rates where the formation of anodic films is unfavorable, only a matte surface results with poor dimensional control. As the applied potential is raised and the flow rate decreased, the surface becomes more bright and the dimensional control improves. Where the most favorable conditions for anodic film formation exist (applied voltage, 30 V; flow rate, 8 liter/min), the machined surface is very bright, and the control of dimensions is good.

It is believed[79,308] that these results conclusively support the validity of the phenomenological model of the ECM process described in this section.

## 8.7. Other ECM Electrolytes

The ECM properties of other electrolytes may now be discussed in terms of this proposed model of the ECM process.

The polarization curve for steel anodes in $Na_2Cr_2O_7$ solution (250 g/liter) is plotted in Figure 29a (inverted triangles) along with the film stripping curve in Figure 29b. In this case, the oxidizing power of the chromate ion was so high that the steel anode was passive at all potentials investigated. Although a transpassive region is evident, the current efficiency for metal removal in this region was found[125,168] to be virtually zero. All of the current is consumed in the evolution of $O_2$ on the surface of a conducting oxide layer, so the ECM of steel cannot be carried out in chromate solutions.

In solutions of $Na_2SO_4$,[39,211,270,309] the polarization curves for steel anodes exhibit active–passive transitions which are shifted to more noble values of potential with increased agitation of the electrolyte.[211] This behavior indicates the presence of a salt film, but current efficiency studies[39] show that virtually all of the transpassive current is consumed in $O_2$ evolution (at 300 A/cm$^2$, 5.5% current efficiency for metal removal). From these data, it is concluded[39] that the protective film begins as a salt film precipitated onto the anode surface. As the potential in the passive region is increased, the salt layer is oxidized to a protective film of an iron oxide of good electrical conductivity. Apparently, the $SO_4^{2-}$ ion exchanges poorly with the oxide ions of the film since the passive layer is still present in the transpassive region[309] with a thickness reduced from the value in the passive region. As a result, most of the current is consumed by the evolution of $O_2$ taking place on the surface of this conducting film. Acceptable ECM of steel cannot be carried out in $Na_2SO_4$ solutions.

The passivation and ECM behavior of steel in $Na_2CO_3$· and $Na_3PO_4$ solutions[270] is similar to that in $Na_2SO_4$. The ECM of steel at industrial rates is not possible in these electrolytes under ordinary conditions.

Solutions of $NaNO_3$ are good ECM electrolytes. The polarization curves obtained for steel anodes in $NaNO_3$ solutions[285,286,305,310] are similar to those obtained for steel anodes in $NaClO_3$ solutions. These curves are independent of the electrolyte agitation rate,[310] indicating that the passive film is formed by the direct oxidation of the metal surface by the $NO_3^-$ ions.

Current efficiency studies[212] show that iron removal is significantly lower for the machining of steel anodes in $NaNO_3$ than in $NaClO_3$ solutions. For example, at 47 $A/cm^2$ in 4.5 $M$ $NaNO_3$, the efficiency for metal removal is 33% and that for $O_2$ evolution is 55.9%. Some $O_2$ may be used in oxidizing $Fe(OH)_2$ to $Fe(OH)_3$. Red sludge is formed, which may account for a total measured efficiency less than 100%. The metal removal efficiency increases with increasing applied potential and with increasing $NaNO_3$ concentration. Film stripping studies show the presence of thick oxide films in the passive region which begin to dissolve in the transpassive region.[286] In all the anodic polarization and ECM studies,[38] the presence of a black, loosely held film or smut is observed on the anode surface. This observation suggests that the passive film on steel anodes in $NaNO_3$ solutions is an $Fe_3O_4$ or ferrous ferrite, $Fe(FeO_2)_2$, film which possesses good electronic conductivity.[297]

When steel bushings were machined in the through-hole ECM test rig with 2 $M$ $NaNO_3$ solution,[38] the machined surface was very rough compared to those machined in $NaClO_3$ or $NaClO_4$ solution (as shown in the Proficorder traces of Figure 33). Apparently, a uniform film is not formed on the steel surface in the transpassive region of $NaNO_3$ solutions as is the case with $NaClO_3$ solutions.

From these observations, it is concluded[38] that the matrix of the protective film on steel anodes in $NaNO_3$ solution is composed of $Fe(FeO_2)_2$ units bonded together in a three-dimensional structure by hydrogen bonding with water molecules. The electronic conductivity of this hydrated ferrous ferrite structure is high, and the highly oxidizing $NO_3^-$ ion exchanges with the oxide ions of the passivating film very inefficiently. A large portion of the transpassive current is consumed in $O_2$ evolution which is in agreement with experimental observation.[38] In practice, the machining of steel by ECM in $NaNO_3$ solution is slower than in $NaClO_3$ solutions because of this poor current efficiency for metal removal. It is possible that the nonuniform dissolution of the passive film in the transpassive region happens because film dissolution can occur only at weak points in the film. These weak points may occur at dislocation sites, at sites where foreign material is occluded, or at other site imperfections. Such behavior can give rise to the surface roughness observed in Figure 33a.

Other research workers[189,276,279,281,282,288,311-314] have recorded polarization curves on steel and ferrous metal anodes in a number of electrolytes. Their results agree in general with the data reported in Figures 28-33.

There has been some concern expressed[168,292] as to whether the results obtained from conventional polarization studies (e.g., Figures 29 and 30) at low current densities, low rates of solution agitation, and large anode-cathode

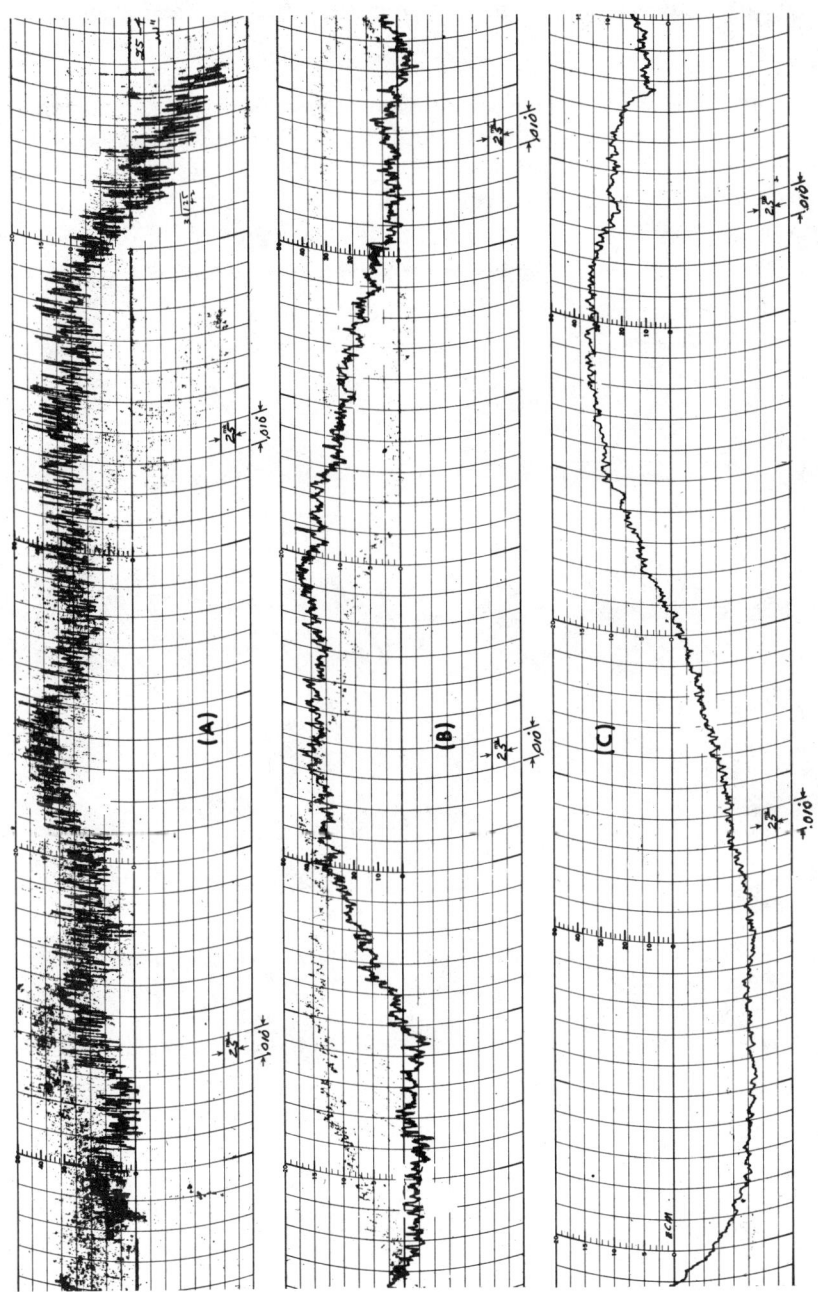

Figure 33. Proficorder traces obtained[38] on the surfaces of fully hardened steel bushings machined in: (A) 4.5 M NaNO₃, (B) 3 M NaClO₄, (C) 4.5 M NaClO₃. x axis, 0.025 cm (0.01 in.)/div.; y axis, 0.625 μm (25 μin.)/div.; RMS roughness estimated by 1/3 peak-to-peak value. (With permission of The Electrochemical Society, Inc.)

separations may be used to form conclusions about the ECM region. In this region current densities and solution flow rates are very high, and the anode–cathode gaps are very narrow. To investigate this problem, the formidable difficulties of designing and debugging the circuits for measuring polarization in a high-rate flow cell at linear flow velocities up to 2000 cm/sec and current densities up to 100 A/cm$^2$ were overcome.[315] High-rate, flow-cell studies were made on steel anodes in NaCl,[315] NaClO$_3$,[303] and NaNO$_3$[316] solutions, and the polarization curves obtained in the ECM region, Figure 34, appear to be extensions of those obtained in conventional cells, Figure 29.

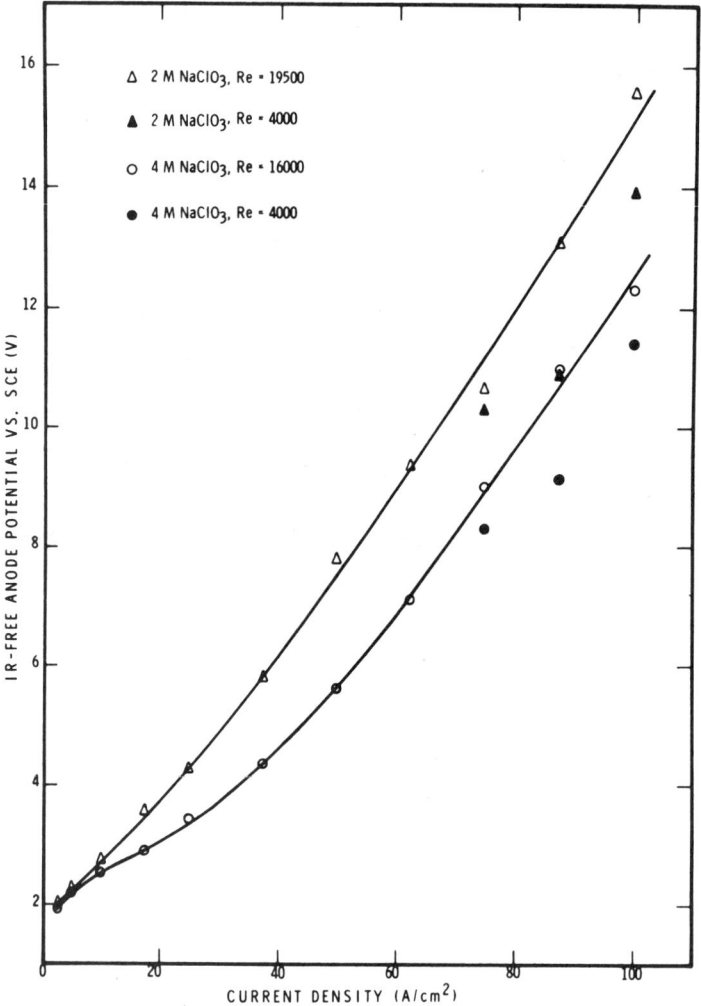

Figure 34. IR-free polarization curves obtained[303] on mild steel anodes in high-rate flow cell as a function of NaClO$_3$ concentration and solution flow rate. (With permission of The Electrochemical Society, Inc.)

These high flow rate investigations confirm the conclusions made based on results from conventional cells. In NaCl solutions, only salt films are detected, and the quality of the surface finish depends on the current density and the solution flow rate.[315] Oxide films are detected on steel anodes in $NaClO_3$ solutions, but these films break down at decreasing anodic potentials with increasing $NaClO_3$ concentration.[303] As the film breaks down, the current efficiency for metal removal increases. In 4 $M$ $NaClO_3$ solution, the passive oxide film disappears at 15 $A/cm^2$ (2.5 V vs. SCE) with a current efficiency for metal removal reaching 100%. Above this point, the passive film is replaced by the thin, uniform, porous oxide film (probably about 1 $\mu$m or less[87] thick) which cannot be detected by the scanning electron microscope (SEM) because of the highly polished nature of the metal surface.

High flow rate cell studies of steel anodes in $NaNO_3$ solutions[316] confirm the presence of the highly conducting film of hydrated $Fe_3O_4$. The uneven dissolution of the oxide film in the transpassive region was also observed, and the severe surface roughness was evident from an SEM micrograph.

Similar studies have been carried out by Franke and Forker.[317]

## 8.8. Mixed Electrolytes

The number of electrolytes which are suitable for the ECM of steel and which are made from a single salt is limited to the salts of a few anions ($Cl^-$, $ClO_3^-$, $ClO_4^-$, and $NO_3^-$). Other than cost considerations and solubility, the choice of cation does not significantly affect the suitability of the ECM electrolyte. When mixtures of salts are considered the number of suitable electrolytes is greatly increased.

It is possible to categorize the anions into two general classifications: (1) those which are good film-forming anions, such as $ClO_3^-$, $NO_3^-$, $SO_4^{2-}$, $CrO_4^{2-}$, etc., and (2) those which are good film-dissolving anions, such as $Cl^-$ and $ClO_4^-$ ions. To obtain a mixed electrolyte in which the ECM of steel proceeds with good control of dimensions and geometry, the solution should contain one of each electrolyte. By varying the relative concentrations of these anions, the passivation properties of the electrolyte may be modified, and a family of polarization curves may be obtained. In general, such curves may be grouped into three idealized patterns[39] sketched in Figure 35: (a) the current in the passive region increases with an increase in the concentration of one of the solution components, (b) the active–passive transition is shifted to higher anodic potentials, and (c) the passive–transpassive transition is shifted to lower anodic potentials.

The family of polarization curves obtained[39] on mild steel anodes in 1 $M$ $Na_2SO_4$ solution to which increasing amounts of NaCl were added is an example of the pattern shown in Figure 36a. As the concentration of $Cl^-$ is increased, the curves demonstrate the progression from a highly passivated system to an active system showing no passivation. Since the passive layer in

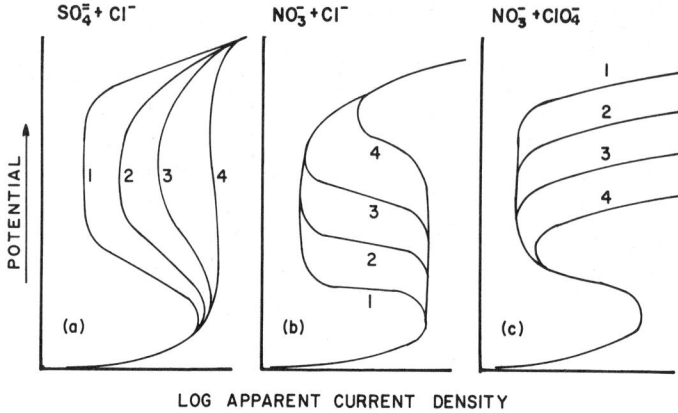

Figure 35. Sketches of families of idealized polarization curves obtained[39] on mild steel anodes in mixed solutions of a film-forming and a film-dissolving anion. Curves are numbered in direction of solutions with increasing concentration of the film-dissolving anion. (With permission of The Electrochemical Society, Inc.)

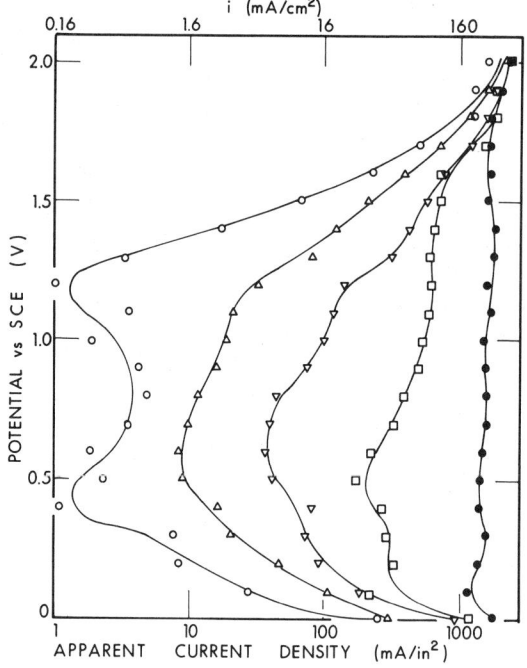

Figure 36. Steady-state polarization curves obtained[39] on mild steel anodes in 1 $M$ $Na_2SO_4$ (○), 1 $M$ $Na_2SO_4$ + 3 × $10^{-3}$ $M$ NaCl (△), 1 $M$ $Na_2SO_4$ + 6 × $10^{-3}$ $M$ NaCl (▽), 1 $M$ $Na_2SO_4$ + $10^{-2}$ $M$ NaCl (□), and 1 $M$ $Na_2SO_4$ + 1 $M$ NaCl (●). (With permission of The Electrochemical Society, Inc.)

*Table 3*
*Current Efficiencies for Metal Removal of Mild Steel Anodes in $Na_2SO_4$–NaCl Solutions*

| Solution composition | Current density, $A/cm^2$ | Current efficiency % Metal removal | Current efficiency % $O_2$ evolution |
|---|---|---|---|
| 1 $M$ $Na_2SO_4$ | 47 | 5.5 | 90.3 |
| 1 $M$ $Na_2SO_4$ + 5 × $10^{-3}$ $M$ NaCl | 37.5 | 40.0 | 54.0 |
| 1 $M$ $Na_2SO_4$ + 1 $M$ NaCl | 37.5 | 95.0 | 0 |
| 4.5 $M$ NaCl | 37.5 | 96.0 | 0 |

1 $M$ $Na_2SO_4$ is formed from a salt film, the presence of $Cl^-$ ions not only increases the solubility of the salt film, thus delaying the point at which a protective oxide film can be formed, but also it increases the ionic conductivity of the passive film as suggested by Hoar,[298] thus lowering the passivating power of the protective film. Only small amounts of $Cl^-$ ion additions are required to affect the polarization curves. The addition of only $3 \times 10^{-3}$ $M$ NaCl to 1 $M$ $Na_2SO_4$ increases the passivation current by nearly an order of magnitude, and the addition of $10^{-2}$ $M$ NaCl virtually destroys the protective properties of the anodic film. Varenko and co-workers[318] have confirmed these observations, and Condit[319] has reported similar results for Fe-Ni alloy anodes in $H_2SO_4$ solutions containing $Cl^-$ ions.

Table 3 contains the current efficiency measurements made[39] on mild steel tube ends in $NaSO_4$–NaCl mixtures. The metal removal efficiency was greatly increased for small NaCl additions, but it was found that the metal surface was badly pitted. At concentrations of NaCl above $10^{-2}$ $M$, the current efficiency for metal removal approaches 100% and the surface is smooth, but large stray cutting is present. Such solutions are not good ECM electrolytes for sizing and finishing steel parts.

When the protective oxide film is formed by the direct oxidation of the electrode surface, the presence of $Cl^-$ affects film formation because the $Cl^-$ is strongly adsorbed on the metal surface and can form soluble complex ions with iron. As a result, the presence of $Cl^-$ ions weakens the oxidizing power of the $NO_3^-$ ion, and causing an increase in the critical potential increases for the active-passive transition with increasing $Cl^-$ ion content of the solution, as shown by the polarization curves[320] plotted in Figure 37. These curves correspond to the pattern in Figure 35b. Similar results have been obtained by others.[321,322]

The results of current efficiency studies[320] are recorded in Table 4. As the $Cl^-$ ion content of the solution increases, the efficiency for metal removal rises while that for $O_2$ evolution falls. However, the surface finish is very poor

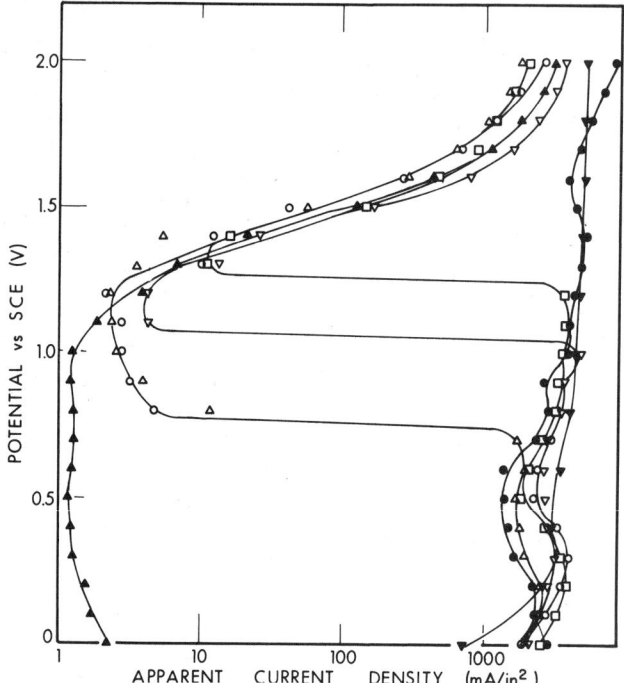

Figure 37. Steady-state polarization curves obtained[320] on mild steel anodes in 3 $M$ NaNO$_3$ (▲), 1 $M$ NaCl + 2 $M$ NaNO$_3$ (○), 2 $M$ NaCl + 1 $M$ NaNO$_3$ (△), 2.25 $M$ NaCl + 0.75 $M$ NaNO$_3$ (▽), 2.4 $M$ NaCl + 0.6 $M$ NaNO$_3$ (□), 2.8 $M$ NaCl + 0.4 $M$ NaNO$_3$ (●), 3 $M$ NaCl (▼). (With permission of Pergamon Press.)

because of uneven metal removal caused by the nonuniform removal of the anodic film in the transpassive region. With respect to ECM, Cl$^-$ ion is a poor choice for a film-removing agent in the transpassive region because of its well-known pitting and uneven metal-removing properties in this range of potentials (see e.g., References 323 and 324).

Table 4

Current Efficiencies for Metal Removal of Mild Steel Anodes in NaNO$_3$–NaCl Solutions

| Solution composition | Current density, A/cm$^2$ | Current efficiency | |
|---|---|---|---|
| | | % Metal removal | % O$_2$ evolution |
| 2 $M$ Na$_2$NO$_3$ | 47 | 11.5 | 84.5 |
| 1 $M$ NaCl + 2 $M$ NaNO$_3$ | 37.5 | 37.0 | Balance |
| 2.8 $M$ NaCl + 0.4 $M$ NaNO$_3$ | 37.5 | 55.0 | Balance |
| 2 $M$ NaCl | 47 | 98.0 | 0 |

Attempts were made to lower the stray cutting of NaCl electrolytes by adding solutions of chromates to them, but two types of behavior have been found that in these mixed electrolytes either (1) no ECM (high chromate content) occurred or (2) strong stray cutting (high chloride content) was observed. There seemed to be no in between behavior.

It was concluded[39] that $NaClO_3$–NaCl mixtures would follow the pattern of $NaNO_3$–NaCl mixtures according to Figure 35b. These solutions are not interesting as ECM electrolytes because it has been our experience that the presence of significant quantities of $Cl^-$ ion has been found to impair the excellent ECM properties of $NaClO_3$ solutions. In a similar manner, additions of NaCl to $NaClO_4$ solutions impair the ECM properties of the $NaClO_4$ electrolyte.

Of course, the other film-removing ion is the $ClO_4^-$. Because it is a symmetrical, unpolarizable ion with low oxidizing power, its presence in a solution has little effect on the film formation, but because it can exchange efficiently with the oxide ions of a passive, iron oxide film, pronounced effects in the transpassive region are expected. The polarization curves on mild steel in $NaNO_3$–$NaClO_4$ electrolytes have been determined[38] and are presented in Figure 38. With increasing $ClO_4^-$ ion concentration, the threshold potential at which the ion-exchange mechanism of anodic film dissolution in the transpassive region takes place is shifted to less noble values. Consequently, the passive–transpassive transition is lowered to less anodic potentials as seen in Figure 38, and the pattern of curves corresponds to Figure 35c.

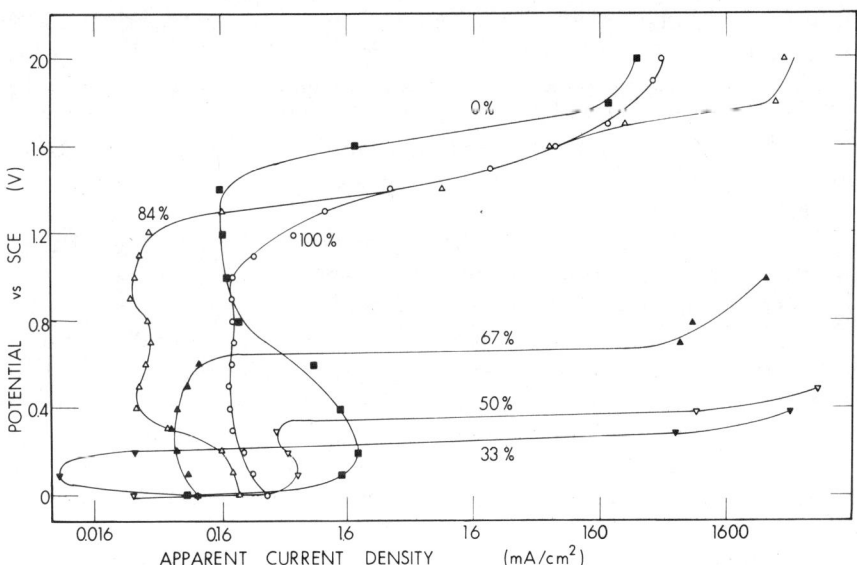

Figure 38. Steady-state polarization curves obtained[38] on mild steel anodes in 3 $M$ solutions of mixtures of $NaNO_3$ and $NaClO_4$ with the percent of $NaNO_3$ indicated on the curves. (With permission of The Electrochemical Society, Inc.)

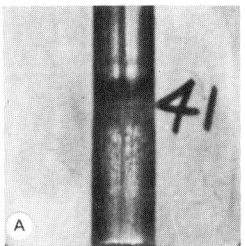 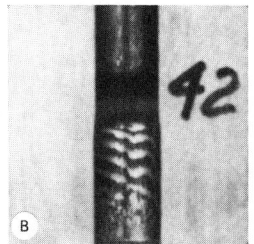 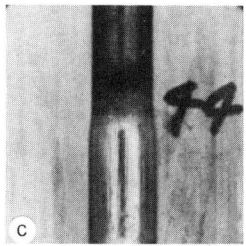

Figure 39a. Open halves of fully hardened steel bushings machined[38] in a solution of 1 $M$ NaClO$_4$ + 2 $M$ NaNO$_3$ at an applied voltage of 10 V (A), 15 V (B), and 25 V (C) at a flow rate of 5.6 liter/min to show breakdown of the protective film with increasing potential. (With permission of the Electrical Society, Inc.)

The black smut characteristic of the presence of a hydrated ferrous ferrite film is found on the machined surface of steel in these mixed electrolytes. Surface finish studies of steel bushings, machined in the through-hole ECM machine in a solution which is 1 $M$ in NaClO$_4$ and 2 $M$ in NaNO$_3$ (Figure 39a) show the breakdown of the passive Fe$_3$O$_4$ film as the applied potential is increased. In Figure 39b, the Proficorder traces correspond to very rough surfaces, and the surface of the machined bushing in Figure 39a(A) is dark and matte at 10 V. The pattern of light and dark bands on the steel surface in Figure 39a(B) at 15 V shows the breaking up of the film. At 25 V applied voltage in Figure 39a(C), the surface is bright, and the Proficorder trace attests to the improved surface smoothness. Filmstripping techniques show that in the transpassive region the oxide film is completely removed and the brightness of the machined surface depends on the solution flow rate, indicating that the polishing of the surface may be traced to a compact salt film on the metal surface in the transpassive region. The film-dissolving properties of the ClO$_4^-$ ion are so great for the removal of iron oxide films that in the presence of NO$_3^-$ ions an oxide film cannot be formed on the steel surface in the transpassive region.

The lowering of the transpassive region in Figure 38 would suggest that for a given applied voltage the metal removal rate would be greater in a mixture of NaNO$_3$ and NaClO$_4$ than in the pure solutions of either NaNO$_3$ or NaClO$_4$. In Table 5 are recorded the metal removal rates as well as the current efficiencies for the solutions of interest. These data support the prediction. Current efficiencies above 100% may result from minute grains of metal dropping out of the metal surface due to grain boundary attack. An advantage in using the NaNO$_3$–NaClO$_4$ electrolyte is the fact that hydrogen is not liberated at the cathode. Closed cell studies[38,212] show that the cathodic current is consumed in nitrate-based electrolytes in the formation of ammonia, nitrate, and hydroxylamine.

Similar results were obtained[305] from studies of a rotating hemispherical electrode in NaNO$_3$–NaClO$_4$ mixtures, but the lowering of the transpassive

# ELECTROCHEMICAL MACHINING

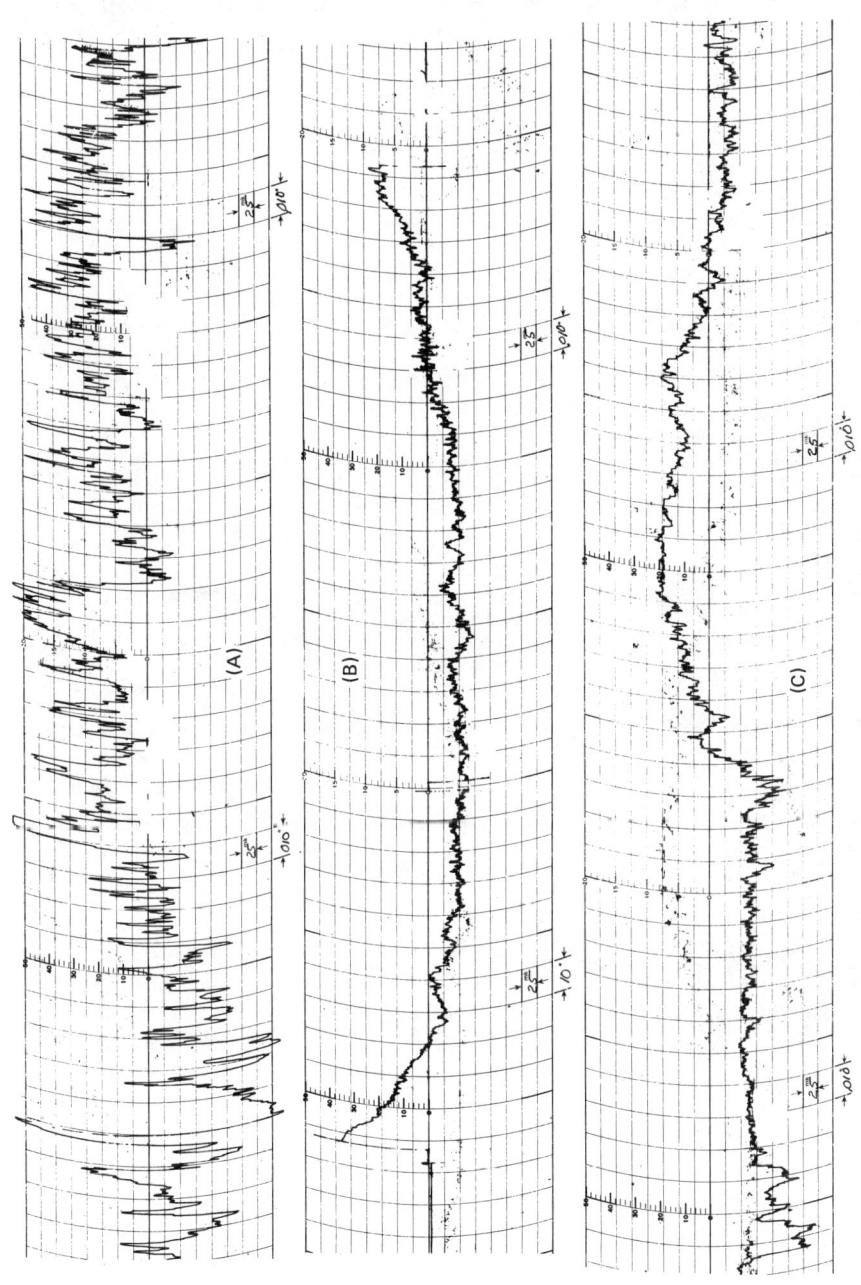

Figure 39b. Proficorder traces obtained[38] on the surfaces of the machined bushings of Fig. 39a. (With permission of The Electrochemical Society, Inc.)

*Table 5*

*Current Efficiencies for Metal Removal of Mild Steel Anodes in $NaClO_4$ Mixed Electrolytes[a]*

| Solution composition | Current density, $A/cm^2$ | Current efficiency % Metal removal | Current efficiency % $O_2$ evolution |
|---|---|---|---|
| 3 M $NaClO_4$ | 47 | 6.0 | 90.0 |
| 2.5 M $NaClO_4$ + 0.5 M $NaNO_3$ | 37.5 | 97.0 | Balance |
| 2 M $NaClO_4$ + 1 M $NaNO_3$ | 37.5 | 100 | 0 |
| 1.5 M $NaClO_4$ + 1.5 M $NaNO_3$ | 47 | 112 | 0 |
| 1 M $NaClO_4$ + 2.0 M $NaNO_3$ | 47 | 112 | 0 |
| 0.5 M $NaClO_4$ + 2.5 M $NaNO_3$ | 47 | 106 | 0 |
| 0.3 M $NaClO_4$ + 2.5 M $NaNO_3$ | 47 | 82 | Balance |
| 2 M $NaNO_3$ | 47 | 11.5 | 84.5 |
| 1.2 M $NaClO_4$ + 0.6 M $Na_2SO_4$ | 47 | 103 | 0 |
| 1.0 M $NaClO_4$ + 1 M $Na_2SO_4$ | 47 | 99.4 | 0 |
| 1 M $Na_2SO_4$ | 47 | 5.5 | 90.3 |
| 1.5 M $NaClO_4$ + 1.5 M $NaClO_3$ | 37.5 | 108 | 0 |
| 0.5 M $NaClO_4$ + 3 M $NaClO_3$ | 37.5 | 114 | 0 |
| 4.5 M $NaClO_3$ | 47 | 79.0 | 25.0 |

[a] Efficiencies greater than 100% may be explained by grain boundary attack.

region is much less than that in Figure 38. The ECM behavior of mixtures of $Na_2SO_4$ and $NaClO_4$[309] and of $NaClO_3$ and $NaClO_4$[309] is very similar to that for $NaNO_3$ and $NaClO_4$ (see Table 5). All of these mixtures with $NaClO_4$ make suitable electrolytes for the ECM of iron and steel although none of these electrolytes surpasses the overall superior ECM properties of $NaClO_3$ solutions. The uniqueness of the $NaClO_3$ electrolyte for the ECM of steels arises due to the formation of the thin, uniform, porous oxide film formed in the transpassive region, for this particular electrolyte alone.

Chikamori and Ito[321] have added a number of salts, including $NaClO_3$, $NH_4SO_4$, $NaBrO_3$, $Na_2SO_4$, and $KBrO_3$, to solutions of $NaNO_3$ attempting to increase the current efficiency for metal removal. They concluded that the addition of O-containing anions causes $O_2$ to be evolved, lowering the pH upon the generation of $H^+$ ions. This lower pH aids in dissolving passive oxide films and increases the efficiency for metal removal. Such a lowering of the pH has been observed[39] in $NaClO_3$ solutions but not in $NaNO_3$, $NaClO_4$, or mixtures of $NaClO_4$ and $NaNO_3$. In $NaClO_3$ solutions, the final steady pH during the ECM of steel was 6.0. This is not the most important basis for determining the metal removal efficiency.

In the early literature,[127,189,325-327] the ECM process was considered to be the reverse of electrodeposition or "deplating." However, only in rare cases (possibly for the $Ag \rightarrow Ag^+ + e$ process) is the cathodic reduction of a metal ion

# ELECTROCHEMICAL MACHINING

the reverse of the oxidation of the metal atom. From the plating industry the term *throwing power*, which is a measure of the ability to deposit a layer of metal over a wide range of current densities, is borrowed. Of course, the reverse is used[280,284,327–329] for ECM in terms of metal removal.

Brook and Iqbal[280] used a modified Haring–Blum cell[330] consisting of two anodes spaced at unequal distances from a common cathode in order to assess the suitability of a given solution as an ECM electrolyte. The throwing power is expressed as a function of the ratio of the distances of the two anodes to the common cathode, but since this ratio may be obtained from more than one empirical formula,[331] a consistent value may not be obtained. The use of a logarithmic function of the ratio of lengths[329] is a possible method for classifying ECM electrolytes. For good dimensional control, the solution of interest is expected to have a low value of throwing power; Landolt[284] has found a negative value for Ni in $NaClO_3$ solutions.

Although the throwing power reflects the passivating behavior of a given metal in a given electrolyte, ECM electrolytes may be better evaluated from a knowledge of the electrochemical properties of the anodic films formed on the given metal in the given electrolyte. By analyzing the results of potentiostatic polarization, film stripping, and current efficiency data for the given system, this knowledge is found. McGeough[261] has criticized the use of quantitative measurements of the throwing power,[284] noting that such calculations depend on electrode configuration and process variables but give little information about the processes that control the ECM properties of passivating and nonpassivating electrolytes.

## 8.9. High-Strength, High-Temperature Alloys

The quality of the electrochemically machined product is directly related to the properties of the anodic film formed on the metal surface in the transpassive region in the given electrolyte. The great advantage in using ECM over conventional machining methods is realized in the machining of the very hard, high-temperature, high-strength alloys. The material presented so far has been concerned only with iron or low-carbon steel anodes. Therefore, the presence of relatively large amounts of metals other than iron in the high-strength-alloy anode and their effect on the properties of the anodic film formed on the electrode surface are of interest.

Samples of seven high-strength alloys and two carbon steels (the composition is given in Table 6) were drilled[273] in the laboratory ECM plunge-cut machine in $NaClO_3$ solution (350 g/liter) at a flow rate of 1.85 liter/min. The recorded feed rate was the highest value which could be maintained without tripping the cutoff switch. After machining, the samples were cut in half longitudinally so that the machined surface of the hole could be analyzed with the SEM and the Proficorder. Steady state, potentiostatic polarization, and cathodic film stripping curves were determined on the pegs or

Table 6
Composition of Some High-Strength Alloys in Weight Per Cent

| Alloy | C | Mn | Fe | Si | Cr | Mo | V | Ti | Al | Ni | Nb | B | Co | W |
|---|---|---|---|---|---|---|---|---|---|---|---|---|---|---|
| 1020 | 0.23 | 0.3 | 99.4 | — | — | — | — | — | — | — | — | — | — | — |
| 5160H | 0.65 | 1.0 | 97.1 | 0.35 | 0.9 | — | — | — | — | — | — | — | — | — |
| H13 | 0.4 | 0.4 | 89.5 | 1.2 | 5.5 | 1.75 | 1.2 | — | — | — | — | — | — | — |
| Inco 901 | 0.05 | 0.5 | 71.5 | 0.35 | 13.0 | 6.0 | — | 2.5 | 2.0 | 4.0 | — | — | — | — |
| Inco 713C | 0.12 | 0.15 | 1.0 | 0.4 | 13.0 | 4.5 | — | 0.6 | 6.0 | 71.9 | 2.25 | — | — | — |
| GMR 235 | 0.15 | 0.25 | 10.0 | 0.6 | 15.5 | 5.25 | — | 2.0 | 3.0 | 62.5 | — | 0.06 | — | — |
| A286 | 0.05 | 1.35 | 51.4 | 0.5 | 15.0 | 1.25 | 0.3 | 2.0 | 2.0 | 26.0 | — | — | — | — |
| HS31 | 0.5 | 0.5 | 1.5 | 0.5 | 25.0 | — | — | — | — | 10.0 | — | — | 54.0 | 8.0 |
| 410 | 0.15 | 1.0 | 84.3 | 1.0 | 13.5 | — | — | — | — | — | — | — | — | — |

slugs trepanned out of the center of the hole. The polarization and stripping curves, the Proficorder trace, and the SEM micrographs are given in Figure 40 for H-13 tool steel, in Figure 41 for Inco 901 nickel–iron alloy, in Figure 42 for Inco 713C nickel-based alloy, in Figure 43 for GMR 235 nickel-based alloy, in Figure 44 for A286 nickel–chromium stainless steel, in Figure 45 for HS31 cobalt-based alloy, and in Figure 46 for 410 chromium stainless steel. In Figure 47, the cutting rate, the weight per cent of Cr, Ni, and Fe, and the surface roughness for the alloys included in Table 6 are plotted.

According to these data in Figure 47, the machining rate is large for those alloys of high Fe content and low Cr content, and the surface finish is very good, in agreement with previous findings (Figure 31). As the Cr content of the alloy increases, the machining rate decreases.

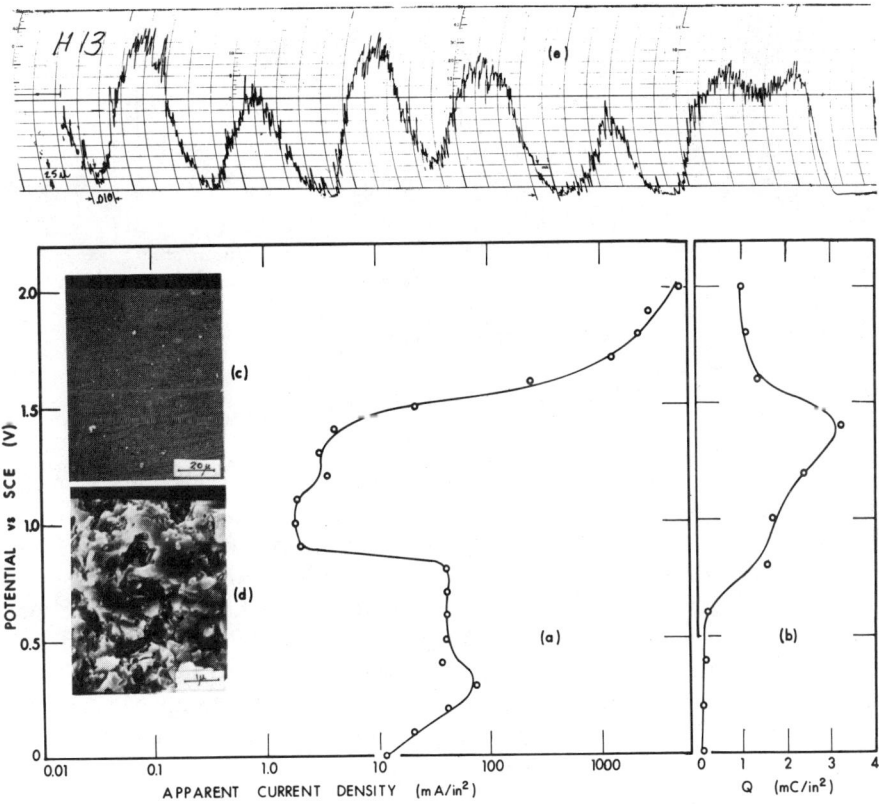

Figure 40. (a) Steady-state polarization curve obtained[273] on H13 tool steel anode in NaClO$_3$ solution (350 g/liter); (b) charge, $Q$, of surface film as a function of potential; (c), (d) SEM micrographs; (e) Proficorder trace: $x$ axis is 0.025 cm (0.01 in.)/div., $y$ axis is 0.625 $\mu$m (25 $\mu$in.)/div. (With permission of The Electrochemical Society, Inc.)

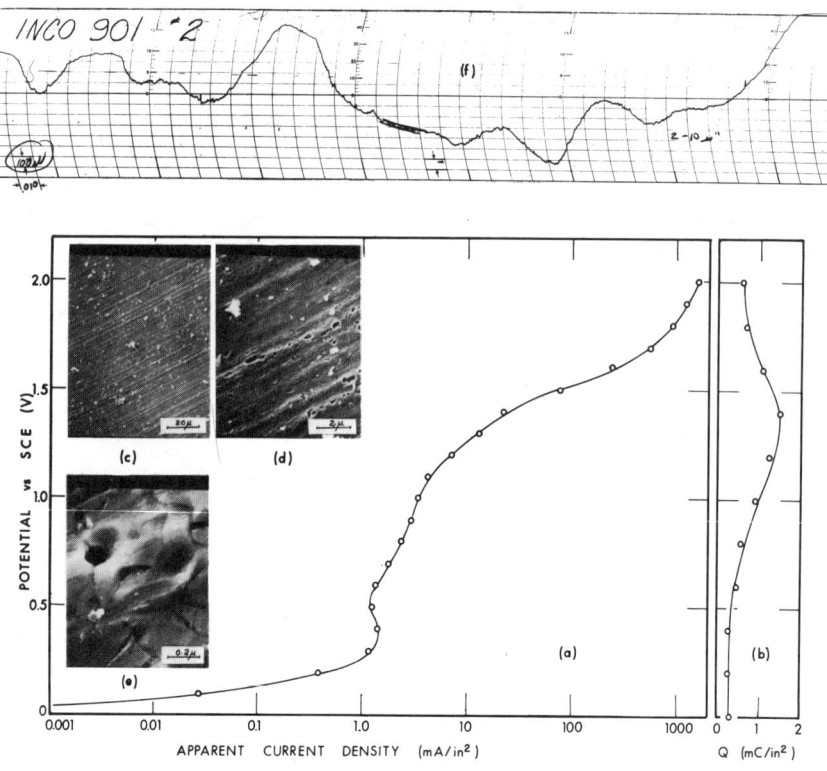

Figure 41. (a) Polarization curve obtained[273] on Inco 901 Ni–Fe alloy anode; (b) $Q$ as $f$ (potential); (c)–(e) SEM micrographs; (f) Proficorder trace: $x$ axis is 0.025 cm (0.01 in.)/div., $y$ axis is 2.5 μm (100 μin.)/div. (With permission of The Electrochemical Society, Inc.)

Hoar and Evans[332] found that the anodic film formed on iron electrodes in NaCl solution to which $CrO_4^{2-}$ ion was added became thinner and more protective as the $CrO_4^{2-}$ ion content of the electrolyte was increased. From tracer studies, Cohen and Beck[333] reported that the chromium oxide content of the protective film on iron anodes in electrolytes containing chromate ions may reach values as high as 25% $Cr_2O_3$. It is reasonable to assume, then, that as the Cr content of the alloy increases, the $Cr_2O_3$ content of the passive film increases, thus increasing the electronic conductivity of the anodic film. Shikata and Ito[334] found a uniform oxide layer on the machined surface of stainless steel anodes in $NaNO_3$ solution composed of iron and chromium oxides. Similar results were reported by Ventsel and Indin.[335] The presence of $Cr_2O_3$ in the film in the transpassive region increases the fraction of current consumed in $O_2$ evolution, thus decreasing the machining rate. In all cases, the thickness of the anodic film is reduced in the transpassive region as expected from ion exchange with adsorbed $ClO_3^-$ ions.

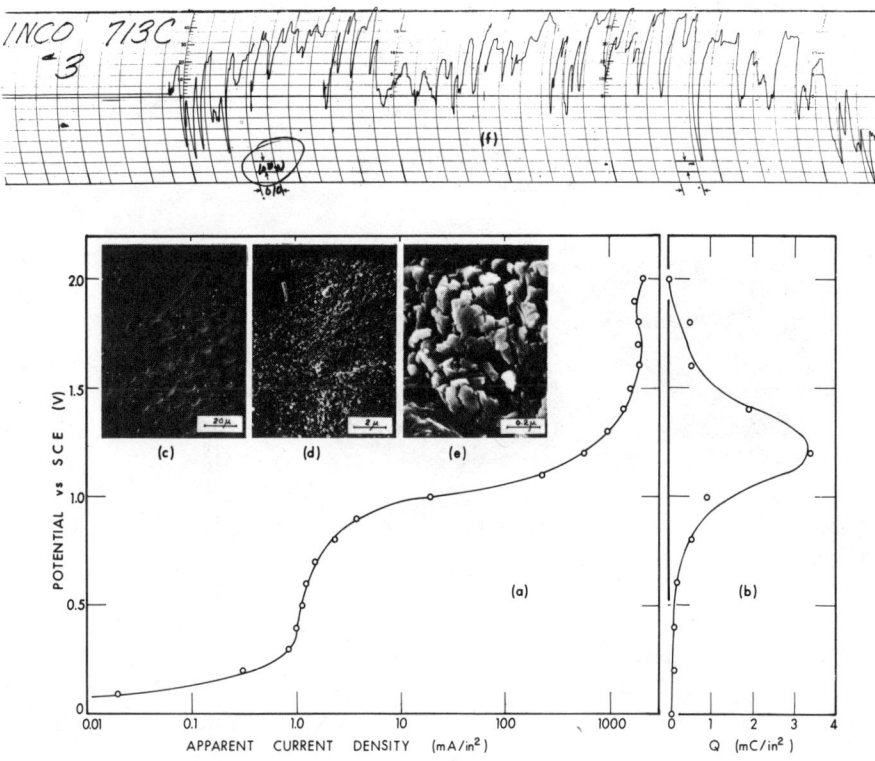

Figure 42. (a) Polarization curve obtained[273] on Inco 713C Ni-based alloy anode; (b) $Q$ as $f$ (potential); (c)–(e) SEM micrographs; (f) Proficorder trace: $x$ axis is 0.025 cm (0.01 in.)/div., $y$ axis is 2.5 μm (100 μin.)/div. (With permission of The Electrochemical Society, Inc.)

For alloys with high Ni content but low Fe content (Inco 713C and GMR 235), the machined surface is very rough. As the Fe content is increased (A286 and 410), the roughness is decreased. A uniform, thin, porous film is not formed on Ni anodes in $NaClO_3$ solutions in the transpassive region, resulting in a relatively rough surface.[336] Morozov and Maksimov[258] have reported that the surface roughness of steels machined in NaCl solution increased with increasing Ni content. Severe grain boundary attack of Nimonic 80A (a Ni-based alloy of Cr) machined[168] in NaCl contributes to surface roughness but can be reduced with lower solution flow rates. It seems that Co acts similarly to Ni from the data obtained on HS31 alloy.

These data support the conclusion that the structure and composition of the passive film on alloy anodes is strongly dependent on metal alloy composition. The properties of this film then influence the quality of the machined product.

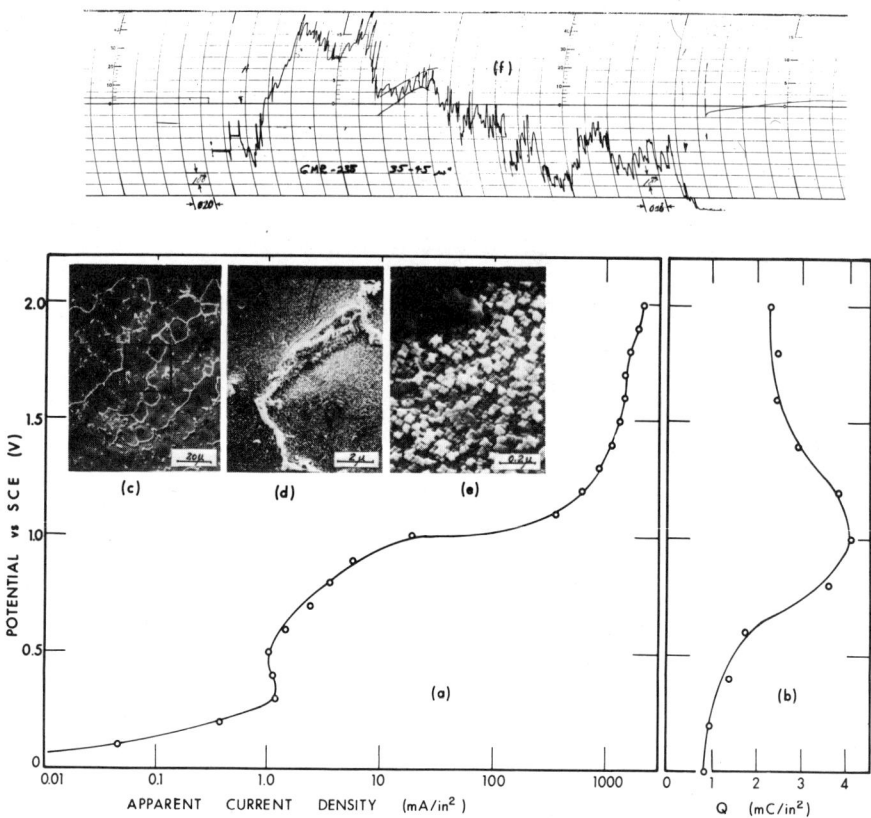

Figure 43. (a) Polarization curve obtained[273] on GMR 235 Ni-based alloy; (b) $Q$ as $f$ (potential); (c)–(e) SEM micrographs; (f) Proficorder trace: $X$ axis is 0.025 cm (0.01 in.)/div., $y$ axis is 2.5 $\mu$m (100 $\mu$in.)/div. (With permission of The Electrochemical Society, Inc.)

A study of the SEM micrographs indicates that the metallurgy of the alloy also greatly influences the quality of ECM. If one phase dissolves more rapidly than another, as in the case of Inco 713C (Figure 42c–e), the surface has a rock-pile appearance. When the material of the grain boundaries is more resistant to attack than that of the grains, the raised network of strands of materials is obtained on GMR 235 (Figure 43c–e). With HS31, Figure 45c–e, the presence of more than one phase with different film-forming and -removing properties gives a machined surface with a dried mud-flat appearance. These situations give rise to surfaces with a high degree of roughness. In contrast, those alloys high in Fe content (H13, Inco 901, A286, 410) are more homogeneous, the metal is removed evenly, and a very smooth surface is obtained.

Figures 44c,d, 46c,d demonstrate the lack of correspondence between reflectivity and surface roughness. In cases of both A286 and 410 stainless

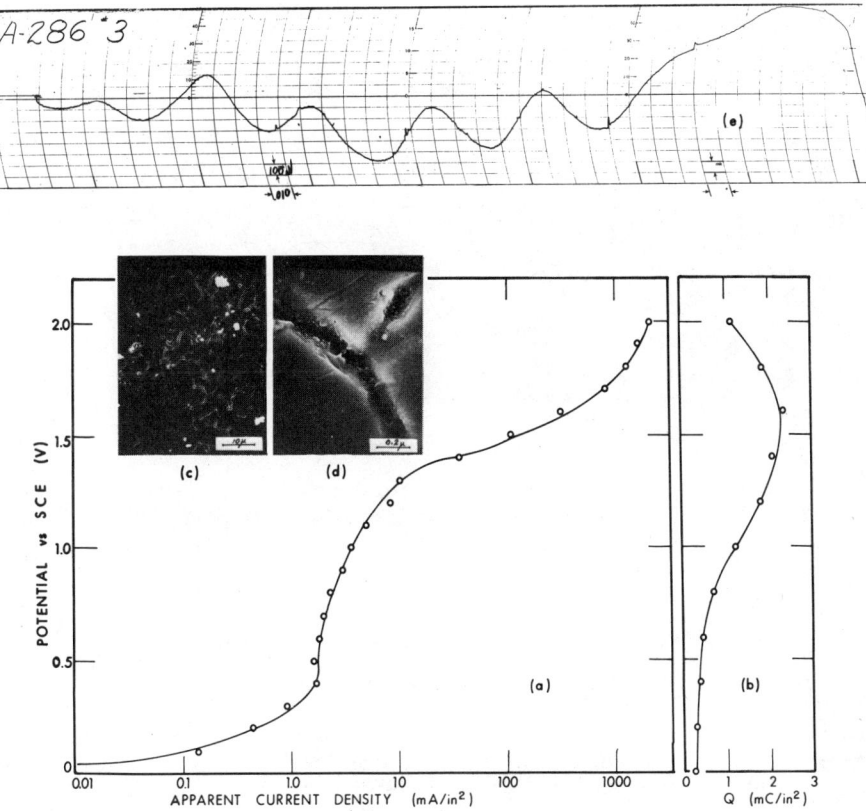

Figure 44. (a) Polarization curve obtained[273] on A286 Ni–Cr stainless steel anode; (b) $Q$ as $f$ (potential); (c), (d) SEM micrographs; (e) Proficorder trace: $x$ axis is 0.025 cm (0.01 in.)/div., $y$ axis is 2.5 μm (100 μin.)/div. (With permission of The Electrochemical Society, Inc.)

steels, the surface roughness is very low (between 0.05 and 0.125 μm) as noted in Figures 44e and 46e, yet the machined surface of A286 is mirror bright, that of 410 is dull. On a microscopic scale, the sponge-type surface of 410 (Figure 46d) scatters the incident light and appears dull, whereas, the slight grain boundary attack on A286 (Figure 44e) produces much less scattering and the surface is bright.

Consequently, in the search for the proper ECM electrolyte to machine a given alloy, one must consider both the electrochemical properties of the anodic films formed and the metallurgical properties of the alloy to be machined. It may be possible to produce a specific electrolyte with ECM properties not found in a single component solution by mixing the solutions of more than one salt in the proper proportions.

Electrochemical machining of inhomogeneous alloys, such as tungsten carbides, W–Co alloys, or cast iron, is characterized by poor machining rates

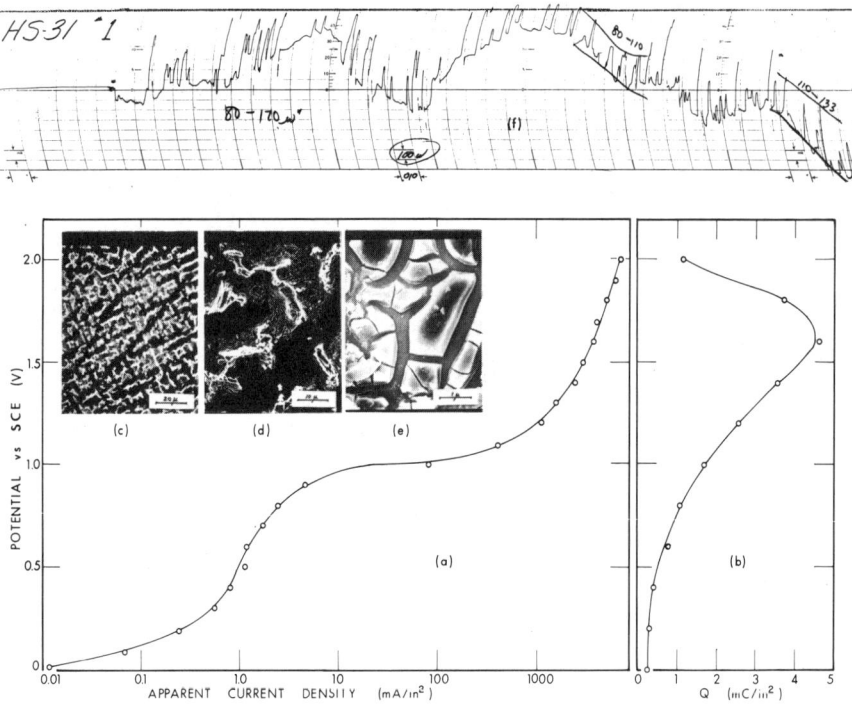

Figure 45. (a) Polarization curve obtained[273] on HS31 Co-based alloy anode; (b) $Q$ as $f$ (potential); (c)–(e) SEM micrographs; (f) Proficorder trace: $x$ axis is 0.025 cm (0.01 in.)/div., $y$ axis is 2.5 μm (100 μin.)/div. (With permission of The Electrochemical Society, Inc.)

and poor finishes.[166,337-340] When the surface was abraded by grinding with ECG, the tungsten carbides[166] and the cast iron[340] were machined efficiently. For such alloys, abrasion of the surface along with anodic corrosion is required to remove the nonconducting carbide particles.

Other studies of the ECM of various alloys[322,341-344] have met with varying degrees of success for NaCl and NaNO$_3$ solutions. Electropolishing of stainless steel anodes has been observed by the author and others and of Nimonic 80A by Evans and Boden (Reference 168, p. 40) in NaCl electrolytes. Possibly, the high Cr content of these alloys generates a thin, uniform, porous oxide film on the anode surface in the transpassive region (similar to that formed on steel anodes in NaClO$_3$ solutions) accounting for the polished metal surface obtained.

## 8.10. The ECM of Other Metals

In the literature,[246-248,284,294,345] there is evidence that the ECM behavior of Ni anodes in various electrolytes may be very different from that of iron or steel. Boden and Evans,[248] using a cell with a segmented anode, reported that good dimensional control could be obtained on Ni samples machined in solutions of NaCl to which the salts of other anions [$Na_2CO_3$, $Na_2SO_4$, $Na_3Fe(CN)_6$, and $Na_3PO_4$] had been added. Those anions which gave insoluble Ni salts reduced the stray cutting to the greatest extent. The best electrolyte for the ECM of Ni had a composition of 15% NaCl plus 2.5% $Na_2CO_3$. Electropolishing of Ni anodes can be obtained in acidified $NaNO_3$ electrolytes,[345] but the degree of brightening depends on the solution flow rate. Apparently, polishing is produced by a compact salt film formed on the Ni-anode surface in the transpassive range of potentials.

Steady-state potentiostatic polarization and cathodic film stripping curves are obtained for Ni anodes in $NaClO_3$ (350 g/liter) solution; the quality of the Ni surface machined using the laboratory plunge-cut machine was determined from Proficorder traces and SEM micrographs.[336] In $NaClO_3$ solutions, the anodic film increases in thickness on a Ni anode as shown in Figure 48 in the

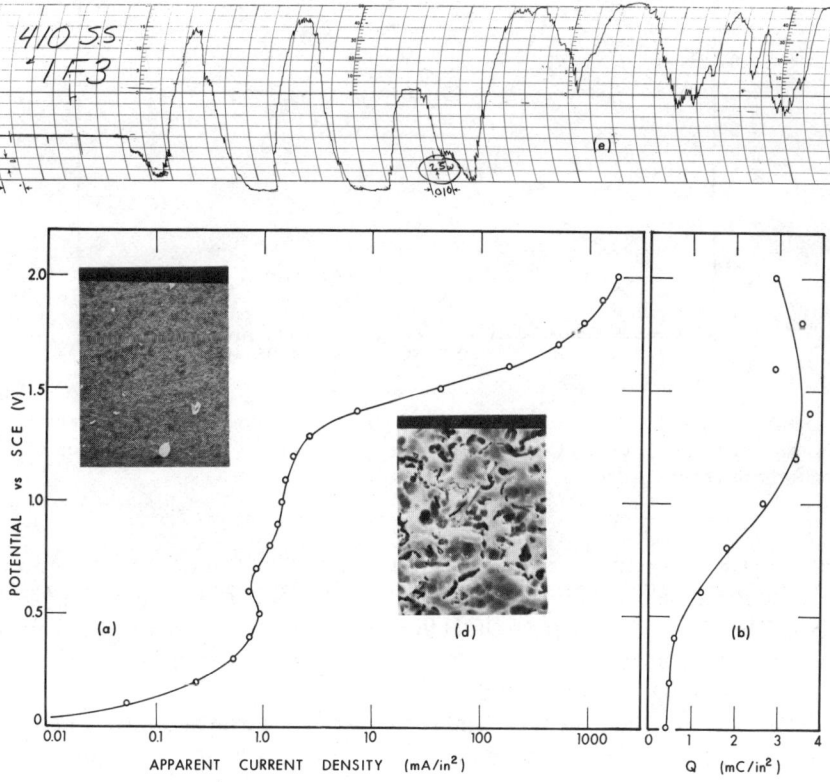

Figure 46. (a) Polarization curve obtained[273] on 410 Cr stainless steel anode; (b) $Q$ as $f$ (potential); (c), (d) SEM micrographs; (e) Proficorder trace: $x$ axis is 0.025 cm (0.01 in.)/div., $y$ axis is 0.625 $\mu$m (25 $\mu$in.)/div. (With permission of The Electrochemical Society, Inc.)

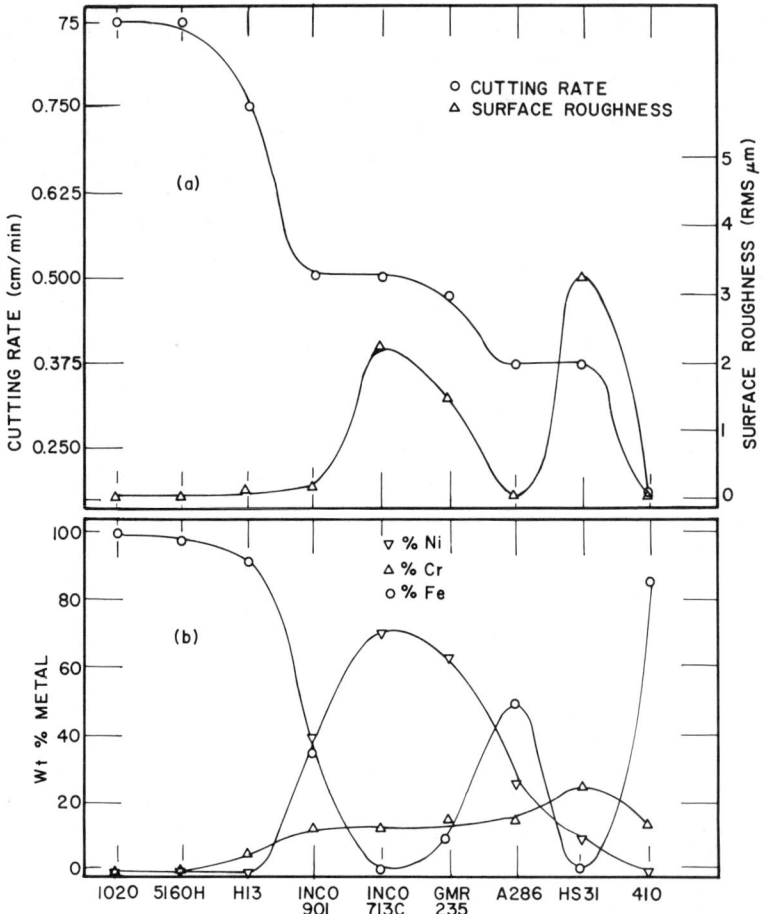

Figure 47. Plots of the plunge-cutting rate in cm/min, the peak-to-valley surface roughness in rms $\mu$m, and the wt% of Ni, Cr, and Fe for the alloys noted.[273] (With permission of The Electrochemical Society, Inc.)

transpassive region in contrast to the decrease in thickness of the film on steel. At potentials below 1 V vs. SCE, the anodic film is hydrated $Ni_2O_3$ or $\beta$-NiOOH.[346] Under the $\beta$-NiOOH film, a thin layer of $NiO_2$ is formed above 1 V. Since these films are good electronic conductors,[87,347] most of the current is consumed in the evolution of $O_2$.

The current efficiency for metal removal of pure iron and pure Ni anodes in a high-rate flow cell[315] was determined[290] and is plotted in Figure 49 for various flow rates. On iron anodes, the efficiency rapidly rises to about 100%, where it is independent of current density and of flow rates. On Ni anodes efficiency is strongly dependent on current density and flow rate. The difference

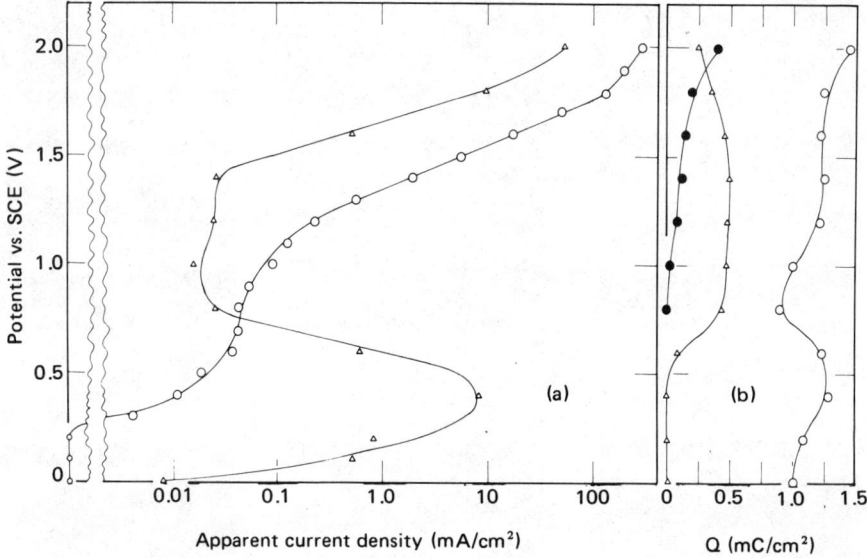

Figure 48. (a) Steady-state polarization curves obtained[336] on a Ni (○) and a mild steel (△) anode in NaClO$_3$ solution (350 g/liter) and (b) the charge, $Q$, associated with the anodic film as a function of potential. In region where O$_2$ is evolved, the charge associated with a higher oxide (●) is shown. (With permission of The Electrochemical Society, Inc.)

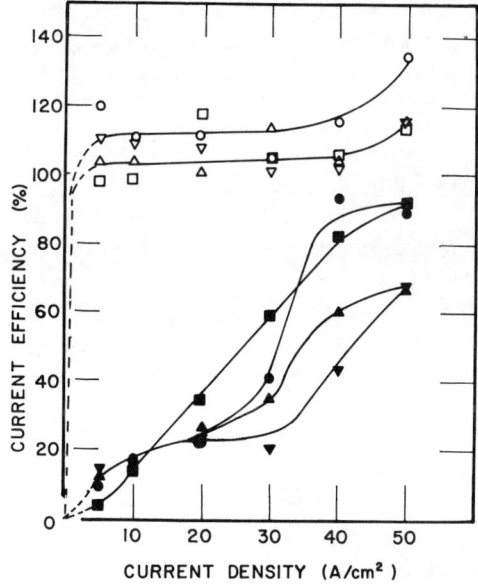

Figure 49. Current efficiency determination[290] for the anodic removal of 99.99% pure iron (open symbols) and 99.9 + % pure nickel (filled symbols) in 4 $M$ NaClO$_3$ solution as a function of current density and flow rate; 500 (○), 1000 (△), 2000 (▽), and 3000 cm/sec (□). Broken lines show behavior in the low-current-density region based on data reported in the literature.[284,303] (With permission of Pergamon Press.)

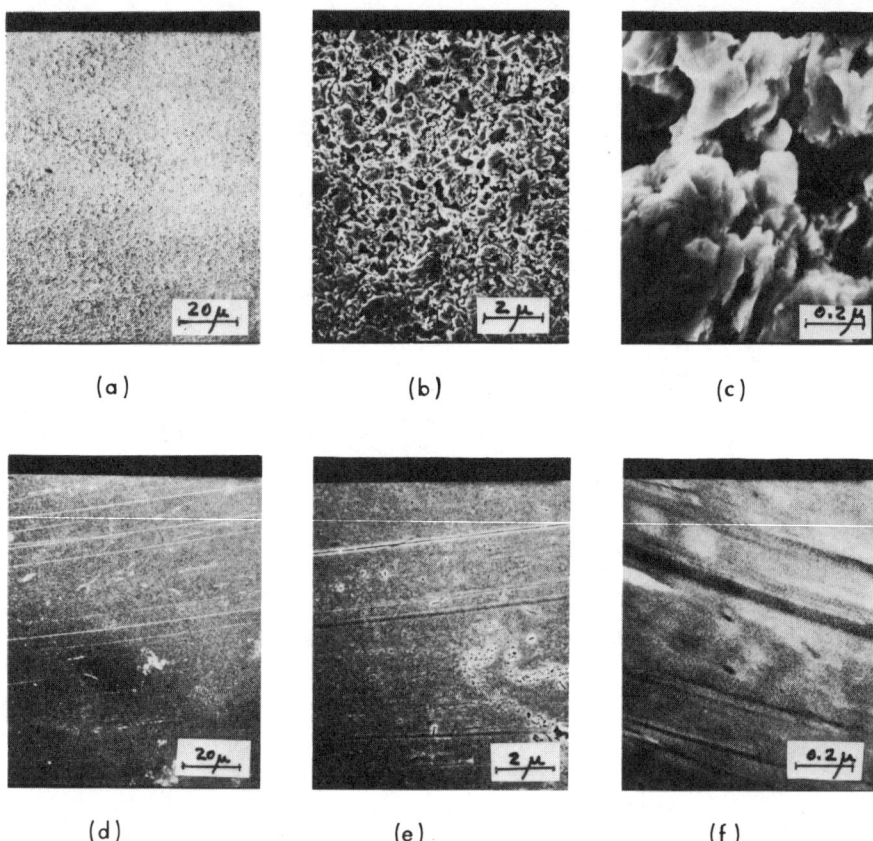

Figure 50. SEM micrographs[336] of the side of a hole plunge-cut in Ni (a,b,c) and steel (d,e,f) samples in NaClO$_3$ solution (350 μg/liter). (With permission of The Electrochemical Society, Inc.)

in ECM behavior is traced to the difference in the properties of the anodic films in the two cases.

Surface roughness determinations[336] indicate that the machined Ni surface is nearly an order of magnitude rougher (0.25–0.38 μm) than the iron surface (0.025–0.05 μm). This result is evident from the SEM micrographs reproduced in Figure 50. At very high anodic potentials, the passive film on Ni is removed unevenly, producing a comparatively poor surface. Because of the conducting properties of these anodic Ni films, which appear to be removed with difficulty in the transpassive region, nickel is machined[336] at about half the rate of steel at an applied voltage of 20 V (0.44 cm/min for nickel vs 0.76 cm/min for steel).

Although one may search[284] for an electrolyte in which most metals will be machined with the same characteristics, applied and fundamental ECM studies show that behavior in a given electrolyte is directly related to the properties of the anodic films formed on the anode surface. Consequently, a universal ECM electrolyte is not available as the properties of anodic films vary among metals in a given electrolyte. It may even be possible to change the phase composition of steels with heat treatments[348] enough to modify the ECM characteristics of the metal. In comparing the ECM properties of different metals in a given electrolyte, one must take care to compare results from the same ECM operation since cell geometry and fixturing influence these results.[290]

Reports of the ECM behavior of a number of metals in various electrolytes appear in the literature,[282,349–354] and an excellent summary of the advantages and disadvantages of a given electrolyte for the ECM of various metals is tabulated by Wilson.[173] A summary of the common electrolytes used to machine a number of metals is recorded in Table 7.

*Table 7*

*Electrolytes for Machining Various Metals*

| Metal | Electrolyte | Remarks | References |
|---|---|---|---|
| Steel and iron-based alloys | $NaClO_3$ (100–600 g/liter) | Excellent dimensional control; brilliant surface finish; high metal removal rate; fire hazard when dry. | 270, 273 |
| | NaCl (50–300 g/liter) | Good surface finish; high metal removal rate; poor dimensional control due to large stray currents; cheap. | 125, 270 |
| | $NaNO_3$ (100–400 g/liter) | Good dimensional control; lower metal removal rates; rougher surface finish; fire hazard when dry. | 38 |
| | $NaClO_4$ (100–400 g/liter) | Good dimensional control; good surface finish; good machining rate; lower fire hazard. | 38, 274 |
| | Potassium salts | Same characteristics as sodium salts but more expensive. $KClO_3$ is less soluble than $NaClO_3$. | |
| Nickel and nickel-based alloys | NaCl (50–300 g/liter) | Good surface finish; high metal removal rate; cheap; large stray currents give poor dimensional control. | 246–248 |
| | $NaNO_3$ (100–400 g/liter) | Good surface finish; better dimensional control; lower metal removal rate. | 345 |
| | $NaClO_3$ (100–600 g/liter) | Good surface finish; good dimensional control; lower metal removal rate. | 273, 336 |

*continued overleaf*

*Table 7 (continued)*

| Metal | Electrolyte | Remarks | References |
|---|---|---|---|
| Titanium and titanium alloys | NaCl (180 g/liter) + NaBr (60 g/liter) + NaF (2.5 g/liter) | Good surface finish; good dimensional control; good machining rate. | 173, 349 |
| | NaCl (50–300 g/liter) | Poor machining rate; strongly passivating films. | 178 |
| | NaClO$_3$ (100–600 g/liter) | Bright surface finish; good machining rate above 24 V. | 349 |
| Cobalt and cobalt alloys | NaCl (50–300 g/liter) | High metal removal rate; poor dimensional control. | 282 |
| | NaNO$_3$ (100–400 g/liter) | Better dimensional control; lower machining rate. | 282 |
| | NaClO$_3$ (100–600 g/liter) | Best dimensional control; excellent surface finish. | 282 |
| Tungsten | NaOH (40–100 g/liter) | NaOH is consumed so must be added continuously. | 172, 173 |
| Molybdenum | NaOH (40–100 g/liter) | NaOH is consumed so must be added continuously. | 173, 354 |
| Uranium | NaCl (20–150 g/liter) | High metal removal rate. | 352 |
| | NaNO$_3$ (20–150 g/liter) | Lower metal removal rate; best surface finish. | 352 |
| Beryllium | NaNO$_3$ (200 g/liter) | Good results. | 351 |
| Aluminum and aluminum alloys | NaNO$_3$ (100–400 g/liter) | Gives best surface finish. | 173 |

## 8.11. Pulsed ECM Methods

To save on rectifier costs, Noble and Shine[355] considered the possibility of applying an ac voltage across the tool and the workpiece. The cathode must be made of a rectifying material such as titanium diboride or an inert material such as high-density graphite to prevent machining of the tool on the positive half-cycle. Machining rates were found to be very low because of the low current efficiency resulting from the loss of anodic current on the workpiece during the negative half-cycle (only gassing without metal dissolution). The use of ac ECM was found to be impractical.

By vibrating the cathode mechanically during ECM, it was found[356,357] that metal dissolution could be increased by as much as 20% (carbon steels in NaNO$_3$ solutions). The vibration of the cathode is thought to accelerate the nucleation of hydrogen bubbles reducing the effect of hydrogen coverage (blanketing effect) of the metal surface.

To increase the machining rate and to improve the surface finish, ultrasonic waves (22,000 Hz) were applied by a transducer directly to the electrolyte entering the gap. Some success was reported.[358,359]

If the current were stopped periodically during ECM, it is conceivable that the flowing electrolyte could carry heat, gases, and sludge away from the machining zone in the gap more efficiently, although lower metal removal rates would result. Russian investigators[360–363] have studied the pulsed ECM system. Saushkin[363] machined stainless steel samples in NaCl electrolyte (80 g/liter) flowing at a rate of 10 m/sec through a 0.3-mm wide gap. Using square-wave pulses, the current was interrupted for 3 msec every 3.5 sec. This current density was 10 A/cm$^2$. At a steady current density, the surface roughness was about 8 $\mu$m, but with pulsed current, the surface roughness was less than 1.5 $\mu$m. Pulsing the current improves the surface finish at the expense of a lower machining rate.

### 8.12. Consensus

Since the central problem in ECM is the determination of the cathode configuration required to generate the desired shaping of the anode, it is highly desirable to formulate a mathematical description of the ECM process from which the required cathode shape can be predicted with confidence. Until the effects of the electrochemical properties of the anodic films formed on the workpiece surface and the effects of solution flow on these properties can be introduced into the mathematical theory of ECM, useful solutions to the central ECM problem will not be obtained. To accomplish this is a formidable task, indeed, but one which promises great rewards, although at the present time the solution appears remote.

## 9. Electrochemical Machining Applications

There are a number of distinct forms of ECM devices used industrially, the most easily recognized of which is plunge cutting. The first working model, made by Anocut Engineering Company in 1954,[149,150] is shown in Figure 51. Other machine-building concerns, such as Cincinnati, Ex-Cell-O, and Sunstrand, offered machines of competitive capacities and cost.

Although some deburring machines were built and demonstrated by Bosch and Wendt in a trade show in Milan in 1963, there was no ECM building activity in Europe before 1966 until the sale of Anocut machines in Great Britain spurred a consortium of firms called PERA (Production Engineering Research Association) to fund a study of ECM.[177] After Anocut announced the sale of its die-sinking machine, ECM experiments were initiated in Japan[156] at Hitachi Ltd. By 1966, there were at least two manufacturers of die-sinking machines in Japan.[364]

The use of ECM plunge-cutting machines was developed in a variety of applications in most ingenious ways. Among these applications are the machining of the following: semicircular slots (Figure 52), a vibration

Figure 51. First plunge-cutting ECM apparatus. (With permission of Anocut Engineering Company.)

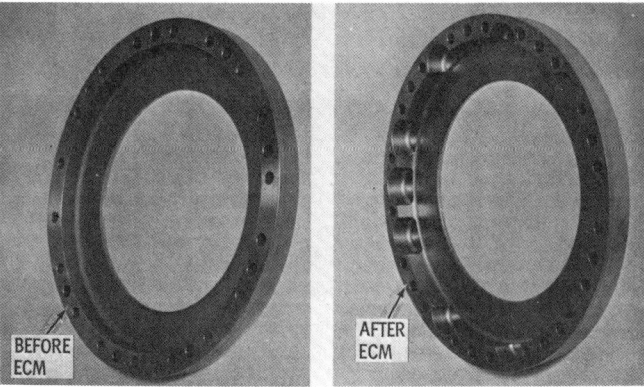

Figure 52. ECM of bearing oil seal retainer. (With permission of the Detroit Diesel Allison Division, General Motors Corporation.)

# ELECTROCHEMICAL MACHINING

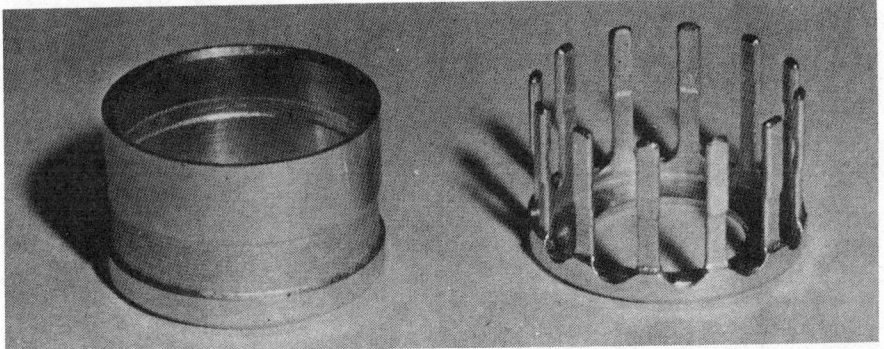

Figure 53. ECM of compressor bearing vibration dampener. (With permission of the Detroit Diesel Allison Division, General Motors Corporation.)

dampener (Figure 53), a turbine shaft bearing sleeve (Figure 54), high pressure turbine wheels (Figure 55), pinion dies (Figure 56), connecting rod die impressions (Figure 57), a steering knuckle forging die (Figure 58), and a drive end of a large steel mill roll (Figure 59).

Because of a saturation of the market, only two producers of plunge cut machines, Anocut Engineering Company and Chem-Form Inc., remain at present. The situation outside the United States does not appear to be much brighter.

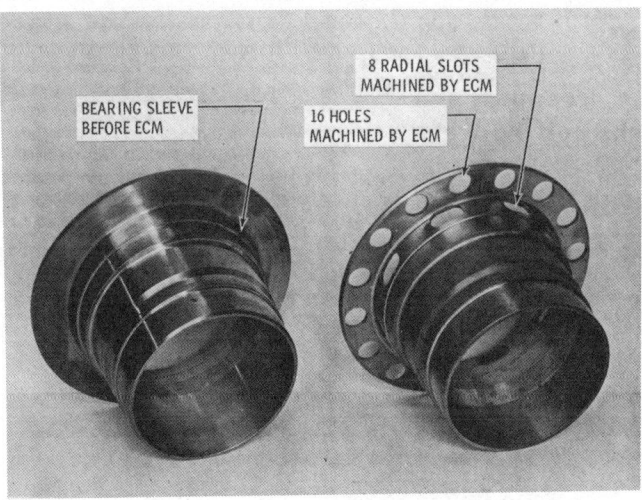

Figure 54. ECM of turbine shaft bearing sleeve. (With permission of the Detroit Diesel Allison Division, General Motors Corporation.)

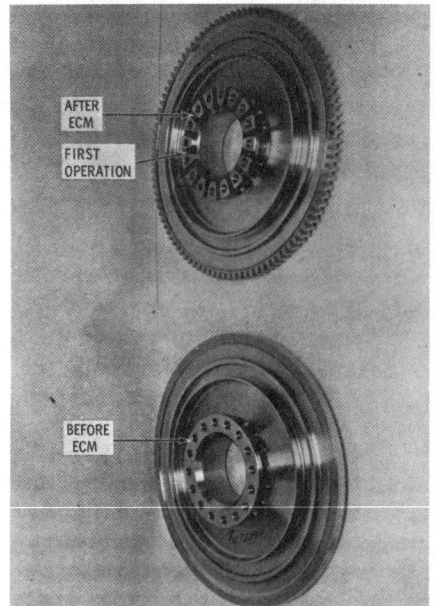
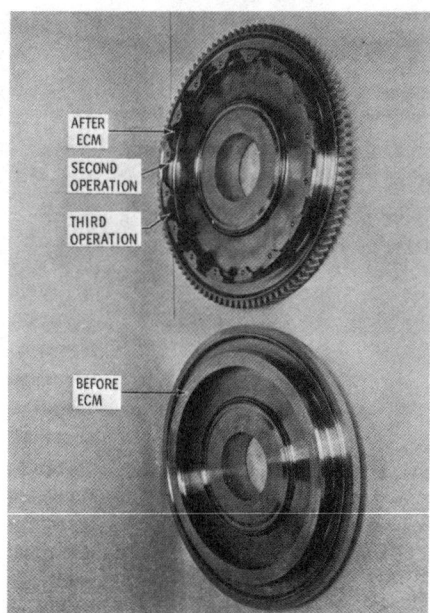

Figure 55. ECM of high-pressure turbine wheels. Front (left), rear (right). (With permission of the Detroit Diesel Allison Division, General Motors Corporation.)

### 9.1. Stem Drilling

Shaped tube electrolyte machining (STEM) is a variation of ECM employing an acid electrolyte.[365] The process was developed from the normal ECM plunge-cut machines to solve the problem of drilling small, deep holes (up to 300 mm deep and down to 0.75 mm diameter) in high-temperature

Figure 56. ECM of large pinion dies. (With permission of Anocut Engineering Company.)

# ELECTROCHEMICAL MACHINING

Figure 57. ECM of connecting rod die cavities. (With permission of Anocut Engineering Company.)

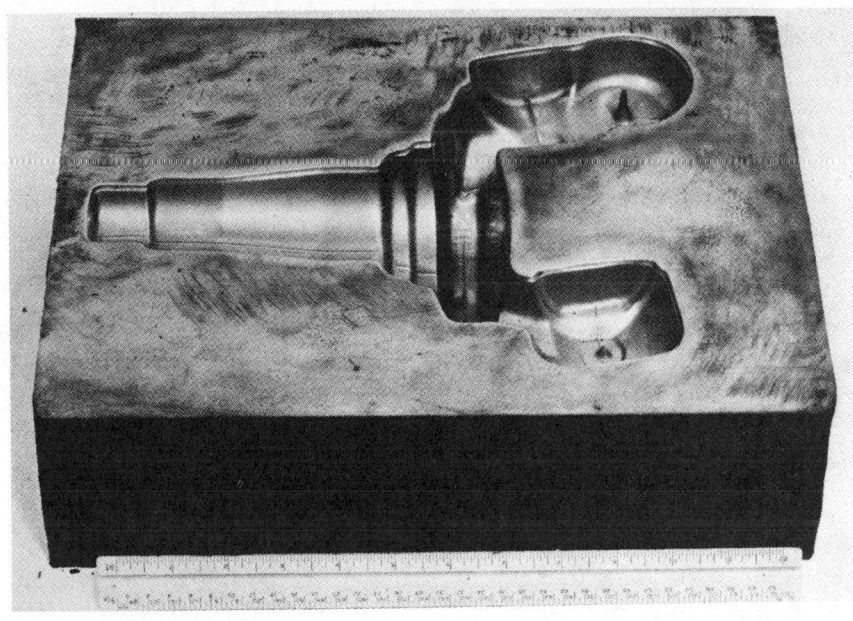

Figure 58. ECM of steering knuckle forging die. (With permission of Anocut Engineering Company.)

Figure 59. ECM of drive end of large steel mill roll. Before ECM (left), after ECM (right). (With permission of Anocut Engineering Company.)

alloys. Holes can be produced with a ±0.05-mm-diameter tolerance and to within 0.03 mm straightness. Surface finish capability is about 0.4 $\mu$m. With depth-to-diameter ratios of 200:1, it is evident that an acid electrolyte is necessary to hold the machined metal in solution so as not to clog the long, finely sized STEM electrodes. The specialization is further carried out in the construction materials of the STEM device; the highly corrosive environment generated by the use of acid electrolytes necessitates the use of corrosion resistant materials throughout. Previously discussed ECM plunge-cut machines are not designed to employ acid electrolytes.

The electrode is a most critical part of the STEM device since the process has several more parameters than the normal plunge-cutting ECM (metal ion content in the electrolyte, shape change of tip due to plating out of metal ions) and since it uses a much more fragile electrode than normal ECM. Electrodes employed in the STEM process are generally titanium tubing, straight, thin-walled, and covered with a smooth, pore-free, concentric dielectric coating about 0.05 mm thick. These must be properly end-dressed and guided to do an effective job.

The STEM process is said to be well suited to drilling materials such as nickel, cobalt, titanium, stainless steels, and superalloys. No cases in which the process was applied to iron or steel alloys were reported, and it appears that the process is not suited in such applications.

## 9.2. Fixture Electrochemical Machining (Cell Type)

In machines in which both the anode and cathode are fixed, only the anodic interface moves because of anodic dissolution on the workpiece. This type of cell (fixture) has not had much use in ECM (only deburring devices) because of the poor characteristics of the electrolytes that have been available. There was no advantage to be gained since the electrolytes, such as acids, sodium chloride, or sodium nitrate, were too corrosive, did not improve the resultant surface finish, or could not remove metal in pertinent areas without maskants of some type. This situation changed significantly when sodium chlorate ECM was discovered.

The fine characteristics of sodium chlorate electrolyte discussed previously introduced a whole new concept to ECM. It was found that finishing (surface finish improvement), sizing, and patterning of various kinds could be performed in cell type fixtures. A number of examples of this type of ECM work follow.

A fixture for performing an embossing-type ECM operation[272] is shown in Figure 60 along with a fully prepared steel blank. The results of the ECM

Figure 60. Fixture for ECM of hydrostatic bearing pads.

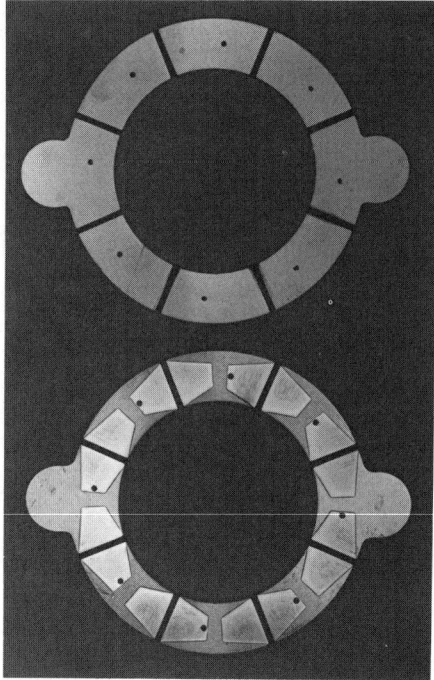

Figure 61. ECM of hydrostatic bearing pads. Initial blank (top) and result of ECM (bottom).

operation are shown in Figure 61. In this case, the hydrostatic bearing pads were machined into the blank to a depth of 6.4 μm in 1 sec. The flatness variation in the bottom of the pads was 0.6 μm. There was no contact of the tool and the work, and no maskant was used on the workpiece. It was as if the cathode face imprinted its image on the anodic face electrochemically.

Other hydrostatic bearing examples were electrochemically machined to the depth of 254 μm with no loss of definition in 15–30 sec. Thus a method for industrial production became entirely feasible in an area where previously none existed. In addition, the machining device is simple and quite adaptable to automation.

In yet another example, a lead screw was electrochemically machined into a blank shaft to a depth of 25–50 μm in 2–3 sec.[308] In Figure 62, the blank is shown on the left, and the finished part with lead screw on the right. The fixture used for ECM with blank inserted is shown in Figure 63.

The requirement in this case was for a triple lead screw cut concentrically (within 2–3 μm with the centerline) in a fully hardened shaft. Because of the shallow, geometrically difficult cutting operation on a fully hardened surface, all mechanical means of making the part failed. The sodium chlorate–fixture ECM made the part easily and is entirely suited to automated production of thousands of replicas daily.

# ELECTROCHEMICAL MACHINING 505

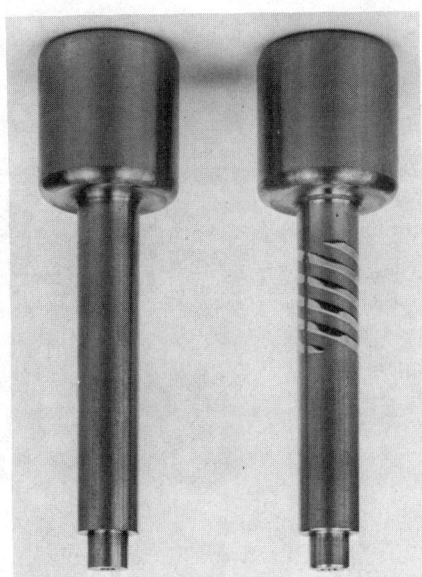

Figure 62. ECM of lead screw into shaft. Initial blank (left) and resultant lead screw (right).

Figure 63. Fixture for the ECM of lead screw into shaft with blank in working position.

Figure 64. Intricate pattern ECM (12.6 lines/cm). ECM depth = 50 $\mu$m. (With permission of The Electrochemical Society, Inc.)

In addition to the relatively gross patterns discussed above, the fixture ECM process can produce very intricate patterns[366] without the use of maskants. The blank and some patterns obtained in this case are shown in Figure 64, and the fixture with a finished part in position is shown in Figure 65. A pattern of 12.6 lines/cm was machined to a depth of 50 $\mu$m in 18 sec. The

Figure 65. Intricate pattern ECM fixture with finished part in position. Cathode on right. (With permission of The Electrochemical Society, Inc.)

# ELECTROCHEMICAL MACHINING

standing dots machined into the soft steel surface measured 0.5 mm in diameter.

This application begins to approach the ultimate of what can be done with an oxide-film forming electrolyte. It shows that not only can an entire steel surface be machined as described previously[270] to a fine finish and accurate dimension, but that it can also be machined with any conceivable gross or fine pattern. It remains to one's ingenuity as to what can be done with the sodium chlorate–fixture ECM.

## 9.3. ECM Broaching

A type of electrolytic broaching machine is shown in Figure 66. The principle of the ECM broaching operation is one in which the inside diameter of an anodic workpiece is electrochemically machined by a limited area cathode passing the anode. By rearranging the elements of the fixture so that the anode moves through a fixed, limited area cathode of torroidal shape, the outside diameter of the workpiece may be machined.

Figure 66. ECM broach device.

The electrolytic broach was developed to offset a severe limitation of the ECM process, the inability to machine large areas. When the average current density of 33 A/cm$^2$ is employed, a 10,000-A machine is capable of machining only 258 cm$^2$ at one time. If the geometry of the anodic workpiece is regular, the area limitation can be overcome, in many cases, by using a moving, limited area cathode in a fixtured ECM broach machine.

The application of ECM broaching by sodium chlorate electrolyte is in its infancy. A number of applications, such as the finishing and sizing of rodlike parts, the skinning of machined porous surfaces, and the finishing of inside and outside diameters of various kinds, have been successful.

### 9.4. Electrolytic Grinding (ELG, ECG)

This method of ECM, which is based to a large extent on the direct application of mechanics by using a grinding wheel, probably has the best opportunity of success in the commercial world. One can obtain an entirely new machine for the process, or existing grinding machines can be converted to

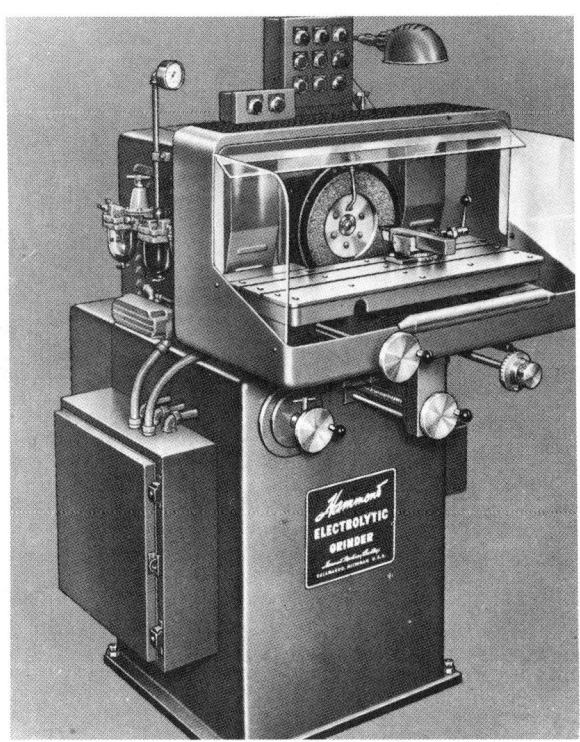

Figure 67. Modern single-table electrolytic grinder. (With permission of Hammond Machine Company.)

Figure 68. Modern duplex-table electrolytic grinder. (With permission of Everite Machine Company.)

ECG. Many manufacturers produce new ECG equipment and some suitable for conversion.[367,368] Some of the machines available are shown in Figures 67 and 68.

The process has gained popularity in many tool rooms throughout the United States. Applications include the following: surface and form grinding of high-speed and carbide form tools, burr-free grinding of all conductive materials, and single point tool sharpening.

Production applications are also promising as indicated by the following: precision form grinding of gears, splines, and other shapes from solid blanks, burr-free cutoff and sharpening of stainless steel hypodermic needles, burr-free slotting operations, form grinding of honeycomb materials, and stress-free grinding to eliminate warping in thin sections.

The weakest link in the ECG system was in the grinding wheels used in early development. Metal-bonded wheels containing diamond or aluminum oxide abrasive had inherent problems. If a complex form was to be ground by ECG, the reverse of this form had to be imparted to the metal-bonded wheel. With the abrasive impregnation of the wheel, this became an almost impossible form-dressing task. There was also a necessary surface preparation or "back etching" step encountered with the metal-bonded wheel. These wheels worked well enough; however, when the form deteriorated past tolerance limits, through general use the wheel had to be removed and returned to the extremely difficult dressing step.

Electrolytic grinding became a viable metal-working technique with the development of resinoid-bonded grinding wheels.[369,370] This type of wheel contains an abrasive, a nonmetallic matrix, and a copper powder. Characteristics of the resinoid bonded wheel are (a) it is easily dressed to desired form, (b) it holds the desired form for economical usage, (c) it has excellent conductivity, and (d) no surface conditioning is necessary.

While the situation for commercial use of plunge-cut ECM is not very encouraging, the future for ECG appears to be better. The coupling of the small mechanical cutting component that scrubs the work surface with the ability to remove fully hardened metals at high rates affords ECG a unique capability. This capability is still largely unexplored, and it would appear that a great economic advantage could be realized if sufficient research and development were focused on industrial use. A dormant bonus to such use would be the ability of the ECG process to be retrofitted to the vast number of grinding and milling machine tools in existence, thus avoiding extensive capital loss.

## Acknowledgments

The authors are indebted to the management of the Research Laboratories of the General Motors Corporation (GMR), in particular to Mr. J. B. Bidwell, Executive Director, Mr. J. C. Holzwarth, Technical Director, and Dr. J. L. Hartman, Head of the Electrochemistry Department, for permission to write this chapter on electrochemical machining. For reading the manuscript critically and authoritatively and for his continued support, we are most grateful to Mr. J. D. Thomas, Assistant Head of the Electrochemistry Department.

We wish to express our thanks to the Detroit Diesel Allison Division of General Motors Corporation, particularly to Mr. J. B. Darling, for making available to us a number of technical publicity photographs; also to Mr. L. Williams, formerly of Anocut Engineering Company, to Mr. Don Ball of Anocut, Incorporated, and to Mr. James Jarvie of Chem-Form Division, KMS Industries, Inc., for supplying data on their equipment.

To Miss G. Sobieska of the Executive Department of GMR, we owe our sincere gratitude for her expert editing and preparation of the manuscript. Our heartfelt thanks go to Mr. J. C. Seifert, Mr. K. O. Wetter, and Mr. R. L. Schulz of the Photographic Section of the Technical Information Department of GMR for their outstanding skill in reproducing the figure content of the manuscript. With pleasure, we offer our thanks to Mr. G. H. Tucker and his associates of the Publications Section of the Technical Information Department of GMR for preparing the manuscript in its final form.

Without the willing cooperation and cheerful help of the Library staff of GMR under Mr. R. W. Gibson, most notably Ms. M. C. Babian, Mr. M. L. Vande Zande, Mr. W. V. Arduino, Mr. N. L. Grow, and Mrs. F. T. Machak, this work would not have been possible, and to them we gratefully give our sincere thanks.

## Auxiliary Notation

| | | | |
|---|---|---|---|
| $A$ | area | $P$ | atmospheric pressure |
| $a$ | atomic weight | $p$ | applied pressure |
| $a_n$ | wave amplitude | $Q$ | charge |
| $B$ | constant | $q$ | quantity of heat energy |
| $b$ | width of exposed anode | $R$ | electrical resistance |
| $b_n$ | wave amplitude | $T$ | temperature |
| $b_n^0$ | initial wave amplitude | $t$ | time |
| $C$ | constant | $t_0$ | initial time |
| $C_p$ | specific heat | $u$ | displacement of movable electrode |
| $c$ | concentration | $V$ | applied voltage |
| $c_m$ | concentration of metal ions | $V_b$ | volume of bubble |
| $c_n$ | wave amplitude | $v_s$ | solution velocity |
| $D$ | diffusion coefficient | $W$ | metal removal (weight loss) |
| $E_p$ | polarization | $W_E$ | equivalent weight |
| $F$ | Faraday constant | $x$ | length on $x$ axis |
| $F_g$ | force exerted by gas | $y$ | length on $y$ axis |
| $F_h$ | horizontal force | $y'$ | $dy/dx$ |
| $F_m$ | force exerted by rotating wheel | $z$ | valence |
| $F_s$ | electrolyte flow rate | $\alpha$ | constant |
| $F_v$ | vertical force | $\delta$ | effective thickness of hydrodynamic boundary layer |
| $f$ | feed rate | | |
| $f_v$ | volume fraction | $\varepsilon$ | amplitude of surface irregularities |
| $f^0$ | coefficient of friction | $\varepsilon_0$ | initial amplitude of irregularities |
| $h$ | gap distance | $\eta$ | overvoltage |
| $h_e$ | equilibrium gap distance | $\theta$ | angle between feed direction and the normal to the anode surface |
| $h_0$ | initial gap distance | | |
| $I$ | current | | |
| $j$ | current density | $\kappa$ | specific conductance |
| $j_0$ | exchange current density | $\kappa_c$ | specific conductance of continuous medium |
| $j^0$ | initial current density | | |
| $\bar{j}$ | average current density | $\kappa_e$ | effective conductance |
| $J$ | mechanical equivalent of heat | $\lambda$ | wavelength |
| $k$ | wave number | $\rho_m$ | density of metal |
| $L$ | length of exposed anode | $\rho_s$ | density of solution |
| $l$ | wave number | $\varphi$ | potential |
| $m$ | mass | $\varphi_1$ | first order perturbation to potential |
| $n$ | index | $\Omega$ | amount of overcut |

## References

1. G. E. Miller, *J. Appl. Phys.* **28**, 149 (1957).
2. W. Pentland and J. A. Ektermanis, *Trans. ASME* **87B**, 46 (1965).
3. T. Vetter and J. Abthoff, *VDI Z.* **108**, 459, 512 (1966).
4. D. Goetze, *J. Acoust. Soc. Am.* **28**, 1053 (1965).
5. M. I. Ross, *Iron Age* **193**(23), 72 (1964).
6. O. Ruediger and A. Winkelmann, *Ind-Anz.* **86**, 1654 (1964).
7. A. L. Lifshits, G. K. Bannikov, and A. M. Sigarev, *Stanki Instrum.* **29**, 23 (1958).
8. T. W. Block, *Tool Manuf. Eng.* **50**, 87 (1963).
9. H. Kurafuji, *Ann. CIRP* **13**, 313 (1966).
10. Anonymous, *Am. Mach.* **103**(23), 99 (1959); *Iron age* **62**(9), 139; **62**(10), 169 (1956).

11. R. F. Duhamel, ASTME Creative Manuf. Seminars, SP 63-29 (1963).
12. T. A. Moore, ASTME Creative Manuf. Seminars, SP 62-30 (1962).
13. R. K. Springborn, Non-traditional machining processes, ASTME Pub., Dearborn, Michigan, 1967.
14. A. B. Meinel, S. Bashkin, and D. A. Loomis, *Appl. Opt.* **4**, 1674 (1965).
15. J. B. Schroeder, S. Bashkin, and J. F. Nester, *Appl. Opt.* **5**, 1031 (1966).
16. J. P. Hoare, *Electrochemistry of Oxygen*, Wiley-Interscience, New York (1968), p. 357ff.
17. C. Wagner and W. Traud, *Z. Elektrochem.* **44**, 391 (1938).
18. U. R. Evans, *The Corrosion and Oxidation of Metals*, St. Martin's, New York (1960).
19. P. M. Aziz, *Corrosion* **9**, 85 (1953); *J. Electrochem. Soc.* **101**, 120 (1954).
20. R. H. Brown and R. B. Mears, *Trans. Electrochem. Soc.* **74**, 495 (1938).
21. C. Edeleanu and U. R. Evans, *Trans. Faraday Soc.* **47**, 1121 (1951).
22. U. R. Evans and D. E. Davies, *J. Chem. Soc.* **1951**, 2607.
23. U. R. Evans, *Corrosion* (*Houston*) **7**, 238 (1951).
24. O. D. Block and L. H. Cutler, *Ind. Eng. Chem.* **50**, 153 (1958).
25. M. Byer, *Mater. Methods* **43**(6), 134 (1956).
26. M. C. Cook, *Product Eng.* **27**(7), 194 (1956).
27. W. C. Hittinger and M. Sparks, *Scient. Am.* **213**(5), 57 (1965).
28. L. H. Sharpe and P. D. Garn, *Ind. Eng. Chem.* **51**, 293 (1959).
29. R. L. Swiggert, *Mod. Plast.* **31**(8), 94 (1954).
30. R. C. Benton and G. D. Woodring, ASTME paper No. 685 (1965).
31. U. R. Evans, L. C. Bannister, and S. C. Britton, *Proc. R. Soc.* **A131**, 355 (1931).
32. M. J. Pryor and D. S. Keir, *J. Electrochem. Soc.* **104**, 269 (1957).
33. S. F. Dorey, *J. Inst. Met.* **82**, 497 (1953).
34. L. Kenworth, *J. Inst. Met.* **69**, 67 (1943).
35. E. C. J. March, *Electroplating* **7**, 88 (1954).
36. N. H. Simpson, *Corrosion* **13**, 151t (1957).
37. T. P. Hoar, *Corros. Sci.* **7**, 341 (1967); *J. Electrochem. Soc.* **117**, 17C (1970).
38. K-W. Mao, M. A. LaBoda, and J. P. Hoare, *J. Electrochem. Soc.* **119**, 419 (1972).
39. J. P. Hoare and K-W. Mao, *J. Electrochem. Soc.* **120**, 1452 (1973).
40. H. Figour and P. A. Jacquet, French Patent 707,526 (1930).
41. P. A. Jacquet, *Sheet Met. Ind.* **24**, 2015 (1947).
42. P. A. Jaquet, *C.R. Acad. Sci.* **205**, 1232 (1937).
43. P. A. Jacquet, *Trans. Electrochem. Soc.* **69**, 629 (1936).
44. M. P. Fedot'ev and S. Ya. Grilikhes, *Electropolishing, Anodizing, and Electrolytic Pickling of Metals*, Robert Draper Ltd., Teddington, England (1959).
45. P. A. Jacquet, *Rev. Metall.* (*Paris*) **37**, 210 (1940).
46. A. W. Hothersall and R. A. F. Hammond, *J. Electrodep. Tech. Soc.* **16**, 83 (1940).
47. O. Zmeskal, *Met. Prog.* **47**, 729 (1945).
48. W. H. J. Vernon and E. G. Stroud, *Nature* **142**, 477 (1938).
49. J. L. Rodda, *Mining Metall.* **24**, 323 (1943).
50. R. M. Imboden and R. S. Sibley, *Trans. Electrochem. Soc.* **82**, 227 (1942).
51. L. A. Hauser, *Iron Age* **153**(3), 48 (1944).
52. C. L. Faust, U.S. Patent 2,334,699 (November 23, 1943); 2,338,321 (January 4, 1944); *Steel* **109**(18), 80 (1941).
53. H. H. Uhlig, *Trans. Electrochem. Soc.* **78**, 265 (1940).
54. C. Schaefer, *Met. Ind.* (*N.Y.*) **38**, 22 (1940).
55. C. B. F. Young and W. L. Brytczuk, *Met. Finish.* **40**, 237, 306 (1942).
56. R. Mondon, *Met. Prog.* **63**(1), 95 (1963).
57. J. Edwards, *J. Electrochem. Soc.* **100**, 189C, 223C (1953).
58. R. Mondon, *Sheet Met. Ind.* **32**, 923 (1955).
59. M. E. Merchant, *Met. Prog.* **37**, 559 (1940).
60. J. K. Higgins, *J. Electrochem. Soc.* **106**, 999 (1959).

61. W. C. Elmore, *J. Appl. Phys.* **10**, 724 (1939); **11**, 797 (1940).
62. T. P. Hoar and T. W. Farthing, *Nature* **169**, 324 (1952).
63. T. P. Hoar and J. A. S. Mowat, *Nature* **165**, 64 (1950).
64. A. Krusenstjern and H. Schlegel, *Metalloberflasche* **10B**, 148 (1955).
65. H. Pray and C. L. Faust, *Iron Age* **145**(15), 33 (1940).
66. H. F. Walton, *J. Electrochem. Soc.* **97**, 219 (1950).
67. T. P. Hoar, *Mod. Aspects Electrochem.* **2**, 262 (1959); *J. Electrodep. Tech. Soc.* **28**, 149 (1952).
68. P. A. Jacquet, *C.R. Acad. Sci.* **201**, 1473 (1935).
69. G. S. Vozdvizhensky and A. I. Turashev, *Dokl. Akad. Nauk SSSR* **114**, 358 (1957).
70. K. Kojima and C. W. Tobias, *J. Electrochem. Soc.* **120**, 1026 (1973).
71. C. Wagner, *J. Electrochem. Soc.* **101**, 225 (1954).
72. J. Edwards, *J. Electrodep. Tech. Soc.* **28**, 137 (1952).
73. M. Zamin, P. Mayer, and M. K. Murthy, *J. Electrochem. Soc.* **123**, 1377 (1976).
74. A. Hickling and J. K. Higgins, *Trans. Inst. Met. Finish.* **29**, 274 (1953).
75. J. F. Nicholas and W. J. M. Tegart, *J. Electrochem. Soc.* **102**, 93C (1955).
76. T. P. Hoar and J. A. S. Mowat, *J. Electrodep. Tech. Soc.* **26**, 7 (1950).
77. T. P. Hoar and G. P. Rothwell, *Electrochim. Acta* **9**, 135 (1964).
78. R. W. Powers, *Electrochem. Tech.* **2**, 274 (1964).
79. J. P. Hoare and M. A. LaBoda, in *Electrochemical Techniques for Corrosion*, R. Baboian, ed., NACE Publishers, Houston (1977), p. 106.
80. P. A. Jacquet and M. Jean, *C.R. Acad. Sci.* **230**, 1862 (1950); *Rev. Met.* **48**, 537 (1951).
81. J. A. Allen, *Trans. Faraday Soc.* **48**, 273 (1952).
82. K. P. Balasev and E. N. Nikitin, *Zh. Prikl. Khim.* **23**, 263 (1950).
83. E. C. Williams, and M. A. Barrett, *J. Electrochem. Soc.* **103**, 364 (1956).
84. D. Laforgue-Kantzer, *C.R. Acad. Sci.* **233**, 547 (1951).
85. S. I. Krichmar and V. P. Galushko, *Zh. Fiz. Khim.* **30**, 577 (1956).
86. K. F. Lorking, *Electrochim. Acta* **7**, 101 (1962).
87. M. Novak, A. K. N. Reddy, and H. Wroblowa, *J. Electrochem. Soc.* **117**, 733 (1970).
88. W. A. Sparks, *J. Electrodep. Tech. Soc.* **21**, 245 (1946).
89. S. Tajima and T. Mori, *C.R. Acad. Sci.* **233**, 160 (1949).
90. K. F. Lorking, *J. Electrochem. Soc.* **102**, 479 (1955).
91. R. R. Cole and Y. Hopenfeld, *Trans. ASME* **85B**, 395 (1963).
92. W. Tegart, *The Electrolytic and Chemical Polishing of Metals*, Pergamon, London (1959).
93. A. Uhlir, *Rev. Sci. Instrum.* **26**, 965 (1955).
94. J. W. Tilley and R. A. Williams, *Proc. IRE* **41**, 1706 (1953).
95. M. H. Farmer and G. H. Glaysher, *J. Sci. Instrum.* **30**, 9 (1953).
96. J. J. Theler, *Metaux* **41**, 463 (1966).
97. S. Sheff, *Electrochem. Tech.* **5**, 47 (1966).
98. P. F. Schmidt and M. Blomgren, *J. Electrochem. Soc.* **106**, 694 (1959).
99. J. A. Eades and K. G. McIntyre, *J. Phys. Sci. Instrum.* **1E**, 491 (1968).
100. P. M. Kelly and J. Nutting, *J. Iron Steel Inst.* **192**, 246 (1959).
101. P. F. Schmidt and D. A. Keiper, *J. Electrochem. Soc.* **106**, 592 (1959).
102. J. C. Bilello, *J. Sci. Instrum.* **44**, 308 (1967).
103. D. G. Brandon and J. Nutting, *Brit. J. Appl. Phys.* **10**, 255 (1959).
104. V. Bulckaen and L. Paganini, *J. Phys. Sci. Instrum.* **4E**, 178 (1971); *Rev. Sci. Instrum.* **42**, 1687 (1971).
105. V. Bulckaen, *Electrochim. Metal.* **4**, 262 (1969).
106. J. L. Bleiweiss and A. J. Fusco, *Met. Alloys* **18**, 1075 (1943).
107. S. G. Ellis, *Phys. Rev.* **100**, 1140 (1955).
108. M. V. Sullivan and J. H. Eigler, *J. Electrochem. Soc.* **103**, 132 (1956).
109. D. R. Turner, *J. Electrochem. Soc.* **103**, 252 (1956).
110. H. Gerischer and F. Beck, *Z. Phys. Chem. N.F.* **13**, 389 (1957).

111. M. Metzger, *Rev. Sci. Instrum.* **29**, 620 (1958).
112. R. A. Peak, *Electronics* **34**(3), 82 (1961).
113. J. A. Muller, *Am. Machinist* **97**(20), 122 (1953).
114. J. R. Duncan, *J. Appl. Electrochem.* **6**, 275 (1976).
115. G. Keeleric, *Steel* **130**(11), 84 (1952); *J. Met.* **4**, 378 (1952).
116. L. H. Metzger and G. Keeleric, *Am. Machinist* **96**(23), 154 (1952).
117. G. F. Keeleric, U.S. Patent 2,359,920 (October 10, 1944); 2,367,286 (January 16, 1945); 2,368,473 (January 30, 1945); 2,370,970 (March 6, 1945).
118. L. A. Williams, *Aircr. Prod.* **22**, 389 (1960).
119. R. R. Cole, *Trans. ASME* **83B**, 194 (1961).
120. G. Pahlitzsch, in *International Research Production Engineering*, ASME, New York (1963), p. 242.
121. A. E. DeBarr and D. A. Oliver, *Electrochemical Machining*, Elsevier, New York (1968), p. 35.
122. A. L. Livshits and V. Ya. Rassakhim, *Stanki Instrum.* **25**, 12 (1954); **26**, 8 (1955); *Eng. Dig. (London)* **16**, 429 (1955).
123. A. G. Ryabinov, *The Electrochemical Processing of Metals and Alloys*, Lenizdat (1965).
124. V. N. Gusev, *Anodic-Mechanical Machining of Metals*, Mashgiz (1952).
125. M. A. LaBoda, J. P. Hoare, and S. E. Beacom, *Coll. Czech. Chem. Commun.* **36**, 680 (1971).
126. C. R. Stroup, *Am. Machinist* **102**(1), 106 (1958).
127. J. A. Gurklis, DDC Rept. AD613261, January 7, 1965.
128. O. W. Storey, *J. Electrochem. Soc.* **100**, 126C (1953).
129. G. Pahlitzsch and A. Visser, *Ann. CIRP* **15**, 229 (1967).
130. R. R. Cole, *J. Eng. Ind.* **1961**, 194.
131. H. Opitz, H. Heitmann, and U. Becker-Barbrock, *Ann. CIRP* **15**, 263 (1967).
132. D. Fishlock, *Metalworking Prod.* **105**, 73 (August 1961).
133. H. Reinhart and W. Grunwald, *Werkstatt Betr.* **1962**, 212.
134. J. S. Spizig, *Werkstattstechnik Maschinenbau* **1958**, 641.
135. A. Geddam and C. F. Noble, *Int. J. Mach. Tool Des. Res.* **11**, 1 (1971).
136. S. I. Zhitnitskii, A. I. Gralevich, V. P. Tselvevskii, and V. I. Kuznetsov, *Sint. Almazy* **5**, 32 (1973); *CA* **80**, 55199d (1974).
137. N. D. G. Mountford, *Trans. Inst. Metal Finish.* **40**, 171 (1960).
137a. J. Frisch and R. R. Cole, *Trans. ASME* **84B**, 483 (1962).
138. G. Comstock, *Aircr. Prod.* **16**, 488 (1954).
139. F. Pearlstein, *Am. Machinist* **102**(1), 110 (1958).
140. K. Beyer, *Metalloberflaeche* **20**, 88 (1966).
141. K. E. Schwartz, *Ind. Anzeiger* **43**, 1357 (1960).
142. H. Reinhart, *Werkstatt Betr.* 1961, 529.
143. J. Frisch and E. G. Thomson, *Trans. ASME* **73**, 337 (1951).
144. J. Just and W. Altgeld, *Prakt. Metallogr.* **7**, 59 (1970).
145. Anonymous, *Iron Age* **186**(23), 115 (1960).
146. V. V. Gostev, *Sint. Almazy* **6**, 24 (1974); *CA* **81**, 130147m (1974).
147. G. Koscholke, *Werkstattstechnik Maschinenbau* **1955**, 562.
147a. Anonymous, *Iron Age* **197**(17), 90 (1966).
148. R. De la Rue and C. W. Tobias, *J. Electrochem. Soc.* **106**, 827 (1959).
149. Anonymous, *Am. Machinist* **103**(23), 99 (1959).
150. Anonymous, *Missiles and Rockets* **5**(45), 29 (1959).
151. V. A. Gusev, British Patent 335,003 (September 18, 1930).
152. R. R. Cole, *Int. J. Prod. Res.* **4**, 75 (1965).
153. C. L. Faust, *Trans. Inst. Metal Finishing* **41**, 1 (1964).
154. W. B. Kleiner, *Tech. Proc. Am. Electroplaters' Soc.* **50**, 147 (1963).
155. H. Kubeth and H. Heitmann, *Ind. Anzeiger* **46**, 975 (1963).
156. K. Kawafune, T. Mekoshiba, K. Noto, and K. Hirata, *Ann. CIRP* **15**, 443 (1967).

157. S. Ito and N. Shikata, *J. Mach. Lab. Japan* **12**, 50 (1966).
158. A. Berecz and V. Horvath, *Gepgyartastech.* **8**, 306 (1968); *CA* **73**, 94034w (1970).
159. O. A. Vodyanitskii, M. Monina, and I. I. Moroz, *Fiz. Khim. Obrab. Mater.* **1968**(5), 45.
160. S. Ito, K. Honda, and F. Sakura, *J. Mech. Lab. Japan* **11**, 67 (1965).
161. A. M. Egorov, *Elektrofiz. Elektrokhim. Metody Obrab. Mater.* **1**, 77 (1967); *CA* **73**, 126337r (1970).
162. J. Bayer, M. A. Cummings, and A. U. Jollis, DDC Rept. AD 450199, September 1964.
163. W. B. Kleiner, SAE paper 618D, January 14, 1963.
164. T. Sato, K. Kodaira, and H. Yamazaki, *Nagoya Kogyo Gijutsu Shikensho Hoboku* **12**, 553 (1963); *CA* **61**, 282 (1964).
165. S. Crisan and M. Ivan, *Constr. Masini* **18**, 583 (1966); *CA* **66**, 71795m (1967).
166. A. D. Davydov, E. Ya. Grodzinski, and A. N. Kamkin, *Elektrokhimya* **9**, 518 (1973).
167. Anonymous, *Chem. Eng. News* **44**(34), 32 (1966).
168. C. L. Faust, ed., *Fundamentals of Electrochemical Machining*, Electrochemical Society, Princeton, New Jersey (1971).
169. W. B. Kleiner, *Trans. SAE* **72**, 123 (1964).
170. V. D. Timashkov, L. Tschirf, and P. I. Yashcheritsyn, *Prog. Tekhnol. Mashostr.* **5**, 93 (1974); *CA* **82**, 65900f (1975).
171. N. H. Cook, G. B. Foote, P. Jordan, and B. N. Kalyani, *Trans. ASME J. Eng. Ind.* **95**, 945 (1975).
172. Anonymous, *Am. Machinist* **111**(22), 149 (1967).
173. J. F. Wilson, *Practice and Theory of Electrochemical Machining*, Wiley Interscience, New York (1971).
174. H. Dietz, K. Otto, and G. Stark, *VDI-Z* **109**, 642 (1967).
175. J. A. McGeough and H. Rasmussen, *Trans. ASME* **92B**, 400 (1970).
176. K. Kawafune, *Bull. Jpn. Soc. Mech. Eng.* **11**, 554, 565 (1968).
177. A. I. W. Moore et al., Electrochemical machining equipment and techniques, PERA Rept. 145, Leicestershire, United Kingdom, 1965.
178. R. F. Rolsten, *J. Appl. Chim.* **18**, 292 (1968).
179. V. D. Myerley, SAE paper 670818, October 2, 1967.
180. A. H. Meleka, *Science J.* **3**(1), 50 (1967).
181. R. G. Dermott, *Met. Progr.* **84**(4), 181 (1963).
182. C. A. Snavely and J. A. Cross, ASTME Creative Manufacturing Seminar, paper SP62-39 (1962).
183. H. Dietz, K. Otto, and G. Stark, *VDI Z.* **109**, 1057 (1967).
184. G. L. Baldwin, D. C. Brown, and J. L. Gulati, *Engineer* **225**, 307 (1968).
185. M. E. Merchant, in *Proceedings of the Third International Congress on Machine Tool Design and Research*, Birmingham, Alabama, Pergamon, New York (1962), p. 93.
186. S. P. Loutrel and N. H. Cook, *Trans. ASME* **95B**, 997 (1973).
187. J. Hopenfeld and R. R. Cole, *Trans. ASME* **88B**, 455 (1966).
188. K. Chikamori and H. E. Freer, *Denki Kagaku* **41**, 275 (1973).
189. J. W. Cuthbertson and T. S. Turner, *Prod. Eng. (London)* **46**, 24 (1967).
190. H. Opitz, in *International Research Production Engineering*, ASME, New York (1963), p. 225.
191. S. Ito, K. Chikamori, and F. Sakurai, *J. Mech. Lab. Japan* **12**, 37 (1966).
192. S. P. Loutrel and N. H. Cook, *Trans. ASME* **95B**, 992 (1973).
193. M. I. Moldavskii, *Elektron. Obrab. Mater.* **1965**(5), 66; *CA* **66**, 81797c (1967).
194. C. W. Tobias, *J. Electrochem. Soc.* **106**, 833 (1959).
195. D. Landolt, R. Acosta, R. H. Muller, and C. W. Tobias, *J. Electrochem. Soc.* **117**, 839 (1970).
196. J. Hopenfeld and R. Cole, *Trans. ASME* **91B**, 755 (1969).
197. S. G. Bankoff, *Trans. ASME* **82C**, 265 (1960).
198. D. Landolt, R. H. Muller, and C. W. Tobias, *J. Electrochem. Soc.* **11**, 1384 (1969).
199. H. H. Kellogg, *J. Electrochem. Soc.* **97**, 133 (1950).

199a. Anonymous, *Aircr. Prod.* **24**, 410 (1962).
200. W. Feitkneckt and G. Keller, *Z. Anorg. Chem.* **262**, 61 (1950).
201. J. O'M. Bockris, D. Drazic, and A. R. Despic, *Electrochim. Acta* **4**, 325 (1961); **7**, 293 (1962).
202. A. D. Davydov, V. D. Kashcheev, and B. N. Kabanov, *Electron. Obrab. Mater.* **1969**(6), 13.
203. M. Minca, *Stud. Circet. Mec. Apl.* **33**, 611 (1974); *CA* **82**, 23399u (1975).
204. V. V. Parshuten and Z. N. Zaidman, *Electron. Obrab. Mater.* **1968**(4), 26; **1968**(5), 41.
205. D. A. G. Bruggemann, *Ann. Phys. (Leipzig)* **24**, 636 (1935).
206. A. J. de Bethune and N. A. S. Loud, *Standard Aqueous Electrode Potentials and Temperature Coefficients*, C. A. Hample, Skokie, Illinois (1964).
207. G. Milazzo, *Electrochemistry*, English transl. by P. J. Mill, Elsevier, Amsterdam (1963), p. 192.
208. G. Kortun and J. O'M. Bockris, *Textbook of Electrochemistry*, Elsevier, Amsterdam (1951), p. 425.
209. J. P. Hoare, M. A. LaBoda, M. L. McMillan, and A. J. Wallace, *J. Electrochem. Soc.* **116**, 199 (1969).
210. D. T. Chin, *J. Electrochem. Soc.* **118**, 1764 (1971); **119**, 1699 (1972); *AIChE J.* **20**, 245 (1974).
211. D. T. Chin, *J. Electrochem. Soc.* **119**, 1043 (1972).
212. K-W. Mao, *J. Electrochem. Soc.* **118**, 1876 (1971).
213. C. F. Carter, *Light Metal Age* **22**(7), 6 (1964).
214. W. Forker and L. Franke, *Technik (Berlin)* **23**(1), 19 (1968).
215. C. A. Weinert, *Auto. Ind.* **123**(11), 60 (1960).
216. T. E. Aaron and R. Wolosewicz, *Machine Design* **41**(28), 160 (1960).
217. S. P. Loutrel and N. H. Cook, *Trans. ASME* **95B**, 1003 (1973).
218. L. A. Williams, *Tool Engineer* **43**(12), 43 (1959).
219. Anonymous, *Iron Age* **195**(15), 106 (1965).
220. W. A. Haggerty, SAE paper 680C (April 1963).
221. W. Konig, D. Pahl, and H. Degenhardt, SME paper MR 70-205 (1970).
222. J. F. Thorpe and R. O. Zerkle, *Int. J. Mach. Tool Des. Res.* **9**, 131 (1969).
223. K. Kawafune, T. Mikoshiba, and K. Noto, *Ann. CIRP* **16**, 345 (1968); **18**, 305 (1970).
224. W. Konig and D. Pahl, *Ann. CIRP* **18**, 223 (1970).
225. Yu. B. Zakharov and L. M. Shcherbakov, *Fiz. Khim. Obrab. Mater.* **1968**(5), 59.
226. J. M. Fitz-Gerald and J. A. McGeough, *J. Inst. Math. Applic.* **5**, 387, 409 (1969); **6**, 102 (1970).
227. J. Kozak, *Arch. Budowy Masz.* **14**, 239 (1967); *CA* **68**, 8628k (1968).
228. V. S. Krylov, A. D. Davydov, and E. Kozak, *Elektrokhimya* **11**, 1155 (1975).
229. A. Haggerty and J. G. Goss, *Am. Machinist* **106**(12), 108 (1962).
230. L. A. Williams, ASTME Creative Manufacturing Seminar, paper SP62-56 (1962).
231. G. A. Alekseev, Yu. S. Volkov, M. A. Monina, and I. I. Moroz, *Fiz. Khim. Obrab. Mater.* **1968**(5), 31.
232. H. J. Huembs, *Ind. Anzeiger.* **96**, 650 (1974).
233. J. Kozak, *Verkstatt Betr.* **106**, 221 (1973).
234. Yu. S. Volkov and I. I. Moroz, *Elektron. Obrab. Mater.* **1965**(5), 59.
234a. H. Tipton, in *Proceedings of the Fifth International Machine Tool and Design Research Conference*, Birmingham, Alabama, Pergamon, New York (1964), p. 509.
234b. H. Tipton, *IEE Conf. Electrical Methods Machining Forming London* **38**, 48 (1967).
234c. J. J. Christie and J. D. Thomas, *Plating* **52**, 855 (1965); **53**, 1207 (1966).
235. S. Ito and N. Shikata, *J. Mech. Lab. Japan* **12**, 29 (1966).
236. A. V. Televnoi *et al.*, *Elektron. Obrab. Mater.* **1974**(1), 23.
237. H. Yamamoto, K. Chikamori, and S. Ito, *Boshoku Gijutsu* **17**, 243 (1968); *CA* **70**, 63347j (1969).

238. V. N. Misra and V. M. Dorkas, *Trans. Indian Inst. Met.* **27**, 156 (1974).
238a. H. Yamamoto and S. Ito, *Denki Kagaku* **40**, 115 (1972); *CA* **77**, 42235a (1972).
239. J. Bannard, *J. Appl. Electrochem.* **4**, 117 (1974).
240. M. A. LaBoda, U.S. Patent 3,429,791 (1969).
241. M. A. LaBoda, U.S. Patent 3,401,104 (1968).
242. M. A. LaBoda, U.S. Patent 3,421,987 (1969).
243. M. A. LaBoda, U.S. Patent 3,389,067 (1968).
244. M. A. LaBoda and W. R. Doty, U.S. Patent 3,389,068 (1968).
245. W. R. Doty and M. A. LaBoda, U.S. Patent 3,404,077 (1968).
246. P. J. Boden and J. M. Evans, *Nature* **222**, 377 (1969).
247. P. J. Boden and J. M. Evans, *J. Electrochem. Soc.* **116**, 1715 (1969).
248. P. J. Boden and J. M. Evans, *Electrochim. Acta* **16**, 1071 (1971).
249. J. P. Hoare, M. A. LaBoda, and A. J. Wallace, *J. Electrochem. Soc.* **116**, 1715 (1969).
250. H. Karafuji and K. Suda, *Ann. CIRP* **14**, 435 (1967).
251. Yu. P. Cherepanov, *Electron. Obrab. Mater.* **1965**(5), 51.
252. S. Ito, K. Honda, and J. Hayashi, *J. Mech. Lab. Japan* **10**, 42 (1964).
253. L. M. Voronenko and I. I. Moroz, *Electron. Obrab. Mater.* **1970**(1), 25.
254. L. M. Voronenko, A. D. Davydov, and V. D. Kashcheev, *Fiz. Khim. Obrab. Mater.* **1972**, 133.
255. J. A. Gurklis, *Cobalt* **39**, 81 (1968).
256. G. Rowden, *Metallurgia* **77**, 188 (1968).
257. V. P. Smolentsev, A. M. Il'chenko, N. Z. Loginov, and I. N. Shkanov, *Stanki Instrum.* **1973**(5), 37; *CA* **79**, 139190m (1973).
258. N. A. Morozov and A. I. Maksimov, *Vestn. Mashinostr.* **47**, 60 (1967); *CA* **67**, 17140b (1967).
258a. V. P. Smolentsev, *Stanki Instrum.* **1973**, 37.
259. B. Cina, *Metallurgia* **55**, 11 (1957).
260. G. V. Kargin, *Vestn. Mashinostr.* **46**, 40 (1966); *CA* **65**, 3449 (1966).
261. J. A. McGeough, *Principles of Electrochemical Machining*, Chapman and Hall, London (1975).
262. C. L. Faust and C. A. Snavely, *Iron Age* **186**(18), 77 (1960).
263. Anonymous, *Aircr. Prod.* **23**, 88 (1961).
264. C. L. Kobrin, *Iron Age* **191**(3), 95 (1963).
265. F. Bergsma, *Ann. CIRP* **16**, 93 (1968).
266. R. H. Eshelman, *Iron Age* **190**(2), 109 (1962).
267. Anonymous, *Aircr. Prod.* **23**, 68 (1961).
268. Anonymous, *Aircr. Prod.* **25**, 2 (1963).
269. E. J. Krabacker, W. A. Haggerty, C. R. Allison, and M. F. Davis, in *International Research and Production Engineering*, ASME, New York (1963), p. 232.
270. M. A. LaBoda and M. L. McMillan, *Electrochem. Tech.* **5**, 340, 346 (1967).
271. M. A. LaBoda, U.S. Patent 3,669,858 (June 13, 1972).
272. J. P. Hoare and M. A. LaBoda, *Sci. Am.* **230**(1), 30 (1974).
273. J. P. Hoare, A. J. Chartrand, and M. A. LaBoda, *J. Electrochem. Soc.* **120**, 1071 (1973).
274. J. P. Hoare, K-W. Mao, and A. J. Wallace, *Corrosion (Houston)* **27**, 211 (1971).
275. K-W. Mao, *J. Electrochem. Soc.* **118**, 1870 (1971).
276. K. Chikamori and S. Ito, *Denki Kagaku* **38**, 492 (1970).
277. P. A. Brook, *Metals Mater.* **3**, 359 (1969).
278. F. Bergsma, *TNO Nieuws* **25**, 125 (1970); *CA* **73**, 126339t (1970); *Ann. CIRP* **16**, 93 (1968).
279. A. G. Atanasyants, A. F. Ivanovskii, D. Ya Dlugach, and Ts. O. Georgiev, *Zh. Fiz. Khim.* **48**, 366 (1974).
280. P. A. Brook and Q. Iqbal, *J. Electrochem. Soc.* **116**, 1458 (1969).
281. M. Kanda and T. Saji, *Kinzoku Hyomen Gijutsu* **24**, 687 (1973).

282. A. D. Davydov, A. D. Romashkan, M. A. Monina, and V. D. Kashcheev, *Elektrokhimya* **10**, 1681 (1974).
283. Yu. S. Volkov, M. A. Monina, and I. I. Moroz, *Elektron. Obrab. Mater.* **1972**(3), 11.
284. D. Landolt, *J. Electrochem. Soc.* **119**, 708 (1972).
284a. Hooker Chemical Co., Bulletin No. 99A, Niagara Falls, New York (1961).
284b. Manufacturing Chemists' Association, Chemical Safety Data Sheet SD-42, Safe handling and use of sodium chlorate, revised 1952.
284c. J. B. Darling and M. A. LaBoda, U.S. Patent 3,718,555 (February 27, 1973).
284d. Federal Register, Part II. Hazardous Materials Regulations, U.S. Department of Transportation, 49 CFR Parts 171-177, January 3, 1977.
285. J. P. Hoare, *Nature* **219**, 1034 (1968).
286. J. P. Hoare, *J. Electrochem. Soc.* **117**, 142 (1970).
287. J. P. Hoare, *Electrochim. Acta* **9**, 599 (1964); R. Thacker and J. P. Hoare, *J. Electroanal. Chem.* **30**, 1 (1971).
288. A. D. Romashkan, A. D. Davydov, V. D. Kashcheev, and B. N. Kabanov, *Elektrokhimya* **10**, 109 (1974).
289. A. D. Davydov, G. N. Korchagin, and V. D. Kashcheev, *Electron. Obrab. Mater.* **1975**(3), 9.
290. J. P. Hoare and C. R. Wiese, *Corros. Sci.* **15**, 435 (1975).
291. K. Chikamori and S. Ito, *Denki Kagaku Oyobi Butsuri Kagaku* **37**, 602 (1969).
292. D. Landolt, R. H. Muller, and C. W. Tobias, *J. Electrochem. Soc.* **118**, 36 (1971).
293. A. D. Davydov, R. A. Mirzoev, V. D. Kashcheev, and B. N. Kabanov, *Elektrokhimya* **8**, 1468 (1972).
294. M. Datta and D. Landolt, *Corros. Sci.* **13**, 187 (1973).
295. A. D. Davydov, *Elektrokhimya* **11**, 809 (1975).
296. T. P. Hoar, *Trans. Faraday Soc.* **33**, 1152 (1937); **45**, 683 (1949).
297. N. V. Sidgwick, *The Chemical Elements and Their Compounds*, Oxford University Press, London (1950), p. 1356.
298. T. P. Hoar, D. C. Mears, and G. P. Rothwell, *Corros. Sci.* **5**, 297 (1965).
299. D. T. Chin, *J. Electrochem. Soc.* **121**, 527 (1974).
300. L. Young, *Anodic Oxide Films*, Academic, New York (1961).
301. D. A. Vermilyea, *Adv. Electrochem. Electrochem. Eng.* **3**, 211 (1963).
302. J. J. Randall, W. J. Bernard, and R. R. Wilkinson, *Electrochim. Acta* **10**, 183 (1965).
303. K-W. Mao and D. T. Chin, *J. Electrochem. Soc.* **121**, 191 (1974).
304. D. T. Chin, *J. Electrochem. Soc.* **118**, 174 (1971).
305. D. T. Chin, *J. Electrochem. Soc.* **122**, 249 (1975).
306. T. Moeller, *Inorganic Chemistry*, Wiley, New York (1951).
307. T. P. Hoar, private communication.
308. J. P. Hoare and M. A. LaBoda, Electrochemical machining, in *Applied and Industrial Electrochemistry*, E. Yeager and A. Salkind, eds., Wiley, New York (1978), Chap. 2, p. 48.
309. J. P. Hoare and K-W. Mao, *Corrosion (Houston)* **29**, 143 (1973).
310. D. T. Chin, *J. Electrochem. Soc.* **119**, 1181 (1972).
311. H. Yamamoto and S. Ito, *Kikai Gijutsu Kenkyusho Shoho* **27**, 53 (1973); *CA* **79**, 37901b (1973).
312. Yu. N. Petrov *et al.*, *Elektron. Obrab. Mater.*, **1969**(2), 87.
313. I. K. Ross, G. C. Wood, and I. Mahmud, *J. Electrochem. Soc.* **113**, 334 (1966).
314. A. D. Davydov, V. D. Kashcheev, and B. N. Kabanov, *Elektrokhimya* **5**, 221 (1969); **6**, 1760 (1970); **8**, 1500 (1972).
315. K-W. Mao, *J. Electrochem. Soc.* **120**, 1056 (1973).
316. D. T. Chin and K-W. Mao, *J. Appl. Electrochem.* **4**, 155 (1974).
317. L. Franke and W. Forker, *Electrochim. Acta* **19**, 27 (1974).
318. E. S. Varenko, V. P. Galushko, and P. P. Kristopa, *Electron. Obrab. Mater.* **1972**(3), 6.
319. D. O. Condit, *Corrosion (Houston)* **28**, 95 (1972).
320. K-W. Mao and J. P. Hoare, *Corros. Sci.* **13**, 799 (1973).

321. K. Chikamori and S. Ito, *Denki Kagaku* **39**, 493 (1971).
322. V. N. Misra and V. M. Dokras, *Trans. Indian Inst. Met.* **27**, 156 (1974); *CA* **81**, 144583n (1974).
323. M. Pourbaix, *Corrosion (Houston)* **26**, 431 (1970).
324. M. G. Fontana and N. D. Greene, *Corrosion Engineering*, McGraw-Hill, New York (1967), pp. 39–58.
325. H. W. Bredin, *Machinery* **66**(11), 113 (1960).
326. A. H. Mekela, *Sci. J. (London)* **3**(1), 50 (1967).
327. F. Bergsma, *TNO Nieuws* **22**, 128 (1967).
328. D. T. Chin and A. J. Wallace, *J. Electrochem. Soc.* **118**, 831 (1971).
329. D. T. Chin, *J. Electrochem. Soc.* **118**, 818 (1971).
330. H. E. Haring and W. Blum, *Trans. Electrochem. Soc.* **44**, 123 (1923).
331. R. V. Jelinek and H. F. David, *J. Electrochem. Soc.* **104**, 279 (1957).
332. T. P. Hoar and U. R. Evans, *Trans. Electrochem. Soc.* **134**, 2476 (1932).
333. M. Cohen and A. F. Beck, *Z. Electrochem.* **62**, 696 (1958).
334. N. Shikata and S. Ito, *Kikai Gijutsu Kenkyusho Shoho* **28**, 8 (1974); *CA* **81**, 20065a (1974).
335. S. V. Ventsel and B. V. Indin, *Elektron. Obrab. Mater.* **1975**(3), 18.
336. M. A. LaBoda, A. J. Chartrand, J. P. Hoare, C. R. Wiese, and K-W. Mao, *J. Electrochem. Soc.* **120**, 643 (1973).
337. M. A. Evseeva, Ya. A. Fedyshkima, and A. I. Levin, *Zh. Prikl. Khim.* **40**, 1380 (1967).
338. Yu. N. Petrov and V. V. Parshutin, *Elektron. Obrab. Mater.* **1972**(4), 22.
339. A. I. Levin, M. A. Evseeva, and I. G. Zhogov, *Zh. Prikl. Khim.* **39**, 584 (1966).
340. G. N. Zaidman, I. N. Verkhovelskii, *Elektron. Obrab. Mater.* **1971**(2), 17.
341. M. A. Monina and I. I. Moroz, *Elektron. Obrab. Mater.* **1972**(3), 11.
342. A. D. Davydov *et al.*, *Elektron. Obrab. Mater.* **1973**(6), 28.
343. A. D. Davydov, V. D. Kashcheev, and R. A. Mirzoev, *Fiz. Khim. Obrab. Mater.* **1973**(6), 32; **1968**(5), 40.
344. N. A. Amirkhanova *et al.*, *Elektron. Obrab. Mater.* **1972**(4), 19.
345. M. Datta and D. Landolt, *J. Electrochem. Soc.* **122**, 1466 (1975).
346. J. P. Hoare, *Nature* **241**, 44 (1973).
347. D. E. Davies and W. Barker, *Corrosion (Houston)*, **20**, 47t (1964).
348. A. D. Davydov, A. Kh. Karimov, and L. M. Varenko, *Elektron. Obrab. Mater.* **1974**(3), 19.
349. Yu. S. Volkov, M. A. Monina, and I. I. Moroz, *Elektron. Obrab. Mater.* **1972**(3), 11.
350. E. Ya. Grodzinskii, *Fiz. Khim. Obrab. Mater.* **1968**(5), 52; *CA* **70**, 25081k (1969).
351. D. R. Pieri, D. A. Christensen, and F. J. Clauss, *Am. Machinist* **115**(8), 113 (1971).
352. J. Van Impe, J. P. Rombaux, and J. Cl. Etienne, *Rev. Universelle Mines Met. Mec.* **116**, 242 (1973); *CA* **80**, 55200x (1974).
353. O. A. Arzhintar *et al.*, *Electron. Obrab. Mater.* **1974**(6), 9.
354. O. M. Tatarinova, N. A. Amirkhanova, and A. G. Akhmadiev, *Electron. Obrab. Mater.* **1974**(3), 12.
355. C. F. Noble and S. J. Shine, *J. Appl. Electrochem.* **5**, 215 (1975).
356. A. G. Atanasyants, D. Ya. Dlugach, and V. G. Reshetnyak, *Zh. Fiz. Khim.* **47**, 2953 (1973).
357. A. N. Amirkhanova *et al.*, *Electron. Obrab. Mater.* **1975**(1), 15.
358. V. N. Karavainikov, *Akust. Ul'trazvukovaya Tekh.* **1972**(7), 30; *CA* **79**, 12767z (1973).
359. B. G. Petrescu and M. Minca, *Constr. Masini* **23**, 29 (1971); *CA* **75**, 93770n (1971).
360. V. D. Kashcheev, N. S. Merkulova, and A. D. Davydov, *Electron. Obrab. Mater. Akad. Nauk Mold. SSR* **1966**(5), 35; *CA* **68**, 26202w (1968).
361. A. L. Vishnitskii, E. A. Drozd, and R. A. Mirzoev, *Nov. Elektrofiz. Elektrokhim. Obrab. Mater.* **1972**, 25; *CA* **78**, 78821y (1973).
362. Yu. N. Petrov, G. N. Zaidman, and B. P. Saushkin, *Electron. Obrab. Mater.* **1974**(5), 17.
363. B. P. Saushkin, *Electron. Obrab. Mater.* **1975**(2), 21; **1975**(3), 14.
364. H. Karafuji, *Ann. CIRP* **13**, 313 (1966).
365. C. Jackson and R. D. Olson, ASTME Engineering Conferences, MR 69-109 (1969).

366. M. A. LaBoda and J. P. Hoare, *J. Electrochem. Soc.* **122**, 1489 (1975).
367. Anonymous, Catalog 965, Setco Industries, Inc., Cincinnati, Ohio (1965).
368. Anonymous, Comprehensive Catalog, Everite Machine Products Co., Philadelphia, Pennsylvania.
369. R. B. Massad, Electro-abrasive machining, in American Society for Abrasive Methods, Fifth National Technical Conference, Pittsburgh, Pennsylvania (1966).
370. Anonymous, *Steel* **155**(25), 42 (1964).

# 9
# Theory of the Structure of Ionomeric Membranes

## ANTON J. HOPFINGER and KENNETH A. MAURITZ

## 1. Introduction

Ionomers are two-phase materials which show considerable promise as ion-selective membranes. Nafion,† in particular, is being used on a limited commercial basis in chloralkali separation processes. Its chemical structure, shown in Figure 1, is typical of ionomers. Pendant side chains, each containing an ionizable group, are essentially spaced uniformly along a linear chain composed of relatively nonpolar groups. The number of nonpolar units is far greater than the number of polar-ionizable groups. This leads to the particular two-phase structure. Polar groups self-associate to form hydrophilic clusters in a sea of nonpolar material. In general, ionomeric structure has been studied by X-ray diffraction, dielectric and mechanical spectroscopies, and by electron microscopy.[1-3] The polar clusters of Nafion have also been investigated by infrared[4] and NMR[5] spectroscopies. Figure 2 shows a transmission electron micrograph of Nafion stained with silver ions. The clusters appear as dark dots.

The two-phase structure of an ionomer is sensitive to its physicochemical state and chemical structure. In particular, exposure to pure solvent and ionic

† Registered trademark of E. I. DuPont de Nemours & Co. (Inc.) for its perfluorosulfonic acid products.

---

**ANTON J. HOPFINGER** • Department of Macromolecular Science, Case Western Reserve University, Cleveland, Ohio 44106.  **KENNETH A. MAURITZ** • T. R. Evans Research Center, Diamond Shamrock Corporation, Painesville, Ohio 44077.

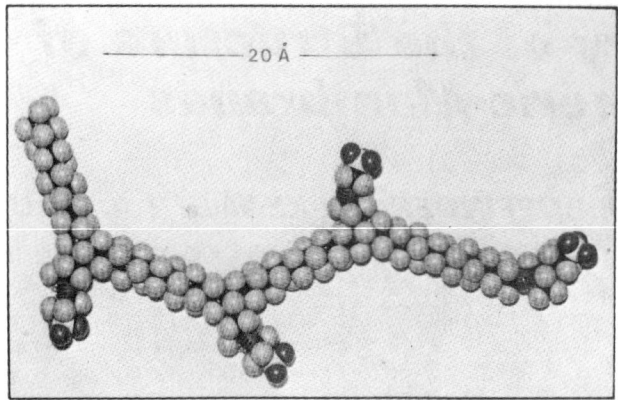

Figure 1. Chemical structure of Nafion.[8]

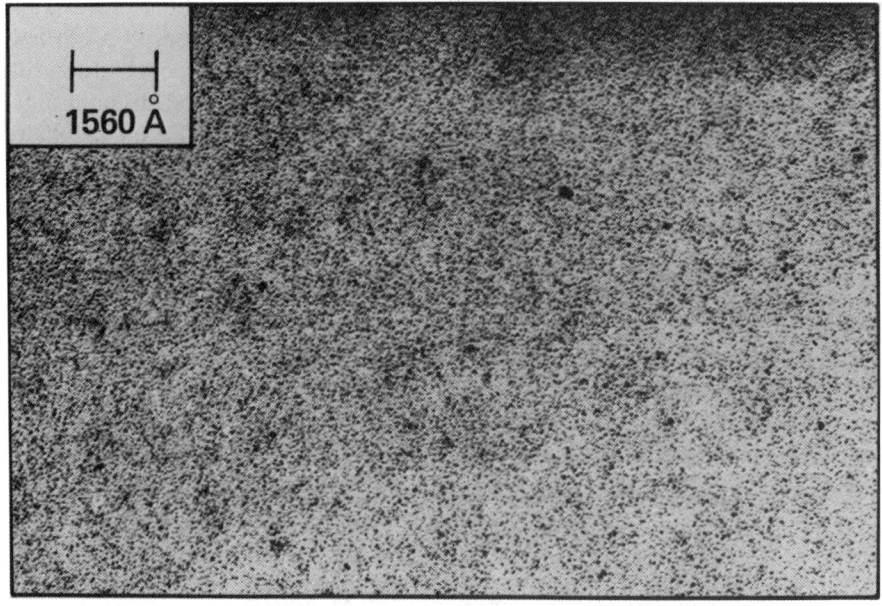

Figure 2. Transmission electron micrograph of Nafion stained with silver ions.

solutions alter both the structural and thermodynamic features of the material. Relatively little theoretical work has been done to understand, or model, these features of ionomers. Eisenberg[6] has put forth an initial theory of ionomer structure which contains some concepts of general merit. His theory has been consulted extensively in our work. Ponomarev and Ionova[7] have attempted to construct a statistical mechanical model to describe the thermodynamic behavior of ionomers. Gierke[8] and Cutler[9] have independently described theories of ion transport for Nafion based upon specific molecular organizations.

In this report we present two ionomer models. The first is a thermomechanical model which describes cluster formation.[10] The second model describes the thermodynamic and structural properties of an existing cluster under varying conditions in which a statistical mechanical formalism is employed. Water is the solvent assumed in both theories.

## 2. Cluster Formation Model

The formalism includes cluster formation under three different physicochemical states: (a) dry, (b) exposed to water, and (c) exposed to an aqueous ionic solution.

### 2.1. General Mechanism of Cluster Formation

For each physicochemical state we restrict ourselves to an equilibrium representation of ionomer structure. On a time average basis, the polar groups on the ionomer side chains are able to "see" one another through their dipolar interactions. Monte Carlo simulation calculations[11] suggest that the configurational dipole–dipole interaction free energy for unrestricted dipolar motion is of the form

$$F = C \ln (n_c - 1) \tag{1}$$

where $C$ is a constant characteristic of the lifetime of the dipoles, strength of the dipoles, and dielectric medium. The number of interacting dipoles is $n_c$.

The self-association of polar groups through dipolar interactions requires expansions and contractions to occur in the backbones of the polymer chain molecules. If we assume this process is random, to the extent that the chain expansions and contractions are equal in number, then the elastic deformation energy per dipole is

$$W = \frac{3kT}{2\langle h^2 \rangle}(\Delta d)^2 \tag{2}$$

where $T$ is the absolute temperature, $\Delta d$ the average chain expansion–contraction, and $\langle h^2 \rangle$ the mean square end-to-end chain dimension. Equation

(2) is valid only for a Gaussian chain. However, we can extend the applicability of Eq. (2) by defining $\Delta d$ in terms of the physicochemical state of the ionomer, and by generalizing the definition of $\langle h^2 \rangle$, for saturated carbon backbone ionomers, to

$$\langle h^2 \rangle = N\left[\left(\frac{n-1}{2}\right)l\right]^\delta \frac{1 - \cos \alpha_0}{1 + \cos \alpha_0} \tag{3}$$

where $N$ is the number of monomer units per chain, $n$ the number of $CX_2$ units per monomer backbone chain, $l$ the trans-distance for the CCC backbone groups, $\alpha_0$ the CCC backbone valence bond angle, and $\Delta$ is a chain-ordering parameter, which is 2 for Gaussian chains. The correct determination of $\langle h^2 \rangle$ is critical to the quantitative implementation of the model, since cluster size is a strong function of $\langle h^2 \rangle$.

The size of a cluster, in terms of $n_c$, for the most simple representation can be found by solving

$$F = W + T \Delta S(n_c) \tag{4}$$

where $\Delta S(n_c)$ is the change in the entropy of the ionomer due to clustering. $\Delta S(n_c)$ can reasonably be equated to the loss in the configurational entropy of a typical chain due to clustering. When normalized to a per dipole scale, the loss in configurational entropy is given by

$$S(n_c) = -\frac{n_c}{N} R \ln[N!(N - n_c)!] \tag{5}$$

where $N$ = (chain molecular weight)/(monomer molecular weight). The right-hand side of Eq. (5) can be simplified using Sterlings' approximation. Unfortunately, additional thermomechanical contributions must be considered in Eq. (4) to describe adequately each physicochemical state of the ionomer.

## 2.2. The Dry State

The polar groups on the pendant side chains, whether neutral or charged, are the interacting dipoles promoting cluster formation. The only additional energy term which needs to be added to Eq. (4) is a destabilizing term on the right to account for the "surface tension" between the cluster and nonpolar medium. Equation (4) becomes, for a spherical cluster,

$$F + \varepsilon r^2(n_c) = W + T \Delta S(n_c) \tag{6}$$

where $r(n_c)$ is the radius of the cluster and $\varepsilon = 4\pi\varepsilon_0$, where, in turn, $\varepsilon_0$ is the surface energy per unit area.

## 2.3. Exposure to Water

Water molecules diffuse into the ionomer upon exposure to an aqueous environment. Clusters may or may not have already formed, depending upon the history of the ionomeric materials as well as the relative rates of aqueous diffusion and cluster formation. We restrict ourselves to the case in which clusters are not formed prior to solvating. In this situation we first must consider the hydrated polar group structure, which is termed a *hydration shell*. The size of the hydration shell is postulated to result from a free energy balance. Polar group$\cdots H_2O$ interactions, coupled with $H_2O \cdots H_2O$ intra-hydration shell interactions, are balanced against the elastic deformation energy needed to move chain segments out of the way of the growing hydration shell. In addition, a surface tension term must again be included to account for the aqueous–nonpolar interface generated by the growing hydration shell. The general form of this energy balance, assuming a spherical hydration shell, is

$$K_e[r_h(n_h)^m - r_p^m] + \varepsilon' r_h^2 = n_b A_1 + (n_h - n_b) A_0 \tag{7}$$

where the first term on the left in Eq. (7) is a generalized deformation energy in which $K_e$ is the elastic force constant, $r_h(n_h)$ is the radius of the hydration shell containing $n_h$ water molecules in the hydration shell, and $r_p$ is the unsolvated radius of the polar group. The second term on the left is the interfacial surface tension term in which $\varepsilon'$ is the $H_2O$–nonpolar surface tension energy per unit area. The first term on the right is the free energy of hydration of the polar group where $n_b$ is the effective number of first layer, hydrating water molecules. $A_1$ is the average free energy of hydration per water. The second term on the right is free energy of hydration per water. The second term on the right is free energy gain due to $H_2O \cdots H_2O$ hydration shell interactions with $A_0$ being the average free energy of interaction per $H_2O$.

The hydration shells are postulated to now come together, through dipolar interactions, to form a hydrated cluster. The form of Eq. (6) can be used once a term on the left is added to account for a possible loss in stabilization free energy due to the expulsion of water molecules from overlapping hydration shells. This loss in energy can, however, be compensated by interhydration shell $H_2O \cdots H_2O$ interactions and a net decrease in the total $H_2O$–nonpolar interfacial area. In total Eq. (6) now becomes

$$F' + \varepsilon' r^2(n_c) + O'(n_c, n_h, F') = W' + T\,\Delta S(n_c) + O'(n_c, n_h, A_0) \tag{8}$$

where the expulsion overlap function is of the form

$$O(n_c, n_h, F') = \phi_1 \Delta R^2(n_c, n_h, F') r_h(n_c) + \phi_2 \Delta R^3(n_c, n_h, F') \tag{9}$$

in which $\phi_1$ and $\phi_2$ are molecular constants characteristic of the chemical groups present, and $\Delta R$ is the radial overlap distance of two hydration shells. $O'(n_c, n_h, A_0)$, the interhydration $H_2O \cdots H_2O$ energy term, is of the same form as Eq. (9), but with different $\phi$'s which depend upon $A_0$.

The primes associated with $F$ and $W$ in Eq. (8) indicate that the numerical forms of these expressions differ from those of Eq. (6). The effective dipole of a hydration shell is due to the polar group plus the net dipolar arrangement of the associated water molecules. Hence, $F'$ is different from $F$. The elastic work, $W$, requires moving hydration shells as well as polymer chains. Thus $W'$ is not equal to $W$.

### 2.4. Exposure to Aqueous Ionic Solutions

Let the ion–counterion interactions of an ionomer in solution be represented by

$$[-G^{(-)} \to I^{(+)}] \underset{k_2}{\overset{k_1}{\rightleftharpoons}} [-G^{(-)} + H^{(+)}], \quad \text{that } k_2 \gg k_1$$

In this model, $-G^{(-)}$ is the anionic form of the polar side-chain unit, and $I^{(+)}$ is the mobile solution cation. The same theory is applicable if a cationic polar side-chain unit and mobile anion are considered. The second theory described in this report eliminates the $k_2 > k_1$ constraint for an *existing* cluster. Nevertheless, the size of the polar clusters for $I$ can be found by solving Eq. (8). However, the numerical constants in the equation will be different from those of the purely aqueous solution situation. The effective hydration shell dipole depends now also upon the type of mobile cation and its solution concentration.

The number of hydration shells which have a bound $I^{(+)}$ should increase with increasing ion solution concentration. The functional relationship will be unique for each choice of ion and/or ionomer. In the case of NaCl solutions and Nafion, each $G^{(-)}$ site has one associated $Na^{+5}$. The presence of $I^{(+)}$ can alter the organization of the hydration shell water molecules. Thus water molecules may be lost or gained, which will change the size of the hydration shell. These factors must be considered in any numerical application of Eq. (8) to describe cluster formation in ionomers exposed to aqueous ionic solutions. A conceptual picture of cluster formation for an ionomer, using Nafion as the example, is shown in Figure 3.

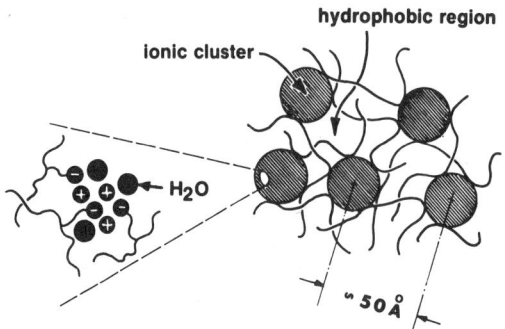

Figure 3. A conceptual picture of cluster formation for an ionomer, using Nafion as the example.

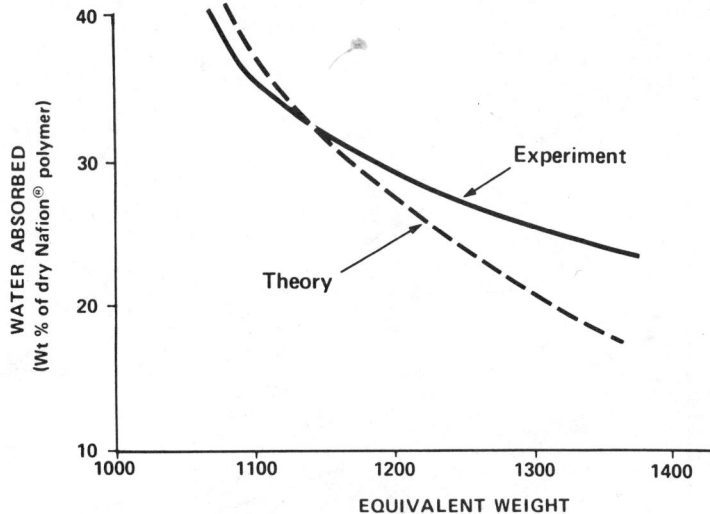

Figure 4. Plots of water absorption for the sulfonic acid form of Nafion in contact with water as a function of ionomer equivalent weight.

## 2.5. Application of the Theory to Nafion

The thermomechanical theory described above has been quantitatively applied to Nafion.[10,11] A generalized computer program to implement the theory has been written. The necessary structural and thermodynamic parameters were obtained from experiments and/or by independent molecular structure calculations and are reported in Reference 12. Results of some of these calculations are presented here.

Figure 4 contains plots of water absorption for the sulfonic acid form of Nafion in contact with water as a function of ionomer equivalent weight.† Both experimentally measured[12] and theoretically predicted water absorption–equiv. wt. curves are presented. Admittedly, the theoretical curve deviates from the experimental data, especially at increasing equivalent weights. However, in the range of major commercial interest (i.e., equiv. wt. = 1000–2000), the agreement between theory and experiment is good ($\pm 6\%$).

Plots of water absorption of Nafion, and wet density of the sodium sulfonate form of the ionomer, as a function of the solution molarity of $Na^+$ ions are shown, respectively, in Figures 5 and 6 for both 1100 and 1200 equiv. wt. materials. As with Figure 4, both theoretical predictions and experimental data are presented to demonstrate the ability of the calibrated model to quantitatively describe the macroscopic equilibrium solution behavior of standard Nafion ionomers.

† Equivalent weight is the average molecular weight of the repeating chemical units in the ionomer chain.

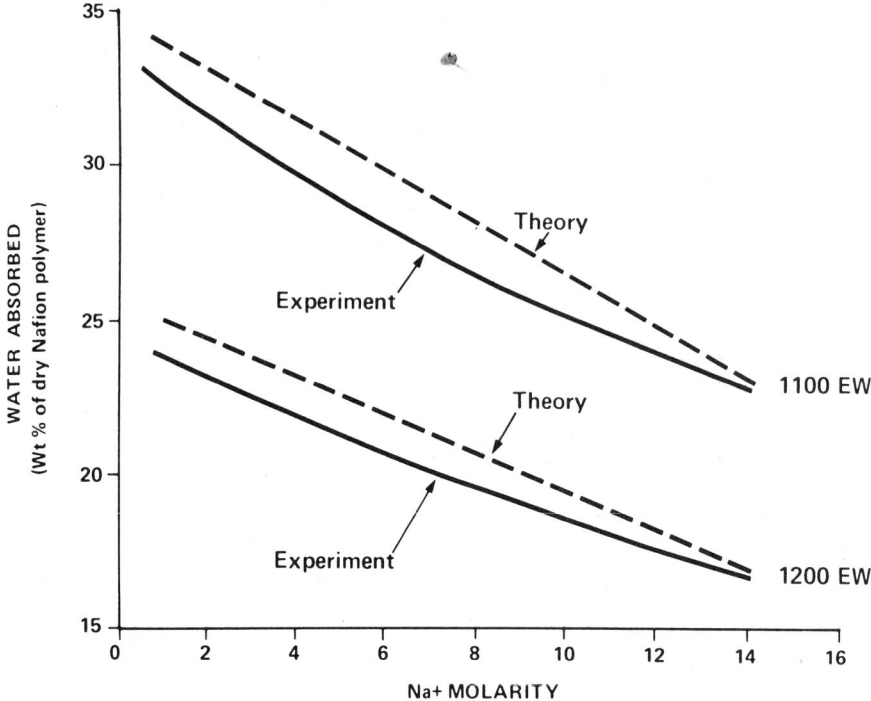

Figure 5. Plots of water absorption of the dry Nafion polymer as a function of the solution molarity of $Na^+$ ions, for both 1100 and 1200 equiv. wt. materials. (EW = equiv. wt.)

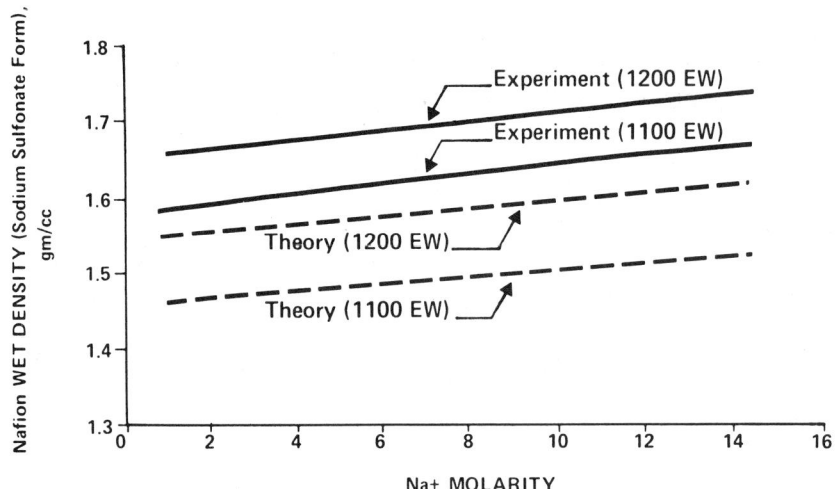

Figure 6. Plots of wet density of the sodium sulfonate form of Nafion as a function of the solution molarity of $Na^+$ ions, for both 1100 and 1200 equiv. wt. materials. (EW = equiv. wt.)

## 3. Cluster Property Model

From a thermodynamic standpoint, the equilibrium membrane structure must depend, in part, upon the internal osmotic pressure which is determined by the water activity, $\bar{a}_w$, within the microscopic cluster regions. $\bar{a}_w$, in turn, should be a function of the relative population of unpaired ions and free water molecules in the cluster solution.

Fourier transform infrared[4] and nuclear magnetic resonance[5] spectroscopies have been used to monitor the phenomenon of counterin association with the negative sulfonate moieties in Nafion membranes in the salt form. Co-ion effects were eliminated by examining membranes containing no excess electrolyte.

The $SO_3^-$ symmetric stretching mode, which occurs around 1060 cm$^{-1}$, is significantly affected by the $H_2O/SO_3^-$ molar ratio as well as monovalent counterion radius.[4] For the series [$Rb^+, K^+, Na^+, Li^+$], this peak has been observed to broaden and shift to higher frequencies with decreasing water content, the magnitudes of the shift decreasing with bare ionic size. In addition, $^{23}Na$ NMR spectra are seen to display an increasing linewidth and chemical shift with decreasing water content.[5] In particular, the onset of both the ir and NMR spectral changes occur at water contents that provide barely enough molecules for complete ionic hydration. This effect is attributed to an increasing population of counterions that are strongly interacting with the sulfonate moiety.[4]

A four-state model, motivated in part by the multistep mechanism for the dissocation of ion pairs in electrolyte solutions, as proposed by Eigen *et al.*,[13,14] is based on the ionic-hydrate associations that are depicted in Figure 7. The states are classified as (1) completely dissociated hydrated ion pairs, (2) ion pairs at the contact of undisturbed primary hydration shells, (3) outer, and (4) inner sphere complexes. The relative populations of these states ($P_i$; $i = 1-4$) have been determined from the use of Boltzmann statistics utilizing specific solvation and interionic energetic formulations. In addition to the direct quantitative incorporation of hydration energetics, the model differs from most

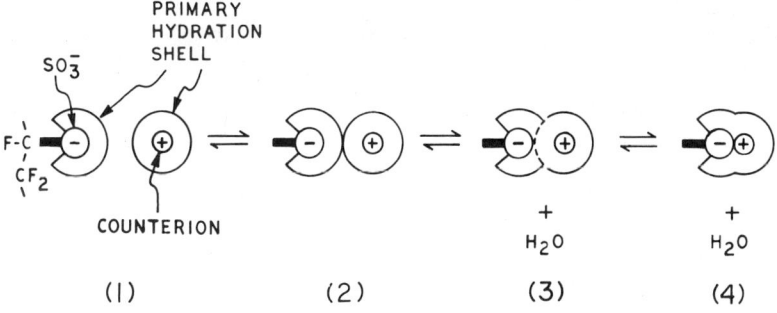

Figure 7. Schematic representation of ionic-hydrate associations.

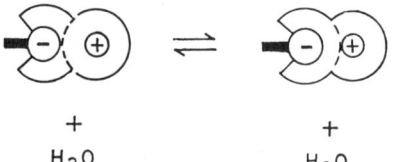

Figure 8. Schematic representation of the two alternate modes of hydration shell overlap.

classical approaches that are based on the theory of associations of Bjerrum[15] insofar as a discrete set of well-specified states of ion–water "complexes" is considered.

The desolvation energetics encountered in passing between states 2 and 3, and 3 and 4, are computed by the utilization of a simple model in which the energy change accompanying the formation of a given ionic association is proportional to the change in the fractional degree of volume overlap between the primary hydration shells of the ionogenic side groups and counterions. The requisite number of ejected water molecules accompanying the overlap is then taken to be the product of this volume ratio and the original molecular population of the penetrated hydration shell. On the average, each displaced water molecule results in a hydrative energy loss of $H_{\pm}^{+}/n_{\pm}^{+}$, where $H_{\pm}^{+}$ and $n_{\pm}^{+}$ are the experimentally determined hydration energies and numbers, respectively, of the given ion. State 3 is considered as a hybridization, that is, energetically weighted mixture of the two alternate modes of hydration shell overlap depicted in Figure 8. The cation–$SO_3^-$ interaction is represented by a simple modification of Coulomb's law for two point charges of opposite sign that are at the energetically favorable separation for the given ionic-hydrate complex. Also, a "molecular" dielectric constant, $K$, that varies linearly with $n$, the average number of water molecules available to a cation–$SO_3^-$ pair, as shown in Figure 9, is utilized in the formulation. Thus, as the water content drops below a level at which completely populated hydration shells can exist, the electrostatic shielding effect of the water becomes less effective, resulting in enhanced cation–$SO_3^-$ attraction.

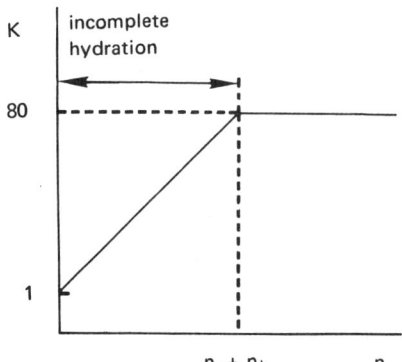

Figure 9. Plot of molecular dielectric constant $K$ as a function of $n$, the average number of water molecules available to a cation–$SO_3^-$ pair.

Finally, the relative population of the various states can be derived from the corresponding energetics by

$$P_i = Q^{-1} \exp(-E_i/RT), \quad i = 1, 2, 3, 4 \tag{10}$$

where

$$Q = \sum_{j=1}^{4} \exp(-E_j/RT) \tag{11}$$

$E_1$ is the combined hydration and coulombic energy per given ion pair for the state $i$ and $T$, the Kelvin temperature. $P_1$ is obviously the theoretical degree of dissociation, and the overall percent of bound cations is $(1 - P_1)$ times 100%.

The calculated shift in the spectrum of states with decreasing water content for the Na$^+$ form is shown in Figure 10. $P_1$ drops steadily after a value of $n$, corresponding to an approximate combined hydration number of a NaSO$_3^-$ pair is reached. State 3, the outer sphere completed, becomes increasingly populated and becomes physically identical to state 2 after the water content has dropped to about 4 mol of water per equivalent of resin. The inner sphere complex, or contact ion pair, while possessing the greatest electrostatic binding, has a population density of practically zero, however, because of the

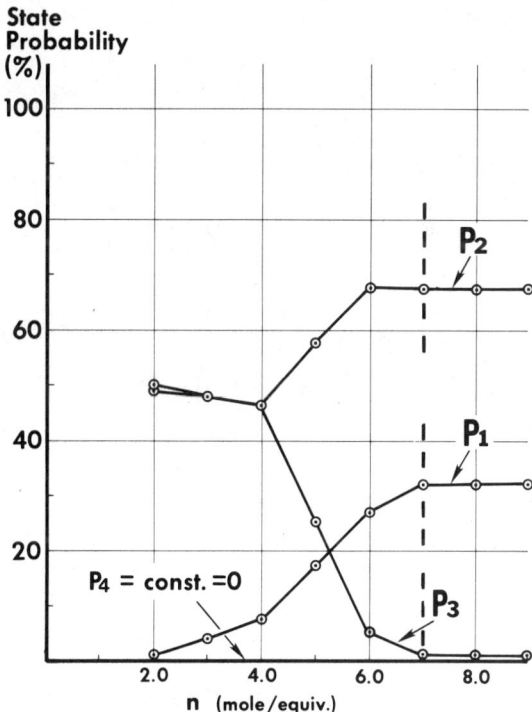

Figure 10. Calculated shift in the spectrum of states with decreasing water content for the Na$^+$ form.

difficulty in removing the final interposed water molecules that must interact with both the $Na^+$ ion and $SO_3^-$ moiety.

Plotted in Figure 11 are (a) the bound percent of $Na^+$ ions and (b) $B$, the overall population percentage of $Na^+$ ions having insufficient thermal kinetic energy necessary to overcome electrostatic binding to the sulfonate group. The situation depicted in Figure 11, i.e., a progressively greater proportion of intimate cation–sulfonate interactions with diminishing internal water content, is at least qualitatively consistent with the physical model arising from spectroscopic interpretation.

If the resin "solution" were ideal, in a thermodynamic sense, then the internal water activity would follow from Raoult's Law, i.e.,

$$\bar{a}_w = n/(n+1) \tag{12}$$

However, since osmotic effects are primarily determined by the mixing of *free water* with *free ions*, $\bar{a}_w$, computed by Eq. (12), must be incorrect because (1) there exists a finite population, $P_2 + P_3 + P_4$, of dissociated or "free" cations, and (2) of the $n$ water molecules per ion exchange site, a certain portion will be constrained in a state of hydrative binding.

Adopting the more realistic definition of the internal water activity as the ratio of the number of moles of free water to the number of moles of free water

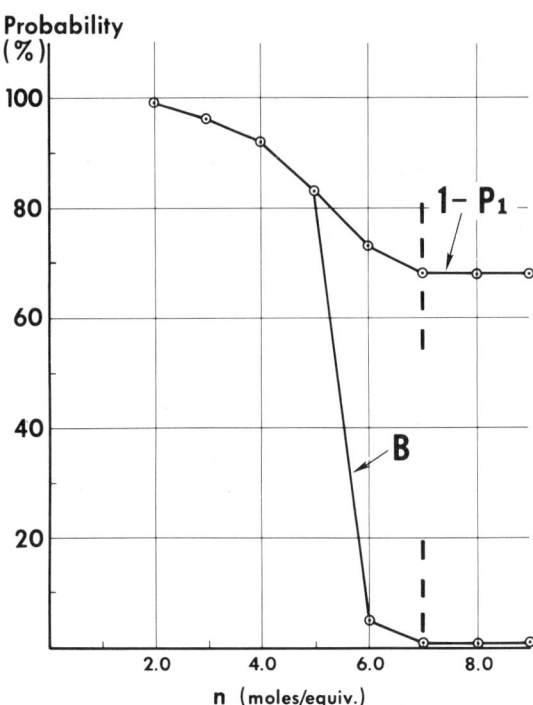

Figure 11. Plots of $1 - P_1$, the bound percent of $Na^+$ ions, and $B$, the overall population percentage of $Na^+$ ions having insufficient thermal kinetic energy necessary to overcome electrostatic binding to the sulfonate group.

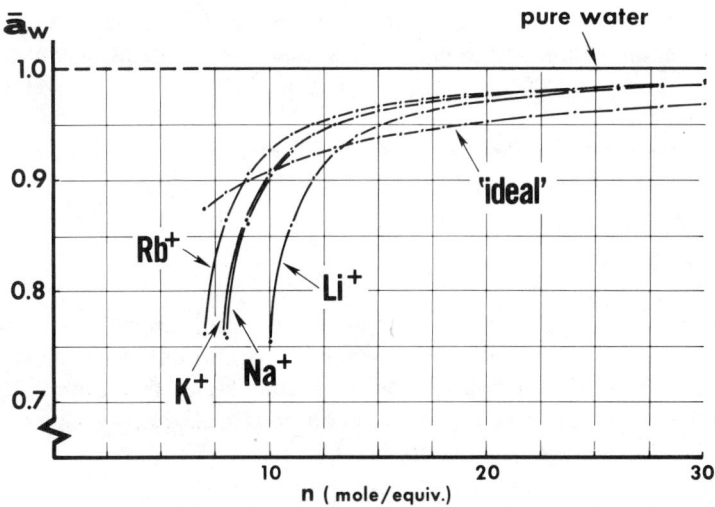

Figure 12. Comparison of theoretically determined internal water activities for membranes in the $Li^+$, $Na^+$, $K^+$, and $Rb^+$ cations over a range of water contents.

plus moles of free cation–sulfonate pairs per equivalent of resin, Eq. (12) becomes

$$\bar{a}_w = n_{\text{free}}/(n_{\text{free}} + P_1) \tag{13}$$

where

$$n_{\text{free}} = n - (n_+ + n_-)(P_1 + P_2) - P_3\langle N_3\rangle - P_4\langle N_4\rangle \tag{14}$$

In addition to $n_+$ and $n_-$, which are the primary hydration numbers of cation and anion, respectively, $\langle N_3\rangle$ and $\langle N_4\rangle$, the average number of water molecules bound in the hydrative associations of states 3 and 4 are also computed using the model for the association–dissociation equilibrium between bound and unbound cations described previously.

A comparison of theoretically determined internal water activities for membranes in the $Li^+$, $Na^+$, $K^+$, and $Rb^+$ forms, over a range of water contents, is provided in Figure 12. For increasingly large $n$, differences between cationic forms becomes negligible, and $\bar{a}_w$ asymptotically approaches 1 (dilution effect). At low $n$, however, a systematic differentiation is apparent wherein the water activities increase in the following progression: $Rb^+ > K^+ > Na^+ > Li^+$. Since, in the process of equilibration with pure water ($\bar{a}_w = 1$), the gradient of water activity across the membrane–solution interface increases in the reverse ionic order; the internal osmotic pressure will be greatest for the $Li^+$ and lowest for the $Rb^+$ form. It is, in fact, observed experimentally that the equilibrium water uptake of Nafion membranes follows the order $Li^+ > Na^+ > K^+ > Rb^+$. The primary cause of this behavior is the ability of the smaller cations to engage a larger number of water molecules in a state of hydration, and thus prevent them from effectively participating in osmotic swelling effects. However, to restore chemical equilibrium, additional water must enter the membrane and result in greater swelling as compared to larger cationic forms. The exact water uptake, of course, also depends on the available limit of Nafion polymer network deformation.

## 4. Summary

A comprehensive theory to explain the equilibrium molecular structures of ionomers under different physicochemical conditions has been presented. Changes in the geometric and thermodynamic states of an ionomer due to exposure to aqueous and aqueous ionic solutions are explicitly considered. The theory has been quantitatively applied to Nafion. A comparison of predicted and observed thermomechanical and spectroscopic properties of Nafion suggest the theory is realistic.

## Auxiliary Notation

$F$ configurational dipole–dipole interaction free energy
$n_c$ number of interacting dipoles
$\Delta d$ average chain expansion–contraction
$\langle h^2 \rangle$ mean square end-to-end chain dimension
$N$ number of monomer units per chain
$n$ number of $CX_2$ units per monomer backbone chain
$l$ trans-distance for the CCC backbone groups
$\alpha_0$ the CCC backbone valence bond angle
$\delta$ a chain-ordering parameter
$W$ elastic deformation energy per dipole
$\Delta S(n_c)$ change in entropy of the ionomer due to clustering
$r(n_c)$ radius of the cluster
$\varepsilon_0$ surface energy per unit area
$\varepsilon$ $4\pi\varepsilon_0$
$K_e$ elastic force constant
$r_h(n_h)$ radius of the hydration shell containing $n_h$ water molecules in the hydration shell

$r_p$ unsolvated radius of the polar group
$\varepsilon'$ $H_2O$–nonpolar surface tension energy per unit area
$n_b$ effective number of first layer, hydrating water molecules
$A_1$ average free energy of hydration per $H_2O$
$A_0$ average free energy of interaction per $H_2O$
$\phi_1, \phi_2$ molecular constants characteristic of the chemical groups present
$\Delta R$ radial overlap distance of two hydration shells
$W'$ elastic work
$G^{(-)}$ anionic form of the polar side-chain unit
$I^{(+)}$ mobile solution cation
$H_{-}^{+}$ experimentally determined hydration energies
$n_{-}^{+}$ experimentally determined hydration numbers
$K$ a molecular dielectric constant
$E_i$ combined hydration and coulombic energy per given ion pair for the state $i$

$P_i$ the theoretical degree of dissociation

$n_+$ primary hydration number of cation

$n_-$ Primary hydration number of anion

$\langle N_x \rangle$ average number of water molecules bound in the hydrative association of state $X$.

## References

1. S. C. Yeo and A. Eisenberg, *J. Appl. Polym. Sci.* **21**, 875 (1977).
2. A. Eisenberg and M. King, in *Ion-Containing Polymers*, Vol. 2. *Physical Properties and Structure*, R. S. Stein, ed., Academic, New York (1977).
3. L. Holliday, ed., *Ionic Polymers*, Applied Science Publishers Ltd., London (1975).
4. K. A. Mauritz and S. R. Lowry, *Polym. Prep. Am. Chem. Soc. Div. Polym. Chem.* **19**(2), 336 (1978).
5. R. A. Komoroski and K. A. Mauritz, *J. Am. Chem. Soc.* **100**, 7487 (1978).
6. A. Eisenberg, *Macromolecules* **3**, 147 (1970).
7. O. A. Ponomarev and I. A. Ionova, *Vysokomol. Soedin. Ser. A* **16**, 1023 (1974).
8. T. D. Gierke, Ionic clustering in Nafion perfluorosulfonic acid membranes and its relationship to hydroxyl rejection and chlor-alkali current efficiency, presented at 152nd National Meeting, The Electrochemical Society, Atlanta, Georgia, October 10–14, 1977.
9. S. G. Cutler, *Polym. Prep. Am. Chem. Soc. Div. Polym. Chem.* **19**(2), 330 (1978).
10. K. A. Mauritz, C. J. Hora, and A. J. Hopfinger, *Polym. Prep. Am. Chem. Soc. Div. Polym. Chem.* **19**(2), 324 (1978).
11. K. A. Mauritz, C. J. Hora, and A. J. Hopfinger, in *Ions in Polymers, Advances in Chemistry Series*, A. Eisenberg, ed., No. 187, American Chemical Society, Washington, D.C. (1980), p. 123.
12. W. G. F. Grot, G. E. Munn, and P. N. Walmsley, Perfluorinated ion exchange membranes, presented at the 141st National Meeting of the Electrochemical Society, Houston, Texas, May 7–11, 1972.
13. H. Diebler and M. Eigen, *Z. Phys. Chem. (Frankfurt)* **20**, 299 (1959).
14. M. Eigen and K. Tamm, *Z. Elektrochem.* **66**(93), 107 (1962).
15. N. Bjerrum, *Kgl. Danske Videnskab. Mat-Fys. Medd.* **7**(9) (1926).

# 10

# Electrodeposition of Paint

*FRITZ BECK*

## 1. Introduction

Electrodeposition of paint (EDP) is a process, in which an electronically conducting substrate is dipped into an aqueous solution of water-borne paint. In most of the cases, the binder is an organic polymer which contains carboxylic groups, thus forming a polyelectrolyte, which provides, at least in the partially neutralized state, some ionic conductivity to the solution. If the substrate is polarized anodically versus a counterelectrode, for which the walls of the tank are usually used (cf. Figure 1), current flow through the system causes electrochemical reactions at the phase boundary substrate–bath, leading to electrocoagulation of the paint.

According to this definition, EDP is an electrochemical surface technique which combines polymer science and electrochemistry. More precisely, it can be classified as a part of organoelectrochemistry, where organic molecules interact with an electrochemical electrode. This process is now applied worldwide to provide the first layer of paint onto car bodies and other industrially manufactured metal mass ware. Today more than 1000 big electrocoat installations are operating for primers and one-coat paint systems. Every year, an area of more than 1000 square miles of paint is now deposited by this process. With an average thickness of about 1 mil, this means an annual turnover of about $10^5$ tons of organic material at electrochemical electrodes. This is the

---

*FRITZ BECK* • University of Duisburg, Gesamthochschule, FB 6 Elektrochemie, D 4100 Duisburg 1, West Germany.

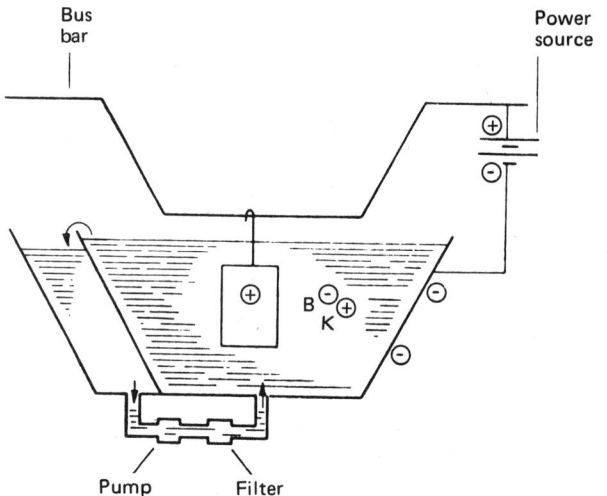

Figure 1. Schematic representation of a tank for anodic EDP process.

same order of magnitude as the largest electro-organic process, the electrosynthesis of adiponitrile.

Another "electric" painting process which has been used for a long time is the electrostatic spray coating. The electric field strength **F** is determined by the dielectric constant $\varepsilon$ in the medium and the surface density of electric charge $Q_A$ in the Gauss equation

$$\mathbf{F}_{\text{stat}} = \frac{1}{\varepsilon \varepsilon_0} Q_A \qquad (1)$$

Thus the field strength in the electrostatically coated film is appreciably lower than in the air gap between both electrodes. On the other hand, in the EDP process, the field strength of the electric "flow fields" is given by the specific electrical conductivity $\kappa$ and the current density $j$ in the layers according to

$$\mathbf{F}_{\text{dyn}} = \frac{1}{\kappa} j \qquad (2)$$

The specific conductivity in the film is lower by 4–5 orders of magnitude in comparison to that of the bath. This causes a very high electric field strength in the film, as is schematically represented in Figure 2.

Processes that can be directly compared with the EDP process are dip painting and electrostatic or nonelectrical spray coating. The technical success of the EDP process is due to certain outstanding advantages, which are summarized in Table 1.[6]

The equipment for the new process is expensive and leads to a continuous operation rather than a batchwise application. Together with the relatively

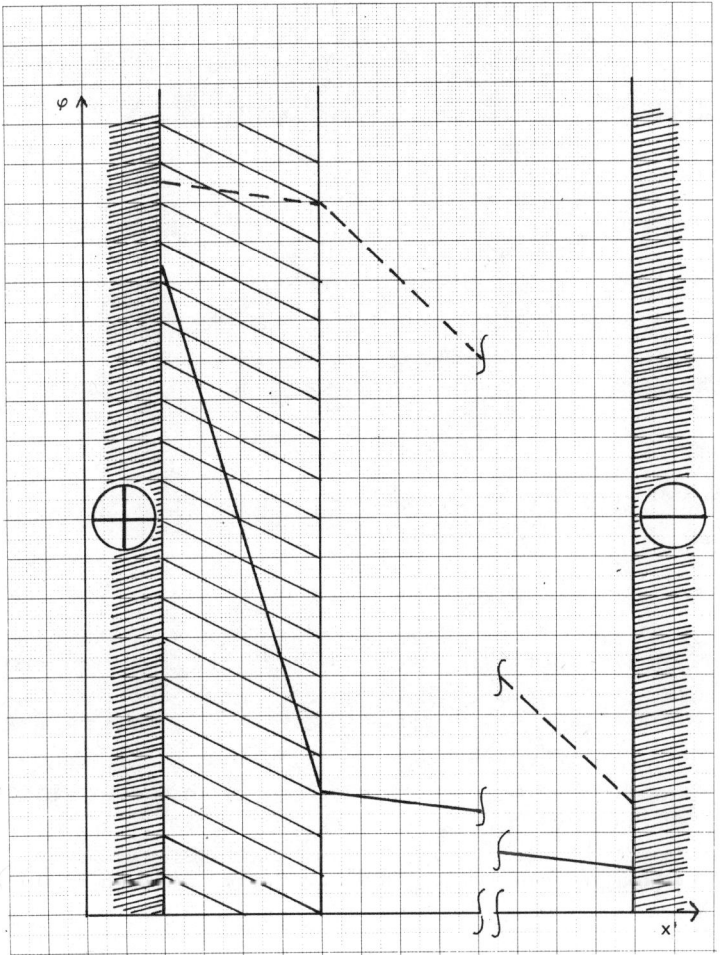

Figure 2. Potential $\varphi$ versus distance $x$ between the electrodes in EDP, ——, and in electrostatic spray, - - -.

short deposition times of about 2 min, all points in Table 1 are highly beneficial for a continuous process.

Some limitations must be mentioned, which are not restrictive at all: Only conducting substrates can be painted, and normally this process can only provide the first layer. Electricity is consumed at a rate of 0.1–0.2 kWh/m². Color changes are difficult and expensive to introduce.

The aim of this chapter is to give a balanced description of the EDP process as a whole, with concentration on the electrochemical features. In the last 15 years, some monographs have been published[1–5] which cover this topic in some detail. Moreover, two more recent review articles on the fundamental,

*Table 1*
*Advantages of EDP Process*

| | |
|---|---|
| Electric process: | Easy to control, to automate and to program, thus leading to minimal labor costs. |
| Aqueous systems: | Minimized air pollution and fire hazards due to minimum concentration of organic solvents. |
| Smooth films: | Of constant thickness, no runs, no drawback from edges; reduction of polish work. |
| Throwing power: | Deep penetration into cracks, flaws, boxes, and other less accessible areas. |
| High utilization: | Of both resin and pigment (95%, in comparison with 80% for other dip procedures and $\approx$50% for spraying). |
| Negligible drag-out: | Short time between deposition and rinsing is possible. |

especially electrochemical aspects[6] as well as on the chemical and preparative background[7] of the EDP process have been published.

## 2. Historical Development

*Electrophoresis* as a migration of colloidal particles dispersed in solutions as well as their deposition on electrodes under the influence of an electrical field was described by Reuss as early as 1808.[8] Helmholtz was the first to derive a quantitative relationship between the velocity of electrophoretic migration and physicochemical properties of the system.[9] "Electrokinetic phenomena" involve in addition the movement of a liquid relative to a charged surface in an electric field and the corresponding electric effects introduced by mechanical forces. However, these topics are only important in connection with transport processes in electrodeposition of dispersed systems (cf. Section 5). The term *electrophoretic painting* for the whole process for this reason is not wholly justified and has been replaced in our time by *electrocoating* or *electrodeposition of paint*.

The beginning of the technical development directly related to the EDP process under discussion occurred at the beginning of this century. An early patent for the painting of conducting articles was granted in 1919.[10] In the twenties and thirties, electrodeposition of lattices was performed on a larger scale.[11] One of the applications was the additional protection of the inside of tin-plate cans.[12] Another was the manufacture of rubber articles like hoses. Oleoresinous layers were deposited by the EDP process as early as 1937.[13] The idea of an acid surface layer as the primary reason for anodic electrocoagulation was disclosed at that time.[14]

In the late 1950s, Brewer and Burnside of Ford Motor Co. in the U.S.A. made the far-sighted proposal to develop a large-scale process of electrophoretic deposition of water-borne paint compositions as a primer for car bodies and other industrially manufactured metal ware. In very tight cooperation with some leading paint manufacturers, a technical process was developed and was ready for use in the early 1960s. This development is published in patents of Ford Motor Co. and in various papers of Brewer and co-workers.[15,16] Due to the disclosure of a considerable amount of basic understanding and technical knowledge prior to this technical development, it was impossible to get any "master" patents. Most of the modern patent literature involves only incremental improvements of paint formulations and process developments, which has been summarized elsewhere.[5,17]

Soon the process was adopted by other car manufacturers. This development proceeded in Europe and in Japan faster than in the U.S. In 1975, 100% of European and Japanese car bodies and 50% of U.S. bodies were primed by the EDP process.[18]

## 3. Anodic EDP Process

### 3.1. General Considerations

In Section 3 only dissolved anionic polyelectrolytes will be discussed, which at the present time have by far the greatest technical importance. In regard to particle size, basically only the first of the following three groups is within the scope of our discussion: $10-10^3$ Å, dissolved polyelectrolytes; $10^3-10^5$ Å, colloidal systems; and $10^5-10^7$ Å, dispersed systems. Systems with small particle size are preferred because of their greater ability to form smooth films.

As has already been mentioned, the anionic polyelectrolytes bear carboxylic groups arranged along the backbone of an organic polymer. These groups are at least partially neutralized with bases like triethylamine or potassium hydroxide. However, the density of ionic groups along the chain is typically one order of magnitude lower than the density in the case of typical polyelectrolytes like sodium acrylate. Thus conclusions from general polyelectrolyte work must be drawn with some caution.[19,20]

The most important property of this "diluted" polyelectrolyte is its solubility in water and the insolubility of the unneutralized acid form in this solvent. This feature is the basis of the EDP process, as will be pointed out in Section 3.3.2. The ionic form of the macromolecule has a stretched structure due to electrostatic repulsion of negative charges; on the other hand, the nonionic, acid form has a random, coil conformation (cf. Figure 3). It must be mentioned that polyacrylic acid—an example of a "concentrated" polyelectrolyte—is soluble in water as an acid as well as an alkali salt.

Some typical data for water-borne EDP systems are collected in Table 2.[6]

The concentration of ionic groups is at a level of about 1 equiv./liter. It is the same as the concentration in typical ion exchange resins and about one

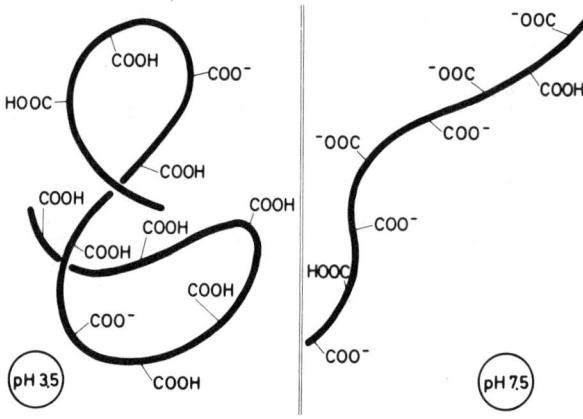

Figure 3. Conformations of a polyelectrolyte.

Table 2
Typical Data of EDP Process

| Process | Parameter | Data |
|---|---|---|
| Chemical | Resin, nature | Polycarboxylic acid |
| | Resin, concentration of ionic groups | ~1 meq/g |
| | Resin, acid number | 40–200 mg KOH/g |
| | Resin, molecular weight | 1000–50,000 |
| | Resin, concentration | ~10 wt % |
| | Solvent | Water, with ~10% org. solv. |
| | Base, nature | $NR_3$ (KOH) |
| | Base, concentration (in terms of degree of neutralization $\alpha$) | $\alpha = 35$–90% |
| | Pigment ($TiO_2$):resin ratio | $70:100^a$ |
| Electrical | Electrode, material | Steel, phosphated steel |
| | Electrode, bias | Anode[b] |
| | Current density | 1–5 mA/$cm^2$ |
| | Voltage | 50–400 V |
| | Specific conductivity in bath | ~$10^{-3}$ mho/cm |
| | Specific conductivity in film | ~$10^{-8}$ mho/cm |
| | Field strength in bath | 1–5 V/cm |
| | Field strength in film | 100–500 kV/cm |
| | Electrochemical equivalent | 10 mg/C = 1 kg/F |
| | Energy consumption | 5 kWh/kg = 0.1 kWh/$m^2$ |
| | Deposition time | 1–3 min |
| | Thickness of film | 20 $\mu$m (~1 mil) |

[a] In conventional systems up to 130:100.
[b] More recently, cathodic systems have gained some technical importance.

order of magnitude smaller than in polyacrylic acid. The data show clearly that the electrochemical process is operated at rather low current densities, and that the electrochemical equivalent is at least ten times higher than that of common electrochemical reactions. A comparison of film and bath shows that the field strength in the film is about $10^5$ times larger due to a specific conductivity which is $10^5$ times lower. Therefore voltages up to several hundred volts are needed to establish a relatively low current density.

Practical baths have a rather complicated composition. They contain a system of polyelectrolytic and nonionic binders, pigments, and additives to promote cross-linking in the baking process and organic solvents from the feed. Baths with 10–20 components are under current use. On the other hand, systems for fundamental research are often restricted to one polyelectrolyte with a relatively simple composition.

### 3.2. Static Properties of the Bath and of the Film

As already mentioned, resinous material can be regarded as a "diluted" polyelectrolyte. Only some of the *counterions* are essentially free, thus contributing to the conductivity of the system. The rest are electrostatically and chemically trapped by the polyion at the following locations (with increasing loss in free energy):[19-21] inside the coil, at the backbone, and at ionic groups of the backbone. Polyvalent ions are fixed more strongly than univalent ones. For the univalent ions, the following order is given according to the increasing chemical interaction[22]:

$$Li^+ < Na^+ < K^+ < NR_3H^+ < NR_4^+$$

Owing to the same forces and the close neighborhood of the ionic groups, the *acidity* of the polycarboxylic acid is weakened. The titration curve is "smeared" over a wider range of pH values.[23] The even more complicated conditions upon neutralizing the weak polyacid with a weak base, e.g., an amine, are discussed elsewhere.[24] The concentration of free amine can be calculated according to these results. The pK value of the polyacid is further lowered due to the presence of organic cosolvent. Experimental results are reported in Reference 5.

At practical concentrations the macroions form *micelles* due to the fact that the critical micelle concentration is about 0.1 wt. %. A maximum micelle diameter of about 1200 Å according to Debye's theory[25] has been confirmed by light-scattering experiments.[26] The micelle diameter decreases with increasing degree of neutralization $\alpha$.

Association of macroions increases strongly with increasing concentration. This leads to a corresponding rise of specific *viscosity*. A higher degree of neutralization leads also to an increased viscosity due to the stretch of polyanion. Examples are given in Table 3. The figures show clearly that at

Table 3

Specific Viscosity (cP) of Polyelectrolyte Solutions: Rise with Concentration

| $C$, wt. % | Acrylic resin[27] | | Styrene maleic acid butylester, 50% degree of esterification[28] |
|---|---|---|---|
| | $\alpha = 45\%$ | $\alpha = 65\%$ | |
| 5 | | | 3.8 |
| 10 | 1.5 | 1.8 | 10 |
| 15 | | | 50 |
| 20 | 2.8 | 5.2 | 800 |
| 30 | 9.2 | 53.7 | |
| 39 | 139 | | |

practical concentrations (~10%), the viscosity of the solution is nearly the same as pure water. This is very important for practical application. The bath liquid is able to run off quickly after the emergence of the coated substrate.

The *stability* of the colloidal system depends among other things on the net charge of the particles. This can be neutralized by addition of a second electrolyte, leading to flocculation. According to the Schulze–Hardy rule, the critical concentration for the second electrolyte decreases rapidly with increasing charge number $z$. With a typical EDP system, this dependence has been found to be not so strong as with other colloidal systems.[29] $H^+$ ions are much more efficient for flocculation than other univalent ions due to the basic character of the $COO^-$ groups.

*Pigments* are an essential component of a practical bath, e.g., titanium dioxide.[30] The net charge of pigment particles is established as a result of wrapping by organic polyelectrolyte and ionization of surface groups. As has been shown in Table 2, the pigment–binder ratio is normally lower in comparison with conventional paint systems.

Some important properties of the *wet film* can be summarized as follows: strong adherence; hydrophobic surface, which is indicative of the presence of –COOH groups[31]; low water concentration (5–15%) in comparison to the high water concentration (80–90%) in the bath;[32] fractionation of bath components with respect to chemical composition and molecular weight (cf. Section 3.3.4); metal content from the substrate, (c.f. Section 3.3.3); and rough, porous surface structure, which can be detected by classical microscopy[33] and by the scanning electron microscope.[27,34,35] In the course of the stoving process, roughness disappears because of passage through a stage of low viscosity.

## 3.3. Electrocoagulation as the Primary Process

### 3.3.1. Transport Processes

The macroions are transported by several processes to the substrate. In the bulk of the solution, convective flow generally predominates. Because of the absence of a supporting electrolyte, the ionic conductivity of the system is directly related to the properties of the polyelectrolyte:

$$\kappa = z c_p f(\mu_p + \mu_c) F \qquad (3)$$

$c_p$ is the molarity of the polyelectrolyte; $\mu_p$ and $\mu_c$ are the mobilities of the polyion and the counterions, respectively. $f$ is an extremely compounded factor, taking into account that only a fraction $f$ of the counterions are free to move in the electric field. In the case of an EDP system, $\kappa$ increases linearly with $\alpha$,[27] while this dependency is nonlinear with polyacrylic acid. Specific conductivities of the EDP bath are on the order of $0.5-2 \times 10^{-3} \, \Omega^{-1} \, cm^{-1}$ at room temperature; they will double by raising the temperature to about 70°C.[36] The molar conductivity of the EDP bath varies: $\Lambda = 10-50 \, \Omega^{-1} \, cm^2$ equiv. wt.$^{-1}$ at a two- to threefold viscosity, thus reaching the same order of magnitude as aqueous solutions of strong electrolytes ($100-150 \, \Omega^{-1} \, cm^2$ equiv. wt.$^{-1}$).[37]

Electrophoresis of colloidal particles can be understood according to Helmholtz as an equilibrium between friction forces and electrical forces. The velocity of migration is given by

$$v = \frac{\varepsilon \zeta}{6 \pi \eta} F \qquad (4)$$

where $\eta$ is the specific viscosity of the medium. Velocities of $\sim 10^{-4}$ cm/sec can be derived from Eq. (4), which is in the same order as the migration velocities of the ions. It must be mentioned that ionic conduction can also be rationalized in terms of a field-enhanced hopping mechanism.[38] This model has been applied to the EDP process.[39]

According to Nernst, a quasistagnant diffusion layer is established at the phase boundary, if the bath is flowing. The mass transfer through this layer is due to diffusion and in our case (ions in the absence of supporting electrolyte) due to migration in addition. A steady state concentration gradient drives a diffusion flux, which is given for the off-transport of protons (cf. Figure 4) by

$$J_{H^+,D} = \frac{j_{H^+}}{F} = D_{H^+} \frac{dc}{dx} \qquad (5)$$

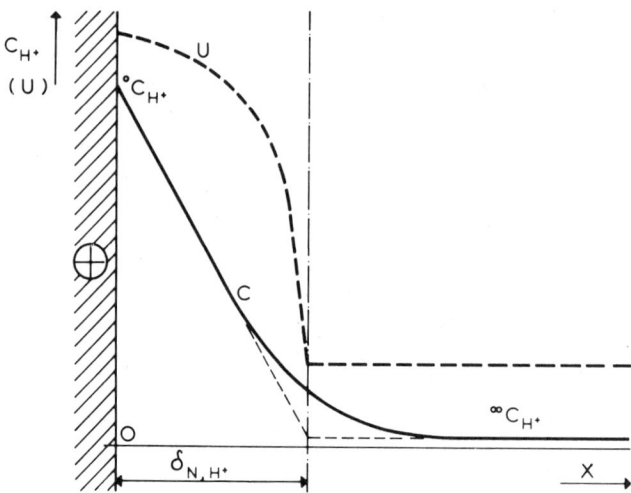

Figure 4. Model of diffusion layer for protons. Proton concentration $C_{H^+}$, ———, and diffusion potential $U$, ---.

From this follows immediately a logarithmic potential curve

$$U = \frac{RT}{zF} \ln c$$

$$\frac{dU}{dx} = \frac{dU}{dc}\frac{dc}{dx} = \frac{RT/zF}{c}\frac{dc}{dx} \qquad (6)$$

which drives a migration flux according to

$$J_{H^+,M} = \mu_{H^+} c \frac{dU}{dx} = \mu_{H^+} \frac{RT}{zF}\frac{dc}{dx} \qquad (7)$$

From the Planck–Einstein relation, this equation becomes essentially identical to Eq. (5). Thus the total flux is

$$J_{H^+} = 2D_{H^+}\frac{dc}{dx} \qquad (8)$$

In regard to polyvalent macroions, transport in the diffusion layer is mainly governed by migration or electrophoresis:[40]

$$J_p = \frac{j_p}{zF} = (1+z)D_p\frac{dc}{dx} \qquad (9)$$

In the EDP process, current density generally increases with increasing voltage. Together with the magnitude of the voltages, these facts indicate that the ion transport through the system is the rate-determining step rather

### 3.3.2. Mechanistic Aspects

The macroions must be discharged in the course of the electrocoating process. This can be done directly via electron transfer or indirectly via neutralization by counterions, e.g., protons:

$$P^{z-} \rightarrow P + ze^- \tag{10a}$$

$$P^{z-} + zH^+ \rightarrow PH_z \tag{10b}$$

However, thickness growth of the film can be understood only with reaction (10b) because of the fact that the organic material has no electronic conductivity. Moreover, chemical evidence like the absence of cross-linkage and the ready solubility of the wet film in organic solvents or in alkaline solutions strongly supports this mechanism. Acid flocculation at pH 3–5 of the binder can be easily demonstrated by adding dilute acid to the bath.

Protons are injected at the phase boundary by anodic decomposition of water at the passivated iron electrode:

$$H_2O \rightarrow \tfrac{1}{2}O_2 + 2H^+ + 2e^- \tag{11}$$

The acidification of the boundary layer, formerly discussed qualitatively,[12,14,41,42] was treated quantitatively by Beck.[6,24,40] According to Eq. (8), the current density is related to the surface concentration of protons $^\circ c_{H^+}$ by

$$j_{H^+} = 2FD_{H^+} \frac{^\circ c_{H^+}}{\delta_{H^+}} \tag{12}$$

Under typical conditions of the EDP process, the pH at the surface is calculated to drop to about 2. This result is not significantly modified if the buffer action of the free amine is taken into account.[24] The macroanion acts as a buffer in addition. Its flocculation occurs at a critical proton concentration $c^*_{H^+}$. As is shown in Figure 5, $c^*_{H^+}$ is generally attained within the diffusion layer for the protons, but outside the diffusion layer of the macroions.† The macroions are transported to this zone, where both the proton and anion concentration drop to virtually zero because of coagulation. The distance $x^*$ from the electrode is calculated to be

$$x^* = \delta_{H^+}\left(1 - \frac{c^*_{H^+}}{^\circ c_{H^+}}\right) \tag{13}$$

---

† The diffusion layer thickness is dependent on the nature of the particle via $D_i$. It is proportional to $D_i^{1/3}D_{H^+}$ and is about $10^{-4}$ cm$^2$/sec, $D_p \sim 10^{-7}$ cm$^2$/sec.[6,40] Thus $\delta_{H^+} : \delta_p \sim 10:1$.

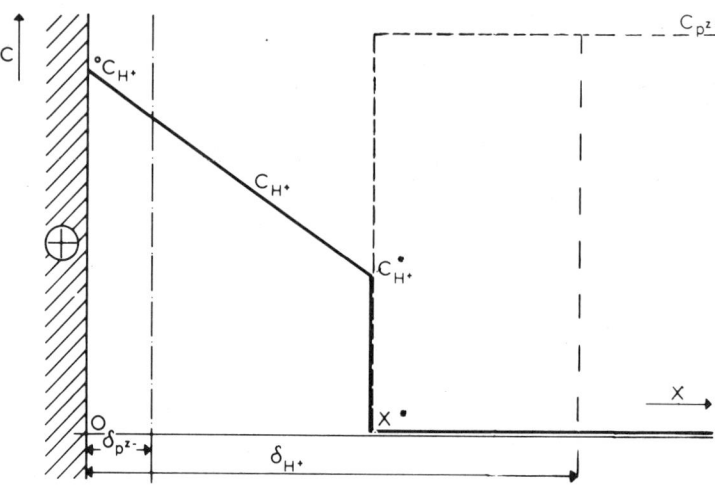

Figure 5. Concentration profiles for protons, ———, and for macroions, – – –, in the diffusion layer.

Thus, $x^*$ increases with increasing $j$ and with decreasing convection and $c_{H^+}$. This may lead to poor adherence of the film. Only at a minimum current density will $x^*$ go to zero and the critical proton concentration be located at the electrode surface. This equilibrium discussion becomes invalid if nucleation phenomena play an important role.

Several experimental proofs show the validity of this concept.

1. Experiments with the *rotating disk electrode* (RDE) have shown that at a critical rotation speed $n^*$, the (galvanostatic) deposition of the layer, indicated by a steep rise of the cell voltage (cf. Figure 9), becomes impossible.[24,40] According to the Levich equation, the diffusion layer thickness at a RDE is proportional to $1/n^{1/2}$. Combining with Eq. (12), the result is that $n^*$ is correlated with the square of a critical current density $j^*$:

$$j^{*2} = 8.20 F^2 D^{4/3} \nu^{-1/3} c_{H^+}^* n^* \qquad (14)$$

This relationship is experimentally verified, as is shown in Figure 6. The influence of $\alpha$ [see Eq. (19)] is demonstrated as well. The influence of other parameters of importance has also been studied with this technique.

2. *Galvanostatic transients* have been measured at EDP electrodes in nonstirred solutions.[37,39,43–45] According to Sands' equation, $1/\tau^{1/2}$ ($\tau$ is transition time) is expected to be proportional to $j$. As is shown in Figure 7, this exact behavior is found in EDP systems. The independency on the polymer concentration indicates that the accumulation of protons rather than the depletion of macroions is the determining step.

3. The *reaction products* according to Eq. (11) have been directly identified. For $H^+$ ions, this was done by a microsonde.[46] Oxygen as the other

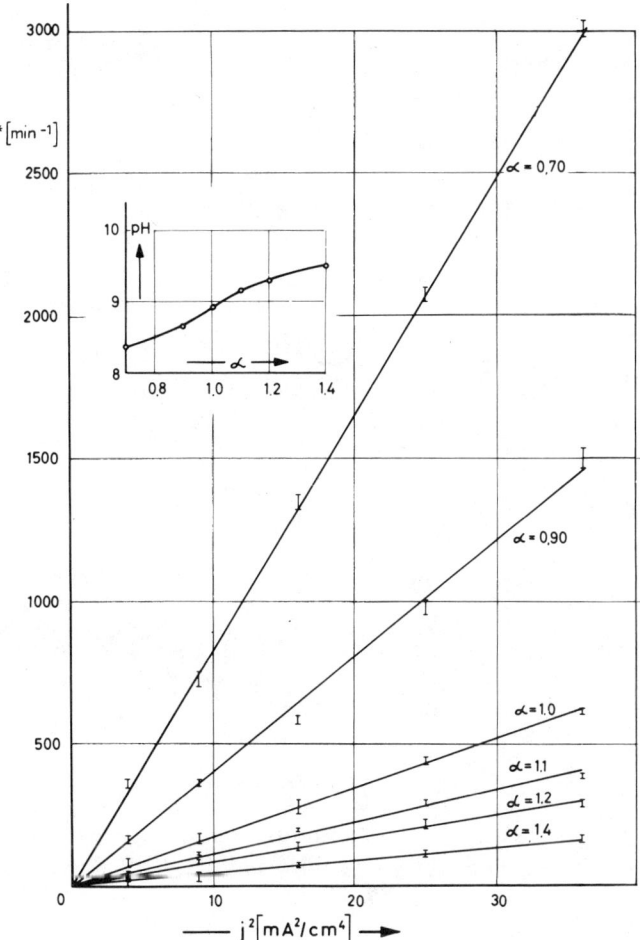

Figure 6. Plot of critical rotation speed $n^*$ vs. square of current density for electrodeposition of acrylic resin on a rotating disk electrode (iron). The parameter is degree of neutralization $\alpha$. A plot of $\alpha$ vs. pH is shown in the inserted diagram.[24]

product was measured volumetrically[45,47] and by flushing the cell with an inert gas which was afterwards conducted through an electrochemical measuring cell for oxygen.[48] The current efficiencies which have been found range from 18% to 95%. In aged EDP baths, this value was found to have increased, which may be due to the oxidative removal of organic material.

### 3.3.3. Relative Importance of Electrode Reactions

The anodic decomposition of water is highly beneficial as a main reaction. It flocculates the polymer in a useful way, and oxygen does not contaminate the

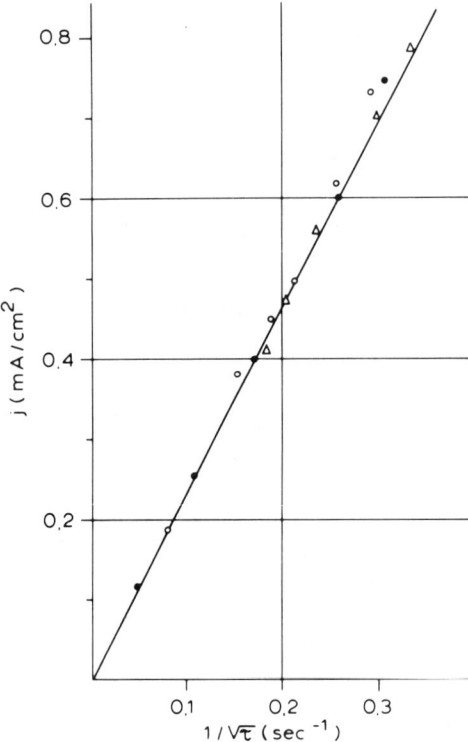

Figure 7. Plot of current density $j$ vs. reciprocal square root of transition time $\tau$ for the galvanostatic electrodeposition of acrylic resins in unstirred solutions.[44] ●, 5 wt. %; △, 10 wt. %; ○, 14 wt. % of polyelectrolyte at pH 8.25.

bath. However, the measured nonquantitative current efficiencies for this reaction indicate that side reactions must play some role. One important side reaction is the *anodic dissolution of the metal*:

$$Me \rightarrow Me^{w+} + we^{-} \qquad (15)$$

According to the Schulze–Hardy rule, polyvalent metal ions are particularly effective in coagulating the polymer. However, metal soaps are unfavorable because of their poor flow behavior in the stoving stage and coloration of the deposit. Normally iron is in the passive state under EDP conditions and the current efficiency does not exceed about 1%.[29,49,50–53] In the presence of chloride ions, this passivity breaks down, and Fe will dissolve with current efficiencies as high as 50–100%.[40,51,52] In Table 4, these data, together with data for other metals, are summarized.

Pt and Au are inert anodes on which electrocoagulation can only proceed according to a pure $H^+$ mechanism. The other metals obviously cannot passivate quantitatively under these conditions. Zinc anodes were found to dissolve with 100% current efficiency in the presence of maleinized oil.[54]

### Table 4
### Result of the Anodic Deposition of a Butyl Acrylate–Acrylic Acid Copolymer onto Different Metals

| Anode | mg/C | Wt. % metal found in film | Current efficiency for metal dissolution, % |
|---|---|---|---|
| Fe + 10 mmol NaCl per liter | 17.0 | 1.5 | 90 |
| Fe | 10 | 0.05 | 1.8 |
| Al | 14.3 | 0.037 | 5.9 |
| Au | 9.7 | 0.005 | 0.1 |
| Zn | 10.2 | 0.34 | 10.5 |
| Ni | 10.3 | 0.73 | 26 |
| Cu | 7.0 | 1.3 | 31 |

Bath concentration = 10%, pH = 7.7.
Deposition time = 1 min, voltage = 100 V.

In spite of these conditions, some authors[54–58] regard the anodic dissolution of iron as a main reaction and oxygen evolution as a "troublesome side reaction."

The anodic attack on the zinc phosphate layer produces metal ions in addition and will be treated in Section 7.

*Electro-organic reactions* at the anode play another important role as possible side reactions. An example is the oxidation of a carbonic acid to an alcohol:

$$R-CH_2-COOH + H_2O \rightarrow R-CH_2OH + CO_2 + 2H^+ + 2e^- \qquad (16)$$

The anodic oxidation of alcohols, amines, and aromatic compounds is very well known in organic electrochemistry.[61,62] Like in water electrolysis, these oxidations are accompanied by the injection of protons at the interface, thus assisting electrocoagulation. However, the organic products contaminate the bath; therefore this type of reaction is highly unwanted. Ionic products of low molecular weight like acetate are very critical. A decrease in acid value of the bath, which was found experimentally, must be interpreted as a fractionation process rather than a decomposition according to Eq. (16).[42,59] On the other hand, $CO_2$ was identified as an anodic product[59] due to reaction (16). The Kolbe coupling does not play any significant role under EDP conditions, in spite of speculations in the literature.[47,55–57,60] This was shown with model experiments (electrolysis of EDP systems with relatively high current densities up to 40 hr[63]). The anodic formation of alcoholic OH groups was found with the sensitive method of reaction gas chromatography.[64]

## 3.4. Thickness Growth of Film as the Secondary Process

### 3.4.1. Laws of Thickness Growth

After the formation of a film, the distribution of voltage in the cell becomes quite uneven according to Figure 1. The thickness growth of the film can be described by the time functions of mass $m$, thickness $l$, voltage $U$, or current density $j$.

According to Faraday's law, the mass of the film should be proportional to the charge $Q$:

$$m = \beta m_e Q \qquad (17)$$

where $\beta$ is the current efficiency for the electrocoagulation process

$$2RCOO^- + H_2O \rightarrow 2RCOOH + \tfrac{1}{2}O_2 + 2e^- \qquad (18)$$

and $m_e$ is the electrochemical equivalent, which can be easily correlated with molecular structure

$$m_e = \frac{56,100}{\alpha \gamma F} \qquad (19)$$

where $\alpha$ is the degree of neutralization and $\gamma$ the acid value of polyelectrolyte (mg KOH/g polyelectrolyte). According to Table 2, $m_e$ is about 1 kg/F. This means that the equivalent weight is essentially the same as in ion exchange resins (1 equiv. wt./kg). With pigmented systems, $m_e$ is appreciably higher (3 kg/F). Owing to the high voltage (200 V on average), the energy consumption is rather high, exceeding 5 kWh/kg.

According to Eq. (19), $m_e$ will increase with decreasing $\alpha$. An example of this behavior is given in Figure 8, which shows in addition linear $m/Q$ curves in agreement with Faraday's law.[65]

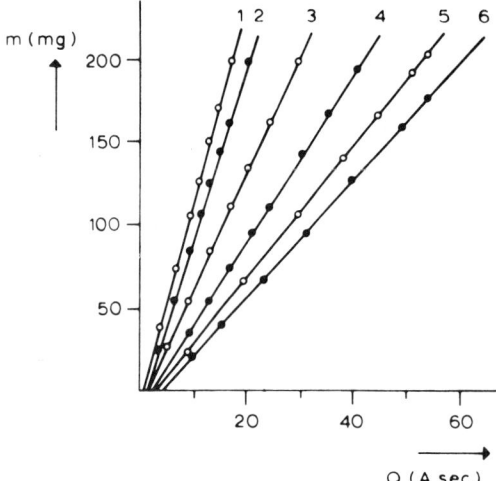

Figure 8. Deposited mass versus charge at various degree of neutralization. 1, 35%; 2, 41.5%; 3, 59%; 4, 79%; 5, 100%; 6, 112%.[65]

# ELECTRODEPOSITION OF PAINT

The *rate of thickness growth* can be obtained by differentiation of Faraday's law [Eq. (17)]:

$$\frac{dl}{dt} = \frac{\beta m_e}{s} j \tag{20}$$

where $s$ is the density of the layer. At *constant current density*, this rate will be time independent. Assuming Ohm's law to be valid, the voltage will rise according to

$$U = \frac{\beta m_e \rho}{s} j^2 t \tag{21}$$

The voltage–time curves in Figure 9 show this behavior after an induction period.[39] However, the proportionality of the slope of $j^2$ is not fulfilled. In other cases, $U/t$ curves which are nonlinear have been measured under galvanostatic conditions.[24,27,29,40] These deviations can be explained on the basis of nonohmic behavior of the film, which will be discussed in the next section. An attempt to modify the $U/t$ curve in this direction has been made in Ref. 6.

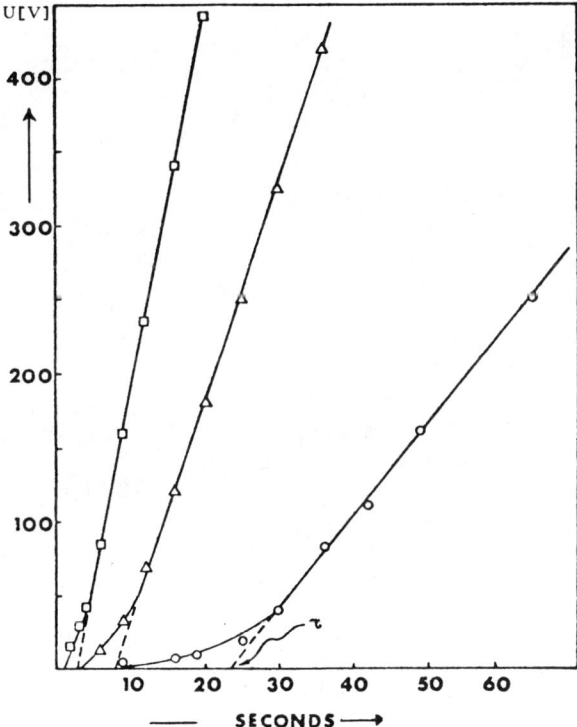

Figure 9. Electrodeposition of oil-modified polyester onto zinc-phosphated steel substrate at constant current densities.[39] (○) $j = 0.963$, $\tau = 23.5$ sec; (△) $j = 1.93$, $\tau = 7.7$ sec; (□) $j = 2.89$, $\tau = 2.8$ sec.

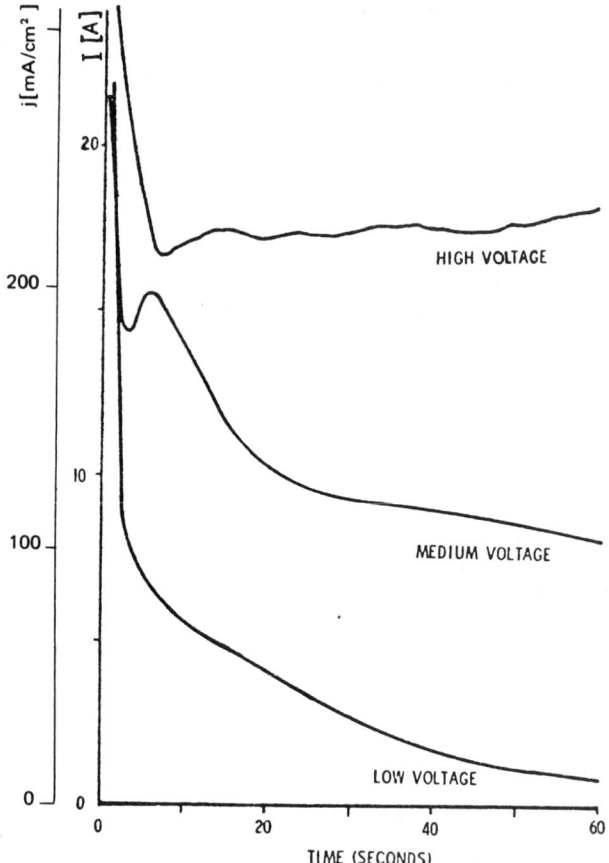

Figure 10. Electrodeposition of maleinized Linseed oil systems onto phosphated steel substrate at constant voltages.[36]

Under technical conditions, films are deposited at a constant voltage. In the primary stage, the current density is relatively high, being controlled by the bath resistance $R_B$. The formation of a film leads to a sharp decrease of current density according to the lower curve of Figure 10. This $I/t$ behavior has been frequently reported in the literature.[39,41,51,58,66–68]

Once again assuming Ohm's law to be valid, the total resistance $R$ after the time $t$ will be

$$R = \frac{U}{I(t)} = \rho' \int_0^t I \, dt + R_B \qquad (22)$$

$\rho'$ is defined as film resistance per A sec. It is related to the specific resistivity $\rho$ according to

$$\rho' = \rho \frac{\beta m_e}{sA^2} \qquad (23)$$

Differentiating Eq. (22), rearranging, and integrating again yields[39,58,67,69]

$$\frac{1}{I^2} = \frac{1}{I_0^2} + \frac{2\rho\beta m_e}{sUA^2}t \qquad (24)$$

$I_0$ is the initial current, determined by the bath resistance alone. A plot of $1/I^2$ vs. $t$ leads to straight lines.[67] However, their slope is not proportional to $1/U$.[6]

Rearrangement of Eq. (24) yields a function $I \sim 1/t^{1/2}$. Integration of this leads to a $Q/t^{1/2}$ relationship. Thus $l$ should grow with $t^{1/2}$. This was experimentally found with various systems.[39,54,56,57]

Another approach starts from a model of ionic conduction through the film according to a hopping mechanism.[39] At the beginning, the field in the thin film is high, and $l$ is expected to grow with log $t$. Later on, when the film has become thicker, Ohm's law will be valid:

$$\frac{dl}{dt} = \frac{\text{const}}{l} \qquad (25)$$

and from this the square root law $l \sim t^{1/2}$ immediately follows.

A third approach to the $t^{1/2}$ law can be obtained by the application of the theory of phase growth to the problem.[54,56,58,66,70]

### 3.4.2. Current–Voltage Behavior

Under the conditions of the EDP process, a very high electric field of $10^5$–$10^6$ V/cm is applied to an organic film of low ionic conductivity. On these grounds alone, the validity of Ohm's law cannot be expected, at least in the initial stage of the process (thin films) and at high voltages. Actually, potentiodynamic current–voltage curves[6,71-74] show clearly that at higher voltages, large deviations from Ohm's law can be observed (cf. Figure 11).[71] The very rapid growth of current density in this region exceeds appreciably the rise, which could be expected according to Child's law[71] or according to field-enhanced dissociation of a weak electrolyte.[71,75] However, it was possible to interpret the curves satisfactorily by a model in which a positive space charge of protons at the surface of the film is opposite to a negative space charge of $COO^-$ groups in the bulk of the film.[102] The following $U/j$ relationship,

$$U = U_0 - \frac{A_0}{(j+j_0)^{1/2}} \qquad (26)$$

fitted very well to the experimental curves. Of the three constants, $U_0$ as a kind of countervoltage is the most important, following from Poisson's equation

$$U_0 = \frac{c_0 F}{2\epsilon\varepsilon_0}d^2 \qquad (27)$$

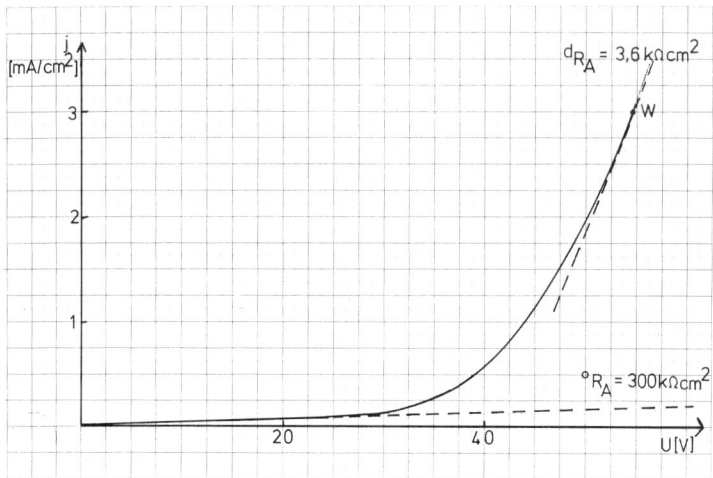

Figure 11. Dynamic current density–voltage curve for the electrodeposition of acrylate resin onto iron. Sweep time 1 sec.[71]

This means that the voltage to separate the charge carriers in the film into two space charge regions is proportional to the proton concentration $c_0$ and to the square of film thickness $d$. $c_0$ should therefore be low in order to keep the voltage low, which is necessary to "hold" a desired film thickness.

Quasilinear $U/j$ curves, which have been reported frequently in the literature,[52,54,67,76–79] apply to the primary part of the curve of Figure 11, where low voltages are applied to thick films. It follows immediately that an Ohmic resistance can be defined only under these conditions. It can be measured with an ac bridge at an unpolarized film (immediately after deposition) or via the residual currents $i_{res}$ flowing after deposition at constant voltage $U$ according to $R = U/i_{res}$. Usually, specific resistivities of about $10^8$ Ω cm are found. The influence of water loss as well as the nature and concentration of pigments in the film have been investigated.[39,51,80–82] Measurement of differential resistance of the film in the course of deposition led to much lower values than those predicted by the general $U/j$ behavior.[6] The current–time behavior of an EDP film after abrupt increase or decrease of voltage can be understood in terms of shifting along the $U/j$ curve from the linear to the nonlinear part and backwards.[6,43,73]

### 3.4.3. Throwing Power

Throwing power is one of the most important features of an EDP system. It means the penetration of current into less accessible regions of the substrate due to the coverage of the other parts in the primary stage of the process. Up to now, no comprehensive theory for throwing power including minimum current density, nonohmic behavior, and rupture voltage is available. Nevertheless, some semiquantitative conclusions can be drawn. They must be based on practical conditions like limited deposition time (2 min) and minimum "inner" film thickness (at least 10 $\mu$m). With randomly shaped anode substrates, the electrolyte must be divided into $n$ sections. In every section, the time-dependent current $i_n$, will be flowing, thus arriving at a total charge after the time $t$ given by

$$\int_0^t I\, dt = \int_0^t \int_1^n i_n\, dt \tag{28}$$

where $I$ is the time-dependent total current.[6]

A calculation in greater detail has been done with defined model configurations. In the case of a pipe electrode perpendicular to a plane counterelectrode, the penetration depth of film formation was found to proceed proportional to $t^{1/2}$ and to $U^{1/2}$.[83] However, the comparability of results obtained with various test configurations is rather poor.[6,84,85]

The following conditions are important in order to obtain a system with high throwing power: (1) high film resistance via low concentration of ionic groups in the film, low concentration of cosolvent, and low temperature; (2) high bath conductivity via mobile counterions and optimized $\alpha$ (40–50%), with the addition of a supporting electrolyte prohibited; (3) low critical current densities and high electrochemical equivalent, and (4) high rupture voltage (see Section 3.4.4). A low concentration of ionic groups in the polymer is highly advantageous for most of these conditions, but must be thoroughly balanced with the need for colloidal stability of the bath.

### 3.4.4. Nonideal Films

Under ideal conditions, a smooth, coherent, well-adhering film without pores will be deposited onto the substrate. In practice, some difficulties arise in approaching this result.

As was pointed out in Section 3.3, anodic *evolution of oxygen* is the main electrode process. However, owing to the low current densities, gas bubbles are not generated under normal conditions. In the diffusion layer, a high degree of supersaturation of the gas is established resulting from an accumulation of organic componennts and the higher viscosity, which retards the rate of nucleation for bubble formation appreciably.[6,86] Some additional precautions to prevent gas bubble formation may be the application of a voltage increasing in steps and rapid stirring. The avoidance of gas bubbles by a slightly increased pressure has been demonstrated.[87]

The formation of an imperfect film is not the result of the gas-producing electrochemical reaction, but is caused by another effect, the *electrical heating of the film*. The electric power, which is dissipated in the film of resistance $R_F$, is given at constant voltage by

$$N_F = I^2 R_F = U^2 \frac{R_F}{(R_F + R_B)^2} \tag{29}$$

As is easily realized, $N_F$ will be at a maximum when $R_F$ is equal to the bath resistance $R_B$.[36] $N_F$ increases very rapidly with the square of the applied voltage. The maximum temperature in the center of the film will be governed by the heat capacity of the metal and by the rate of heat transport through the liquid boundary layer.[88] The situation is similar to that in a diaphragm of an electrolysis cell where the resistivity is lower, but the current density is higher.[89] The rise of temperature may be as high as 100°C.[36] Eventually, film rupture and even blowing off may occur if the voltage is raised to a critical point called the rupture voltage, which is normally above 200 V.[77–79] Current–voltage curves, which are representative for this perturbation, are shown in the upper part of Figure 10.

### 3.4.5. Transport Mechanisms in the Film

The film is an ionic conductor of high resistivity due to the low concentration of free protons (about 1 $\mu$mol/liter). A schematic representation of

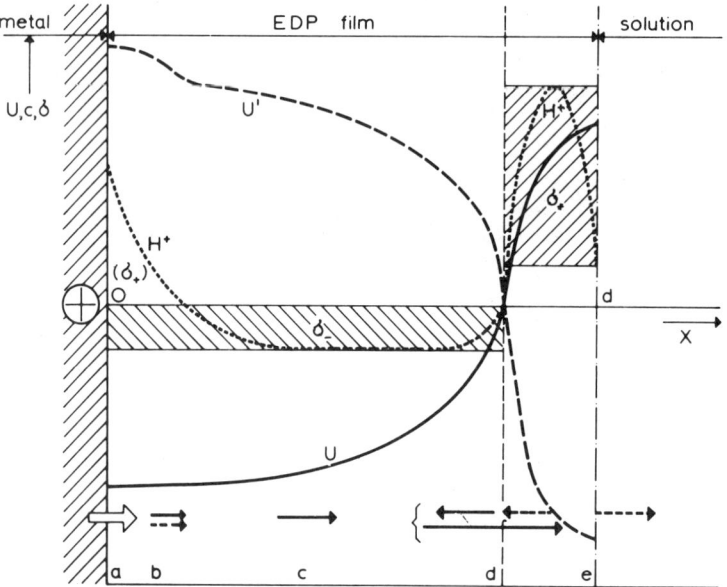

Figure 12. Schematic representation of positive and negative space charge regions in an anodic EDP film. ——, Voltage drop in space charge regions; ---, applied voltage; ⇨, proton injection; ——▶, migration; ---▶, diffusion.

positive and negative space charge regions in the film, arising from the high field strength across the film, is given in Figure 12.[6,71] Protons are injected via the electrochemical reaction at the metal interface and they are transported via migration and diffusion through the film. The transport of alkyl ammonium ions[43,72,73] and even of the macroanions[42,90] was discussed elsewhere. On the basis of the above-mentioned model, it was possible to analyze the current–voltage curve.[71] Another approach to the problem was tried by the application of a field-activated hopping mechanism for protons.[39]

In the course of electrodeposition, a strong dehydration process occurs which is predominantly a squeezing effect due to electrocoagulation[91] and only to a minor extent due to electro-osmosis.[51] The final water concentration in the film is about 10%. During thickness growth of the film, only nonsteady states are established. However, after a decay of current density below the critical current density, thickness growth ceases and a constant residual current is flowing. Water must then be transported towards the metal with the same rate as protons and oxygen are transported out of the film. This residual current is a property of the homogeneous film in the absence of cracks and imperfections of the film; these imperfections are assumed erroneously as the reason for it.[92]

## 4. Cathodic EDP Process

The mechanism of cathodic deposition of cationic resins is analogous to the anodic process. The cathodic water decomposition is once again the main reaction

$$H_2O + e^- \rightarrow \tfrac{1}{2}H_2 + OH^- \qquad (30)$$

Generated $OH^-$ ions will accumulate in the cathodic diffusion layer, and cationic species of resinous material, coming in from the solution, will be electrocoagulated in the alkaline boundary layer at the cathode:

$$R-\underset{R''}{\overset{R'}{N^\oplus}}-H + OH^- \rightarrow R-\underset{R''}{\overset{R'}{N}} + H_2O \qquad (31)$$

Analogously to the anodic mechanism, discharge proceeds indirectly via acid–base reactions rather than directly. One of the inherent advantages of cationic systems is the impossibility of electrochemical oxidation of resin at the substrate and the avoidance of electrochemical dissolution of metal. Thus deposition curves have been found to be independent of the nature of the cathode, even under a wide variation of hydrogen overvoltages (Fe vs. Pb).[51]

## 5. Deposition of Dispersions and Latices

In the early days of electrodeposition of organic coatings, most of the work concentrated on dispersions and latices[5,10-12,14] owing to the nearly total lack of water-soluble resinous material. A very thorough investigation in this field, mainly on aqueous dispersions of copolymers of vinylidene chloride and acrylonitrile (Saran) of 0.1-$\mu$m particle size was published in 1948.[66]

Generally, the particles carry some unneutralized charge at the surface due to the presence of ionizable groups or the specific adsorption of ions, but even neutral particles can be charged by charge injection at the counterelectrode.[93] Some typical results for aqueous dispersions are compiled in Table 5. Because of the larger electrochemical equivalent and the very low deposition voltage, the energy consumption by this process is much lower than in the case of dissolved systems (Table 2). Current densities and deposition times are usually larger, leading to a film thickness of up to several millimeters. The deposits are spongy and not well adhering. The moisture content is as high as 40%; it must be removed by long drying (30 min at 85°C). This point is a severe restriction for practical application. The low voltages lead to poor throwing power behavior.

Again, the mechanism for electrodeposition is a combined electrocoagulation by electrochemically generated ions ($H^+$, $Fe^{2+}$ at anode, $OH^-$ at the cathode).[11,12,14,66] Observations of depositions even at very low voltages led to the conclusion that accumulation of particles at the electrode by electrophoresis, leading to a stronger interaction and thus stimulating coagulation, may also play an important role ("concentration coagulation").

A growth rate of the film proportional to $t^{1/2}$ is discussed in terms of a rate-determining diffusion of metal ions through the film.[58] In nonaqueous

Table 5

*Some Typical Results for Aqueous Dispersions*

| Process | Parameter | Data |
|---|---|---|
| Chemical | Nature of latex | Particles with ionogenic surface groups, e.g., $\bigcirc\!\!\sim\!\!COOH$ |
|  | Particle size | 0.1–0.5 $\mu$m |
|  | Concentration of latex | 10–50 wt. % |
| Electrical | Electrode bias | Anode or cathode |
|  | Electrode, material | Iron, zinc |
|  | Current density | 5–25 mA/cm$^2$ |
|  | Voltage | 2–20 V |
|  | Electrochemical equivalent | 25–35 mg/C |
|  | Energy consumption | 0.01 kWh/kg |
|  | Deposition time | 3–30 min |
|  | Thickness of film | Up to several millimeters |

## 6. Chemistry of Resins and of the Bath

In present practical applications, very numerous EDP resins are available,[2,4,5,7] although none of these are universally employable. *Anodic* binders carry nearly exclusively COOH groups, the amount of which governs the solubility of the polyelectrolyte in a given solvent at a given pH value.

The classical anodic binders have been oleoresins, e.g., maleinized oils, which are prepared via maleinization or Diels–Alder addition:

$$\text{ROOC}\cdots\diagup\!\!\diagdown\!\!=\!\!\diagup\!\!\diagdown\!\!\diagup\!\!\cdots\text{COOR}^1 \quad \text{with side group } \text{CH(COOH)CH}_2\text{COO}^\ominus \tag{32}$$

The importance of this low-cost material has diminished because of some drawbacks such as yellowing tendency and low resistance to hydrolysis. Similar properties are found in maleinized or styrenized alkyds.

Medium-cost material like polyesters and epoxyesters

$$-\text{O}-\text{CH}_2-\underset{\text{OH}}{\text{CH}}-\text{R}-\underset{\text{OH}}{\text{CH}}-\text{CH}_2-\text{OCO}\cdots \tag{33}$$

are of great practical interest according to their excellent hardness, adhesion, and flexibility. The corrosion resistance is comparable to the above-mentioned low-cost material. The only polymer with a backbone, which contains no hydrolyzable ester groups, thus preventing the formation of low-molecular-weight products, is the polyacrylic ester

$$\cdots\text{CH(COOH)}-\text{CH(COOR)}-\text{CH(COO}^\ominus)-\text{CH(COOR)}\cdots \tag{34}$$

This material has excellent whiteness, overbake resistance, and hardness properties. Adhesion and corrosion resistance are fairly good. However, the relatively high cost is a limiting factor for a broad application.

More recently, *cathodically* depositable cationic resins are used in practice according to their excellent corrosion protection. Examples are ammonium- or amino-modified epoxy resins, in special epoxy-modified Mannich bases of polyphenols[94] and sulfonium and phosphonium groups containing binders. Another advantage of these polyelectrolytes is the stability of the metal substrate under cathodic polarization.

In most of the cases, the binders are self-cross-linking. They need no additional polymers to perform the cross-linking reaction in the course of stoving (baking). *Bases* used to neutralize the polyacid at least partially are ammonia, amines, and KOH. The polybase is neutralized with acetic acid or phosphoric acid. The idea was advanced in order to improve the corrosion resistance via corrosion inhibitors introduced as the neutralizing agent.[95]

## 7. Pretreatment of the Substrate

The substrate, which is normally iron, is pretreated in order to improve corrosion resistance and adhesion. The most common method is the deposition of a layer of zinc- and iron phosphate a few microns thick. Zinc or zinc-coated steel is phosphated as well, or it is chromated. The current flowing in the EDP process thus concentrates in the pores of these inorganic layers. The acid, which is generated inside the pores at high current density, attacks to some extent the phosphate film. Weight losses up to 60% have been found.[96,97] It is possible to identify the pores via copper cementation or by the formation of gas bubbles. In the primary stage of the EDP process, the pores are filled with electro-coagulated resin.[6]

Dissolution of zinc phosphate layers may also be caused by $OH^-$ ions which are generated at the surface by normal corrosion[97] or by the action of localized cells.[98] Delamination of organic coatings can be promoted by corrosion.[99] The strong influence of water concentration in phosphate layers was investigated by potentiodynamic current–voltage curves.[100]

## 8. Practical Aspects of the EDP Process

As is shown in Table 2, the *energy consumption* in the EDP process is rather high. This figure can be lowered by adjusting the electrochemical equivalent and the degree of neutralization.

Usually the components of the EDP system are not found in the film with the same relation as in the bath. A *fractionation* of the polyelectrolyte according to its molecular weight (preferential deposition of low-molecular-weight fractions,[101] and in other cases high-molecular-weight fractions,[102] as well as according to the structure[103]) has been found. The pigment is enriched in the film. Retention of amine in the electrodeposited layer was found to approach 20%.[104,105]

This means that most of the amine tends to accumulate in the bath. The amine ("solubilizer") balance is maintained according to the following two methods:[106]

1. *The amine-deficient method.* In the replenishment feed, only that part of the amine is introduced which is removed by the deposition process. Homogenizing the feed is somewhat difficult.

2. *Continuous removal of unbalanced amine by dialysis, cathodic electrodialysis, ion exchange, and ultrafiltration.*[107] The feed contains the total amount of amine and it is readily homogenized.

In most cases, both strategies are combined. As has been shown in Section 3.3, a *minimum current density* exists for the EDP process which is on the order of $1 \text{ mA/cm}^2$. Approaching this critical value leads to an increase of an induction period which is due to a nucleation phenomenon.[6,24,40] A *maximum critical current density* on the order of $10 \text{ mA/cm}^2$ must be considered, as well. Exceeding this value leads to excessive gassing and blow off of the film.[40,86] Both limits depend on convection.

The practical *polyelectrolyte concentration* is about 10 wt %. Higher concentrations lead to viscous solutions with substantial drag-out losses. At low concentrations, brittle films have been observed.

Most electrodepositions are performed with a voltage program[36,113,114] to optimize *throwing power* and to minimize gassing. Auxiliary electrodes in less accessible parts of car bodies are sometimes necessary, but should be avoided as much as possible. The *pH value* of anodic baths is slightly alkaline, thus allowing the use of mild steel tanks and pipes. Cathodic baths can offer some problems due to their slightly acid pH values. Some systems, however, show pH values above 6, thus avoiding the corrosion of steel.

Usually only the first layer of paint can be provided by the EDP process. However, a *second layer* can be deposited on top of the first, if this first layer is made conducting by a pigment with good electronic conductivity.[108]

## 9. Follow-Up Steps after Electrodeposition

As is shown in Figure 13, a step preceding the EDP process is surface pretreatment, generally the formation of a zinc phosphate layer by spray or dip procedure.

Follow-up steps after the electrochemical stage are as follows:

1. *Rinsing* removes the carry out on the surface and in hollow sections of the workpiece when emerging from the tank. The adherence of the film is such that even a powerful jet of water does not damage the workpiece.

2. *Stoving* (baking) is the last step in the overall process. Temperatures range from 160–200°C with stoving times of 10–40 min. The surface of the film

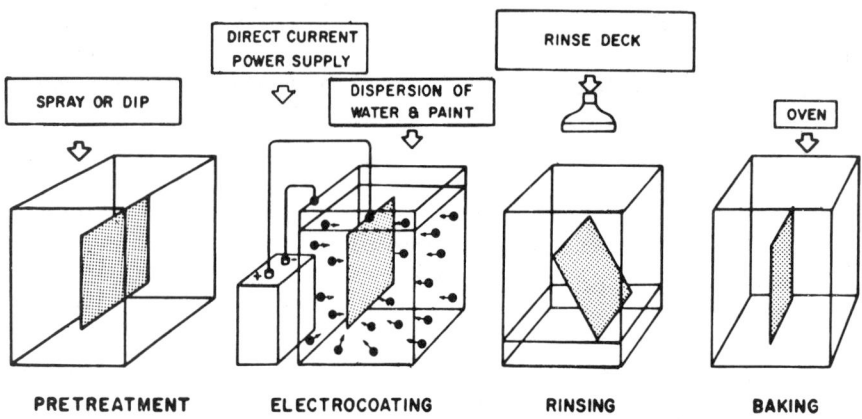

Figure 13. Steps of the EDP process after Brewer.[4]

is smoothed in the primary stage of the step due to increasing flow caused by a decrease in viscosity. As soon as cross-linking sets in at elevated temperatures, the viscosity rises again. Time and measure of the mentioned low viscosity phase must be thoroughly controlled in order to avoid a retreat of the layer from corners and edges. Solvent molecules and other organics escape during stoving and must be retained from escaping into the environment.

## 10. Electrochemical Alternative Processes

Until now, none of the electrochemical alternative processes which have been developed have gained practical importance. In the following, these alternatives are described briefly.

*Chemiphoresis* or *autophoresis* is a new process in which an acid solution, containing HCl, $H_2O_2$, surfactants, and other materials, corrodes the inner and outer surface of the metal body, thus providing the change in concentration of $H^+$ and $Fe^{2+}$ in the diffusion layer:[109,110]

$$Fe + \tfrac{3}{2}H_2O_2 + 3H^+ \rightarrow Fe^{3+}(aq) + 3H_2O \tag{35}$$

No external field is applied. A virtually 100% throwing power has been claimed, for there is no need for current lines to penetrate into slots or boxes. The unconsumed iron ions are absorbed by ion exchange resins. Diffusion of ions through the film seems to be the rate-determining step, according to a square root law for thickness growth and metal dissolution (cf. Figure 14).[111]

A similar improvement in throwing power can be realized in "*reverse EDP*," where the outer surfaces are precoated by classical paint deposition, preferentially powder coating.[112]

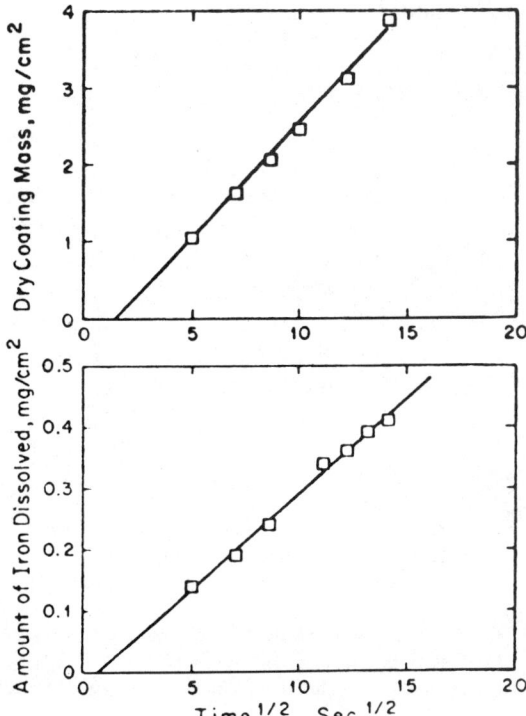

Figure 14. Extent of deposited material and metal dissolution vs. square root of time.

An EDP process with *nonionic binders* is possible in the presence of ionic resins, acting as protective colloids. This type of process has been known for a long time (cf. Reference 12). The electrodeposition of Teflon particles of ~ 0.5 nm diameter from an aqueous dispersion stabilized by perfluoroctanic acid is a more recent example.[115] The films are cured at 380°C by sintering. Dispersions of particles of polyesters, polyamides, polypropylene, and other nonionic polymers are electrodeposited in a similar way, leading to thick films.[116] In the so-called EPLA process, solid particles of thermoplastic resins are dispersed in water and stabilized by anion-active emulsifying agents. The current density is as high as 80 mA/cm$^2$, thus generating Joule's heat which smoothens the thick layers.[117,118] The deposition time is only 10–20 sec.

Finally, the possibility of a film-building *electrochemical polymerization* of monomers dissolved in the electrolyte must be mentioned. From a preparative point of view, the product should not adhere at the electrode.[61,62] However, under some conditions the polymers form a film. The polymerization is initiated by electrochemically generated radicals or radical ions. One of the examples studied in detail is the electropolymerization of diacetone acrylamide at a cathode†:

$$CH_2=CH-CO-NH-\underset{\underset{CH_3}{|}}{\overset{\overset{CH_3}{|}}{C}}-CH_2-CO-CH_3 = \text{monomer} \quad (36)$$

The current densities are rather high (above 20 mA/cm$^2$), and the deposition times are extended.[61,119,120] Because of this fact, this process is currently not of great practical importance in spite of the inherent possibility of reducing the sequence of 12 steps in the original EDP process to four or five.[6] Other examples are the electropolymerization of acrylic esters *in situ* with tetrabutylammonium perchlorate[121] or dissolved in dimethylformamide[122] and *p*-xylilene-bis(trimethylamonium) nitrate in DMSO.[123]

## References

1. R. L. Yeates, *Electropainting*, Draper, Teddington England (1966).
2. K. Weigel, *Elektrophorese Lacke*, Wissenschaftliche Verlagsgesellschaft, Stuttgart (1967).
3. M. W. Ranney, *Electrodeposition and Radiation Curing*, Noyes Data Corp., Park Ridge, New Jersey (1970).
4. G. E. F. Brewer, ed., *Electrodeposition of Coatings* Advances in Chemistry Series Vol. 119, American Chemical Society, Washington, D.C. (1973).
5. W. Machu, *Elektrotauchlackierung*, Verlag Chemie, Weinheim (1974).
6. F. Beck, *Prog. Org. Coat.* **4**, 1 (1976).
7. H. U. Schenck, H. Spoor, and M. Marx, *Prog. Org. Coat.* **7**, 1 (1979).
8. Reuss, *Mem. Soc. Imper. Nat. Moscow* **2**, 327 (1808).
9. H. Helmholtz, *Wiss. Ann.* **7**, 337 (1879).
10. W. P. Davey, U.S. Patent 1,294,627 (1919).
11. B. S. E. Shepard and L. W. Eberlin, *Ind. Eng. Chem.* **17**, 711 (1925).
12. G. C. Sumner, *Trans. Faraday Soc.* **36**, 272 (1940).
13. Crosse and Blackwell Ltd., British Patent 455,810; 496,945 (1937).
14. C. L. Beal, *Ind. Eng. Chem.* **25**, 609 (1933).
15. G. L. Burnside and G. E. F. Brewer, *J. Paint Technol.* **38**, 96 (1966).
16. G. E. F. Brewer, *J. Paint Technol.* **45**, 37 (1973).
17. W. Brushwell, various papers in *Am. Paint J.* and *Farbe + Lack* (1970–1979).
18. *Chem. Eng. News*, 10 (26 August 1974); *Farbe + Lack* **80**, 38 (1974).
19. F. Oosawa, *Polyelectrolytes*, Marcel Dekker, New York (1971).
20. A. Katchalski, Z. Alexandrowicz, and O. Kedem, in *Chemical Physics of Ionic Solutions*, B. E. Conway and R. G. Barradas, eds., Wiley, New York (1966).
21. F. T. Wall, H. Terrayama, and S. S. Techakumpuch, *J. Polym. Sci.* **20**, 477 (1956).
22. W. J. Blank and G. G. Parekh, *Ind. Lackier Betr.* **40**, 142 (1972).
23. W. Kern, *Z. Phys. Chem.* **181**, 249 (1937).
24. F. Beck, *Chem. Ing. Tech.* **40**, 575 (1968).
25. P. Debye, *Ann. N.Y. Acad. Sci.* **51**, 575 (1949).
26. P. E. Pierce and C. E. Cowan, *J. Paint Technol.* **44**, 61 (1972).
27. D. Saatweber and B. Vollmert, *Angew. Makromol. Chem.* **8**, 1 (1969).
28. H. Schuller, cited in F. Beck, *Chem. Ing. Tech.* **40**, 575 (1968).

---

† Cathodic electropolymerization is preferred in general to avoid dissolution of a metal.

29. J. P. Giboz and J. Lahaye, *J. Paint Technol.* **43**, 79 (1971).
30. H. Rechmann, *Farbe + Lack* **71**, 717, 797 (1965).
31. G. E. F. Brewer, *J. Paint Technol.* **45**, 37 (December 1973).
32. A. Kühnert and H. Feichtenbeiner, *Ind. Lackier Betr.* **36**, 3 (1968).
33. W. Jettmar, in Proceedings of the Xth FATIPEC Congress, Montreux, 1970, p. 493.
34. D. T. Smith, L. Landauer, and G. J. Ebben, *Appl. Polym. Symp.* **23**, 1 (1974).
35. G. Rudolph and H. Hansen, *Trans. Inst. Met. Finish* **50**, 33 (1972); cf. Bonder Post, No. 16, May 1969.
36. W. B. Brown and C. A. Campbell, in *Electrodeposition of Coatings*, G. E. F. Brewer, ed., American Chemical Society, Washington, D.C. (1973).
37. B. A. Cooke, *Paint Technol.* **34**, 12 (1970).
38. J. O'M. Bockris and A. K. Reddy, *Modern Electrochemistry*, Vol. I, Plenum Press, New York (1970), p. 391.
39. Z. Kovac-Kalko, in *Electrodeposition of Coatings*, G. E. F. Brewer, ed., American Chemical Society, Washington, D.C. (1973).
40. F. Beck, *Farbe + Lack* **72**, 218 (1966).
41. S. R. Finn and C. C. Mell, *J. Oil Colour Chem. Assoc.* **47**, 219 (1964).
42. A. R. H. Tawn and J. R. Berry, *J. Oil Colour Chem. Assoc.* **48**, 790 (1965).
43. B. A. Cooke, N. M. Ness, and A. L. L. Palluel, in *Industrial Electrochemical Processes*, A. T. Kuhn, ed., Elsevier, Amsterdam (1971).
44. J. C. Catonne and J. Royon, in *Proceedings Interfinish*, Basel, 1972, Forster Verlag, Zürich (1973), p. 258.
45. Nakamura, S. Ando, and H. Nozaki, *J. Electrochem. Soc. Jpn.* **37**, 13 (1969).
46. Y. Nakamura, K. Komata, and H. Nozaki, *Bull. Chem. Soc. Jpn.* **43**, 663 (1970).
47. L. R. Le Bras, *J. Paint Technol.* **38**, 85 (1966).
48. F. Beck and H. Guthke, *Farbe + Lack* **77**, 299 (1971).
49. W. Rausch, personal communication, 1971.
50. D. Saatweber and B. Vollmert, *Angew. Makromol. Chem.* **10**, 143 (1970).
51. F. Beck, H. Pohlemann, and H. Spoor, *Farbe + Lack* **73**, 298 (1967).
52. S. Mercouris and W. F. Graydon, *J. Electrochem. Soc.* **117**, 717 (1970).
53. J. P. Giboz and J. Lahaye, *J. Paint Technol.* **42**, 371 (1970).
54. A. E. Rheineck and A. M. Usmani, *J. Paint Technol.* **41**, 597 (1969); cf. G. E. F. Brewer, ed., *Electrodeposition of Coatings*, American Chemical Society, Washington, D.C. (1973).
55. C. A. May and G. Smith, *J. Paint Technol.* **40**, 526 (1968).
56. D. A. Olsen, *J. Paint Technol.* **38**, 429 (1966).
57. W. J. V. Westrenen, J. R. Weber, and C. A. May, in Proceedings of the VIIIth FATIPEC Congress, Scheveningen, 1966.
58. S. L. Philips and E. P. Damm, Jr., *J. Electrochem. Soc.* **118**, 1916 (1971).
59. J. P. Giboz and J. Lahaye, *J. Paint Technol.* **42**, 501 (1970).
60. G. Smith and C. A. May, *Am. Chem. Soc. Div. Org. Coat. Plast. Chem.* **28**(1), 480 (1968).
61. M. M. Baizer, ed., *Organic Electrochemistry*, Marcel Dekker, New York (1973).
62. F. Beck, *Elektroorganische Chemie*, Verlag Chemie, Weinheim (1974).
63. H. Junek and H. Rauch-Puntigam, *Farbe + Lack* **78**, 1162 (1972).
64. V. Novak, M. Dlaskowa, J. Kasc, and J. Franc, *Farbe + Lack* **81**, 1109 (1975).
65. D. Saatweber and B. Vollmert, *Angew. Makromol. Chem.* **9**, 61 (1969).
66. C. G. Fink and M. Feinleib, *Trans. Electrochem. Soc.* **94**, 309 (1948).
67. M. Elmas, *Ind. Lackier Betr.* **39**, 49 (1971).
68. E. Hahn, *Ind. Lackier Betr.* **32**, 234 (1964).
69. W. Maisch, *Ind. Lackier Betr.* **35**, 3 (1967).
70. L. R. Munson, *J. Paint Technol.* **44**, 83 (July 1972).
71. F. Beck, *Ber. Bunsenges. Phys. Chem.* **72**, 445 (1968).
72. B. A. Cooke and T. A. Strivens, *J. Oil Colour Chem. Assoc.* **51**, 344 (1968).
73. B. A. Cooke, *Paint Technol.* **34**, 12 (1970).

74. J. P. Netillard, *Double Liaison* **17**, 225 (1970).
75. L. Onsager, *J. Chem. Phys.* **2**, 599 (1934).
76. K. H. Frangen, *Farbe + Lack* **70**, 271 (1964); *Tenside* **3**, 253 (1966).
77. A. Gemant, *Ind. Eng. Chem.* **31**, 1233 (1939).
78. W. Machu and L. Steinbrecher, *Werkst. Korros.* **21**, 439 (1970).
79. D. B. Bruce, *Trans. Inst. Met. Finish.* **45**, 93 (1967).
80. H. Clarke and G. Asten, *Paint Manuf.* **37**, 21 (July 1967).
81. F. Holzinger, *Paint Technol.* **30**(9), 26 (1966).
82. M. Bono and D. Pagani, *J. Paint Technol.* **40**, 123 (1968).
83. D. A. Olsen, P. J. Boardman, and St. Prager, *J. Electrochem. Soc.* **114**, 445 (1967).
84. B. A. Cooke, in Proceedings of the Electropaint Conference, London, 1971, paper III.
85. H. Schene, *Metalloberflaeche* **22**, 299 (1968).
86. F. Beck, in *Proceedings Interfinish*, Basel, 1972, Forster Verlag, Zürich (1973), p. 263; *Surface Oberflaeche* **14**, 79 (1973).
87. W. Göring, B. Ancykowski, and H. Noak, *Farbe + Lack* **75**, 327 (1969).
88. G. A. Campbell and W. B. Brown, in *Electrodeposition of Coatings*, G. E. F. Brewer, ed., American Chemical Society, Washington, D.C. (1973).
89. P. Javet, in Proceedings of the Electrochemical Engineering Meeting, Southampton, April 17–19, 1979.
90. A. E. Gilchrist and D. O. Schuster, in *Electrodeposition of Coatings*, G. E. F. Brewer, ed., American Chemical Society, Washington, D.C. (1973).
91. W. Funke and G. Handloser, *Farbe + Lack* **72**, 850 (1966).
92. D. R. Hays and C. S. White, *J. Paint Technol.* **41**, 461 (1969).
93. W. F. Pickard, *J. Electrochem. Soc.* **115**, 105C (1968).
94. F. E. Kempter, E. Gulbins, and F. Ceasar, Proceedings of the XIV FATIPEC Congress, Budapest, 1978.
95. BASF, U.S. Pat. 3,819,548 (1970).
96. W. Menzer, *Prod. Finish. (London)*, 21 (June 1968).
97. R. R. Wiggle, A. G. Smith, and J. V. Petrocelli, *J. Paint Technol.* **40**, 174 (1968).
98. F. Beck and H. Kaiser, *Metalloberflaeche* **29**, 281 (1975).
99. H. Leidheiser, Jr. and M. W. Kendig, *Ind. Eng. Chem. Prod. Res. Dev.* **17**, 1 (1978).
100. C. A. May, in *Electrodeposition of Coatings*, G. E. F. Brewer, ed., American Chemical Society, Washington, D.C. (1973).
101. D. Saatweber and B. Vollmert, *Angew. Makromol. Chem.* **25**, 131 (1972).
102. T. Nakamura et al., *Chem. Abstr.* **67**, 101088 (1968); J. Motoyama et al., *J. Jpn. Soc. Col. Mat.* **41**, 275 (1968).
103. A. Hilt, G. Handloser, and W. Funke, *Angew. Makromol. Chem.* **1**, 174 (1967).
104. W. J. Blank, J. N. Koral, and J. C. Petropoulos, *J. Paint Technol.* **42**, 609 (1970).
105. W. J. Blank, in *Electrodeposition of Coatings*, G. E. F. Brewer, ed., American Chemical Society, Washington, D.C. (1973).
106. G. L. Burnside, G. E. F. Brewer, and G. G. Strosberg, *J. Paint Technol.* **41**, 431 (1969).
107. ICI, U.S. Patent 3,419,488 (1962).
108. S. E. Stromberg, *Ind. Finish.*, 22 (October 1969).
109. ACHEM Prod., U.S. Patent 3,585,084 (1971).
110. ACHEM Prod., U.S. Patent 3,791,431; 3,839,097 (1972).
111. D. C. Prieve, H. L. Gerhart, and R. E. Smith, *Ind. Eng. Chem. Prod. Res. Dev.* **17**(1), 32 (1978).
112. A. Krüger, German patent 1,905,270.
113. G. E. F. Brewer and R. F. Hines, *Am. Chem. Soc. Div. Org. Coat. Plast. Chem.* **30**, 249 (1970).
114. G. E. F. Brewer, *J. Paint Technol.* **43**, 71 (1971).
115. A. Oswald, *Plaste Kautsch.* **15**, 498 (1968).
116. Mitsubishi Electric Co., Japanese Patent 1/51275 (1975).

117. J. Weigel, *Galvanotechnik* **57**, 832 (1966).
118. L. Winkler, *Aluminium (Düsseldorf)* **47**, 390 (1971).
119. Grace & Co, British Patent 1,134,387 (1966).
120. A. Bogenschütz, J. L. Jostan, and W. Krusemark, *Kunststoffe* **60**, 127 (1970).
121. G. Mengoli and B. M. Tidswell, *Polymer* **16**, 881 (1975).
122. Toyota, Japanese Patent 48/56235 (1971).
123. S. D. Ross and D. J. Kelley, *J. Appl. Polym. Sci.* **11**, 1209 (1967).

# 11
# Mineral Flotation

**R. WOODS**

## 1. Introduction

In the extraction of metals from their ores, the process of mineral flotation plays a most important role. Flotation provides the means of separating and concentrating the valuable components of an ore to produce a grade of mineral concentrate suitable for feeding to efficient pyrometallurgical or hydrometallurgical operations. The flotation process involves crushing the ore to liberate separate grains of the various valuable minerals and gangue components, pulping the ore particles with water, and then selectively rendering hydrophobic the surface of the mineral of interest. A stream of air bubbles is then passed through the pulp; the bubbles attach to and levitate the hydrophobic particles, which collect in a froth layer which flows over the weir of the flotation cell.

The froth flotation process came into prominence early in this century, replacing gravity methods as the single most important concentration process. It is employed in the treatment of ores of metals such as Pb, Zn, Cu, Ni, Mo, Sn, and W, the recovery of minerals such as sulfur, graphite, and coal, and in the concentration or purification of minerals such as phosphates and quartz sand. Fleming and Kitchener[1] pointed out that, in light of the very large tonnages treated annually by this process ($>10^9$ tons), flotation must be considered as one of the major practical applications of surface chemistry.

---

**R. WOODS** • CSIRO, Institute of Earth Resources, Division of Mineral Chemistry, Port Melbourne, Victoria 3207, Australia.

The mechanisms of the many processes involved in flotation cover a wide range of disciplines. Of these, the conditioning of the interface between the mineral and the aqueous solution to render the particles hydrophobic is, as this chapter will endeavour to show, the province of the electrochemist. First, the thermodynamic and kinetic aspects of particle–bubble adhesion is discussed. Then the manner in which the requirements for adhesion are met in flotation systems is illustrated by considering two mineral types—oxides and sulfides. Flotation of these minerals exemplifies two different mechanisms—ion adsorption for oxides, and oxidation and chemisorption for sulfides.

## 2. Adhesion of Particles to Gas Bubbles

### 2.1. Thermodynamics of Bubble Attachment

From a consideration of the flotation process, it is clear that an air bubble must be able to adhere to the mineral surface for a particle to be floatable. The attachment of air bubbles to treated galena particles is illustrated in Figure 1, taken from the work of Sutherland and Wark.[2]

The general thermodynamic condition for three-phase contact between particle, solution, and gas bubble is a finite contact angle ($\theta$ in Figure 2) at the three-phase boundary. The Young equation,[3] modified to take into account the influence of water vapor in the gas phase, defines the contact angle in terms of the three interfacial tensions,

$$\cos \theta = \frac{\gamma_{sv} - \gamma_{sl}}{\gamma_{lv}} \qquad (1)$$

where $\gamma_{sv}$, $\gamma_{sl}$, and $\gamma_{lv}$ are the surface tensions of the solid/vapor, solid/liquid and liquid/vapor interfaces.

The application of contact angle measurements to determine solution conditions under which flotation would be efficient was developed, in particular, by Wark and co-workers.[2]

The Dupré relationship gives the free energy change accompanying the replacement of a unit area of the solid/liquid interface by a solid/vapor interface:

$$\Delta G = \gamma_{sv} + \gamma_{lv} - \gamma_{sl} \qquad (2)$$

Combining Eqs. (1) and (2) gives

$$\Delta G = \gamma_{lv}(\cos \theta - 1) \qquad (3)$$

Thus, for any finite contact angle, there will be a free energy decrease upon attachment of a mineral particle to an air bubble and hence adhesion is thermodynamically favorable.

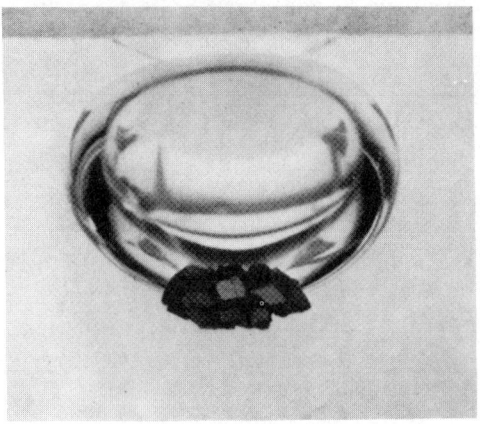

Figure 1. The adhesion of galena particles to gas bubbles. (From reference 2.)

The criterion for a finite contact angle is that

$$\gamma_{sv} - \gamma_{sl} < \gamma_{lv} \qquad (4)$$

In general, the condition in Eq. (4) is not met by mineral surfaces in contact with water. Thus the key to flotation lies in treating the mineral particle with a surface-active agent which will adsorb at the mineral–solution interface. It is clear that, for Eq. (4) to be fulfilled, the adsorption of collector molecules must result in $\gamma_{sv}$ being decreased to a greater extent than $\gamma_{sl}$, assuming $\gamma_{lv}$ remains

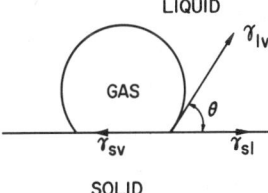

Figure 2. Equilibrium relationships for a gas bubble in contact with a solid immersed in a liquid.

unchanged. Thus a complete analysis of collector action in flotation involves a consideration of all three interfaces. However, as pointed out by Fuerstenau and Raghavan,[5] the collector must adsorb at the interface between the solid and liquid phases for flotation to occur, and hence investigation of adsorption at this interface is of prime importance in understanding flotation. It is not only the coverage but also the nature of the adsorbed molecules which determines the influence of $\gamma_{sl}$ and $\gamma_{sv}$ and hence $\theta$. Adsorption at the liquid/vapor interface is also important in practical flotation since it determines frothing properties. However, control of the froth layer is not normally the role of the collector; frothing agents are generally added for this purpose.

The conclusion, that it is the solid/liquid interface which is important in determining flotation properties, is more readily understood from consideration of the work of adhesion of water to the solid surface, $W_a$, in relation to the work of cohesion of water, $W_c$. Since $W_c = 2\gamma_{lv}$ and $W_a = \gamma_{sv} + \gamma_{lv} - \gamma_{sl}$ then, by substituting the Young equation,

$$W_a - W_c = \gamma_{lv}(\cos \theta - 1) \tag{5}$$

Thus $\theta$ is finite and flotation possible if $W_a < W_c$. The work of adhesion, $W_a$, is the work required to separate liquid water from a solid/water interface, leaving behind an adsorbed layer in equilibrium with saturated water vapor.[6] The action of the collector is to influence the water structure at the interface such that $W_a < W_c$, i.e., $W_a < 144$ erg cm$^{-2}$.

The work of adhesion, $W_a$, is made up of contributions from London–van der Waals (dispersion) forces, $W_d$, hydrogen bonding of water to polar sites, $W_h$, and interactions with ionic sites, $W_i$. $W_d$ for all solids is less than $W_c$, and hence the requirement for bubble contact can be met if $W_h$ and $W_i$ can be diminished sufficiently. This can be achieved by, for example, adsorption of a collector to eliminate or shield polar sites and replace them with sites that do not take part in hydrogen bonding or ionize.

Kitchener and co-workers[7–9] replaced the hydroxyl groups on the surface of silica by methylation with trimethyl chlorosilane. The zeta potential remained the same as for untreated silica, but the contact angle became finite. The silica had been converted from a hydrophilic to a hydrophobic condition by replacing polar hydroxyl sites by nonpolar methyl groups. This replacement diminishes $W_h$ and $W_i$ and hence $W_a$.

These considerations lead to the conclusion that flotation collectors must contain groups which interact unfavorably with water and that these groups must be directed towards the solution side of the solid/solution interface. Hydrocarbon chains possess the required property, displaying a strong hydrophobic effect. The influence of hydrocarbon species on the structure of neighboring water molecules is important in determining a number of properties—the solubility of organic molecules, micelle formation, biological membrane phenomena, etc.[10] The conclusions of hydrophobicity derived from studies of such systems are equally relevant to the mechanism of mineral flotation.

## 2.2. Kinetics of Bubble Attachment

The kinetics of attachment of an air bubble to a mineral surface in the flotation situation depends on the rate of replacement of water from the solid/solution interface. It was pointed out by Frumkin[11] that consideration of the rate of thinning of the water layer at the mineral surface is important in developing flotation theory. Frumkin[12] regarded the difference between wetting and nonwetting systems as being due to the difference in the dependence of the total energy of the liquid film at the solid surface on its thickness. Figure 3 presents the expected dependence of surface energy on thickness for three situations.

Curve A represents a hydrophilic solid in which the water layer "wets" the surface and is stable. Curve C describes a strongly hydrophobic surface in which the surface energy decreases with film thickness. If a gas bubble approaches such a surface, the water layer will spontaneously thin and bubble contact take place rapidly. In curve B, there is an energy barrier which must be overcome before the film will thin spontaneously and bubble adhesion will occur. Curve B represents the case in which there is moderate hydration of the solid surface.

Derjaguin and co-workers[13,14] considered the phenomena associated with film thinning in terms of a "disjoining pressure" which acted either to aid or resist film thinning. Rehbinder[15] derived curves of the form of Figure 3 using this approach. The criterion for flotation is that the disjoining pressure is negative at some distance close to the mineral surface.

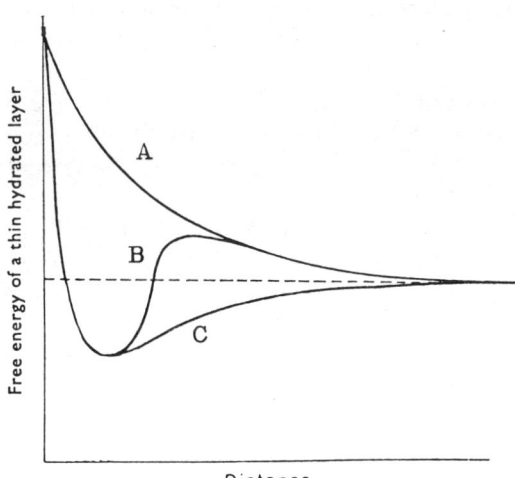

Figure 3. Dependence of the free energy of the water film between a solid surface and a gas bubble on film thickness. A, strongly hydrated; B, moderate hydration; C, very weak hydration. (From Reference 15.)

In practical flotation, the mineral surface is usually not rendered strongly hydrophobic, since selectivity of flotation between different constituents of an ore is more important than the actual flotation rate. Thus the situation generally encountered in real systems is described more by curve B in Figure 3. There is an "induction time" for bubble adhesion which could arise from the presence of an energy barrier, hence the mechanics of particle–bubble approach is important.

## 3. Flotation of Oxide Minerals

### 3.1. Surface Charge

Oxide minerals are hydrophilic because there are polar ionic groups on their surfaces which interact strongly with water. The ionic groups arise from reactions at surface sites of the type

$$M - OH \rightleftharpoons M - O^- + H^+ \tag{6}$$

$$M - OH + H^+ \rightleftharpoons M \cdot OH_2^+ \tag{7}$$

where M denotes a surface site on the mineral. These reactions lead to a finite surface charge under most conditions, which can be either positive or negative depending on the pH of the solution phase.

The electrical properties of the oxide/solution interface are characterized by the surface potential and the change in potential with distance from the surface. These parameters cannot be determined directly, and double-layer properties are generally deduced from measurements of the zeta potential, i.e., the potential across the diffuse double layer from the plane of shear which, in kinetic situations, is where the double layer divides between the inner part that moves with the particle and the outer layer that remains with the solution. The zeta potential can be obtained from determinations of streaming potential or of electrophoretic mobility. A knowledge of zeta potentials has proved most valuable in understanding the flotation properties of nonconducting minerals.

The zeta potential of oxides varies with pH (Figure 4), being positive at low pH due to reaction (7) and negative at high pH where reaction (6) is operative. The zeta potential passes through zero at a pH value characteristic of the particular mineral. This value is the isoelectric point, iep; hydrogen and hydroxide ions are potential determining for oxide systems. Fuerstenau and Palmer have reported[16] the iep values for various oxide minerals.

Other inorganic ions in solution will be present in the double layer at the oxide/solution interface and will influence the magnitude of the zeta potential (Figure 4).

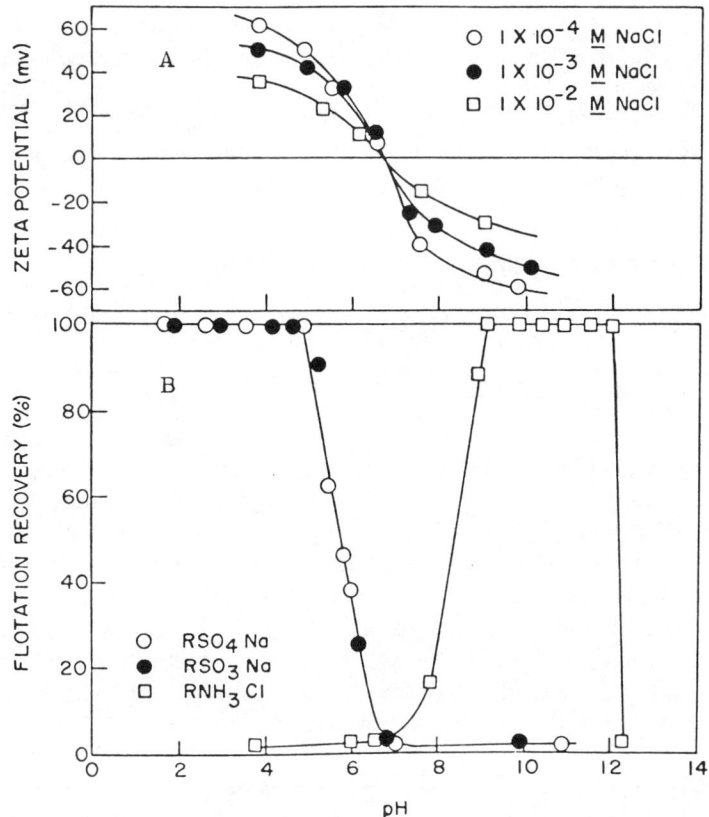

Figure 4. The dependence of zeta potential and flotation response of goethite (FeOOH) on pH. (A) The zeta potential at different concentrations of sodium chloride; (B) the flotation recovery in $10^{-3}$ $M$ solutions of sodium dodecyl sulfate, sodium dodecyl sulfonate, and dodecyl ammonium chloride. (From Reference 17.)

## 3.2. Collector Adsorption

Specific adsorption of ions can take place on oxide surfaces due to electrostatic attraction between dissolved ions and ionized surface sites. Flotation collectors for oxide minerals are, in general, organic ions which contain a terminal charged group which promotes specific adsorption and a hydrocarbon chain which influences the neighboring water structure through hydrophobic interactions.

Reactions (6) and (7) suggest that, if the oxide is to be floated at a pH below its iep, it will be positively charged and require an anionic collector, while at a pH above this value it will be negatively charged and a cationic collector needed. This behavior is demonstrated by the results of Iwasaki, Cooke, and Colombo[17] which are presented in Figure 4. Efficient flotation occurs at low pH in the presence of $C_{12}H_{25}SO_4^-$ and $C_{12}H_{25}SO_3^-$ and at high pH with $C_{12}H_{25}NH_3^+$. Close to the iep, flotation is inefficient with both collectors since the surface charge is small in this region.

There are a number of factors which complicate the situation such as the influence of other inorganic ions in the double layer and chain–chain interactions which favor the concentration of collector molecules at the interface.

### 3.3. Influence of Inorganic Ions in Solution

The presence of inorganic ions in solution can result in *activation*, i.e., enhancement, or *depression*, i.e., inhibition of flotation. It is clear from the discussion in the preceding section that changes in the concentration of hydrogen ions, i.e., in pH, influence flotation recovery. Other inorganic ions can have a similar effect. For example, multivalent ions such as $SO_4^{2-}$ or $SiF_5^{2-}$ can adsorb on a positively charged mineral surface and lead to a reversal of the sign of the zeta potential. Modi and Fuerstenau[18] demonstrated that the addition of $10^{-2} M$ $Na_2SO_4$ results in the surface of corundum $Al_2O_3$ at pH 6 becoming negatively charged. This promotes the adsorption of a cationic collector, dodecyl ammonium chloride, and efficient flotation results.

Multivalent metal cations can also adsorb at the oxide/solution interface and activate flotation of the mineral. Figure 5 shows the results of Fuerstenau[19] for the pH dependence of the flotation response of quartz, $SiO_2$, with a sulfonate collector in the presence of ferric, plumbous, and manganous

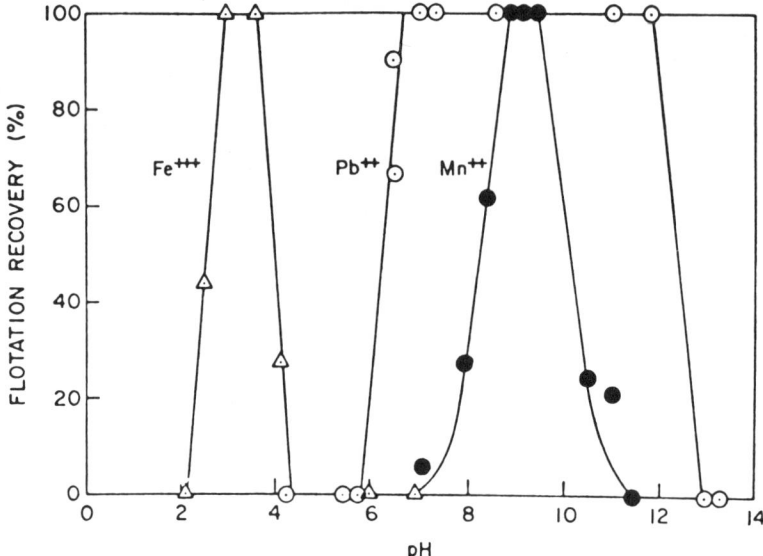

Figure 5. Flotation response of quartz ($SiO_2$) as a function of pH in the presence of $10^{-4} M$ sulfonate and $10^{-4} M$ metal ions. (From Reference 19.)

species in solution. For each metal, flotation is activated over the region of pH in which the predominant ion in solution is the hydroxy complex. The ions $Fe^{3+}$, $Pb^{2+}$, and $Mn^{2+}$ are subject to only weak diffuse layer adsorption, probably because they are too strongly hydrated to approach the surface closely enough to adsorb in the inner layer. The species $FeOH^{2+}$ [or $Fe(OH)_2^+$], $PbOH^+$, and $MnOH^+$ are strongly specifically adsorbed and change the sign of the zeta potential from negative to positive, promoting the adsorption of the cationic collector. The upper pH limit of flotation for each activating species corresponds to the precipitation of the uncharged hydroxide.

## 3.4. Interaction between Hydrocarbon Chains

The hydration of hydrocarbon chains is thermodynamically unfavorable due to a large negative entropy of hydration.[10] The association of hydrocarbon chains of collector molecules adsorbed on the mineral surface leads to an increase in the free energy of adsorption since it results in a decrease in $CH_2$ to $H_2O$ interactions and replaces them with favorable $CH_2$ to $CH_2$ interactions.

Association of hydrocarbon chains occurs in solution when the concentration of long-chain organic ions is above a critical value and this leads to the formation of micelles. The similarity between the association of collector molecules in the adsorbed layer and in micelles led Gaudin and Fuerstenau[20] to refer to the surface phenomenon as "hemi-micelle" formation.

Figure 6 presents the schematic representation of Gaudin and Fuerstenau[20] for the composition of the interfacial region between a negatively charged solid and a solution containing low, medium, and high concentrations of an amine acetate collector. In Figure 6a, they suggest that the adsorption of

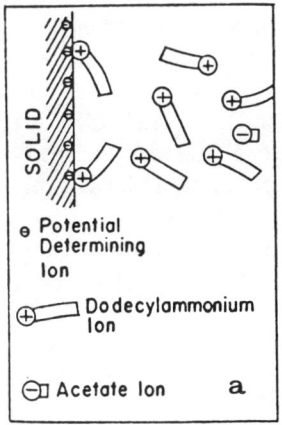
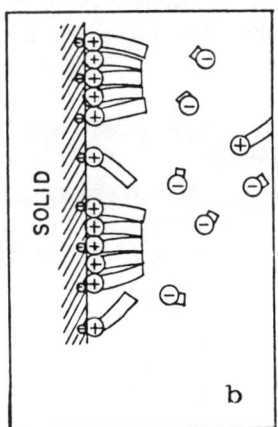
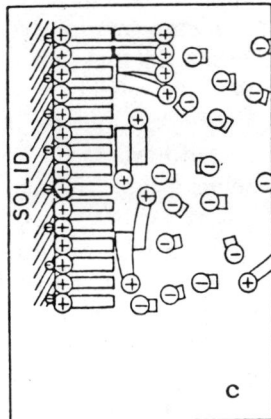

Figure 6. Schematic representation of the double layer at a negatively charged mineral surface in the presence of dodecyl ammonium acetate. (a) Individual ion adsorption, (b) hemi-micelle formation, and (c) multilayer formation. (From Reference 20.)

organic ions in the Stern plane results from electrostatic forces alone. In this case, the surface excess of the collector, $\Gamma$, is given by

$$\Gamma = 2rC \exp\left(-\frac{vF\psi}{RT}\right) \tag{8}$$

where $r$ is the radius of the ion, $C$ the concentration in the bulk solution, $v$ is the valence of the ion, and $\psi$ is the surface potential. In Figure 6b, they consider the concentration to be sufficiently large for chain–chain interaction to occur and hence hemi-micelles form on the surface of the mineral. In this case the adsorption density relationship becomes

$$\Gamma = 2rC \exp\left(-\frac{vF\psi + n\phi}{RT}\right) \tag{9}$$

where $\phi$ is the free energy of adsorption due to chain–chain interaction per $CH_2$ group and $n$ is the number of $CH_2$ groups in the collector molecule. According to Eq. (9), chain–chain interaction will lead to a surface coverage greater than that due to electrostatic attraction alone. Furthermore, the coverage will become greater as the length of the hydrocabon chain of the collector is increased for constant collector concentration in solution. Thus the effect of lengthening the hydrocarbon chain is not only to increase the hydrophobicity of the adsorbed molecule itself, but also to increase the concentration of hydrophobic molecules at the solid/solution interface.

The magnitude of $\phi$ has been evaluated from adsorption measurements to be about 0.6 kcal/mol of $CH_2$ groups.[5] This is comparable with values of the dependence on chain length of the free energy of transfer of organic species from a water to a hydrocarbon environment derived from solubility measurements and of the free energy of micelle formation.

Figure 6c represents the situation in which there is excess collector in solution. In this case, it is concluded[20] that chain–chain interaction has a deleterious effect on flotation, since it results in the adsorption of a second ad-layer with reverse orientation. The polar groups of this second layer are directed towards the solution phase and this leads to a decrease in hydrophobicity of the surface. Wark and co-workers[2] pointed out that high concentrations of collector ions can also influence flotation through their adsorption at the gas–solution interface. They demonstrated that strongly adsorbed ions "armor" the gas bubble, inhibiting adhesion between the bubble and a hydrophobic surface.

The influence of hemi-micelle formation on properties of the solid/solution interface is illustrated by the results of Wakamatsu and Fuerstenau[21] presented in Figure 7. These authors studied the interaction of dodecyl sulfonate with alumina in considerable detail. At low concentrations, region I in Figure 7, the adsorption density increases slowly with increasing bulk concentration. The presence of the collector at the interface creates a finite

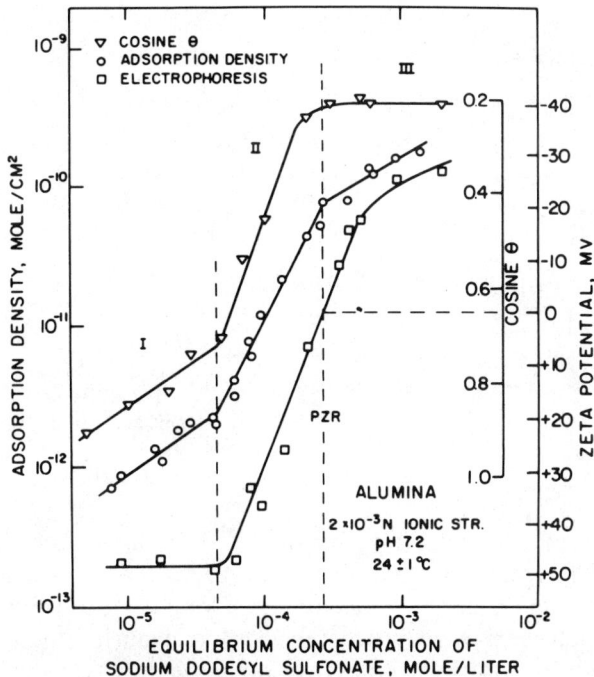

Figure 7. The adsorption density, zeta potential, and contact angle at an alumina ($Al_2O_3$) surface as a function of the equilibrium concentration of sodium dodecyl sulfonate at pH 7.2 with sodium chloride added to maintain the ionic strength at $2 \times 10^{-3}$ M. (From Reference 21.)

contact angle (i.e., $\cos \theta$ becomes <1). The angle increases with coverage (i.e., $\cos \theta$ decreases). In this region, adsorption does not influence the zeta potential because the adsorption of the organic molecules replaces chloride ions adsorbed at the interface.

In region II of Figure 7, it is concluded that hemi-micelles are formed, resulting in a steeply rising adsorption isotherm and an increase in the contact angle. Since more collector ions are now adsorbed than chloride ions displaced, the zeta potential decreases in magnitude. At high concentrations the zeta potential changes sign (PZR, Figure 7) and the coverage of collector ions becomes greater than that corresponding to the surface charge on the mineral. In region III, further increase in coverage is opposed by the excess positive charge, and the slope of the isotherm decreases accordingly.

## 3.5. Influence of Neutral Organic Molecules

The favorable free energy change resulting from the interaction of hydrocarbon chains of organic molecules can result in the coadsorption of neutral organic species with the collector ions. Figure 8 shows the schematic

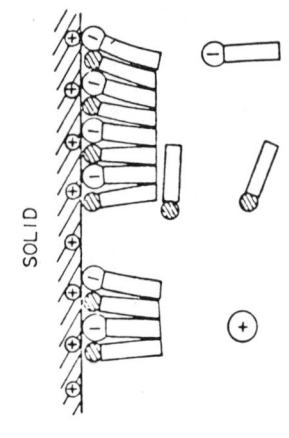

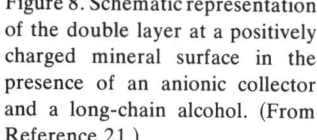

Figure 8. Schematic representation of the double layer at a positively charged mineral surface in the presence of an anionic collector and a long-chain alcohol. (From Reference 21.)

representation of Aplan and Fuerstenau[22] for the constitution of the solid/solution interface at a positively charged surface in contact with a solution containing an anionic collector and a long-chain alcohol. The coadsorption, through chain–chain interaction, leads to a much greater coverage of the surface with hydrocarbon chains than would be the case for adsorption of ions alone. It was shown[22] that the addition of decyl alcohol reduced the concentration of dodecyl sulfate necessary for flotation by about one order of magnitude. Of course, the presence in solution of the alcohol alone would not be effective, since ion adsorption is necessary to anchor the organic species to the mineral surface.

## 4. Flotation of Sulfide Minerals

### 4.1. Mixed Potential Mechanism of Collector Adsorption

Sulfide minerals can be floated in the same manner as oxides with long-chain collector ions which adsorb in the double layer. However, short-chain xanthates (dithiocarbonates), which were introduced as collectors for sulfides in 1925[23] are much more selective since they are most efficient for the flotation of sulfides but quite inert for common gangue minerals. Ever since their collecting properties were first recognized, xanthates have been the major collector species for sulfides although other similar thiol molecules have found application in certain circumstances. The mechanism of flotation with thiol collectors differs from that of ion adsorption discussed previously for oxides and involves oxidative adsorption processes at the mineral surface.

Sulfides, in general, display good electronic conductivity and hence can sustain coupled anodic and cathodic processes on their surfaces. It was recognized as early as 1830[24] that corrosion-type reactions could play an important role in determining the behavior of sulfide ores in nature. However, an electrochemical mechanism for sulfide–collector interaction was not established until after the middle of this century. The similarity between corrosion and flotation systems led Salami and Nixon[25] to suggest that collector interaction proceeded by a mixed potential mechanism at the sulfide surface, involving (i) the anodic oxidation of the collector to form a hydrophobic species on the surface and (ii) the cathodic reduction of oxygen. For example, the oxidation of xanthate to the disulfide, dixanthogen,

$$2ROCS_2^- + \tfrac{1}{2}O_2 + 2H^+ \rightarrow ROCS_2-S_2COR + H_2O \qquad (10)$$

which is very slow in solution, can be catalyzed at a conducting surface through the separate electrochemical steps,

$$2ROCS_2^- \rightarrow ROCS_2-S_2COR + 2e \qquad (11)$$

and

$$\tfrac{1}{2}O_2 + 2H^+ + 2e \rightarrow H_2O \qquad (12)$$

This mechanism is supported by investigations[26] which demonstrate that oxygen plays an important role in the flotation of metal sulfides with thiol collectors.

The mixed potential concept leads to an understanding of collector action in terms of well-established electrochemical principles. Since electrodes can be prepared from sulfide minerals, it also means that the kinetics and mechanisms of the reactions taking part can be studied by conventional electrochemical techniques.

Sulfides are, in fact, semiconductors and this property has been considered as significant in determining flotation properties.[27] However, most sulfides of metallurgical interest have narrow energy gaps and possess a very high surface concentration of conducting electrons and positive holes. Hence these sulfides can be efficient electrocatalysts for both anodic and cathodic reactions, and the rates of electrochemical processes are not limited by the supply of either positive or negative charge carriers at the surface.

### 4.2. Anodic Oxidation of Sulfide Minerals

Eh–pH diagrams for sulfide minerals[28] show that the region of stability is bounded (i) at low pH by dissolution to metal cations and $H_2S$, (ii) at high pH by dissolution to oxymetal anions and $HS^-$, (iii) at low Eh by reduction to the metal with release of $H_2S$ or $HS^-$, and (iv) at high Eh by oxidation to metal cations, metal hydroxide, or oxymetal anions and sulfur or sulfate ions. The most important processes in the flotation environment relate to oxidation since oxidation of the mineral in the aerated pulp can influence flotation response.

Except at low pH, the most favorable oxidation process from a thermodynamic viewpoint is the formation of sulfate. However, the sulfur–sulfate equilibrium is exceedingly slow and a large overpotential is required for sulfate to be formed at a significant rate. Thus, in practical situations, sulfides can exhibit an extensive region of metastability which extends to the boundary for the formation of other oxysulfur anions or sulfur. Only in geological situations, e.g., ore genesis or supergene alterations, are conventional Eh–pH diagrams for sulfides an accurate representation of the reactions and phases actually observed.

Possible oxidation and reduction reactions for galena (PbS) are presented in Figure 9 in the form of an Eh–pH diagram. It has been demonstrated[29] that the cathodic reduction of galena produces Pb and $H_2S$ in accordance with the diagram. Cyclic voltammograms[30] in which the potential traverses the equilibrium values for oxidation of galena to sulfate, to thiosulfate, and to sulfur are shown in Figure 10. Comparison with Figure 9 indicates that oxidation of galena over the entire pH range results in the formation of sulfur since the initial increase in anodic current occurs at the potential, and has the pH dependence, expected for this process. This conclusion is confirmed by the resulting cathodic peaks on the negative-going scans which arise from reduction of the sulfur to PbS, $H_2S$, or $HS^-$ depending on pH and stirring rate. Analysis of the cyclic voltammograms[30] indicated that thiosulfate was formed

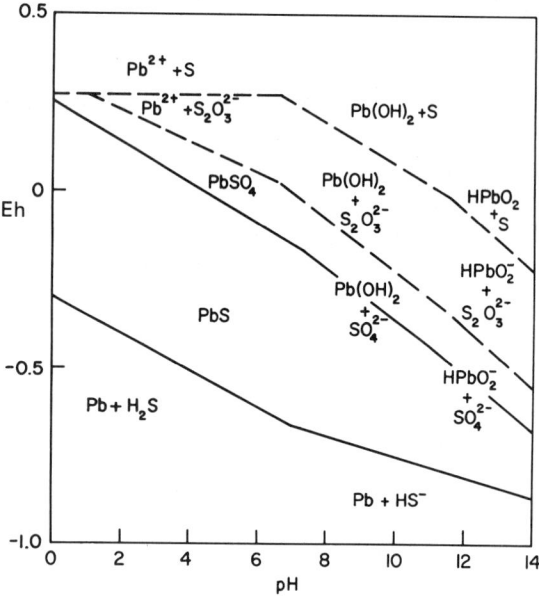

Figure 9. Eh–pH diagram for galena (PbS) showing the region of stability (solid lines) and of metastability (dashed lines) for the mineral. Equilibrium lines correspond to dissolved species at $10^{-3}$ M.

in addition to sulfur, the quantity increasing with increase in pH. However, the major anodic process at all pH values in Figure 10 was found to be the formation of sulfur. Thus galena is metastable up to the upper dashed lines in Figure 9. This is important in understanding the anodic oxidation of flotation collectors on galena surfaces.

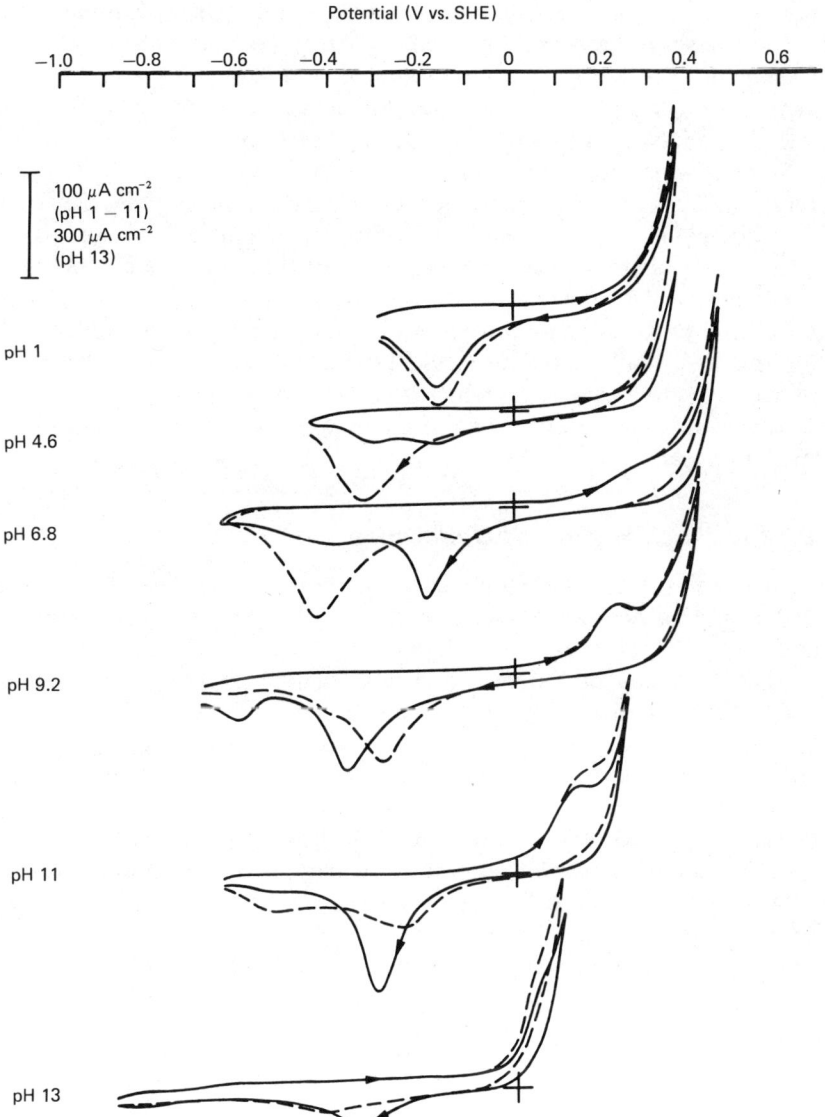

Figure 10. Cyclic voltammograms for galena (PbS) in solutions of different pH. Triangular potential cycles at 20 mV sec$^{-1}$. ———, Quiescent solution; – – –, stirred solution. (From Reference 30.)

Investigations on chalcopyrite ($CuFeS_2$) using cyclic voltammetry[31] have shown that this mineral oxidizes in neutral and alkaline media according to the reaction

$$CuFeS_2 + 3H_2O \rightarrow CuS + Fe(OH)_3 + S + 3H^+ + 3e \qquad (13)$$

It has been demonstrated[32] that chalcopyrite will float in the absence of any collector species if the mineral is in an oxidizing, but not if it is in a reducing, environment. The mineral begins to float at approximately the potential at which reaction (13) occurs[31,32] suggesting that the formation of sulfur on the surface is the factor responsible for inducing flotation.

Sulfides which have layer structures, such as molybdenite ($MoS_2$), are naturally hydrophobic. This characteristic arises from the fact that they cleave along sulfur planes, breaking only weak sulfur–sulfur bonds. Thus the surface is similar to that of sulfur itself. The behavior of these sulfides can be understood[33] in terms of the work of adhesion of water to the surface being largely determined by dispersion forces, with hydrogen bonding and ionic interactions being small. A similar situation could exist at the surface of oxidized chalcopyrite, but the manner in which the elemental sulfur counteracts the influence of the hydrophilic iron hydroxide, also formed by reaction (13), is difficult to explain.

### 4.3. Anodic Oxidation of Collectors

Nearly all sulfide minerals require the presence of a collector for efficient flotation, and the flotation of minerals which can float naturally is substantially improved by the addition of collectors.

The first study of the rate of oxidation of thiol collectors at sulfide electrodes was carried out by Tolun and Kitchener[34] using a polarographic technique. Later workers[35-39] employed linear potential sweep voltammetry in order to study the formation of the initial submonolayer quantities of oxidized collector species, i.e., the quantities relevant to practical flotation systems. With proper design of the electrode, contact angles can be measured simultaneously with electrochemical investigations of the anodic processes.

At noble metal electrodes, xanthate collectors are oxidized to the corresponding dixanthogens [reaction (11)]. The kinetics of this process indicates[35] that it proceeds through an adsorbed xanthate radical:

$$ROCS_2^- \rightarrow (ROCS_2)_{ads} + e \qquad (14)$$

$$2(ROCS_2)_{ads} \rightarrow ROCS_2{-}S_2COR \qquad (15)$$

or

$$(ROCS_2)_{ads} + ROCS_2^- \rightarrow ROCS_2{-}S_2COR + e \qquad (16)$$

where R denotes an alkyl radical. Xanthates used in practical flotation range from the ethyl to the pentyl compounds.

Xanthate ions are specifically adsorbed at low potentials on platinum[35] but the contact angle is zero. A finite angle is only observed in the presence of xanthate when the potential is above that at which oxidation to dixanthogen takes place. It has been suggested[40] that adsorbed xanthate ions are orientated at the interface with the polar group directed towards the solution phase, and that this is the reason the xanthate ion ad-layer does not render the surface hydrophobic. An alternative explanation is that, in short-chain xanthate ions, the charge counteracts the influence of the hydrocarbon chain. The strong attraction of charged polar head groups for water is known to influence the neighboring $CH_2$ groups such that their contribution to the hydrophobic effect is substantially diminished.[10] This could result in short-chain xanthate ions themselves being insufficiently hydrophobic to induce flotation. However, when the charge is removed by oxidation, and the resulting xanthate radical is chemisorbed at a surface site, or bonded to a base-metal atom or to another xanthate radical in dixanthogen, the polarity of the head group is greatly decreased. Hydrophobic interactions between the hydrocarbon chain and water are then unaffected by the head group and become the dominant factor, and the presence of those oxidized species on a mineral surface can render it floatable. This explanation is supported by the fact that alkali xanthates are readily soluble, while dixanthogens are only slightly soluble and behave as typical hydrophobic molecules.[41]

The oxidation of xanthates on pyrite ($FeS_2$)[38,39] and chalcopyrite ($CuFeS_2$)[42] also results in the formation of dixanthogen. The Tafel slope observed for this process, ~80 mV, suggests that either reaction (15) or (16) is rate determining.[38,42] Figure 11 presents cyclic voltammograms and contact angle measurements for pyrite in the presence of methyl, ethyl, and butyl xanthates. It can be seen that, for each xanthate homolog, oxidation to dixanthogen commences close to the reversible potential, $E_r$, of the particular xanthate/dixanthogen redox couple. The decrease in $E_r$ with increase in alkyl chain length arises from an enhanced hydrophobic effect arising from the greater number of $CH_2$ groups which diminishes the solubility of the dixanthogen. The formation of dixanthogen on pyrite gives rise to a finite contact angle which increases with increase in dixanthogen coverage to a limiting value. The limiting contact angle is that for the dixanthogen/water interface itself.

Figure 12 shows the results of analogous experiments with galena (PbS).[39] For this mineral, a current peak appears at potentials below the formation of dixanthogen. This peak is not evident in the absence of xanthate (cf. Figures 10 and 12), appearing in a potential region where galena is metastable (Figure 9). The peak arises from the chemisorption of xanthate, reaction (14), which, on galena, gives rise to monolayer coverage before further oxidation takes place. Continued oxidation leads to the formation of lead xanthate in addition to dixanthogen:

$$PbS + 2ROCS_2^- \rightarrow Pb(ROCS_2)_2 + S + 2e \qquad (17)$$

$$2PbS + 4ROCS_2^- + 3H_2O \rightarrow 2Pb(ROCS_2)_2 + S_2O_3^{2-} + 6H^+ + 8e \qquad (18)$$

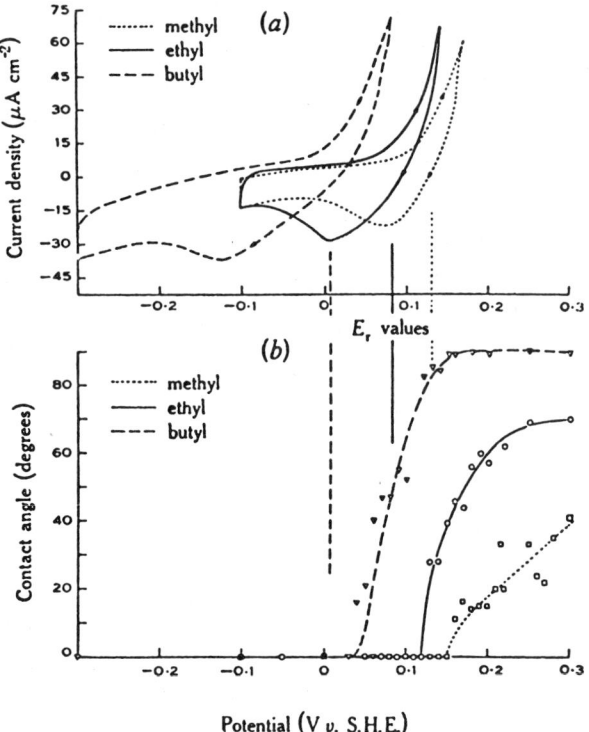

Figure 11. Pyrite (FeS$_2$) electrodes in pH 9.2 solutions containing 1000 ppm of three potassium alkyl xanthates. (a) Cyclic voltammograms at 4 mV sec$^{-1}$, (b) contact angles measured after holding the potential at each value for 30 sec. Vertical lines are the reversible potentials, $E_r$, of the xanthate–dixanthogen couples. (From Reference 39.)

It can be seen from Figure 12 that the chemisorbed xanthate layer, for ethyl and butyl xanthates, is hydrophobic and gives rise to a significant contact angle. This suggests that chemisorbed xanthate is the important surface species in inducing the floatability of galena. The formation of dixanthogen enhances the contact angle, but lead xanthate reduces hydrophobicity. In Figure 12, the decrease in contact angle at high potentials for methyl, and the lower angles in curve A than curve B for ethyl, result from lead xanthate formation.

The chemisorption of xanthate on galena was found[35] to obey the Elovich adsorption rate equation. The initial adsorption current, derived from analysis of current–time curves, has a Tafel slope of 120 mV, which is the value expected for reaction (14).

Current peaks arising from the chemisorption processes have also been identified on cyclic voltammograms of chalcocite (Cu$_2$S) in xanthate[36] and dithiophosphate[37] solutions and of galena in dithiocarbamate solution.[42]

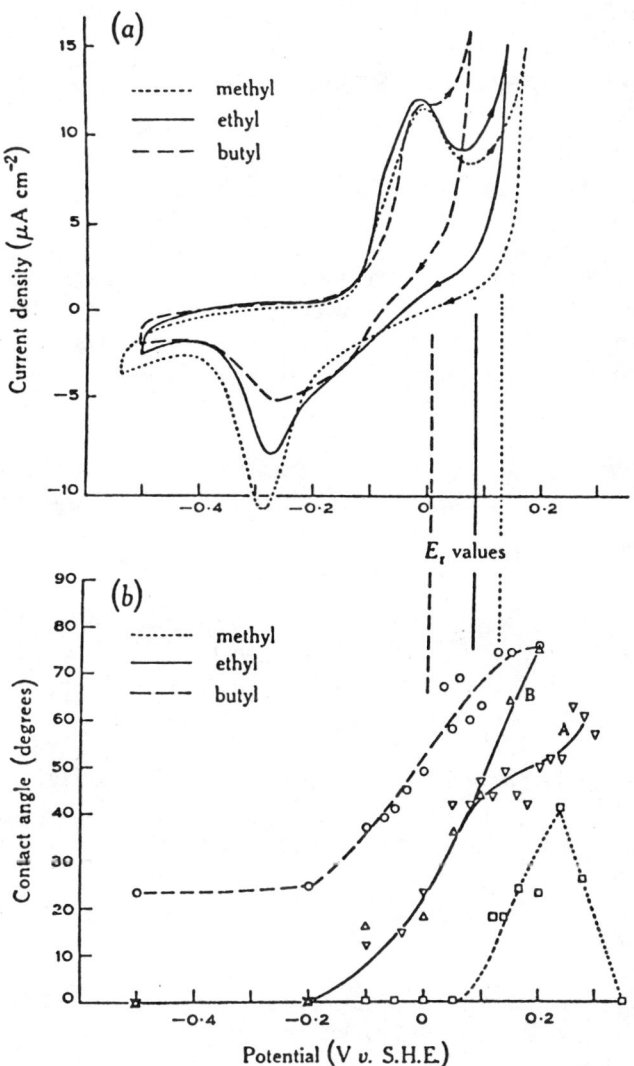

Figure 12. Galena (PbS) electrodes in pH 9.2 solution containing 1000 ppm of three potassium alkyl xanthates. (a) Cyclic voltammograms at 4 mV sec$^{-1}$, (b) contact angles measured after holding the potential at each value for 30 sec. Vertical lines are the reversible potentials, $E_r$, of the xanthate/dixanthogen couples. (From Reference 39.)

## 4.4. Cathodic Reduction of Oxygen at Sulfide Surfaces

The reduction of oxygen is the most important cathodic reaction in a number of corrosion-type processes involving sulfide minerals. In addition to flotation, it plays a role in the weathering of sulfides in nature,[43,44] in oxidation during mining, storage, and transport,[45] and in metal extraction by a variety of leaching techniques.[46]

Oxygen reduction has been studied in detail on pyrite ($FeS_2$)[47] and the reaction shown to proceed through the intermediate formation of hydrogen peroxide:

$$O_2 \xrightarrow{2H^+ + 2e} H_2O_2 \xrightarrow{2H^+ + 2e} 2H_2O \qquad (19)$$

Recent rotated ring-disk electrode studies[48] have confirmed this conclusion.

The Tafel slope for the initial stage on pyrite was determined[47] from analysis of the stirring-independent region at the foot of the reduction wave. In acid solution, the slope is 120 mV, which is indicative of the first electron transfer step,

$$O_2 + e \rightarrow O_2^- \qquad (20)$$

being rate determining. The Tafel slope was found to decrease with increase in pH due to an approach to quasiequilibrium of reaction (20) with the following step,

$$O_2^- + 2H^+ + e \rightarrow H_2O_2 \qquad (21)$$

also being important in determining the kinetics. Analysis of the complete wave indicated that the reduction of hydrogen peroxide has a Tafel slope of 245 mV.

The reduction of oxygen on a range of sulfide minerals[49,50] is illustrated in Figure 13. It can be seen that the electrocatalytic activity for oxygen reduction varies considerably between the different sulfides. In alkaline solution, the medium most relevant to sulfide flotation, pyrite is only slightly less active than gold. The thiospinel $Co_3S_4$ has been found[40] to be more active than pyrite.

The reduction of hydrogen peroxide has been studied on a number of sulfide minerals[40,45,51] and oscillatory behavior observed, arising from interaction with the sulfide surface. To date, the formation of hydrogen peroxide during reduction of oxygen has not been determined directly on any sulfide other than pyrite, but there is no reason to believe that the mechanism will be different. The significance of hydrogen peroxide formation on flotation theory and practice has not been considered in any detail. However, there are two mechanisms by which peroxide could influence flotation response. First, it reacts with xanthates to form other compounds with different collector properties.[52] Second, it could lead to oxidation of the surface of the sulfide to sulfate[45] in a potential region where the mineral is metastable in the absence of peroxide.

### 4.5. Modulation of Sulfide Flotation

Control of the solution pH is one of the most widely employed methods for achieving selectivity between flotation of sulfide minerals. The influence of pH

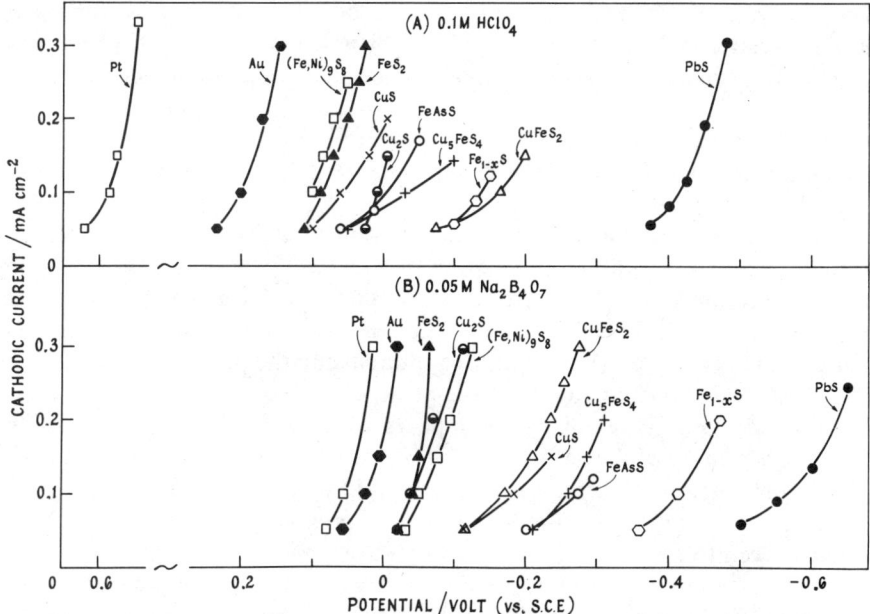

Figure 13. Activity for oxygen reduction of various sulfide mineral and noble metal electrodes in solutions of (A) pH 1 (0.1 $M$ HClO$_4$) and (B) pH 9.2 (0.05 $M$ Na$_2$B$_4$O$_7$). Curves are the activation controlled currents at the foot of the oxygen wave in oxygen saturated solution. (From Reference 50.)

on the flotation characteristics of sulfides is not, as it is in the case of oxide minerals, one of changing the sign or magnitude of the surface charge. Rather, its effect is to change the relative ease of collector and mineral oxidation.

The different ways in which the rate of oxidation of thiol collectors at sulfide surfaces can be enhanced or inhibited has been discussed[53] in terms of the anodic and cathodic processes which can participate in a mixed potential system. The mechanism of the influence of pH on the flotation of pyrite provides a good example of this concept.

The kinetics of oxidation of xanthate to dixanthogen on pyrite is independent of pH,[38] but the anodic wave for the oxidation of pyrite itself shifts to lower potentials by ~70 mV for each increase in pH by one unit. At pH values below about 11, xanthate oxidation is more rapid than that of pyrite and constitutes the anodic process in the mixed potential system. Above this pH, pyrite oxidation is favored, the mixed potential system involves pyrite oxidation and oxygen reduction, dixanthogen is not formed, and the mineral does not float.

The shift in the xanthate oxidation wave for a tenfold increase in concentration is approximately the same as the shift in pyrite oxidation due to pH. Thus the effect of increasing the pH by one unit can be circumvented by increasing the xanthate concentration by an order of magnitude, i.e., for equal flotation response, $[ROCS_2^-]/[OH^-]$ is a constant. This is the Barsky relationship[2] established empirically for alkali depression of sulfide minerals. Selective flotation of one sulfide mineral from another exploits the differences in the upper pH limit for flotation of the two minerals which arises from the differences in ease of oxidation, e.g., chalcopyrite and galena can be floated at high pH where pyrite is depressed.

The influence of cyanide ion on the flotation of chalcopyrite results from a similar mechanism to alkali depression of pyrite, enhanced oxidation of the mineral probably resulting from dissolution to form complex copper cyanides.[42] Indeed, Wark and Cox[2,54] established in the 1930s that there was a direct correlation between electrode potential and the prevention of adsorption of xanthate by chalcopyrite in the presence of cyanide.

Cyanide depression of pyrite appears to be due to a surface cyanide blocking oxidation of the collector.[42,55] The formation of surface compounds with depressant ions occurs at galena in the presence of sulfite and chromate.[56]

In practical flotation, combinations of depressants, e.g., $CN^-$ and $OH^-$, are employed to fine-tune the system to give maximum selectivity.

One of the most important flotation activation processes is that of sphalerite (ZnS) with copper sulfate. Sphalerite is the only economic sulfide with a large band gap and low electronic conductivity. It does not float with the shorter-chain thiol collectors alone but this does not appear to be due to a low electrocatalytic activity arising from its semiconducting properties. The explanation lies in sphalerite oxidation occurring at lower potentials than collector oxidation. When cupric ions are present, they exchange with zinc in the surface of the sphalerite to produce a surface which has the flotation characteristics of copper sulfides.[57] Thus, in practice, sulfides such as galena or chalcopyrite can be floated first from a mixed sulfide ore with a thiol collector. The sphalerite content is then activated with copper, collector added, and the sphalerite floated away from the gangue.

### 4.6. Investigations of the Flotation of Mineral Particles as a Function of Potential

Flotation response is not determined solely by the magnitude of the equilibrium contact angle, but also by the kinetics of particle–bubble contact. In order to carry out electrochemical investigations on systems more closely related to the practical flotation situation, a particulate bed electrode cell has been employed.[58] By using nitrogen bubbles instead of air, and controlling the potential across the interface between the bed of mineral particles by means of an external circuit, it is possible to investigate, in a controlled manner, the surface conditions required for the mineral particles to float.

In a model system using gold spheres, the quantity of charge passed in oxidizing xanthate to form dixanthogen on the gold surface before the particles leave the bed through flotation was measured.[58,59] The dixanthogen coverage when the gold commenced to float varied from multilayers for the dimethyl to 4% of a monolayer for the dipentyl compound. Thus, increasing the length of the alkyl chain in xanthates has two effects which reinforce one another to enhance flotation response. First, the redox potential of the xanthate/dixanthogen couple is lowered and dixanthogen is formed at less positive potentials and hence at a higher rate at the mixed potential. Second, a smaller quantity of dixanthogen is required for flotation.

The flotation characteristics of galena (PbS) and pyrite ($FeS_2$) have been studied[39] using the particulate bed electrode system. The investigations confirm the conclusions reached from the contact angle studies discussed previously and also demonstrate that excess xanthate oxidation has a deleterious effect on flotation. The latter arises from the hydrophobic particles forming into aggregates too large to float.

Heyes and Trahar[32] controlled the potential of mineral particles in a modified laboratory flotation cell through the addition of reducing or oxidizing agents to the solution phase. The potential was monitored with a platinum electrode inserted in the mineral pulp. This experimental system has the

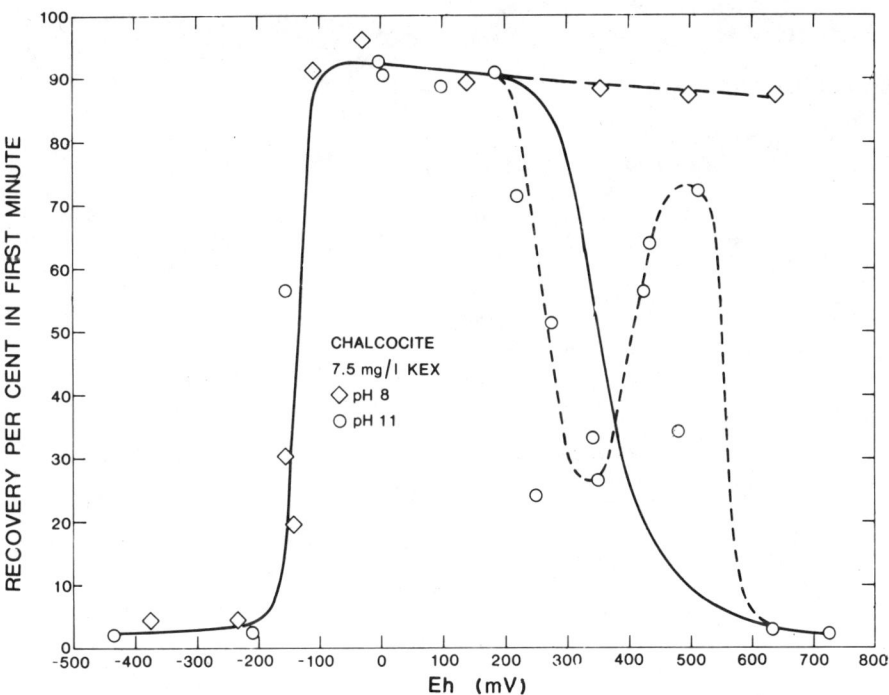

Figure 14. The flotation response of chalcocite ($Cu_2S$) at pH 8 and 11 as a function of potential. The potential was controlled by addition of dithionite or hypochlorite to the solution. Alternative curves for pH 11 are shown. (From Reference 60.)

advantage that the particle size, the pulp density, and the particle–bubble contact mechanism are all close to those evident in practical flotation. Figure 14 presents results of recent investigations[60] on the flotation of chalcocite ($Cu_2S$) with ethyl xanthate at pH 8 and 11. The potential of onset of flotation is independent of pH and is close to that found for the commencement of chemisorption of xanthate on this mineral.[36] Flotation is inhibited at high potentials at pH 11 but not at pH 8. This is explained by oxidation of the surface xanthate to copper oxide by a reaction of the type

$$Cu(C_2H_5OCS_2) + H_2O \rightarrow CuO + C_2H_5OCS_2^- + 2H^+ + e \quad (22)$$

The equilibrium potential of reaction (22) decreases by 118 mV for each increase in pH unit and hence the reaction is favored at high pH.

It is interesting to compare Figures 4 and 14. In the former the flotation response of an oxide is seen to be a function of pH. In the latter, flotation of a sulfide is determined by the Eh. This exemplifies the different mechanisms operative for the two mineral types, viz. collector ion adsorption on charged surface sites for oxides and collector oxidation at an electrocatalytic surface for sulfides.

## References

1. M. G. Fleming and J. A. Kitchener, *Endeavour* **24**, 101 (1965).
2. K. L. Sutherland and I. W. Wark, *Principles of Flotation*, Australian Institute of Mining and Metallurgy, Melbourne (1955).
3. T. Young, *Philos. Trans. Royal Soc. (London)* **95**, 65 (1805).
4. A. Dupré, *Theorie Mechanique de la Chaleur*, Gauthiers-Villars, Paris (1869), p. 369.
5. D. W. Fuerstenau and S. Raghavan, in *Flotation—A. M. Gaudin Memorial Volume*, Vol. 1, M. C. Fuerstenau, ed., AIME, New York (1976), Chap. 3.
6. O. Mellgren, R. J. Gochim, H. L. Shergold, and J. A. Kitchener, in Proceedings of the 10th International Mineral Processing Congress, London, 1973, p. 451.
7. J. Laskowski and J. A. Kitchener, *J. Colloid Interface Sci.* **29**, 670 (1969).
8. A. D. Reed and J. A. Kitchener, *J. Colloid Interface Sci.* **30**, 391 (1969).
9. T. D. Blake and J. A. Kitchener, *J. Chem. Soc. Faraday Trans. 1* **68**, 1435 (1972).
10. C. Tanford, *The Hydrophobic Effect*, Wiley and Sons, New York (1973).
11. A. N. Frumkin, *Usp. Khim.* **2**, 1 (1933).
12. A. N. Frumkin, *Acta Physicochim. URSS* **9**, 313 (1938).
13. B. V. Derjaguin, *J. Phys. Chem. (USSR)* **3**, 29 (1932).
14. B. V. Derjaguin and E. Obuchov, *Acta Physicochim. URSS* **5**, 1 (1936).
15. P. A. Rehbinder, *General Course in Colloid Chemistry*, Moscow University (Moscow), 1949.
16. M. C. Fuerstenau and B. R. Palmer, in *Flotation—A. M. Gaudin Memorial Volume*, Vol. 1, M. C. Fuerstenau, ed., AIME, New York (1976), Chap. 7.
17. I. Iwasaki, S. R. B. Cooke, and A. F. Colombo, Report of Investigation No. 5593, U.S. Bureau of Mines, 1960.
18. H. J. Modi and D. W. Fuerstenau, *Trans. AIME* **217**, 381 (1960).
19. M. C. Fuerstenau, in *AIChE Symposium Series 150*, Vol. 71, P. Somasundaran and R. B. Grieves, eds., AIChE, New York (1975), p. 16.
20. A. M. Gaudin and D. W. Fuerstenau, *Trans. AIME* **202**, 958 (1955).
21. T. Wakamatsu and D. W. Fuerstenau, *Trans. AIME* **254**, 123 (1973).

22. F. F. Aplan and D. W. Fuerstenau, in *Froth Flotation, 50th Anniversary Volume*, D. W. Fuerstenau, ed., AIME, New York (1962), p. 170.
23. C. H. Keller and C. P. Lewis, U.S. Patent 1,554,216, 1,554,220 (1925).
24. R. W. Fox, *Roy. Soc. (London) Philos. Trans.*, 399 (1830).
25. S. G. Salami and J. C. Nixon, in *Recent Developments in Mineral Dressing, Institute of Mining and Metallurgy*, London (1953), p. 503; J. C. Nixon, *Proc. Int. Congr. Surface Activ.* 2nd, **3**, 369 (1957).
26. I. N. Plaksin and S. V. Bessonov, *Proc. Int. Congr. Surface Activ.*, 2nd, **3**, 361 (1957).
27. I. N. Plaksin and R. Sh. Shafeev, *Dokl. Akad. Nauk SSSR* **128**, 777 (1957).
28. R. M. Garrels and C. L. Christ, *Solutions, Minerals and Equilibria*, Harper and Row, New York (1965).
29. E. Peters in *Trends in Electrochemistry*, J. O'M. Bockris, D. A. J. Rand, and B. J. Welch, eds., Plenum, New York (1977), p. 167.
30. J. R. Gardner and R. Woods, *J. Electroanal. Chem.* **100** (1979).
31. J. R. Gardner and R. Woods, *Int. J. Miner. Process.* **6**, 1 (1979).
32. G. W. Heyes and W. T. Trahar, *Int. J. Miner. Process.* **4**, 317 (1977).
33. S. Chander and D. W. Fuerstenau, *Trans. AIME* **252**, 62 (1972).
34. R. Tolun and J. A. Kitchener, *Trans. I.M.M.* **73**, 313 (1964).
35. R. Woods, *J. Phys. Chem.* **75**, 354 (1971).
36. A. Kowal and A. Pomianowski, *J. Electroanal. Chem.* **46**, 411 (1973).
37. S. Chander and D. W. Fuerstenau, *J. Electroanal. Chem.* **56**, 217 (1974).
38. N. D. Janetski, S. I. Woodburn, and R. Woods, *Int. J. Miner. Process.* **4**, 227 (1977).
39. J. R. Gardner and R. Woods, *Aust. J. Chem.* **30**, 981 (1977).
40. S. M. Ahmed, *Int. J. Miner. Process*, **5**, 168 (1978).
41. I. C. Hamilton and R. Woods, *Aust. J. Chem.* **32**, 2171 (1979).
42. R. Woods, in *Flotation—A. M. Gaudin Memorial Volume*, Vol. 1, M. C. Fuerstenau, ed., AIME, New York (1976), Chap. 10.
43. M. Sato, *Econ. Geol.* **55**, 1202 (1960).
44. E. H. Nickel, J. R. Ross, and M. R. Thornber, *Econ. Geol.* **69**, 93 (1974).
45. H. Tributsch and H. Gerischer, *J. Appl. Chem. Biotechnol.* **26**, 247 (1976).
46. H. Majima and E. Peters, in *Proceedings of the 8th Mineral Processing Congress*, Leningrad, 1968, Vol. II, Institut Mekhanobr., Leningrad (1969), p. 5.
47. T. Biegler, D. A. J. Rand, and R. Woods, *J. Electroanal. Chem.* **60**, 151 (1975).
48. M. Aston, D. A. J. Rand, and R. Woods, unpublished work.
49. T. Biegler, D. A. J. Rand, and R. Woods, in *Trends in Electrochemistry*, J. O'M. Bockris, D. A. J. Rand, and B. J. Welch, eds., Plenum Press, New York (1977), p. 291.
50. D. A. J. Rand, *J. Electroanal. Chem.* **83**, 19 (1977).
51. H. Tributsch, *Ber. Bunsenges. Phys. Chem.* **79**, 571, 580 (1975).
52. M. H. Jones and J. T. Woodcock, *Int. J. Miner. Process.* **5**, 285 (1978).
53. R. Woods, in *Avances en Flotacion*, Vol. 3, S. Castro and J. Alvarez, eds., Universidad de Concepcion, Concepcion, Chile (1977), p. 1.
54. I. Wark and A. B. Cox, *Trans. AIME* **112**, 245 (1934).
55. B. Ball and R. S. Rickard, in *Flotation—A. M. Gaudin Memorial Volume*, Vol. 1, M. C. Fuerstenau, ed., AIME, New York (1977), Chap. 15.
56. J. Shimoiizaku, S. Usui, J. Matsuoka, and H. Sasaki, in *Flotation—A. M. Gaudin Memorial Volume*, Vol. 1, M. C. Fuerstenau, ed., AIME, New York (1976), Chap. 13.
57. N. P. Finkelstein and S. A. Allison, in *Flotation—A. M. Gaudin Memorial Volume*, Vol. 1, M. C. Fuerstenau, ed., AIME, New York (1976), Chap. 14.
58. J. R. Gardner and R. Woods, *Aust. J. Chem.* **26**, 1635 (1973).
59. J. R. Gardner and R. Woods, *Aust. J. Chem.* **30**, 981 (1974).
60. G. W. Heyes and W. J. Trahar, *Int. J. Miner. Process.* **6**, 229 (1979).

# Index

Acheson and Castner, and synthetic graphite, 109
Addition agents, 330
  and electrochemical machining, 444
  and electrorefining, 330
Adipodinitrile, Baizer synthesis, 294
Adsorption, and hydrogen evolution, 26
Alcoa, and electrolysis of aluminum chloride, 323
Alkaline electrolysis, 517
Alkaline electrolyzers, novel, 49
Alkaline solution
  and parameters for hydrogen evolution, 21
  and parameters for oxygen evolution, 31
Allen, and electropolishing, 412
Alloy plating, 385
Alloys, electrodepositions, 393
Alumina, dissolution of, 306
Aluminum
  density of, 304
  electrolysis cell with prebaked electrodes, 301
  electrometallurgy, 301
  high purity, production of, 321
Aluminum chloride, electrolysis of, 322
Aluminum production
  and cathodic reaction, 306
  and reaction rate, factors, controlling, 317
  and the Tafel slopes, interpreted by Blyholder and Eyring, 310
Annealing temperature, of gold deposits, 390

Anode coatings, and the Dow Chemical Company, 126
Anode effect
  in aluminum electrolysis, 312
  and Haupin, 313
Anode life cycles, tabulated, 336
Anode material, and chlorate electrolysis, 195
Anode preparation, and lead refining, 356
Anodes
  and flowsheet, 352
  inert, 313
  insoluble, for electrowinning, 338
Anodic corrosion, 401
Anodic deposition processes, described, 541
Anodic electrodeposition, tank for, 538
Anodic generation, of radicals, 279
Anodic overvoltage, in the collection of aluminum, 310
Anodic oxidation, 267
Anodic oxidation
  of naphthalene, 267
  of oliphenes, 267
Anodic preparation, in silver refineries, 360
Anodic process, and cryolite-alumina, 308
Anodic processes
  and electrochemical synthesis, 291
  in refining, 328
Anodic solution, of nickel, 329
Antony and Martinson, and anode coatings for chlorine, 126
Arrhenius plots, of exchange current density, in hydrogen evolution, 29

**597**

Aryl halides, and radical cations, 270
Asahi cells, 155
Asahi Chemical Company, and glass membrane, 138
Asahi chemical membrane electrolyzer, 155
Asahi chemicals, and perfluorocarboxylic acid, competitor for Nafion, 75
Asahi Glass Company, and perfluorocarboxylic acid, 138
Asbestos
  binders and hydrogen evolution, 33
  and chlorine electrolysis, 132
  and separated material, 32
Asbestos diaphragms, in chlorine electrolysis, 132
Askenasy and Revai, and diaphragms with chromic acid formation, 236
Aswan, and Brown Boveri cells, 42
Aswan dam, its electrolyzer, 42
Atomic number, and hydrogen evolution, 25
Automobile industry, chromium layers in, 382

Baizer
  and cathodic carboxylation, 276
  and electrochemical synthesis, 1965, 252
  his famous synthesis of adipodinitrile, 294
  his well-known book, 253
  two of his processes, 271, 273
Balej and Kaderavek, and peroxy compounds, in the preparation of periodates, 223
Barber-Webb, and thermoplastic materials, 343
Baths, for electrodeposition with paint, 543
Beck
  and the capillary-gap cell, 253
  his contributions to the electrodepositions of paint, 546
  his treatise on electro-organic synthesis, 253
Beer, Henri
  DSA patents, 130
  thanks to, 126
Benzoquinone synthesis, 293
Berl and Berl, and hydrogen peroxide preparation by direct cathodic reduction, 227
Bertollet, and the bleaching properties of chlorine, 107
Bilello, and shaped tungsten rods, 416
Bimetalic corrosion, 401

Bipolar cell, 3
  arrangements in series-parallel, 7
Blyholder and Eyring, and Tafel slopes in aluminum production, 310
Bockris
  and constitution of cryolite alumina melts, 305
  his mechanism for the dissolution of iron, 432
  and quantum efficiency in water electrolysis, 92
Bockris and Razumney, and role of inhibitors during metal deposition, 331
Bockris and Uosaki, and water photo-electrolysis, 92
Boden and Evans
  effect of additions of carbonate in electrochemical machining, 445
  and electrochemical machining, 491
Borisoglebsky, and one-percent electronic conductivity in aluminum melts, 317
Born-Haber cycle, and electrochemical reactions, 256
Bowman, and voltammetry in aluminum deposition, 307
Brenner, his famous book on alloy deposition, 393
Brewer and Burnside, and the electrophoretic deposition of paint on cars, 540
Brey, his work on the preparation of bromate, 216
Bright deposits, and crystal faces, 389
Brine, purification of, 158
Broach device, for electrochemical machine, 507
Bromate, industrial significance of, 216
Bromate technology, reviewed by Redford, 219
Brook and Iqbal, and the modified Herring-Blum cell, 483
Brown Boveri
  detailed description of cell, 40
  their bipolar cell, 13
  their cell, 34
    flow diagram of, 38
  their technology, 39
  and various cell types, 40
Brown and Walker, and Kolbe synthesis, 293
Bruggman, his equation for chlorine electrolysis, 195

# INDEX

Burgess, and the shaving of metals, 1941, 423
Bubble attachment, kinetics of, 574
Bubbles
  and Haupin, 312
  and Landau, 311
  and Tobias, 311
Bunsen, and electrolysis of aluminum chloride, 323

Capillary-gap cell, and Beck, 253
Capital recovery, for several interest rates, 82
Carbon dioxide, on carbon surfaces, 309
Carbon frame, for electrochemical machines, 455
Carbon–halogen bonds, 277
  cleavage of, 277
Carbon–heteroatom bond formation, 285
Carbon monoxide–carbon dioxide ratio, during evolution of aluminum, 309
Carbon radical cations, and carbanions, 275
Carbon radicals, ease of oxidation, 273
Carboxylate anions, their anodic oxidation, 280
Carlson, and patent for perchloric acid, 210
Caro's acid, on route to hydrogen peroxide, 227
Carrous Chemical Company, its cell for permanganate reduction, 233
Case Western Reserve, its contract with Diamond Shamrock, 131
Cassenova and Eberson, and their work on carbon–halogen bonds, 277
Castner cell, for chlorine evolution, 110
Castner and Kellner, their cell for chlorine evolution, 110
Cathode banks, in electrorefining, 340
Cathode potential, in chlorine cells, 113
Cathodes
  depolarized, 130
  for chlorine evolution, 130
  for electrorefining, 340
  and flowsheet, 352
  for hydrogen evolution, 127
  low overvoltage, 129
Cathodic generation, of carbon radicals, by cleavage of carbon–heteroatom bonds, 277
Cathodic losses, during the electrolysis of brine, 173
Cathodic processes, and impurities, 329

Cathodic products, of optically active compounds, 286
Cathodic reaction, in aluminum production, 306
Cathodic reduction
  of carbon–heteroatom double bonds, 275
  of olephine, 284
  of onium compounds, 278
Cauquis and Tabacovic, and the electrochemistry of heterocycles, 282
Caustic solar processing, 159
Cell activity, electrochemical, 328
Cell, bipolar, 3
Cell components, in chlorine cells, 121
Cell configuration, and water electrolyzers, 63
Cell
  construction of, for hydrogen, 5
  for hydrogen evolution, detailed description, 40
  monopolar, 3
  tank type, 4
  unipolar, diagramed, 4
Cell potential, in SPE water electrolyzers, 67
Cell separators, and DuPont, 135
Cell technology, and diaphragm cell, 138
Cell types, for electrochemical machining, 503
Cell voltage
  in chlorine evolution, 112
  components of, 16
  for hydrogen oxygen cells, 16
  in Norsk-hydro plants, 45
Cells, monopolar, electrical arrangements, 6
Chalcocite, its flotation, 594
Chalcopyrite, and flotation of minerals, 587
Chalk River Nuclear Labs, and work on electrocatalysis, 93
Chalk River Tritium Recovery Plant, 93
Chao, and diffusion coefficients, 185
Characteristics, for 5-megawatt SPE electrolysis system, 76
Chemical reactions
  of radical anions, 271
  of radical cations and radical anions, 261
Chemical structure of Nafion membrane, 523
Chikamori and Ito, their work on salts in electrochemical machining, 482
Chloralkali cells, 111
Chloralkali electrolysis cells, and asbestos electrodes, 131

Chloralkali plant auxiliaries, 156
Chlorate
  industrial significance, 169
  its effects on electrochemical machining, 464
Chlorate cells, their operation, 194
Chlorate conversion to perchlorite, 212
Chlorate electrolysis
  factors relevant for industrial production, 189
  and flow velocity, 193
  as a function of volume and temperature, 191
  and interelectrode distance, 194
  and pH, 195
Chlorate formation
  anodic, 173
  parameters for, 183
Chlorate production, current efficiency of, 179
Chloride, and intermediate formation, 177
Chloride formation, at anode, stoichiometry, 177
Chlorine
  direct oxidation to, 213
  futuristic technology, 160
  and Hooker cells, 139
  and hypochlorite, 182
  nonelectrolytic processes, 107
  processing, 158
  production of, 105
Chlorine–caustic, the balance, 105
Chlorine cells
  and cell components, 121
  and current efficiency, 117
  and energy consumption, 120
  and energy efficiency, 120
  and resistivity, 117
Chlorine discharge potential, at 1 atmosphere, 113
Chlorine efficiency, and oxygen anions, 119
Chlorine overvoltage, 121
  tabulated, 121
Chlorine production, global, 105
Chlorite cells, 197
  their design, 198
  unipolar, 199
  with integrated chemical reaction volumes, 200
  with low-carbon seals, 199
Chlorite formation, at anode, 181
Chlorite production, industrial significance, 209

Choudhuri, and aluminum droplets in quenched electrolyte, 316
Chromic acid
  industrial cells, 235
  its industrial significance, 234
  preparation of, 234
Chromium, electrowinning, 375
Chromium layers, in automobile industry, 382
Chrysotile fibers, and chlorine electrolysis, 132
Cladding, and perchlorate cells, 214
Cleavage, of carbon–halogen bonds, 277
Cluster formation models, and Nafion membranes, 523
Cluster property, 529
Clusters, and membranes, 526
Coal, hydrogen from, 2
Coatings
  on anodes, in chlorine evolution, 125
  nonplatinum, 126
Cobalt deposition, and gas content, 387
Cobalt electrorefining, and new developments, 375
Cobalt electrowinning, 374
Cobalt oxide spinel and chlorine coatings, 126
Cobaltic oxide, and coatings, 126
Cole, and the dynemometer, 421
Collector absorption, of minerals, 577
Collectors, and anodic oxidation, 585
Complex cyanide, and copper deposition, 392
Complexing, and nickel–zinc deposit, 394
Components
  of cell voltage, 16
  of water electrolyzer cells, 16
Composition, of anode catalysts, in water electrolyzers, 69
Computer programs, for electrochemical machining, 443
Comstock, and electronic detector, 420
Concentration profile, in the electrodeposition of paint, 548
Conducting sheets, and electrochemical grinding, 419
Contact systems, in electrorefining, 346
Contamination, in electrochemical machining, 432
Convection, in electrowinning cells, 333
Copper
  and the complex cyanide, 392
  purification of, 359

# INDEX

Copper electrowinning
 tabulated, 368
 typical requirements for, 370
Copper refineries, tabulated, 351
Copper refining, 330
 detailed cell description, 343
 flowsheet for, 352
 novel approaches, 370
 in operating practice, 351
 requirements, tabulated, 356
Copper refining tank house, and flowsheet, 352
Corrosion
 anodic, 401
 bimetalic, 401
 local cell, 400
 and spacecraft, 65
Corrosion processes, 400
Corrosion products, and flow rates in solution, 432
Corrosion-type reactions, and flotation, recognized relevant in 1830, 583
Cost, reduction of, with anode catalysts, 70
Cost reduction, and performance characteristics, with new anodes, 70
Credit, and deuterium oxide, 94
Cruickshank, and decomposition of salt water, 1800, 108
Cryolite
 in the aluminum process, 303
 ionic constitution, 305
Cryolite-alumina
 the anodic process, 308
 the ionic constitution, 305
Cryolite constitution, work of Paucirova, 305
Current, distribution in cells, for hydrogen production, 6
Current collector, developments, 71
Current consumption, for chlorine cells, theoretical, 117
Current density
 in chlorate electrolysis, 193
 and flow rates in solution, 432
 and transition time, in the electrodeposition of paint, 550
Current density distribution in electrorefining, 347
Current density and flow velocity, 193
Current density–potential curve, summary, in electrode plating practice, 394
Current density–voltage curve, and electrodeposition of paint, 555

Current efficiencies
 in electrochemical machining, 477, 482
Current efficiency
 and caustic strength, 138
 in chlorate production, 211
 in chlorine cells, 117
 in chlorite formation, and temperature, 192
 of the losses in, 317
 and mechanism of loss, 314
 for oxygen production, and hypochlorite, 180
 of stainless steel, 493
 in electrochemical machining, 472
Current-voltage behavior, in electrodeposition of paint, 555
Currey and Pumplin, and reversible chlorine potentials, 113
Curve, electropolishing, 403

Darling and LaBoda, and fire danger in electrochemical machining, 461
DeBarr and Oliver, and electrochemical machining, 447
Deburring, in electrochemical machining, 442
Deburring machine, diagramed, 452
Decomposition voltage, and electrolytic cell, 14
Dehumidifiers and hydrogen electrolyzers, 53
Deionization, and water electrolyzers, 74
Deionized water, essential, for electrolytic cells, 11
Deisser, and the avoidance of diaphragm use in permanganate production, 232
De la Rue, and conductivity effects in electrochemical machining, 434
De la Rue and Tobias
 and bubbles, 311
 the effect of gas bubbles, 115
Demisters, and electrolytic cells, 10
DeNora
 and bipolar cells, marketing of, 139
 cells, and double-woven asbestos, 47
 and chlorate production, 210
 and diaphragm cells, 147
 electrolyzer, diagram of, 49
 his cell, 34
 and the influence of hypochlorite, 211
 and multicomponent coatings, 124
 and overpotential on platinum and lead dioxide, 211
DeNora and Nangal, Northern India, 47

Deodorization, and chlorite cells, 202
Deposition of rutile acrylate–acrylic acid copolymers, in different metals, 550
Deposits, in electroplating, properties of, 387
Derjaguin, and film thinning in kinetics of bubble attachment, 574
Design, features of, for hydrogen evolution, 33
Despic, and electrochemical oxidation of $ClO^-$, 181
Deuterium oxide credit, 94
Deuterium recovery, at Chalk River, 93
Deuterium–tritium exchange, at Chalk River, 94
deValera, and the viewpoint of Foerster and Muller, 181
Deville, and electrolysis of aluminum chloride, 323
Dewing and Yoshida, and the lack of electronic conductivity in aluminum melts, 317
Diagram
  of deburring machine, 452
  of electrochemical machining, 448
Dialkyl sulfides, and radical cations, 269
Diamond alkali and cell separators, 134
Diamond Alkali Company, and their cells, 143
Diamond diaphragm cells, 146
Diamond Shamrock
  cells, for hyprochlorite, 207
  and coating of cathodes, 130
  and diaphragm membrane, 139
  and DuPont Chemical Company, 135
  hydrogen recovery, 159
  and membrane cells, 150
  its monopoly on metal anodes, 124
  their many patents on DSA electrodes, 124
Diaphragm, in chlorine cell, 108
Diaphragm cells
  and Dow Chemical Company, 148
  and steel cathodes, 129
  their characteristics, 140
Diaphragm membranes, 139
Diaphragms, 32
  in cells, 139
  part played in chlorine electrolysis, 132
  and the preparation of bromate, 218
Dielectric constant, as a function of concentration, 530
Diels–Alder addition, in the electrodeposition of paint, 561

Die-sinking, by electrochemical machining, 449
Die-sinking model, 425
  for fixed anode, 435
Diffusion layer, and the electrodeposition of paint, 546
Dihydrophthalic acid, its preparation, 294
Dimerization, 275
Direct current electric power, 157
Disinfection, of seawater, 202
Dispersion, aqueous, in the electrodeposition of paint, 558
Dispersions and latices in the electrodeposition of paint, 560
Dissolution, of alumina, 306
DMSO in the electrodeposition of paint, 565
Dobos, and membrane cell electrolytes, in chlorine evolution, 116
Dornier system, and electrolytic cells for hydrogen production, 88
Dotson, his negative evaluation of depolarized cathodes, 131
Double layer, in flotation of minerals, portrayed, 582
Dow cell, largest chloralkali plant, at Freeport, 145
Dow Chemical Company, and anode coatings, 126
Drossbach, and impedance measurement, 311
DSA
  and electrocatalysis, 125
  electrode for chlorine, its composition a secret, 124
  use in electrorefining, 339
Duhamel du Monceau, and common salt, his discovery, 107
Duisburger Kupferhütte, and processes used in leaching copper, 370
DuPont
  and cell separators, 135
  and Diamond Shamrock, 135
  and Nafion, 135
Dupré equation, and mineral flotation, 571
Duval Corporation, its novel approach for copper refining, 370

Economics, of hydrogen production, 78
Edwards
  data, in electropolishing theory, 409
  and diffuse acceptor species, 410
  and the electrode polishing of grooved copper disks, 409
  and the smoothing mechanism, 407

# INDEX

Efficiency, caustic, in chlorine cells, 119
Eisenberg, and theory of ionomer structure, 523
Electrical circuits, and the Onohama refinery, 345
Electrical connection, of tanks in series, 345
Electrical discharge machining, 399
Electrical energy, in cells, 12
Electrocatalysis
  and DSA, 125
  and hydrogen evolution, 25
  on metals and alloys, 28
Electrocatalyst, for hydrogen evolution, 27
Electrocatalysts
  high activity, 67
  for SPE water electrolyzers, 67
Electrochemical broaching, 507
Electrochemical cell activity, 328
Electrochemical generation, of carbon radicals, their reactions, 277
Electrochemical grinder, apparatus for, 418
Electrochemical grinding rates, 421
Electrochemical machines, 453
Electrochemical machining, 399
  and addition agents, 444
  applications of, 497
  and bubble effects, 429
  cell types for, 503
  of compressor-bearing vibration dumper, 499
  current efficiency studies, 472
  electrolytes for, 471
    tabulated, 495
  and fully hardened steel bushings, 478
  and the gap, 423
  of inhomogeneous alloys, 489
  and large steel mill roll, 502
  in mixed electrolytes, 475
  operations, diagramed, 448
  of other metals, 489
  and polarization studies, 462
  and pressure effects, 429
  principles of, 422
  and Russian investigations, 496
  and shape of cathode, 435
  solution flow effects, 468
  of steering knuckle forging dye, 501
  and stray currents, 443
  and surface finish considerations, 446
  theory of, 405, 424
Electrochemical organic synthesis, technical, 291
Electrochemical rates, for metals, 422

Electrochemical synthesis
  and future developments, 253
  mediation, 253
Electrocoagulation, as a primary process, 544
Electrocontact systems in electrorefining, 347
Electrode configuration, and solid-polymer electrolytes, 61
Electrode kinetics, as a function of temperature, 16
Electrode-membrane assembly, and water electrolyzers, 63
Electrode reactions, their relative importance in the electrodeposition of paint, 549
Electrodeposition
  of maleinized linseed oil, 554
  of oil-modified polyester, 553
  of paint, 537, 538
    advantages, 540
    baths, 543
    cathodic processes, 558
    and colloids, 545
    and current–voltage behavior, 555
    follow-up states, 563
    and Levich's equation, 548
    and mechanistic aspects, 546
    in 1937, 540
    practical aspects, 562
    and reaction products, 548
    and thickness of film, 552
    typical data, 543
  and X-ray patents, 389
Electrolysis
  of aluminum chlorite, 322
  of brine, 173
  cell, for aluminum, 301
  at Chalk River, 94
  of concentrated brine, and reactions occurring, 186
  early history, 108
  and flowsheet, 352
  at high pressure, 7
  of hypochlorite, 202
Electrolyte, 354
Electrolyte purification, 364
Electrolyte studies, in electrochemical machining, 456
Electrolyte surface, and electrorefining, 347
Electrolytes
  and electrochemical machining, 471, 495
  for electropolishing metals, 404

Electrolytic grinder, 509
Electrolytic grinding, 508
Electrolytic production, of hydrogen, 84
Electrolytic purification, of aluminum, 322
Electrolyser Corporation, and Stewart cells, 35
Electrolyzer
  detailed diagrams, 48
  solid polymer, 61
Electrolyzers, novel, 49
Electromagnetic effects, in Hall–Heroult cells, 320
Electrometallurgy, of aluminum, 301
Electron beam welding, 399
Electron micrograph, of Nafion membrane, 522
Electron transfer, from benzene, 290
Electro-organic synthesis, 251
  and electrochemical engineering, 252
  a treatise by Beck, 253
Electrophilic groups, their part in electro-organic reactions, 262
Electroplating, 381
  solution used in, 384
  in thin layers, 384
Electropolishing
  and Jacquet, 403
  of metals, and electrolytes, 404
  and smoothing mechanism, 405
Electrorefining
  and electrolyte surface, 347
  and electrowinning, 328
  of lead, 356
  and metal anodes, 358
  new developments, 355
  of silver, 360
  and solid anodes, 335
  and sulfide anodes, 358
Electrosorption
  and electro-organic synthesis, 285
  and mass transfer, in electrosynthesis, 287
  of optically active cations, 286
Electrostatic calculations, in electrosynthesis, 257
Electrosynthesis
  historical, 168
  of hydrogen peroxide, 226
  and industrial cells, 224
  inorganic, 167
  of iodate and periodate, 219
  reaction types, 283

Electrowinning, 331, 348, 364
  of cadmium, 375
  of chromium, 375
  and circulation, 348
  of copper, 367
  in copper electrorefining, 369
  of gallium, 375
  of manganese, 375
  of other materials, 375
  process, for copper, 369
  and refining, principles, 335
  of selenium, 370
  of zinc, 341
Elmore, and the smoothing mechanism, 406
Endurance test, for water electrolyzers, 70
Energy considerations, 318
Energy consumption, in chlorine cells, 120
Energy efficiency, in chlorine cells, 120
Entrapment, of anode-produced lead sulfate, 339
Equations, for chlorate electrolysis, 190
  efficiency of, 189
Equilibrium gap, in electrochemical machining, 426
Esshoney, and fluorides in aluminum production, 321
Evans, susceptible sites, 401
Evans and Boden, and electrochemical machining, 489
Evolution of oxygen, in the electrodeposition of paint, 558
Exposure to aqueous ionic solutions, and membranes, 526
Exposure to water, and membrane theory, 525
Exchange current density, and hydrogen evolution, 26
Exchange current density, and temperature, in hydrogen evolution, 29
External shaping, by electrochemical machining, 447

Faita, Longhi, and Mussini, the potential of cells, 113
Faraday
  describing electro-organic synthesis in 1834, 251
  and laws of electrolysis, 1832, 108
Fedom'ev, and overvoltage on steel, 129
Ferric oxide, and electrosynthesis, 288
Films, and electrodeposition of paint, 543

# INDEX

Filter press cells
  bipolar, 37
  construction, 5
  their differences, 34
Fioshin, and a modern review of electrosynthesis, 171
Fitz-Gerald and McGeough, theory of electrochemical machining, 436
Fleischmann, invention of the fluidized bed electrode, 253
Fleming and Kitchner, their estimate of flotation, 571
Flemion, and the Asahi Glass Company, 138
Flis and Bynayaeva
  and hypochlorate formation, 176
  and rate constant in chlorate electrolysis, 176
Floating slime, and cathodic contamination, 330
Flotation
  of chalcocyte, 593
  its inhibition at pH 11, 594
  of mineral particles, as a function of potential, 592
  of minerals, 571
  of oxide minerals, 575
  of sulfide minerals, 582
Fluidized bed electrode
  and Fleishmann, 253
  and Goodrich, 253
Flow diagram, for Brown Boveri cell, 38
Flow sheet
  for copper refining, 352
  for adipodinitrile process, 294
  of the Nalco process, 294
  for propylene oxide production, 292
  for water electrolysis plant, 12
Flow velocity, in chlorate electrolysis, 193
Foerster
  and chlorate production, 210
  his mechanism for periodate formation, 222
  and perchlorate electrolysis, 178, 179
  and the preparation of iodate, 219
  and primary chlorate formation, 188
  his remarkable book, 171
  stoichiometry, in chlorate electrolysis, 189
  his work on the preparation of bromate, 217
Foerster and Muller, and the synthesis of chlorate, 173

Ford's patent, on electrodeposition, 540
Fourier integral, in electrochemical machining, 440
Frary and Edwards, and three-layer electrolytic refining process, 321
Frumkin, and water layer at mineral surface, 574
Fuerstenau, and effects of addition of sodium sulfate on surface of corundum, 578
Fukuda, and titanium supported ruthenium dioxide anodes, in ammonium sulfate electrolysis, 223
Fully hardened steel bushes, made by electrochemical machining, 478
Fusion, and possible application in electrolysis, 88
Fusion reactor, and hydrogen electrolysis, 88
Future trends, in chlorine industry, 160
Futuristic concepts, for hydrogen production, 84

Galena
  and flotation, 584
  and flotation of minerals, 587
Gallium, electrowinning of, 375
Gallone, his unpublished calculations on chlorate electrolysis, 193
Galvanostatic transients, and the electrodeposition of paint, 548
Gap and feed rate, in electrochemical machining, 423
Gasometers, and electrolytic cells, 13
Gaudian and Fuerstenau, and double layer at negatively charged mineral surfaces, 579
Gerach, and aluminum droplets in cryolyte aluminum, 316
Gerkler and Kuckler, and Nafion membranes, 523
Germanium, etching of, 418
General Electric, and scale-up program, 76
Ginsberg and Wrigge, permeable diaphragm in alumina cells, 309
Goals, in water electrolysis, 76
Gold-nickel deposits, and stress, 386
Gold refining, 362
Goodrich, invention of the fluidized bed, 253
Gremshaw, and optically active compounds, 286
Grignard reagent, oxidation, 279

Grinder, mechanical, 419
Grjotheim, and fluoride emissions, 305
Grube, and chromate content, 173
Gurklis, and electrochemical machining, 446
Gusev, and the first electrochemical patent, 423

Haber and Grinberg, and the electrosynthesis of chlorate, 171
Half-wave potentials
  for a series of saturated hydrocarbons, 255
  of vinyl, inide, and carbonyl compounds, tabulated, 259
Hall, and the first patent for inert electrodes, 313
Hall–Heroult cell, for alumina reduction, described, 301
Hall–Heroult cells, 319
  and dispersed alumina, 316
  and electrode magnetic effects, 320
Hall and Van Gemert, and molybdenum-tungsten alloys, 130
Hammar and Wranglén
  work on chlorine electrolysis, 191
  work on the oxygen–hypochlorite relation, 181
Hardness, and various ion deposits, 387
Haring–Blum, and Hull cells, 391
Harver, his mechanism for periodate formation, 222
Haupin
  and bubbles, 312
  and increase in current density when cell goes into anode effect, 313
Haussermann, and the development chromic acid production, 234
Hazelett caster for continuous anode refining, 339
Heal, Kohn, and Lartey, chlorate formation, chemical and anodic, 173
Heat balance, in Hall–Heroult cell, 320
Heavy water, as by-product, 92
Heine and Murakami, and verification of equation for gas bubbles, 115
Heisinger and Berzelius, and sodium perchlorate, 170
Helmholtz, and the first quantitative relations in electrophoresis, 540
Helmholtz plane, and electrosynthesis, 288
Heyes and Traher, and control of potential on mineral particles, 593

Hickling and Huggins, and electropolished laminates, 409
Hickling and Jones, and oxygen evolution, 223
High-strength, high-temperature alloys, in electrochemical machining, 482
High-strength steel alloys, 484
High-temperature capability, 72
Hills, and electrochemical machining, 439
Hine
  and resistance in cells, 116
  and streaming potential in asbestos, 133
  and verification of Mukaibo's model, 133
History, of electrodeposition of paint, 540
Hitachi, and electrochemical machining, 497
Hoar
  and the current–voltage curve in electropolishing, 410
  and dual anodic film theory, 413
  and electrochemical machining, 465
  and passive films, 465
Hoar and Evans, and anodic films in ion electrodes, 486
Hoar and Frawling, and electropolishing, 411
Hoar and Rothwell, and solution flow in electropolishing, 410
Holmes, and high-temperature chlorate electrolysis, 192
Hoopes, and high-purity electrodes, for aluminum, 321
Hooker
  and diaphragm membrane, 139
  and future cells for chlorine, 160
  and ion-selective chlorine separators, 134
  and membrane cells, 150, 154
Hooker cells
  for chlorine, 139
  and their current–voltage curves, 143
Hooker Chemical Company, and steel cathodes, 128
Hooker diaphragm cells, and voltages, 142
Hooker membrane cell, 154
Hooker MX cell, for chlorine, 153
Hooker-type diaphragm cells, their characteristics, 141
Hopenfeld and Cole, and solution temperature rise in gap, 421
Hückel molecular orbital (HMO), in oxidation reduction, 258
Hückel $\pi$-electron MO systems, for electrode reactions, 256

# INDEX

Hull cells, 391
Hybrid cycles, and water electrolyzers, 90
Hydration, 530
Hydrocarbon chains, and interaction, 579
Hydrogen
  and alloy deposition, 393
  catalyst for, 70
  consumption of, in USA, 2
  and electrolysis, 2
  in nickel–zinc deposits, 394
  production of, electrolytic, 1
Hydrogen bubble effects, in electrochemical machining, 429
Hydrogen cells, and electrolysis, 8
Hydrogen evolution
  cathodes for, 127
  and cathodic electrolysis for chlorine, 127
  and parameters, 22
    for alkaline solutions, 18
  and the solid-polymer electrolyte, 61
  from various metals, 128
Hydrogen overvoltage, 17
  on steel, 129
Hydrogen peroxide
  production by anodic oxidation, 226
  production by direct cathodic reduction, 227
  production by direct reduction, 229
  methods of preparation of, 226
  reactions in the preparation of, 227
  reduction of, and sulfide minerals, 590
Hydrogen plants, by various manufacturers, tabulated, 42
Hydrogen production, economics of, 78
Hydrostatic bearing paths, and electrochemical machining, 504
Hypobromite, and current efficiency, 218
Hypochlorate concentration, and current efficiency in brine cells, 187
Hypochlorite
  in diffusion layer, 184
  industrial significance, 201
  and sewage, 202
  theory of production, 203
  typical operation data, 206
Hypochlorite cell, high velocity, 207
Hypochlorite cells, their reaction efficiency, 204

Ibl and Landolt, and the kinetic model for chlorate formation, 181

Imagawa
  and chlorate formation, 176
  and the activity coefficient of $ClO^-$, 185
Impurities in refining processes, 329
Industrial cells
  and chlorate electrolysis, 196
  for hypochlorite, 205
  for permanganate reduction, 232
Industrial significance of chlorite formation, 201
Industrial significance of chlorite production, 209
Industrial significance of chromic acid, 234
Industrial significance of permanganate, 231
Inert anodes, 313
Inhibitors, of gold deposits, 390
Inorganic ions, and flotation processes, 577
Insoluble anodes, 340
Internal grooving, by electrochemical machining, 451
Iodate, electrolytic, formation of, 219
Ion-exchange membranes, and electrolysis, 134
Ionic constitution of cryolite alumina, 305
Ionic-hydrate association, 528
Ionics, and cell separators, 134
Ions present in aluminum–cryolite mixtures, 307
IR-free polarization studies, in electrochemical machining, 474
Ito and Shikata, and centrifuging sludges away, 434

Jacquet
  and electropolishing, 403
  his ideas on smoothing mechanisms, 405
Jaksic
  and chlorate electrolysis, 178
  and foreign anions, 187
  and optimum flow velocity for chlorate electrolysis, 189
  his modification of the treatment of Ibl and Landolt for electrochemical chlorate formation, 185
Japan
  and chlorine production, 105
  faster development than in the United States, 541
Japanese refinery, at Onahama, 338
Jet etching, 413
  conditions for, tabulated, 415
  modifications and procedures for, 417

Johnson Matthey Company and electrowinning, 377
Jumbo cathodes, and zinc electrowinning, 343

Kako refinery, in Japan, 357
Kariv, and optically active compounds, 286
Kastening, and pause gas electrodes in hydrophobizing cell, for direct reduction of oxygen, 228
Kavance, and activities of aluminum fluoride, 307
Kellog process, 292
Kelly and Nutting, and uniform foils, 414
Kennecott Copper Corporation, 367
Kershaw
  and early cells for chlorate electrolysis, 196
  and the survey of technology in 1886, 171
Kheifets and Goldberg, theory of anolyte circulation in diaphragm cells, 116
Kinetics, of hydrogen-oxygen cells, 14
Kirkendall and voids, 386
Kitchener and co-workers, and mineral flotation theory, 574
Klonowski, and his production of permanganate, 232
Knibbs, and chlorate production, 210
Kojima and Tobias, and test of electropolishing theory, 412
Kokoulina and Krishtalik, and activity coefficient for chlorate, 176
Kolbe, his electrosynthetic formation of chlorate, 170
Kolbe electrolysis, and vinyl and halene compounds, 282
Kolbe reaction, 251
Kolbe synthesis, of sebacinic acid, 293
Krasilova, her studies of chlorine oxidation to perchloric acid, 213
Krebskosmo technology, 37, 97
Krebskosmo, water electrolyzer, 95
Kretzchmar, and the hydrolysis work in the preparation of bromate, 216, 218
Krylov, and review of Russian investigations, 435
Kubik, and ions present, 307
Kuhn, and tubular cells, his analysis, 206
Kvande, and vapor pressure of cryolyte-alumina cells, 316

Landsberg, and the kinetics of electrolytic iodate formation, 219
Laplace equation, and electrochemical machining, 437
Laryman refinery, in Greece, 357
Leaching, 363
  in copper electrorefining, 369
  and zinc refining, 363
Leaching copper, 369
Leaching-electrowinning process, 367
Lead dioxide
  as an anode for iodate, 219
  as an anode material, 173
  and electrosynthesis, 288, 290
  superior to platinum or graphite, in periodate preparation, 221
Lead refining, 331, 356
  and anode preparation, 356
Lead tetraethyl, and Kolbe synthesis, 293
LeGendre, his review of perchlorate cells, 214
LeSueur, and percolating diaphragm, 1897, 108
Leveling, 393
Levich equation, in the electrodeposition of paint, 548
Life systems
  detailed diagrams, 52
  major advantages, 52
  practical cells, 51
  their cells, 50
Light scattering, and the science of micelles, 543
Lister, and intermediate formation of chloride, 177
Local cell corrosion, 400
Loss, of current efficiency, in alumina cells, 314
Losses
  in current efficiency, 317
  of heat in electrolytic cells, 14
Lowenstein-Riedel process, and hydrogen peroxide, 227
Luilu, and copper electrorefining, 374
Lurgi cell, 34
Lurgi electrolyzer, 46
  detailed view, 48
  detailed view of installation, 48
  exploded view, 47
  their size, 49
Lurgi technology, and differentiating features, 44

Machining, electrochemical, 399

# INDEX

Machu, and technical processes for perborate formation, 231
MacIntosh, and invention of calcium hypochloride, 107
MacMullin's paper, and the vapor pressure of water, 114
Magnetite
 and anode material, 172
 as a chlorine electrode, 126
 as a spinel, 126
Manganese, electrowinning of, 375
Marsh, and interleaved anodes and cathodes, 1913, 109
Mass balance, in cells, 11
Mass and energy balance
 in hydrogen–oxygen cell, diagramed, 10
 in water electrolyzers, 9
Mass transfer, and refining, 331
Mass transport, and refining cells, 331
Material handling, 349
Mathieson, alkali works, and their patent, 1894, 110
Mavroides, and photoelectrolysis of water, 92
Maxwell, and steam enriched bubbles, in chlorate electrolysis, 192
McGeough and Rasmussen, temperature effects, in electrochemical machining, 426
McKee, his work on quaternary alkali ammonium cations, 274
McKee and Leo, regeneration of chromic acid, 236
Mechanism
 of cluster formation, in Nafion membrane, 523
 of periodate formation, 222
Mechanistic aspects of the electrodeposition of paint, 546
Mediator systems
 anodic, 209
 cathodic, 291
Mehltretter and Weiss, and the preparation of periodate, 220
Membrane, large-scale, 76
Membrane cells, 150
 operating characteristics, 153
Membrane theory, and exposure to water, 526
Membranes
 in cluster formation, 523
 structure of, 521
Mercury cathode electrolyzers, 150
Mercury cell, 109
Mercury cells
 and Dow Chemical Company, 149
 operating characteristics, 152
Metal anode preparation, and nickel refining, 357
Metal anodes, electrorefining, 358
Metal dissolution, and thickness of electrodeposited material, 565
Metal metering ladels, at Outokumpu Oy, 337
Metal removal
 and electrochemical rates, 422
 rate considerations, 434
Metals, and flowsheet, 352
Metzger, and stress-free slices, 418
Metzger and Keeleric, 419
Micelle, and its diameter, 543
Michael addition, 273
Micromachining, 414
Microthrowing power, and electrodeposition, 391
Mineral flotation, 571
Mirror bright finishes in electrochemical machining, 467
Mist, in electrowinning, 349
Mitsubishi, and the Onohama refinery, 345
Mixed electrolytes, in electrochemical machining, 474
Mixed potential mechanism, on collector adsorption, 582
MO presentation, for carbon–heteroatom double bond, 258
Model for concentrated sodium chloride electrolysis, 185
Module, of Teledyne cell, 55
Molecular orbital presentation, in electrosynthesis, 254
Monopolar–bipolar cells, comparison, 8
Moore, and electrochemical machining theory, 444
Morozov and Maxisov, and surface roughness, 487
Muller, his work on the preparation of bromate, 216
Müller and Suter, and industrial cells for permanganate reduction, 232
Multicell, diagramed, 9

Nafion advanced studies, 73
 and cell separators, 134
 and DuPont Chemical Company, 135
 its cost, 75

Nafion advanced studies (*cont.*)
  membrane, 9, 292
    and Asahi cells, 155
    membranes, their characteristics, tabulated, 74
    and water absorption, 135
Nafion membrane, electron micrograph of, 522
Nafion polymer, 534
Nagalingam, and chloric production, 209
Nagy
  and rate constants in perchlorate production, 119
  and stereoelectrochemical reduction on steel cathode, 119
Nalco process, 279, 293
Naphthalene, anodic oxidation, 267
Natural convection, and refining cells, 331
Nernst–Einstein equation, 133
Nickel
  and hydrogen deposition, 393
  and hydrogen evolution, 17
  sulfur-depolarized, 360
  typical requirements of, 358
Nickel-alumina, as catalyst for hydrogen evolution, 27
Nickel boride, as catalyst for hydrogen evolution, 27
Nickel cobaltate
  and hydrogen evolution, 29
  and time dependence of oxygen evolution, 32
Nickel electrorefining, 371
  typical requirements, 359
Nickel refinery at Thompson, flow sheet, 360
Nickel refining, 357
  new developments, 373
Nickel sulfide, as catalyst for hydrogen evolution, 27
Nickel-zinc, as catalyst for hydrogen evolution, 27
Nickel-zinc deposits, and hydrogen, 394
Niihama refinery, in Japan, 357
Nippon, light metals and electrowinning, 376
Nitro compounds, and electrosynthesis, 291
Nitrogen-radical anions, 276
Nitrogen-radical cations, reactions of, 268
Nitrogen–nitrogen bond formation, 285
Noble and Shine, and rectifier costs, 496
Nohe, and the preparation of dihydrophthalic acid, 294

Nonideal films in the electrodeposition of paint, 557
Noranda, and starter sheets, 341
Norsk Hydro
  exploded view of one cell, 43
  front view, 45
  generating plant, 44
  their cells, 34, 41
Notches, and bioelectrochemical machining, 460
Novel approaches to copper refining, 370

Oettell
  and the electrosynthesis of chlorate, 171
  and the use of diaphragms, 171
Ohmic drop
  across cathode and anode, and hydrogen evolution, 32
  in chlorine cell, 114
  and structure of cells, 116
Ohmic effects, in electrochemical machining, 438
Ohm's law, in electrochemical machining, 441
Oil fields, and seawater injection, 203
Oil-seal retainer, 498
O'Leary, and his patent, 124
Oliphenes
  anodic oxidation, 267
  and electrosynthesis, 288
Onahama, and Japanese refinery, 338
Onium compounds, their cathodic reduction, 278
Operating costs, in electrorefining, 354
Operating practice, in copper refining, 350
Optically active compounds, and cathodic products, 286
Optimization, of cells for peroxodisulfate, 225
Optimum, economic, and Lurgi, 59
Organic molecules
  influence of, in flotation of minerals, 581
  oxidation–reduction, 253
Organic substrates, and electrochemical conversion, 290
Organic sulfides, and radical cations, 269
Organoelectrochemistry, typical reaction types, 283
Organoelectrosynthetic reactions, 281
Outokumpu
  and a flow sheet for its electrorefining, 371

Outokumpu (*cont.*)
  its nickel refinery in Finland, 371
  its refineries in Africa, 371
  starter-sheet stripping material, 341
Outokumpu process, diagram of, 337
Outokumpu system, and electrorefining, 336
Overvoltage
  of chlorine, 121
  of hydrogen, 17
  of oxygen, 27
Overvoltage effects, in electrochemical machining, 442
Oxidation
  anodic
    of carbanions, 279
    of nitrogen anions, 280
  of nickelous to nickelic, and oxygen evolution, 29
  of thiol collectors, 591
Oxidation and reduction, of organic molecules, 253
Oxidizing salt electrolytes, 458
Oxygen evolution
  in chlorate cells, 211
  in electrochemical machining, 463
  and parameters in alkaline solution, 29
  reaction, and solid-polymer electrolyte, 61
Oxygen overpotential, and time dependence on nickel cobaltate, 32
Oxygen overvoltage, 27

Paint, electrodeposition of, 537
Paint films, and potential versus distance, 539
Parameters
  for cell operation, tabulated, 36
  for hydrogen evolution, in alkaline solution, 20
  of various cells, in hydrogen electrolysis, 78
  for various water electrolyzers, 81
Parker, and the midpoint potential, 258
Passivation, in electrorefining, 333
Passivation processes, steady-state polarization, 402
Patent, English, for production of chlorine, 1851, 108
Patents on chlorine and cobalt oxide spinels, 126
Paucirova, and cryolite constitution, 305
Pauli, his work in the preparation of bromate, 216

Perborate, its preparation, 230
Perchlorate behavior, explained by Hoar, 464
Perchlorate cells, 214
Perchlorate manufacture, industrial anodes for, 225
Perchloric acid
  cell for manufacture, 214
  its electrochemical production, 207
  and Riche, 210
Perchloric cells, operational data, 215
Perchlorite cells, reactions in, 213
Performance, of two SPE water electrolyzers, compared with platinum black, 69
Periodate
  its preparation, 220
  and the use of lead anodes, 221
Permanganate
  its electrolytic preparation, 231
  its industrial significance, 231
  loss in reduction of, 232
  preparation, reaction fundamentals, 231
  reduction
    and the Carus Chemical Company, 233
    its technology, 232
Permeation, and hydrogen evolution, 26
Peroxodisulfate
  improvement of cells for, 225
  its preparation, 221
Peroxodisulfate cells, characteristics of, tabulated, 224
Phelps-Dodge refinery, 342
Photoelectrolysis of water, 91
Photon layer, and the electrodeposition of paint, 546
Physicochemical data for chloride, bromide, and iodide, tabulated, 217
Pietzsch-Adolph process, for hydrogen peroxide, 227
Pigments, and the electrodeposition of paint, 544
Piontelli, and industrial cells in electrochemistry, 312
Pipe elbow, by electrochemical machining, 448
Plastic separators, and electrolysis, 134
Plating, of alloys, 385
Platinum and electrosynthesis, 288
Platinum group, their application in chlorine electrolysis, 124
Plunge cutting
  electrochemical machining apparatus, 497

Plunge cutting (*cont.*)
  by electrochemical sheeting, 450
  rate, of stainless steel, 492
Polarization, and hydrogen evolution, 28
Polarization curve
  of 410 Cr stainless steel, 491
  in electrochemical machining, 488
Polarization curves
  and $n$- and $p$-type germanium electrodes, 417
  for stainless steel, 493
  steady-state, 486
Polarization studies, in electrochemical machining, 462, 463
Polarogram, of styrene, 265
Polarograms, for unsaturated hydrocarbons, 265
Polorography, of 1$e$ reduction of oxidation of anthracene and pyrylene, 254
Polyelectrolyte, confirmation of, 542
Polysulfone, and thin fibers, 33
Ponomarev and Ionova, statistical mechanical theory of electrochemical machining, 523
Porous gas electrodes, and carbonaceous materials, 228
Potassium iodate, operation data, tabulated, 220
Potential-distance curves in the electrodeposition of paint, 540
Potential versus distance for paint films, 539
Power, and rectifiers, in electrolytic cells, 13
Powers, and electropolishing, 410
Practical aspects of the electrodeposition of paint, 563
Practical cells, by life systems, 51
Preparation, of dihydrophthalic acid, 294
Pressure, and electrolysis, 7
Pressure effects, in electrochemical machining, 429
Principles
  of electrochemical machining, 422
  of refining and electrowinning, 335
Processes, alternative, electrochemical, for the electrodeposition of paint, 564
Processing, of chlorine, 158
Production, of chlorine, 105
Proficocorder traces, in electrochemical machining, 473
Profitability index, and water electrolyzers, 80
Properties, of deposits in electroplating, 387
Protonation, of radical cations, 272

Pulse electrolysis, for alkaline metal persulfate reactions, 225
Pulsed electrochemical machining methods, 496
Pulver, and current efficiency as a function of time, 135
Purification, of electrolyte, 364
Purifications, electrolytic, of copper, 359
Pyrite, and flotation of minerals, 587, 588

Radical addition reactions, 281
Radical anions, and unsaturated hydrocarbons, 271
Radical cations
  from aliphatic and aromatic hydroxy compounds, 268
  from alkali and halides, 270
  from aryl amines, their reactions, 268
  and dialkyl sulfides, 269
  from diaryl sulfides, 270
  of aliphatic and aromatic amines, 267
  from organic sulfides, 269
  from phenyls, 268
  and some organic reactions, 264
Radical formation, by anodic oxidation of carboxylate anions, 280
Radicals, anodic generation of, 279
Raman spectroscopy
  and cryolite-alumina, 305
  and dissolution of alumina, 306
Rate considerations, in metal removal, 434
Reaction fundamentals
  for chromic acid, 234
  for permanganate preparation, 231
Reaction paths, for radical cations of alkali aromates, 266
Reaction products, in the electrodeposition of paint, 548
Reaction rate, in aluminum production, factors controlling, 316
Reaction routes, in electro-organic reactions, 263
Reactions
  electro-organic, in the electrodeposition of paint, 551
  in hypochlorite cells, fundamental, 203
  in perchlorite cells, 212
  in the preparation of diborate, 230
  in the preparation of hydrogen peroxide, 227
  in the preparation of persulfuriodate, 221
  in the production of chromic acid, 235

# INDEX

Reactions (*cont.*)
  of radical anions, and cathodic reduction, 275
  using radical cations, 264
  radicalic, 263
Reactivity, 259
Rebazyan, and reactions of $CO_2$, 210
Reduction of oxygen, on sulfide surfaces, 589
Refining
  of aluminum, cell for, 323
  anodic process in, 328
  and winning of metals, 327
Refining cells, and natural convection in, 331
Refining processes, and impurities, 329
Rehbinder, and flotation theory, 575
Resins, their chemistry with respect to the electrodeposition of paint, 561
Resistivities in chlorine cells, 115, 117
Reuss
  and electrodeposition of paint, 540
  and the first electrophoresis, 540
Reversible cell potential in chlorine evolution, 112
Reynolds numbers, in electrochemical machining, 446
Rhenium, in oxygen cells, 59
Richards, and Tafel plots, 308
Riche, and perchloric acid, 210
Rifi, and carbon–halogen bonds, 277
Ring closure, in electro-organic chemistry, 284
Ring closure reactions, tabulated, 284
Rius and Garcia, and oxygen evolution, 223
Rius and Lopis, and perchlorate electrolysis, 178
Roasting
  in copper electrorefining, 368
  and zinc refining, 363
Rogers, and the conductivity of ruthenium oxide, 124
Rohm and Haas, and ion-selective separators, 134
Rolin, and physicochemical properties of cryolite, 305
Rotating disk electrode experiments, in the electrodeposition of paint, 548
Rotating disk studies, of reductional oxygen, in flotation, 590
Russian investigations, and pulsed electrochemical machining, 496
Ruthenium, in oxygen cells, 59

Ruthenium dioxide
  as an anode material, 173
  and chlorate electrolysis, 196
Ruthenium oxide, extensive literature, 124
Ruthenium oxide films, in chlorine evolution, 125
Ruthenium oxide–titanium oxide, DSA, a standard anode material, 123

Salami and Nixon, connection between corrosion and flotation, 583
Sarghel, his work on the preparation of bromate, 219
Sarto, and the ionic constitution in cryolite, 305
Sasaki and Matsui, and patents for water electrolysis cathodes, 130
Saushkin, and steel samples, 497
Scanning electron microscope, and micrographs for alloys, 488
Schafer
  and reactions involving azides, 281
  and the synthesis of some pheromones, 281
  and vinyl compounds, 287
Scheele, and first investigation of chlorine, 107
Schulze–Hardy rule, and the electrodeposition of paint, 550
Schumacher, and his review of perchlorate cells, 214
Schumann–Leclercq, and separators, 171
Selenium, electrowinning, 370
Seleniun removal, and electrowinning, 370
Semiconductor materials, machined, 416
Sen's equation, and the electrodeposition of paint, 548
Separated materials and hydrogen cells, 60
Separators, and hydrogen evolution, tabulated, 33
Sewage, and hypochlorite, 202
Shituru, and cobalt electrorefining, 374
Shlyapnikov, and chlorite electrolysis, 191
Shlyapnikov and Filippov, and perchlorate electrolysis, 178
Silicide, and catalysis of hydrogen evolution reaction, 70
Silicon, and rubber gasket, 65
Silver, massive deposition, 383
Silver cyanide, and silver deposition, 392
Silver refining, anodic preparation, 360
Simmrock process, 292
Sine-wave profile, for anodic surface, 408

Skrabal, his work on the preparation of bromate, 216
Slime handling, 348
Smit and Hoogland, and overpotential in oxygen evolution, 223
Smoothing
  and the theory of Edwards, 407
  and the theory of Wagner, 407
Smoothing mechanism, of metal, during electropolishing, 405
Snavley and Cross, calculation of temperature rise in electrochemical machining, 427
Soderberg anode, 302
  in Hall–Heroult cell, 320
Sodium chlorate production, graphs of change with time, 169
Sodium ions, bound in Nafion membrane, 532
Sodium sulfate electrolysis, 236
Solid polymer, water electrolyzer, 61
Solid-polymer electrolyte
  and the hydrogen evolution reaction, 61
  and the oxygen evolution reaction, 61
Soluble anodes, and electrorefining, 335
Solution flow, and electropolishing, 411
Solution flow effects, in electrochemical machining, 468
Solutions, used in electroplating, 384
Solvay, and experiments with stationary cells, 1898, 110
Solvent systems and electrosynthesis, 289
South Africa, the Nchanga Tailings leach plant, 369
Spacecraft, and water electrolyzers, 65
Spalek, and carbonaceous materials used in porous gas electrodes, 228
Spark-detecting device, 420
SPE cells in series, 66
SPE membranes
  advanced studies, 73
  and dimensional stability, 67
SPE water electrolysis system, and scale-up, 75
SPE water electrolyzer, 61
SPE water electrolyzers
  and cell performance, 73
  small, 76
  and transition metals, 72
Spectrum of states, in the Nafion membrane, 531
Stadion, the first production of chlorate, 210

Stainless steel, and polarization curves, 493
Starter sheets, 341
Steady-state polarization, in electrochemical machining, 476, 478
Steady-state polarization curve, in passivation processes, 402
Steady-state polarization curves, 486
Steam, electrolysis of, 87
Steel bushing, electrochemically machined, 470
Steel cathodes, their service life, 128
Stem drilling, and electrochemical machining, 500
Stender, Ksenzhek, and Lazarev, minimum diagram thickness in electrolyzers, 133
Sterlings' approximation, used in theory of membranes, 524
Stray currents, in electrochemical machining, 443
Stress-free slices of gold, and titanium, 418
Stress, in gold–nickel deposits, 386
Stuart
  founder of the Electrolyser Corporation, 34
  1928, and a Marsh-type cell, 109
Stuart cell, 34
  photograph, 35
  tank-type, 35
Substrate, pretreatment of, in the electrodeposition of paint, 562
Sulfamic acid process, in lead refining, 356
Sulfide anode preparation, 357
Sulfide flotation, reduction of, 590
Sulfide minerals, and anodic oxidation, 583
Sulfides, and electronic conductivity, 582
Sulfonamide membranes, 135
Sulfur–oxygen bond formation, 285
Summary of reaction types, for electrochemical conversion, 283
Surface brightening, in electrochemical machining, 467
Surface charge, of oxide minerals, 576
Surface finish considerations, in electrochemical machining, 446
Surface roughness, in electrochemical machining, 494
Sutherland and Wark, and early theories of mineral flotation, 571
Swann, and his computation on electro-organic synthesis, 1936–1952, 252

Tafel equation, for hydrogen evolution, 17

Tafel plots
  by Richards, 308
  in SPE water electrolyzers, 67
  and Thonstad, for aluminum-cryolite, 308
Tafel slope on DSA electrodes, 125
Taniguchi and Sekine, and the chlorate electrolysis, 176
Tank, for anodic electrodeposition, 538
Tank-type cell, 4
  monopolar, 33
Tank-type Stewart cell, 35
Tarnishing in electrowinning, 383
Technical electrochemical organic synthesis, 291
Technology for chromic acid, 234
Technology
  developed for permanganate reduction, 232
  of electroplating, 381
  futuristic, for chlorine, 160
  for hydrogen peroxide preparation, 229
  of Krebskomos, 98
  for permanganate reduction, 232
  of water electrolysis, general aspects, 3
Teflon, and potassium titanate, in hydrogen evolution, 33
Teflon gaskets, and Asahi cells, 155
Teledyne
  detailed diagrams, 58
  isotopes Electra cell, general diagram, 56
  their cell, 54
  their technology, diagram, 52
Teledyne systems, and the Brookhaven Laboratory, 59
Tellurium, and electrowinning, 377
Temperature coefficient, for hydrogen evolution reaction, 114
Temperatures, in electrochemical machining, 428
Tensile strength, of gold deposits, 390
Texas City, and the tin refining plant, 363
Theile and Matschiner, and industrial cells for peroxydisulfate preparation, 225
Theory of chlorite production, 210
Theory of electrochemical machining, 405, 435, 440
Thermochemical cycles, 91
Thermodynamic decomposition voltage, and electrolytic cells, 14
Thermodynamics
  and decomposition potential, 16
  of hydrogen–oxygen cells, 14
  of water electrolyzers, 13, 61

Thermoneutral voltage, potential, as a function of temperature, 15
Thickness of films, in the electrodeposition of paint, 552
Thin layers, in electroplating, 384
Thompson nickel refinery, in Canada, 357
Thonstad
  and current in alumina cells, 312
  and impedance measurements, 311
  and Tafel plots, 308
Thonstad and Hove, and impedance measurements, 311
Thonstad and Rolseth, and limiting current of platinum anode, 315
Throwing power
  and electrochemical machining, 483
  and electrodeposition, 391
  and the electrodeposition of paint, 556
  and lamellar structure, 391
Tilley and Williams, and jet etching, 413
Tin refining, 362
  in Texas City, 363
Titanium anodes, coated, 339
Titanium electrodes, bipolar, 201
Titanium oxide, its relationship to ruthenium oxide, 124
Tobias, and bubble effects in electrochemical machining, 429
Tobias and De La Rue, bubbles, 115
Tobias and Kojima, test of electropolishing theory, 412
Tolun and Kitchener, and rate of oxidation of thiol collectors, 586
Transpassive dissolution of anodic films, 464
Transport mechanisms, in films, 558
Transition metals, and SPE water electrolyzers, 72
Traube, and hydrogen peroxide, 227
Trepanning, by electrochemical machining, 451
Tritium recovery, at Chalk River, 93
Tumanow, and the use of diaphragms, 232
Turning, by electrochemical machining, 451

Uhlir, and jet etching, 413
Unsaturated hydrocarbons, 281
  charge density of, 260

Venkatachalaphathy, and lead dioxide in the preparation of iodate, 220
Ventsel and Indin, chromous chloride films in electrochemical machining, 486

Vetter
  and his 1960 discussion of electro-organic synthesis, 252
  and impedance measurements, 311
Vetyukov and Vinokurov, and replacing the sodium with bromobutane, in alumina electrolysis, 315
Virtual cathode, and jet etching, 414
Viscosity, of polyelectrolyte solutions, and concentration, 544
Void fraction, and hydrogen evolution, 32
Voids, from the Kirkendall effect, 386
Volmer–Tafel mechanisms, and chlorine electrodes, 125
Voltages, of Hooker diaphragm cells, 143
Volume and temperature, their influence on chlorate electrolysis, 191

Wagner
  and the theory of electrolysis through membranes, 171
  and the theory of smoothing, 407
Wagner model for electropolishing, test of, 412
Wakamatsu and Fuerstenau, and Hemi-micelle, formation, 580
Wark, and effect of high concentration on flotation, 580
Wark and co-workers, with respect to mineral flotation, 572
Wark and Cox, direct correlation between electrode potential and prevention of adsorption of xanthate, 592
Water absorption
  of dry Nafion membrane, 528
  of Nafion polymers, 135
Water activities, in membranes, 533
Water electrolyzer module, full-scale, 76
Water electrolyzers
  mass and energy balance, 9
  thermodynamics, 13
  types of, 3
Water vapor electrolysis, 86
Watt, and chlorate manufacture, 170

Wavelength, concert in electrochemical machining, 439
Weill, Janssen, and Hoogland, and kinetics of anodic oxidation reaction, 231
Weissenstein, and hydrogen peroxide preparation, 229
Westinghouse
  and the improvement of performance in their cell, 91
  their solid electrolyte fuel cell, 88
Wet films, and the electrodeposition of paint, 544
Wettability, and hydrogen evolution, 32
Williams, and the wheel for electrochemical grinding, 422
Wilson, and book on electrochemical machining, 446
Wire cutting, by electrochemical machining, 452
Woolard and Ralston, and the preparation of periodate, 220
Wranglen and Hammar, and steel cathodes in chlorate formation, 173

Xanthates and flotation of minerals, 587, 588
X-ray, patents in electrodeposition, 389

Yoshida and Dewing
  and mass transfer coefficient in alumina cells, 315
  and saturation melt with aluminum, 314
Yurkevich and Vrevskii, and ion-exchange membranes in hychlorite manufacture, 207

Zamin, and the verification of the Wagner model, 409
Zholudev and Stender, and the Tafel equation, 127
Zinc, a flowsheet for, 366
Zinc electrowinning, 340
  typical requirements of, 367
Zinc plants, tabulated, 365
Zinc refining, 363

# RAYMOND H. FOGLER LIBRARY
## DATE DUE

**BOOKS ARE SUBJECT TO RECALL AFTER TWO WEEKS**